郭柏灵院士

# 郭柏灵论文集

（第八卷）

郭柏灵　著

华南理工大学出版社
·广州·

图书在版编目(CIP)数据

郭柏灵论文集. 第八卷/郭柏灵著. —广州：华南理工大学出版社，2010.11
ISBN 978-7-5623-3341-8

Ⅰ.①郭… Ⅱ.①郭… Ⅲ.①郭柏灵-文集 ②非线性方程-文集 Ⅳ.①O175-53

中国版本图书馆 CIP 数据核字（2010）第 208398 号

总 发 行：华南理工大学出版社（广州五山华南理工大学 17 号楼，邮编 510640）
营销部电话：020-87113487 87111048（传真）
E-mail：scutc13@scut.edu.cn http://www.scutpress.com.cn
责任编辑：张 颖 王建洲
印 刷 者：广东省农垦总局印刷厂
开 本：787mm×1092mm 1/16 印张：32 插页：2 字数：805 千
版 次：2010 年 11 月第 1 版 2010 年 11 月第 1 次印刷
定 价：96.00 元

# 序

今年是恩师郭柏灵院士70寿辰，华南理工大学出版社决定出版《郭柏灵论文集》。郭老师的弟子，也就是我的师兄弟，推举我为文集作序。这使我深感荣幸。我于1985年考入北京应用物理与计算数学研究所，师从郭柏灵院士和周毓麟院士。研究生毕业后我留在研究所工作，继续跟随郭老师学习和研究偏微分方程理论。老师严谨的治学作风和对后学的精心培养与殷切期望，给我留下了深刻的印象，同时老师在科研上的刻苦精神也一直深深地印在我的脑海中。

郭老师1936年生于福建省龙岩市新罗区龙门镇，1953年从福建省龙岩市第一中学考入复旦大学数学系，毕业后留校工作。1963年，郭老师服从祖国的需要，从复旦大学调入北京应用物理与计算数学研究所，从事核武器研制中有关的数学、流体力学问题及其数值方法研究和数值计算工作。他全力以赴地做好了这项工作，为我国核武器的发展作出了积极的贡献。1978年改革开放以后，他又在非线性发展方程数学理论及其数值方法领域开展研究工作，现为该所研究员、博士生导师，中国科学院院士。迄今他共发表学术论文300余篇、专著9部，1987年获国家自然科学三等奖，1994年和1998年两度获得国防科工委科技进步一等奖，为我国的国防建设与人才培养作出了巨大贡献。

郭老师的研究方向涉及数学的多个领域，其中包括非线性发展方程的数学理论及其数值解、孤立子理论、无穷维动力系统等，其研究工作的主要特点是紧密联系数学物理中提出的各种重要问题。他对力学及物理学等应用学科中出现的许多重要的非线性发展方程进行了系统深入的研究，其中对Landau－Lifshitz方程和Benjamin－Ono方程的大初值的整体可解性、解的唯一性、正则性、渐近行为以及爆破现象等建立了系统而深刻的数学理论。在无穷维动力系统方面，郭老师研究了一批重要的无穷维动力系统，建立了有关整体吸引子、惯性流形和近似惯性流形的存在性和分形维数精细估计等理论，提出了一种证明强紧吸引子的新方法，并利用离散化等方法进行理论分析和数值计算，展示了吸引子的结构和图像。下面我从这几个方面介绍郭老师的一些学术成就。

Landau－Lifshitz方程（又称铁磁链方程）由于其结构的复杂性，特别是强耦合性和不带阻尼时的强退化性，在20世纪80年代之前国内外几乎没有从数学上进行理论研究的成果出现。最先进行研究的，当属周毓麟院士和郭老师，他们在1982年到1986年间，采用Leray－Schauder不动点定理、离散方法、Galerkin方法证明了从一维到高维的各种边值问题整体弱解的存在性，比国外在1992年才出现的同类结果早了将近10年。

20世纪90年代初期，周毓麟、郭柏灵和谭绍滨，郭柏灵和洪敏纯得到了两个在国内外至今影响很大的经典结果。第一，通过差分方法结合粘性消去法，利用十分巧妙的先验估计，证明了一维Landau－Lifshitz方程光滑解的存在唯一性,对于一维问题给出了完整的答案,解决了长期悬而未决的难题。第二,系统分析了带阻尼的二维 Landau－Lifshitz 方程弱解的奇性,发现了 Landau－Lifshitz 方程与调和映照热流的联系,其弱解具有与调和映照热流完全相同的奇性。现在,国内外这方面的文章基本上都引用这个结果。调和映照的 Landau－Lifshitz 流的概念,即是源于此项结果。

20 世纪 90 年代中期,郭老师对于 Landau－Lifshitz 方程的长时间性态、Landau－Lifshitz 方程耦合 Maxwell 方程的弱解及光滑解的存在性问题进行了深入的研究,得到了一系列的成果。铁磁链方程的退化性以及缺少相应的线性化方程解的表达式,对研究解的长时间性态带来很大困难。郭老师的一系列成果克服了这些困难,证明了近似惯性流形的存在性、吸引子的存在性,给出了其 Hausdorff 和分形维数的上、下界的精细估计。此外,我们知道,与调和映照热流比较,高维铁磁链方程的研究至今还很不完善。其中最重要的是部分正则性问题,其难点在于单调不等式不成立,导致能量衰减估计方面的困难。另外一个是 Blow-up 解的存在性问题,至今没有解决;而对于调和映照热流来说,这样的问题的研究是比较成熟的。

对于高维问题,20 世纪 90 年代后期至今,郭老师和陈韵梅、丁时进、韩永前、杨干山一道,得出了许多成果,大大地推动了该领域的研究。首先,证明了二维问题的能量有限弱解的几乎光滑性及唯一性,这个结果类似于 Freire 关于调和映照热流的结果。第二,得到了高维 Landau－Lifshitz 方程初边值问题的奇点集合的 Hausdorff 维数和测度的估计。第三,得到了三维 Landau－Lifshitz－Maxwell 方程的奇点集合的 Hausdorff 维数和测度的估计。第四,得到了一些高维轴对称问题的整体光滑解和奇性解的精确表达式。郭老师还开创了一些新的研究领域。例如,关于一维非均匀铁磁链方程光滑解的存在唯一性结果后来被其他数学家引用并推广到一般流形上。其次,率先讨论了可压缩铁磁链方程测度值解的存在性。最近,在 Landau－Lifshitz 方程耦合非线性 Maxwell 方程方面,也取得了许多新的进展。

多年来,郭老师还对一大批非线性发展方程解的整体存在唯一性、有限时刻的爆破性、解的渐近性态等开展了广泛而深入的研究,受到国内外同行的广泛关注。研究的模型源于数学物理、水波、流体力学、相变等领域,如含磁场的 Schrödinger 方程组、Zakharov 方程、Schrödinger－Boussinesq 方程组、Schrödinger－KdV 方程组、长短波方程组、Maxwell 方程组、Davey－Stewartson 方程组、Klein－Gordon－Schrödinger 方程组、波动方程、广义 KdV 方程、Kadomtsev－Petviashvili(KP)方程、Benjamin－Ono 方程、Newton－Boussinesq 方程、Cahn－Hilliard 方程、Ginzburg－Landau 方程等。其中不少耦合方程组都是郭老师得到了第一个结果,开创了研究的先河,对国内外同行的研究产生了深远的影响。

郭老师在无穷维动力系统方面也开展了广泛的研究,取得了丰硕的成果。对耗散非线性发展方程所决定的无穷维动力系统,研究了整体吸引子的存在性、分形维数估计、惯性流形、近似惯性流形、指数吸引子等问题。特别是在研究无界域上耗散非线性发展方程的强紧整体吸引子存在性时所提出的化弱紧吸引子成为强紧吸引子的重要方法和技巧,颇受同行关注并广为利用。对五次非线性 Ginzburg－Landau 方程,郭老师利用空间离散化方法将无限维问题化为有限维问题,证明了该问题离散吸引子的存在性,并考虑五次 Ginzburg－Landau 方程的定态解、慢周期解、异宿轨道等的结构。利用有限维动力系统的理论和方法,结合数值计算得到具体的分形维数(不超过4)和结构以及走向混沌、湍流的具体过程和图像,这是一种寻求整体吸引子细微结构新的探索和尝试,对其他方程的研究也是富有启发性的。1999 年以来,郭老师集中于近可积耗散的和 Hamilton 无穷维动力系统的结构性研究,利用孤立子理论、奇异摄动理论、Fenichel 纤维理论和无穷维 Melnikov 函数,对于具有小耗散的三次到五次非线性 Schrödinger 方程,证明了同宿轨道的不变性,并在有限维截断下证明了 Smale 马蹄的存在性,目前,正把这一方法应用于具小扰动的 Hamilton 系统的研究上。他对于非牛顿流无穷维动力系统也进行了系统深入的研究,建立了有关的数学理论,并把有关结果写成了专著。以上这些工作得到国际同行们的高度评价,被称为“有重大的国际影响”、“对无穷维动力系统理论有重要持久的贡献”。最近,郭老师及其合作者又证明了具耗散的 KdV 方程 $L^2$ 整体吸引子的存在性,该结果也是引人注目的。

郭老师不仅自己辛勤地搞科研,还尽心尽力培养了大批的研究生(硕士生、博士生、博士后),据不完全统计,有 40 多人。他根据每个人不同的学习基础和特点,给予启发式的具体指导,其中的不少人已成为了该领域的学科带头人,有些人虽然开始时基础较差,经过培养,也得到了很大提高,成为该方向的业务骨干。

《郭柏灵论文集》按照郭老师在不同时期所从事的研究领域,分成多卷出版。文集中所搜集的都是郭老师正式发表过的学术成果。把这些成果整理成集出版,不仅系统地反映了他的科研成就,更重要的是对于从事这方面学习、研究的学者无疑大有裨益。这本文集的出版得到了多方面的帮助与支持,特别要感谢华南理工大学校长李元元教授、华南理工大学出版社范家巧社长和华南理工大学数学科学学院吴敏院长的支持。还要特别感谢华南理工大学的李用声教授、华南师范大学的丁时进教授、北京应用物理与计算数学研究所的苗长兴研究员等人在论文的搜集、选择与校对等工作中付出了辛勤的劳动。感谢华南理工大学出版社的编辑对文集的精心编排。

谭绍滨

2005 年 8 月于厦门大学

# 目　录

## 2008 年



## 2009 年

# Attractor for Nonlinear Schrödinger Equation Coupling with Stochastic Weakly Damped, Forced KdV Equation *

Guo Boling(郭柏灵) Wang Guolian(王国联)

**Abstract**

The nonlinear Schrödinger equation coupling with stochastic weakly damped, forced KdV equation with additive noise can be solved pathwise, and the unique solution generates a random dynamical system. Then we prove that the system possesses a global weak random attractor.

**Keywords** Korteweg-de Vries (KdV) equation; nonlinear Schrödinger equation; additive noise

## 1 Introduction

In this paper, we are concerned with the long time behavior of the system of the nonlinear Schrödinger equation

$$\mathrm{i}m_t + m_{xx} - bum + \mathrm{i}\gamma m = g_1(x), \tag{1.1}$$

coupling with a weakly damped forced Korteweg-de Vries (KdV) equation with a random potential

$$u_t + \frac{\beta}{2}u_{xxx} + \nu u + \frac{1}{2}|m|_x^2 = g_2(x) + \Phi\frac{\partial^2 B}{\partial t\partial x}, \tag{1.2}$$

where

$$m(x,t) = (m^1(x,t), m^2(x,t), \cdots, m^k(x,t))$$

is a complex valued vector process, $u(x,t)$ is a real scalar process defined on $(t,x)\in \mathbf{R}_+\times\mathbf{R}$, $b$, $\beta$, $\gamma$, $\nu$ are real constants, and $\Phi$ is a linear operator. Also, $B$ is a two-parameter Brownian motion on $\mathbf{R}_+\times\mathbf{R}$, that is, a zero mean Gaussian process whose correlation function is given by

$$E(B(t,x)B(s,y)) = (t\wedge s)(x\wedge y),$$

for $t,s\geqslant 0, x,y\in\mathbf{R}$.

The stochastic KdV equation arises when modeling the propagation of weakly nonlinear waves in a noisy plasma[4,14,17]. It is also of interest in any circumstance when the KdV equation is used, since the stochastic may represent terms that have been neglected in the derivation of this ideal model. Note also that many articles have been devoted to the derivation of a forced KdV equation (see Ref. [11] and reference therein). In that case it can sometimes

* Front Math China, 2008, 3(4): 495-510.

be reasonable to assume that the forcing contains a stochastic part. In Refs. [7,8], the well-posedness of a stochastic KdV equation with an additive noise was studied.

The periodic initial value problem for the deterministic system of a nonlinear Schrödinger equation coupling with a weakly damped, forced KdV equation (i.e., system (1.1)~(1.2) as $\Phi=0$) was raised in plasma physics. The soliton solutions for the system when dissipation is zero have been studied in Refs. [1,10,15], and the existence and uniqueness of the global solution was studied in Ref. [13].

Note that the covariance operator $\Phi$ can, in general, be described by a kernel $\mathscr{K}(x,y)$. The correction function of the noise is then given by

$$E\left(\Phi\frac{\partial^2 B}{\partial t\partial x}(t,x)\Phi\frac{\partial^2 B}{\partial t\partial x}(s,y)\right)=c(x,y)\delta_{t-s},$$

where $\delta$ is the Dirac function and

$$c(x,y)=\int_{\mathbf{R}}\mathscr{K}(x,z)\mathscr{K}(y,z)\mathrm{d}z.$$

Alternatively, we can consider a cylindrical Wiener process $W$ by setting

$$W(t)=\frac{\partial B}{\partial x}=\sum_{i=0}^{\infty}\beta_i e_i,$$

where $(e_i)_{i\in\mathbf{N}}$ is an orthonormal basis of $L^2(\mathbf{R})$ and $(\beta_i)_{i\in\mathbf{N}}$ is a sequence of mutually independent real Brownian motions in a fixed probability space $(\Omega,\mathscr{F},P)$ adapted to a filtration $\{\mathscr{F}_t\}_{t\geqslant 0}$.

Let us write the Itô form of (1.2):

$$\mathrm{d}u+\left(\frac{\beta}{2}u_{xxx}+\nu u+\frac{1}{2}|m|_x^2\right)\mathrm{d}t=g_2(x)\mathrm{d}t+\Phi\mathrm{d}W. \tag{1.3}$$

We consider (1.1) and (1.3) for $x\in\mathbf{R}$ with the initial conditions

$$m(0)=m_0,\quad u(0)=u_0. \tag{1.4}$$

The aim of this paper is to study the long time behavior of the stochastic systems (1.1), (1.3) and (1.4). We know that attractors have been quite well investigated so as to describe the long time behavior for deterministic systems. Temam[18] gives a comprehensive presentation. In Ref. [12], the author studied the long time behavior of the periodic initial value problem for the nonlinear Schrödinger equation coupling with a weakly damped, forced KdV equation

$$\mathrm{i}m_t+m_{xx}-bum+\mathrm{i}\gamma m=g_1(x),$$

$$u_t+\frac{1}{2}f(u)_x+\frac{\beta}{2}u_{xxx}+\nu u+\frac{1}{2}|m|_x^2=g_2(x),$$

and the finite dimensional attractor was obtained. However, all the analysis breaks down as soon as one wants to take random influences on the system under investigation into account. In particular, when subjecting the system to additive noise, there is no chance that bounded subsets of the state space remain invariant. White noise pushes the system out of every bounded set with probability one.

Recently, the notion of random attractors for a stochastic dynamical system has been introduced in Refs. [5,6]. Random attractors are, in fact, compact invariant sets, however, they are not fixed, they depend on chance, and they move with time (in a coherent, 'stationary' manner). This seems to be a good generalization of the classical concept of global attractors for deterministic dynamical systems, and this has been successfully applied to many infinite dimensional stochastic dynamical systems (see Refs. [5,6,19]).

This paper is arranged as follows. In Sect. 2, we first recall some relevant concepts and results of random dynamical systems. Then we construct a random dynamical system for problems (1.1), (1.3) and (1.4). In Sect. 3, we first give some *a priori* estimates. Using them, we obtain the main result, i.e., there exists a weak random attractor for system (1.1), (1.3) and (1.4).

# 2 Preliminaries

## 2.1 Preliminaries on Random Dynamical Systems

We now recall some concepts and results from Refs. [2,5,6].

Let $(\Omega, \mathcal{F}, P)$ be a probability space and $\{\theta_t: \Omega \to \Omega, t \in \mathbf{R}\}$ a family of measure-preserving transformations, such that $(t, \omega) \mapsto \theta_t\omega$ is measurable, $\theta_0 = \mathrm{id}$, and $\theta_{t+s} = \theta_t \circ \theta_s$ for all $s, t \in \mathbf{R}$. The flow $\theta_t$ together with the corresponding probability space $(\Omega, \mathcal{F}, P)$ is called a metric dynamical system.

A continuous random dynamical system (RDS) on a polish space $(X, d)$ with Borel sigma-algebra $\mathcal{B}$ over $\theta_t$ on $(\Omega, \mathcal{F}, P)$ is by definition a measurable map

$$\phi: \mathbf{R}_+ \times \Omega \times X \to X,$$
$$(t, \omega, x) \mapsto \phi(t, \omega)x,$$

such that $P$-a.s.:

(i) $\phi(0, \omega) = \mathrm{id}$ on $X$;

(ii) $\phi(t+s, \omega) = \phi(t, \theta_s\omega)\phi(s, \omega)$ for all $s, t \in \mathbf{R}_+$; (cocycle property)

(iii) $\phi(t, \omega): X \to X$ is continuous.

A set-valued map $K: \Omega \to 2^X$, the set of all subsets of $X$, is called a random compact set if $K(\omega)$ is a compact $P$-almost surely and if $\omega \mapsto d(x, K(\omega))$ is measurable with respect to $\mathcal{F}$ for every $x \in X$, where

$$d(x, M) := \inf_{y \in M} d(x, y).$$

Let $\mathcal{A}(\omega)$ be a random set and $B \subset X$. One says that $\mathcal{A}(\omega)$ attracts $B$ if

$$\lim_{t \to \infty} d(\phi(t, \theta_{-t}\omega)B, \mathcal{A}) = 0, \quad P\text{-a.s.}$$

A random set $\mathcal{A}(\omega)$ is said to be a random attractor for the RDS $\phi$ if $P$-a.s.:

(i) $\mathcal{A}(\omega)$ is a random compact set;

(ii) $\mathcal{A}(\omega)$ is invariant, that is, $\phi(t, \omega)\mathcal{A}(\omega) = \mathcal{A}(\theta_t\omega)$ for all $t > 0$;

(iii) $\mathcal{A}(\omega)$ attracts all deterministic bounded sets $B \subset X$.

Similar to the deterministic theory, the existence result of random attractors can be stated as follows (see Refs. [5,6]).

**Theorem 2.1** If there exists a random compact set absorbing every bounded set $B\subset X$, then the RDS $\phi$ possesses a random attractor $\mathscr{A}(\omega)$,

$$\mathscr{A}(\omega)=\overline{\bigcup_{B\subset X}\Lambda_B(\omega)},$$

where

$$\Lambda_B(\omega):=\bigcap_{s\geqslant 0}\overline{\bigcup_{t\geqslant s}\phi(t,\theta_{-t}\omega)B}$$

is the omega-limit set of $B$.

## 2.2 Solve Equations and Generate an RDS

For the mathematical setting of the problem, let

$$IH^m=H^m(\mathbf{R})\times\cdots\times H^m(\mathbf{R})$$

denote the product of $k$ complex Hilbert spaces $H^m(\mathbf{R})$ endowed with the norm

$$\|v\|_m=\Big(\sum_{j=1}^{k}\|v_j\|_m^2\Big)^{1/2},\quad v=(v^1,v^2,\cdots,v^k)\in IH^m.$$

Let

$$X_k=IH^k\times H^k(\mathbf{R}),\quad k=0,1,2,\cdots,m.$$

The first factor in the product is complex and the second is real with the usual norm

$$\|\xi\|_k=(\|\xi_1\|_k^2+\|\xi_2\|_k^2)^{1/2},\quad \xi=(\xi_1,\xi_2)\in X_k,\quad k=1,2,\cdots,m.$$

Denote by $(\cdot,\cdot)$ and $\|\cdot\|_0$ the inner product and the norm in $L^2(\mathbf{R})$, respectively. And denote by $\|\cdot\|_X$ the norm of the Banach space $X$.

Given two separable Hubert spaces $\overline{H}$ and $\widetilde{H}$, we denote by $L_2^0(\overline{H};\widetilde{H})$ the space of Hibert-Schmidt operators from $\overline{H}$ into $\widetilde{H}$. Its norm is given by

$$\|\Phi\|^2_{L_2^0(\overline{H};\widetilde{H})}=\sum_{i\in\mathbf{N}}\|\Phi e_i\|^2_{\widetilde{H}},$$

where $(e_i)_{i\in\mathbf{N}}$ is any orthonormal basis of $\overline{H}$. When

$$\overline{H}=L^2(\mathbf{R}),\quad \widetilde{H}=H^m(\mathbf{R}),$$

$L_2^0(L^2(\mathbf{R});H^m(\mathbf{R}))$ is simply denoted by $L_2^{0,m}$.

Throughout this paper, $C$ represents different constants from line to line.

We consider the following linear problem (P):

$$\begin{cases}\mathrm{d}z+\Big(\dfrac{\beta}{2}z_{xxx}+\nu z\Big)\mathrm{d}t=\Phi\mathrm{d}W\\ z(0)=0\end{cases},$$

whose solution is given by the stochastic integral

$$z(t)=\int_0^t U(t-s)\Phi\mathrm{d}W(s),$$

where

$$U(t)=\mathscr{F}_x^{-1}\mathrm{e}^{t\left(\mathrm{i}\frac{\beta}{2}\xi^3-\nu\right)}\mathscr{F}_x$$

is the unitary operator associated with the linear problem (P).

Let us introduce the following space to study the solution of problem (P):

$$X_\sigma(T) = \{u \in C(0,T;H^\sigma(\mathbf{R})) \cap L^2((\mathbf{R});L^\infty([0,T])),$$
$$D^\sigma \partial_x u \in L^\infty((\mathbf{R});L^2([0,T])), \partial_x u \in L^4([0,T];L^\infty(\mathbf{R}))\}$$

for some $\sigma < 1$.

We now give some properties about the solution of the linear problem (P), which will be useful in the next section to get some *a priori* estimates.

**Lemma 2.2**[7] Assume that $\Phi \in L_2^{0,\bar\sigma}$ for some constant $\bar\sigma > 3/4$. Then $z$ is almost surely in $X_\sigma(T)$ for any $T>0$ and any $\sigma \in (3/4, \bar\sigma)$. More precisely,

$$E\Big(\sup_{t\in[0,T]} \|z\|_{H_x^\sigma}^2\Big) \leqslant C\|\Phi\|_{L_2^{0,\sigma}}^2 \quad (3/4 < \sigma \leqslant \bar\sigma), \tag{2.1}$$

$$E\left(\left(\int_0^T \sup_{x\in\mathbf{R}} |\partial_x z|^4 \mathrm{d}t\right)^{1/2}\right) \leqslant C\|\Phi\|_{L_2^{0,\bar\sigma}}^2, \tag{2.2}$$

$$E\left(\int_{\mathbf{R}} \sup_{t\in[0,T]} |z|^2 \mathrm{d}t\right) \leqslant C(\bar\sigma)\|\Phi\|_{L_2^{0,\bar\sigma}}^2. \tag{2.3}$$

Let $0 < \varepsilon < \inf\{\bar\sigma, 2\}$. Then

$$E\left(\sup_{x\in\mathbf{R}} \int_0^T |D^{\bar\sigma-\varepsilon}\partial_x z|^2 \mathrm{d}t\right) \leqslant C(\varepsilon)\|\Phi\|_{L_2^{0,\bar\sigma}}^2. \tag{2.4}$$

We also have

$$E\left(\left(\int_0^T \sup_{x\in\mathbf{R}} |\partial z|^4 \mathrm{d}t\right)^{1/2}\right) \leqslant \|\Phi\|_{L_2^{0,\bar\sigma}}^2, \tag{2.5}$$

where $C(\varepsilon)$ and $C(\bar\sigma)$ depend on $\varepsilon$ and $\bar\sigma$, respectively.

**Remark** We have to impose a stronger condition on $\Phi \in L_2^{0,2}$ in the present paper than that on $\Phi \in L_2^{0,1}$ in Ref. [7]. Thus, the solution $z$ of the linear problem (P) is more regular, which will be used in proving the boundedness of $u$ in $H^1(\mathbf{R})$. More precisely, we can get

$$E\left(\left(\int_0^T \sup_{x\in\mathbf{R}} |\partial_x^2 z|^4 \mathrm{d}t\right)^{1/2}\right) \leqslant C\|\Phi\|_{L_2^{0,\bar\sigma}}^2, \tag{2.6}$$

$$E\left(\sup_{x\in\mathbf{R}} \int_0^T |D\partial_x z|^2 \mathrm{d}t\right) \leqslant C\|\Phi\|_{L_2^{0,\bar\sigma}}^2. \tag{2.7}$$

Let us study systems (1.1), (1.3) and (1.4) by means of the change of variable

$$n(t) = u(t) - z(t). \tag{2.8}$$

Then $(m,u)$ satisfies (1.1), (1.3) and (1.4) if and only if $(m,n)$ is a solution of

$$\mathrm{i}m_t + m_{xx} - b(n+z)m + \mathrm{i}\gamma m = g_1(x), \tag{2.9}$$

$$n_t + \frac{\beta}{2} n_{xxx} + \nu n + \frac{1}{2}|m|_x^2 = g_2(x). \tag{2.10}$$

$$m(0) = m_0, \quad n(0) = n_0. \tag{2.11}$$

This system with random parameters can be solved pathwise. For the existence of the solution of (2.9)~(2.11), we have the following theorem.

**Theorem 2.3** Suppose that

$$\Phi \in L_2^{0,m}, \quad g = (g_1, g_2) \in X_m, \quad (m_s, n_s) \in X_m, \quad m \geqslant 2.$$

Then for every $T\in(0,\infty)$, there exists a unique solution $(m(t),n(t))\in C(0,T;X_n)$ of problem (2.9)~(2.11) and the mapping

$$X_m\to C(0,T;X_m),$$
$$(m_s,n_s)\mapsto(m(t),n(t))$$

restrict to $[0,T]$ is continuous and bounded.

Exploiting the Galerkin approximation and *a priori* estimates similar to Refs. [3,13], for $P$-a.s. $\omega\in\Omega$, we can prove Theorem 2.3.

We summarize the above existence results for $P$-a.s. $\omega\in\Omega$ of system (2.9)~(2.10) with the initial conditions

$$m(s,x)=m_s,\quad n(s,x)=n_s=u_s-z(s),\quad s\in\mathbf{R}$$

as follows:

(ⅰ) Under the assumption of Theorem 2.3, for $s<T$ (any $T\in\mathbf{R}$) and $(m_s,n_s)\in X_2$, there exists a unique solution $v\in C([s,T];X_2)$;

(ⅱ) Denoting such a solution $(m(t,\omega;s,m_s),n(t,\omega;s,v_s))$, the mapping $(m_s,n_s)\mapsto(m(t,\omega;s,m_s),n(t,\omega;s,v_s))$ is continuous for all $s\leqslant T$.

Now we construct an RDS modeling nonlinear Schrödinger equation coupling with a stochastic weakly damped, forced KdV equation. For example, consider the set of continuous functions with value 0 at 0,

$$\Omega=\{\omega\in C(\mathbf{R},\mathbf{R}):\ \omega(0)=0\}.$$

Let $\mathscr{F}$ be the Borel $\sigma$-algebra induced by the compact-open topology of $\Omega$, and let $P$ be a Wiener measure on $(\Omega,\mathscr{F})$. we write

$$(\beta_1(t,\omega),\cdots,\beta_k(t,\omega),\cdots)=\omega(t).$$

The time shift is simply defined by

$$\theta_s\omega(t)=\omega(t+s)-\omega(s),\quad t,s\in\mathbf{R}. \tag{2.12}$$

Then $(\Omega,\mathscr{F},P,\theta_t)$ is an ergodic metric dynamical system which models white noise.

Let

$$(m(t,\omega),n(t,\omega))=\sum(t,s;\omega)(m_s,n_s),$$
$$(m(t,\omega),u(t,\omega))=S(t,s;\omega)(m_s,n_s),$$

where $(m(t,\omega),n(t,\omega))$ is a solution to (2.9)~(2.11) and $u(t)$ satisfies (1.1), (1.3) and (1.4). Obviously, for $s\leqslant r\leqslant t$, we have

$$S(t,s;\omega)=S(t,r;\omega)S(r,s;\omega).$$

By (2.12), for any $s,t\in\mathbf{R}_+,(m_0,u_0)\in X_m$, we have $P$-a.s.

$$S(t+s,0;\omega)(m_0,u_0)=S(t,0;\theta_s\omega)S(s,0;\omega)(m_0,u_0).$$

Therefore, the process $\phi$: $\mathbf{R}_+\times\Omega\times X_m\to X_m$, defined by

$$\phi(t,\omega)(m_0,u_0)=S(t,0;\omega)(m_0,u_0),$$

is a cocycle. It is continuous on $X_m$ over $(\Omega,\mathscr{F},P,(\theta_t)_{t\in\mathbf{R}}$ and models the continuous random dynamical system associated with problems (1.1), (1.3) and (1.4).

# 3 Random Attractor

In the following computations, $\omega \in \Omega$ is fixed. The results will hold $P$-a. s. We usually denote $L^P(\mathbf{R})$ $(1 \leqslant p \leqslant \infty)$ by $L_x^p$ in this section.

**Proposition 3.1** Assume that $(m_s, u_s) \in X_0$. Given any ball of $X_0, B(0,\rho)$, centered at 0 of radius $\rho$, there exist $r_0$ and $\bar{t}_0(\rho) \leqslant -1$, such that for any $s \leqslant \bar{t}_0(\rho) \leqslant -1$,

$$\| m(0) \|_0 \leqslant r_0(\omega), \quad \| m(-1) \|_0 \leqslant r_0(\omega), \quad P\text{-a.s.}, \tag{3.1}$$

where $(m(t), n(t))$ is a solution to system (2.9)~(2.11).

**Proof** Taking the inner product of (2.9) with $m$ in $IH = L^2(\mathbf{R}) \times \cdots \times L^2(\mathbf{R})$ of $k$ complex Hilbert spaces $L^2(\mathbf{R})$, we have

$$(\mathrm{i}m_t, m) - \| m_x \|_0^2 - b((n+z)m, m) + \mathrm{i}\gamma \| m \|_0^2 = (g_1, m). \tag{3.2}$$

The imaginary part of (3.2) reads

$$\frac{1}{2}\frac{\mathrm{d}}{\mathrm{d}t}\| m \|_0^2 + \gamma \| m \|_0^2 = \mathrm{Im}\int g_1 \bar{m}\,\mathrm{d}x.$$

Here and below, whenever the limits of integration are omitted, spatial integration over $\mathbf{R}$ is understood.

Using Hölder inequality to the above equality, we have

$$\frac{\mathrm{d}}{\mathrm{d}t}\| m \|_0^2 + \gamma \| m \|_0^2 \leqslant \gamma \| g_1 \|_0^2.$$

For $0 \geqslant t \geqslant s$, by Gronwall's Lemma,

$$\| m(t) \|_0^2 \leqslant \| m(s) \|_0^2 \mathrm{e}^{-\gamma(t-s)} + \frac{1}{\gamma}\int_s^t \mathrm{e}^{-\gamma(t-\tau)} \| g_1 \|_0^2 \mathrm{d}\tau. \tag{3.3}$$

Let $t = 0, -1$. For $(m_s, u_s) \in B(0,\rho) \subset X_0$, there exists $\bar{t}_0(\rho) \leqslant -1$, such that for all $s \leqslant \bar{t}_0(\rho)$,

$$\mathrm{e}^{\gamma s}\rho^2 < 1, \quad \mathrm{e}^{\gamma(1+s)}\rho^2 < 1.$$

Choosing

$$r_0^2 = 1 + \frac{C}{\gamma^2}\| g_1 \|_0^2,$$

we get (3.1).

**Proposition 3.2** Suppose that $\Phi \in L_2^{0,1}, (m_s, n_s) \in X_1$ and $\nu > \gamma$. Then, for any ball of $X_1, B(0,\rho)$, centered at 0 of radius $\rho$, and for any solution $(m(t), n(t))$ of (2.9)~(2.11), there exist constants $C_1, C_2, C_3$ (given in the proof), such that for any $s \leqslant t \leqslant 0$,

$$\| n(t) \|_0^2 \leqslant C_1 + C_2 \mathrm{e}^{-2\gamma(t-s)} + C_3(1 - \mathrm{e}^{-8\gamma(t-s)/3})^{3/4}. \tag{3.4}$$

**Proof** Taking the inner product of (2.9) with $m_x$ in $IH$, the real part reads

$$\mathrm{Im}(m_t, m_x) + \mathrm{Re}(b(n+z)m, m_x) + \mathrm{Im}(\mu m, m_x) = -\mathrm{Re}(g_1, m_x). \tag{3.5}$$

Taking the inner product of (2.10) with $n$, we get

$$\frac{1}{2}\frac{\mathrm{d}}{\mathrm{d}t}\| n \|_0^2 + \nu \| n \|_0^2 + \frac{1}{2}(| m |_x^2, n) = (g_2, n). \tag{3.6}$$

Applying (3.5), we get

$$\frac{1}{2}|b|(|m|_x^2,n)=\operatorname{sgn}b\operatorname{Re}(bnm,m_x)$$
$$=\operatorname{sgn}b\{-\operatorname{Im}(m_t,m_x)-\operatorname{Im}(\gamma m,m_x)-\operatorname{Re}(zm,m_x)-\operatorname{Re}(g_1,m_x)\}.$$

We also have

$$\operatorname{Im}(m_t,m_x)=\frac{1}{2}\frac{\mathrm{d}}{\mathrm{d}t}\operatorname{Im}\int m\bar{m}_x\mathrm{d}x,\quad -\operatorname{Re}(zm,m_x)=\frac{1}{2}(z_xm,m).$$

Combining the above two equalities with (3.6), we get

$$\frac{1}{2}\frac{\mathrm{d}}{\mathrm{d}t}\int\{|b|n^2-\operatorname{sgn}b\operatorname{Im}(m\bar{m}_x)\}\mathrm{d}x+\nu|b|\|n\|_0^2-\gamma\int\operatorname{sgn}b\operatorname{Im}(m\bar{m}_x)\mathrm{d}x$$
$$=|b|(g_2,n)+\operatorname{sgn}b\{\operatorname{Re}(g_1,m_x)-\frac{1}{2}(z_xm,m)\}.$$

Noting that $\nu>\gamma$ and using Hölder inequality, we deduce

$$\frac{\mathrm{d}}{\mathrm{d}t}\int\{|b|n^2-\operatorname{sgn}b\operatorname{Im}(m\bar{m}_x)\}\mathrm{d}x+2\gamma\int\{|b||n|^2-\operatorname{sgn}b\operatorname{Im}(m\bar{m}_x)\}\mathrm{d}x$$
$$\leqslant\frac{|b|^2|g_2|_0^2}{2(\nu-\gamma)}+\|g_1x\|_0^2+(\|z_x\|_{L_x^\infty}+1)\|m\|_0^2.$$

By Gronwall's Lemma, we deduce from the above inequality that

$$|b|\|n(t)\|_0^2-\int\operatorname{sgn}b\operatorname{Im}(m(t)\bar{m}_x(t))\mathrm{d}x$$
$$\leqslant\left\{|b|\|n(s)\|_0^2-\int\operatorname{sgn}b\operatorname{Im}(m(s)\bar{m}_x(s))\mathrm{d}x\right\}\mathrm{e}^{-2\gamma(t-s)}+$$
$$\int_s^t\left\{\frac{|b|^2|g_2|_0^2}{2(\nu-\gamma)}+\|g_{1x}\|_0^2+(\|z_x\|_{L_x^\infty}+1)\|m\|_0^2\right\}\mathrm{e}^{-2\gamma(t-\tau)}\mathrm{d}\tau.$$

Noting (2.5), let

$$C_1=\|n_s\|_0^2+2|b|^{-1}\|m_s\|_0\|m_s\|_1,$$
$$C_2=|b|^{-1}\frac{1}{2\gamma}\left(\frac{|b|^2|g_2|_0^2}{2(\nu-\gamma)}+\|(g_1)_x\|_0^2+\|m\|_0^2\right),$$
$$C_3=\|\Phi\|_{L_2^{0,\sigma}}\left(\frac{3}{8\gamma}\right)^{3/4}.$$

Then we get (3.4).

**Proposition 3.3** Suppose that $b\beta<0,\Phi\in L_2^{0,2}$, and $(m_s,u_s)\in X_1$. Then, for any ball of $X_1$, $B(0,\rho)$, centered at 0 of radius $\rho$, and for any solution $(m(t),u(t))$ of (1.1), (1.3) with initial conditions $m(s)=m_s$, $u(s)=u_s$, there exist constants $r_1(\omega)$ and $\bar{t}_2(\rho)\leqslant-1$, such that for any $s\leqslant\bar{t}_2(\rho)\leqslant-1$,

$$\|(m(-1),u(-1))\|_1\leqslant r_1(\omega).\tag{3.7}$$

**Proof** Multiplying (2.9) by $m_t$ and integrating on $\mathbf{R}$ and taking the real part, we get

$$\frac{\mathrm{d}}{\mathrm{d}t}\|m_x\|_0^2+\int bn\frac{\mathrm{d}}{\mathrm{d}t}|m|^2\mathrm{d}x+2\operatorname{Im}(\gamma m,m_t)=-2\operatorname{Re}(g_1,m_t)-\operatorname{Re}(zm,m_t).\tag{3.8}$$

Reading the real part of (3.2), we get

$$\operatorname{Im}(m_t, m) + \| m_x \|_0^2 + \int bn \,|m|^2 \mathrm{d}x = -\operatorname{Re}(g_1, m) - \int bz \,|m|^2 \mathrm{d}x. \tag{3.9}$$

Taking the inner product of (2.10) with $|m|^2$ and $n_{xx}$, respectively, we obtain

$$\int |m|^2 n_t \mathrm{d}x + \frac{\beta}{2}\int |m|^2 n_{xxx} \mathrm{d}x + \nu \int |m|^2 n \mathrm{d}x = \int |m|^2 g_2 \mathrm{d}x, \tag{3.10}$$

$$-\frac{1}{2}\frac{\mathrm{d}}{\mathrm{d}t}\| n_x \|_0^2 - \nu \| n_x \|_0^2 + \frac{1}{2}(|m|_x^2, n_{xx}) = (g_2, n_{xx}). \tag{3.11}$$

From (3.8)~(3.11), we have

$$\begin{aligned}&\frac{\mathrm{d}}{\mathrm{d}t}\left\{ \| m_x \|_0^2 + b\int n\,|m|^2 \mathrm{d}x - \frac{b\beta}{2}\| n_x \|_0^2 \right\} + \\ &2\gamma\left\{ \| m_x \|_0^2 + \operatorname{Re}(zg_1, m) + b\int (n+z)|m|^2 \mathrm{d}x \right\} - \\ &\nu b\beta \| n_x \|_0^2 + b\nu \int n\,|m|^2 \mathrm{d}x - b(g_2, |m|^2) + \\ &b\beta(g_{2x}, n_x) + 2\operatorname{Re}(g_1, m_t) + \operatorname{Re}(zm, m_t) = 0.\end{aligned} \tag{3.12}$$

Notice that

$$\operatorname{Re}(g_1, m_t) = -\operatorname{Im}((g_1)_x, m_x) - \operatorname{Im}(g_1, b(n+z)m) - \operatorname{Re}(g_1, \gamma m), \tag{3.13}$$

$$\operatorname{Re}(zm, m_t) = -\operatorname{Im}(z_x m, m_x) - \gamma(zm, m) - \operatorname{Re}(zm, g_1). \tag{3.14}$$

Combining (3.6), (3.13) and (3.14) with (3.12), we find

$$\frac{\mathrm{d}}{\mathrm{d}t}H(m,n) + K(m,n) = 0, \tag{3.15}$$

where

$$H(m,n) = \| m_x \|_0^2 + b\int n\,|m|^2 \mathrm{d}x - \frac{b\beta}{2}\| n_x \|_0^2 + \lambda \| n \|_0^2, \tag{3.16}$$

$$\begin{aligned}K(m,n) = &2\gamma\left\{ \| m_x \|_0^2 + \operatorname{Re}(g_1, m) + b\int (n+z)|m|^2 \mathrm{d}x \right\} - \\ &\nu b\beta \| n_x \|_0^2 + b\nu \int n\,|m|^2 \mathrm{d}x - b(g_2, |m|^2) + b\beta(g_{2x}, n_x) + \\ &2(-\operatorname{Im}((g_1)_x, m_x) - \operatorname{Im}(g_1, b(n+z)m) - \operatorname{Re}(g_1, \gamma m)) + \\ &2b(-\operatorname{Im}(z_x m, m_x) - \gamma(zm, m) - \operatorname{Re}(zm, g_1)) + \\ &2\nu\lambda \| n \|_0^2 + \lambda \int n\,|m|_x^2 \mathrm{d}x - 2\lambda(g_2, n).\end{aligned} \tag{3.17}$$

By Agmon inequality and Hölder inequality, we have

$$\left| b\int n\,|m|^2 \mathrm{d}x \right| \leqslant \frac{2|b|^2}{3}\| n \|_0^2 + \frac{1}{4}\| m_x \|_0^2 + \frac{9}{64}\| m \|_0^6. \tag{3.18}$$

Therefore,

$$H(m,n) \geqslant \frac{3}{4}\| m_x \|_0^2 + \frac{|b\beta|}{2}\| n_x \|_0^2 + \left(\lambda - \frac{2|b|^2}{3}\right)\| n \|_0^2 - \frac{9}{64}\| m \|_0^6, \tag{3.19}$$

$$H(m,n) \leqslant \frac{5}{4}\| m_x \|_0^2 + \frac{|b\beta|}{2}\| n_x \|_0^2 + \left(\lambda + \frac{2|b|^2}{3}\right)\| n \|_0^2 + \frac{9}{64}\| m \|_0^6. \tag{3.20}$$

For $K(m,n)$ and $H(m,n)$, we have the following lemma.

**Lemma 3.4** Suppose that conditions of Proposition 3.3 hold. Then we have

$$K(m,n)-\gamma H(m,n)\geqslant 2(\nu-\gamma)\|n\|_0^2-\widetilde{K}_0, \tag{3.21}$$

where

$$\widetilde{K}_0=\widetilde{K}_1+\widetilde{K}_2+\widetilde{K}_3,$$

$$\widetilde{K}_1=C(\|g_{2x}\|_0^2+\|g_2\|_0^2+\|g_{1x}\|_0^2),$$

$$\widetilde{K}_2=C(\|m\|_0^6+\|g_2\|_0^2\|m\|_0^3+\|g_1\|_0^4\|m\|_0^2+\|m\|_0^2),$$

$$\widetilde{K}_3=C(\|z\|_{L_x^\infty}^2\|g_1\|_0^2+\|z_x\|_{L_x^\infty}^2\|m\|_0^2+\|z\|_{L_x^\infty}\|m\|_0^2+\|z\|_{L_x^\infty}^2\|g_1\|_0^2).$$

The proof of Lemma 3.4 is simple. Applying Agmon inequality and Hölder inequality on $-\gamma(3.16)+(3.17)$, and computing carefully, we can get Lemma 3.4.

Now, if we choose $\lambda$ large enough such that

$$\lambda>\frac{2|b|^2}{3}+|b|,$$

then from (3.15) and (3.21) we have

$$\frac{\mathrm{d}}{\mathrm{d}t}H(m,n)+\gamma H(m,n)\leqslant\widetilde{K}_0.$$

Therefore, by Gronwall's Lemma, for $s\leqslant t\leqslant 0$, we get

$$H(m(t),n(t))\leqslant H(m(s),n(s))\mathrm{e}^{-\gamma(t-s)}+\int_s^t\widetilde{K}_0\mathrm{e}^{-\gamma(t-\tau)}\mathrm{d}\tau. \tag{3.22}$$

Let $t=-1$. For $(m_s,u_s)\in B(0,\rho)\subset X_1$, there exists $\bar{t}_2(\rho)\leqslant -1$, such that for all $s\leqslant\bar{t}_0(\rho)$,

$$\mathrm{e}^{-\gamma(-1-s)}\rho^2<1.$$

It is easy to see that

$$\int_s^{-1}\widetilde{K}_1\mathrm{e}^{\gamma(1+\tau)}\mathrm{d}\tau\leqslant\int_\infty^{-1}\widetilde{K}_1\mathrm{e}^{\gamma(1+\tau)}\mathrm{d}\tau\leqslant C.$$

By (3.3), we have

$$\int_s^{-1}\|m(\tau)\|_0^2\mathrm{e}^{\gamma(1+\tau)}\mathrm{d}\tau$$

$$\leqslant\int_s^{-1}\left(\rho^2\mathrm{e}^{-\gamma(\tau-s)}+\frac{1}{\gamma}\int_s^\tau\mathrm{e}^{-\gamma(\tau-\delta)}\|g_1\|_0^2\mathrm{d}\delta\right)\mathrm{e}^{\gamma(1+\tau)}\mathrm{d}\tau\leqslant C.$$

By Lemma 2.2 and (3.3), we have

$$\int_s^{-1}\|z_x\|_{L_x^\infty}^2\|m(\tau)\|_0^2\mathrm{e}^{\gamma(1+\tau)}\mathrm{d}\tau$$

$$\leqslant\|\Phi\|_{L_2^{0,\sigma}}^2\left(\int_s^{-1}\|m(\tau)\|_0^4\mathrm{e}^{2\gamma(1+\tau)}\mathrm{d}\tau\right)^{1/2}\leqslant C.$$

Then we deduce that

$$\int_s^{-1}(\widetilde{K}_1+\widetilde{K}_2+\widetilde{K}_3)\mathrm{e}^{\gamma(1+\tau)}\mathrm{d}\tau\leqslant C.$$

By (3.19), (3.22) and the above inequality, we can choose

$$r_3^2(\omega)=1+C+\frac{|b\beta|}{2}\|z_x(-1)\|_0^2+\frac{9}{64}\|m(-1)\|_0^6.$$

Thus, we get (3.7).

**Proposition 3.5** Suppose that

$$\nu > \gamma, \quad b\beta < 0, \quad \Phi \in L_2^{0,2}, \quad (m_s, n_s) \in X_2.$$

Then for any ball of $X_2, B(0,\rho)$, centered at 0 of radius $\rho$, and for any solution $(m(t), n(t))$ of (2.9) ~ (2.10) with initial data $m(s) = m_s$, $n(s) = n_s$, there exist constants $r^3(\omega)$ and $\bar{t}_3(\rho) \leqslant 0$, such that for any $s \leqslant \bar{t}_3(\rho) \leqslant 0$,

$$\|(m(0), u(0))\|_2 \leqslant r_3(\omega). \tag{3.23}$$

**Proof** It suffices to estimate the uniform bound of $\|m_{xx}\|_0$ and $\|n_{xx}\|_0$. Let $A = -\partial_{xx}$. Taking the inner product of (2.9) with $Am_t + \gamma Am$, we read the real part

$$\frac{1}{2}\frac{\mathrm{d}}{\mathrm{d}t}\|Am\|_0^2 + b\gamma \mathrm{Re}\int (nm)_x \overline{m}_x \mathrm{d}x + \gamma\|Am\|_0^2 + b\mathrm{Re}(nm, Am_t)$$
$$= -\mathrm{Re}(zg_1, Am_t + \gamma Am) - \mathrm{Re}(zm, Am_t + \gamma Am). \tag{3.24}$$

Using invariants of the equations, we deduce that

$$\frac{1}{2}\beta\frac{\mathrm{d}}{\mathrm{d}t}\|An\|_0^2 + (|m|_{xx}^2, n_t)$$
$$= \beta(g_{2xx}, n_{xx}) - (|m|_{xx}^2, \nu n - g_2) - \nu\beta\|An\|_0^2. \tag{2.25}$$

Since

$$\mathrm{Re}(nm, Am_t) = -2\mathrm{Re}(n_x m_x, m_t) - \mathrm{Re}(n_{xx}m, m_t) - \mathrm{Re}(nm_{xx}, m_t),$$

and

$$(|m|_{xx}^2, n_t) = \frac{\mathrm{d}}{\mathrm{d}t}(|m|_{xx}^2, n) - (|m|_t^2, n)$$
$$= \frac{\mathrm{d}}{\mathrm{d}t}(|m|_{xx}^2, n) - 2\mathrm{Re}\int m\overline{m}_t n_{xx}\mathrm{d}x,$$

we deduce that

$$\frac{\mathrm{d}}{\mathrm{d}t}\widetilde{H}(m(t), n(t)) + 12\gamma\|Am\|_0^2 + 6\nu|b\beta|\,\|An\|_0^2 +$$
$$b(6\beta(g_{2xx}, n_{xx}) - 6(|m|_{xx}^2, \nu n - g_2)) + h_1 + h_2 = 0,$$

where

$$\widetilde{H}(m(f), n(t)) = 6\|Am\|_0^2 + 3|b\beta|\,\|An\|_0^2 + 6b(|m|_x^2, n_x),$$
$$h_1 = 12b\gamma\mathrm{Re}\int (mn)_x \overline{m}_x \mathrm{d}x - 24b\mathrm{Re}(n_x m_x, m_t),$$
$$h_2 = -12b\mathrm{Re}(nm_{xx}, m_t) + 12\mathrm{Re}(g_1, Am_t + \gamma A_m) +$$
$$12\mathrm{Re}(zm, Am_t + \gamma A_m).$$

Then we deduce that

$$\frac{\mathrm{d}}{\mathrm{d}t}\widetilde{H}(m(t), n(t)) + \gamma\widetilde{H}(m(t), n(t)) \leqslant \widetilde{K}_4,$$

where $\widetilde{K}_4$ depends on

$$\|m\|_0, \quad \|m_x\|_0, \quad \|n\|_0, \quad \|n_x\|_0,$$
$$\|g_1\|_0, \quad \|g_{1x}\|_0, \quad \|g_{1xx}\|_0, \quad \|g_2\|_0, \quad \|g_{2x}\|_0, \quad \|g_{2xx}\|_0$$

polynomially.

By Gronwall's Lemma, we deduce that

$$\widetilde{H}(m(t),n(t)) \leqslant \widetilde{H}(m(s),n(s))e^{-\gamma(t-s)} + \int_s^t \widetilde{K}_4 e^{-\gamma(t-\tau)} d\tau.$$

Let $t=0$. Then there exists $\bar{t}_3(\rho) \leqslant -1$, such that for all $s \leqslant \bar{t}_3(\rho)$,

$$e^{-\gamma(-s)}\rho^2 < 1.$$

The boundedness of

$$\int_s^0 \widetilde{K}_4 e^{\gamma\tau} d\tau$$

can be obtained as that of

$$\int_s^{-1} \widetilde{K}_0 e^{\gamma(1+\tau)} d\tau.$$

Let

$$r_3^2(w) = 1 + C + \| z_{xx}(0) \|_0^2 + \| m_x(0) \|_0 \| m(0) \|_0^3.$$

Then we complete the proof of Proposition 3.5. □

Before we construct the random attractor, it is worthwhile to point out that the domain about the spatial variable is unbounded. Fortunately, we can get weak compactness in the space $X_2$ by Lemma 3.6 below. Motivated by the definition of a weak attractor for the deterministic system in Ref. [9], we extend this definition to the RDS. Then we get a weak random global attractor for systems (1.1), (1.3) with initial conditions $m(s)=m_s$, $u(s,x)=u_s$.

**Lemma 3.6**[16] The solution operator $\{S(t)\}_{t\in\mathbf{R}}$ is weakly continuous in $X_2$, in the sense that if $(m_{0n}, n_{0n})$ converges weakly in $X_2$ to some $(m_0, n_0)$ as $n\to\infty$, then $S(t)(m_{0n}, n_{0n})$ weakly converges to $S(t)(m_0, n_0)$ in $X_2$ for all $t\in\mathbf{R}$.

Having Proposition 3.3, 3.5 and Lemma 3.6 in hand, we begin to construct the weak random attractor in space $X_2$.

From Subsect. 2.2, we get an RDS for the nonlinear Schrödinger equation coupling with a stochastic weakly damped, forced KdV equation, i.e.,

$$(m(t),\ u(t)) = S(t,s;\omega)(m_s, u_s).$$

We recall in Ref. [6] that if $B$ and $A$ are random sets such that for $P$-a.s. $\omega$, there exists a time $t_B(\omega)$ such that for all $t \leqslant t_B(\omega)$,

$$\phi(t,\theta_{-t}\omega)B(\theta_{-t}\omega) \subset A(\omega),$$

then $A$ is said to absorb $B$. Of course, if $A$ absorbs $B$, then $A$ attracts $B$.

Denote

$$A_1 = \{\|(m(0),u(0))\|_2 \leqslant r_3(\omega)\}.$$

It is easy to see that $A_1$ is an absorbing set for RDS $\{\phi(t,\omega)\}$ in $X_2$. Furthermore, it has weak compactness according to Lemma 3.6.

**Proposition 3.7** Let $B$ be a bounded set in $X_2$. Let

$$\mathscr{A}(\omega) = \overline{\bigcup_{B\subset X} \Lambda_B(\omega)},$$

where

$$\Lambda_B(\omega) := \bigcap_{s\geqslant 0}\overline{\bigcup_{t\geqslant s}\phi(t,\theta_{-t}\omega)B}$$

is the omega-limit set of $B$, where the closures are taken with respect to the weak topology of $X_2$. Then $\mathscr{A}(\omega)$ is included in $A_1$ and is nonempty. It is invariant by $\phi(t,\omega)$, i.e.,

$$\phi(t,\omega)\mathscr{A}(\omega) = \mathscr{A}(\theta_t\omega), \quad \forall t > 0.$$

The proof of Proposition 3.7 is easy using the following facts.

(a) $\phi(t,\omega)$ is weakly continuous on $X_2$, which is an obvious corollary of Lemma 3.6.

(b) A point $b\in\mathscr{A}(\omega)$ if and only if there exist two sequences $(s_n)_{n\in\mathbf{N}}\in\mathbf{R}$, $(x_n)_{n\in\mathbf{N}}\in\mathscr{A}(\omega)$, such that $s_n\to-\infty$ and $\phi(s_n)x_n\to b$.

Then $\mathscr{A}(\omega)$ has the following properties:

(i) $\mathscr{A}(\omega)$ is a random weakly compact set;

(ii) $\phi(t,\omega)\mathscr{A}(\omega)=\mathscr{A}(\theta_t\omega)$ for all $t\geqslant 0$;

(iii) $\mathscr{A}(\omega)$ attracts all deterministic bounded sets in the sense that

$$\lim_{t\to\infty} d^w(\phi(t,\theta_t\omega)B,\mathscr{A}) = 0,$$

where $d^w$ denotes the distance in the weak topology of $X_2$.

Thus, we get the main result of this paper.

**Theorem 3.8** Let the assumptions of Proposition 3.5 hold. Then there exists a global weak random attractor $\mathscr{A}$ for system (1.1), (1.3) with initial data $m(s)=m_s$, $u(s)=u_s$ in $X_2$.

## References

[1] Appert K, Vaclavik J. Dynamics of coupled solitons. Physics Fluids, 1977, 20: 1845–1949.

[2] Arnold L. Random dynamical system. Springer Monographs in Mathematics. Berlin: Springer, 1998.

[3] Bona J L, Smith R. The initial-value problem for the Korteweg-de Vries equation. Philos Trans Roy Soc London, Ser A, 1975, 278: 555–604.

[4] Chang H Y, Lien Ch, Sukarto S, et al. Propagation of ion-acoustic solitons in a non-quiescent plasma. Plasma Phys Control Fusion, 1986, 28: 675–681.

[5] Crauel H, Debussche A, Franco F. Random attractors. J Dyn Diff Eq, 1997, 9(2): 307–341.

[6] Crauel H, Flaudoli F. Attractors for random dynamical systems. Probab Th Rel Fields, 1994, 100: 365–393.

[7] De Bouard A, Debussche A. On the stochastic Korteweg-de Vries equation. J Funct Anal, 1998, 154: 215–251.

[8] De Bouard A, Debussche A, Tsutsumi Y. White noise driven Korteweg-de Vries equation. J Funct Anal, 1999, 169: 532–558.

[9] Ghidaglia J M. Finite dimensional behavior for weakly damped driven Schrödinger equations. Ann Inst Henri Poincaré, Analyse Non Linéaire, 1988, 5(4): 365–405.

[10] Gibbous J. On the theory of Langmuir solitons. J Plasma Phys, 1977, 17(2): 153–170.

[11] Grimshaw R, Pelinovsky E, Tian X. Interaction of a solitary wave with an external force. Physica D, 1994, 77: 405–433.

[12] Guo B, Chen F. Finite dimensional behavior of global attractors for weakly damped and forced KdV

equations coupling with nonlinear Schrödinger equations. Nonlinear Analysis TMA, 1997, 29(5): 569-584.

[13] Guo B, Shen L. The periodic initial value problem and the initial value problem for the system of KdV equation coupling with nonlinear Schrödinger equations// Proceedings of DD-3 Symposium, Changchun, 1982: 417-435.

[14] Herman R. The stochastic, damped Korteweg-de Vries equation. J Phys A, 1990, 23: 1063-1084.

[15] Makhankov V G. Dynamics of classical solitons. Physics Reports (Section C of Physics Letters), 1978, 35(1): 1-128.

[16] Rosa R. The global attractor of a weakly damped, forced Korteweg-de Vries equation in $H^1(\mathbf{R})$. Mat Contemp, 2000, 19: 129-152.

[17] Scalerandi M, Romano A, Condat C A. Korteweg-de Vries solitons under additive stochastic perturbations. Phys Rev E, 1998, 58: 4166-4173.

[18] Temam R. Infinite-dimensional systems in mechanics and physics. Applied Math Sciences 68. New York: Springer, 1988.

[19] Yang D. The asymptotic behavior of the stochastic Ginzburg-Landau equation with multiplicative noise. J Math Phys, 2004, 45(11): 4064-4076.

# Existence of the Global Smooth Solution to the Period Boundary Value Problem of Fractional Nonlinear Schrödinger Equation *

Guo Boling(郭柏灵) Han Yongqian(韩永前) Xin Jie(辛杰)

**Abstract**

In this paper, we prove the existence and uniqueness of the global smooth solution to the period boundary value problem of fractional nonlinear Schrödinger equation by using energy method.

**Keywords** fractional nonlinear Schrödinger equation; global smooth solution; Gagliardo-Nirenberg inequality

## 1 Introduction

The Schrödinger equation is the fundamental equation of physics for describing nonrelativistic quantum mechanical behavior. It is well known that Feynman and Hibbs[1] used path integrals over Brownian paths to derive the standard (non-fractional) Schrödinger equation. Recently, Laskin[5,6] showed that the path integral over the Lévy-like quantum mechanical paths allows to develop the generalization of the quantum mechanics. Namely, if the path integral over Brownian trajectories leads to the well known Schrödinger equation, then the path integral over Lévy trajectories leads to the fractional Schrödinger equation. The fractional Schrödinger equation includes the space derivative of order $\alpha$ instead of the second ($\alpha=1$) order space derivative in the standard Schrödinger equation. Laskin[7] showed the Hermiticity of the fractional Hamilton operator and established the parity conservation law. Guo and Xu[4] studied some physical applications of the fractional Schrödinger equation.

We consider the following fractional nonlinear Schrödinger equation:

$$\mathrm{i}u_t + (-\Delta)^{\alpha}u + \beta|u|^{\rho}u = 0, \quad x \in \mathbf{R}^n, \quad t > 0 \tag{1.1}$$

with initial condition and periodic boundary condition:

$$u(x,0) = u_0(x), \quad x \in \mathbf{R}^n, \tag{1.2}$$

$$u(x + 2\pi e_i, t) = u(x,t), \quad x \in \mathbf{R}^n, \quad t > 0 \tag{1.3}$$

where $e_i = (0,\cdots,0,1,0,\cdots,0)$ $(i = 1,\cdots,n)$ is an orthonormal basis of $\mathbf{R}^n$. In (1.1), i is the imaginary unit, $\alpha$ is a positive fraction, $\beta \in \mathbf{R}, \beta \neq 0, \rho > 0$.

When $\alpha = 1$, (1.1) becomes the standard nonlinear Schrödinger equation which was studied by many works. The existence and uniqueness of the weak solution to the initial-

* Applied Mathematics and Computation, 2008, 204: 468 – 477.

boundary value problem can be found in [2]. The existence of the global smooth solution to the initial-boundary value problem was studied in [3]. However, there is few theory analysis for the fractional nonlinear Schrödinger equation (1.1).

In this paper, we prove the existence and uniqueness of the global smooth solution to the period boundary value problem of fractional nonlinear Schrödinger equation (1.1) by using the energy method. Theorems 2.1 and 4.1 are the new fundamental results of the fractional nonlinear Schrödinger equation. We obtain the global existence of smooth solution and weak solution to the period boundary value problem of the fractional nonlinear Schrödinger equation in Theorems 2.1 and 4.1, respectively.

## 2 Preliminaries and Main Result

Since $u$ is a periodic function, we can express $u$ by a Fourier series:

$$u = \sum_{k\in \mathbf{Z}^n} a_k e^{i\langle k,x\rangle}.$$

Then,

$$\partial_{x_j} u = \sum_{k\in \mathbf{Z}^n} i k_j a_k e^{i\langle k,x\rangle}.$$

So $(-\Delta)^\alpha u$ is defined by

$$(-\Delta)^\alpha u = \sum_{k\in \mathbf{Z}^n} |k|^{2\alpha} a_k e^{i\langle k,x\rangle}.$$

Let,

$$A = \left\{ u \,\middle|\, u = \sum_{k\in \mathbf{Z}^n} a_k e^{i\langle k,x\rangle},\quad \sum_{k\in \mathbf{Z}^n} |k|^{4\alpha} a_k^2 < \infty,\quad \sum_{k\in \mathbf{Z}^n} a_k^2 < \infty \right\}.$$

Let $H^\alpha$ be a complete space of the set $A$ under the norm:

$$\| u \|_{H^\alpha} = \Big(\sum_{k\in \mathbf{Z}^n} |k|^{4\alpha} a_k^2\Big)^{\frac{1}{2}} + \Big(\sum_{k\in \mathbf{Z}^n} a_k^2\Big)^{\frac{1}{2}}.$$

Then $H^\alpha$ is a Banach space. It is easy to show that $H^\alpha$ is a Hilbert space with the inner product:

$$(u,v)_{H^\alpha} = ((-\Delta)^\alpha u, (-\Delta)^\alpha v) = \sum_{k\in \mathbf{Z}^n} |k|^{4\alpha} a_k b_k.$$

Let $\Omega = (0,2\pi)\times(0,2\pi)\times\cdots\times(0,2\pi)\subset\mathbf{R}^n$. Throughout this paper, we denote by $\|\cdot\|$ the norm of $H = L^2(\Omega)$ with usual inner product $(\cdot,\cdot)$, denote by $\|\cdot\|_{L^p(\Omega)}$ the norm of $L^p(\Omega)$ for all $1\leqslant p\leqslant\infty$ ($\|\cdot\|_{L^2(\Omega)} = \|\cdot\|$). We denote by $H^{-\alpha}(\Omega)$ the dual space to $H^\alpha(\Omega)$. In order to study the problem (1.1)~(1.3), we introduce the Banach space $V = H^\alpha(\Omega)\cap L^{\rho+2}(\Omega)$ with norm:

$$\| v \|_{H^\alpha(\Omega)\cap L^{\rho+2}(\Omega)} = \| v \|_{H^\alpha(\Omega)} + \| v \|_{L^{\rho+2}(\Omega)}.$$

Let $X$ denote a Banach space, with norm $\|\cdot\|_X$. We need the following definitions.

**Definition 2.1** The space $L^p(0, T; X)$ consists of all measurable functions $f:[0,T]\to$

$X$ with

$$\|f\|_{L^p(0,T;X)} = \left(\int_0^T \|f\|_X^p \mathrm{d}t\right)^{\frac{1}{p}} < \infty$$

for $1 \leqslant p < \infty$, and

$$\|f\|_{L^\infty(0,T;X)} = \lim_{0\leqslant t\leqslant T} \sup \|f(t)\|_X < \infty.$$

**Definition 2.2** The space $C([0,T];X)$ comprises all continuous functions $f:[0,T] \to X$ with

$$\|f\|_{C([0,T];X)} = \max_{0\leqslant t\leqslant T} \|f(t)\| < \infty.$$

We have the following main result.

**Theorem 2.1** Let $\alpha > \frac{n}{2}$. If $\rho$ is an even, suppose that $\rho > 0$ when $\beta > 0$ and $0 < \rho < \frac{4\alpha}{n}$ when $\beta < 0$. If $\rho$ is not an even, suppose that $p > 2[\alpha] + 1$ when $\beta > 0$ and $2[\alpha] + 1 < p < \frac{4\alpha}{n}$ when $\beta < 0$. Then for all $u_0 \in H^{4\alpha}(\Omega)$, there exists a unique global smooth solution $u(x,t)$ of the problem 1.1, 1.2, 1.3, such that

$$u \in L^\infty(0,T;H^{4\alpha}(\Omega)), \quad u_t \in L^\infty(0,T;H^{2\alpha}(\Omega)).$$

Theorem 2.1 generalizes the result of the standard nonlinear Schrödinger equation in [3].

## 3 *A Priori* Estimates

We can construct the global approximate solution by Galërkin's method. In order to prove the convergence of these approximate solutions, we need to obtain *a priori* estimates of these approximate solutions. Therefore, we first prove *a priori* estimates of the problem (1.1) ~ (1.3), these estimates are same as *a priori* estimates of Galërkin approximate solutions.

**Lemma 3.1** Suppose that $\alpha > 0$ and $\rho > 0$, $u = u(x,t)$ solves the problem 1.1, 1.2, 1.3. Then

$$\sup_{0\leqslant t\leqslant\infty} \|u\| = \|u_0\|. \tag{3.1}$$

**Proof** Multiply (1.1) by $\bar{u}$, integrate with respect to $x$, and then take the imaginary part to get

$$\frac{\mathrm{d}}{\mathrm{d}t}\|u\|^2 = 0. \qquad \square$$

Here, and throughout, we use $T$ to denote an arbitrary positive constant, and use $C$ to denote different positive constants, which depends only on initial data and positive constant $T$.

**Lemma 3.2** Let $\alpha > 0$. Suppose that $\rho > 0$ when $\beta > 0$ and $0 < \rho < \frac{4\alpha}{n}$ when $\beta < 0$, $u = u(x,t)$ solves the problem 1.1, 1.2, 1.3. Then

$$\sup_{0\leqslant t\leqslant\infty} \left(\|(-\Delta)^{\frac{\alpha}{2}} u\| + \|u\|_{L^{\rho+2}(\Omega)}\right) \leqslant C\left(\|u_0\|_{H^\alpha(\Omega)}, \|u_0\|_{L^{\rho+2}(\Omega)}\right). \tag{3.2}$$

**Proof** Multiply (1.1) by $\bar{u}_t$, and then integrate with respect to $x$ to obtain

$$(iu_t, u_t) + ((-\Delta)^\alpha u, u_t) + (\beta|u|^\rho u, u_t) = 0.$$

Take the real part of the above equality to get

$$\frac{\mathrm{d}}{\mathrm{d}t}\int_{\Omega}\left(|(-\Delta)^{\frac{\alpha}{2}}u|^2+\frac{2\beta}{\rho+2}|u|^{\rho+2}\right)\mathrm{d}x=0.$$

By (3.1) and the above equality, we get

$$\begin{aligned}&\|(-\Delta)^{\frac{\alpha}{2}}u\|^2+\frac{2\beta}{\rho+2}\|u\|^{\rho+2}_{L^{\rho+2}(\Omega)}\\&=\|(-\Delta)^{\frac{\alpha}{2}}u_0\|^2+\frac{2\beta}{\rho+2}\|u_0\|^{\rho+2}_{L^{\rho+2}(\Omega)}=E(u_0).\end{aligned}\tag{3.3}$$

If $\beta>0$, by (3.3),

$$\begin{aligned}&\|(-\Delta)^{\frac{\alpha}{2}}u\|^2\leqslant E(u_0)\leqslant C(\|u_0\|_{H^{\alpha}(\Omega)},\|u_0\|_{L^{\rho+2}(\Omega)}),\\&\|u\|_{L^{\rho+2}(\Omega)}\leqslant C(\|u_0\|_{H^{\alpha}(\Omega)},\|u_0\|_{L^{\rho+2}(\Omega)}).\end{aligned}$$

In the following, we consider the case that $\beta<0$. Let $\theta=\frac{n\rho}{2\alpha(\rho+2)}<1$. Then

$$\frac{1}{\rho+2}=\theta\left(\frac{1}{2}-\frac{\alpha}{n}\right)+(1-\theta)\frac{1}{2}.$$

So by Gagliardo-Nirenberg inequality, we have

$$\|u\|^{\rho+2}_{L^{\rho+2}(\Omega)}\leqslant C\|(-\Delta)^{\frac{\alpha}{2}}u\|^{\theta(\rho+2)}\|u\|^{(1-\theta)(\rho+2)}\leqslant C\|(-\Delta)^{\frac{\alpha}{2}}u\|^{\frac{n\rho}{2\alpha}}.$$

Since $\rho<\frac{4\alpha}{n}$, $\frac{n\rho}{2\alpha}<2$. Then from the above inequality, we get

$$\frac{2|\beta|}{\rho+2}\|u\|^{\rho+2}_{L^{\rho+2}(\Omega)}\leqslant\frac{1}{2}\|(-\Delta)^{\frac{\alpha}{2}}u\|^2+C.\tag{3.4}$$

By (3.3) and (3.4), when $\beta<0$, we get

$$\begin{aligned}&\|(-\Delta)^{\frac{\alpha}{2}}u\|^2\leqslant C(\|u_0\|_{H^{\alpha}(\Omega)},\|u_0\|_{L^{\rho+2}(\Omega)}),\\&\|u\|_{L^{\rho+2}(\Omega)}\leqslant C(\|u_0\|_{H^{\alpha}(\Omega)},\|u_0\|_{L^{\rho+2}(\Omega)}).\end{aligned}$$

□

**Lemma 3.3** Let $\alpha>\frac{n}{2}$. Suppose that $\rho$ satisfy the conditions of Lemma 3.2, $u=u(x,t)$ solves the problem (1.1),(1.2),(1.3). Then,

$$\sup_{0\leqslant t<\infty}(\|u_t\|+\|(-\Delta)^{\alpha}u\|)\leqslant C(\|u_0\|_{H^{2\alpha}(\Omega)}).\tag{3.5}$$

**Proof** Differentiate (1.1) with respect to $t$, multiply by $\bar{u}_t$, and then integrate with respect to $x$ to obtain

$$(\mathrm{i}u_{tt},u_t)+((-\Delta)^{\alpha}u_t,u_t)+\left(\frac{\mathrm{d}}{\mathrm{d}t}(\beta|u|^{\rho}u),u_t\right)=0.$$

By taking the imaginary part of the above equality, we get

$$\frac{1}{2}\frac{\mathrm{d}}{\mathrm{d}t}\|u_t\|^2+\mathrm{Im}\left(\frac{\mathrm{d}}{\mathrm{d}t}(\beta|u|^{\rho}u),u_t\right)=0.\tag{3.6}$$

But

$$\begin{aligned}\mathrm{Im}\left(\frac{\mathrm{d}}{\mathrm{d}t}(\beta|u|^{\rho}u),u_t\right)&=\mathrm{Im}\int_{\Omega}\frac{\mathrm{d}}{\mathrm{d}t}(\beta|u|^{\rho}u)\bar{u}_t\mathrm{d}x\\&=\mathrm{Im}\int_{\Omega}\beta|u|^{\rho}|u_t|^2\mathrm{d}x+\mathrm{Im}\int_{\Omega}\frac{\rho\beta}{2}|u|^{\rho-2}(|u_t|^2|u|^2+u^2\bar{u}_t^2)\mathrm{d}x\end{aligned}$$

$$= \mathrm{Im}\int_\Omega \frac{\rho\beta}{2}|u|^{\rho-2}(u^2\bar{u}_t^2)\mathrm{d}x. \tag{3.7}$$

Then by using (3.6) and (3.7), we obtain

$$\frac{1}{2}\frac{\mathrm{d}}{\mathrm{d}t}\|u_t\|^2 + \mathrm{Im}\int_\Omega \frac{\rho\beta}{2}|u|^{\rho-2}(u^2\bar{u}_t^2)\mathrm{d}x = 0.$$

Integrating the above equality from 0 to $t$, we have

$$\begin{aligned}\|u_t\|^2 &= -\int_0^t \mathrm{Im}\int_\Omega \rho\beta|u|^{\rho-2}(u^2\bar{u}_t^2)\mathrm{d}x\mathrm{d}s + \|u_t(x,0)\|^2\\ &\leqslant C\int_0^t\int_\Omega |u|^\rho|u_t|^2\mathrm{d}x\mathrm{d}s + \|u_t(x,0)\|^2.\end{aligned} \tag{3.8}$$

By applying Eq. (1.1), we have

$$\|u_t(x,0)\| \leqslant C\|(-\Delta)^\alpha u_0\| + C\|\beta|u_0|^\rho u_0\| \leqslant C(\|u_0\|_{H^{2\alpha}(\Omega)}).$$

Since $\alpha > \frac{n}{2}$, $\|u\|_{L^\infty(\Omega)} \leqslant C\|u\|_{H^\alpha(\Omega)} \leqslant C$. Then from (3.8), we get

$$\|u_t\|^2 \leqslant C\int_0^t \|u\|^\rho_{L^\infty(\Omega)}\|u_t\|^2\mathrm{d}s + \|u_t(x,0)\|^2 \leqslant C\int_0^t\|u_t\|^2\mathrm{d}s + C(\|u_0\|_{H^{2\alpha}(\Omega)}).$$

In term of Gronwall inequality, we deduce that

$$\|u_t\|^2 \leqslant C(\|u_0\|_{H^{2\alpha}(\Omega)}).$$

By (1.1) and the above inequality, we obtain

$$\begin{aligned}\|(-\Delta)^\alpha u\| &\leqslant \|u_t\| + \|\beta|u|^\rho u\| \leqslant C(\|u_0\|_{H^{2\alpha}(\Omega)}) + C\|u\|^\rho_{L^\infty(\Omega)}\|u\|\\ &\leqslant C(\|u_0\|_{H^{2\alpha}(\Omega)}).\end{aligned}$$

□

**Lemma 3.4** Let $\alpha > \frac{n}{2}$. Suppose that $\rho$ satisfies the conditions of Lemma 3.2 if $\rho$ is an even. If $\rho$ is not an even, suppose that $\rho > [\alpha]$ when $\beta > 0$ and $[\alpha] < p < \frac{4\alpha}{n}$ when $\beta < 0$. Assume that $u = u(x,t)$ solves the problem 1.1,1.2,1.3. Then

$$\sup_{0\leqslant t<\infty}\|(-\Delta)^{\frac{\alpha}{2}}u_t\| \leqslant C(\|u_0\|_{H^{3\alpha}(\Omega)}). \tag{3.9}$$

**Proof** Differentiate (1.1) with respect to $t$, multiply by $\bar{u}_{tt}$, and then integrate with respect to $x$ to obtain

$$(\mathrm{i}u_{tt}, u_{tt}) + ((-\Delta)^\alpha u_t, u_{tt}) + \left(\frac{\mathrm{d}}{\mathrm{d}t}(\beta|u|^\rho u), u_{tt}\right) = 0.$$

By taking the real part of the above equality, we get

$$\frac{\mathrm{d}}{\mathrm{d}t}\|(-\Delta)^{\frac{\alpha}{2}}u_t\|^2 + 2\mathrm{Re}\left(\frac{\mathrm{d}}{\mathrm{d}t}(\beta|u|^\rho u), u_{tt}\right) = 0. \tag{3.10}$$

But

$$\begin{aligned}&2\mathrm{Re}\left(\frac{\mathrm{d}}{\mathrm{d}t}(\beta|u|^\rho u), u_{tt}\right)\\ &= \int_\Omega\left(\frac{\rho}{2}+1\right)\beta|u|^\rho\frac{\mathrm{d}}{\mathrm{d}t}|u_t|^2\mathrm{d}x + \int_\Omega\frac{\rho\beta}{4}|u|^{\rho-2}\left(u^2\frac{\mathrm{d}}{\mathrm{d}t}\bar{u}_t^2 + \bar{u}^2\frac{\mathrm{d}}{\mathrm{d}t}u_t^2\right)\mathrm{d}x.\end{aligned} \tag{3.11}$$

Then from (3.10) and (3.11), we have

$$\frac{\mathrm{d}}{\mathrm{d}t}\|(-\Delta)^{\frac{\alpha}{2}}u_t\|^2+\int_\Omega\left(\frac{\rho}{2}+1\right)\beta|u|^\rho\frac{\mathrm{d}}{\mathrm{d}t}|u_t|^2\mathrm{d}x+\int_\Omega\frac{\rho\beta}{4}|u|^{\rho-2}\left(u^2\frac{\mathrm{d}}{\mathrm{d}t}\bar{u}_t^2+\bar{u}^2\frac{\mathrm{d}}{\mathrm{d}t}u_t^2\right)\mathrm{d}x=0.$$

The above equality implies that

$$\begin{aligned}&\frac{\mathrm{d}}{\mathrm{d}t}\|(-\Delta)^{\frac{\alpha}{2}}u_t\|^2+\frac{\mathrm{d}}{\mathrm{d}t}\int_\Omega\left(\frac{\rho}{2}+1\right)\beta|u|^\rho|u_t|^2\mathrm{d}x+\frac{\mathrm{d}}{\mathrm{d}t}\int_\Omega\frac{\rho\beta}{4}|u|^{\rho-2}(u^2\bar{u}_t^2+\bar{u}^2u_t^2)\mathrm{d}x\\&=-\left(\frac{\rho}{2}+1\right)\beta\int_\Omega\frac{\mathrm{d}}{\mathrm{d}t}(|u|^\rho)|u_t|^2\mathrm{d}x-\frac{\rho\beta}{4}\int_\Omega\frac{\mathrm{d}}{\mathrm{d}t}(|u|^{\rho-2}u^2)\bar{u}_t^2\mathrm{d}x-\frac{\rho\beta}{4}\int_\Omega\frac{\mathrm{d}}{\mathrm{d}t}(|u|^{\rho-2}\bar{u}^2)u_t^2\mathrm{d}x\\&\leqslant C\int_\Omega|u|^{\rho-1}|u_t|^3\mathrm{d}x\leqslant C\|u\|_{L^\infty(\Omega)}^{\rho-1}\|u_t\|_{L^3(\Omega)}^3\leqslant C\|u_t\|_{L^3(\Omega)}^3.\end{aligned}\tag{3.12}$$

Let $\theta=\frac{n}{6\alpha}<\frac{1}{3}$. Then

$$\frac{1}{3}=\theta\left(\frac{1}{2}-\frac{\alpha}{n}\right)+(1-\theta)\frac{1}{2}.$$

Thus, by Gagliardo-Nirenberg inequality and (3.5), we have

$$\begin{aligned}\|u_t\|_{L^3(\Omega)}^3&\leqslant C\|u_t\|^{3(1-\theta)}\|(-\Delta)^{\frac{\alpha}{2}}u_t\|^{3\theta}\leqslant C\|(-\Delta)^{\frac{\alpha}{2}}u_t\|^{3\theta}\\&\leqslant C\|(-\Delta)^{\frac{\alpha}{2}}u_t\|^2+C.\end{aligned}\tag{3.13}$$

Then from (3.12) and (3.13), we obtain

$$\begin{aligned}&\|(-\Delta)^{\frac{\alpha}{2}}u_t\|^2+\int_\Omega\left(\frac{\rho}{2}+1\right)\beta|u|^\rho|u_t|^2\mathrm{d}x+\int_\Omega\frac{\rho\beta}{4}|u|^{\rho-2}(u^2\bar{u}_t^2+\bar{u}^2u_t^2)\mathrm{d}x\\&\leqslant\|(-\Delta)^{\frac{\alpha}{2}}u_t(x,0)\|^2+\int_\Omega|\beta|\left(\frac{\rho}{2}+1\right)|u_0|^\rho|u_t(x,0)|^2\mathrm{d}x+\\&\int_\Omega\frac{\rho|\beta|}{4}|u_0|^{\rho-2}(u_0^2\bar{u}_t(x,0)^2+\bar{u}_0^2u_t(x,0)^2)\mathrm{d}x+C\int_0^t\|(-\Delta)^{\frac{\alpha}{2}}u_t\|^2\mathrm{d}s+C\\&\leqslant C+C\int_0^t\|(-\Delta)^{\frac{\alpha}{2}}u_t\|^2\mathrm{d}s.\end{aligned}\tag{3.14}$$

In fact, by (1.1), we have

$$\begin{aligned}\|(-\Delta)^{\frac{\alpha}{2}}u_t(x,0)\|&\leqslant\|(-\Delta)^{\frac{3\alpha}{2}}u_0\|+\|(-\Delta)^{\frac{\alpha}{2}}(\beta|u_0|^\rho u_0)\|\\&\leqslant C\|u_0\|_{H^{3\alpha}(\Omega)}+C\||u_0|^\rho u_0\|_{H^{[\alpha]+1}(\Omega)}\leqslant C(\|u_0\|_{H^{3\alpha}(\Omega)}),\end{aligned}$$

where we use the condition that $\rho>[\alpha]$ if $\rho$ is not an even.

$$\begin{aligned}&\int_\Omega|\beta|\left(\frac{\rho}{2}+1\right)|u_0|^\rho|u_t(x,0)|^2\mathrm{d}x+\int_\Omega\frac{\rho|\beta|}{4}|u_0|^{\rho-2}(u_0^2\bar{u}_t(x,0)^2+\bar{u}_0^2u_t(x,0)^2)\mathrm{d}x\\&\leqslant C\|u_0\|_{L^\infty(\Omega)}^\rho\|u_t(x,0)\|^2\leqslant C(\|u_0\|_{H^{2\alpha}(\Omega)}).\end{aligned}$$

By (3.13), we have

$$\begin{aligned}&\int_\Omega\left(\frac{\rho}{2}+1\right)\beta|u|^\rho|u_t|^2\mathrm{d}x+\int_\Omega\frac{\rho\beta}{4}|u|^{\rho-2}(u^2\bar{u}_t^2+\bar{u}^2u_t^2)\mathrm{d}x\\&\leqslant C\int_\Omega|u|^\rho|u_t|^2\mathrm{d}x\leqslant C\left(\int_\Omega|u|^{3\rho}\mathrm{d}x\right)^{\frac{1}{3}}\left(\int_\Omega|u_t|^3\mathrm{d}x\right)^{\frac{2}{3}}\\&\leqslant C\|(-\Delta)^{\frac{\alpha}{2}}u_t\|^{2\theta}\leqslant\frac{1}{2}\|(-\Delta)^{\frac{\alpha}{2}}u_t\|^2+C.\end{aligned}\tag{3.15}$$

Then by using (3.14), (3.15) and Gronwall inequality, we deduce that

$$\|(-\Delta)^{\frac{\alpha}{2}} u_t\|^2 \leqslant C + C\int_0^t \|(-\Delta)^{\frac{\alpha}{2}} u_t\|^2 ds \leqslant C(\|u_0\|_{H^{3\alpha}(\Omega)}). \qquad \square$$

**Lemma 3.5** Let $\alpha > \frac{n}{2}$. Suppose that $\rho$ satisfies the conditions of Lemma 3.2 if $\rho$ is an even. If $\rho$ is not an even, suppose that $\rho > 2[\alpha]+1$ when $\beta > 0$ and $2[\alpha]+1 < \rho < \frac{4\alpha}{n}$ when $\beta < 0$. Assume that $u = u(x,t)$ solves the problem 1.1, 1.2, 1.3. Then

$$\sup_{0\leqslant t<\infty}(\|u_{tt}\| + \|(-\Delta)^{\alpha} u_t\|) \leqslant C(\|u_0\|_{H^{4\alpha}(\Omega)}). \tag{3.16}$$

**Proof** Differentiate (1.1) with respect to $t$ two times, multiply by $\bar{u}_{tt}$, and then integrate with respect to $x$ to obtain

$$(\mathrm{i} u_{ttt}, u_{tt}) + ((-\Delta)^{\alpha} u_{tt}, u_{tt}) + \left(\frac{d^2}{dt^2}(\beta|u|^{\rho}u), u_{tt}\right) = 0.$$

By taking the imaginary part of the above equality, we get

$$\frac{1}{2}\frac{d}{dt}\|u_{tt}\|^2 + \mathrm{Im}\left(\frac{d^2}{dt^2}(\beta|u|^{\rho}u), u_{tt}\right) = 0. \tag{3.17}$$

But

$$\begin{aligned}
&\mathrm{Im}\left(\frac{d^2}{dt^2}(\beta|u|^{\rho}u), u_{tt}\right)\\
&= \mathrm{Im}\left(\frac{\rho^2}{2} + \rho\right)\beta(|u|^{\rho-2}|u_t|^2 u, u_{tt}) + \mathrm{Im}\left(\frac{\rho^2}{4} + \frac{\rho}{2}\right)\beta(|u|^{\rho-2}u_t^2\bar{u}, u_{tt}) +\\
&\quad \mathrm{Im}\left(\frac{\rho^2}{4} - \frac{\rho}{2}\right)\beta(|u|^{\rho-4}u_t^2 u^3, u_{tt}) + \mathrm{Im}\frac{\rho\beta}{2}(|u|^{\rho-2}u^2\bar{u}_{tt}, u_{tt}).
\end{aligned} \tag{3.18}$$

For the first term to the right hand side of (3.18), we get

$$\begin{aligned}
\mathrm{Im}\left(\frac{\rho^2}{2} + \rho\right)\beta(|u|^{\rho-2}|u_t|^2 u, u_{tt}) &\leqslant C\int_{\Omega}|u|^{\rho-1}|u_t|^2|u_{tt}|\,dx \leqslant C\|u\|_{L^{\infty}(\Omega)}^{\rho-1}\|u_t\|_{L^4(\Omega)}^2\|u_{tt}\|\\
&\leqslant C\|u_t\|_{L^4(\Omega)}^4 + C\|u_{tt}\|^2.
\end{aligned} \tag{3.19}$$

As the same reason, for the second term and the third term to the right hand side of (3.18), we have

$$\begin{aligned}
&\mathrm{Im}\left(\frac{\rho^2}{4} + \frac{\rho}{2}\right)\beta(|u|^{\rho-2}u_t^2\bar{u}, u_{tt}) + \mathrm{Im}\left(\frac{\rho^2}{4} - \frac{\rho}{2}\right)\beta(|u|^{\rho-4}u_t^2 u^3, u_{tt})\\
&\leqslant C\|u_t\|_{L^4(\Omega)}^4 + C\|u_{tt}\|^2.
\end{aligned} \tag{3.20}$$

For the last term to the right hand side of (3.18), we get

$$\mathrm{Im}\frac{\rho\beta}{2}(|u|^{\rho-2}u^2\bar{u}_{tt}, u_{tt}) \leqslant C\|u_{tt}\|^2. \tag{3.21}$$

Then by (3.17)~(3.21), we conclude that

$$\|u_{tt}\|^2 \leqslant C\int_0^t \|u_t\|_{L^4(\Omega)}^4 ds + C\int_0^t \|u_{tt}\|^2 ds + \|u_{tt}(x,0)\|^2. \tag{3.22}$$

Let $\theta = \frac{n}{8\alpha} < \frac{1}{4}$. Then

$$\frac{1}{4} = \theta\left(\frac{1}{2} - \frac{\alpha}{n}\right) + (1-\theta)\frac{1}{2}.$$

By (3.5), (3.9) and Gagliardo-Nirenberg inequality, we have

$$\|u_t\|_{L^4(\Omega)} \leqslant C\|u_t\|^{1-\theta}\|(-\Delta)^{\frac{\alpha}{2}}u_t\|^{\theta} \leqslant C(\|u_0\|_{H^{3\alpha}(\Omega)}). \tag{3.23}$$

While, by (1.1) and (3.5), we get

$$\begin{aligned}\|u_{tt}(x,0)\| &\leqslant \|(-\Delta)^{\alpha}((-\Delta)^{\alpha}u_0+\beta|u_0|^{\rho}u_0)\|+\left\|\frac{\mathrm{d}}{\mathrm{d}t}(\beta|u|^{\rho}u)\right\| \\ &\leqslant C\|(-\Delta)^{2\alpha}u_0\|+C\|(-\Delta)^{\alpha}(|u_0|^{\rho}u_0)\|+C\||u_0|^{\rho}|u_t(x,0)|\| \\ &\leqslant C(\|u_0\|_{H^{4\alpha}(\Omega)})+C\|(-\Delta)^{\alpha}(|u_0|^{\rho}u_0)\|+C\|u_t(x,0)\| \\ &\leqslant C(\|u_0\|_{H^{4\alpha}(\Omega)})+C\|(-\Delta)^{\alpha}(|u_0|^{\rho}u_0)\|.\end{aligned} \tag{3.24}$$

If $\alpha \geqslant \max\left\{\frac{n}{2},1\right\}$,

$$\|(-\Delta)^{\alpha}(|u_0|^{\rho}u_0)\| \leqslant C\|(-\Delta)^{[\alpha]+1}(|u_0|^{\rho}u_0)\| \leqslant C(\|u_0\|_{H^{4\alpha}(\Omega)}), \tag{3.25}$$

where we use the condition $\rho>2[\alpha]+1$ if $\rho$ is not an even.

If $n=1$ and $\frac{1}{2}<\alpha<1$,

$$\|(-\Delta)^{\alpha}(|u_0|^{\rho}u_0)\| \leqslant C\|\Delta(|u_0|^{\rho}u_0)\| \leqslant C(\|u_0\|_{H^{4\alpha}(\Omega)}). \tag{3.26}$$

Hence, from (3.24)~(3.26), we derive that

$$\|u_{tt}(x,0)\| \leqslant C(\|u_0\|_{H^{4\alpha}(\Omega)}). \tag{3.27}$$

From (3.22) and (3.27), we deduce that

$$\|u_{tt}\|^2 \leqslant C\int_0^t \|u_{tt}\|^2 \mathrm{d}s + C(\|u_0\|_{H^{4\alpha}(\Omega)}).$$

Therefore, Gronwall inequality implies that

$$\|u_{tt}\|^2 \leqslant C(\|u_0\|_{H^{4\alpha}(\Omega)}).$$

By (1.1) and the above inequality, we have

$$\|(-\Delta)^{\alpha}u_t\| \leqslant C\|u_{tt}\|+C\left\|\frac{\mathrm{d}}{\mathrm{d}t}(|u|^{\rho}u)\right\| \leqslant C(\|u_0\|_{H^{4\alpha}(\Omega)})+C\left\|\frac{\mathrm{d}}{\mathrm{d}t}(|u|^{\rho}u)\right\|. \tag{3.28}$$

But,

$$\left\|\frac{\mathrm{d}}{\mathrm{d}t}(|u|^{\rho}u)\right\| = \left\|\frac{\rho}{2}|u|^{\rho-2}(u\bar{u}_t+\bar{u}u_t)u+|u|^{\rho}u_t\right\| \leqslant C\|u\|^{\rho}_{L^{\infty}(\Omega)}\|u_t\| \leqslant C(\|u_0\|_{H^{2\alpha}(\Omega)}).$$

Then from (3.28) and the above inequality, we get

$$\sup_{0\leqslant t<\infty}\|(-\Delta)^{\alpha}u_t\| \leqslant C(\|u_0\|_{H^{4\alpha}(\Omega)}). \quad \square \tag{3.29}$$

**Lemma 3.6** Suppose that $\alpha$ and $\rho$ satisfy the conditions of Lemma 3.5, $u=u(x,t)$ solves the problem 1.1,1.2,1.3. Then

$$\sup_{0\leqslant t<\infty}\|(-\Delta)^{2\alpha}u\| \leqslant C(\|u_0\|_{H^{4\alpha}(\Omega)}). \tag{3.30}$$

**Proof** When $\alpha \geqslant \max\left\{\frac{n}{2},1\right\}$, by (1.1),(3.5) and (3.29), we obtain

$$\begin{aligned}\|(-\Delta)^{2\alpha}u\| &\leqslant C\|(-\Delta)^{\alpha}u_t\|+C\|(-\Delta)^{\alpha}(|u|^{\rho}u)\| \\ &\leqslant C\|(-\Delta)^{\alpha}u_t\|+C\|(-\Delta)^{[\alpha]+1}(|u|^{\rho}u)\| \leqslant C(\|u_0\|_{H^{4\alpha}(\Omega)}).\end{aligned} \tag{3.31}$$

When $n=1$ and $\frac{1}{2}<\alpha<1$, by (1.1) and (3.16), we have

$$\begin{aligned}\|(-\Delta)^{2\alpha}u\| &\leqslant C\|(-\Delta)^{\alpha}u_t\| + C\|(-\Delta)^{\alpha}(|u|^{\rho}u)\| \\ &\leqslant C(\|u_0\|_{H^{4\alpha}(\Omega)}) + C\|\Delta(|u|^{\rho}u)\| \\ &\leqslant C(\|u_0\|_{H^{4\alpha}(\Omega)}) + C\|u\|^{\rho}_{L^{\infty}(\Omega)}\|\Delta u\| + C\|u\|^{\rho-1}_{L^{\infty}(\Omega)}\||\nabla u|^2\| \\ &\leqslant C(\|u_0\|_{H^{4\alpha}(\Omega)}) + C\|\Delta u\| + C\|\nabla u\|^2_{L^4(\Omega)}. \end{aligned} \tag{3.32}$$

Let $\theta=\frac{2}{4\alpha}<1$. Then

$$\frac{1}{2} = 2+\theta\left(\frac{1}{2}-4\alpha\right)+(1-\theta)\frac{1}{2}.$$

By Gagliardo-Nirenberg inequality and (3.1), we have

$$C\|\Delta u\| \leqslant C\|(-\Delta)^{2\alpha}u\|^{\theta}\|u\|^{1-\theta} \leqslant \frac{1}{4}\|(-\Delta)^{2\alpha}u\| + C. \tag{3.33}$$

Let $\delta=\frac{1}{16\alpha-4}<\frac{1}{4}$. Then

$$\frac{1}{4} = \delta\left(\frac{1}{2}-(4\alpha-1)\right)+(1-\delta)\frac{1}{2}.$$

By Gagliardo-Nirenberg inequality and (3.5), we have

$$\begin{aligned} C\|\nabla u\|^2_{L^4(\Omega)} &\leqslant C\|(-\Delta)^{2\alpha}u\|^{2\delta}\|\nabla u\|^{2(1-\delta)} \leqslant C\|(-\Delta)^{2\alpha}u\|^{2\delta}\|(-\Delta)^{\alpha}u\|^{2(1-\delta)} \\ &\leqslant \frac{1}{4}\|(-\Delta)^{2\alpha}u\| + C. \end{aligned} \tag{3.34}$$

Then, by (3.32)~(3.34), we conclude that

$$\|(-\Delta)^{2\alpha}u\| \leqslant C(\|u_0\|_{H^{4\alpha}(\Omega)})$$

when $n=1$ and $\frac{1}{2}<\alpha<1$. Therefore, from (3.31) and the above inequality, we complete the proof of (3.30). □

The *a priori* estimates in 3.1~3.6, when $\alpha=1$ include the *a priori* estimates related to the standard nonlinear Schrödinger equations in [3].

## 4 Proof of the Main Result

Before proving Theorem 2.1, we first prove the existence of the weak solution to the problem (1.1)~(1.3) by using Faedo-Galërkin's method. We need the following lemmas.

**Lemma 4.1** Let $B_0$, B and $B_1$ be three Banach spaces. Assume that $B_0 \subset B \subset B_1$ and $B_i, i=0,1$ are reflective. Suppose also that $B_0$ is compactly embedded in $B$. Let

$$W = \left\{ v \mid v\in L^{p_0}(0,T;B_0), v' = \frac{\mathrm{d}v}{\mathrm{d}t} \in L^{p_1}(0,T;B_1) \right\},$$

where $T$ is finite and $1<p_i<\infty$, $i=0,1$. $W$ is equipped with the norm

$$\|v\|_{L^{p_0}(0,T;B_0)} + \|v'\|_{L^{p_1}(0,T;B_1)}.$$

Then $W$ is compactly embedded in $L^{p_0}(0,T;B)$.

**Lemma 4.2** Suppose that $Q$ is a bounded domain in $\mathbf{R}^n_x\times\mathbf{R}_t$, $g_\mu, g\in L^q(Q)(1<q<\infty)$ and $\|g_\mu\|_{L^q(Q)}\leqslant C$. Furthermore, suppose that

$$g_\mu \to g \text{ a.e. in } Q.$$

Then

$$g_\mu \rightharpoonup g \text{ weakly in } L^q(Q).$$

**Lemma 4.3** X is a Banach space. Suppose that $g \in L^p(0,T;X), \frac{\partial g}{\partial t} \in L^p(0,T;X)(1 \leqslant p \leqslant \infty)$. Then $g \in C([0,T],X)$ (after possibly being redefined on a set of measure zero).

In the following, we prove the existence of weak solution to the problem (1.1)~(1.3).

**Theorem 4.1** Let $\alpha > 0$. When $\beta > 0$, suppose that $\rho > 0$ if $\alpha \geqslant \frac{n}{2}$ and $0 < \rho < \frac{4\alpha}{n-2\alpha}$ if $\alpha < \frac{n}{2}$. When $\beta < 0$, suppose that $0 < \rho < \frac{4\alpha}{n}$. $u_0 \in H^\alpha(\Omega)$. Then there exists a function $u = u(x,t)$ satisfying 1.1,1.2,1.3, such that

$$u \in L^\infty(0,T;H^\alpha(\Omega) \cap L^{\rho+2}(\Omega)), \quad u_t \in L^\infty(0,T;H^{-\alpha}(\Omega)). \tag{4.1}$$

**Proof** We prove Theorem 4.1 by three steps:

**Step 1** Construct of approximate solutions by Faedo-Galërkin's method.

Fix now a positive integer $m$, we will look for a function $u_m = u_m(t)$ of the form:

$$u_m(t) = \sum_{|j|=0}^{m} g_{jm}(t) w_j, \quad w_j = e^{i\langle j,x\rangle}, \quad j \in \mathbf{Z}^n \tag{4.2}$$

where $g_{jm}(t)$ ($|j| = 0,1,\cdots,m$) are selected by the following conditions:

$$(iu_{m,t}, w_j) + ((-\Delta)^\alpha u_m, w_j) + (\beta |u_m|^\rho u_m, w_j) = 0, \quad 0 \leqslant |j| \leqslant m \tag{4.3}$$

and

$$u_m(0) - u_{0m} \in [w_j, 0 \leqslant |j| \leqslant m], \quad u_{0m} \to u_0 \ (m \to \infty) \text{ in } H^\alpha(\Omega). \tag{4.4}$$

Then (4.3) becomes the system of nonlinear ODE subject to the initial condition (4.4). According to standard existence theory for nonlinear ordinary differential equations, there exists a unique solution of (4.3) and (4.4) for a.e. $0 \leqslant t \leqslant t_m$. By *a priori* estimates we obtain that $t_m = T$.

**Step 2** *A priori* estimates.

As the proof of Lemmas 3.1 and 3.2, we have

$$u_m \in L^\infty(0,T;H^\alpha(\Omega) \cap L^{\rho+2}(\Omega)). \tag{4.5}$$

For $\forall \varphi \in H^\alpha(\Omega)$, we have

$$(iu_{m,t}, \varphi) + ((-\Delta)^\alpha u_m, \varphi) + (\beta |u_m|^\rho u_m, \varphi) = 0. \tag{4.6}$$

Then

$$\begin{aligned} |(u_{m,t}, \varphi)| &\leqslant |((-\Delta)^\alpha u_m, \varphi)| + |(\beta |u_m|^\rho u_m, \varphi)| \\ &\leqslant C\|(-\Delta)^{\frac{\alpha}{2}} u_m\| \|(-\Delta)^{\frac{\alpha}{2}} \varphi\| + C\|u_m\|_{L^{\rho+2}(\Omega)}^{\rho+1} \|\varphi\|_{L^{\rho+2}(\Omega)} \\ &\leqslant C\|(-\Delta)^{\frac{\alpha}{2}} \varphi\| + C\|\varphi\|_{L^{\rho+2}(\Omega)}. \end{aligned} \tag{4.7}$$

By Sobolev imbedding theorem, we have

$$\|\varphi\|_{L^{\rho+2}(\Omega)} \leqslant C\|(-\Delta)^{\frac{\alpha}{2}} \varphi\|.$$

So by (4.6) and (4.7), we have

$$|(u_{m,t},\varphi)| \leqslant C\|(-\Delta)^{\frac{\alpha}{2}}\varphi\|, \quad \forall \varphi \in H^{\alpha}(\Omega).$$

Therefore

$$u_{m,t} \in L^{\infty}(0,T;H^{-\alpha}(\Omega)). \tag{4.8}$$

**Step 3** Passing to the limit.

By applying (4.5) and (4.8), we deduce that there exists a subsequence $u_{\mu}$ from $u_m$, such that

$$u_{\mu} \rightharpoonup u \quad * - \text{weakly in } L^{\infty}(0,T;H^{\alpha}(\Omega)), \tag{4.9}$$

$$u_{\mu,t} \rightharpoonup u_t \quad * - \text{weakly in } L^{\infty}(0,T;H^{-\alpha}(\Omega)). \tag{4.10}$$

By (4.5), we have

$$u_m \text{ is bounded in } L^2(0,T;H^{\alpha}(\Omega)). \tag{4.11}$$

By (4.8), we have

$$u_{m,t} \text{ is bounded in } L^2(0,T;H^{-\alpha}(\Omega)). \tag{4.12}$$

Let

$$W = \{v \mid v \in L^2(0,T;H^{\alpha}(\Omega)), \quad v_t \in L^2(0,T;H^{-\alpha}(\Omega))\}.$$

We equip $W$ with the norm

$$\|v\|_w = \|v\|_{L^2(0,T;H^{\alpha}(\Omega))} + \|v_t\|_{L^2(0,T;H^{-\alpha}(\Omega))}.$$

Since $H^{\alpha}(\Omega)$ is compactly embedded in $L^2(\Omega)$, $W$ is compactly embedded in $L^2(0,T;L^2(\Omega))$ by Lemma 4.1. By (4.11) and (4.12), $u_m \in W$. Then, there exists the subsequence $u_{\mu}$ (not rebelled) which satisfies

$$u_{\mu} \to u \text{ strongly in } L^2(0,T;L^2(\Omega)) \text{ and a.e.} \tag{4.13}$$

By using (4.5), (4.13) and Lemma 4.2, we have

$$|u_{\mu}|^{\rho}u_{\mu} \rightharpoonup |u|^{\rho}u \quad * - \text{weakly in } L^{\infty}(0,T;L^{\frac{\rho+2}{\rho+1}}(\Omega)). \tag{4.14}$$

Fixing $j$, by (4.3), we get

$$(\mathrm{i}u_{\mu,t},w_j) + ((-\Delta)^{\alpha}u_{\mu},w_j) + (\beta|u_{\mu}|^{\rho}u_{\mu},w_j) = 0. \tag{4.15}$$

By applying (4.9), (4.10) and (4.14), we deduce that there exists a subsequence $u_{\mu}$ from $u_m$, such that

$$((-\Delta)^{\alpha}u_{\mu},w_j) \rightharpoonup ((-\Delta)^{\alpha}u,w_j) \quad * - \text{weakly in } L^{\infty}(0,T),$$

$$(u_{\mu,t},w_j) \rightharpoonup (u_t,w_j) \quad * - \text{weakly in } L^{\infty}(0,T),$$

$$(\beta|u_{\mu}|^{\rho}u_{\mu},w_j) \rightharpoonup (\beta|u|^{\rho}u,w_j) \quad * - \text{weakly in } L^{\infty}(0,T).$$

Then from (4.15), we have

$$(\mathrm{i}u_t,w_j) + ((-\Delta)^{\alpha}u,w_j) + (\beta|u|^{\rho}u,w_j) = 0.$$

The above equality holds for any fixed $j$. By the density of the basis $w_j$ ($j \in \mathbf{Z}^n$), we have

$$(\mathrm{i}u_t,v) + ((-\Delta)^{\alpha}u,v) + (\beta|u|^{\rho}u,v) = 0, \quad \forall v \in H^{\alpha}(\Omega).$$

Hence $u$ satisfies (1.1) and (4.1).

By (4.5), (4.8) and Lemma 4.3, we obtain that $u_{\mu} \in C([0,T],H^{-\alpha}(\Omega))$. Then,

$$u_{\mu}(0) \rightharpoonup u(0) \text{ weakly in } H^{-\alpha}(\Omega).$$

But from (4.4), we have

$$u_\mu(0)\rightarrow u_0 \text{ in } H^\alpha(\Omega).$$

Therefore, $u(0)=u_0$. □

Theorem 4.1 generalizes the result of the global existence of weak solution to the standard nonlinear Schrödinger equation in [2]. Now we prove our main results.

By *a priori* estimates in Lemmas 3.1 ~ 3.6 and Theorem 4.1, we derive a global smooth solution $u(t,x)$ of the problem (1.1)~(1.3), such that

$$u\in L^\infty(0,T;H^{4\alpha}(\Omega)),\quad u_t\in L^\infty(0,T;H^{2\alpha}(\Omega)).$$

We prove the uniqueness of the solution to the problem (1.1)~(1.3) in the following.

Suppose that $u$ and $v$ are two solutions which satisfy the problem (1.1)~(1.3). Then $w=u-v$ satisfies

$$\mathrm{i}w_t+(-\Delta)^\alpha w+\beta(|u|^\rho u-|v|^\rho v)=0 \tag{4.16}$$

with $w(0)=0$. Take the inner product of (4.16) with $w$ to get

$$\mathrm{i}(w_t,w)+((-\Delta)^\alpha w,w)+\beta(|u|^\rho u-|v|^\rho v,w)=0. \tag{4.17}$$

Consider the imaginary part of (4.17) to obtain

$$\frac{1}{2}\frac{\mathrm{d}}{\mathrm{d}t}\|w\|^2+\mathrm{Im}\beta(|u|^\rho u-|v|^\rho v,u-v)=0. \tag{4.18}$$

But

$$\begin{aligned}\mathrm{Im}\beta(|u|^\rho u-|v|^\rho v,u-v)&\leqslant C|(|u|^\rho(u-v)+(|u|^\rho-|v|^\rho)v,u-v)|\\&\leqslant C|u|^\rho\|u-v\|^2+C|(v(|u|^\rho-|v|^\rho),u-v)|\\&\leqslant C\|u\|_{L^\infty(\Omega)}\|u-v\|^2+C\|v\|_{L^\infty(\Omega)}\||u|^\rho-|v|^\rho\|\|u-v\|\\&\leqslant C\|w\|^2.\end{aligned}$$

By the above inequality and Gronwall's inequality, (3.13) implies that $\|w(0)\|^2=0$. So $w=0$.

Therefore, we complete the proof of Theorem 2.1.

**References**

[1] Feynman R P, Hibbs A R. Quantum mechanics and path integrals. New York: McGraw-Hill, 1965.

[2] Guo Boling, Wang Lireng. Some methods of nonlinear boundary value problem (Chinese). Zhongshan University Publishing House, 1989.

[3] Guo Boling. Nonlinear evolution equations (Chinese). Shanghai: Shanghai Scientific and Technological Education Publishing House, 1995.

[4] Guo Xiaoyi, Xu Mingyu. Some physical applications of fractional Schrödinger equation. J Math Phys, 2006, 47: 082104.

[5] Laskin N. Fractional quantum mechanics and Lévy integrals. Phys Lett A, 2000,268: 298 – 305.

[6] Laskin N. Fractional quantum mechanics. Phys Rev E, 2000, 63: 3135.

[7] Laskin N. Fractional Schrödinger equation. Phys Rev E, 2002, 66: 056108.

# 液晶模型和 Landau-Lifshitz 方程强解的正则性*

樊继山　郭柏灵

**摘　要**

对液晶模型和 Landau-Lifshitz 方程的初值问题的强解给出一些临界空间 Serrin 类正则性准则.

**关键词**　液晶；Landau-Lifshitz 方程；临界空间；Besov 空间

## 1　引　言

本文研究液晶和铁磁学中非线性问题的正则性问题. 林芳华于 1989 年提出如下液晶模型[1]：

$$\begin{cases}\partial_t u+(u\cdot\nabla)u+\nabla p-\Delta u=-\nabla\cdot(\nabla d\odot\nabla d), & \nabla\cdot u=0\\ \partial_t d+(u\cdot\nabla)d=\Delta d-f(d), & |d|\leqslant 1\\ (u,d)|_{t=0}=(u_0,d_0), & |d_0|\leqslant 1,\text{在 }\mathbf{R}^3\text{ 内}\end{cases}. \tag{1.1}$$

这里，$u$ 和 $p$ 分别表示速度和压力. $d$ 是描述液晶分子方向的一个向量函数. $f(d):=\frac{1}{\varepsilon^2}(|d|^2-1)d$. $(\nabla d\odot\nabla d)_{ij}=\sum_{k=1}^{3}\partial_{x_i}d_k\partial_{x_j}d_k,\ \forall\, i,j=1,2,3.$

Lin-Liu[2,3]和 Coutand-Shkoller[4]研究模型 (1.1) 的适定性问题，得到二维时的经典解的存在唯一性和三维问题局部古典解的存在唯一性，但三维问题是否存在整体古典解目前还不知道.

当小参数 $\varepsilon\to 0_+$ 时，模型 (1.1) 形式上收敛于下述问题：

$$\begin{cases}\partial_t u+(u\cdot\nabla)u+\nabla p-\Delta u=-\nabla\cdot(\nabla d\odot\nabla d), & \nabla\cdot u=0\\ \partial_t d+(u\cdot\nabla)d=\Delta d+d|\nabla d|^2, & |d|=1\\ (u,d)|_{t=0}=(u_0,d_0), & |d_0|=1,\text{在 }\mathbf{R}^3\text{ 内}\end{cases}. \tag{1.2}$$

当 $u\equiv 0$ 时，$(1.2)_2$ 式即为映到球上的调和映照热流方程[5]. 与此微分几何中重要方程相关的是如下 Landau-Lifshitz 方程[6]：

$$\begin{cases}d_t=\Delta d+d|\nabla d|^2+d\times\Delta d, & |d|=1\\ d|_{t=0}=d_0, & |d_0|=1,\text{在 }\mathbf{R}^3\text{ 内}\end{cases}. \tag{1.3}$$

这里，“×”表示 $\mathbf{R}^3$ 中两个向量的叉积. 利用公式 $a\times(b\times c)=(a\cdot c)b-(a\cdot b)c$ 与 $|d|=1$ 可得 $d\Delta d=-|\nabla d|^2$，方程 (1.3) 有如下等价的方程：

* 本文刊于中国科学：A辑 数学，2008，38(1)：31－40.

$$\begin{cases}\frac{1}{2}d_t-\frac{1}{2}d\times d_t=\Delta d+d|\nabla d|^2, & |d|=1\\ d|_{t=0}=d_0 \text{ 在 } \mathbf{R}^3 \text{ 内}, & |d_0|=1\end{cases}. \tag{1.4}$$

近年来，关于方程 (1.4) 或 (1.3) 已有许多研究，可见文献[7－15].

本文将用 Morrey 空间和 Besov 空间研究上述问题解的正则性条件.

这一方向关于 Navier-Stokes 方程的 Leray 弱解[16]的正则性条件已有许多. Prodi[17], Serrin[18]和 Giga[19]得到 Leray 弱解的正则性条件为

$$u\in L^{\theta}(0,T;L^p(\mathbf{R}^n)),\quad \frac{2}{\theta}+\frac{n}{p}=1,\quad n<p\leqslant\infty. \tag{1.5}$$

da Veiga[20]得到正则性准则

$$\omega:=\mathrm{curl}u\in L^{\theta}(0,T;L^p(\mathbf{R}^n)),\quad \frac{2}{\theta}+\frac{n}{p}=2,\quad \frac{n}{2}<p\leqslant\infty. \tag{1.6}$$

Kozono-Ogawa-Taniuchi 证明如下正则性条件：

$$\omega\in L^{\theta}(0,T;\dot{B}^0_{p,\infty}(\mathbf{R}^n)),\quad \frac{2}{\theta}+\frac{n}{p}=2,\quad n\leqslant p\leqslant\infty, \tag{1.7}$$

这里，$\dot{B}^s_{p,q}$是均匀的 Besov 空间[21－23].

以后，Kozono-Shimada[24]得到如下条件：

$$u\in L^{\frac{2}{1-\alpha}}(0,T;\dot{F}^{-\alpha}_{\infty,\infty}),\quad 0<\alpha<1, \tag{1.8}$$

这里，$\dot{F}^s_{p,q}$是均匀的 Triebel-Lizorkin 空间[22－24].

关于调和映照热流问题，Kozono-Ogawa-Taniuchi[21]证明了如下正则性条件：

$$\nabla d\in L^{\theta}(0,T;\dot{B}^0_{p,\frac{2}{n}p}(\mathbf{R}^n)),\quad \frac{2}{\theta}+\frac{n}{p}=1,\quad n<p<\infty. \tag{1.9}$$

让 $1<q\leqslant p<\infty$，定义均匀 Morrey 空间

$$\dot{M}_{p,q}:=\left\{f\in L^q_{\mathrm{loc}}(\mathbf{R}^3)\,\middle|\,\|f\|_{\dot{M}_{p,q}}:=\sup_{R>0}\sup_{x}R^{3\left(\frac{1}{p}-\frac{1}{q}\right)}\left(\int_{B(x,R)}|f(y)|^q\mathrm{d}y\right)^{1/q}<\infty\right\}.$$

让 $1\leqslant p'\leqslant q'<\infty$，定义均匀空间

$$\dot{N}_{p',q'}:=\left\{\begin{array}{l}f\in L^{p'}/f=\sum_{k\in\mathbf{N}}g_k,\quad (g_k)\subset L^{q'}_{\mathrm{comp}}(\mathbf{R}^3),\\ \sum_{k\in\mathbf{N}}d_k^{3\left(\frac{1}{p}-\frac{1}{q}\right)}\|g_k\|_{L^{q'}}<+\infty,\quad \forall k,d_k=\mathrm{diam}(\mathrm{supp}g_k)<\infty\end{array}\right\}.$$

令 $1<p'\leqslant q'<\infty$ 且 $\frac{1}{p}+\frac{1}{p'}$，$\frac{1}{q}+\frac{1}{q'}=1$，则 $\dot{M}_{p,q}$是 $\dot{N}_{p',q'}$的对偶空间[25].

令 $1\leqslant p'\leqslant q'<2$ 且 $r=\frac{3}{p}$，则存在 $C>0$，使得[25]

$$\|uv\|_{\dot{N}_{p',q'}}\leqslant C\|u\|_{L^2}\|v\|_{\dot{H}^2}. \tag{1.10}$$

本文的主要结果是：

**定理 1.1** 设 $u_0\in H^1$，$d_0\in H^2$ 且 $|d_0|\leqslant 1$. 如果下列两条件之一成立：

$$u\in L^s(0,T;\dot{M}_{p,q}(\mathbf{R}^3)),\quad \frac{2}{s}+\frac{3}{p}=1,\quad 3<p\leqslant\infty,\quad 1\leqslant q\leqslant p; \tag{1.11}$$

$$\nabla u \in L^s(0,T;\dot{M}_{p,q}(\mathbf{R}^3)),\quad \frac{2}{s}+\frac{3}{p}=2,\quad \frac{3}{2}<p\leqslant\infty,\quad 1\leqslant q\leqslant p; \tag{1.12}$$

则问题 (1.1) 的解$(u,\ d)$满足

$$u\in L^\infty(0,T;H^1)\cap L^2(0,T;H^2),\quad d\in L^\infty(0,T;H^2)\cap L^2(0,T;H^3). \tag{1.13}$$

**注 1.1** 这里并没有对 $d$ 加条件，本定理表明液晶中流体速度占主要因素.

**定理 1.2** 设 $u_0\in H^1$，$d_0\in H^2$ 且$|d^0|=1$，则问题 (1.2) 的强解$(u,\ d)$满足如下正则性条件：

$$u\in L^s(0,T;\dot{M}_{p,q}(\mathbf{R}^3)),\quad \frac{2}{s}+\frac{3}{p}=1,\quad p>3,\quad p\geqslant q\geqslant 1, \tag{1.14}$$

及

$$\nabla d\in L^\theta(0,T;L^r(\mathbf{R}^3)),\quad \frac{2}{\theta}+\frac{3}{r}=1,\quad 3<r\leqslant\infty. \tag{1.15}$$

换言之，若解$(u,\ d)$在有限时刻 $T$ 爆破，则有

$$\int_0^T \|u(t)\|_{\dot{M}_{p,q}}^s+\|\nabla d(t)\|_{L^r}^\theta \mathrm{d}t=\infty. \tag{1.16}$$

**定理 1.3** 设$(u_0,\ d_0)\in H^3\times H^4$，则问题 (1.2) 的强解满足如下正则性条件：

$$(\mathrm{curl}u,\nabla d)\in L^{\frac{2p}{2p-3}}(0,T;\dot{B}^0_{p,\infty})\times L^{\frac{2p}{2p-3}}(0,T;\dot{B}^0_{p,\frac{2}{3}p}),\quad 3<p\leqslant\infty. \tag{1.17}$$

换言之，若解$(u,\ d)$在有限时刻 $T$ 爆破，则有

$$\int_0^T \|\mathrm{curl}u(t)\|_{\dot{B}^0_{p,\infty}}^{\frac{2p}{2p-3}}+\|\nabla d(t)\|_{\dot{B}^0_{p,\frac{2}{3}p}}^{\frac{2p}{2p-3}}\mathrm{d}t=\infty. \tag{1.18}$$

**注 1.2** 如果$(u,\ d)$是问题 (1.2) 的解，则$(u_\lambda,d_\lambda):=(\lambda u(\lambda^2t,\lambda x),d(\lambda^2t,\lambda x))$也是问题 (1.2) 的解. 在这种意义下，条件 (1.14) 和 (1.15) 及 (1.17) 是最优的.

**定理 1.4** 设 $d_0\in H^4$，$|d_0|=1$，则问题 (1.3) 的强解 $d$ 满足如下正则性条件：

$$\nabla d\in L^{\frac{2p}{p-3}}(0,T;\dot{B}^0_{p,\frac{2}{3}p}(\mathbf{R}^3)),\quad 3<p\leqslant\infty. \tag{1.19}$$

换言之，若解 $d$ 在有限时刻 $T$ 爆破，则有

$$\int_0^T \|\nabla d(t)\|_{\dot{B}^0_{p,\frac{2}{3}p}}^{\frac{2p}{p-3}}\mathrm{d}t=\infty. \tag{1.20}$$

**注 1.3** 设 $d$ 是方程 (1.3) 的一个解，则 $d_\lambda:=d(\lambda^2t,\ \lambda x)$也是方程 (1.3) 的一个解 . 在这种意义下，条件 (1.19) 是最优的.

另外，在这里还要指出：

(i) 由于耦合项 $-\nabla\cdot(\nabla d\odot\nabla d)$与 $u\cdot\nabla d$ 的存在，文献[21]中的方法不能用来证明定理 1.1.

(ii) 文献[21]中用来处理困难项 $d|\nabla d|^2$ 的方法是：对问题 $(1.2)_2$ 作用 $\Delta$,然后乘以$|\Delta d|^{q-2}\Delta d$ $(q>3)$并分部积分. 但是他们的方法在这里也不管用，因为耦合项 $-\nabla\cdot(\nabla d\odot\nabla d)$和 $u\cdot\nabla d$ 的存在.

(iii) 由于非线性项 $d\times\Delta d$ 的存在，文献[21]中的方法也不能用来证明定理 1.4.

# 2 定理1.1的证明

本节证明定理1.1成立.

模型 $(1.1)_1$ 两端乘以 $-\Delta u$，分部积分并利用不可压条件,得

$$\frac{1}{2}\frac{\mathrm{d}}{\mathrm{d}t}\int|\nabla u|^2\mathrm{d}x+\int|\Delta u|^2\mathrm{d}x=-\sum_{i,j,k}\int\partial_k u_j\cdot\partial_j u_i\cdot\partial_k u_i\mathrm{d}x-\sum_{i,k}\int\Delta d_k\cdot\partial_i\nabla d_k\cdot\nabla u_i\mathrm{d}x-\sum_{i,k}\int\partial_i d_k\cdot\nabla\Delta d_k\cdot\nabla u_i\mathrm{d}x. \tag{2.1}$$

模型 $(1.1)_2$ 两端作用 $\Delta$，然后乘以 $\Delta d$，分部积分并利用不可压条件，有

$$\begin{aligned}&\frac{1}{2}\frac{\mathrm{d}}{\mathrm{d}t}\int|\Delta d|^2\mathrm{d}x+\int|\nabla\Delta d|^2\mathrm{d}x+\int\Delta f(d)\cdot\Delta d\mathrm{d}x\\&=\sum_{i,k}\int\partial_i d_k\cdot\nabla\Delta d_k\cdot\nabla u_i\mathrm{d}x-\sum_{i,k}\int\partial_i\partial_j d_k\cdot\partial_j\nabla d_k\cdot\nabla u_i\cdot\mathrm{d}x.\end{aligned} \tag{2.2}$$

联立式 (2.1) 和式 (2.2) 并舍弃无害项

$$\int\Delta f(d)\cdot\Delta d\mathrm{d}x=-\int\nabla f(d):\nabla\Delta d\mathrm{d}x,$$

得

$$\begin{aligned}&\frac{1}{2}\frac{\mathrm{d}}{\mathrm{d}t}\int|\nabla u|^2+|\Delta d|^2\mathrm{d}x+\int|\Delta u|^2+|\nabla\Delta d|^2\mathrm{d}x\\&\leqslant-\sum_{i,j,k}\int\partial_k u_j\cdot\partial_j u_i\cdot\partial_k u_i\mathrm{d}x-\sum_{i,k}\int\Delta d_k\cdot\partial_i\nabla d_k\cdot\nabla u_i\mathrm{d}x-\\&\quad\sum_{i,j,k}\int\partial_i\partial_j d_k\cdot\partial_j\nabla d_k\cdot\nabla u_i\mathrm{d}x=:I_1+I_2+I_3.\end{aligned} \tag{2.3}$$

在以下计算中，我们将经常利用如下 Gagliardo-Nirenberg 不等式:

$$\|w\|_{\dot{H}^{3/p}}\leqslant C\|w\|_{L^2}^{1-3/p}\|\nabla w\|_{L^2}^{3/p},\quad 3<p\leqslant\infty. \tag{2.4}$$

现在设条件 (1.12) 成立. 每项 $I_i(i=1,2,3)$ 的估计如下:

$$\begin{aligned}I_1&\leqslant\sum_{i,j,k}\|\partial_k u_j\|_{\dot{M}_{p,q}}\|\partial_j u_i\cdot\partial_k u_i\|_{\dot{N}_{p',q'}}\\&\leqslant C\|\nabla u\|_{\dot{M}_{p,q}}\|\nabla u\|_{L^2}\|\nabla u\|_{\dot{H}^{3/p}}\\&\leqslant C\|\nabla u\|_{\dot{M}_{p,q}}\|\nabla u\|_{L^2}^{2-3/p}\|\Delta u\|_{L^2}^{3/p}\quad(\text{由式}(2.4),w=\nabla u)\\&\leqslant\varepsilon\|\Delta u\|_{L^2}^2+C\|\nabla u\|_{\dot{M}_{p,q}}^{\frac{2p}{2p-3}}\|\nabla u\|_{L^2}^2,\quad\forall\varepsilon>0.\end{aligned}$$

类似地,

$$\begin{aligned}I_2,I_3&\leqslant C\|\nabla u\|_{\dot{M}_{p,q}}\|\Delta d\|_{L^2}\|\Delta d\|_{\dot{H}^{3/p}}\\&\leqslant C\|\nabla u\|_{\dot{M}_{p,q}}\|\Delta d\|_{L^2}^{2-3/p}\|\nabla\Delta d\|_{L^2}^{3/p}\quad(\text{由式}(2.4),w=\Delta d)\\&\leqslant\varepsilon\|\nabla\Delta d\|_{L^2}^2+C\|\nabla u\|_{\dot{M}_{p,q}}^{\frac{2p}{2p-3}}\|\Delta d\|_{L^2}^2,\quad\forall\varepsilon>0.\end{aligned}$$

将上述估计代入式 (2.3) 并让 $\varepsilon$ 充分小，然后由 Gronwall 不等式知式 (1.13) 成立.

下面设条件 (1.11) 成立. 类似地, 有式 (2.3) 成立. 但是, 对每项 $I_i(i=1,2,3)$做如下估计:

$$\begin{aligned}I_1 &= \sum_{i,j,k}\int u_i\cdot\partial_k u_j\cdot\partial_k\partial_j u_i\mathrm{d}x \leqslant \sum_{i,j,k}\|u_i\|_{\dot{M}_{p,q}}\|\partial_k u_j\cdot\partial_k\partial_j u_i\|_{\dot{N}_{p',q'}}\\ &\leqslant C\|u\|_{\dot{M}_{p,q}}\|\nabla u\|_{\dot{H}^{3/p}}\|\Delta u\|_{L^2}\\ &\leqslant C\|u\|_{\dot{M}_{p,q}}\|\nabla u\|_{L^2}^{1-3/p}\|\Delta u\|_{L^2}^{1+3/p}\quad(\text{由式}(2.4), w=\nabla u)\\ &\leqslant \varepsilon\|\Delta u\|_{L^2}^2+C\|u\|_{\dot{M}_{p,q}}^{\frac{2p}{p-3}}\|\nabla u\|_{L^2}^2,\quad\forall\varepsilon>0.\end{aligned}\tag{2.5}$$

$$\begin{aligned}I_2 &= \sum_{i,k}\int u_i\cdot\partial_i\nabla d_k\cdot\Delta\nabla d_k\mathrm{d}x \leqslant \sum_{i,k}\|u_i\|_{\dot{M}_{p,q}}\|\partial_i\nabla d_k\cdot\Delta\nabla d_k\|_{\dot{N}_{p',q'}}\\ &\leqslant C\|u\|_{\dot{M}_{p,q}}\|\Delta d\|_{\dot{H}^{3/p}}\|\nabla\Delta d\|_{L^2}\\ &\leqslant C\|u\|_{\dot{M}_{p,q}}\|\Delta d\|_{L^2}^{1-3/p}\|\nabla\Delta d\|_{L^2}^{1+3/p}\quad(\text{由式}(2.4), w=\Delta d)\\ &\leqslant \varepsilon\|\nabla\Delta d\|_{L^2}^2+C\|u\|_{\dot{M}_{p,q}}^{\frac{2p}{p-3}}\|\Delta d\|_{L^2}^2,\quad\forall\varepsilon>0.\end{aligned}\tag{2.6}$$

$$\begin{aligned}I_3 &= \sum_{i,j,k}\int \partial_i\partial_j d_k\cdot\partial_j\Delta d_k\cdot u_i\mathrm{d}x \leqslant \sum_{i,j,k}\|u_i\|_{\dot{M}_{p,q}}\|\partial_i\partial_j d_k\cdot\partial_j\Delta d_k\|_{\dot{N}_{p',q'}}\\ &\leqslant C\|u\|_{\dot{M}_{p,q}}\|\Delta d\|_{\dot{H}^{3/p}}\|\nabla\Delta d\|_{L^2}\\ &\leqslant \varepsilon\|\nabla\Delta d\|_{L^2}^2+C\|u\|_{\dot{M}_{p,q}}^{\frac{2p}{p-3}}\|\Delta d\|_{L^2}^2,\quad\forall\varepsilon>0.\end{aligned}\tag{2.7}$$

将式 (2.5)~式(2.7) 代入式 (2.3), 取 $\varepsilon$ 充分小并利用 Gronwall 不等式可得.

定理证毕.

## 3 定理1.2 的证明

本节证明定理 1.2. 与式 (2.3) 类似可得

$$\begin{aligned}&\frac{1}{2}\frac{\mathrm{d}}{\mathrm{d}t}\int|\nabla u|^2+|\Delta d|^2+\int|\Delta u|^2+|\nabla\Delta d|^2\mathrm{d}x\\ &=\sum_{i,j,k}\int u_i\cdot\partial_k u_j\cdot\partial_k\partial_j u_i\mathrm{d}x+\sum_{i,k}\int u_i\cdot\partial_i\nabla d_k\cdot\Delta\nabla d_k\mathrm{d}x+\\ &\quad\sum_{i,j,k}\int u_i\cdot\partial_i\partial_j d_k\cdot\partial_j\Delta d_k\mathrm{d}x-\sum_{i,k}\int\nabla d_k\cdot|\nabla d_i|^2\cdot\nabla\Delta d_k\mathrm{d}x-\\ &\quad\sum_{i,j,k}2\int d_k\cdot\partial_j d_i\cdot\nabla\partial_j d_i\cdot\nabla\Delta d_k\mathrm{d}x=:I_1+I_2+I_3+I_4+I_5.\end{aligned}\tag{3.1}$$

与式 (2.5)~式(2.7) 完全一样地估计 $I_1$, $I_2$, $I_3$. 下面估计 $I_5$ 与 $I_4$.

$$\begin{aligned}I_5&\leqslant 2\sum_{i,j,k}\|\partial_j d_i\|_{L^r}\|\nabla\partial_j d_i\|_{L^{\frac{2r}{r-2}}}\|\nabla\Delta d_k\|_{L^2}\\ &\leqslant C\|\nabla d\|_{L^r}\|\Delta d\|_{L^{\frac{2r}{r-2}}}\|\nabla\Delta d\|_{L^2}\end{aligned}$$

$$\leqslant C\|\nabla d\|_{L^r}\|\Delta d\|_{L^2}^{1-3/r}\|\nabla\Delta d\|_{L^2}^{1+3/r}$$

$$\leqslant \varepsilon\|\nabla\Delta d\|_{L^2}^2 + C\|\nabla d\|_{L^r}^{\frac{2r}{r-3}}\|\Delta d\|_{L^2}^2,\quad \forall\,\varepsilon>0,$$

这里用了 Gagliardo-Nirenberg 不等式

$$\|\Delta d\|_{L^{\frac{2r}{r-2}}} \leqslant C\|\Delta d\|_{L^2}^{1-3/r}\|\nabla\Delta d\|_{L^2}^{3/r}. \tag{3.2}$$

为估计 $I_4$，将利用 Gagliardo-Nirenberg 不等式

$$\|\nabla d\|_{L^{\frac{4r}{r-2}}}^2 \leqslant C\|d\|_{L^\infty}\|\Delta d\|_{L^{\frac{2r}{r-2}}}. \tag{3.3}$$

$$I_4\leqslant \sum_{i,k}\|\nabla d_k\|_{L^r}\|\nabla\Delta d_k\|_{L^2}\|\nabla d_i\|_{L^{\frac{4r}{r-2}}}^2 \leqslant C\|\nabla d\|_{L^r}\|\nabla\Delta d\|_{L^2}\|\Delta d\|_{L^{\frac{2r}{r-2}}},$$

余下的估计与 $I_5$ 完全一样，有

$$I_4, I_5\leqslant \varepsilon\|\nabla\Delta d\|_{L^2}^2 + C\|\nabla d\|_{L^r}^{\frac{2r}{r-3}}\|\Delta d\|_{L^2}^2,\quad \forall\,\varepsilon>0.$$

将这些估计代入式 (3.1)，取 $\varepsilon$ 充分小，用 Gronwall 不等式即得.

定理证毕.

## 4 定理 1.3 的证明

本节证明定理 1.3. 在下面的证明中，我们将用到如下的对数 Sobolev 不等式[21]：

$$\|\nabla u\|_{L^p}\leqslant C(1+\|\mathrm{curl}u\|_{\dot B^0_{p,\infty}}(\log(3+\|u\|_{\dot H^3}))^{1/2}), \tag{4.1}$$

$$\|\nabla d\|_{L^p}\leqslant C(1+\|\nabla d\|_{\dot B^0_{p,\frac{2}{3}p}}(\log(3+\|d\|_{\dot H^4}))^{\frac{p-3}{2p}}), \tag{4.2}$$

及 David-Journé[26]与 Kato-Ponce[27]的换位子估计.

**引理 4.1**[26,27] 设 $1<p<\infty, f,g\in W^{s,p}$，则有

$$\|\partial^\alpha(fg)-f\partial^\alpha g\|_{L^r}\leqslant C(\|\nabla f\|_{L^q}\|\partial^{\alpha-1}g\|_{L^r}+\|g\|_{L^q}\|\partial^\alpha f\|_{L^r}), \tag{4.3}$$

其中，$1\leqslant\alpha\leqslant s$，$\dfrac{1}{p}=\dfrac{1}{q}+\dfrac{1}{r}$，$1<q\leqslant\infty$，$1<r<\infty$.

问题 $(1.2)_1$ 两端作用$\partial^3$，然后乘以$\partial^3 u$，分部积分且利用不可压条件，得

$$\frac{1}{2}\frac{\mathrm{d}}{\mathrm{d}t}\int|\partial^3 u|^2\mathrm{d}x+\int|\partial^4 u|^2\mathrm{d}x=-\sum_i\int(\partial^3(u\cdot\nabla u_i)-u\partial^3\nabla u_i)\partial^3 u_i\mathrm{d}x+$$
$$\sum_{i,j,k}\int\partial^3(\partial_i d_k\partial_j d_k)\partial^3\partial_j u_i\mathrm{d}x=:I_1+I_2. \tag{4.4}$$

利用换位子估计 (4.3)，估计 $I_1$ 如下：

$$I_1\leqslant\sum_i\|\partial^3(u\cdot\nabla u_i)-u\partial^3\nabla u_i\|_{L^2}\|\partial^3 u_i\|_{L^2}$$

$$\leqslant C\|\nabla u\|_{L^p}\|\partial^3 u\|_{L^{\frac{2p}{p-2}}}\|\partial^3 u\|_{L^2}$$

$$\leqslant C\|\nabla u\|_{L^p}\|\partial^3 u\|_{L^2}^{2-3/p}\|\partial^4 u\|_{L^2}^{3/p}$$

$$\leqslant \varepsilon\|\partial^4 u\|_{L^2}^2+C\|\nabla u\|_{L^p}^{\frac{2p}{2p-3}}\|\partial^3 u\|_{L^2}^2,\quad\forall\,\varepsilon>0.$$

这里用了如下的 Gagliardo-Nirenberg 不等式：

$$\|u\|_{L^{\frac{2p}{p-2}}} \leqslant C\|u\|_{L^2}^{1-3/p}\|\nabla u\|_{L^2}^{3/p}. \tag{4.5}$$

$$I_2 \leqslant C\int |\nabla d|\cdot|\partial^4 d|\cdot|\partial^4 u| + |\partial^3 d|\cdot|\partial^2 d|\cdot|\partial^4 u|\,\mathrm{d}x =: I_3 + I_4. \tag{4.6}$$

$$\begin{aligned} I_3 &\leqslant C\|\nabla d\|_{L^p}\|\partial^4 d\|_{L^{\frac{2p}{p-2}}}\|\partial^4 u\|_{L^2} \\ &\leqslant C\|\nabla d\|_{L^p}\|\partial^4 d\|_{L^2}^{1-3/p}\|\partial^5 d\|_{L^2}^{3/p}\|\partial^4 u\|_{L^2} \quad (\text{由式}(4.5), u = \partial^4 d) \\ &\leqslant \varepsilon\|\partial^4 u\|_{L^2}^2 + \varepsilon\|\partial^5 d\|_{L^2}^2 + C\|\nabla d\|_{L^p}^{\frac{2p}{p-3}}\|\partial^4 d\|_{L^2}^2, \quad \forall\,\varepsilon > 0. \end{aligned} \tag{4.7}$$

为估计 $I_4$，我们将利用如下两个有趣的 Gagliardo-Nirenberg 不等式：

$$\|\partial^2 d\|_{L^{\frac{6p}{p+2}}} \leqslant C\|\nabla d\|_{L^p}^{2/3}\|\partial^4 d\|_{L^{\frac{2p}{p-2}}}^{1/3}, \tag{4.8}$$

$$\|\partial^3 d\|_{L^{\frac{3p}{p-1}}} \leqslant C\|\nabla d\|_{L^p}^{1/3}\|\partial^4 d\|_{L^{\frac{2p}{p-2}}}^{2/3}. \tag{4.9}$$

$$I_4 \leqslant C\|\partial^2 d\|_{L^{\frac{6p}{p+2}}}\|\partial^3 d\|_{L^{\frac{3p}{p-1}}}\|\partial^4 u\|_{L^2} \leqslant C\|\nabla d\|_{L^p}\|\partial^4 d\|_{L^{\frac{2p}{p-2}}}\|\partial^4 u\|_{L^2}.$$

余下的估计与 $I_3$ 完全一样，因此，有

$$I_4 \leqslant \varepsilon\|\partial^4 u\|_{L^2}^2 + \varepsilon\|\partial^5 d\|_{L^2}^2 + C\|\nabla d\|_{L^p}^{\frac{2p}{p-2}}\|\partial^4 d\|_{L^2}^2, \quad \forall\,\varepsilon > 0. \tag{4.10}$$

问题 $(1.2)_2$ 两端作用$\partial_4$，然后乘以$\partial^4 d$，分部积分并利用不可压条件，有

$$\begin{aligned} &\frac{1}{2}\frac{\mathrm{d}}{\mathrm{d}t}\int|\partial^4 d|^2\mathrm{d}x + \int|\partial^5 d|^2\mathrm{d}x \\ &= -\sum_i\int(\partial^4(u\cdot\nabla d_i) - u\partial^4\nabla d_i)\partial^4 d_i\mathrm{d}x + \sum_{i,j,k}\int\partial^4(d_k(\partial_j d_i)^2)\cdot\partial^4 d_k\mathrm{d}x \\ &=: J + L. \end{aligned} \tag{4.11}$$

再次利用换位子估计 (4.3)，估计 $J$ 如下：

$$\begin{aligned} J &\leqslant \sum_i\|\partial^4(u\cdot\nabla d_i) - u\partial^4\nabla d_i\|_{L^{\frac{2p}{p+2}}}\|\partial^4 d_i\|_{L^{\frac{2p}{p-2}}} \\ &\leqslant C(\|\nabla u\|_{L^p}\|\partial^4 d\|_{L^2} + \|\nabla d\|_{L^p}\|\partial^4 u\|_{L^2})\|\partial^4 d\|_{L^{\frac{2p}{p-2}}} \\ &\leqslant C\|\nabla u\|_{L^p}\|\partial^4 d\|_{L^2}^{2-3/p}\|\partial^5 d\|_{L^2}^{3/p} + C\|\nabla d\|_{L^p}\|\partial^4 u\|_{L^2}\|\partial^4 d\|_{L^2}^{1-3/p}\|\partial^5 d\|_{L^2}^{3/p} \quad (\text{由式}(4.5)) \\ &\leqslant \varepsilon\|\partial^4 u\|_{L^2}^2 + \varepsilon\|\partial^5 d\|_{L^2}^2 + C\|\nabla u\|_{L^p}^{\frac{2p}{2p-3}}\|\partial^4 d\|_{L^2}^2 + C\|\nabla d\|_{L^p}^{\frac{2p}{p-3}}\|\partial^4 d\|_{L^2}^2, \quad \forall\,\varepsilon > 0. \end{aligned} \tag{4.12}$$

下面估计 $L$.

$$\begin{aligned} L &\leqslant C\int|\nabla d|^2\cdot|\partial^4 d|^2 + |\nabla d|\cdot|\partial^4 d|\cdot|\partial^5 d| + |\partial^3 d|^2\cdot|\partial^4 d| + \\ &\quad |\partial^2 d|^3\cdot|\partial^4 d| + |\partial^2 d|\cdot|\partial^4 d|^2 + |\nabla d|\cdot|\partial^2 d|\cdot|\partial^3 d|\cdot|\partial^4 d|\,\mathrm{d}x \\ &=: \sum_{i=1}^6 l_i. \end{aligned} \tag{4.13}$$

$$\begin{aligned} l_1 &\leqslant C\|\nabla d\|_{L^p}^2\|\partial^4 d\|_{L^{\frac{2p}{p-2}}}^2 \\ &\leqslant C\|\nabla d\|_{L^p}^2\|\partial^4 d\|_{L^2}^{2(1-3/p)}\|\partial^5 d\|_{L^2}^{6/p} \quad (\text{由式}(4.5)) \end{aligned}$$

$$\leqslant \varepsilon\|\partial^5 d\|_{L^2}^2 + C\|\nabla d\|_{L^2}^{\frac{2p}{p-3}}\|\partial^4 d\|_{L^2}^2,\quad \forall\, \varepsilon > 0;$$

$$l_2 \leqslant C\|\nabla d\|_{L^p}\|\partial^4 d\|_{L^{\frac{2p}{p-2}}}\|\partial^5 d\|_{L^2}$$

$$\leqslant \varepsilon\|\partial^5 d\|_{L^2}^2 + l_1$$

$$\leqslant C\varepsilon\|\partial^5 d\|_{L^2}^2 + C\|\nabla d\|_{L^2}^{\frac{2p}{p-3}}\|\partial^4 d\|_{L^2}^2,\quad \forall\, \varepsilon > 0;$$

$$l_3 \leqslant C\|\partial^3 d\|_{L^{\frac{4p}{p+2}}}^2\|\partial^4 d\|_{L^{\frac{2p}{p-2}}}$$

$$\leqslant C\|\nabla d\|_{L^p}\|\partial^5 d\|_{L^2}\|\partial^4 d\|_{L^{\frac{2p}{p-2}}},$$

这里用了 Gagliardo-Nirenberg 不等式

$$\|\partial^3 d\|_{L^{\frac{4p}{p+2}}}^2 \leqslant C\|\nabla d\|_{L^p}\|\partial^5 d\|_{L^2}, \tag{4.14}$$

因此 $l_3$ 与 $l_2$ 完全一样可估计为

$$l_3 \leqslant C\varepsilon\|\partial^5 d\|_{L^2}^2 + C\|\nabla d\|_{L^p}^{\frac{2p}{p-3}}\|\partial^4 d\|_{L^2}^2,\quad \forall\, \varepsilon > 0.$$

用如下的 Gagliardo-Nirenberg 不等式估计 $l_4$:

$$\|\partial^2 d\|_{L^{\frac{6p}{p+2}}}^3 \leqslant C\|d\|_{L^\infty}\|\partial^3 d\|_{L^{\frac{4p}{p+2}}}^2, \tag{4.15}$$

$$l_4 \leqslant C\|\partial^2 d\|_{L^{\frac{6p}{p+2}}}^3\|\partial^4 d\|_{L^{\frac{2p}{p-2}}} \leqslant C\|d\|_{L^\infty}\|\partial^3 d\|_{L^{\frac{4p}{p+2}}}^2\|\partial^4 d\|_{L^{\frac{2p}{p-2}}}.$$

余下的计算与 $l_3$ 完全一样，因此有

$$l_4 \leqslant C\varepsilon\|\partial^5 d\|_{L^2}^2 + C\|\nabla d\|_{L^p}^{\frac{2p}{p-3}}\|\partial^4 d\|^2 L^2,\quad \forall\, \varepsilon > 0.$$

为估计 $l_5$，用如下的 Gagliardo-Nirenberg 不等式：

$$\|\partial^4 d\|_{L^{5/2}}^{5/2} \leqslant C\|d\|_{L^\infty}^{1/2}\|\partial^5 d\|_{L^2}^2. \tag{4.16}$$

$$l_5 \leqslant C\int |\partial^2 d|\cdot|\partial^4 d|^2\mathrm{d}x = C\int |\partial^2 d|\cdot|\partial^4 d|^{1/3}\cdot|\partial^4 d|^{5/3}\mathrm{d}x$$

$$\leqslant C\int |\partial^2 d|^3\cdot|\partial^4 d|\mathrm{d}x + \varepsilon\int|\partial^4 d|^{5/2}\mathrm{d}x$$

$$\leqslant l_4 + \varepsilon\int|\partial^4 d|^{5/2}\mathrm{d}x$$

$$\leqslant l_4 + C\varepsilon\|d\|_{L^\infty}^{1/2}\|\partial^5 d\|_{L^2}^2$$

$$\leqslant C\varepsilon\|\partial^5 d\|_{L^2}^2 + C\|\nabla d\|_{L^p}^{\frac{2p}{p-3}}\|\partial^4 d\|_{L^2}^2,\quad \forall\, \varepsilon > 0.$$

为估计 $l_6$，用如下的 Gagliardo-Nirenberg 不等式：

$$\|\partial^2 d\|_{L^{\frac{4p}{p-2}}}^2 \leqslant C\|d\|_{L^\infty}\|\partial^4 d\|_{L^{\frac{2p}{p-2}}}. \tag{4.17}$$

$$l_6 \leqslant C\int(|\nabla d|^2\cdot|\partial^4 d|^2 + |\partial^2 d|^2\cdot|\partial^3 d|^2)\mathrm{d}x$$

$$\leqslant l_1 + C\|\partial^2 d\|^2_{L^{\frac{4p}{p-2}}}\|\partial^3 d\|^2_{L^{\frac{4p}{p+2}}}$$

$$\leqslant l_1 + C\|d\|_{L^\infty}\|\partial^4 d\|_{L^{\frac{2p}{p-2}}}\|\partial^3 d\|^2_{L^{\frac{4p}{p+2}}}.$$

余下的计算与 $l_1$ 和 $l_3$ 完全一样，因此有

$$l_6 \leqslant C\varepsilon\|\partial^5 d\|^2_{L^2} + C\|\nabla d\|^{\frac{2p}{p-3}}_{L^p}\|\partial^4 d\|^2_{L^2}, \quad \forall\, \varepsilon > 0.$$

将上述估计代入式 (4.4) 与式 (4.11)，将它们相加并取 $\varepsilon$ 充分小，由式 (4.1) 和式 (4.2) 可得

$$\frac{\mathrm{d}}{\mathrm{d}t}\int|\partial^3 u|^2 + |\partial^4 d|^2\mathrm{d}x + \int|\partial^4 u|^2 + |\partial^5 d|^2\mathrm{d}x$$

$$\leqslant C\left(\|\nabla u\|^{\frac{2p}{2p-3}}_{L^p} + \|\nabla d\|^{\frac{2p}{p-3}}_{L^p}\right)\int|\partial^3 u|^2 + |\partial^4 d|^2\mathrm{d}x$$

$$\leqslant C\left(\|\mathrm{curl}u\|^{\frac{2p}{2p-3}}_{\dot{B}^0_{p,\infty}} + \|\nabla d\|^{\frac{2p}{p-3}}_{\dot{B}^0_{p,\frac{2}{3}}}\right)(\|\partial^3 u\|^2_{L^2} + \|\partial^4 d\|^2_{L^2})\log(3 + \|\partial^3 u\|^2_{L^2} + \|\partial^4 d\|^2_{L^2}),$$

由此立知定理成立.

## 5 定理1.4的证明

本节证明定理 1.4.

方程 $(1.3)_1$ 两端作用$\partial^4$，然后乘以$\partial^4 d$，利用$(a\times b)\cdot a=0$，有

$$\frac{1}{2}\frac{\mathrm{d}}{\mathrm{d}t}\int|\partial^4 d|^2\mathrm{d}x + \int|\partial^5 d|^2\mathrm{d}x$$

$$\leqslant\sum_{i,j,k}\int\partial^4(d_k(\partial_j d_i)^2)\cdot\partial^4 d_k\mathrm{d}x +$$

$$C\int|\nabla d|\cdot|\partial^4 d|\cdot|\partial^5 d| + |\partial^2 d|\cdot|\partial^4 d|^2 + |\partial^3 d|^2\cdot|\partial^4 d|\mathrm{d}x$$

$$\leqslant\sum_{i=1}^{6} l_i,$$

余下的计算与式 (4.13) 完全一样，因此定理成立. 我们略去此处的证明.

### 参 考 文 献


[1] Lin F H. Nonlinear theory of defects in nematic liquid crystals: phase transition and flow phenomena. Comm Pure Appl Math, 1989, 42: 789－814.

[2] Lin F H, Liu C. Nonparabolic dissipative systems modelling the flow of liquid crystals. Comm Pure Appl Math, 1995, 48: 501－537.

[3] Lin F H, Liu C. Existence of solutions for the Ericksen-Leslie system. Arch Ration Mech Anal, 2000, 154: 135－156.

[4] Coutand D, Shkoller S. Well posedness of the full Ericksen-Leslye model of nematic liquid crystals. C R Acad Sci Ser I, 2001, 333: 919－924.

[5] Chen Y, Struwe M. Existence and partial regularity results for the heat flow of harmonic maps. Math Z, 1989, 201: 83-103.

[6] Landau L D, Lifshitz E M. On the theory of the dispersion of magnetic permeability in ferromagnetic bodies. Phys Z Sowj, 1935, 8: 153-169.

[7] Chen Y, Ding S, Guo B L. Partial regularity for two-dimensional Landau-Lifshitz equations. Acta Math Sin, New Ser, 1998, 14(3): 423-432.

[8] Guo B L, Hong M. The Landau-Lifshitz equations of the ferromagnetic spin chain and harmonic maps. Calc Var PDE, 1993, 1: 311-334.

[9] Liu X. Partial regularity for the Landau-Lifshitz systems. Calc Var PDE, 2003, 20(2): 153-173.

[10] Zhou Y, Guo B L. Existence of weak solution for boundary problems of systems of ferromagnetic chain. Sci China Ser A-Math, 1984, 27: 799-811.

[11] Zhou Y, Guo B L. Weak solution systems of ferromagnetic chain with several variables. Sci China Ser A-Math, 1987, 30: 1251-1266.

[12] Zhou Y, Guo B L, Tan S. Existence and uniqueness of smooth solution of systems of ferromagnetic chain. Sci China Ser A-Math, 1991, 34: 257-266.

[13] Zhou Y, Sun H, Guo B L. Multidimensional system of ferromagnetic chain type. Sci China Ser A-Math, 1993, 36: 1422-1434.

[14] Alouges F, Soyeur A. On global weak solutions for Landau-Lifshitz equations: existence and nonuniqueness. Nonlinear Anal, 1992, 18(11): 1084-1096.

[15] Carbou G, Fabrie P. Regular solutions for Landau-Lifshitz equation in a bounded domain. Differential Integral Equations, 2001, 14: 213-229.

[16] Leray J. Sur le mouvement d'un liquide visqeux emplissant l'espace. Acta Math, 1934, 63: 193-248.

[17] Prodi G. Un teorema di unicità per le equazioni di Navier-Stokes. Ann Mat Pura Appl, 1959, 48: 173-182.

[18] Serrin J. On the interior regularity of weak solutions of the Navier-Stokes equations. Arch Ration Mech Anal, 1962, 9: 187-195.

[19] Giga Y. Solutions for semilinear parabolic equations in $L^p$ and regularity of weak solutions of the Navier-Stokes system. J Differential Equations, 1986, 62: 186-212.

[20] Da Veiga H B. A new regularity class for the Navier-Stokes equations in $\mathbf{R}^n$. Chinese Ann Math Ser B, 1995, 16: 407-412.

[21] Kozono H, Ogawa T, Taniuchi Y. The critical Sobolev inequalities in Besov spaces and regularity criterion to some semilinear evolution equations. Math Z,2002, 242: 251-278.

[22] Miao C X. Harmonic analysis and its applications in PDE. 2nd ed (in Chinese). Beijing: Science Press, 2004.

[23] Triebel H. Theory of function spaces Ⅱ. Boston: Birkhäuser Basel, 1992.

[24] Kozono H, Shimada Y. Bilinear estimates in homogeneous Triebel-Lizorkin spaces and the Navier-Stokes equations. Math Nachr, 2004, 276:63-74.

[25] Lemarié-Rieusaet P G. Recent developments in the Navier-Stokes problem. London: Chapman & Hall/CRC, 2002.

[26] David G, Journé J L. A boundedness criterion for generalized Calderón Zygmund operators. Ann of Math, 1984, 120: 371-397.

[27] Kato T, Ponce G. Commutator estimates and the Euler and Navier-Stokes equations. Comm Pure Appl Math, 1988, 41: 891-907.

# Random Attractors for a Quasi-geostrophic Dynamical System under Stochastic Forcing *

Huang Daiwen(黄代文) Guo Boling(郭柏灵) Han Yongqian(韩永前)

**Abstract**

We consider a dissipative two-dimensional quasi-geostrophic equation, which model a class of large-scale geophysical flows under a stochastic external forcing (the forcing is a Gaussian space-time random field, white noise in time). First, by Faedo-Galerkin method, we prove the existence and uniqueness of the global solution to the initial boundary value problem of the stochastic equation. Second, by studying the asymptotic behaviour of the solution, we obtain the existence of random attractors for the stochastic quasi-geostrophic dynamical system.

**Keywords** stochastic equation; quasi-geostrophic dynamics; white noise; random attractors

## 1 Introduction

The barotropic model, also called the rotating shallow-water model in fluid mechanics, is derived from the primitive equations of the atmosphere or the ocean under the assumption that the velocity and pressure fields are independent of the vertical coordinate[1]. Experience has shown that the barotrophic model is capable of describing important aspects of atmospheric and oceanic motions. Charney et al[2] used the vorticity form of the barotrophic model to successfully make the first numerical weather prediction.

Since the inviscid barotrophic model is a hyperbolic system, the numerical solutions of the model develop high-frequency gravity wave solutions which can seriously distort short-term forecasts (on the order of hours to days, typically 12 hours). In order to filter out the high-frequency waves, from the physical point of view, one can derive the two-dimensional quasi-geostrophic equation as an approximation of the rotating shallow-water equations by a conventional asymptotic expansion for small Rossby number (the ratio of the horizontal acceleration to the Coriolis force)[1]. Recently, the papers [3-5] have shown by rigorous mathematical proofs that the two-dimensional inviscid quasi-geostrophic equation is a valid approximation of the rotating inviscid shallow water equations without friction and external forcing in the limit of zero Rossby number, for some appropriate initial data.

Atmospheric or oceanic motions exhibit fluctuations over space and time scales ranging from centimeters to thousands of kilometers and from seconds to decades and beyond. Some of these fluctuations are thought to be caused by internal instability processes, others by external forcing. For

* Int J Dynamical Systems and Differential Equations, 2008, 1(3): 147-154.

the atmosphere, the main external forcing coming from the solar radiation is stochastic[6] The external forcing of the ocean comes mainly from the atmosphere. As usual, the atmospheric forcing field must be regarded as random[7-11], for instance. Therefore, in study the models of large-scale ocean or atmosphere, taking the stochastic external factors into account is reasonable and necessary. There are many works about mathematical study of several stochastic climate models[8,12-15].

In the present paper, we study the dissipative two-dimensional quasi-geostrophic equation with a stochastic external forcing (the explicit definition of the equation will be given in Section 2):

$$\left(\frac{\partial}{\partial t}+\frac{\partial\psi}{\partial x}\frac{\partial}{\partial y}-\frac{\partial\psi}{\partial y}\frac{\partial}{\partial x}\right)(\Delta\psi-F\psi+\beta_0 y)=\frac{1}{Re}\Delta^2\psi-\frac{r}{2}\Delta\psi+f(x,y,t) \tag{1.1}$$

on a regular enough bounded open domain $D\subset\mathbf{R}^2$, where $\psi$ is the stream function, $\frac{1}{Re}\Delta^2\psi$ the viscosity term, $-\frac{r}{2}\Delta\psi$ the friction term, $f(x,y,t)$ the external forcing form, cf. [1] or [16]. In this paper, we take $f(x, y, t)$ as a stochastic forcing given later.

The paper [12-14] considered the two-dimensional non-divergent quasi-geostrophic equation

$$\left(\frac{\partial}{\partial t}+\frac{\partial\psi}{\partial x}\frac{\partial}{\partial y}-\frac{\partial\psi}{\partial y}\frac{\partial}{\partial x}\right)(\Delta\psi+\beta_0 y)=\frac{1}{Re}\Delta^2\psi-\frac{r}{2}\Delta\psi+f(x,y,t) \tag{1.2}$$

on a bounded regular enough domain in $\mathbf{R}^2$, where $f(x,y,t)$ is a stochastic forcing. In [13], by the theory for semigroup of operators and Banach Fixed Pointed Theorem, Brannan etc. obtained the global well-posedness of equation (1.2). However, they did not consider the asymptotic behaviour of the solutions. Blömker et al[12] was devoted to studying enstrophy dynamics of stochastically forced large-scale geophysical flows which is modelled by (1.2). In [14], Duan etc. have proved the exponential stability of equation (1.2) with random boundary conditions.

Our paper is devoted to understanding the asymptotic behaviour of the dynamical system

$$\begin{cases}\left(\frac{\partial}{\partial t}+\frac{\partial\psi}{\partial x}\frac{\partial}{\partial y}-\frac{\partial\psi}{\partial y}\frac{\partial}{\partial x}\right)(\Delta\psi-F\psi+\beta_0 y)=\frac{1}{Re}\Delta^2\psi-\frac{r}{2}\Delta\psi+f(x,y,t), & \text{in } D\\ \psi(x,y,t)=0, \quad \text{on } \partial D\\ \Delta\psi(x,y,t)=0, \quad \text{on } \partial D\end{cases} \tag{1.3}$$

The understanding of the asymptotic behaviour of dynamical systems is one of the most important problems of modern mathematical physics. One way to attack the problem is to consider its global attractors[17-18]. By taking advantage of the geostrophic balance and other geophysical consideration, several intermediate models have been the subject of studying the long-time dynamics and global attractors in order to describe the mechanism of long-term weather prediction and climate dynamics[19-21]. We aim to study the existence of random attractors of the system (1.3) under some appropriate assumptions of $f$. Our main results are Theorems 3.1 and 5.1. First, applying the properties of an auxiliary Ornstein-Uhlenbeck process, we prove the existence and uniqueness of the global solution to the initial boundary value problem of the stochastic equation (1.1) by Faedo-Galerkin method. Second, by studying the asymptotic behaviour of the solutions, we obtain the existence of random attractors for the stochastic quasi-geostrophic dynamical system.

The paper is organised as follows. In Section 2, we shall introduce the model of a dissipative two

-dimensional quasi-geostrophic equation with a stochastic forcing term. In Section 3, we shall prove global well-posedness of the initial boundary value problem for the stochastic equation (1.1). We are going to give some preliminaries for random attractors in Section 4. Section 5 is devoted to proving the existence of random attractors of dissipative quasi-geostrophic dynamical system under stochastic forcing.

## 2 The Model

The two-dimensional stochastic quasi-geostrophic equation in $D$ is written as[1]

$$\left(\frac{\partial}{\partial t}+\frac{\partial \psi}{\partial x}\frac{\partial}{\partial y}-\frac{\partial \psi}{\partial y}\frac{\partial}{\partial x}\right)(\Delta\psi - F\psi + \beta_0 y) = \frac{1}{Re}\Delta^2\psi - \frac{r}{2}\Delta\psi + f(x,y,t), \tag{2.1}$$

where $\Delta$ is the Laplacian, $x$ and $y$ the eastward and northward directed coordinates respectively, $F$ Froude number, $Re$ Reynolds number, $\beta_0$ a positive constant, $r$ the Ekman dissipation constant, $f(x,y,t) = -\frac{\mathrm{d}W}{\mathrm{d}t}$ (the definition of $W$ will be given later) a Gaussian space-time random field, white noise in time, subject to the restrictions imposed below. Here we assume that the topography is equal to zero.

Let $A = -\frac{1}{Re}\Delta$, $A: L^2(D)\rightarrow L^2(D)$ with domain $D(A) = H^2(D)\cap H_0^1(D)$, where $L^2(D)$ is usual Lebesgue space and $H^2(D)$, $H_0^1(D)$ are Sobolev spaces. $|\cdot|_p$ is the usual norm of the space $L^p(D)$ $(1\leqslant p\leqslant +\infty)$ and $\|\cdot\|$ is the usual norm in $H_0^1(D)$. The operator $A$ is positive, self-adjoint with the compact inverse operator $A^{-1}$. We denote by $0<\lambda_1<\lambda_2\leqslant\cdots$ the eigenvalues of $A$ and by $e_1$, $e_2$, $\cdots$ a corresponding complete orthonormal system of eigenvectors. And we remark that $\frac{1}{Re}\|u\|^2\geqslant\lambda_1|u|_2^2$ for any $u\in H_0^1(D)$. We denote by $e^{t(-A)}, t\geqslant 0$ the semigroup on $L^2(D)$ generated by $-A$.

Here we assume that the random process $W$ be a two-sided in time Wiener process with the form

$$W(t) = \sum_{i=1}^{+\infty}\mu_i\omega_i(t)e_i,$$

where $\omega_1, \omega_2, \cdots$ is a sequence of independent standard Brownian motions on a complete probability space $(\Omega, \mathcal{F}, P)$ (with expectation denoted by $E$), and the coefficients $\mu_i$ satisfy the condition

$$\sum_{i=1}^{+\infty}\frac{\mu_i^2}{\lambda_i^{\frac{1}{2}-2\beta_1}} < +\infty, \quad \text{for some } \beta_1 > 0.$$

We shall study equation (2.1) with the no-penetration and free-slip boundary conditions[16,22].

$$\psi(x,y,t) = 0, \quad \text{on } \partial D,$$
$$\Delta\psi(x,y,t) = 0, \quad \text{on } \partial D.$$

For any $u\in L^2(D)$, by solving the elliptic equation with Dirichlet boundary condition

$$\begin{cases} F\psi - \Delta\psi = u \\ \psi|_{\partial D} = 0 \end{cases},$$

we obtain $\psi=(F-\Delta)^{-1}u=B(u)$. By elliptic regular theory[23], we know that $B: L^2(D)\to H_0^1(D)\cap H^2(D)$. Therefore the equation (2.1) can be rewritten as

$$u_t+J(\psi,u)-\beta_0\psi_x=\frac{1}{Re}\Delta u+\left(\frac{F}{Re}-\frac{r}{2}\right)u-F\left(\frac{F}{Re}-\frac{r}{2}\right)\psi+\frac{\mathrm{d}W}{\mathrm{d}t}, \tag{2.2}$$

where $W$ is differential in a weak sense $\left(\frac{\mathrm{d}W}{\mathrm{d}t}\text{ is defined by the usual Itô integration, cf. [24-25]}\right)$, $\psi=B(u)$, $J$ the Jacobian operator which is defined by $J(\psi,u)=\frac{\partial\psi}{\partial x}\frac{\partial u}{\partial y}-\frac{\partial\psi}{\partial y}\frac{\partial u}{\partial x}$.

Define

$$G(u)=-J(\psi,u)+\beta_0\psi_x+\left(\frac{F}{Re}-\frac{r}{2}\right)u-F\left(\frac{F}{Re}-\frac{r}{2}\right)\psi.$$

In the following, we shall consider

$$u_t-\frac{1}{Re}\Delta u=G(u)+\frac{\mathrm{d}W}{\mathrm{d}t} \tag{2.3}$$

with boundary condition

$$u|_{\partial D}=0, \tag{2.4}$$

and initial condition

$$u(x,y,s)=u_s, \tag{2.5}$$

where $s\in\mathbf{R}$. We remark that the boundary value problem (2.3) and (2.4) coincides with the dynamical system (1.3).

Now we rewrite equations (2.3)~(2.5) as the abstract differential stochastic equation together with the initial-boundary condition

$$\begin{cases}\mathrm{d}u=-Au\mathrm{d}t+G(u)\mathrm{d}t+\mathrm{d}W\\ u(s,\omega)=u_s\end{cases}. \tag{2.6}$$

## 3 Existence, Uniqueness and Regularity of Global Solutions

In order to obtain the existence, uniqueness and regularity of global solutions for the problem (2.6), we should consider an auxiliary Ornstein-Uhlenbeck process. For any $\alpha>0$, let

$$z(t)=W_A^\alpha(t)=\int_{-\infty}^{t}\mathrm{e}^{(t-s)(-A-\alpha)}\mathrm{d}W(s),$$

where the Ornstein-Uhlenbeck process $W_A^\alpha(t)$ is the solution of the initial value problem

$$\begin{cases}\mathrm{d}z=(-Az-\alpha z)\mathrm{d}t+\mathrm{d}W(t)\\ z(0)=\int_{-\infty}^{0}\mathrm{e}^{-s(-A-\alpha)}\mathrm{d}W(s)\end{cases}.$$

$W_A^\alpha(t)$ has a continuous version with values in $D(A^{\frac{1}{4}+\beta})$ $(\beta<\beta_1)$ for $P$-a.e. $\omega\in\Omega$ (here "almost everywhere about probability" is denoted by $P$-a.e.), especially in $C_0(D)$ (see [25-27]), where $C_0(D):=\{u; u\in C(D)$, supp $u$ is a compact subset in $D\}$. For the definition of differential operator $A^{\frac{1}{4}+\beta}$, one can refer to [28]. Assume that $u$ be the solution of the initial boundary value problem

$$
(\mathrm{P}_1)\quad \begin{cases} du - \dfrac{1}{Re}\Delta u dt = G(u)dt + dW(t) \\ u(s,w) = u_s \\ u|_{\partial D} = 0 \end{cases}.
$$

Let $v(t)=u(t)-z(t)$. Then $v(t)$ is a solution for the problem

$$
(\mathrm{P}_2)\quad \begin{cases} dv = -Av + G(v+z) + \alpha z \\ v(s,w) = u_s - z(s) \end{cases}. \tag{3.1}
$$

One of our main results is the following theorem.

**Theorem 3.1** (i) For any $T>s$, each initial condition $u_s\in L^2(D)$ for $P$-a.e. $\omega\in\Omega$, there exists a unique solution $v\in C(s,T;L^2(D))\cap L^2(s,T;H_0^1(D))$ to the problem $(\mathrm{P}_2)$ in the weak sense for $P$-a.e. $\omega\in\Omega$ and the solution $u=v+z$ for the problem $(\mathrm{P}_1)$ satisfies $u\in C(s,T;L^2(D))\cap L^2(s,T;D(A^{\min\{\frac{1}{4}+\beta,\frac{1}{2}\}}))$ a.s. (almost surely) for all $\beta<\beta_1$, where $\beta_1$ is given in the definition of the random process $W(t)$ in the Section 2.

(ii) If $u_s\in D(A^\theta)$ for $P$-a.e. $\omega\in\Omega$ and some $\theta\in(0,2\beta_1)\cap\left(0,\frac{1}{2}\right]$, then $v\in C(s,T;D(A^\theta))\cap L^2(s,T;D(A^{\frac{1}{2}+\theta}))$ for $P$-a.e. $\omega\in\Omega$.

The proof of Theorem 3.1 is classically obtained by Faedo-Galerkin method[29]. Since this method is now standard, we only give the *a priori* estimates.

**Energy estimates on** $v$. In the sequel, $\omega\in\Omega$ is fixed. By choosing $v$ as a test function in equation (3.1), we have

$$
\frac{1}{2}\frac{d|v|_2^2}{dt} + \frac{1}{Re}\|v\|^2 = \int_D G(v+z)v + \alpha\int_D zv, \tag{3.2}
$$

where

$$
\int_D G(v+z)v = -\int_D [J(B(v+z),v+z)v + \beta_0(B(v+z))_x v] + \int_D\left[\left(\frac{F}{Re}-\frac{r}{2}\right)(v+z)v - F\left(\frac{F}{Re}-\frac{r}{2}\right)B(v+z)v\right]. \tag{3.3}
$$

By integration by parts, Hölder inequality and Young inequality,

$$
\begin{aligned}
-\int_D J(B(v+z),v+z)v &= \int_D J(B(v+z),v+z)z \\
&= \int_D J(B(v+z),v)z \\
&\leqslant c|v|_2^2|z|_4^4 + c|z|_4^4 + 2\varepsilon\|v\|^2,
\end{aligned} \tag{3.4}
$$

$$
\left|\beta_0\int_D (B(z))_x v\right| \leqslant c|(B(z))_x|_2|v|_2 \leqslant c|z|_2|v|_2 \leqslant c|z|_2^2 + \varepsilon|v|_2^2. \tag{3.5}
$$

Similarly to equation (3.5),

$$
\left|\left(\frac{F}{Re}-\frac{r}{2}\right)\int_D zv\right| \leqslant c|z|_2^2 + \varepsilon|v|_2^2, \tag{3.6}
$$

$$
\left|-F\left(\frac{F}{Re}-\frac{r}{2}\right)\int_D vB(z)\right| \leqslant c|z|_2^2 + \varepsilon|v|_2^2. \tag{3.7}
$$

By the definition of the operator $B$, we prove

$$\beta_0\int_D (B(v))_x v + \left(\frac{F}{Re}-\frac{r}{2}\right)\int_D v^2 - F\left(\frac{F}{Re}-\frac{r}{2}\right)\times\int_D vB(v)\leqslant 0. \tag{3.8}$$

From equations (3.2)~(3.8), by choosing small enough $\varepsilon>0$, we obtain

$$\frac{d|v|_2^2}{dt}+\frac{1}{Re}\|v\|^2\leqslant c|z|_4^4|v|_2^2+c|z|_2^2+c|z|_4^4. \tag{3.9}$$

By Poincaré inequality $\frac{1}{Re}\|v\|^2\geqslant\lambda_1|v|_2^2$,

$$\frac{d|v|_2^2}{dt}\leqslant\left(c|z|_4^4-\frac{\lambda_1}{Re}\right)|v|_2^2+c|z|_2^2+c|z|_4^4.$$

So, by Gronwall inequality, for all $t\in[s,T]$, we have

$$|v(t)|_2^2\leqslant e^{\int_s^t(-\lambda_1+c|z(\tau)|_4^4)d\tau}|v(s)|_2^2+\int_s^t e^{\int_\sigma^t(-\lambda_1+c|z(\tau)|_4^4)d\tau}(c|z|_2^2+c|z|_4^4)d\sigma. \tag{3.10}$$

Integrating equation (3.9) from $t_1$ to $t_2$ ($[t_1,t_2]\subset[s,T]$),

$$\frac{1}{Re}\int_{t_1}^{t_2}\|v(\tau)\|^2d\tau\leqslant|v(t_1)|_2^2+\int_{t_1}^{t_2}(c|z|_4^4|v|_2^2+c|z|_2^2+c|z|_4^4)d\tau. \tag{3.11}$$

**Lemma 3.2**[26] For any two arbitrary real valued functions $f,g\in H^{\frac{1}{2}+\theta}(D), 0<\theta<\frac{1}{2}$,

$$\|fg\|_{H^{2\theta}}\leqslant\|f\|_{H^{\frac{1}{2}+\theta}}\|f\|_{H^{\frac{1}{2}+\theta}}.$$

**Energy estimates on $A^\theta v$.** As above, $\omega\in\Omega$ is fixed. Assume $v$ is a solution to the problem $(P_2)$, with the regularity $v\in C(s,T;D(A^\theta))\cap L^2(s,T;D(A^{\frac{1}{2}+\theta}))$. Since the proof of the case $\theta=\frac{1}{2}$ is classical[18], here we only prove the case $0<\theta<\frac{1}{2}$.

Then we get

$$\frac{1}{2}\frac{d|A^\theta v|_2^2}{dt}+\frac{1}{Re}|A^{\frac{1}{2}+\theta}v|_2^2=\int_D A^\theta G(v+z)A^\theta v+\alpha\int_D A^\theta zA^\theta v. \tag{3.12}$$

By the definition of $A^\theta$, the interpolation inequalities and Lemma 3.2,

$$\begin{aligned}
&-\int_D A^\theta J(B(v+z),v+z)A^\theta v\\
&=\int_D\left(\frac{\partial B(v+z)}{\partial x}\frac{\partial A^{2\theta}v}{\partial y}-\frac{\partial B(v+z)}{\partial y}\frac{\partial A^{2\theta}v}{\partial x}\right)(v+z)\\
&\leqslant c\left\|\frac{\partial A^{2\theta}v}{\partial y}\right\|_{H^{-2\theta}}\left\|\frac{\partial B(v+z)}{\partial x}(v+z)\right\|_{H^{2\theta}}+c\left\|\frac{\partial A^{2\theta}v}{\partial x}\right\|_{H^{-2\theta}}\left\|\frac{\partial B(v+z)}{\partial y}(v+z)\right\|_{H^{2\theta}}\\
&\leqslant c|A^{\frac{1}{2}+\theta}v|_2|A^{\frac{1}{2}\theta}(v+z)|_2|A^{\frac{1}{4}+\frac{\theta}{2}}(v+z)|_2\\
&\leqslant\varepsilon|A^{\frac{1}{2}+\theta}v|_2^2+c|A^\theta v|_2^2|v|_2^2+c|A^{\frac{1}{4}+\frac{\theta}{2}}v|_2^4+c|A^{\frac{1}{4}+\frac{\theta}{2}}z|_2^4.
\end{aligned} \tag{3.13}$$

By interpolation inequality and Young inequality,

$$\begin{aligned}
|A^{\frac{1}{4}+\frac{\theta}{2}}v|_2^4&\leqslant c|A^{\frac{1}{4}}v|_2^2|A^{\frac{1}{4}+\theta}v|_2^2\\
&\leqslant c|v|_2|A^{\frac{1}{2}}v|_2|A^\theta v|_2|A^{\frac{1}{2}+\theta}v|_2\\
&\leqslant\varepsilon|A^{\frac{1}{2}+\theta}v|_2^2+c|v|_2^2|A^{\frac{1}{2}}v|_2^2|A^\theta v|_2^2.
\end{aligned} \tag{3.14}$$

Combining equation (3.13) with (3.14), we get

$$-\int_D A^\theta J(B(v+z),v+z)A^\theta v$$

$$\leqslant 2\varepsilon|A^{\frac{1}{2}+\theta}v|_2^2+c|v|_2^2|A^\theta v|_2^2+c|v|_2^2|A^{\frac{1}{2}}v|_2^2|A^\theta v|_2^2+c|A^{\frac{1}{4}+\frac{\theta}{2}}z|_2^4. \tag{3.15}$$

By Hölder inequality and Cauchy-Schwarz inequality,

$$\beta_0\int_D A^\theta(B(v+z))_x A^\theta v\leqslant c|A^\theta(B(v+z))_x|_2|A^\theta v|_2$$

$$\leqslant c|A^\theta v|_2^2+c|z|_2^2. \tag{3.16}$$

Similarly to equation (3.13),

$$\left(\frac{F}{Re}-\frac{r}{2}\right)\int_D A^\theta(v+z)A^\theta v-F\left(\frac{F}{Re}-\frac{r}{2}\right)\int_D A^\theta B(v+z)A^\theta v$$

$$\leqslant c|A^\theta v|_2^2+\varepsilon|A^{\frac{1}{2}+\theta}v|_2^2+c|z|_2^2. \tag{3.17}$$

By equations (3.12), (3.15)~(3.17), we obtain

$$\frac{1}{2}\frac{\mathrm{d}|A^\theta v|_2^2}{\mathrm{d}t}+\frac{1}{Re}|A^{\frac{1}{2}+\theta}v|_2^2$$

$$\leqslant 3\varepsilon|A^{\frac{1}{2}+\theta}v|_2^2+c|v|_2^2|A^{\frac{1}{2}}v|_2^2|A^\theta v|_2^2+c|v|_2^2|A^\theta v|_2^2+c|A^\theta v|_2^2+|A^{\frac{1}{4}+\frac{\theta}{2}}z|_2^4+c|z|_2^2.$$

Therefore, by choosing small enough $\varepsilon>0$, we get

$$\frac{\mathrm{d}|A^\theta v|_2^2}{\mathrm{d}t}+\frac{1}{Re}|A^{\frac{1}{2}+\theta}v|_2^2$$

$$\leqslant c|v|_2^2|A^{\frac{1}{2}}v|_2^2|A^\theta v|_2^2+c|A^\theta v|_2^2+c|v|_2^2|A^\theta v|_2^2+c|A^{\frac{1}{4}+\frac{\theta}{2}}z|_2^4+c|z|_2^2. \tag{3.18}$$

By Gronwall inequality, for all $t\in[s,T]$,

$$|A^\theta v(t)|_2^2\leqslant \mathrm{e}^{\int_s^t(c|v|_2^2|A^{\frac{1}{2}}v|_2^2+c|v|_2^2+c)\mathrm{d}\tau}|A^\theta v(s)|_2^2+$$

$$\int_s^t \mathrm{e}^{\int_\sigma^t(c|v|_2^2|A^{\frac{1}{2}}v|_2^2+c|v|_2^2+c)\mathrm{d}\tau}(c|z|_2^2+c|A^{\frac{1}{4}+\frac{\theta}{2}}z|_2^2)\mathrm{d}\sigma. \tag{3.19}$$

Integrating equation (3.18) from $t_1$ to $t_2$ ($[t_1,t_2]\subset[s,T]$), we have

$$\frac{1}{Re}\int_{t_1}^{t_2}|A^{\frac{1}{2}+\theta}v|_2^2\mathrm{d}\tau$$

$$\leqslant|A^\theta v(t_1)|_2^2+\int_{t_1}^{t_2}c(|v|_2^2|A^{\frac{1}{2}}v|_2^2|A^\theta v|_2^2+|v|_2^2|A^\theta v|_2^2+|A^\theta v|_2^2+|A^{\frac{1}{4}+\frac{\theta}{2}}z|_2^4+|z|_2^2)\mathrm{d}\tau.$$

## 4 Preliminaries for Random Attractor

We recall some results about random attractors in [30].

Let $(X, d)$ be a complete separable metric space and $(\Omega,\mathscr{F},P)$ be a complete probability space. A family of mappings $S(t,s;\omega):X\to X$, $-\infty<s\leqslant t<\infty$, parametrised by $\omega\in\Omega$, satisfy for $P$-a.e. $\omega$ the two properties:

(i) $S(t,r;\omega)S(r,s;\omega)x=S(t,s;\omega)x$ for all $s\leqslant r\leqslant t$ and $x\in X$;

(ii) $S(t,s;\omega)$ is continuous in $X$, for all $s\leqslant t$.

**Definition 4.1** Given $t\in\mathbf{R}$ and $\omega\in\Omega$, a set $K(t,\omega)\subset X$ is called an attracting set if, for all bounded sets $B\subset X$,

$$d(S(t,s;\omega)B,K(t,\omega))\to 0,\quad \text{as } s\to-\infty.$$

**Definition 4.2** $(S(t,s;\omega))_{t\geqslant s,\omega\in\Omega}$ is said to be asymptotically compact if there exists a measurable set, with measure one, $\Omega_0\subset\Omega$ such that, for all $t\in\mathbf{R}$ and $\omega\in\Omega_0$, there exists a compact attracting set $K(t,\omega)$.

**Proposition 4.3** Assume the stochastic dynamical system $(S(t,s;\omega))_{t\geqslant s,\omega\in\Omega}$ is asymptotically compact. Then, for $P$-a.e. $\omega$ the following results hold true. For all $t\in\mathbf{R}$, the set $A(t,\omega)$ is a nonempty compact subset of $K(t,\omega)$, it attracts all bounded sets from $-\infty$($A(t,\omega)$ is usually called the random attractor), and it is the minimal closed set with this property, where

$$A(t,\omega)=\overline{\bigcup_{B\subset X}A(B,t,\omega)},$$

$$A(B,t,\omega)=\bigcap_{T<t}\overline{\bigcup_{s<T}S(t,s;\omega)B}.$$

Moreover, $A(t,\omega)$ is invariant, in the sense that, for all $s\leqslant t$,

$$S(t,s;\omega)A(s,\omega)=A(t,\omega).$$

**Proposition 4.4** Under assumptions (ⅰ), (ⅱ),

(ⅲ) for all $s<t$ and $x\in X$, the mapping

$$\omega\mapsto S(t,s;\omega)x$$

is measurable from $(\Omega,\mathscr{F})$ to $(X,\mathscr{B}(X))$, and

(ⅳ) for all $t,x\in X$, and $P$-a.e. $\omega$, the mapping $s\mapsto S(t,s;\omega)$ is right continuous at any point.

And suppose that there exists a group $\theta_t$, $t\in\mathbf{R}$, of measure preserving transformations of $(\Omega,\mathscr{F},P)$, with the property that for all $s<t$ and $x\in X$,

$$S(t,s;\omega)x=S(t-s,0;\theta_s\omega)x,\quad P\text{-a.s.},\tag{4.1}$$

and that for $P$-a.e. $\omega\in\Omega$ there exists a compact attracting set $K(\omega)$ at time 0, i.e., such that for all bounded sets $B\subset X$,

$$d(S(0,s;\omega)B,K(\omega))\to 0,\quad \text{as } s\to-\infty.$$

Then the stochastic dynamical system $(S(t,s;\omega))_{t\geqslant s,\omega\in\Omega}$ is asymptotically compact.

Applying Propositions 4.3 and 4.4, one can have the following theorem.

**Theorem 4.5** Let $(S(t,s;\omega))_{t\geqslant s,\omega\in\Omega}$ be a stochastic dynamical system satisfying (ⅰ), (ⅱ), (ⅲ), and (ⅳ). Assume that there exists a group $\theta_t$, $t\in\mathbf{R}$, of measure preserving mapping such that (4.1) holds and that, for $P$-a.e. $\omega\in\Omega$, there exists a compact attracting set $K(\omega)$ at time 0. For $P$-a.e. $\omega\in\Omega$, we set

$$A(\omega)=\overline{\bigcup_{B\subset X}A(B,\omega)}$$

where the union is taken over all the bounded subsets of $X$ and $A(B,\omega)$ is given by

$$A(B,\omega)=\bigcap_{\tau<0}\overline{\bigcup_{s<\tau}S(0,s;\omega)B}.$$

Then we have for $P$-a.e. $\omega\in\Omega$:

1. $A(\omega)$ is a nonempty compact subset of $X$, and if $X$ is connected, it is a connected subset of $K(\omega)$.

2. The family $A(\omega),\omega\in\Omega$ is measurable (A family $A(\omega),\omega\in\Omega$ of closed subsets of $X$ is measurable if, for all $x\in X$, the function $\omega\mapsto d(A(\omega),x)$ is measurable).

3. $A(\omega)$ is invariant in the sense that

$$S(t,s;\omega)A(\theta_s\omega) = A(\theta_t\omega), \quad s \leqslant t.$$

4. It is the minimal closed set such that for $t \in \mathbf{R}, B \subset X$ bounded

$$d(S(t,s;\omega)B,\ A(\theta_t\omega)) \to 0, \quad \text{when } s \to -\infty.$$

5. For any bounded set $B \subset X$, $d(S(t,s;\omega)B,\ A(\theta_t\omega)) \to 0$ in probability when $t \to \infty$.

And if the time shift $\theta_t, t \in \mathbf{R}$ is ergodic, then

6. There exists a bounded $B \subset X$ such that

$$A(\omega) = A(B,\omega).$$

7. $A(\omega)$ is the largest compact measurable set which is invariant in the sense of equation (3).

## 5 Existence of Random Attractors

In this section, by studying the dissipativity property of the dynamical system (1.3), we obtain the existence of random attractors for the system.

**Theorem 5.1** The two dimensional quasi-geostrophic equation with a general white noise

$$\mathrm{d}u - \frac{1}{Re}\Delta u \mathrm{d}t = G(u)\mathrm{d}t + \mathrm{d}W(t), \quad u|_{\partial D} = 0$$

has a random attractor $\mathscr{A}(\omega)$ in $L^2(D)$.

Before the proof of Theorem 5.1, we give some notations.

Let

$$\Omega = \{\omega : \omega \in C(\mathbf{R}, D(A^{-\frac{1}{4}+\beta_1})), \omega(0) = 0\},$$

$\mathscr{F}$ the Borel sigma-algebra induced by the compact open-topology of $\Omega$, $P$ a Wiener measure on $(\Omega,\mathscr{F})$. Writing $W(t,\omega) = \omega(t)$, we define

$$\vartheta_t\omega(s) = \omega(t+s) - \omega(t).$$

Then $\vartheta_t$ satisfies $\vartheta_{t+s} = \vartheta_t \circ \vartheta_s$ for all $t, s \in \mathbf{R}$, and $\vartheta_t$ is ergodic under $P$. By Theorem 3.1, let

$$u(t,\omega) = S(t,s;\omega)u_s,$$

where $u(t,\omega)$ is a unique solution to equations (2.9)~(2.15) on $[s,t]$ with the initial value $u(s) = u_s$. Then, for $s \leqslant r \leqslant t$, we have $S(t,s;\omega) = S(t,r;\omega)S(r,s;\omega)$. For any $s, t \in \mathbf{R}_+$, $u_0 \in L^2(D)$, we have $P$-a.s.

$$S(t+s,0;\omega)u_0 = S(t,0;\vartheta_s\omega)S(s,0;\omega)u_0.$$

**Remark 5.2** Define $\psi : \mathbf{R}_+ \times \Omega \times L^2(D) \to L^2(D), \psi(t,\omega)u_0 = S(t,0;\omega)u_0$. By Theorem 3.1, $\psi$ is a continuous random dynamical system (in the usual sense) on $L^2(D)$ over $(\Omega, \mathscr{F}, \mathscr{P}, \{\vartheta_t\}_{t\in\mathbf{R}})$ and models the dynamical system generated by equation (1.3). $\mathscr{A}(\omega)$ attracts all deterministic bound sets $B \subset L^2(D)$, i.e.,

$$\lim_{t\to+\infty} d(\psi(t,\vartheta_{-t}\omega)B, A(\omega)) = \lim_{s\to-\infty} d(S(0,s;\omega)B, K(0,\omega)) = 0, \quad P\text{-a.s.}$$

where $\psi(t,\vartheta_{-t}\omega)x$ can be interpreted as the position at $t=0$ of the trajectory which was $x$ at time $-t$, that is, when $t$ is moving , the trajectory $\psi(t,\vartheta_{-t}\omega)x$ is always at the position at $t=0$, $x \in L^2(D)$. Therefore, the random attractor $\mathscr{A}(\omega)$ is also called the random pull-back attractor.

**Proof** From the ergodicity of the process $z$ with values in $D(A^{\frac{1}{4}+\beta})(\beta<\beta_1)$, and $D(A^{\frac{1}{4}+\beta})\hookrightarrow L^4$ by Sobolev embedding, we have (cf. [27])

$$\lim_{s_0\to-\infty}\frac{1}{-s_0}\int_{s_0}^{0}|z(\tau)|_4^4\mathrm{d}\tau=E(|z(0)|_4^4).$$

From [24], it is known that $E(|z(0)|_4^4)$ is arbitrary small for $\alpha$ sufficiently large.

Therefore, by choosing large enough $\alpha$, we have

$$\lim_{s\to-\infty}\frac{1}{-s}\int_{s}^{0}(-\lambda_1+c|z|_4^4)\mathrm{d}\tau=-\lambda_1+cE(|z(0)|_4^4)\leqslant-\frac{\lambda_1}{2}.$$

This implies the existence of $S_0(\omega)$ such that $s<S_0(\omega)$,

$$\int_{s}^{0}(-\lambda_1+c|z|_4^4)\mathrm{d}\tau\leqslant-\frac{\lambda_1}{4}(-s). \tag{5.1}$$

Since $|z(s)|_2^2$ and $|z(s)|_4^4$ have at most polynomial growth as $s\to-\infty$ (See [26]), (3.10) and (5.1) imply the existence of a.s. finite random variable $r_1(\omega)$ such that a.s.

$$|v(t)|_2^2\leqslant r_1(\omega),\quad \text{for any } t\in[-1,0], \tag{5.2}$$

when $u_s$ is in a bounded set in $L^2(D)$.

Define $t_1=-1, t_2=0$. By equations (3.11) and (5.2), we know that there exists an a.s. finite random variable $r_2(\omega)$ such that a.s.

$$\int_{-1}^{0}\|v(\tau)\|^2\mathrm{d}\tau\leqslant r_2(\omega). \tag{5.3}$$

From equation (3.19), defining $t=0$, $s\in[-1,0]$, we have

$$\begin{aligned}|A^\theta v(0)|_2^2\leqslant{}&\mathrm{e}^{\int_s^0(c|v|_2^2|A^{\frac{1}{2}}v|_2^2+c)\mathrm{d}\tau}|A^\theta v(s)|_2^2+\\&\int_s^0\mathrm{e}^{\int_\sigma^0(c|v|_2^2|A^{\frac{1}{2}}v|_2^2+c)\mathrm{d}\tau}(c|z|_2^2+c|A^{\frac{1}{4}+\frac{\theta}{2}}z|_2^2)\mathrm{d}\sigma.\end{aligned} \tag{5.4}$$

Integrating equations (5.4) about $s$ from $-1$ to $0$, we obtain

$$\begin{aligned}|A^\theta v(0)|_2^2\leqslant{}&\mathrm{e}^{\int_{-1}^0(c|v|_2^2|A^{\frac{1}{2}}v|_2^2+c)\mathrm{d}\tau}\int_{-1}^0|A^\theta(s)|_2^2\mathrm{d}s+\\&\int_{-1}^0\mathrm{e}^{\int_\sigma^0(c|v|_2^2|A^{\frac{1}{2}}v|_2^2+c)\mathrm{d}\tau}(c|z|_2^2+c|A^{\frac{1}{4}+\frac{\theta}{2}}z|_2^2)\mathrm{d}\sigma.\end{aligned} \tag{5.5}$$

Combining equations (5.2) and (5.3), we know that equation (5.5) implies the existence of an a.s. finite random variable $r_3(\omega)$ such that a.s.

$$|A^\theta v(0)|_2^2\leqslant r_3(\omega).$$

Let $K(\omega)$ be the ball in $D(A^\theta)$ of radius $r_3(\omega)+|A^\theta z(0,\omega)|_2^2$. $K(\omega)$ is an attracting set at time 0 which is compact (since the embedding $H^{2\theta}(D)\hookrightarrow L^2(D)$ is compact). Theorem 4.5 applies. The proof of Theorem 5.1 is complete.

**References**


[1] Pedlosky J. Geophysical fluid dynamics. 2nd ed. Berlin: Springer-Verlag, 1987.

[2] Charney J G, Fjortaft R, Von Neumann J. Numerical integration of the barotropic vorticity equation. Tellus, 1950, 2: 237-254.

[3] Bourgeois A J, Beale J T. Validity of the quasi-geostrophic model for large-scale flow in the atmosphere and ocean. SIAM J Math Anal, 1994, 25: 1023-1068.

[4] Embid P F, Majda A J. Averaging over fast gravity waves for geophysical flows with arbitrary potential vorticity. Comm in PDE, 1996, 21: 619-658.

[5] Schochet S H. Singular limits in bounded domains for quasilinear symmetric hyperbolic systems having a vorticity equations. J Diff Equ, 1987, 68: 400-428.

[6] Zhou X. Atmospheric stochastic dynamics and predictability. Acta Meteo Sinica (in Chinese), 2005, 63(5): 806-811.

[7] Frankignoul C, Hasselmann K. Stochastic climate models, part II: application to sea-surface temperature anomalies and thermocline variability. Tellus, 1997, 29: 289-305.

[8] Müller P. Stochastic forcing of quasi-geostrophic eddies//Adler R J, Müller P, Rozovskii B. Stochastic modelling in physical oceanography. Basel: Birkhäuser, 1996.

[9] Mikolajewicz U, Maier-Reimer E. Internal secular variability in an OGCM. Climate Dyn, 1990, 4:145-156.

[10] Phillips O M. On the generation of waves by turbulent winds. J Fluid Mech, 1957, 2: 417-445.

[11] Rubenstein D. A spectral model of wind-forced internal waves. J Phys Oceanogr, 1994, 24: 819-831.

[12] Blömker D, Duan J, Wanner T. Enstrophy dynamics of stochastically forced large-scale geophysical flow. J Math Phys, 2002, 43(5): 2616-2626.

[13] Brannan J, Duan J, Wanner T. Dissipative quasi-geostrophic dynamics under random forcing. J Math Anal, 1998, 228:221-233.

[14] Duan J, Kloeden P E, Schmalfuss B. Exponential stability of the quasi-geostrophic equation under random perturbations. Prog in Probability, 2001, 49: 241-256.

[15] Majda A, Eijnden E V. A mathematical framework for stochastic climate models. Comm Pure Appl Math, 2001, 54:891-974.

[16] Pedlosky J. Ocean circulation theory. Berlin: Springer-Verlag, 1996.

[17] Leith C. Nonlinear normal mode initialization and quasi-geostrophic theory. J Atmospheric Sci, 1980, 37: 958-968.

[18] Temam R. Infinite-dimensional dynamical systems in mechanics and physics. Appl Math Ser, Vol 68. New York: Springer-Verlag, 1988.

[19] Guo B, Huang D. Existence of weak solutions and trajectory attractors for the moist atmospheric equations in geophysics. J Math Phys, 2006, 47: 083508, 1-23.

[20] Li J, Chou J. Existence of atmosphere attractors. Science in China, Series D, 1997, 40(2): 215-224.

[21] Wang S. On the 2-D model of large-scale atmospheric motion: well-posedness and attractors. Nonlinear Anal TMA, 1992, 18(1): 17-60.

[22] Cessi P, Ierley G R. Symmetry-breaking multiple equilibria in quasi-geostrophic wind-driven flows. J Phys Ocean, 1995, 25: 1196-1205.

[23] Gilbarg D, Trudinger N S. Elliptic partial differential equations of the second order. 2nd ed. Berlin: Springer-

Verlag, 1983.
[24] Friedman A. Stochastic differential equations and application, Vol 1. New York: Academic Press, 1975.
[25] Da Prato G, Zabczyk J. Stochastic equations in infinite dimensions, encyclopedia of mathematics and its application. Cambridge: Cambridge Univ Press, 1992.
[26] Flandoli F. Dissipativity and invariant measures for stochastic Navier-Stokes equations. Nonlinear Diff Eq Appl, 1994, 1: 403-423.
[27] Da Prato G, Zabczyk J. Ergodicity for infinite dimensional systems. London Mathematical Society Lecture Note Series, Vol 229. Cambridge: Cambridge Univ Press, 1996.
[28] Pazy A. Semigroup of linear operator and applications to partial differential equations. Berlin: Springer-Verlag, 1983.
[29] Lions J L. Quelques méthodes de résolutions des problèmes aux limites nonlinéaires. Paris: Dunod, 1969.
[30] Crauel H, Debussche A, Flandoli F. Random attractors. J Dyn and Diff Eq, 1997, 29(2): 307-341.

# Random Attractors for Quasi-continuous Random Dynamical Systems and Applications to Stochastic Reaction-diffusion Equations *

Li Yangrong(李杨荣) Guo Boling(郭柏灵)

**Abstract**

This paper first introduces the so-called quasi-continuous random dynamical system (RDS) on a separable Banach space. The quasi-continuity is weaker than all the usual continuities and thus is easier to check in practice. We then establish a necessary and sufficient condition for the existence of random attractors for the quasi-continuous RDS. We also give a general method to obtain the random attractors for the RDS on the Banach space $L^q(D)$ for $q \geqslant 2$. As an application, it is shown that the RDS generated by the stochastic reaction-diffusion equation possesses a finite-dimensional random attractor in $L^q(D)$ for any $q \geqslant 2$, a comparison result of fractal dimensions under the different $L^q$-norms is also obtained.

**Keywords** random dynamical systems; random attractors; quasi-continuity; stochastic reaction-diffusion equations; wiener processes; fractal dimensions

## 1 Introduction

Random attractors for random dynamical systems (RDS) were first introduced by Crauel and Flandoli[8] and Schmalfuss[19], with notable developments given in [2 – 9, 11, 12, 17] and in the references there among many others. They are compact invariant random sets attracting all the orbits. To apply it to the certain stochastic partial differential equations (SPDE), one must rely on some sufficient or necessary and sufficient conditions for the existence of random attractors. For example, it is well known that there exists a random attractor if there is a compact absorbing random set (see [7, 8]), while Crauel[6] has proved that there exists a random attractor if and only if there exists a compact attracting set.

As we know, the above general results have a basic assumption that the RDS is norm-to-norm continuous in some Banach spaces. However, there are some important RDS which are not norm-to-norm continuous. For example, consider the following stochastic reaction-diffusion equations (see [16, 22])

$$du - (\Delta u - f(u))dt = dW(t) \tag{1.1}$$

where $f$ is a polynomial of odd degree $2p-1$ and $W(t)$ is a Wiener process, the further assumptions and details will be placed in Section 6. Even in the deterministic case of Eq. (1.1), the norm-to-norm continuity of the solutions semigroup in $L^q(D)$ $(q \neq 2)$ or in $H_0^1(D)$ is unknown as pointed out by

* J Differential Equations, 2008, 245: 1775 – 1800.

Robinson[18] and Zhong et al[25], the norm-to-weak continuity of the semigroup in whole space $L^q(D)$ $(q\neq 2)$ is also not clear as pointed out by Zhong et al[25], they have only proved the norm-to-weak continuity on some bounded sets of $L^q(D)$.

To overcome the continuity difficulty, we introduce in present paper the so-called quasi-continuity of a RDS. The quasi-continuity is weaker than all continuities mentioned above and thus is easier to check in practice. In fact, we will see that the RDS is always quasi-continuous in $L^q(D)(q>2)$ if it is (norm-to-norm or norm-to-weak) continuous in $L^2(D)$ (see Proposition 3.3).

We then establish a criterion of the existence of random attractors for the quasi-continuous RDS (see Theorem 4.1). Such a criterion in the deterministic case for the (norm-to-norm or norm-to-weak) continuous semigroup has been established by Zhong et al[25] and Ma et al[13], in which a very useful tool of the Kuratowski measure of non-compactness and the related notion of omega-compactness were involved. Of course, these notions have been used by other papers in the continuous cases when the phase space is the Hilbert space $L^2(D)$ (see [3, 20, 21, 23, 24]). Recently, Kloeden and Langa[11] gave a generalization of the Ma et al method to the RDS. In present paper, by using the above tool, we generalize their results to the quasi-continuous RDS case. Some difficulty arose in proving the compactness, but it is shown that an ergodicity argument can solve the problem.

In comparison with the criterion of random attractors obtained by Crauel[6] (as mentioned above), our criterion has the weaker continuity assumption and uses omega-compactness instead of the ordinary compactness. In particular, we can apply it to the wider range when the phase space is the Banach space $L^q(D)$ for $q>2$ rather than the Hilbert space $L^2(D)$.

It is well known that many RDS generated by the concrete SPDEs possess a random attractor in the Hilbert space $L^2(D)$. Under the assumption of this fact, we will give in this paper a criterion for the existence of random attractors in $L^q(D)$ for $q>2$ (see Theorem 5.3).

Our theoretical result can be applied to the stochastic reaction-diffusion equation (1.1) when the phase space is $L^q(D)$ for any $q>2$. It is well known that the RDS generated by the above equation possesses a random attractor in $L^2(D)$ (see [4, 6 − 9, 12] for different noise assumptions). We will prove that the RDS has a random attractor in $L^q(D)$ for any $q>2$ under some bounded noise assumptions (see Theorem 6.7). It is worth pointing out that, even in the non-random case, it seems to be the best result that there is an attractor in $L^q(D)$ for $q=4p-2$ and thus for all $2\leqslant q\leqslant 4p-2$ (see [25]).

We also prove that the random attractor for Eq. (1.1) has finite fractal dimensions in $L^q(D)$ (Theorem 6.9). We emphasize here that the fractal dimension is defined by the $L^q$-norm rather than the $L^2$-norm, and thus the same set may enjoy different fractal dimensions under different $L^q$-norms. A relationship of fractal dimensions under different $L^q$-norms will be given. This subject seems not to be discussed upon now even in the deterministic case.

## 2 Preliminaries

In this section, we first recall some basic notions of the theory of random dynamical systems

(RDS) (see [1,6 – 9]). We then review briefly the Kuratowski measure of non-compactness (see [10]), which is a useful tool to study the attractor (see [13,20,21,23 – 25]).

### 2.1 Random Attractors

Throughout this paper, $X$ is a separable Banach space equipped with the norm $\|\cdot\|$ and the Borel $\sigma$-algebra $\mathscr{B}(X)$. $(\Omega, \mathscr{F}, P)$ is a complete probability space and $\{\theta_t : \Omega \to \Omega, t \in \mathbf{R}\}$ a group of measure preserving transformations of $\Omega$. We assume that $\theta_t$ is ergodic.

**Definition 2.1** A random dynamical system (RDS) $\varphi$ on $X$ over $(\Omega, \mathscr{F}, P, \theta_t)$ is a measurable map

$$\varphi: \mathbf{R}_+ \times \Omega \times X \to X, \quad (t, \omega, x) \mapsto \varphi(t, \omega)x$$

such that $P$-a.s. $\varphi(0, \omega) = \mathrm{id}$ on $X$ and the cocycle property holds, that is,

$$\varphi(t+s, \omega) = \varphi(t, \theta_s\omega)\varphi(s, \omega), \quad s, t \in \mathbf{R}_+. \tag{2.1}$$

A random set $\{K(\omega)\}_{\omega \in \Omega}$ is a family of subsets indexed by $\omega$ such that for every $x \in X$ the mapping $\omega \mapsto \mathrm{dist}(x, K(\omega))$ is measurable with respect to $\mathscr{F}$, where we denote $\mathrm{dist}(\cdot, \cdot)$ the Hausdorff semi-distance in $X$

$$\mathrm{dist}(A, B) = \sup_{a \in A} \inf_{b \in B} \|a - b\|, \quad A, B \subset X. \tag{2.2}$$

A random set $\mathscr{A}(\omega)$ is said to be an attracting set if $\mathscr{A}(\omega)$ attracts all deterministic bounded sets $B \subset X$, that is,

$$\lim_{t \to \infty} \mathrm{dist}(\varphi(t, \theta_{-t}\omega)B, \mathscr{A}(\omega)) = 0, \quad P\text{-a.s. } \omega \in \Omega. \tag{2.3}$$

A random set $\mathscr{A}(\omega)$ is said to be an absorbing set if for every deterministic bounded sets $B \subset X$ there exists a time $t_B(\omega)$ such that

$$\varphi(t, \theta_{-t}\omega)B \subset \mathscr{A}(\omega), \quad t \geqslant t_B(\omega), \quad P\text{-a.s. } \omega \in \Omega. \tag{2.4}$$

**Definition 2.2** A random set $\mathscr{A}(\omega)$ is said to be a random attractor for the RDS $\varphi$ if $\mathscr{A}(\omega)$ is compact, attracting (as (2.3)) and invariant, that is,

$$\varphi(t, \omega)\mathscr{A}(\omega) = \mathscr{A}(\theta_t\omega), \quad \forall t \geqslant 0, \quad P\text{-a.s. } \omega \in \Omega. \tag{2.5}$$

The following result on the existence of random attractors can be found in [6 – 8].

**Theorem 2.3** A continuous RDS possesses a random attractor if there exists a compact absorbing set, while a continuous RDS possesses a random attractor if and only if there exists a compact attracting set.

### 2.2 The Kuratowski Measure of Non-compactness

**Definition 2.4** Let $B$ be a bounded set in a Banach space $X$. The Kuratowski measure of non-compactness $\kappa(B)$ of $B$ is defined by

$$\kappa(B) = \inf\{d > 0 \mid B \text{ admits a finite cover by sets of diameter} \leqslant d\}. \tag{2.6}$$

We define $\kappa(B) = \infty$ if $B$ is unbounded. The properties of $\kappa(B)$, which we need to use in this paper, are given in the following lemma (see [9], the conclusion (ⅴ) due to Zhong et al[25]).

**Lemma 2.5** (ⅰ) $\kappa(B) = 0$ iff $\bar{B}$ is compact, where $\bar{B}$ is the norm closure of $B$.

(ⅱ) $\kappa(\overline{\mathrm{co}\, B}) = \kappa(B)$, where $\overline{\mathrm{co}\, B}$ is the closed convex hull of $B$.

(iii) $\kappa(B_1 \cup B_2) \leqslant \max\{\kappa(B_1), \kappa(B_2)\}$ and $\kappa(B_1 + B_2) \leqslant \kappa(B_1) + \kappa(B_2)$.

(iv) For a decreasing family $\{B_t\}_{t>r}$ of nonempty, closed and bounded sets, if $\kappa(B_t) \to 0$ as $t \to \infty$, then $\bigcap B_t$ is a nonempty, compact set.

(v) $\kappa(B) = \kappa(\bar{B}^{WS})$, where $\bar{B}^{WS}$ is the weakly sequential closure of $B$ defined by

$$\bar{B}^{WS} = \{x \in X \mid \exists x_n \in B, \text{ such that } x_n \rightharpoonup x\} \tag{2.7}$$

where $\rightharpoonup$ means the weak convergence.

## 3 Norm-to-weak Continuity and Quasi-continuity

**Definition 3.1** A RDS $\varphi$ on a Banach space $X$ is said to be norm-to-weak continuous if $P$-a.s. $\omega \in \Omega$, $\varphi(t_n, \omega)x_n \rightharpoonup \varphi(t, \omega)x$ whenever $(t_n, x_n) \to (t, x)$ in $\mathbf{R}_+ \times X$, where $\rightharpoonup$ means weak convergence. A RDS $\varphi$ is called to be quasi-continuous if $P$-a.s. $\omega \in \Omega$, $\varphi(t_n, \omega)x_n \rightharpoonup \varphi(t, \omega)x$ whenever $\{(t_n, x_n)\}$ is a sequence in $\mathbf{R}_+ \times X$ such that $\{\varphi(t_n, \omega)x_n\}$ is bounded and $(t_n, x_n) \to (t, x)$ as $n \to \infty$.

Both kinds of continuity above are weaker than the usual one. In fact we have: norm (resp. weak) continuity $\Rightarrow$ norm-to-weak continuity $\Rightarrow$ quasi-continuity.

As pointed out in Section 1, it is difficult to verify that the RDS is continuous (or weak continuous) in stronger norm spaces than $L^2(D)$, e.g. $L^q(D)$ for $q > 2$. The following results present some easy-to-check criteria for norm-to-weak continuity and quasi-continuity in the stronger norm space.

Let $X$, $Y$ be two Banach spaces with the dual spaces $X^*$, $Y^*$, respectively, and assume also

(i) the embedding $i: X \to Y$ is densely continuous;

(ii) the adjoint operator $i^*: Y^* \to X^*$ is dense, i.e. $i^*(Y^*)$ is dense in $X^*$.

We remark here that $i^*: Y^* \to X^*$ is always injective and continuous if the assumption (i) holds (comparing this fact to the assumptions given in [25]). Indeed, if $i^*(y^*) = 0$ for $y^* \in Y^*$, then obviously $\langle i(X), y^* \rangle = 0$. Since $i(X)$ is dense in $Y$, it follows from the Hahn-Banach theorem that $y^* = 0$. Thus, $i^*$ is injective. The continuity of the linear operator $i^*$ follows from the continuity of $i$. We also note that the assumption (ii) holds if $X$, $Y$ satisfy the assumption (i) and, in addition, $X$ is reflexive.

**Proposition 3.2** Let $X$, $Y$ satisfy the assumptions above, and let $\varphi$ be a RDS on $X$, $Y$ respectively. Suppose $\varphi$ is continuous in $Y$, then $\varphi$ is norm-to-weak continuous in $X$ if and only if for $P$-a.s. $\omega \in \Omega$, $\varphi(\cdot, \omega)$ maps compact subsets of $\mathbf{R}_+ \times X$ into bounded sets of $X$.

**Proof** It is the same as in the non-random case (see [25, Theorem 3.2]). □

It is also difficult to verify the norm-to-weak continuity by using the above proposition. However, we will see from the following result that the quasi-continuity holds automatically in stronger norm spaces.

**Proposition 3.3** Let $X$, $Y$ satisfy the assumptions (i) and (ii) above. Let $\varphi$ be a RDS on $X$, $Y$ respectively. If $\varphi$ is continuous or norm-to-weak continuous in $Y$, then $\varphi$ is quasi-continuous in $X$.

**Proof** We work for every fixed $\omega \in \Omega_0$ with $P(\Omega_0)=1$. Assume $\varphi$ is continuous or norm-to-weak continuous in $Y$, then $\varphi$ is quasi-continuous in $Y$. We must show that $\varphi$ is quasi-continuous in $X$. To do this, we take $(t_n, x_n) \in \mathbf{R}_+ \times X$ such that $\{\varphi(t_n,\omega)x_n\}$ is bounded in $X$ and $(t_n, x_n) \to (t, x)$. We need to show that $\varphi(t_n,\omega)x_n \rightharpoonup \varphi(t,\omega)x$, i.e.

$$\langle \varphi(t_n,\omega)x_n - \varphi(t,\omega)x, x^* \rangle_X \to 0, \quad \text{as } n \to \infty \text{ for every } x^* \in X^*. \tag{3.1}$$

Indeed, let $M$ be the positive constant such that

$$\| \varphi(t_n,\omega)x_n - \varphi(t,\omega)x \|_X \leqslant M. \tag{3.2}$$

For any $\varepsilon > 0$ and $x^* \in X^*$, by the assumption (ii) we can choose $y^* \in Y^*$ such that

$$\| i^*(y^*) - x^* \|_{X^*} < \frac{\varepsilon}{2M}. \tag{3.3}$$

Since $\varphi$ is norm or norm-to-weak continuous in $Y$, there exists $N>0$ such that for all $n \geqslant N$,

$$|\langle i(\varphi(t_n,\omega)x_n - \varphi(t,\omega)x), y^* \rangle_Y| < \frac{\varepsilon}{2},$$

which, together with (3.1)~(3.2), implies that

$$\begin{aligned} &|\langle \varphi(t_n,\omega)x_n - \varphi(t,\omega)x, x^* \rangle_X| \\ \leqslant &|\langle \varphi(t_n,\omega)x_n - \varphi(t,\omega)x, i^*(y^*) - x^* \rangle_X| + |\langle i(\varphi(t_n,\omega)x_n - \varphi(t,\omega)x), y^* \rangle_Y| \\ \leqslant &M \cdot \frac{\varepsilon}{2M} + \frac{\varepsilon}{2} = \varepsilon. \end{aligned}$$

Therefore (3.3) holds and thus $\varphi$ is quasi-continuous in $X$. □

**Remark 3.4** For the concrete SPDEs, one can choose $Y=L^2(D)$, $X=L^q(D)$ $(q>2)$ or $H_0^m(D)$, and conclude that $\varphi$ is quasi-continuous in $L^q(D)$ or $H_0^m(D)$ if $\varphi$ is continuous in $L^2(D)$.

## 4 Random Attractors for the Quasi-continuous RDS

In this section, we present an existence criterion of random attractors.

**Theorem 4.1** Suppose $\varphi(t,\omega)$ be a quasi-continuous RDS on a separable Banach space $X$ over an ergodic system $(\Omega, \mathscr{F}, P; \theta_t)$. Then $\varphi$ possesses a random attractor if and only if

(i) $\varphi$ has a bounded absorbing set $\mathscr{B}(\omega)$;

(ii) $\varphi$ is omega-limit compact, that is, for every bounded non-random set $B \subset X$, we have

$$\lim_{T \to +\infty} \kappa\Big( \bigcup_{t \geqslant T} \varphi(t, \theta_{-t}\omega)B \Big) = 0, \quad P\text{-a.s. } \omega \in \Omega. \tag{4.1}$$

Under the assumptions in Theorem 4.1, we will prove one of the random attractors is

$$\mathscr{A}(\omega) = \bigcup_{B \subset X} A(B,\omega) \tag{4.2}$$

where the union is taken over all bounded subsets of $X$, and $A(B,\omega)$ is the weakly sequence omega-limit set, that is,

$$A(B,\omega) = \bigcap_{T \geqslant 0} \overline{\bigcup_{t \geqslant T} \varphi(t, \theta_{-t}\omega)B}^{WS} \tag{4.3}$$

for every bounded set $B \subset X$. Using the same method as in [25], one can prove that if $\varphi$ is omega-limit compact then $A(B,\omega)$ has the following character:

$$x \in A(B,\omega) \quad \text{iff} \quad \exists\, x_n \in B,\ t_n \to \infty \quad \text{such that} \quad \varphi(t_n, \theta_{-t_n}\omega)x_n \rightharpoonup x. \tag{4.4}$$

**Proof of Theorem 4.1** Necessity. Suppose $\varphi$ has a random attractor $\widetilde{\mathscr{A}}(\omega)$. Then it is easy to prove that the $\varepsilon$-neighborhood $N_\varepsilon(\widetilde{\mathscr{A}}(\omega))$ is a bounded absorbing set. But $\widetilde{\mathscr{A}}(\omega)$ is compact, we know that $\kappa(\widetilde{\mathscr{A}}(\omega))=0$ and, thus, for large $T$,

$$\kappa(\bigcup_{t\geq T}\varphi(t,\theta_{-t}\omega)B)\leqslant\kappa(N_\varepsilon(\widetilde{\mathscr{A}}(\omega)))\leqslant 2\varepsilon,$$

which proves (4.1). Thus, $\varphi$ is omega-limit compact.

Sufficiency. We will show that $\mathscr{A}(\omega)$, defined by (4.2), is a random attractor. The proof can be accomplished by the following 4 steps.

**Step 1** We show that $\mathscr{A}(\omega)$ is invariant. It is obvious that $\mathscr{A}(\omega)$ is invariant if every $A(B,\omega)$, defined by (4.3), is invariant. To prove $A(B,\omega)$ is invariant, we let $x\in A(B,\omega)$. Then, by the character (4.4), there exist $t_n\to\infty$ and $x_n\in B$ such that

$$\varphi(t_n,\theta_{-t_n}\omega)x_n\rightharpoonup x.$$

By the omega-limit compactness of $\varphi$, we can choose a subsequence $\{n_k\}$ such that

$$y_k:=\varphi(t_{n_k},\theta_{-t_{n_k}}\omega)x_{n_k}\to x. \tag{4.5}$$

Since, by the assumption (i), there exists a bounded absorbing set $\mathscr{B}(\omega)$, it follows from the cocycle property (2.1) that, for fixed $t>0$ and large $k$,

$$\varphi(t,\omega)y_k=\varphi(t+t_{n_k},\theta_{-(t+t_{n_k})}\theta_t\omega)x_{n_k}\in\mathscr{B}(\theta_t\omega).$$

In particular, $\{\varphi(t,\omega)y_k,k\in\mathbf{N}\}$ is bounded in $X$. Thus, by the quasi-continuity of $\varphi$, it follows from (4.5) that

$$\varphi(t,\omega)y_k\rightharpoonup\varphi(t,\omega)x.$$

Therefore, by the character (4.4) again, $\varphi(t,\omega)x\in A(B,\theta_t\omega)$, which shows that

$$\varphi(t,\omega)A(B,\omega)\subset A(B,\theta_t\omega),\quad t\geqslant 0.$$

Conversely, let $x\in A(B,\theta_t\omega)$, where $t$ is a fixed positive number. Then, by (4.4), there exist $x_n\in B$ and $t_n$ with $t<t_n\to\infty$ such that

$$\varphi(t_n,\theta_{-t_n}\theta_t\omega)x_n\rightharpoonup x.$$

Since $\varphi$ is omega-limit compact, there exist a $z\in X$ and a subsequence $\{n_k\}$ such that

$$z_k:=\varphi(t_{n_k}-t,\theta_{-(t_{n_k}-t)}\omega)x_{n_k}\to z. \tag{4.6}$$

Thus, $z\in A(B,\omega)$ in view of (4.4). Using the cocycle property (2.1), we know that the sequence

$$\varphi(t,\omega)z_k=\varphi(t_{n_k},\theta_{-t_{n_k}}\theta_t\omega)x_{n_k}\rightharpoonup x.$$

Noting that $\varphi(t,\omega)z_k$ is bounded in $X$ since a weakly convergent sequence is norm bounded, we see, by the quasi-continuity of $\varphi$ and by (4.6), that

$$\varphi(t,\omega)z_k\rightharpoonup\varphi(t,\omega)z.$$

Thus, $x=\varphi(t,\omega)z\in\varphi(t,\omega)A(B,\omega)$, which proves the converse inclusion and thus $A(B,\omega)$ is invariant as required.

**Step 2** We show that there exists a bounded deterministic set $B_0\subset X$ such that

$$\mathscr{A}(\omega)=A(B_0,\omega),\quad P\text{-a.s. }\omega\in\Omega. \tag{4.7}$$

This step is a key to prove the compactness of $\mathscr{A}(\omega)$ and needs the ergodicity assumption of $\theta_t$.

Follows the idea of [6, Proposition 2.3], we set

$$R(\omega) = \inf\{r \geqslant 0 \mid \mathscr{A}(\omega) \subset B(O,r)\} \tag{4.8}$$

where $B(O,r)$ is the ball of center $O$ and radius $r$. We must verify that $R(\omega)$ is well defined. Indeed, for every bounded set $B \subset X$, there exists $t_B$ such that

$$\overline{\mathrm{co}\Big(\bigcup_{t \geqslant t_B} \varphi(t,\theta_{-t}\omega)B\Big)^{WS}} = \overline{\mathrm{co}\Big(\bigcup_{t \geqslant t_B} \varphi(t,\theta_{-t}\omega)B\Big)} \subset \overline{\mathrm{co}\ \mathscr{B}(\omega)}.$$

Then it follows from (4.2) and (4.3) that $\mathscr{A}(\omega) \subset \overline{\mathrm{co}\ \mathscr{B}(\omega)}$. Since the absorbing set $\mathscr{B}(\omega)$ is bounded, it follows that $\overline{\mathrm{co}\ \mathscr{B}(\omega)}$ and thus $\mathscr{A}(\omega)$ is bounded. We conclude that $R(\omega)$ is finite for all $\omega \in \Omega_0$ with $P(\Omega_0) = 1$. We set also

$$R(\omega) = 0 \quad \text{if} \quad \omega \in \Omega \setminus \Omega_0.$$

Then the function $R(\omega)$ is measurable, and it follows from the ergodic theorem (e.g. see [1]) that there exists a sequence $t_n \to \infty$ such that

$$R(\theta_{-t_n}\omega) \leqslant R_0 + 1, \quad P\text{-a.s. } \omega \in \Omega,$$

where

$$R_0 = \mathrm{ess\ inf}(R(\omega))_{\omega \in \Omega}.$$

Let now $B_0 = B(O, R_0 + 1)$, we claim that $\mathscr{A}(\omega) = A(B_0,\omega)$. Indeed, let $x \in \mathscr{A}(\omega)$ and let $t_n$ be the sequence defined above. Then, by the invariant property (which has been proved in Step 1), there exists $x_n \in \mathscr{A}(\theta_{-t_n}\omega)$ such that

$$\varphi(t_n, \theta_{-t_n}\omega)x_n = x. \tag{4.9}$$

Since $\mathscr{A}(\theta_{-t_n}\omega) \subset B_0$, it follows from (4.9) and the character (4.4) that $x \in A(B_0,\omega)$. We have proved $\mathscr{A}(\omega) \subset A(B_0,\omega)$. Conversely, it is obvious that $A(B_0,\omega) \subset \mathscr{A}(\omega)$ and thus (4.7) holds.

**Step 3** We show that $\mathscr{A}(\omega)$ is compact. By (4.7) in Step 2, we need only to prove that every $A(B,\omega)$ is compact. Indeed, by Lemma 2.5(v) and the omega-limit compactness of $\varphi$, we have

$$\kappa\Big(\overline{\bigcup_{t \geqslant T} \varphi(t,\theta_{-t}\omega)B^{WS}}\Big) = \kappa\Big(\bigcup_{t \geqslant T} \varphi(t,\theta_{-t}\omega)B\Big) \to 0 \quad \text{as} \quad T \to \infty.$$

Since $\overline{\bigcup_{t \geqslant T} \varphi(t,\theta_{-t}\omega)B^{WS}}$ is norm-closed in $X$, thanks to the property (iv) in Lemma 2.5, we know that $A(B,\omega)$ is nonempty and compact as required.

**Step 4** We show that $\mathscr{A}(\omega)$ is an attracting set. For every bounded set $B \subset X$, we claim that $A(B,\omega)$ attracts $B$, that is,

$$\mathrm{dist}(\varphi(t,\theta_{-t}\omega)B, A(B,\omega)) \to 0 \quad \text{as} \quad t \to \infty, \quad P\text{-a.s. } \omega \in \Omega. \tag{4.10}$$

Indeed, if not, then there exist $\delta > 0$, $t_n \to \infty$ and $x_n \in B$ such that

$$\mathrm{dist}(\varphi(t,\theta_{-t_n}\omega)x_n, A(B,\omega)) \geqslant \delta. \tag{4.11}$$

By the omega-limit compactness of $\varphi$, there exist a subsequence $\{n_k\}$ and an $x \in X$ such that

$$\varphi(t_{n_k}, \theta_{-t_{n_k}}\omega)x_{n_k} \to x.$$

But by (4.4) $x \in A(B,\omega)$, which contradicts (4.11). Thus (4.10) holds true and implies that

$$\mathrm{dist}(\varphi(t,\theta_{-t}\omega)B, \mathscr{A}(\omega)) \leqslant \mathrm{dist}(\varphi(t,\theta_{-t}\omega)B, A(B,\omega)) \to 0$$

as $t \to \infty$. That is, $\mathscr{A}(\omega)$ attracts $B$, which completes the proof of Theorem 4.1. □

As a consequence, we obtain the sufficient condition for the existence of random attractor given in

[7,8] under the weaker continuity assumption.

**Corollary 4.2** A quasi-continuous RDS $\varphi$ has a random attractor $\mathscr{A}(\omega)$ if $\varphi$ has a compact absorbing set $\mathscr{B}(\omega)$.

**Proof** For any bounded set $B\subset X$, there exists a $t_B$ such that

$$\bigcup_{t\geqslant t_B}\varphi(t,\theta_{-t}\omega)B\subset\mathscr{B}(\omega).$$

But $\mathscr{B}(\omega)$ is compact, $\kappa(\mathscr{B}(\omega))=0$, and so

$$\kappa\Big(\bigcup_{t\geqslant t_B}\varphi(t,\theta_{-t}\omega)B\Big)=0, \tag{4.12}$$

which implies that $\varphi$ is omega-limit compact. □

**Remark 4.3** The formulation (4.2) of the random attractor seems to differ from the random attractor obtained by the usual method in [7,8]. However, under the assumptions of Theorem 4.1, we can show that they are the same one. That is,

$$\mathscr{A}(\omega)=\bigcup_{B\subset X}A(B,\omega)=\overline{\bigcup_{B\subset X}A_B(\omega)} \tag{4.13}$$

where $A(B,\omega)$ is defined by (4.3), and $A_B(\omega)$ is defined by (see [7,8])

$$A_B(\omega)=\bigcap_{T\geqslant 0}\overline{\bigcup_{t\geqslant T}\varphi(t,\theta_{-t}\omega)B}. \tag{4.14}$$

Indeed, By (4.4) and the omega-limit compactness of $\varphi$, we can prove that $A(B,\omega)\subset A_B(\omega)$ and thus $A(B,\omega)=A_B(\omega)$. But by the proof of Theorem 4.1, $\bigcup_{B\subset X}A(B,\omega)$ is compact and thus norm closed particularly, which implies that

$$\bigcup_{B\subset X}A(B,\omega)=\overline{\bigcup_{B\subset X}A(B,\omega)}=\overline{\bigcup_{B\subset X}A_B(\omega)}.$$

**Remark 4.4** Using the same argument given in [7,8], one can show that the random attractor $\mathscr{A}(\omega)$, defined by (4.2), is the largest compact invariant set and the minimal closed attracting set.

**Remark 4.5** Of course, Theorem 4.1 still holds if quasi-continuity of $\varphi$ is replaced by the norm continuity, weak continuity or norm-to-weak continuity. Even in the deterministic case, Theorem 4.1 seems to be a slightly stronger result than [25, Theorem 4.2] since we need the weaker continuity assumption (than norm-to-weak continuity).

## 5 Random Attractors on the $q$-times Integrated Spaces

It is well known that many RDS generated by the concrete SPDEs possess a random attractor on the Hilbert space $L^2(D)$, $D\subset\mathbf{R}^n$ is bounded. Assuming this fact, we will give in this section a general method to prove the existence of random attractors on the Banach space $L^q(D)$ for $q>2$.

We denote by $\|\cdot\|_q$ the norm of $L^q(D)$, $|\cdot|$ the usual modular (or the absolute value), $m(\cdot)$ the Lebesgue measure and $D(|u|\geqslant M)=\{x\in D\mid |u(x)|\geqslant M\}$.

**Lemma 5.1** Suppose a RDS $\varphi$ has a bounded absorbing set $\mathscr{B}(\omega)$ in $L^q(D)$. Then for $P$-a.s. $\omega\in\Omega$, $\varepsilon>0$ and bounded $B\subset L^q(D)$, there exist $T=T(B,\omega)$ and $M=M(\varepsilon,\omega)$ such that

$$m(D(|\varphi(t,\theta_{-t}\omega)u|\geqslant M))<\varepsilon,\quad \forall t\geqslant T,\ u\in B. \tag{5.1}$$

**Proof** We let $\psi(t):=\varphi(t,\theta_{-t}\omega)$ for convenience and work for every fixed $\omega\in\Omega$. By the absorbing assumption, there exists a random variable $M_0=M_0(\omega)$ such that, for bounded

$B\subset L^q(D)$, we can find a constant $T=T(B,\omega)$ such that

$$\|\psi(t)u\|_q^q\leqslant M_0,\quad \text{whenever}\quad t\geqslant T,\ u\in B.$$

On the other hand

$$\|\psi(t)u\|_q^q=\int_D|\psi(t)u|^q\geqslant\int_{D(|\psi(t)|\geqslant M)}|\psi(t)|^q\geqslant M^q\cdot m(D|\psi(t)u|\geqslant M).$$

Thus $m(D(|\psi(t)u|\geqslant M))<\varepsilon$ if we choose $M$ large enough such that $M>(M_0/\varepsilon)^{1/q}$. □

We also need the following basic lemma given in [25].

**Lemma 5.2**[25] If a bounded set $B\subset L^q(D)$ has a finite $\varepsilon$-net in $L^q(D)$, then there is a number $M$ such that

$$\sup_{u\in B}\int_{D(|u|\geqslant M)}|u|^q\leqslant q^{q+1}\varepsilon^q. \tag{5.2}$$

**Theorem 5.3** Let $\varphi(t,\omega)g$ be a RDS on $L^q(D)$ and a continuous RDS on $L^r(D)$ for some $1<r\leqslant q$, where $D$ is bounded. Suppose $\varphi$ has a random attractor in $L^r(D)$. Then $\varphi$ has a random attractor in $L^q(D)$ if and only if

(i) $\varphi$ has a bounded absorbing set in $L^q(D)$;

(ii) for every bounded deterministic set $B\subset L^q(D)$, we have

$$\lim_{t,M\to\infty}\sup_{u\in B}\int_{D(|\varphi(t,\theta_{-t}\omega)u|\geqslant M)}|\varphi(t,\theta_{-t}\omega)u|^q=0,\quad P\text{-a.s. }\omega\in\Omega. \tag{5.3}$$

**Proof** Taking $X=L^q(D)$, $Y=L^r(D)$ in Proposition 3.3, we know that $\varphi$ is quasi-continuous in $L^q(D)$. Then by Theorem 4.1 we have only to consider the omega-limit compactness of $\varphi$ in $L^q(D)$. We work for every fixed $\omega\in\Omega_0$ with $P(\Omega_0)=1$ and let $\psi(t)=\varphi(t,\theta_{-t}\omega)$ for convenience.

Necessity. Since $\varphi$ is by Theorem 4.1 omega-limit compact in $L^q(D)$, it follows that for every $\varepsilon>0$ and bounded $B\subset L^q(D)$, there exists a $T=T(\varepsilon,B,\omega)$ such that the set

$$\tilde{B}=\bigcup_{t\geqslant T}\psi(t)B$$

admits a finite $\varepsilon$-net in $L^q(D)$. Applying Lemma 5.2 to the bounded set $\tilde{B}$, we can also choose an $M_0=M_0(B,\varepsilon,\omega)$ such that if $M\geqslant M_0$ and $t\geqslant T$ then

$$\sup_{u\in B}\int_{D(|\psi(t)u|\geqslant M)}|\psi(t)u|^q\leqslant 2^{q+1}\varepsilon^q,$$

which proves (5.3).

Sufficiency. For any $\varepsilon>0$ and bounded $B\subset L^q(D)$ $(\subset L^r(D))$, by the assumption (5.3) there exist $t_1=t_1(B,\varepsilon,\omega)$ and $M=M(B,\varepsilon,\omega)$ such that

$$\sup_{u\in B}\int_{D(|\psi(t)u|\geqslant M)}|\psi(t)u|^q<\frac{\varepsilon^q}{2^{q+3}},\quad\forall\, t\geqslant t_1. \tag{5.4}$$

Since by Theorem 4.1 $\varphi$ is omega-limit compact in $L^r(D)$, there exists a $t_2\geqslant t_1$ such that the set

$$\hat{B}=\bigcup_{t\geqslant t_2}\psi(t)B$$

has a finite $(2M)^{(r-q)/r}(\varepsilon^q/2)^{1/r}$-net $\{u_k\}\subset\hat{B}$ in $L^r(D)$. Thus, for any $u\in\hat{B}$ we can choose some $u_k\in\hat{B}$ such that

$$\| u - u_k \|_r^r < (2M)^{r-q} \cdot \frac{\varepsilon^q}{2}. \tag{5.5}$$

It is obvious that $D = D_1 \cup D_2 \cup D_3 \cup D_4$, where

$$\begin{aligned} D_1 &= D(|u| \leqslant M, |u_k| \leqslant M), \quad D_2 = D(|u| \geqslant M, |u_k| \leqslant M), \\ D_3 &= D(|u| \leqslant M, |u_k| \geqslant M), \quad D_4 = D(|u| \geqslant M, |u_k| \geqslant M). \end{aligned} \tag{5.6}$$

Then (5.5) implies that

$$\int_{D_1} |u - u_k|^q \leqslant \int_{D(|u-u_k| \leqslant 2M)} |u - u_k|^q \leqslant (2M)^{q-r} \| u - u_k \|_r^r < \frac{\varepsilon^q}{2}. \tag{5.7}$$

Since $|u - u_k| \leqslant 2|u|$ in $D_2$, it follows from (5.4) that

$$\int_{D_2} |u - u_k|^q \leqslant 2^q \int_{D(|u| \geqslant M)} |u|^q < \frac{\varepsilon^q}{8}. \tag{5.8}$$

Similarly, since $|u - u_k| \leqslant 2|u_k|$ in $D_3$,

$$\int_{D_3} |u - u_k|^q \leqslant 2^q \int_{D(|u_k| \geqslant M)} |u_k|^q < \frac{\varepsilon^q}{8}. \tag{5.9}$$

Using the inequality $(a+b)^q \leqslant 2^q(|a|^q + |b|^q)$ and (5.4), we also find

$$\int_{D_4} |u - u_k|^q \leqslant 2^q \left( \int_{D(|u| \geqslant M)} |u|^q + \int_{D(|u_k| \geqslant M)} |u_k|^q \right) < \frac{\varepsilon^q}{4}. \tag{5.10}$$

It follows from (5.7)~(5.10) that

$$\| u - u_k \|_q^q \leqslant \left( \int_{D_1} + \int_{D_2} + \int_{D_3} + \int_{D_4} \right) |u - u_k|^q < \varepsilon^q, \tag{5.11}$$

which proved $\hat{B}$ has an $\varepsilon$-net in $L^q(D)$ and thus $\varphi$ is omega-limit compact in $L^q(D)$ as needed. □

# 6 Applications to the Stochastic Reaction-diffusion Equation

## 6.1 Preliminaries for the Equation

Let $D \subset \mathbf{R}^n$, $n \in \mathbf{N}$, be a bounded open set with regular boundary $\partial D$. We consider the following stochastic reaction-diffusion equation

$$\begin{cases} du - (\Delta u - f(u))dt = \sum_{j=1}^{m} g_j \, dW_j(t) \\ u = 0 \quad \text{on} \quad \partial D \end{cases}. \tag{6.1}$$

The function $f$ is polynomial defined by

$$f(s) = \sum_{j=1}^{2p-1} a_j s^j, \quad a_{2p-1} > 0, \ p > 1. \tag{6.2}$$

We also assume that

$$g_j \in L^\infty(D), \quad j = 1, 2, \cdots, m. \tag{6.3}$$

The random functions $W_j$, $j = 1, \cdots, m$, are independent two-sided real-valued Wiener processes on a probability space $(\Omega, \mathscr{F}, P)$, where $\Omega = \{\omega \in C(\mathbf{R}, \mathbf{R}^n) \mid \omega(0) = 0\}$, with $P$ being a product measure of two Wiener measures on the negative and the positive time parts of $\Omega$. We have

$$(W_1(t,\omega), W_2(t,\omega), \cdots, W_m(t,\omega)) = \omega(t), \quad t \in \mathbf{R}.$$

The time shift is simply defined by

$$\theta_s \omega(t) = \omega(t+s) - \omega(s), \quad t, s \in \mathbf{R}. \tag{6.4}$$

It is a family of ergodic transformations.

Let $A_q = -\Delta$ with $D(A_q) = W^{2,q}(D) \cap W_0^{1,q}(D)$ and write $A = A_2$. Then $A_q$ generates an analytic semigroup $e^{tA_q}$ on $L^q(D)$ (see [15]). The Ornstein-Uhlenbeck process defined by

$$z(t) = \sum_{j=1}^{m} \int_{-\infty}^{t} e^{A(t-s)} g_j dW_j(s) \tag{6.5}$$

is $P$-a.s. pathwise continuous and is the unique stationary solution of the linear equation

$$dz + Az dt = \sum_{j=1}^{m} g_j dW_j(t). \tag{6.6}$$

It is well known (see [8,9]) that for any $s \in \mathbf{R}$, $u_0 \in L^2(D)$ and $\omega \in \Omega$, there exists a unique solution

$$v \in C([s,t]; L^2(D)) \cup L^2(s,t; H_0^1(D)) \cup L^{2p}(s,t; L^{2p}(D))$$

for any $t \geqslant s$ of the equation

$$\begin{cases} \dfrac{dv}{dt} + Av + f(v+z) = 0 \\ v(s) = u_0 - z(s) \end{cases}, \tag{6.7}$$

and $u = v + z$ is a solution of (6.1) such that $u(s) = u_0$. Denote this solution by $u(t) = u(t; s, u_0; \omega)$ (denoted sometimes by $u(t; s, u_0)$, $u(t; s)$, $u(t)$ or even $u$ if no confusions). Then

$$\varphi(t,\omega) u_0 := u(t; 0, u_0; \omega), \quad t \geqslant 0, \tag{6.8}$$

is a continuous RDS on $L^2(D)$ with the following fact

$$\varphi(s, \theta_{-s}\omega) u_0 = u(0; -s, u_0; \omega), \quad s \geqslant 0. \tag{6.9}$$

The random attractor in $L^2(D)$ has been obtained by [8,9].

**Theorem 6.1**[8,9] The RDS defined by (6.8) possesses a random attractor in $L^2(D)$.

### 6.2 Absorption in $L^q(D)$

In order to prove the RDS $\varphi$ defined by (6.8) has a random attractor in $L^q(D)(q>2)$, by Theorem 5.3, we must show that $\varphi$ has a bounded absorbing set and some asymptotic *a priori* estimates in $L^q(D)$ $(q>2)$.

In this subsection, we will prove that the RDS $\varphi$ generated by Eq. (6.1) has a bounded absorbing set $\mathscr{B}(\omega) \subset L^q(D), q>2$, which absorbs in fact all bounded sets $B \subset L^2(D)$.

We need the properties of the polynomial $f$ defined by (6.2). The proof is elementary and so is omitted. In the sequel, the large positive constant $C$ may be different from line to line.

**Lemma 6.2** For the polynomial $f$ defined by (6.2), there exist $c_1, c_2$ and large number $C$ such that

(i) $f'(x) \geqslant -C$, for all $x \in \mathbf{R}$;

(ii) $f(x+y)x^{2k+1} \geqslant c_1 x^{2p+2k} - C(y^{2p+2k} + 1)$, for all $k \in \mathbf{N}$ and $x, y \in \mathbf{R}$;

(iii) there exists an $M>0$ such that $f(x)\geqslant c_2x^{2p-1}$ if $x\geqslant M$, while $f(x)\leqslant c_2x^{2p-1}$ if $x\leqslant -M$.

We now prove the existence of bounded absorbing sets in $L^q(D)$. The idea of the induction proof follows from the proof of differentiability of the system in $L^2(D)$ given in [9, Lemma 4.3] or [14, Lemma 2.3]. Recall that $u(t)=u(t;s,u_0;\omega)$ is the solution of (6.1) with $u(s)=u_0$ and $v(t)=v(t;s,u_0-z(s);\omega)$ is the solution of (6.7) with $v(s)=u_0-z(s)$.

**Lemma 6.3** For every fixed $q>2$, there exists a random variable $\gamma(\omega)$ such that, for any $R>0$ there exists an $s_0=s_0(R,\omega)<-1$ satisfying

$$\|u(0,s,u_0,\omega)\|_q\leqslant\gamma(\omega),\quad \text{whenever}\quad s\leqslant s_0,\ \|u_0\|_2\leqslant R. \tag{6.10}$$

**Proof** We have only to prove that (6.10) holds for all

$$q_k=2k(p-1)+2,\quad k\in\mathbf{N}, \tag{6.11}$$

since for every $q$, $L^q(D)\supset L^{q_k}(D)$ for some $q_k$. To do this, we shall prove, recursively in $k$, that there exist random numbers $\gamma_1^k(\omega),\gamma_2^k(\omega)$ such that, for any $R>0$, there exists an $s_k=s_k(R,\omega)\leqslant -1$ satisfying

$$\int_{-\frac{1}{k}}^{0}\|v(\tau;s,u_0-z(s);\omega)\|_{q_k}^{q_k}\mathrm{d}\tau\leqslant\gamma_1^k(\omega), \tag{6.12}$$

$$\|u(t;s,u_0;\omega)\|_{q_k}\leqslant\gamma_2^k(\omega),\quad -\frac{1}{k+1}\leqslant t\leqslant 0, \tag{6.13}$$

whenever $s\leqslant s_k$ and $\|u_0\|_2\leqslant R$.

We first prove (6.12)~(6.13) for $k=1$, that is, $q_1=2p$. Taking the scalar product of (6.7) with $v$ on $L^2(D)$ and using Lemma 6.2(ii), we have

$$\frac{\mathrm{d}}{\mathrm{d}t}\|v\|_2^2+\|v\|_{H^1}^2+c_1\|v\|_{q_1}^{q_1}\leqslant C\|z\|_{q_1}^{q_1}+C. \tag{6.14}$$

Integrating (6.14) over $t\in[-1,0]$ we find

$$\int_{-1}^{0}\|v(\tau)\|_{q_1}^{q_1}\mathrm{d}\tau\leqslant C\int_{-1}^{0}\|z(\tau)\|_{q_1}^{q_1}\mathrm{d}\tau+C+\|v(-1)\|_2^2. \tag{6.15}$$

But it has been proved in [7] that there exist $\gamma_0=\gamma_0(\omega)$ and $s_0=s_0(R,\omega)$ such that

$$\sup_{-1\leqslant t\leqslant 0}\|v(t;s,u_0-z(s))\|_2\leqslant\gamma_0 \tag{6.16}$$

whenever $s\leqslant s_0$ and $\|u_0\|_2\leqslant R$. It note also that by the assumption (6.3) $g_j\in L^\infty(D)$ and thus $z$ has continuous path on $L^q(D)$ for any $q\geqslant 2$, which, together with (6.15)~(6.16), implies that

$$\int_{-1}^{0}\|v(\tau)\|_{q_1}^{q_1}\mathrm{d}\tau\leqslant\gamma_1^1:=C\int_{-1}^{0}|z(\tau)|_{q_1}^{q_1}\mathrm{d}\tau+C+\gamma_0^2<+\infty, \tag{6.17}$$

which proved (6.12) for $k=1$. Multiplying (6.7) by $v^{q_1-1}$ and using Lemma 6.2 (ii) again we obtain

$$\frac{\mathrm{d}}{\mathrm{d}t}\|v\|_{q_1}^{q_1}+c_2\|v\|_{q_2}^{q_2}\leqslant C\|z\|_{q_2}^{q_2}+C. \tag{6.18}$$

Integrating (6.18) from $\tau(-1\leqslant\tau\leqslant -1/2)$ to $t(-1/2\leqslant t\leqslant 0)$ we get

$$\|v(t)\|_{q_1}^{q_1}\leqslant C\int_{\tau}^{t}\|z(\sigma)\|_{q_2}^{q_2}\mathrm{d}\sigma+C(t-\tau)+\|v(\tau)\|_{q_1}^{q_1},\quad -1\leqslant\tau\leqslant-\frac{1}{2}\leqslant t\leqslant 0.$$

Integrating the above inequality from $\tau=-1$ to $\tau=-1/2$ and using (6.17), we have

$$\|v(t)\|_{q_1}^{q_1}\leqslant \gamma_2^1:=2C\int_{-1}^{0}\|z(\sigma)\|_{q_2}^{q_2}d\sigma+2C+2\gamma_1^1<\infty,\quad -\frac{1}{2}\leqslant t\leqslant 0,$$

which proved (6.13) for $k=1$.

Assume now (6.12) ~ (6.13) hold for $k$. Using a similar argument as in the derivation of (6.18), we can prove the existence of $c_1^k$ and $C$ such that

$$\frac{d}{dt}\|v\|_{q_k}^{q_k}+c_1^k\|v\|_{q_{k+1}}^{q_{k+1}}\leqslant C\left(\|z\|_{q_{k+1}}^{q_{k+1}}+1\right).\tag{6.19}$$

Integrating (6.19) from $t=-\frac{1}{k+1}$ to $t=0$ yields

$$c_1^k\int_{-\frac{1}{k+1}}^{0}\|v(t)\|_{q_{k+1}}^{q_{k+1}}dt\leqslant C\left(\int_{-\frac{1}{k+1}}^{0}\|z(t)\|_{q_{k+1}}^{q_{k+1}}dt+1\right)+\left\|v\left(-\frac{1}{k+1}\right)\right\|_{q_k}^{q_k}.$$

Thanks to the induction assumption (6.13) for $k$, we see that

$$\left\|v\left(-\frac{1}{k+1}\right)\right\|_{q_k}^{q_k}\leqslant 2^{q_k}\left[\left\|u\left(-\frac{1}{k+1}\right)\right\|_{q_k}^{q_k}+\left\|z\left(-\frac{1}{k+1}\right)\right\|_{q_k}^{q_k}\right]\leqslant 2^{q_k}(\gamma_2^k+C).$$

Thus we have proved (6.12) for $k+1$. Integrating from $\tau\left(-1\leqslant\tau\leqslant-\frac{1}{k+2}\right)$ to $t\left(-\frac{1}{k+2}\leqslant t\leqslant 0\right)$ in (6.19) with $k$ replaced by $k+1$, we obtain

$$\|v(t)\|_{q_{k+1}}^{q_{k+1}}\leqslant C+\|v(\tau)\|_{q_{k+1}}^{q_{k+1}},\quad -1\leqslant\tau\leqslant-\frac{1}{k+2}\leqslant t\leqslant 0.$$

Integrating the above inequality from $\tau=-\frac{1}{k+1}$ to $\tau=-\frac{1}{k+2}$ we know

$$\left(\frac{1}{k+1}-\frac{1}{k+2}\right)\|v(t)\|_{q_{k+1}}^{q_{k+1}}\leqslant C+\int_{-\frac{1}{k+1}}^{0}\|v(\tau)\|_{q_{k+1}}^{q_{k+1}}d\tau,\quad -\frac{1}{k+2}\leqslant t\leqslant 0,$$

which, together with (6.12) for $k+1$, implies that (6.13) holds for $k+1$ and completes the induction proof. □

### 6.3 Asymptotic Estimates

This subsection presents some asymptotic *a priori* estimates for the unbounded part of the modular $|u|$ in $L^q(D)(q>2)$. We need an auxiliary lemma which gives the asymptotic estimate in $L^2(D)$ at all time $t\in[-1,0]$. (The special case that $t=0$ is immediate from Theorems 5.3 and 6.1.)

**Lemma 6.4** For any $\varepsilon>0$ and bounded set $B\subset L^2(D)$, there exist $s_0=s_0(\varepsilon,B,\omega)$, $M_1=M_1(\varepsilon,B,\omega)$, $M_2=M_2(\varepsilon,B,\omega)$ such that, for all $s\leqslant s_0$, $u_0\in B$, $P$-a.s. $\omega\in\Omega$

$$\int_{D(|u(t)|\geqslant M_1)}|u(t;s,u_0;\omega)|^2<\varepsilon,\quad -1\leqslant t\leqslant 0,\tag{6.20}$$

$$\int_{D(|v(t)|\geqslant M_2)}|v(t;s,u_0-z(s);\omega)|^2<\varepsilon,\quad -1\leqslant t\leqslant 0.\tag{6.21}$$

**Proof** Using a similar argument as given in [7], one can prove that there exists a compact absorbing set $\mathscr{B}(\omega)\subset L^2(D)$, which absorbs every bounded set $B$ in $L^2(D)$ at all time $t\in[-1,0]$ (rather than $t=0$ only). More precisely, there exists $s_0\leqslant -1$ such that if $s\leqslant s_0$ then

$$u(t;s,B) := \bigcup_{u_0\in B}\{u(t;s,u_0)\}\subset \mathscr{B}(\omega), \quad -1\leqslant t\leqslant 0. \tag{6.22}$$

Since $\mathscr{B}(\omega)$ is compact, for any $\varepsilon>0$, the set

$$\bigcup_{-1\leqslant t\leqslant 0}\ \bigcup_{s\leqslant s_0} u(t;s,B)$$

has an $(\varepsilon/8)^{1/2}$-net. Thus by Lemma 5.2 there exists an $M_1$ such that

$$\sup_{-1\leqslant t\leqslant 0}\int_{D(|u(t)|\geqslant M_1)}|u(t;s,u_0)|^2 < 2^{2+1}\varepsilon/8 = \varepsilon,$$

whenever $u_0\in B$ and $s\leqslant s_0$, which proved (6.20).

To prove (6.21), we let

$$E = \sup_{-1\leqslant t\leqslant 0}\|z(t)\|_\infty, \tag{6.23}$$

which is finite by the bounded assumption (6.3) of $g_j$. By Lemma 5.1, we can choose $s_1\leqslant s_0\leqslant -1$ and $\widetilde{M}\geqslant M_1$ such that, for all $s\leqslant s_1$ and $u_0\in B$,

$$m(D(|u(t;S,u_0)|\geqslant \widetilde{M}))\geqslant \frac{\varepsilon}{E^2}, \quad -1\leqslant t\leqslant 0. \tag{6.24}$$

Taking now $M_2=\widetilde{M}+E$ and noting that

$$D(|v(t)|\geqslant M_2)\subset D(|u(t)|\geqslant \widetilde{M})\subset D(|u(t)|\geqslant M_1),$$

we get from (6.20) and (6.23)~(6.24) that

$$\int_{D(|v|\geqslant M_2)}|v(t)|^2\leqslant 2\int_{D(|u|\geqslant M_1)}|u(t)|^2 + 2\int_{D(|u|\geqslant \widetilde{M})}|z(t)|^2\leqslant 4\varepsilon,$$

which completes the proof of (6.21). □

Using the auxiliary lemma above, we can now give the asymptotic estimates in $L^q(D)$ for $q=q_1=2p$.

**Lemma 6.5** For any bounded set $B\subset L^2(D)$ and $P$-a.s. $\omega\in\Omega$, we have

$$\lim_{-s,M\to+\infty}\sup_{u_0\in B}\int_{D(|u(t;s,u_0;\omega)|\geqslant M)}|u(t;s,u_0;\omega)|^{2p} = 0, \quad t\in\left[-\frac{1}{2},0\right]. \tag{6.25}$$

**Proof** For any $\varepsilon>0$, let $s_0, M_1, M_2$ be the constants defined in Lemma 6.4. By Lemma 5.1 we choose $s_1\leqslant s_0\leqslant -1$ and $M_3$ such that, for all $s\leqslant s_1, u_0\in B$,

$$m(D(|u(t;s,u_0)|\geqslant M_3))\leqslant \frac{\varepsilon}{E^{2p}}, \quad -1\leqslant t\leqslant 0, \tag{6.26}$$

where $E=\sup\{\|z(t)\|_\infty; -1\leqslant t\leqslant 0\}$ defined as in (6.23). By Lemma 6.2 (ⅲ), we can choose $M_4$ such that

$$f(x)\geqslant 0 \quad \text{if} \quad x\geqslant M_4. \tag{6.27}$$

Let now $M=E+\max\{M_1,M_2,M_3,M_4\}$. Multiplying Eq. (6.7) by $(v-M)_+$ and integrating over $D$ we find

$$\frac{\mathrm{d}}{\mathrm{d}t}\|(v-M)_+\|_2^2 + 2\int_D|\nabla(v-M)_+|^2 + 2\int_{D(v\geqslant M)}f(v+z)(v-M) = 0 \tag{6.28}$$

where

$$(v-M)_+ = \begin{cases} v-M & \text{if } v \geqslant M \\ 0 & \text{if } v \leqslant M \end{cases}.$$

By (6.27) we know that $f(v+z)(v-M) \geqslant 0$ on $D(v \geqslant M)$ (since $v+z \geqslant M-E \geqslant M_4$). Integrating (6.28) from $t=-1$ to $t=0$ and using (6.21) we obtain

$$2\int_{-1}^{0}\int_{D(v\geqslant M)} f(v+z)(v-M) \leqslant \|(v(-1)-M)_+\|_2^2 < \varepsilon. \tag{6.29}$$

Noting that

$$\int_{D(v\geqslant 2M)} f(v+z)v \leqslant 2\int_{D(v\geqslant M)} f(v+z)(v-M)$$

and

$$v^{2p} \leqslant c_1 f(v+z)v + C(|z|^{2p}+1),$$

we get from (6.26) and (6.29) that

$$\int_{-1}^{0}\int_{D(v\geqslant 2M)} v^{2p} \leqslant C\varepsilon \tag{6.30}$$

where $C$ does not depend on $\varepsilon$ and $B$. We then take the product of (6.7) with $(v-2M)_+^{2p-1}$ to find

$$\frac{1}{2p}\frac{\mathrm{d}}{\mathrm{d}t}\|(v-2M)_+\|_{2p}^{2p} + \int_{D(v\geqslant 2M)} f(v+z)(v-2M)^{2p-1} \leqslant 0. \tag{6.31}$$

In particular

$$\frac{\mathrm{d}}{\mathrm{d}t}\|(v-2M)_+\|_{2p}^{2p} \leqslant 0, \tag{6.32}$$

which implies obviously that

$$\|(v(t)-2M)_+\|_{2p}^{2p} \leqslant \|(v(\tau)-2M)_+\|_{2p}^{2p}, \quad -1 \leqslant \tau \leqslant -\frac{1}{2},\ -\frac{1}{2} \leqslant t \leqslant 0.$$

Integrating it from $\tau=-1$ to $\tau=-1/2$ and using (6.30) we get

$$\|(v(t)-2M)_+\|_{2p}^{2p} \leqslant 2\int_{-1}^{-1/2}\int_{D(v\geqslant 2M)} v^{2p} < C\varepsilon, \quad -\frac{1}{2} \leqslant t \leqslant 0,$$

which implies that

$$\int_{D(v\geqslant 4M)} v^{2p}(t) \leqslant 2^{2p+1}\int_{D(v\geqslant 2M)} (v(t)-2M)^{2p} \leqslant C\varepsilon, \quad -\frac{1}{2} \leqslant t \leqslant 0.$$

This, together with (6.26), implies that

$$\int_{D(u\geqslant 4M+E)} u^{2p}(t) \leqslant C\varepsilon, \quad -\frac{1}{2} \leqslant t \leqslant 0. \tag{6.33}$$

Taking $(v+M)_-$ and $(v+2M)_-^{2p-1}$ instead of $(v-M)_+$ and $(v-2M)_+^{2p-1}$ in the preceding proof, we deduce similarly that

$$\int_{D(u\leqslant -4M-E)} u^{2p}(t) \leqslant C\varepsilon, \quad \frac{1}{2} \leqslant t \leqslant 0,$$

which completes the proof of (6.25). □

In what follows we give the asymptotic estimate in $L^q(D)$ for any $q>2$.

**Lemma 6.6** Let $q>2$ be fixed. For every bounded set $B \subset L^2(D)$, we have

$$\lim_{-s, M\to+\infty} \sup_{u_0\in B} \int_{D(|u|\geqslant M)} |u(0;s,u_0;\omega)|^q = 0, \quad P\text{-a.s. } \omega \in \Omega. \tag{6.34}$$

**Proof** Let $q_k = 2k(p-1)+2$ defined as in (6.11). We will prove by induction that for any $\varepsilon > 0$ there exist $s_0^k = s_0^k(\varepsilon, B, \omega)$, $M_1^k = M_1^k(\varepsilon, B, \omega)$ and $M_2^k = M_2^k(\varepsilon, B, \omega)$ such that, for all $s < s_0^k$, $u_0 \in B$,

$$\int_{-\frac{1}{k}}^{0}\left(\int_{D(|v| \geqslant M_1^k)} |v(\tau; s, u_0 - z(s))|^{q_k}\right)\mathrm{d}\tau < C\varepsilon, \tag{6.35}$$

$$\int_{D(|u| \geqslant M_2^k)} |v(t; s, u_0)|^{q_k} < C\varepsilon, \quad -\frac{1}{k+1} \leqslant t \leqslant 0. \tag{6.36}$$

The case that $k=1$ has been proved by Lemma 6.5. Assume now that (6.35)~(6.36) hold for $k$. By Lemma 5.1 we can choose $s_1^k \leqslant s_0^k$ and $M_3^k$ such that, for $s \leqslant s_1^k$ and $u_0 \in B$,

$$m(D(|u(t; s, u_0)| \geqslant M_3^k)) \leqslant \frac{\varepsilon}{E^{q_{k+1}}}, \quad -1/k \leqslant t \leqslant 0. \tag{6.37}$$

Let $M = E + \max\{M_1^k, M_2^k, M_3^k, M_4\}$, where $M_4$ is the constant such that $f(x) \geqslant 0$ whenever $x \geqslant M_4$. Using similar arguments as in the derivation of (6.31) we can prove the existence of $c_k$ such that

$$\frac{\mathrm{d}}{\mathrm{d}t}\|(v-M)_+\|_{q_k}^{q_k} + c_k\int_{D(v \geqslant M)} f(v+z)(v-M)^{q_k-1} \leqslant 0. \tag{6.38}$$

Using (6.37) and the induction assumption (6.36) we know that, for $-\frac{1}{k+1} \leqslant t \leqslant 0$,

$$\|(v(t)-M)_+\|_{q_k}^{q_k} \leqslant C\left(\int_{D(u \geqslant M_2^k)} u^{q_k}(t) + \int_{D(u \geqslant M_3^k)} z^{q_k}(t)\right) < C\varepsilon, \tag{6.39}$$

which implies by integrating (6.38) over $\left[-\frac{1}{k+1}, 0\right]$ that

$$\int_{-\frac{1}{k+1}}^{0}\int_{D(v \geqslant M)} f(v+z)(v-M)^{q_k-1} < C\varepsilon. \tag{6.40}$$

It is easy to prove that

$$\int_{D(v \geqslant 2M)} f(v+z)v^{q_k-1} \leqslant 2^{q_k}\int_{D(v \geqslant M)} f(v+z)(v-M)^{q_k-1}.$$

This, together with (6.40), (6.37), Lemma 6.2 (ii), implies that

$$\int_{-\frac{1}{k+1}}^{0}\int_{D(v \geqslant 2M)} v^{q_{k+1}} < C\varepsilon. \tag{6.41}$$

We also multiply Eq. (6.7) by $(v-2M)_+^{q_{k+1}-1}$ to find

$$\frac{\mathrm{d}}{\mathrm{d}t}\|(v-2M)_+\|_{q_{k+1}}^{q_{k+1}} \leqslant 0. \tag{6.42}$$

Then

$$\|(v(t)-2M)_+\|_{q_{k+1}}^{q_{k+1}} \leqslant \|(v(\tau)-2M)_+\|_{q_{k+1}}^{q_{k+1}}, \quad -\frac{1}{k+1} \leqslant \tau \leqslant -\frac{1}{k+2} \leqslant t \leqslant 0.$$

Integrating it from $\tau = -\frac{1}{k+1}$ to $\tau = -\frac{1}{k+2}$ and using (6.41) we have

$$\left(\frac{1}{k+1} - \frac{1}{k+2}\right)\|(v(t)-2M)_+\|_{q_{k+1}}^{q_{k+1}} \leqslant \int_{-\frac{1}{k+1}}^{0}\int_{D(v \geqslant 2M)} (v-2M)^{q_{k+1}} < C\varepsilon, \tag{6.43}$$

whenever $-\frac{1}{k+2} \leqslant t \leqslant 0$, which implies easily that

$$\int_{D(v\geq 4M)} v^{q_{k+1}}(t) \leq C\varepsilon, \quad -\frac{1}{k+2} \leq t \leq 0,$$

and thus

$$\int_{D(u\geq 4M+E)} u^{q_{k+1}}(t) \leq C\varepsilon, \quad -\frac{1}{k+2} \leq t \leq 0. \tag{6.44}$$

On the other hand, we have the similarly asymptotic estimates for the negative part and thus (6.35)~(6.36) hold for $k+1$, which completes the induction proof of (6.35) ~ (6.36). In particular, (6.36) with $t=0$ implies that (6.34) holds for all $q_k$. Since for every $q>2, L^q(D)\supset L^{q_k}(D)$ for some large $q_k$, (6.34) holds for all $q>2$. □

## 6.4 Existence and the Fractal Dimension of Random Attractors

By using the preceding results, we can now state our conclusions about the existence of random attractors for Eq. (6.1) in $L^q(D)$ for any $q\geq 2$.

**Theorem 6.7** Assume that $D\subset \mathbf{R}^n$ is bounded and $g_j\in L^\infty(D), j=1,\cdots,m$. Then the RDS $\varphi$ generated by the stochastic reaction-diffusion equation (6.1) possesses a random attractor $\mathscr{A}_q(\omega)$ in $L^q(D)$ for any $q\geq 2$. $\mathscr{A}_q(\omega)$ is an invariant and compact random set which attracts bounded sets of $L^2(D)$ under the $L^q$-norm topology. Furthermore, $\mathscr{A}_q(\omega)=\mathscr{A}_2(\omega):=\mathscr{A}(\omega)$, for all $q\geq 2$, where $\mathscr{A}(\omega)$ is the usual attractor in $L^2(D)$.

**Proof** By Lemmas 6.3, 6.6 and Theorem 6.1, we know that the hypotheses of Theorem 5.3 are satisfied, which proved the RDS $\varphi$ has a random attractor $\mathscr{A}_q(\omega)$ in $L^q(D)$. It notes that by Lemma 6.3 $\mathscr{A}_q(\omega)$ attracts in fact all bounded sets in $L^2(D)$. By this fact and a similar argument as in the deterministic case, it is easy to prove that all $\mathscr{A}_q(\omega)$ $(q\geq 2)$ are the same set $\mathscr{A}(\omega)$. □

Finally we are concerned with the fractal dimension of $\mathscr{A}(\omega)$ under the $L^q$-norm topology. Recall that, for a bounded set $B\subset L^q(D)$, the fractal dimension of $B$ under $L^q$-norm is defined by (see [18,20])

$$\dim_q(B) = \limsup_{\varepsilon\to 0} \frac{\ln M_q(\varepsilon)}{-\ln\varepsilon} \tag{6.45}$$

where $M_q(\omega)$ denotes the minimum number of balls $B_q(u_i,\varepsilon):=\{u\in L^q(D) \mid \|u-u_i\|_q<\varepsilon\}$ of radius $\varepsilon$ required to cover $B$. By [12, Lemma 2.1], the centers $u_i$ can be taken in $B$, that is,

$$\dim_q(B) = \limsup_{\varepsilon\to 0} \frac{\ln N_q(\varepsilon)}{-\ln\varepsilon} \tag{6.46}$$

where $N_q(\varepsilon)$ denotes the minimum number of balls of radius $\varepsilon$ with centers in $B$ that required to cover $B$ under the $L^q$-norm.

In order to prove the finiteness of the fractal dimension of the random attractor $\mathscr{A}(\omega)$ under any $L^q$-norm $(q\geq 2)$, we need a key lemma, which gives a comparison result of fractal dimensions under the different $L^q$-norm.

**Lemma 6.8** Suppose a set $B$ is bounded in $L^\delta(D)$ for all $1<\delta<+\infty$, where $D$ is bounded. Then we have

$$\dim_r(B) \leqslant \dim_q(B) \leqslant \frac{2q-r}{r}\dim_r(B), \quad \text{for } q \geqslant r > 1. \tag{6.47}$$

**Proof** It is relatively easy to prove the first inequality. Indeed, by the Hölder inequality we know that

$$\| u \|_r \leqslant c_1 \| u \|_q, \quad \forall\, u \in L^q(D), \tag{6.48}$$

where $c_1 = |D|^{(q-r)/qr}$. Consider an $\varepsilon$-net $B_q(u_i, \varepsilon)$ in $L^q(D), i = 1,2,\cdots, N_q(\varepsilon)$. By (6.48) we have

$$B_q(u_i, \varepsilon) \subset B_r(u_i, c_1\varepsilon),$$

it follows that $N_r(c_1\varepsilon) \leqslant N_q(\varepsilon)$, and so

$$\frac{\ln N_r(c_1\varepsilon)}{-\ln(c_1\varepsilon)} \leqslant \frac{\ln N_q(\varepsilon)}{-\ln c_1 - \ln\varepsilon}$$

yields $\dim_r(B) \leqslant \dim_q(B)$.

In order to prove the second inequality in (6.47), we first show that for any fixed $\delta > 1$ and $q \geqslant r > 1$ there exists a number $C_1 = C_1(\delta, B, q)$ such that

$$\sup_{u \in B} \int_{D(|u| \geqslant C_1\varepsilon^{-\delta})} |u|^q < \varepsilon^q, \quad \forall\, \varepsilon > 0, \tag{6.49}$$

where $C_1$ does not depend on $\varepsilon$. Indeed, by the assumptions, $B$ is bounded in $L^{q\delta'}(D)$, where $\delta'$ satisfies $1/\delta + 1/\delta' = 1$, then

$$c_2 := \sup_{u \in B}\{\| u \|_{q\delta'}, \| u \|_q\} < +\infty. \tag{6.50}$$

By the Hölder inequality we have

$$\begin{aligned}\int_{D(|u| \geqslant C_1\varepsilon^{-\delta})} |u|^q &\leqslant \| u \|_{q\delta'}^q \cdot (m(D(|u| \geqslant C_1\varepsilon^{-\delta})))^{\frac{1}{\delta}} \\ &\leqslant c_2^q (m(D(|u| \geqslant C_1\varepsilon^{-\delta})))^{\frac{1}{\delta}}.\end{aligned} \tag{6.51}$$

But by (6.50)

$$m(D(|u| \geqslant C_1\varepsilon^{-\delta})) \leqslant \frac{1}{(C_1\varepsilon^{-\delta})^q}\int_{D(|u| \geqslant C_1\varepsilon^{-\delta})} |u|^q \leqslant \frac{c_2^q}{C_1^q}\varepsilon^{q\delta}, \tag{6.52}$$

which, together with (6.51), implies that

$$\int_{D(|u| \geqslant C_1\varepsilon^{-\delta})} |u|^q \leqslant \frac{c_2^{q+q/\delta}}{C_1^{q/\delta}}\varepsilon^q. \tag{6.53}$$

Thus (6.49) holds if we choose $C_1$ such that $C_1^{q/\delta} > c_2^{q+q/\delta}$.

We then show that for any $\delta > 1$ there exists a constant $C_2$ such that

$$N_q(C_2\varepsilon) \leqslant N_r(\varepsilon^{(\delta(q-r)+q)/r}), \quad \forall\, \varepsilon > 0. \tag{6.54}$$

To this end, let $B$ have a net $B_r(u_k, \varepsilon^{(\delta(q-r)+q)/r})$ in $L^r(D)$, $k = 1,2,\cdots, N_r(\varepsilon^{(\delta(q-r)+q)/r})$. Then for any $u \in B$ there exists some $u_k$ such that

$$\| u - u_k \|_r^r \leqslant \varepsilon^{\delta(q-r)+q}. \tag{6.55}$$

Let

$$D_1 = D(|u| \leqslant C_1\varepsilon^{-\delta}, |u_k| \leqslant C_1\varepsilon^{-\delta}), \quad D_2 = D(|u| \geqslant C_1\varepsilon^{-\delta}, |u_k| \leqslant C_1\varepsilon^{-\delta}),$$

$$D_3 = D(|u| \leqslant C_1\varepsilon^{-\delta}, |u_k| \geqslant C_1\varepsilon^{-\delta}), \quad D_4 = D(|u| \geqslant C_1\varepsilon^{-\delta}, |u_k| \geqslant C_1\varepsilon^{-\delta}).$$

By (6.55) we have

$$\int_{D_1} |u - u_k|^q \leqslant (2C_1\varepsilon^{-\delta})^{q-r}\|u - u_k\|_r^r \leqslant (2C_1)^{q-r}\varepsilon^q.$$

Similar to the proof of (5.8)~(5.10), it follows from (6.49) that

$$\left(\int_{D_2} + \int_{D_3} + \int_{D_4}\right)|u - u_k|^q \leqslant 4 \cdot 2^q\varepsilon^q.$$

Thus $\|u - u_k\|_q \leqslant C_2\varepsilon$ if we choose $C_2$ such that $C_2^q \geqslant (2C_1)^{q-r} + 2^{q+2}$. That is, $B$ has a $C_2\varepsilon$-net in $L^q(D)$, where the number of balls equals to $N_r(\varepsilon^{(\delta(q-r)+q)/r})$, which proved (6.54).

Finally, it follows from (6.54) that

$$\frac{\ln N_q(C_2\varepsilon)}{\ln C_2 - \lg(C_2\varepsilon)} \leqslant \frac{\ln N_r(\varepsilon^{(\delta(q-r)+q)/r})}{-\ln \varepsilon^{(\delta(q-r)+q)/r}} \cdot \frac{\delta(q-r)+q}{r}.$$

Taking the sup-limit as $\varepsilon \to 0$ we find

$$\dim_q(B) \leqslant \frac{\delta(q-r)+q}{r}\dim_r(B), \quad \forall \delta > 1,\ q \geqslant r > 1.$$

Letting $\delta \to 1$ we obtain

$$\dim_q(B) \leqslant \left(\frac{2q}{r} - 1\right)\dim_r(B), \quad q \geqslant r > 1,$$

which completes the proof of the lemma. □

**Theorem 6.9** Let $\mathscr{A}(\omega) = \mathscr{A}_q(\omega)$ be the same random attractor obtained in Theorem 6.7. Then $\mathscr{A}(\omega)$ has finite fractal dimensions under $L^q$-norm for any $q \geqslant 2$. Furthermore, we have

$$\dim_2\mathscr{A}(\omega) \leqslant \dim_q\mathscr{A}(\omega) \leqslant (q-1)\dim_2\mathscr{A}(\omega), \quad q \geqslant 2, \tag{6.56}$$

where $\dim_q\mathscr{A}(\omega)$ denotes the fractal dimension under the $L^q$-norm and $\dim_2\mathscr{A}(\omega)$ is the usual fractal dimension under the $L^2$-norm.

**Proof** Taking $r = 2$ in Lemma 6.8 we obtain (6.56) immediately. But it has been proved by Langa and Robinson[12] that $\dim_2\mathscr{A}(\omega) < +\infty$, then (6.56) implies that $\dim_q\mathscr{A}(\omega) < +\infty$ for all $q > 2$. □

**Remark 6.10** Since the Hausdorff dimension is less than the fractal dimension, the attractor $\mathscr{A}(\omega)$ has finite Hausdorff dimensions under any $L^q$-norm ($q \geqslant 2$). It is also possible to derive a similar estimate of (6.56) for the Hausdorff dimensions.

**References**

[1] Arnold L. Random dynamical system. Springer Monogr Math. Berlin: Springer-Verlag, 1998.
[2] Arnold L, Schmalfuss B. Lyapunov's second method for random dynamical systems. J Differential Equations, 2001, 177:235-265.
[3] Brzezniak Z, Li Y H. Asymptotic compactness and absorbing sets for 2D stochastic Navier-Stokes equations on some unbounded domains. Trans Amer Math Soc, 2006, 358: 5587-5629.
[4] Caraballo T, Langa J A. Stability and random attractors for a reaction-diffusion equation with multiplicative noise. Discrete Contin Dyn Syst, 2000, 6: 875-892.
[5] Caraballo T, Langa J A, Robinson J C. Upper semi-continuity of attractors for random perturbations of dynamical

systems. Comm Partial Differential Equations, 1998, 23: 1557 – 1581.

[6] Crauel H. Random point attractors versus random set attractors. J London Math Soc, 2001, 63: 413 – 427.

[7] Crauel H, Debussche A, Flandoli F. Random attractors. J Dynam Differential Equations, 1997, 9: 307 – 341.

[8] Crauel H, Flandoli F. Attractors for random dynamical systems. Probab Theory Related Fields, 1994, 100: 365 – 393.

[9] Debussche A. Hausdorff dimension of a random invariant set. J Math Pures Appl, 1998, 77: 967 – 988.

[10] Deimling K. Nonlinear functional analysis. Berlin: Springer-Verlag, 1985.

[11] Kloeden P E, Langa J A. Flattening, squeezing and the existence of random attractors. Proc R Soc Lond Ser A, 2007, 463: 163 – 181.

[12] Langa J A, Robinson J C. Fractal dimension of a random invariant set. J Math Pures Appl, 2006, 85: 269 – 294.

[13] Ma Q F, Wang S H, Zhong C K. Necessary and sufficient conditions for the existence of global attractors for semigroups and applications. Indiana Univ Math J, 2002, 51: 1541 – 1559.

[14] Marion M. Attractors for reaction-diffusion equations: existence and estimate of their dimension. Appl Anal, 1987, 25: 101 – 147.

[15] Pazy A. Semigroups of linear operators and applications to partial differential equations. New York: Springer-Verlag, 1983.

[16] Da Prato G, Zabczyk J. Stochastic equations in infinite dimensions. Encyclopedia Math Appl. Cambridge: Cambridge Univ Press, 1992.

[17] Robinson J C. Stability of random attractors under perturbation and approximation. J Differential Equations, 2002, 186: 652 – 669.

[18] Robinson J C. Infinite-dimensional dynamical systems: an introduction to dissipative parabolic PDEs and the theory of global attractors. Cambridge: Cambridge Univ Press, 2001.

[19] Schmalfuss B. Backward cocycle and attractors of stochastic differential equations // Reitmann V, Riedrich T, Koksch N. International seminar on applied mathematics—nonlinear dynamics: attractor approximation and global behavior. Dresden: Technische Universität, 1992: 185 – 192.

[20] Sun C Y, Zhong C K. Attractors for the semi-linear reaction-diffusion equation with distribution derivatives in unbounded domains. Nonlinear Anal, 2005, 63: 49 – 65.

[21] Sun C Y, Yang M H, Zhong C K. Global attractors for the wave equation with nonlinear damping. J Differential Equations, 2006, 227: 427 – 443.

[22] Temam R. Infinite-dimensional dynamical systems in mechanics and physics. 2nd ed. New York: Springer-Verlag, 1997.

[23] Wang Y J, Zhong C K, Zhou S F. Pullback attractors of non-autonomous dynamical systems. Discrete Contin Dyn Syst, 2006, 16: 587 – 614.

[24] Wu D L, Zhong C K. The attractors for the non-homogeneous non-autonomous Navier-Stokes equations. J Math Anal Appl, 2006, 321: 426 – 444.

[25] Zhong C K, Yang M H, Sun C Y. The existence of global attractors for the norm-to-weak continuous semigroup and application to the nonlinear reaction-diffusion equations. J Differential Equations, 2006, 223: 367 – 399.

# The Global Solution of the (2+1)-dimensional Long Wave-short Wave Resonance Interaction Equation *

Xin Jie(辛杰) Guo Boling(郭柏灵) Han Yongqian(韩永前) Huang Daiwen(黄代文)

**Abstract**

In this paper, we prove the existence and uniqueness of the global smooth solution to the (2+1)-dimensional long wave-short wave resonance interaction equation.

## 1 Introduction

The interactions of long waves and short waves play an important role in fluid mechanics. It is well known that a strong interaction between long and short waves occurs when the phase velocity of the long waves equals (or is close to) the component of the group velocity of the short wave in the direction of propagation of the long wave.

For one-dimensional propagation of waves, there are many studies on this interaction. Boling Guo[4,5] obtained the existence of global solution for long wave-short wave LS-type equations and generalized LS-type equations. The approximation inertial manifolds for LS-type equations have been studied in Ref. [6]. The well posedness of the Cauchy problem for the long wave-short wave resonance equations were studied in Refs. [1,2,7,9,12,13,15].

In Ref. [10], Oikawa et al obtained the interaction equations describing the resonance interaction between a long interfacial wave and a short surface wave of (2+1)-dimensions (2 spatial and 1 temporal) in a two-layer fluid to the case in which they propagate obliquely each other. If $n(x,y,t)$ is a long interfacial wave and $u(x,y,t)$ is a short surface wave packet, their interaction on the $(x,y)$ plane is described by the system

$$\mathrm{i}(u_t + u_y) - \Delta u + nu = 0, \tag{1.1}$$

$$n_t - 2|u|_x^2 = 0, \tag{1.2}$$

where $u(x,y,t)$ is a complex valued unknown function and $n(x,y,t)$ is a real valued unknown function. In Refs. [8,11,14], the authors constructed the solutions of the (2+1)-dimensional long wave-short wave resonance interaction equations.

We consider the long wave-short wave resonance interaction equations (1.1) and (1.2) with the following initial condition:

$$u(x,y,0) = u_0(x,y), \quad n(x,y,0) = n_0(x,y), \quad -\infty < x < \infty, \quad -\infty < y < \infty. \tag{1.3}$$

* Journal of Mathematical Physics, 2008, 49: 073504-1-13.

Throughout this paper, we denote by $\|\cdot\|$ the norm of $H=L^2(\mathbf{R}^2)$ with usual inner product $(\cdot,\cdot)$, denote by $\|\cdot\|_p$ the norm of $L^p(\mathbf{R}^2)$ for all $1\leqslant p\leqslant\infty$ ($\|\cdot\|_2=\|\cdot\|$), and denote by $\|\cdot\|_{H^k}$ the norm of a usual Sobolev space $H^k(\mathbf{R}^2)$ for all $1\leqslant k<\infty$.

In this paper, we prove the existence and uniqueness of the global smooth solution to the (2+1)-dimensional long wave-short wave resonance interaction equation.

## 2 *A Priori* Estimates

**Lemma 2.1** If $u_0\in L^2$, then for the solution of the problem (1.1)~(1.3), we have

$$\|u\|=\|u_0\|. \tag{2.1}$$

**Proof** Taking the inner product of (1.1) with $u$, we get

$$(\mathrm{i}u_t+\mathrm{i}u_y-\Delta u+nu,u)=0. \tag{2.2}$$

Taking the imaginary part of (2.2), we find that

$$\frac{\mathrm{d}}{\mathrm{d}t}\|u\|^2=0.$$

Then we derive (2.1) □

Here, and throughout this section, we use $T$ to denote an arbitrary positive constant and use $C$ to denote different positive constants, which depends only on initial data and positive constant $T$.

**Lemma 2.2** Suppose that $u_0\in H^1$ and $n_0\in L^2$, $u$ and $n$ solve the problem (1.1)~(1.3), Then

$$\sup_{0\leqslant t\leqslant T}(\|u\|_{H^1}+\|n\|)\leqslant C(\|u_0\|_{H^1},\|n_0\|). \tag{2.3}$$

**Proof** Taking the inner product of (1.1) with $u_t$, we have

$$(\mathrm{i}u_t+\mathrm{i}u_y-\Delta u+nu,u_t)=0. \tag{2.4}$$

Taking the real part of (2.4), we get

$$\mathrm{Re}\,\mathrm{i}(u_y,u_t)+\frac{1}{2}\frac{\mathrm{d}}{\mathrm{d}t}\|\nabla u\|^2+\frac{1}{2}\int n|u|_t^2\mathrm{d}x\mathrm{d}y=0. \tag{2.5}$$

By (1.2),

$$\begin{aligned}\int n|u|_t^2\mathrm{d}x\mathrm{d}y&=\frac{\mathrm{d}}{\mathrm{d}t}\int n|u|^2\mathrm{d}x\mathrm{d}y-\int n_t|u|^2\mathrm{d}x\mathrm{d}y=\frac{\mathrm{d}}{\mathrm{d}t}\int n|u|^2\mathrm{d}x\mathrm{d}y-2\int|u|_x^2|u|^2\mathrm{d}x\mathrm{d}y\\&=\frac{\mathrm{d}}{\mathrm{d}t}\int n|u|^2\mathrm{d}x\mathrm{d}y,\end{aligned} \tag{2.6}$$

$$\mathrm{Re}\,\mathrm{i}(u_y,u_t)=-\mathrm{Im}(u_y,u_t)=-\mathrm{Im}(u_{yt},u)=-\frac{1}{2}\frac{\mathrm{d}}{\mathrm{d}t}\mathrm{Im}(u_y,u). \tag{2.7}$$

By (2.5)~(2.7), we derive

$$\frac{\mathrm{d}}{\mathrm{d}t}\left(\|\nabla u\|^2+\int n|u|^2\mathrm{d}x\mathrm{d}y-\mathrm{Im}(u_y,u)\right)=0,$$

i.e.,

$$\|\nabla u\|^2+\int n|u|^2\mathrm{d}x\mathrm{d}y-\mathrm{Im}(u_y,u)=\|\nabla u_0\|^2+\int n_0|u_0|^2\mathrm{d}x\mathrm{d}y-\mathrm{Im}(u_{0y},u_0). \tag{2.8}$$

Then by (2.8),

$$\|\nabla u\|^2 \leqslant C\|n\|\|u\|_4^2 + C\|\nabla u\|\|u\| + C(\|u_0\|_{H^1}, \|n_0\|)$$

$$\leqslant C\|n\|\|\nabla u\|\|u\| + C\|\nabla u\|\|u\| + C \leqslant C\|n\|^2 + \frac{1}{2}\|\nabla u\|^2 + C.$$

So

$$\|\nabla u\|^2 \leqslant C\|n\|^2 + C. \tag{2.9}$$

Taking the inner product of (1.2) with $n$, we have

$$(n_t - 2|u|_x^2, n) = 0. \tag{2.10}$$

Then by (2.10) and (1.1),

$$\frac{\mathrm{d}}{\mathrm{d}t}\|n\|^2 = 4\int n|u|_x^2 \mathrm{d}x\mathrm{d}y = 4\int(nu\bar{u}_x + n\bar{u}u_x)\mathrm{d}x\mathrm{d}y$$

$$= 4\int[(-\mathrm{i}u_t - \mathrm{i}u_y + \Delta u)\bar{u}_x + (\mathrm{i}\bar{u}_t + \mathrm{i}\bar{u}_y + \Delta\bar{u})u_x]\mathrm{d}x\mathrm{d}y$$

$$= 4\mathrm{i}\int(\bar{u}_t u_x - u_t\bar{u}_x)\mathrm{d}x\mathrm{d}y + 4\mathrm{i}\int(\bar{u}_y u_x - u_y\bar{u}_x)\mathrm{d}x\mathrm{d}y$$

$$= 4\mathrm{i}\int(\bar{u}_t u_x - u_t\bar{u}_x)\mathrm{d}x\mathrm{d}y. \tag{2.11}$$

By (2.11) and

$$\frac{\mathrm{d}}{\mathrm{d}t}\int(\bar{u}u_x - u\bar{u}_x)\mathrm{d}x\mathrm{d}y = \int(\bar{u}_t u_x + \bar{u}u_{xt} - u_t\bar{u}_x - u\bar{u}_{xt})\mathrm{d}x\mathrm{d}y$$

$$= \int(\bar{u}_t u_x - u_t\bar{u}_x - \bar{u}_x u_t + u_x\bar{u}_t)\mathrm{d}x\mathrm{d}y$$

$$= 2\int(\bar{u}_t u_x - u_t\bar{u}_x)\mathrm{d}x\mathrm{d}y,$$

we have

$$\frac{\mathrm{d}}{\mathrm{d}t}\left(\|n\|^2 - 2\mathrm{i}\int(\bar{u}u_x - u\bar{u}_x)\mathrm{d}x\mathrm{d}y\right) = 0,$$

i.e.,

$$\|n\|^2 - 2\mathrm{i}\int(\bar{u}u_x - u\bar{u}_x)\mathrm{d}x\mathrm{d}y = \|n_0\|^2 - 2\mathrm{i}\int(\bar{u}_0 u_{0x} - u_0\bar{u}_{0x})\mathrm{d}x\mathrm{d}y. \tag{2.12}$$

By (2.12),

$$\|n\|^2 \leqslant C\|\nabla u\|\|u\| + C(\|u_0\|_{H^1}, \|n_0\|) \leqslant C\|\nabla u\| + C. \tag{2.13}$$

By (2.9) and (2.13),

$$\|\nabla u\|^2 \leqslant C\|\nabla u\| + C.$$

Then

$$\|\nabla u\|^2 \leqslant C(\|u_0\|_{H^1}, \|n_0\|). \tag{2.14}$$

By (2.13) and (2.14),

$$\|n\|^2 \leqslant C(\|u_0\|_{H^1}, \|n_0\|).$$ □

**Lemma 2.3** Suppose that $u_0 \in H^2$ and $n_0 \in H^1$, $u$ and $n$ solve the problem (1.1)~(1.3). Then

$$\sup_{0 \leqslant t < T}(\|u\|_{H^2} + \|u_t\| + \|n\|_{H^1} + \|n_t\|_{H^1}) \leqslant C(\|u_0\|_{H^2}, \|n_0\|_{H^1}). \tag{2.15}$$

**Proof** By (1.1) and Brezis-Gallouet inequality[3],

$$\|u\|_\infty \leqslant C(1+\sqrt{\ln(1+\|u\|_{H^2})}), \tag{2.16}$$

for every $u \in H^2$ with $\|u\|_{H^1} \leqslant C$, we have

$$\|\Delta u\| \leqslant \|u_t\| + \|u_y\| + \|nu\| \leqslant C + \|u_t\| + \|u\|_\infty \|n\| \leqslant C + \|u_t\| + C(1+\sqrt{\ln(1+\|\Delta u\|)})$$
$$\leqslant C + \|u_t\| + C(1+(1+\|\Delta u\|)^{1/2}) \leqslant C + \|u_t\| + \frac{1}{2}\|\Delta u\|.$$

So

$$\|\Delta u\| \leqslant C + C\|u_t\|. \tag{2.17}$$

Differentiating (1.1) with respect to $t$ and then taking the inner product with $u_t$, we obtain

$$(\mathrm{i}u_{tt} + \mathrm{i}u_{yt} - \Delta u_t + n_t u + n u_t, u_t) = 0. \tag{2.18}$$

Taking the imaginary part of (2.18), by (1.2) and (2.17), and Brezis-Gallouet inequality, we derive

$$\frac{\mathrm{d}}{\mathrm{d}t}\|u_t\|^2 = -2\mathrm{Im}(n_t u, u_t) = -4\mathrm{Im}(|u|_x^2 u, u_t)$$
$$\leqslant C\|u\|_\infty^2 \|u_x\| \|u_t\| \leqslant C(1+\sqrt{\ln(1+\|\Delta u\|)})^2 \|u_t\|$$
$$\leqslant C(1+(1+\|\Delta u\|)^{1/2})\|u_t\| \leqslant C\|u_t\| + C\|\Delta u\|^{1/2}\|u_t\|$$
$$\leqslant C\|u_t\| + C\|u_t\|^{3/2} \leqslant C + C\|u_t\|^2.$$

Applying Gronwall inequality and (1.1), we deduce

$$\|u_t\| \leqslant C(\|u_t(x,y,0)\|) = C(\|u_0\|_{H^2}, \|n_0\|).$$

By (2.17),

$$\|u\|_{H^2} \leqslant C(\|u_0\|_{H^2}, \|n_0\|).$$

By the Gagliardo-Nirenberg inequality and the above inequality,

$$\|u\|_\infty \leqslant C(\|u_0\|_{H^2}, \|n_0\|). \tag{2.19}$$

Combining (1.2) with (2.19), we get

$$\|n_t\| = 2\|u\bar{u}_x + \bar{u}u_x\| \leqslant C\|u\|_\infty |u_x| \leqslant C(\|u_0\|_{H^2}, \|n_0\|).$$

By (1.2),

$$\|\nabla n_t\| = 2\|\nabla(|u|_x^2)\| \leqslant C\|u\|_\infty \|\nabla u_x\| + C\|\nabla u\|_4 \|u_x\|_4 \leqslant C(\|u_0\|_{H^2}, \|n_0\|).$$

Taking the inner product of (1.2) with $\Delta n$, we obtain

$$(n_t - 2|u|_x^2, \Delta n) = 0.$$

Then

$$\frac{\mathrm{d}}{\mathrm{d}t}\|\nabla n\|^2 \leqslant C\|\nabla(uu_x)\| \|\nabla n\| \leqslant C\|\nabla u\|_4 \|u_x\|_4 \|\nabla n\| + C\|u\|_\infty \|\nabla u_x\| \|\nabla n\|$$
$$\leqslant C\|\nabla u\|^{1/2}\|\nabla u\|^{1/2}\|\nabla u_x\|^{1/2}\|u_x\|^{1/2}\|\nabla n\| + C\|\nabla n\| \leqslant C + \|\nabla n\|^2.$$

By Gronwall inequality,

$$\|\nabla n\|^2 \leqslant C(\|u_0\|_{H^2}, \|n_0\|_{H^1}).$$

□

**Lemma 2.4** Suppose that $u_0 \in H^3$ and $n_0 \in H^2$, $u$ and $n$ solve the problem (1.1)~(1.3). Then

$$\sup_{0\leqslant t<T}(\|u\|_{H^3}+\|u_t\|_{H^1}+\|n\|_{H^2}+\|n_{tt}\|)\leqslant C(\|u_0\|_{H^3},\|n_0\|_{H^2}). \quad (2.20)$$

**Proof** By (1.1),

$$(\mathrm{i}u_{tt}+\mathrm{i}u_{yt}-\Delta u_t+n_t u+nu_t,\Delta u_t)=0. \quad (2.21)$$

Taking the imaginary part of (2.21), we have

$$\begin{aligned}\frac{\mathrm{d}}{\mathrm{d}t}\|\nabla u_t\|^2&=-2\mathrm{Im}(n_t u+nu_t,\Delta u_t)=2\mathrm{Im}(\nabla(2|u|_x^2u+nu_t),\nabla u_t)\\&=2\mathrm{Im}(\nabla(2|u|_x^2)u+2|u|_x^2\nabla u+\nabla nu_t,\nabla u_t)\\&\leqslant C|(\nabla uu_xu,\nabla u_t)|+C|(u^2\nabla u_x,\nabla u_t)|+C|(uu_x\nabla u,\nabla u_t)|+C\|\nabla n\|_4\|u_t\|_4\|\nabla u_t\|\\&\leqslant C(\|u\|_\infty\|\nabla u\|_4^2+\|u\|_\infty^2\|\nabla u_x\|+\|\nabla n\|^{1/2}\|\Delta n\|^{1/2}\|u_t\|^{1/2}\|\nabla u_t\|^{1/2})\|\nabla u_t\|\\&\leqslant C\|\nabla u_t\|+C\|\Delta n\|^{1/2}\|\nabla u_t\|^{3/2}\leqslant C+C(\|\nabla u_t\|^2+\|\Delta n\|^2). \quad (2.22)\end{aligned}$$

By (1.2),

$$(\Delta n_t-2\Delta(|u|_x^2),\Delta n)=0. \quad (2.23)$$

Then by (2.25) and (1.1), and Brezis-Gallouet inequality,

$$\begin{aligned}\frac{\mathrm{d}}{\mathrm{d}t}\|\Delta n\|^2&=4(\Delta(|u|_x^2),\Delta n)\\&=4(\Delta u\bar{u}_x+u\Delta\bar{u}_x+2\nabla u\cdot\nabla\bar{u}_x+\Delta\bar{u}u_x+\bar{u}\Delta u_x+2\nabla\bar{u}\cdot\nabla u_x,\Delta n)\\&\leqslant C\|\Delta u\|_4\|\nabla u\|_4\|\Delta n\|+C\|u\|_\infty\|\Delta u_x\|\|\Delta n\|\\&\leqslant C(\|u_t\|_4+\|u_y\|_4+\|u\|_\infty\|n\|_4)\|\Delta n\|+C(\|u_{tx}\|+\|u_{xy}\|+\|n_xu+nu_x\|)\|\Delta n\|\\&\leqslant C(\|u_t\|^{1/2}\|\nabla u_t\|^{1/2}+C+\|\nabla u_t\|+\|u\|_\infty\|n_x\|+\|n\|_\infty\|u_x\|)\|\Delta n\|\\&\leqslant C\|\nabla u_t\|^2+C\|\Delta n\|^2+C(1+\sqrt{\ln(1+\|\Delta n\|)})\|\Delta n\|+C\\&\leqslant C\|\nabla u_t\|^2+C\|\Delta n\|^2+C(1+\|\Delta n\|)^{1/2}\|\Delta n\|+C\\&\leqslant C(\|\nabla u_t\|^2+\|\Delta n\|^2)+C. \quad (2.24)\end{aligned}$$

By (2.22) and (2.24),

$$\frac{\mathrm{d}}{\mathrm{d}t}(\|\nabla u_t\|^2+\|\Delta n\|^2)\leqslant C(\|\nabla u_t\|^2+\|\Delta n\|^2)+C.$$

Using Gronwall inequality, we have

$$\|\nabla u_t\|^2+\|\Delta n\|^2\leqslant C(\|\nabla u_t(x,y,0)\|,\|\Delta n(x,y,0)\|)=C(\|u_0\|_{H^3},\|n_0\|_{H^2}).$$

By (1.1),

$$\|\nabla\Delta u\|\leqslant C\|\nabla u_t\|+C\|\nabla u_y\|+C\|\nabla(nu)\|\leqslant C(\|u_0\|_{H^3},\|n_0\|_{H^2}).$$

So

$$\|u\|_{H^3}\leqslant C(\|u_0\|_{H^3},\|n_0\|_{H^2}).$$

By Gagliardo-Nirenberg inequality,

$$\|\nabla u\|_\infty\leqslant C\|\nabla u\|^{1/2}\|\nabla\Delta u\|^{1/2}\leqslant C(\|u_0\|_{H^3},\|n_0\|_{H^2}).$$

Therefore, by (1.2), we get

$$\|\Delta n_t\| = 2\|\Delta(|u|_x^2)\| \leqslant C\|u_x\|_\infty\|\Delta u\| + C\|u\|_\infty\|\Delta u_x\| + C\|\nabla u\|_\infty\|\nabla u_x\|$$
$$\leqslant C(\|u_0\|_{H^3}, \|n_0\|_{H^2})$$

and

$$\|n_{tt}\| = 2\|u_{xt}\bar{u} + u_x\bar{u}_t + \bar{u}_x u_t + u\bar{u}_{xt}\|$$
$$\leqslant C\|u\|_\infty\|u_{xt}\| + C\|u_x\|_4\|u_t\|_4 \leqslant C(\|u_0\|_{H^3}, \|n_0\|_{H^2}).$$ □

**Lemma 2.5** Suppose that $u_0 \in H^4$ and $n_0 \in H^3$, $u$ and $n$ solve the problem (1.1)~(1.3). Then

$$\sup_{0\leqslant t<T}(\|u\|_{H^4} + \|u_{tt}\| + \|n\|_{H^3} + \|n_{tt}\|_{H^1}) \leqslant C(\|u_0\|_{H^4}, \|n_0\|_{H^3}). \tag{2.25}$$

**Proof** By (1.1),

$$(\mathrm{i}u_{ttt} + \mathrm{i}u_{ytt} - \Delta u_{tt} + n_{tt}u + u_{tt}n + 2n_t u_t, u_{tt}) = 0. \tag{2.26}$$

Taking the imaginary part of (2.26), by (2.20), we have

$$\frac{\mathrm{d}}{\mathrm{d}t}\|u_{tt}\|^2 = -2\mathrm{Im}(n_{tt}u + 2n_t u_t, u_{tt}) \leqslant C\|u\|_\infty\|n_{tt}\|\|u_{tt}\| + C\|n_t\|_4\|u_t\|_4\|u_{tt}\| \leqslant C + \|u_{tt}\|^2.$$

By Gronwall inequality,

$$\|u_{tt}\|^2 \leqslant C(\|u_{tt}(x,y,0)\|^2, \|n_0\|_{H^2}) = C(\|u_0\|_{H^4}, \|n_0\|_{H^2}).$$

By (1.1) and the above inequality,

$$\|u_t\|_{H^2} \leqslant C\|\Delta u_t\| \leqslant C\|u_{tt}\| + C\|u_{ty}\| + C\|n_t u + n u_t\| \leqslant C(\|u_0\|_{H^4}, \|n_0\|_{H^2})$$

and

$$\|u\|_{H^4} \leqslant C\|\Delta u\|_{H^2} \leqslant C\|u_t\|_{H^2} + C\|u_y\|_{H^2} + C\|u\|_\infty\|n\|_{H^2} \leqslant C(\|u_0\|_{H^4}, \|n_0\|_{H^2}).$$

By (1.2),

$$\|\nabla n_{tt}\| = 2\|\nabla(|u|_{xt}^2)\|$$
$$\leqslant C\|u_x\|_\infty\|\nabla u_t\| + C\|u_t\|_\infty\|\nabla u_x\| + C\|\nabla u\|_\infty\|u_{xt}\| + C\|u\|_\infty\|\nabla u_{xt}\|$$
$$\leqslant C(\|u_0\|_{H^4}, \|n_0\|_{H^2}).$$

From (1.2), we notice that

$$(\Delta n_t - 2\Delta(|u|_x^2), \Delta^2 n) = 0,$$

then

$$\frac{\mathrm{d}}{\mathrm{d}t}\|\nabla\Delta n\|^2 = C(|(\nabla\Delta u u_x, \nabla\Delta n)| + |(\Delta u\,\nabla u_x, \nabla\Delta n)| + |(\nabla u\Delta u_x, \nabla\Delta n)| +$$
$$|(u\,\nabla\Delta u_x, \nabla\Delta n)| + |(\nabla(\nabla u\cdot\nabla u_x), \nabla\Delta n)|)$$
$$\leqslant C(\|\nabla u\|_\infty\|u\|_{H^3} + \|\Delta u\|_4^2 + \|u\|_\infty\|\nabla\Delta u_x\|)\|\nabla\Delta n\|$$
$$\leqslant C\|\nabla\Delta n\| + C\|\nabla(\mathrm{i}u_{tx} + \mathrm{i}u_{xy} + n_x u + n u_x)\|\|\nabla\Delta n\|$$
$$\leqslant C\|\nabla\Delta n\| + C(\|u_t\|_{H^2} + \|u\|_{H^3} + \|u\|_\infty\|\Delta n\| + \|n\|_\infty\|\Delta u\| + \|\nabla u\|_4\|\nabla n\|_4)\|\nabla\Delta n\|$$
$$\leqslant C\|\nabla\Delta n\| \leqslant C + \|\nabla\Delta n\|^2.$$

Therefore, by Gronwall inequality,

$$\|\nabla\Delta n\| \leqslant C(\|u_0\|_{H^4}, \|n_0\|_{H^3}).$$ □

**Lemma 2.6** Suppose that $u_0 \in H^5$ and $n_0 \in H^4$, $u$ and $n$ solve the problem (1.1)~(1.3). Then

$$\sup_{0\leqslant t<T}(\|u\|_{H^5}+\|u_{tt}\|_{H^1}+\|n\|_{H^4}+\|n_{ttt}\|)\leqslant C(\|u_0\|_{H^5},\|n_0\|_{H^4}). \tag{2.27}$$

**Proof** By (1.1),

$$(\mathrm{i}u_{ttt}+\mathrm{i}u_{ytt}-\Delta u_{tt}+n_{tt}u+u_{tt}n+2n_tu_t,\Delta u_{tt})=0. \tag{2.28}$$

Taking the imaginary part of (2.28), we have

$$\begin{aligned}\frac{\mathrm{d}}{\mathrm{d}t}\|\nabla u_{tt}\|^2&=-2\mathrm{Im}(n_{tt}u+nu_{tt}+2n_tu_t,\Delta u_{tt})\\&\leqslant C(\|u\|_\infty\|\nabla n_{tt}\|+\|\nabla u\|_\infty\|n_{tt}\|+\\&\quad\|\nabla n\|_\infty\|u_{tt}\|+\|u_t\|_\infty\|\nabla n_t\|+\|n_t\|_\infty\|\nabla u_t\|)\|\nabla u_{tt}\|\\&\leqslant C+\|\nabla u_{tt}\|^2.\end{aligned}$$

Then by Gronwall inequality,

$$\|\nabla u_{tt}\|^2\leqslant C(\|\nabla u_{tt}(x,y,0)\|^2,\|n_0\|_{H^3})=C(\|u_0\|_{H^5},\|n_0\|_{H^3}).$$

From (1.1), we derive

$$\begin{aligned}\|u\|_{H^5}&\leqslant C\|\Delta u\|_{H^3}\leqslant C\|u_t\|_{H^3}+C\|u_y\|_{H^3}+C\|u\|_\infty\|n\|_{H^3}\\&\leqslant C\|\Delta u_t\|_{H^1}+C(\|u_0\|_{H^4},\|n_0\|_{H^2})\\&\leqslant C\|u_{tt}\|_{H^1}+C\|u_{ty}\|_{H^1}+C\|n_tu+nu_t\|_{H^1}+C\\&\leqslant C(\|u_0\|_{H^5},\|n_0\|_{H^3}).\end{aligned}$$

By (1.2),

$$\begin{aligned}\|n_{ttt}\|&=2\|\,|u|^2_{xtt}\|\\&\leqslant C(\|u_{tt}u_x\|+\|uu_{xtt}\|+\|u_tu_{xt}\|)\\&\leqslant C(\|u_x\|_\infty\|u_{tt}\|+\|u\|_\infty\|u_{xtt}\|+\|u_t\|_\infty\|u_{xt}\|)\\&\leqslant C(\|u_0\|_{H^5},\|n_0\|_{H^3}).\end{aligned}$$

From (1.2), we get

$$(\Delta^2n_t-2\Delta^2(|u|^2_x),\Delta^2n)=0.$$

Then

$$\frac{\mathrm{d}}{\mathrm{d}t}\|\Delta^2n\|^2\leqslant C\|\Delta^2uu_x+u\Delta^2u_x+\nabla\Delta u\,\nabla u_x+\nabla u\,\nabla\Delta u_x+\Delta u\Delta u_x\|\,\|\Delta^2n\|$$

$$\leqslant C(\|u_x\|_\infty\|\Delta^2u\|+\|u\|_\infty\|\Delta^2u_x\|+\|\nabla u_x\|_\infty\|\nabla\Delta u\|+\|\nabla u\|_\infty\|\nabla\Delta u_x\|+\|\Delta u\|_\infty\|\Delta u_x\|)\|\Delta^2n\|$$

$$\leqslant C\|u\|_{H^5}\|\Delta^2n\|\leqslant C(\|u_0\|_{H^5},\|n_0\|_{H^3})+\|\Delta^2n\|^2.$$

Then by Gronwall inequality,

$$\|\Delta^2n\|\leqslant C(\|u_0\|_{H^5},\|n_0\|_{H^4}).$$ □

From Lemma 2.5 and Lemma 2.6, we obtain the following lemma by induction.

**Lemma 2.7** Suppose that $u_0\in H^{2k}\,(k\geqslant2)$ and $n_0\in H^{2k-1}$, $u$ and $n$ solve the problem (1.1)~(1.3). Then

$$\sup_{0\leqslant t<T}(\|u\|_{H^{2k}}+\|D_t^{k-1}u\|_{H^1}+\|D_t^k u\|+\|n\|_{H^{2k-1}}+\|D_t^{k-1}n\|_{H^1}+\|D_t^k n\|)$$
$$\leqslant C(\|u_0\|_{H^{2k}},\|n_0\|_{H^{2k-1}}). \tag{2.29}$$

where $D_t^k u$ is the $k$th order derivative of $u$ with respect of $t$.

Suppose also that $u_0\in H^{2k+1}(k\geqslant 2)$ and $n_0\in H^{2k}$, $u$ and $n$ solve the problem (1.1)~(1.3). Then

$$\sup_{0\leqslant t<T}(\|u\|_{H^{2k+1}}+\|D_t^k u\|_{H^1}+\|n\|_{H^{2k}}+\|D_t^k n\|_{H^1}+\|D_t^{k+1}n\|)$$
$$\leqslant C(\|u_0\|_{H^{2k+1}},\|n_0\|_{H^{2k}}). \tag{2.30}$$

**Proof** The proof of (2.30) is similar to the proof of (2.29), so we only prove (2.29) in the following.

By Lemma 2.5, (2.29) is true for $k=2$. Now we suppose that (2.29) is also true as $k-1(k\geqslant 3)$. By (1.2),

$$\begin{aligned}\|\nabla D_t^{k-1}n\| &= 2\|\nabla D_t^{k-2}|u|_x^2\|\\ &\leqslant C\|u_x\|_\infty\|\nabla D_t^{k-2}u\|+C\|u\|_\infty\|\nabla D_t^{k-2}u_x\|+C\left\|\nabla\sum_{l=1}^{k-3}C_{k-2}^l D_t^l u D_t^{k-2-l}\bar{u}_x\right\|\\ &\leqslant C+\|D_t^{k-2}\Delta u\|+C\leqslant C+C\|D_t^{k-2}u_t\|+C\|D_t^{k-2}u_y\|+C\|D_t^{k-2}(nu)\|\leqslant C.\end{aligned}$$

Here, and throughout the proof of this lemma, the positive constant $C$ depends on $\|u_0\|_{H^{2k}}$ and $\|n_0\|_{H^{2k-1}}$.

By (1.1),

$$(\mathrm{i}\,\nabla D_t^{k-1}u_t+\mathrm{i}\,\nabla D_t^{k-1}u_y-\Delta\nabla D_t^{k-1}u+\nabla D_t^{k-1}(nu),\nabla D_t^{k-1}u)=0. \tag{2.31}$$

Taking the imaginary part of (2.31), we have

$$\begin{aligned}\frac{\mathrm{d}}{\mathrm{d}t}\|\nabla D_t^{k-1}u\|^2 &= -2\mathrm{Im}(\nabla D_t^{k-1}(nu),\nabla D_t^{k-1}u)\\ &= -2\mathrm{Im}\left(u\,\nabla D_t^{k-1}n+\nabla\sum_{l=1}^{k-2}C_{k-1}^l D_t^l u D_t^{k-1-l}n,\nabla D_t^{k-1}u\right)\\ &\leqslant C\|u\|_\infty\|\nabla D_t^{k-1}n\|\,\|\nabla D_t^{k-1}u\|+C\|\nabla D_t^{k-1}u\|\leqslant C+\|\nabla D_t^{k-1}u\|^2.\end{aligned}$$

Then by Gronwall inequality,

$$\|\nabla D_t^{k-1}u\|^2\leqslant C.$$

From (1.2), we derive

$$\begin{aligned}\|D_t^k n\| &= 2\|D_t^{k-1}(|u|_x^2)\|\\ &\leqslant C\|u_x\|_\infty\|D_t^{k-1}u\|+C\|u\|_\infty\|\nabla D_t^{k-1}u\|+C\left\|\sum_{l=1}^{k-2}C_{k-1}^l D_t^l u D_t^{k-1-l}\bar{u}_x\right\|\leqslant C.\end{aligned}$$

By (1.1),

$$(\mathrm{i}D_t^k u_t+\mathrm{i}D_t^k u_y-\Delta D_t^k u+D_t^k(nu),D_t^k u)=0. \tag{2.32}$$

Taking the imaginary part of (2.32), we have

$$\frac{\mathrm{d}}{\mathrm{d}t}\|D_t^k u\|^2=-2\mathrm{Im}(D_t^k(nu),D_t^k u)=-2\mathrm{Im}\left(uD_t^k n+\sum_{l=1}^{k-1}C_k^l D_t^l n D_t^{k-l}u,D_t^k u\right)$$

$$\leqslant C\|u\|_\infty\|D_t^k u\|^2 + C\|D_t^k u\|\leqslant C + \|D_t^k u\|^2.$$

Then by Gronwall inequality,

$$\|D_t^k u\|^2\leqslant C.$$

$$\|u\|_{H^{2k}}\leqslant C\|\Delta u\|_{H^{2k-2}}\leqslant C\|u_t\|_{H^{2k-2}} + C\|u_y\|_{H^{2k-2}} + C\|nu\|_{H^{2k-2}}\leqslant C\|\Delta u_t\|_{H^{2k-4}}$$
$$\leqslant C\|u_{tt}\|_{H^{2k-4}} + C\|u_{ty}\|_{H^{2k-4}} + C\|n_t u + nu_t\|_{H^{2k-4}}\leqslant C.$$

By (1.2),

$$(\nabla^{2k-1}n_t - 2\nabla^{2k-1}(|u|_x^2), \nabla^{2k-1}u) = 0. \tag{2.33}$$

Then

$$\frac{d}{dt}\|\nabla^{2k-1}n\|^2\leqslant C\|\nabla^{2k-1}(|u|_x^2)\|\,\|\nabla^{2k-1}n\|$$
$$\leqslant C\Big(\|u\|_\infty\|\nabla^{2k}u\| + \|u_x\|_\infty\|\nabla^{2k-1}u\| +$$
$$\Big\|\sum_{l=1}^{2k-2}C_{2k-1}^l\nabla^l u\,\nabla^{2k-1-l}\bar{u}_x\Big\|\Big)\|\nabla^{2k-1}n\|$$
$$\leqslant C + \|\nabla^{2k-1}n\|^2.$$

Therefore, by Gronwall inequality,

$$\|\nabla^{2k-1}n\|^2\leqslant C.$$ □

## 3 The Existence of the Problem (1.1)~(1.3)

We use the following period initial value problem (1.1), (1.2), and (3.1) to approximate the initial value problem (1.1) ~ (1.3) by using the uniform estimates in Sec. Ⅱ and the method in Ref. [16].

$$u(x,y,0) = \tilde{u}_0(x,y),\quad n(x,y,0) = \tilde{n}_0(x,y),\quad (x,y)\in \mathbf{R}^2,$$
$$u(x+2D,y,t) = u(x,y+2D,t) = u(x,y,t),$$
$$n(x+2D,y,t) = n(x,y+2D,t) = n(x,y,t),\quad D>0,\quad (x,y)\in\mathbf{R}^2,\quad t\geqslant 0. \tag{3.1}$$

Let $\Omega = (-D,D)\times(-D,D)$. At first, we give the following definitions of solutions to the problem (1.1)~(3.1).

**Definition 3.1** Suppose that $\tilde{u}_0\in H^2_{\text{per}}(\Omega)$, $\tilde{n}_0\in L^2_{\text{per}}(\Omega)$. We say that $u, n$ is a strong solution to the problem (1.1)~(3.1) in $(0,T)$ if $u, n$ satisfies

$$\int_0^T\int_\Omega - iu(\bar{v}_{1t} + \bar{v}_{1y}) + \nabla u\,\nabla\bar{v}_1 + nu\bar{v}_1 = \int_\Omega \tilde{u}_0\bar{v}_1(0),$$
$$\int_0^T\int_\Omega - n\bar{v}_{2t} + 2|u|^2\bar{v}2x = \int_\Omega\tilde{n}_0\bar{v}_2(0),$$

where $v_1, v_2\in C^1(0,T;C^\infty_{\text{per}}(\Omega))$, $v_1(T) = v_2(T) = 0$; moreover

$$u\in L^\infty(0,T;H^2_{\text{per}}(\Omega)),\ u_t\in L^\infty(0,T;L^2_{\text{per}}(\Omega)),$$
$$n\in L^\infty(0,T;L^2_{\text{per}}(\Omega)),\ n_t\in L^\infty(0,T;H^1_{\text{per}}(\Omega)).$$

**Definition 3.2** Suppose that $\tilde{u}_0\in H^2_{\text{per}}(\Omega)$, $\tilde{n}_0\in L^2_{\text{per}}(\Omega)$. We say that $u, n$ is a local (global) strong solution to the problem (1.1)~(3.1) if there exists $t_0(\|\tilde{u}_0\|_{H^2}, \|\tilde{n}_0\|_{L^2})>0$ ($\forall T>0$) such that $u, n$ is a strong solution to (1.1)~(3.1) in $(0,t_0)$ ($(0,T)$).

**Proposition 3.3** If $\tilde{u}_0\in H^4_{\text{per}}(\Omega)$, $\tilde{n}_0\in H^3_{\text{per}}(\Omega)$, then for $\forall T>0$, there exists a unique global solution $u(x,y,t)$, $n(x,y,t)$ of the problem (1.1)~(3.1), such that

$$u(x,y,t)\in L^\infty(0,T;H^4_{\text{per}}(\Omega)),\quad n(x,y,t)\in L^\infty(0,T;H^3_{\text{per}}(\Omega)).$$

Moreover, if for $k\geqslant 5$, $\tilde{u}_0\in H^k_{\text{per}}(\Omega)$, $\tilde{n}_0\in H^{k-1}_{\text{per}}(\Omega)$, then for $\forall T>0$, there exists a unique global solution $u(x,y,t)$, $n(x,y,t)$ of the problem (1.1)~(1.2), and (3.1), such that

$$u(x,y,t)\in L^\infty(0,T;H^k_{\text{per}}(\Omega)),\quad n(x,y,t)\in L^\infty(0,T;H^{k-1}_{\text{per}}(\Omega)).$$

**Proof** We prove Proposition 3.3 by two steps.

**Step 1** Existence.

We construct the approximate solution of the periodic initial value problem (1.1)~(3.1) by using Galërkin's method. We apply the following approximate solution of the problem (1.1)~(3.1):

$$u^l(x,y,t)=\sum_{j=1}^{l}a_j^l(t)w_j(x,y),\quad n^l(x,y,t)=\sum_{j=1}^{l}b_j^l(t)w_j(x,y),$$

to approach the solution $u(x,y,t)$, $n(x,y,t)$ of the problem (1.1)~(3.1), where $\{w_j(x,y)\}$ is a basis, which is a system of the orthogonal eigen functions with period $2D$ satisfying $-\Delta w_j(x,y)=\lambda_j w_j$ $(j=1,2,\cdots)$. By the Galërkin method, the unknown functions $u^l(x,y,t)$, $n^l(x,y,t)$ satisfy

$$(\mathrm{i}u_t^l+\mathrm{i}u_y^l-\Delta u^l+n^l u^l, w_j)=0,$$
$$(n_t^l-2|u^l|_x^2, w_j)=0,\quad j=1,2,\cdots,l,$$
$$(u^l(x,y,0),w_j)=(\tilde{u}_0,w_j),\quad (n^l(x,y,0),w_j)=(\tilde{n}_0,w_j). \tag{3.2}$$

It is easy to see that the system (3.2) is an initial value problem of nonlinear ordinary system of the equations for the unknown coefficients $a_i^l(t)$, $b_j^l(t)$. By the theory of ordinary differential equations and the *a priori* estimates in Sec. Ⅱ, for any $l$, there exists $u^l, n^l$, such that

$$\sup_{0\leqslant t<\infty}(\|u^l\|_{H^{2k+1}}+\|D_t^k u^l\|_{H^1}+\|n^l\|_{H^{2k}}+\|D_t^k n^l\|_{H^1}+\|D_t^{k+1}n^l\|)$$
$$\leqslant C(\|\tilde{u}_0\|_{H^{2k+1}},\|\tilde{n}_0\|_{H^{2k}})\ (k=1,2),$$

where $C$ is independent of $l$.

There is a subsequence $\{u^{l_k}, n^{l_k}\}$ of $\{u^l, n^l\}_{l=1}^\infty$ and two functions $u, n$ such that

$$u_{tt}^{l_k}\to u_{tt}\quad *-\text{weakly in } L^\infty(0,T;L^2_{\text{per}}(\Omega)),$$
$$n_t^{l_k}\to n_t\quad *-\text{weakly in } L^\infty(0,T;H^1_{\text{per}}(\Omega)),$$
$$u^{l_k}\to u\quad *-\text{weakly in } L^\infty(0,T;H^4_{\text{per}}(\Omega)),$$
$$n^{l_k}\to n\quad *-\text{weakly in } L^\infty(0,T;H^3_{\text{per}}(\Omega)),$$

where $v^l\to v$ $*-$weakly in $L^\infty(0,T;Z)$ means that for any $f\in L^1(0,T;Z^*)$, $\int_0^T(v^l,f)\to\int_0^T(v,f)$, $Z^*$ is the dual space of $Z$. By Definition 3.1, $(u, n)$ is a solution to (1.1)~

(3.1), in $(0, T)$.

Due to the *a priori* estimates in Sec. Ⅱ and the continuous extension theorem, we obtain the global solution of the period initial value problem (1.1)~(3.1).

**Step 2** Uniqueness.

Suppose that there are two solutions, $u_1(x,y,t)$, $n_1(x,y,t)$ and $u_2(x,y,t)$, $n_2(x,y,t)$, to the problem (1.1)~(3.1). Let

$$u(x,y,t) = u_1(x,y,t) - u_2(x,y,t), \quad n(x,y,t) = n_1(x,y,t) - n_2(x,y,t).$$

Thus from (1.1) and (1.2),

$$\mathrm{i}(u_t + u_y) - \Delta u + n_1 u_1 - n_2 u_2 = 0, \tag{3.3}$$

$$n_t - 2(|u_1|^2 - |u_2|^2)_x = 0, \tag{3.4}$$

$$u(x,y,0) = 0, \quad n(x,y,0) = 0. \tag{3.5}$$

Note that $n_1 u_1 - n_2 u_2 = n u_1 + n_2 u$, $|u_1|_x^2 - |u_2|_x^2 = u_x \bar{u}_1 + u_{2x}\bar{u} + u\bar{u}_{1x} + u_2 \bar{u}_x$. Taking the inner product of (3.3) with $u$, it follows that

$$(\mathrm{i}u_t + \mathrm{i}u_y - \Delta u + n u_1 + n_2 u, u) = 0. \tag{3.6}$$

Taking the imaginary part of (3.6), we have

$$\frac{\mathrm{d}}{\mathrm{d}t}\|u\|^2 \leqslant C\|u_1\|_\infty \|n\| \|u\| \leqslant C(\|n\|^2 + \|u\|^2). \tag{3.7}$$

Taking the inner product of (3.4) with $n$,

$$(n_t - 2(u_x \bar{u}_1 + u_{2x}\bar{u} + u\bar{u}_{1x} + u_2\bar{u}_x), n) = 0. \tag{3.8}$$

Then by (3.8),

$$\begin{aligned}\frac{\mathrm{d}}{\mathrm{d}t}\|n\|^2 &\leqslant C(\|u_1\|_\infty \|u_x\| + \|u_{2x}\|_\infty \|u\| + \|u_{1x}\|_\infty \|u\| + \|u_2\|_\infty \|u_x\|)\|n\| \\ &\leqslant C(\|u_x\|^2 + \|n\|^2 + \|u\|^2).\end{aligned} \tag{3.9}$$

Taking the inner product of (3.3) with $u_t$, we get

$$(\mathrm{i}u_t + \mathrm{i}u_y - \Delta u + n u_1 + n_2 u, u_t) = 0. \tag{3.10}$$

Taking the real part of (3.10),

$$\begin{aligned}\frac{\mathrm{d}}{\mathrm{d}t}\|\nabla u\|^2 &\leqslant C(\|u_y\| + \|u_1\|_\infty \|n\| + \|n_2\|_\infty \|u\|)\|u_t\| \\ &\leqslant C(\|u_y\|^2 + \|u_t\|^2 + \|n\|^2 + \|u\|^2).\end{aligned} \tag{3.11}$$

By (3.3),

$$(\mathrm{i}u_{tt} + \mathrm{i}u_{ty} - \Delta u_t + n_t u_1 + n u_{1t} + n_{2t} u + n_2 u_t, u_t) = 0. \tag{3.12}$$

Taking the imaginary part of (3.12), we have

$$\begin{aligned}\frac{\mathrm{d}}{\mathrm{d}t}\|u_t\|^2 &\leqslant C(\|u_1\|_\infty \|n_t\| + \|u_{1t}\|_\infty \|n\| + \|n_{2t}\|_\infty \|u\|)\|u_t\| \\ &\leqslant C(\|u_x\bar{u}_1 + u_{2x}\bar{u} + u\bar{u}_{1x} + u_2\bar{u}_x\| + \|n\| + \|u\|)\|u_t\| \\ &\leqslant C(\|u_1\|_\infty \|u_x\| + \|u_{2x}\|_\infty \|u\| + \|u_{1x}\|_\infty \|u\| + \|u_2\|_\infty \|u_x\| + \|n\| + \|u\|)\|u_t\| \\ &\leqslant C(\|u_x\| + \|n\| + \|u\|)\|u_t\| \leqslant C(\|u_x\|^2 + \|n\|^2 + \|u\|^2 + \|u_t\|^2).\end{aligned} \tag{3.13}$$

Therefore, by (3.7), (3.9), (3.11), and (3.13), we conclude that

$$\frac{\mathrm{d}}{\mathrm{d}t}(\|u\|^2+\|n\|^2+\|\nabla u\|^2+\|u_t\|^2)\leqslant C(\|u\|^2+\|n\|^2+\|u_x\|^2+\|u_y\|^2+\|u_t\|^2).$$

From Gronwall inequality and zero initial conditions (3.5),

$$u(x,y,t)\equiv n(x,y,t)\equiv 0.\qquad\square$$

**Definition 3.4** Suppose that $u_0\in H^2(\mathbf{R}^2)$, $n_0\in L^2(\mathbf{R}^2)$. We say that $u, n$ is a strong solution to the problem (1.1)~(3.1) in $(0,T)$ if $u, n$ satisfies

$$\int_0^T\int_{\mathbf{R}^2}-\mathrm{i}u(\bar{v}_{1t}+\bar{v}_{1y})+\nabla u\,\nabla\bar{v}_1+nu\bar{v}_1=\int_{\mathbf{R}^2}u_0\bar{v}_1(0),$$

$$\int_0^T\int_{\mathbf{R}^2}-n\bar{v}_{2t}+2|u|^2\bar{v}_{2x}=\int_{\mathbf{R}^2}n_0\bar{v}_2(0),$$

where $v_1, v_2\in C^1(0,T;C_0^\infty(\mathbf{R}^2))$, $v_1(T)=v_2(T)=0$; moreover

$$u\in L^\infty(0,T;H^2(\mathbf{R}^2)),\quad u_t\in L^\infty(0,T;L^2(\mathbf{R}^2)),$$

$$n\in L^\infty(0,T;L^2(\mathbf{R}^2)),\quad n_t\in L^\infty(0,T;H^1(\mathbf{R}^2)).$$

Suppose that $u_0\in H^2(\mathbf{R}^2)$, $n_0\in L^2(\mathbf{R}^2)$. We say that $u, n$ is a local (global) strong solution to the problem (1.1)~(3.1) if there exists $t_0(\|u_0\|_{H^2},\|n_0\|_{L^2})>0$ ($\forall T>0$) such that $u, n$ is a strong solution to (1.1)~(3.1) in $(0,t_0)$ ($(0,T)$).

In fact, the uniform estimates in Sec. Ⅱ are independent of the period $D$ and only depend on the norm of initial data. Therefore, by using the *a priori* estimates of the solution to the problem (1.1)~(3.1) for $D$, as in Ref. [16], we derive the global smooth solution of the problem (1.1)~(3.1) as $D\to\infty$. Thus we have the following theorem.

**Theorem 3.5** If $u_0\in H^k(\mathbf{R}^2)$, $n_0\in H^{k-1}(\mathbf{R}^2)$ $(k\geqslant 4)$, then for $\forall T>0$, there exists a global solution $u(x,y,t)$, $n(x,y,t)$ for the initial value problem (1.1)~(3.1) such that

$$u(x,y,t)\in L^\infty(0,T;H^k(\mathbf{R}^2)),\quad n(x,y,t)\in L^\infty(0,T;H^{k-1}(\mathbf{R}^2)).$$

**Proof** Let $D_m=m$, $\Omega_m=(-m,m)\times(-m,m)$, $u_0^m(x,y)$, $n_0^m(x,y)$ be two period functions with period $2m$, $u_0^m(x,y)=u_0(x,y)$, $n_0^m(x,y)=n_0(x,y)$ in $\left(-m+\frac{1}{2},m-\frac{1}{2}\right)\times\left(-m+\frac{1}{2},m-\frac{1}{2}\right)$; moreover $\|u_0^m(x,y)-u_0(x,y)\|_{H^{2k}_{\mathrm{per}}}\to 0$, $\|n_0^m(x,y)-n_0(x,y)\|_{H^{2k-1}_{\mathrm{per}}}\to 0\ (k\geqslant 2)$. $u_m(x,y,t)$, $n_m(x,y,t)$ is the global solution to the problem (1.1)~(3.1) when $\Omega=\Omega_m$; moreover $u_m(x,y,0)=u_0^m(x,y)$, $n_m(x,y,0)=n_0^m(x,y)$. Similarly to *a priori* estimates in Sec. Ⅱ, we can obtain

$$\sup_{0\leqslant t<T}(\|u_m\|_{H^{2j}}+\|D_t^{j-1}u_m\|_{H^1}+\|D_t^ju_m\|+\|n_m\|_{H^{2j-1}}+\|D_t^{j-1}n_m\|_{H^1}+\|D_t^jn_m\|)\leqslant C,\quad \text{if } k=2j,$$

$$\sup_{0\leqslant t<T}(\|u_m\|_{H^{2j+1}}+\|D_t^ju_m\|_{H^1}+\|n_m\|_{H^{2j}}+\|D_t^jn_m\|_{H^1}+\|D_t^{j+1}n_m\|)\leqslant C,\quad \text{if } k=2j+1,$$

where $C$ is independent of $m$. Going if necessary to a subsequence, there exist $u(x,y,t)\in L^\infty(0,T;H^k(\mathbf{R}^2))$, $n(x,y,t)\in L^\infty(0,T;H^{k-1}(\mathbf{R}^2))$ such that

$$u_m(x,y,t)\to u(x,y,t)\neq 0\quad *-\text{weakly in } L^\infty(0,T;H^k(\mathbf{R}^2)),$$

$$n_m(x,y,t)\to n(x,y,t)\neq 0\quad *-\text{weakly in } L^\infty(0,T;H^{k-1}(\mathbf{R}^2)),$$

and for any $L>0$,

$$u_m(x,y,t) \to u(x,y,t) \quad \text{in } H^{k-1}(\Omega_L) \text{ uniformly for } 0 \leqslant t < T,$$
$$n_m(x,y,t) \to n(x,y,t) \quad \text{in } H^{k-2}(\Omega_L) \text{ uniformly for } 0 \leqslant t < T.$$

Therefore, by Definition 3.4, $u$, $n$ is the global solution to the problem (1.1)~(3.1). □

### References


[1] Bekiranov D, Ogawa T, Ponce G. On the well-posedness of Benney's interaction equation for short and long waves. Adv Differ Equ, 1996, 1: 919.

[2] Bekiranov D, Ogawa T, Ponce G. Interaction equation for short and long dispersive waves. J Funct Anal, 1998, 158: 357.

[3] Brezis H, Gallouet T. Nonlinear Schrödinger evolution equations. Nonlinear Anal Theory, Methods Appl, 1980, 4: 677.

[4] Guo B. The global solution for one class of the system of LS nonlinear wave interaction. J Math Res Exposition, 1987, 1: 69.

[5] Guo B. The period initial value problems and initial value problems for one class of generalized LS type equations. J Eng Math, 1991, 8: 47.

[6] Guo B, Wang B. The global solution and its longtime behavior for a class of generalized LS type equations. Prog Nat Sci, 1996, 5: 533.

[7] Huo Z, Guo B. Local well-posedness of interaction equations for short and long dispersive waves. J Partial Differ Equ, 2004, 17: 137.

[8] Lai W C, Chow K W. Positon and dromion solutions of the (2 + 1) dimensional long wave-short wave resonance interaction equations. J Phys Soc Jpn, 1999, 68: 1847.

[9] Li Y. Long time behavior for the weakly damped driven long-wave-short-wave resonance equations. J Differ Equations, 2006, 223: 261.

[10] Oikawa M, Okamura M, Funakoshi M. Two-dimensional resonant interaction between long and short waves. J Phys Soc Jpn, 1989, 58: 4416.

[11] Tang X, Lou S. Periodic and localized solutions of the long wave-short wave resonance interaction equation. J Phys A, 2005, 38: 9649.

[12] Tsutsumi M, Hatano S. Well-posedness of the Cauchy problem for Benney's first equations of long wave short wave interactions. Finkcialaj Ekvacioj, 1994, 37: 289.

[13] Tsutsumi M, Hatano S. Well-posedness of the Cauchy problem for the long-wave-short-wave resonance equations. Nonlinear Anal Theory, Methods Appl, 1994, 22: 155.

[14] Yurova M. Application of dressing method for long wave-short wave resonance interaction equation. J Math Phys, 2007, 48: 053516.

[15] Zhang R, Guo B. Global solution and its long time behavior for the generalized long-short wave equations. J Partial Differ Equ, 2005, 18: 206.

[16] Zhou Y, Guo B. Period boundary value problem and initial value problem for the generalized Korteweg-de Vries equations of high order. Acta Math Sinica, 1984, 27: 154.

# The Decay Rates of Solutions of Generalized Benjamin-Bona-Mahony Equations in Multi-dimensions *

Fang Shaomei(房少梅) Guo Boling(郭柏灵)

**Abstract**

In this paper, we study the decay rates of the generalized Benjamin-Bona-Mahony equations in multi-dimensional space. By using Fourier analysis for low frequencies, and by applying the energy method for high frequencies, we obtain the $L_2$ convergence rates of the solution when the initial data is in a bounded subset of the phase space $H^1(\mathbf{R}^n)$ $(n \geqslant 3)$. The optimum decay rate is obtained in our results since it is the same as for the heat equation.

**Keywords** generalized Benjamin-Bona-Mahony; Fourier analysis; energy method; optimum convergence rates

## 1 Introduction

In this paper, we consider the system defined on $\mathbf{R}^n (n \geqslant 3)$

$$u_t - \Delta u_t - \eta \Delta u + (\beta \cdot \nabla) u + \operatorname{div}(f(u)) = 0, \tag{1.1}$$

where $\eta$ is a positive constant, $\beta$ is a real constant vector, $f(u) = (f_1(u), \cdots, f_n(u))^\tau$. The initial data is given by

$$u|_{t=0} = u_0(x). \tag{1.2}$$

The BBM equation was proposed and studied in [3] by Benjamin, Bona and Mahony for specific physical situations in the long wave limit for nonlinear dispersive media. In [7], Guo et al proposed and studied a generalized BBM equation in multi-dimensions (GBBM). The existence and uniqueness of solutions for GBBM equations has been proved by many authors (see [2,3,6,7]). The decays of solutions were also studied in [1,4,5,8]. But these studies do not give a clear picture of the decay rate for multi-dimensions. In this paper, the goal is to give the optimum decay rates of solutions to Cauchy problems of the generalized Benjamin-Bona-Mahony equations in multi-dimensional space.

As we know, it is difficult to get the optimum decay rates of solutions just from energy estimates. Some new techniques are needed. In this paper, we give the frequency of decomposition for solutions. Then by using Fourier analysis for low frequencies and by applying the energy method for high frequencies to solutions, the $L_2$ convergence rates of the solution

* Nonlinear Analysis, 2008, 69: 2230-2235.

were obtained for when the initial data is in a bounded subset of the phase space $H^1(\mathbf{R}^n)$ ($n \geq 3$). This is the main difference between the study in our paper and related results in previous works.

Throughout this paper we denote the generic constants by $C$. $W^{s,p}(\mathbf{R}^n)$, $s \in \mathbf{Z}_+$, $p \in [1,\infty]$, is the usual Sobolev space with the norm

$$\|f\|_{W^{s,p}} := \sum_{|\alpha|=0}^{s} \|\partial^\alpha f\|_{L_p}.$$

In particular, $W^{s,2} = H^s$.

By using the above notation, we can now state the main result in this paper as follows.

**Theorem 1.1** Given a ball $\mathscr{B} = \{v \in H^1(\mathbf{R}^n) \cap L^1(\mathbf{R}^n) : \max(\|v\|_{L^1}, \|v\|_{H^1}) \leq R\}$, if the initial data $u_0(x) \in \mathscr{B}$, and $f_j(u) = u^2 (j = 1, \cdots, n)$, then there exists a solution $u(x, t) \in C([0, \infty), H^1)$ to Eq. (1.1) defined globally in time. Moreover, for $|\alpha| \leq 1$,

$$\|\partial_x^\alpha u(\cdot, t)\|_{L^2} \leq C(1+t)^{-\frac{n}{4}-\frac{|\alpha|}{2}}. \tag{1.3}$$

**Remark** The decay rate of the heat equations is also $(1+t)^{-\frac{n}{4}-\frac{|\alpha|}{2}}$ in $L^2$-norm.

The rest of the paper is arranged as follows. In Section 2, we will study the estimates for low frequencies of solutions by using Fourier analysis. The estimates for high frequencies of solutions obtained by using the energy method are given in Section 3. Then the time-asymptotic behavior of the solutions to (1.1) and (1.2) follow from these estimates.

## 2 The Estimates for Low Frequencies

In this section, we will study the estimates for low frequencies of solutions, using Fourier analysis. As usual, the Fourier transformation to the variable $x \in \mathbf{R}^n$ is

$$\hat{f}(\xi, t) \equiv (\mathscr{F}f)(\xi, t) = \int_{\mathbf{R}^n} f(x, t) e^{-\sqrt{-1}x\xi} dx,$$

and the inverse Fourier transform to the variable $\xi$ is

$$f(x, t) \equiv (\mathscr{F}^{-1}\hat{f})(x, t) = (2\pi)^{-n} \int_{\mathbf{R}^n} \hat{f}(\xi, t) e^{\sqrt{-1}x\xi} d\xi.$$

Let

$$\chi(\xi) = \begin{cases} 1, & |\xi| < \varepsilon \\ 0, & |\xi| > 2\varepsilon \end{cases}, \tag{2.1}$$

be a smooth cut-off function, with $\varepsilon$ being sufficiently small. Set $\chi(D)$ as a pseudo-differential operator with symbol $\chi(\xi)$. Define

$$u_L(x, t) = \chi(D) u(x, t), \quad u_H(x, t) = (1 - \chi(D)) u(x, t).$$

We will first recall some basic results for Fourier analysis.

**Lemma 2.1** (Bernstein Inequality) For all $1 \leq p \leq q \leq \infty$, we have

$$\|\chi(D) D_x^\alpha f\|_{L^q(\mathbf{R}^n)} \leq C_{p,q} \varepsilon^{|\alpha| + n(1/p - 1/q)} \|f\|_{L^p(\mathbf{R}^n)},$$

where $C_{p,q}$ is a constant depending only on $p$ and $q$.

Now we come back to considering (1.1). Applying the operator $\chi(D)$ to (1.1), we have

$$(u_L)_t - \Delta(u_L)_t - \eta\Delta u_L + (\beta\cdot\nabla)u_L + \mathrm{div}(\chi(D)f(u)) = 0. \tag{2.2}$$

Defining $\Phi(\mu) = -\mathrm{div}\varphi(u)$ with $\varphi(u) = -\chi(D)f(u)$, we can write (2.2) as

$$(u_L)_t - \Delta(u_L)_t - \eta\Delta u_L + (\beta\cdot\nabla)u_L = \Phi(u).$$

For the low frequencies of the solutions, we first consider the Cauchy problem of the linear part of (2.2) as follows:

$$\begin{cases}(E_L)_t - \Delta(E_L)_t - \eta\Delta E_L + (\beta\cdot\nabla)E_L = 0\\ E|_{t=0} = \chi(D)\delta(x)\end{cases}, \tag{2.3}$$

where $\delta(x)$ is the Dirac function. After making the Fourier transformation to the variable $x\in\mathbf{R}^n$ in (2.3), we obtain the following ordinary differential equation for $\hat{E}_L$:

$$\begin{cases}(\hat{E}_L)_t + |\xi|^2(\hat{E}_L)_t + \eta|\xi|^2\hat{E}_L + \sqrt{-1}(\beta\cdot\xi)\hat{E}_L = 0\\ \hat{E}|_{t=0} = \chi(\xi)\end{cases}, \tag{2.4}$$

where $\xi$ corresponds to $D_x$. By direct calculation, we have

$$\hat{E}_L(\xi,t) = \chi(\xi)\mathrm{e}^{\lambda(\xi)t}, \tag{2.5}$$

with

$$\lambda(\xi) \equiv \frac{-\eta|\xi|^2 - \sqrt{-1}(\beta\cdot\xi)}{1+|\xi|^2}. \tag{2.6}$$

In the following, we are going to obtain some properties of the solutions $E_L(x,t)$.

**Lemma 2.2** For any fixed $\varepsilon$, there exists a positive constant $C$ such that

$$|\xi^\alpha\hat{E}_L(\xi,t)| \leqslant C(1+t)^{-\frac{|\alpha|}{2}},$$

$$\int_{\mathbf{R}^n}|\xi^\alpha(\hat{E}_L)(\xi,t)|^2\mathrm{d}\xi \leqslant C(1+t)^{-|\alpha|-\frac{n}{2}}. \tag{2.7}$$

**Proof** For $|\xi|\leqslant\varepsilon$, we have

$$\begin{aligned}|\xi^\alpha\hat{E}_L(\xi)| &\leqslant |\xi|^{|\alpha|}\exp\left(-\eta\frac{|\xi|^2}{1+\varepsilon^2}t\right)\\ &\leqslant C(1+t)^{-|\alpha|/2}\left|(|\xi|\sqrt{t})^{|\alpha|}\exp\left(-\eta\frac{(|\xi|\sqrt{t})^2}{2}\right)\right|\\ &\leqslant C(1+t)^{-|\alpha|/2}.\end{aligned}$$

Moreover

$$\begin{aligned}\|\xi^\alpha\hat{E}_L(\cdot,t)\|_{L^2}^2 &\leqslant C(1+t)^{-|\alpha|}\int_{|\xi|\leqslant\varepsilon}(|\xi|\sqrt{t})^{2|\alpha|}\exp(-\eta(|\xi|\sqrt{t})^2)\mathrm{d}\xi\\ &\leqslant C(1+t)^{-|\alpha|-\frac{n}{2}}.\end{aligned}$$

Thus the lemma is proved. □

The solution of (2.2) can be formulated by the Duhamel principle as

$$u_L(x,t) = (E_L*(\chi(D)u_0))(x,t) + \int_0^t E_L(\cdot,t-s)*\Phi(u(\cdot,s))\mathrm{d}s. \tag{2.8}$$

Since $\| D_x^\alpha(E_L * (\chi(D)u_0))\|_{L^2} \leqslant \| D_x^\alpha E_L\|_{L^2} \|(\chi(D)u_0)\|_{L^1}$, by using Lemma 2.2, one has

$$\| D_x^\alpha(E_L * (\chi(D)u_0))\|_{L^2} \leqslant C(1+t)^{-\frac{|\alpha|}{2}-\frac{n}{4}} \| u_0\|_{L^1}. \tag{2.9}$$

For the second term of (2.8), first

$$\begin{aligned}
&\int_0^t \| D_x^\alpha E_L(\cdot, t-s) * \Phi(u(\cdot,s))\|_{L^2} ds \\
&= \int_0^{t/2} \| D_x^\alpha E_L(\cdot, t-s) * \Phi(u(\cdot,s))\|_{L^2} ds + \\
&\quad \int_{t/2}^t \| E_L(\cdot, t-s) * D_x^\alpha \Phi(u(\cdot,s))\|_{L^2} ds \\
&\leqslant C\int_0^{t/2} \| D_x^\alpha E_L(\cdot, t-s)\|_{L^2} \| \Phi(u(\cdot,s))\|_{L^1} ds + \\
&\quad C\int_{t/2}^t \| E_L(\cdot, t-s)\|_{L^2} \| D_x^\alpha \Phi(u(\cdot,s))\|_{L^1} ds.
\end{aligned}$$

Assume that

$$I(t) = \sup_{0\leqslant s\leqslant t, |\alpha|\leqslant 1} \| D^\alpha u(\cdot,s)\|_{L^2} (1+s)^{\frac{|\alpha|}{2}+\frac{n}{4}}.$$

Then by using the Bernstein inequality, we have

$$\begin{aligned}
\| D_x^\alpha \Phi(u(\cdot,s))\|_{L^1} &\leqslant C\varepsilon \| D_x^\alpha \varphi(u(\cdot,s))\|_{L^1} \\
&\leqslant C\varepsilon \sum_{|\alpha_1|+|\alpha_2|=|\alpha|} \| D_x^{\alpha_1} u(\cdot,s)\|_{L^2} \| D_x^{\alpha_2} u(\cdot,s)\|_{L^2} \\
&\leqslant C\varepsilon (1+s)^{-\frac{|\alpha|}{2}-\frac{n}{2}} I^2(t).
\end{aligned} \tag{2.10}$$

Thus by using Lemma 2.1 and (2.10) we can get

$$\begin{aligned}
&\int_0^t \| D_x^\alpha E_L(\cdot, t-s) * \Phi(u(\cdot,s))\|_{L^2} ds \\
&\leqslant C\varepsilon \left( \int_0^{t/2} (1+t-s)^{-\frac{|\alpha|}{2}-\frac{n}{4}} (1+s)^{-\frac{n}{2}} ds + \right. \\
&\quad \left. \int_{t/2}^t (1+t-s)^{-\frac{n}{4}} (1+s)^{-\frac{|\alpha|}{2}-\frac{n}{2}} ds \right) I^2(t) \\
&\leqslant C\varepsilon (1+t)^{-\frac{|\alpha|}{2}-\frac{n}{4}} I^2(t).
\end{aligned} \tag{2.11}$$

Summing (2.8), (2.9) and (2.11), we have

$$\| D^\alpha u_L(\cdot,t)\|_{L^2} \leqslant C(1+t)^{-\frac{|\alpha|}{2}-\frac{n}{4}} (\| u_0\|_{L^1} + \varepsilon I^2(t)). \tag{2.12}$$

## 3 The Estimates for High Frequencies

In this section, we will establish some $L_2$ estimates for high frequencies by using the energy method. By applying the operator $1-\chi(D)$ to (1.1), we obtain

$$(u_H)_t - \Delta(u_H)_t - \eta\Delta u_H + (\beta\cdot\nabla)u_H + \mathrm{div}(((1-\chi(D))f(u)) = 0. \tag{3.1}$$

By integrating its product with $e^{\mu t}u_H$ over $\mathbf{R}^n\times[0,t]$, we find

$$\frac{1}{2}\frac{\mathrm{d}}{\mathrm{d}t}(e^{\mu t}(\|u_H(\cdot,t)\|_{L^2}^2+\|\nabla u_H(\cdot,t)\|_{L^2}^2))-\mu(e^{\mu t}(\|u_H(\cdot,t)\|_{L^2}^2+\|\nabla u_H(\cdot,t)\|_{L^2}^2))+$$

$$\eta e^{\mu t}\|\nabla u_H(\cdot,t)\|_{L^2}^2+\int_{\mathbf{R}^n}e^{\mu t}((\beta\cdot\nabla)u_H+\mathrm{div}(1-\chi(D))f(u)))u_H\mathrm{d}x=0. \tag{3.2}$$

Since

$$\int_{\mathbf{R}^n}((\beta\cdot\nabla)u_H)u_H\mathrm{d}x=0,$$

and

$$\int_{\mathbf{R}^n}(\mathrm{div}(f(u)))u\,\mathrm{d}x=0,$$

one has

$$\begin{aligned}&\int_{\mathbf{R}^n}((\beta\cdot\nabla)u_H+\mathrm{div}((1-\chi(D))f(u)))u_H\mathrm{d}x\\&=\int_{\mathbf{R}^n}(\mathrm{div}(\chi(D))f(u))u_L\mathrm{d}x-\int_{\mathbf{R}^n}(\mathrm{div}((1-\chi(D))f(u)))u_L\mathrm{d}x-\int_{\mathbf{R}^n}(\mathrm{div}(\chi(D))f(u))u_H\mathrm{d}x\\&=:R_1+R_2+R_3.\end{aligned}$$

By the Bernstein inequality,

$$\begin{aligned}|R_1|&\leqslant C\|\mathrm{div}(\chi(D))f(u)\|_{L^2}\|u_L\|_{L^2}\\&\leqslant C\varepsilon^{n/2}\|\mathrm{div}(\chi(D))f(u)\|_{L^1}\|u_L\|_{L^2}\\&\leqslant C\varepsilon^{n/2}\|\mathrm{div}u\|_{L^2}\|u\|_{L^2}^2\leqslant C\varepsilon^{n/2}(1+t)^{-\frac{1}{2}-\frac{3n}{4}}I^3(t).\end{aligned}$$

For $R_2$, set

$$\chi_1(\xi)=\begin{cases}1, & |\xi|<2\varepsilon\\ 0, & |\xi|>3\varepsilon\end{cases},$$

as a smooth cut-off function. It is easy to see that $(1-\chi(\xi))\chi(\xi)=\chi_1(\xi)\chi(\xi)$. Thus

$$\begin{aligned}|R_2|&\leqslant C\|\mathrm{div}(1-\chi(D))f(u)\|_{L^2}\|u_L\|_{L^2}\\&\leqslant C\varepsilon^{n/2}\|\mathrm{div}(\chi_1(D))f(u)\|_{L^1}\|u_L\|_{L^2}\\&\leqslant C\varepsilon^{n/2}\|\mathrm{div}u\|_{L^2}\|u\|_{L^2}^2\\&\leqslant C\varepsilon^{n/2}(1+t)^{-\frac{1}{2}-\frac{3n}{4}}I^3(t).\end{aligned}$$

For $R_3$, like for $R_1$, one has

$$|R_3|\leqslant C\varepsilon^{n/2}(1+t)^{-\frac{1}{2}-\frac{3n}{4}}I^3(t).$$

In summary, by above inequalities, we obtain

$$\begin{aligned}&\left|\int_{\mathbf{R}^n}((\beta\cdot\nabla)u_H+\mathrm{div}((1-\chi(D))f(u)))u_H\mathrm{d}x\right|\\&\leqslant C\varepsilon^{n/2}(1+t)^{-\frac{1}{2}-\frac{3n}{4}}I^3(t).\end{aligned} \tag{3.3}$$

Taking $\mu\leqslant\min\{1,\varepsilon^2\}\frac{\eta}{4}$, and by using Plancherel's theorem,

$$\|\nabla u_H(\cdot,t)\|_{L^2}^2 = \|\xi(1-\chi(\xi))\hat{u}(\cdot,t)\|_{L^2}^2 \geqslant \varepsilon^2\|u_H(\cdot,t)\|_{L^2}^2,$$

one has

$$-\mu(e^{\mu t}(\|u_H(\cdot,t)\|_{L^2}^2 + \|\nabla u_H(\cdot,t)\|_{L^2}^2)) + \eta e^{\mu t}\|\nabla u_H(\cdot,t)\|_{L^2}^2 \geqslant 0. \quad (3.4)$$

Combining (3.2) ~ (3.4) gives

$$\frac{1}{2}\frac{d}{dt}(e^{\mu t}(\|u_H(\cdot,t)\|_{L^2}^2 + \|\nabla u_H(\cdot,t)\|_{L^2}^2)) \leqslant C\varepsilon^{n/2}e^{\mu t}(1+t)^{-\frac{1}{2}-\frac{3n}{4}}I^3(t).$$

Thus

$$\begin{aligned}&\|u_H(\cdot,t)\|_{L^2}^2 + \|\nabla u_H(\cdot,t)\|_{L^2}^2\\ &\leqslant Ce^{-\mu t}(\|u_0(\cdot)\|_{L^2}^2 + \|\nabla u_0(\cdot)\|_{L^2}^2) +\\ &\quad C\varepsilon^{n/2}\left(\int_0^t e^{-\mu(t-s)}(1+s)^{-\frac{1}{2}-\frac{3n}{4}}ds\right)I^3(t).\end{aligned} \quad (3.5)$$

Setting $\theta(t) = e^{-\mu t}(1+t)^{1+\frac{n}{2}}$, we have

$$\begin{aligned}&(1+t)^{\frac{n}{2}}\|u_H(\cdot,t)\|_{L^2}^2 + (1+t)^{1+\frac{n}{2}}\|\nabla u_H(\cdot,t)\|_{L^2}^2\\ &\leqslant C\theta(t)((\|u_0(\cdot)\|_{L^2}^2 + \|\nabla u_0(\cdot)\|_{L^2}^2) + \varepsilon^{n/2}I^3(t)).\end{aligned} \quad (3.6)$$

Here $\theta(0) = 1$, and it is easy to see that there exists $t_1 > 0$ such that $\theta(t) \leqslant 1$ when $t > t_1$.

Now we come back to proving Theorem 1.1 by using (2.12) and (3.6). In fact

$$\begin{aligned}I^2(t) &\leqslant (1+t)^{-n/2}\|u\|_{L^2}^2 + (1+t)^{-1-n/2}\|\nabla u\|_{L^2}^2\\ &\leqslant 2((1+t)^{-n/2}\|u_L\|_{L^2}^2 + (1+t)^{-1-n/2}\|\nabla u_L\|_{L^2}^2 +\\ &\quad (1+t)^{-n/2}\|u_H\|_{L^2}^2 + (1+t)^{-1-n/2}\|\nabla u_H\|_{L^2}^2).\end{aligned} \quad (3.7)$$

By combining (3.7), (2.12) and (3.6), we have

$$I^2(t) \leqslant C(\theta(t)\|u_0\|_{H^1}^2 + \|u_0\|_{L^1}^2 + \varepsilon(I^4(t) + \theta(t)I^3(t))). \quad (3.8)$$

Let $M^2 = 4C(\|u_0\|_{H^1}^2 + \|u_0\|_{L^1}^2) > 0$; since $C$ is a constant independent of $\varepsilon$ and $t$, and the above inequality holds for any $\varepsilon > 0$, taking $\varepsilon = (2C(M^2+M))^{-1}$, we see that

$$I^2(0) \leqslant 2C(\theta(0)\|u_0(\cdot)\|_{H^1}^2 + \|u_0(\cdot)\|_{L^1}^2) = \frac{1}{2}M^2.$$

We will prove that $I^2(t) \leqslant M^2$ for all $t > 0$. For this, we define

$$T = \sup\{t \geqslant 0 : I^2(t) \leqslant M^2\}.$$

Without loss of generality, let $T > t_1 > 0$. If $T < \infty$, then by the definition we find $I(T) = M^2$; then

$$M^2 \leqslant \frac{1}{4}M^2 + \varepsilon(M^4 + M^3) \leqslant \frac{3}{4}M^2. \quad (3.9)$$

It then follows that $M^2 < 0$; the contradiction tells us that $T = \infty$ and therefore

$$I^2(t) \leqslant 4C(\|u_0(\cdot)\|_{H^1}^2 + \|u_0(\cdot)\|_{L^1}^2). \quad (3.10)$$

From (3.10) and the definition of $I(t)$, we know that Theorem 1.1 is valid.

### References

[1] Albert J. On the decay of solutions of the generalized Benjamin-Bona-Mahony equation. J Math Anal Appl, 1989, 141: 527-537.

[2] Avrin J, Goldstein J A. Global existence for the Benjamin-Bona-Mahony equations. Nonlinear Anal, 1985, 9: 861-865.

[3] Benjamin T B, Bona J L, Mahony J J. Model equations for long waves in nonlinear dispersive systems. Philos Trans R Soc Lond, 1972, 272: 47-78.

[4] Biler P. Long time behavior of solutions of the generalized Benjamin-Bona-Mahony equation in two-space dimensions. Differential Integral Equations, 1992, 5: 891-901.

[5] Fang S M, Guo B L. Long time behavior for solution of initial-boundary value problem for one class of system with multidimensional inhomogeneous GBBM equations. Appl Math Mech, 2005, 26(6): 665-675.

[6] Goldstein J A, Wichnoski B J. On the Benjamin-Bona-Mahony equation in higher dimensions. Nonlinear Anal, 1980, 4: 861-865.

[7] Guo B L. Initial boundary value problem for one class of system of multidimensional inhomogeneous GBBM equations. Chinese Ann Math Ser B, 1987, 8(2): 226-238.

[8] Zhang L. Decay of solutions of generalized Benjamin-Bona-Mahony equation in $n$-space dimensions. Nonlinear Anal, 1995, 25: 1343-1369.

# Blow up and Instability of Solitary Wave Solutions to a Generalized Kadomtsev-Petviashvili Equation and Two-dimensional Benjamin-Ono Equations *

Chen Jianqing(陈建清) Guo Boling(郭柏灵) Han Yongqian(韩永前)

**Abstract**

Let $p \geqslant 2$ with $p$ being the ratio of an even to an odd integer. For the generalized Kadomtsev-Petviashvili equation, coupled with Benjamin-Ono equations, in the form

$$(u_t + u_{xxx} + \beta\mathscr{H}u_{xx} + u^p u_x)_x = u_{yy}, \quad (x, y) \in \mathbf{R}^2, \quad t \geqslant 0,$$

it is proved that the solutions blow up in finite time even for those initial data with positive energy. As a by-product, it is proved that for all $c > \beta/2$, the solitary waves $\varphi(x - ct, y)$ are strongly unstable if $2 \leqslant p < 4$. This result, even in a special case $\beta = 0$, improves a previous work by Liu[1] where the instability of solitary waves was proved only in the case of $2 < p < 4$.

**Keywords** Kadomtsev-Petviashvili equation; Benjamin-Ono equations; blow-up solutions; strong instability of solitary waves

## 1 Introduction

This paper is concerned with the following generalized Kadomtsev-Petviashvili (KP) equation and two-dimensional Benjamin-Ono (BO) equations

$$\left.\begin{aligned} &u_t + u_{xxx} + \beta\mathscr{H}u_{xx} - v_y + u^p u_x = 0, \\ &u_y = v_x, \quad (x, y) \in \mathbf{R}^2, \quad t \in \mathbf{R}_+, \end{aligned}\right\} \tag{1.1}$$

$$u(x, y, t)\Big|_{t=0} = u_0, \tag{1.2}$$

where $\beta \geqslant 0$, $p > 0$ and $\mathscr{H}$ is the Hilbert operator defined by

$$\mathscr{H}u(x, y, t) = \text{p.v.}\frac{1}{\pi}\int_{\mathbf{R}} \frac{u(z, y, t)}{z - x}\mathrm{d}z.$$

Here, p.v. denotes the Cauchy principal value.

System (1.1) has been raised in many branches of physics. For example, when $\beta = 0$ and $p = 1$, (1.1) is the well known KP-I equation[2,3]. When $\beta = 0$ and $p > 0$, (1.1) is called the generalized KP (GKP) equation. Mathematical interests are concentrated on the existence of solitary waves (i.e. solutions of (1.1) with the form $u(x, y, t) = \varphi(x - ct, y)$), existence of solutions of Cauchy problem (1.1) and (1.2) as well as the possible blow-up solutions of (1.1) and (1.2). But one of the most important and challenging issues is to investigate the existence

* Proc R Soc A, 2008, 464: 49 - 64.

of solutions that blow up in a finite time.

Recently, Guo & Han[4] proved the existence of blow-up solutions of (1.1) and (1.2) if $p \geqslant 4$ and the initial energy

$$E_\beta(u_0) = \int\left(\frac{u_{0x}^2 + v_0^2 + \beta u_{0x}\mathscr{H}u_0}{2} - \frac{u_0^{p+2}}{(p+2)(p+1)}\right) < 0, \quad \text{where } u_{0y} = v_{0x}.$$

This is one motivation of the present paper. Another motivation is the recent results for (1.1) and (1.2) in the case of $\beta = 0$. When $\beta = 0$, Liu & Wang[5] and Liu[1] proved the existence of solitary waves of (1.1) for $0 < p < 4$ and the existence of blow-up solutions of (1.1) and (1.2) for $4/3 \leqslant p < 4$, as well as the stability of solitary waves of (1.1) for $0 < p < 4/3$ and the instability of solitary waves of (1.1) for $2 < p < 4$. The main ingredients used in Liu & Wang[5] and Liu[1] are based on solving several complicated minimization problems and constructing several invariant sets (e. g. [5], Theorem 1 and 3; [1], Lemma 3.1). In particular, they use essentially the property of invariance of the problem under scaling $c^{1/p}u(\sqrt{c}\,\cdot, c\,\cdot)$ and $u(\cdot/\mu)$. But the case of $\beta \neq 0$ is quite different from those of $\beta = 0$ since the scaling invariance does not hold any more due to the term $\int u_x Hu\,\mathrm{d}x\mathrm{d}y$. The method used in [1] seems not to be applied in the case of $\beta \neq 0$.

There are two main purposes of the present paper. One is to prove the existence of solitary waves $u(x, y, t) = \varphi(x - ct, y)$ of (1.1) in the case of $\beta \geqslant 0$ and $0 < p < 4$, which covers and generalizes the previous works in [5] and [6]. In particular, we do not require the wave speed $c$ satisfying $c > \beta/2$, which is usually assumed in finding solitary waves of the BO equations. The other is to prove that (1.1) and (1.2) has a solution that has positive initial energy and blows up in a finite time. As a by-product, it is proved that the solitary wave is strongly unstable for $2 \leqslant p < 4, \beta \geqslant 0$ and $c > \beta/2$. This result improves a previous work by Liu[1] where the strong instability of a solitary wave is proved only in the case of $2 < p < 4$ and $\beta = 0$.

This paper is organized as follows. In Section 2, we use the Nehari method to prove that for any $c > 0$, (1.1) has a solitary wave solution $\varphi(x - ct, y)$. In Section 3, we use several minimization problems with multiple constraints to construct sets $\mathscr{R}_\pm$ and $\mathscr{Q}_\pm$ that are invariant under the flow generated by the Cauchy problem (1.1) and (1.2). In Section 4, we prove that the solutions of (1.1) and (1.2) with initial data in $\mathscr{R}_+ \cap \mathscr{Q}_-$ must blow up in a finite time. In Section 5, we prove that for any $c > \beta/2$, the solitary wave $\varphi(x - ct, y)$ is strongly unstable in the sense of Definition 5.1.

We end this introduction by notations where $|\cdot|_q$ and $\|\cdot\|_s$ will denote the norm in $L^q(\mathbf{R}^2)$ and Sobolev space $H^S(\mathbf{R}^2)$. Throughout this paper, $\hat{f}$ denotes the Fourier transform of $f$, defined as $\hat{f}(\xi) = \int \exp(-\mathrm{i}(\xi_1 x + \xi_2 y)) f(x, y)\mathrm{d}x\mathrm{d}y$. All integrals are taken over $\mathbf{R}^2$ unless stated otherwise and $\mathrm{d}x\mathrm{d}y$ will be omitted if no confusion occurs. Let $Y$ be the closure of $\partial_x(C_0^\infty(\mathbf{R}^2))$ under the norm

$$\|u\|_Y = \|\partial_x\psi\|_Y = (\|\nabla\psi\|_{L^2}^2 + \|\partial_x^2\psi\|_{L^2}^2)^{1/2},$$

for $u \in Y$ and $u = \partial_x \psi$, where $\psi \in L^q_{loc}(\mathbf{R}^2)$, $\forall 2 \leqslant q < \infty$. $Y'$ is the dual space of $Y$. We also have $v = \partial_y \psi$ by a choice of $\psi \in L^q_{loc}$. Let

$$X_s = \{f \in H^s(\mathbf{R}^2); (\xi_1^{-1}\hat{f}(\xi))^\vee \in H^s(\mathbf{R}^2)\},$$

with the norm $\|u\|_{X_s} = \|u\|_s + \|(\xi_1^{-1}\hat{u})^\vee\|_s$, where "$\vee$" is the Fourier inverse transform. We define the operator $D_x^{-k}$ by

$$\widehat{D_x^{-k}f}(\xi) = (\mathrm{i}\xi_1)^{-k}\hat{f}(\xi), \quad k > 0.$$

$D^s = (-\partial_x^2)^{s/2}$ is the Riesz potential of order $s$ defined by $\widehat{D^s f}(\xi) = |\xi_1|^s \hat{f}(\xi)$. $S(\mathbf{R}^2)$ is the Schwartz class in $\mathbf{R}^2$.

Throughout this paper, we only consider the case of $\beta \geqslant 0$ and $p = n_1/n_2$, where $n_1$ is any even integer and $n_2$ any odd integer so that $\int u^{p+2} = \int |u|^{p+2}$. Different positive constants might be denoted by the same letter $C$ or $C_j$. If necessary, by $C(\cdot, \cdot)$ we denote the constant depending only on the quantities appearing in parentheses.

## 2 Solitary Wave Solutions

In this section, we will use the Nehari method to prove that (1.1) possesses a solitary wave solution. Before doing this, we recall the following well-known properties of the Hilbert operator $\mathcal{H}$[7]:

$$\widehat{\mathcal{H}f}(\xi) = \mathrm{i}\ \mathrm{sgn}\xi_1 \hat{f}(\xi); \quad (\mathrm{I})$$

$$|\mathcal{H}f|_2 = |f|_2; \quad (\mathrm{II})$$

$$\mathcal{H}f_x = \frac{\partial}{\partial x}(\mathcal{H}f); \quad (\mathrm{III})$$

$$\int f\mathcal{H}g = -\int g\mathcal{H}f; \quad (\mathrm{IV})$$

$$\int f_x \mathcal{H}f \mathrm{d}x \geqslant 0; \quad \text{and} \quad (\mathrm{V})$$

$$|D^{1/2}f|_2 \leqslant C|f|_2^{1/2}|f_x|_2^{1/2}, \quad \text{for any } f \in S(\mathbf{R}^2). \quad (\mathrm{VI})$$

For any $c > 0$, a solitary wave of (1.1) with wave speed $c$ is a solution of (1.1) of the form $u(x, y, t) = \varphi(x - ct, y)$ and decaying to zero at infinity. More precisely, $\varphi$ is a solution of

$$-c\varphi + \varphi_{xx} + \beta\mathcal{H}\varphi_x - D_x^{-2}\varphi_{yy} - \frac{u^{p+1}}{p+1} = 0, \quad \text{in } Y'. \quad (2.1)$$

We say that $\varphi \in Y$ is a solution of (2.1) if, and only if, $L_c'(\varphi) = 0$, where $L_c(u) = E_\beta(u) + cV(u)$ with $V(u) = 1/2\int u^2$ and

$$E_\beta(u) = \int\left(\frac{u_x^2 + v^2 + \beta u_x \mathcal{H}u}{2} - \frac{u^{p+2}}{(p+2)(p+1)}\right).$$

Let

$$N(u) = \int\left(u_x^2 + v^2 + \beta u_x \mathcal{H}u + cu^2 - \frac{u^{p+2}}{p+1}\right).$$

Then $N(\varphi)=\langle L_c'(\varphi),\varphi\rangle=0$ whenever $\varphi$ is a solution of (2.1). Here, and after, $\langle\cdot,\cdot\rangle$ denotes the dual product between $Y'$ and $Y$.

**Definition 2.1** We say that $\varphi$ is a groundstate solution of (2.1) if $L_c'(\varphi)=0$ and for any $\psi$ satisfying $L_c'(\psi)=0$ there holds $L_c(\varphi)\leqslant L_c(\psi)$.

Next we will prove that the following minimum

$$d=\inf_{u\in\mathcal{N}}L_c(u),\quad \mathcal{N}=\{u\in Y;\ u\neq 0,\ N(u)=0\},\tag{2.2}$$

is achieved by some $\varphi$, which is a groundstate solution of (2.1).

**Lemma 2.2** Suppose that both $\{u_n\}$ and $\{u_{nx}\}$ are bounded in $L^2(\mathbf{R}^2)$. Assuming $u_n\to u$ a.e. in $\mathbf{R}^2$, then

$$\int u_{nx}\mathcal{H}u_n=\int(u_n-u)_x\mathcal{H}(u_n-u)+\int u_x\mathcal{H}u+o(1).\tag{2.3}$$

Here, and after, $o(1)$ goes to zero as $n$ goes to infinity.

**Proof** We firstly claim that if $u_n\rightharpoonup u$ in $L^2(\mathbf{R}^2)$, then $\mathcal{H}u_n\rightharpoonup\mathcal{H}u$ in $L^2(\mathbf{R}^2)$. Indeed, for any $w\in L^2(\mathbf{R}^2)$,

$$\int w\mathcal{H}u_n=-\int u_n\mathcal{H}w\to-\int u\mathcal{H}w=\int w\mathcal{H}u,$$

proving the claim.

Since $u_n\rightharpoonup u$ in $L^2(\mathbf{R}^2)$ and $u_{nx}\rightharpoonup u_x$ in $L^2(\mathbf{R}^2)$, we have, from the properties of $\mathcal{H}$, that

$$\begin{aligned}\int(u_n-u)_x\mathcal{H}(u_n-u)&=\int u_{xx}\mathcal{H}u_n-\int u_x\mathcal{H}u_n-\int u_{nx}\mathcal{H}u+\int u_x\mathcal{H}u\\&=\int u_{nx}\mathcal{H}u_n-\int u_x\mathcal{H}u+o(1)\quad(n\to\infty).\end{aligned}$$

The proof is complete. □

**Lemma 2.3** If $c>0$ and $0<p<4$, then for any $0\neq u\in Y$, there exists a unique $\theta_u>0$ such that $\theta_u u\in\mathcal{N}$. Moreover if $N(u)<0$, then $0<\theta_u<1$.

**Proof** For any $0\neq u\in Y$, define

$$\gamma(\theta)=L_c(\theta u)=\int\left(\frac{1}{2}\theta^2(u_x^2+v^2+\beta u_x\mathcal{H}u+cu^2)-\frac{\theta^{p+2}u^{p+2}}{(p+2)(p+1)}\right).$$

Direct computations show that

$$\theta_u=\left(\frac{\int(u_x^2+v^2+\beta u_x\mathcal{H}u+cu^2)}{\int\frac{u^{p+2}}{p+1}}\right)^{1/p}.$$

Clearly, from the expression of $\theta_u$ we know that if $N(u)<0$, then $0<\theta_u<1$. □

**Lemma 2.4** Suppose that $c>0$ and $0<p<4$. Then

(i) $\mathcal{N}\neq\emptyset$ and $\mathcal{N}$ is manifold. Moreover, there exists $\alpha>0$ such that for any $u\in\mathcal{N}$, $\|u\|_Y\geqslant\alpha>0$ and

(ii) if $u$ achieves the minimum $d$, then $u$ is a critical point of $L_c$ in $Y$, i.e. $L_c'(u)=0$.

**Proof** $\mathcal{N}\neq\emptyset$ follows from Lemma 2.3. For any $u\in\mathcal{N}$,

$$\langle N'(u),u\rangle=\int\left(2(u_x^2+v^2+\beta u_x\mathcal{H}u+cu^2)-\frac{p+2}{p+1}u^{p+2}\right)=-\frac{p}{p+1}\int u^{p+2}<0,$$

this implies that $\mathcal{N}$ is manifold. For any $u\in\mathcal{N}$, denote $\rho=\int(u_x^2+v^2+cu^2)$, then using the anisotropic Sobolev embedding theorem[8] we have

$$\rho+\int\beta u_x\mathcal{H}u=\int\frac{u^{p+2}}{p+1}\leqslant C_2|u|_2^{(4-p)/2}|v|_2^{p/2}|u_x|_2^p\leqslant C_3\rho^{(2+p)/2},$$

which implies that $\rho\geqslant C_3^{-2/p}:=\alpha>0$. Thus (i) is proved.

Since $u$ achieves the minimum $d$, we know that there exists $\theta\in\mathbf{R}$ such that $L_c'(u)=\theta N'(u)$. Noticing that $\langle L_c'(u),u\rangle=N(u)=0$ and

$$\langle N'(u),u\rangle=-\frac{p}{p+1}\int u^{p+2}<0,$$

we have that $\theta=0$. It follows that $L_c'(u)=0$. Thus (ii) is proved. □

Next we prove that $d$ is achieved.

**Theorem 2.5** Suppose that $c>0$ and $0<p<4$. Then the minimum $d$ is achieved by some $\varphi\in Y$, which is a groundstate solution of (2.1).

**Remark 2.6** If $u\neq0$ is a solution of (2.1), then $L_c'(u)=0$ and hence $N(u)=0$. The definition of $d$ implies that $L_c(u)\geqslant L_c(\varphi)$ for any $\varphi$ satisfying $\varphi\in\mathcal{N}$ and $L_c(\varphi)=d$.

**Proof of Theorem 2.5** From remark 2.6 and (ii) of Lemma 2.4, it suffices to prove that $d$ is achieved by some $\varphi\in\mathcal{N}$. Let $(u_n)\subset Y$ be such that $N(u_n)=0$ and $d+o(1)=L_c(u_n)$. We know from (i) of Lemma 2.4 that

$$\liminf_{n\to\infty}\int u_n^{p+2}>0.$$

Since the functionals $L_c$ and $N$ are invariant under translation, i.e. for any $(x_1,y_1)\in\mathbf{R}^2$, $L_c(u(\cdot+x_1,\cdot+y_1))=L_c(u)$ and $N(u(\cdot+x_1,\cdot+y_1))=N(u)$, we obtain from concentration compactness lemma of Lions[9,10] (see also [11] and [12]) that there exist $(x_n,y_n)\in\mathbf{R}^2$ such that $\varphi_n=u_n(\cdot+x_n,\cdot+y_n)$ satisfies $N(\varphi_n)=0$ and $d+o(1)=L_c(\varphi_n)$. Moreover, $\varphi_n\to\varphi$ a.e. in $\mathbf{R}^2$ and $\varphi\neq0$.

If $N(\varphi)<0$, then by Lemma 2.3 there exists $0<\theta_\varphi<1$ such that $\theta_\varphi\varphi\in\mathcal{N}$. Using $\varphi_n\in\mathcal{N}$ and the Fatou lemma, we have that

$$\begin{aligned}d+o(1)=L_c(\varphi_n)&=\left(\frac{1}{2}-\frac{1}{p+2}\right)\int\frac{\varphi_n^{p+2}}{p+1}\\&\geqslant\frac{p}{2(p+2)}\int\frac{\varphi^{p+2}}{p+1}+o(1)\\&=\frac{p\theta_\varphi^{-(p+2)}}{2(p+2)}\int\frac{(\theta_\varphi\varphi)^{p+2}}{p+1}+o(1).\end{aligned}\tag{2.4}$$

Since $0<\theta_\varphi<1$, (2.4) implies that $d>L_c(\theta_\varphi\varphi)$, which is a contradiction because $\theta_\varphi\varphi\in\mathcal{N}$.

If $N(\varphi)>0$, then using Lemma 2.2 and the Brezis-Lieb lemma[13] we have that

$$0=N(\varphi_n)=N(\varphi)+N(w_n)+o(1),$$

with $w_n=\varphi_n-\varphi$ and hence $\limsup_{n\to\infty}N(w_n)<0$. From Lemma 2.3 we know that there are $\theta_n:=\theta_{w_n}$, such that $\theta_nw_n\in\mathcal{N}$. Moreover, we claim that $\limsup_{n\to\infty}\theta_n\in(0,1)$. Indeed, if otherwise there exists a subsequence still denoted by $\theta_n$ such that $\theta_n \to 1$ $(n \to \infty)$, then $N(w_n) = N(\theta_n w_n) + o(1) = o(1)$. That is a contradiction, since

$$\begin{aligned} d + o(1) = L_c(\varphi_n) &= \frac{p}{2(p+2)(p+1)} \int \varphi_n^{p+2} \\ &\geqslant \frac{p}{2(p+2)(p+1)} \int w_n^{p+2} + o(1) \\ &\geqslant \frac{p\theta_n^{-(p+2)}}{2(p+2)(p+1)} \int (\theta_n w_n)^{p+2} + o(1), \end{aligned} \tag{2.5}$$

one has again $d > L_c(\theta_n w_n)$ for large $n$.

Thus $N(\varphi) = 0$. Using Lemma 2.2 and the Brezis-Lieb lemma again, we easily get that $\| w_n \|_Y \to 0$, i.e. $\varphi_n \to \varphi \neq 0$ in $Y$. The proof is complete. □

**Remark 2.7**

(ⅰ) When $\beta = 0$, [6] proved that any solitary wave solutions of (1.1) belong to $C^\infty(\mathbf{R}^2)$ for $p = 1,2,3$ (see [6], Theorem 4.1.).

(ⅱ) Observing the proofs of [6] (p. 227, Theorem 4.1), we know that if we replace the $|\xi|^2 + \xi_1^4$ by $c\xi_1^2 + \xi_2^2 + \xi_1^4 - \beta|\xi_1|^3$ in the multipliers $\Phi_1$, $\Phi_2$ and $\Phi_3$, then all the proofs contained in [6] (p. 227, Theorem 4.1) hold provided $c > \beta/2$. Therefore if $c > \beta/2$, then the solitary wave $\varphi(x - ct, y)$ of (1.1) belongs to $C^\infty(\mathbf{R}^2)$ for $p = 1,2,3$.

We end this section by another characterization of the minimum $d$.

**Proposition 2.8** Let $s \geqslant 3$ and

$$\tilde{d} = \inf_{u \in \tilde{\mathcal{N}}} L_c(u), \quad \tilde{\mathcal{N}} = \{u \in X_s;\ u \neq 0,\ N(u) = 0\}. \tag{2.6}$$

Then $\tilde{d} = d$.

**Proof** Clearly $\tilde{d} \geqslant d$. To see $d \geqslant \tilde{d}$, it suffices to prove that for any $\varepsilon > 0$ and any $u^* \in \mathcal{N}$ there holds

$$L_c(u^*) \geqslant \inf_{u \in \tilde{N}} L_c(u) - \varepsilon. \tag{2.7}$$

Since $X_s$ is dense in $Y$, we find a sequence $(u_n) \subset X_s$ such that

$$u^* = u_n + w_n \quad \text{with} \quad w_n \to 0 \quad \text{in} \quad Y \quad \text{as} \quad n \to \infty.$$

From $u^* \neq 0, N(u^*) = 0$ and $w_n \to 0$ in $Y$, we know that $\liminf_{n\to\infty} \int |u_n|^{p+2} \neq 0$ and $\lim_{n\to\infty} N(u_n) = 0$. The proof of Lemma 2.3 implies that there exists a sequence $(\rho_n)$ such that $\rho_n u_n \in \widetilde{N}$ and $\lim_{n\to\infty} \rho_n = 1$.

Next, denoting $u^* = \rho_n u_n + (1 - \rho_n) u_n + w_n$, we compare $L_c(u^*)$ with $L_c(\rho_n u_n)$. Since $1 - \rho_n \to 0$ and $w_n \to 0$ strongly in $Y$, we have immediately that

$$\frac{1}{2} \| u^* \|_Y^2 \geqslant \frac{1}{2} \| \rho_n u_n \|_Y^2 - \frac{\varepsilon}{3}$$

and

$$-\frac{1}{(p+2)(p+1)} \int (u^*)^{p+2} \geqslant -\frac{1}{(p+2)(p+1)} \int (\rho_n u_n)^{p+2} - \frac{\varepsilon}{3}.$$

We now calculate the term

$$\frac{\beta}{2}\int(\rho_n u_n + (1-\rho_n)u_n + w_n)_x \mathscr{H}(\rho_n u_n + (1-\rho_n)u_n + w_n).$$

Indeed, using the fact that $(w_n)_x \to 0$ in $L^2(\mathbf{R}^2)$, $w_n \to 0$ in $L^2(\mathbf{R}^2)$, $1-\rho_n \to 0$ and interpolation inequality (VI), we know that all other terms go to zero as $n\to\infty$ but $\frac{\beta}{2}\int(\rho_n v_n)_x \mathscr{H}(\rho_n v_n)$. Hence we get that for large $n$

$$\frac{\beta}{2}\int(u^*)_x \mathscr{H}u^* \geqslant \frac{\beta}{2}\int(\rho_n u_n)_x \mathscr{H}(\rho_n u_n) - \frac{\varepsilon}{3}.$$

Therefore

$$L_c(u^*) \geqslant L_c(\rho_n u_n) - \varepsilon \geqslant \inf_{u\in\widetilde{\Lambda}} L_c(u) - \varepsilon,$$

and equation (2.7) holds. The proof is complete. □

# 3 Invariant Sets

In this section, we will use the minimization problem defined in (2.1) and the property of the groundstate solution $\varphi$ to construct several sets that are invariant under the flow generated by the Cauchy problem (1.1) and (1.2). We emphasize that the method used in [1] seems not to be applied here due to the term $\int u_x \mathscr{H}u\,dx dy$. Hence we need to develop further the techniques in [14] to get the invariant sets. First, we have the following local well-posed result.

**Lemma 3.1**([4], Theorem 2.1) Suppose $u_0\in X_s(\mathbf{R}^2)$ and $s\geqslant 3$. Then $T>0$ such that (1.1) and (1.2) have a unique solution

$$u\in C(0,T;X_s)\cap C^1(0,T;H^{s-3}(\mathbf{R}^2)),\quad v\in C(0,T;H^{s-1}(\mathbf{R}^2)),$$

and if $\partial_x^{-2}u_{0yy}\in L^2(\mathbf{R}^2)$, one has

$$u_t\in L^\infty(0,T;X_0),\quad v_t\in L^\infty(0,T;H^{-1}(\mathbf{R}^2)).$$

Moreover there holds

$$\frac{d}{dt}V(u) = \frac{d}{dt}\int\frac{1}{2}|u(x,y,t)|^2 dx dy = 0, \tag{3.1}$$

$$E_\beta(u) = \int\left(\frac{u_x^2+v^2+\beta u_x\mathscr{H}u}{2} - \frac{u^{p+2}}{(p+2)(p+1)}\right)dx dy \equiv E_\beta(u_0). \tag{3.2}$$

Next, we define another two functionals

$$Q(u) = \int\left(u_x^2 + v^2 + \frac{\beta}{2}u_x\mathscr{H}u - \frac{3p}{2(p+2)(p+1)}u^{p+2}\right), \tag{3.3}$$

$$R(u) = \int\left(u_x^2 + \frac{\beta}{2}u_x\mathscr{H}u - \frac{p}{(p+2)(p+1)}u^{p+2}\right). \tag{3.4}$$

We denote that

$$\mathscr{R} = \{u\in X_s; u\neq 0, N(u)<0, R(u)=0\},$$
$$\mathscr{Q} = \{u\in X_s; u\neq 0, N(u)<0, Q(u)=0\}.$$

We set that

$$d_{\mathscr{R}} = \inf\{L_c(u); u \in \mathbf{R}\}, \quad d_{\mathscr{Q}} = \inf\{L_c(u); u \in \mathbf{Q}\}.$$

**Lemma 3.2** If $c>0$ and $2\leqslant p<4$, then $d_{\mathscr{R}}\geqslant d$.

**Proof** Keep the definition of $L_c(u)$, $N(u)$ and $R(u)$ in mind. For any $u\in\mathscr{R}$, we want to find $w\in\widetilde{\mathscr{N}}$ such that $L_c(u)\geqslant L_c(w)$. Firstly, for any $u\in\mathscr{R}, N(u)<0$ and $R(u)=0$. We denote that $u_\lambda(x)=\lambda u(\lambda x,\lambda y)$, then

$$N(u_\lambda) = \int\left(\lambda^2 u_x^2 + v^2 + \lambda\beta u_x\mathscr{H}u + cu^2 - \frac{\lambda^p u^{p+2}}{p+1}\right).$$

Noticing that $N(u_\lambda)\to\int(v^2+cu^2)>0$ as $\lambda\to 0$ and $N(u_\lambda)\to N(u)<0$ as $\lambda\to 1$, we find a $\lambda_1\in(0,1)$ such that $N(u_{\lambda_1})=0$, i.e. $u_{\lambda_1}\in\widetilde{\mathscr{N}}$. Therefore $L_c(u_{\lambda_1})\geqslant d$.

Secondly, we want to prove that $L_c(u)\geqslant L_c(u_\lambda)$ for any $\lambda\in(0,1)$. In fact, from

$$L_c(u_\lambda) = \int\left(\frac{\lambda^2 u_x^2 + v^2 + \lambda\beta u_x\mathscr{H}u + cu^2}{2} - \frac{\lambda^p u^{p+2}}{(p+2)(p+1)}\right),$$

we get that

$$L_c(u) - L_c(u_\lambda) = \int\left(\frac{(1-\lambda^2)u_x^2 + (1-\lambda)\beta u_x\mathscr{H}u}{2} - \frac{(1-\lambda^p)u^{p+2}}{(p+2)(p+1)}\right).$$

$R(u)=0$ implies that

$$L_c(u) - L_c(u_\lambda) = \int(g_1(\lambda)u_x^2 + g_2(\lambda)\beta u_x\mathscr{H}u), \tag{3.5}$$

where $g_1(\lambda)=\frac{1}{2}(1-\lambda^2)-\frac{1}{p}(1-\lambda^p)$ and $g_2(\lambda)=\frac{1}{2}(1-\lambda)-\frac{1}{2p}(1-\lambda^p)$. Since $g_1(\lambda)\to\frac{1}{2}-\frac{1}{p}\geqslant 0$ for $p\geqslant 2$ as $\lambda\to 0$, $g_1(\lambda)\to 0$ as $\lambda\to 1$ and $g'_1(\lambda)=-\lambda(1-\lambda^{p-2})<0$ for $\lambda\in(0,1)$, we know that $g_1(\lambda)\geqslant 0$ for $\lambda\in(0,1)$. Similar arguments arrive at $g_2(\lambda)\geqslant 0$ for $\lambda\in(0,1)$. Thus $L_c(u)\geqslant L_c(u_\lambda)$ for any $\lambda\in(0,1)$. In particular, $L_c(u)\geqslant L_c(u_{\lambda_1})$. Taking $w=u_{\lambda_1}$, we conclude the proof. □

**Lemma 3.3** If $c>0$ and $2\leqslant p<4$, then $d_{\mathscr{Q}}\geqslant d$.

**Proof** The idea is similar to those used in the proof of Lemma 3.2, but we need a quite different scaling argument. Our purpose is to show that for any $u\in\mathscr{Q}$, there is $w\in\widetilde{\mathscr{N}}$ such that $L_c(u)\geqslant L_c(w)$. Firstly, for any $u\in\mathscr{Q}$, $N(u)<0$ and $Q(u)=0$. We denote that $u^\lambda(x)=\lambda^{\frac{3}{2}}u(\lambda x,\lambda^2 y)$, then

$$N(u^\lambda) = \int\left(\lambda^2 u_x^2 + \lambda^2 v^2 + \lambda\beta u_x\mathscr{H}u + cu^2 - \frac{\lambda^{\frac{3p}{2}} u^{p+2}}{p+1}\right).$$

Noticing that $N(u^\lambda)\to c\int u^2>0$ as $\lambda\to 0$ and $N(u^\lambda)\to N(u)<0$ as $\lambda\to 1$, we find a $\lambda_2\in(0,1)$ such that $N(u^{\lambda_2})=0$, i.e. $u^{\lambda_2}\in\widetilde{\mathscr{N}}$. Therefore $L_c(u^{\lambda_2})\geqslant d$.

Secondly, we prove that $L_c(u)\geqslant L_c(u^\lambda)$ for any $\lambda\in(0,1)$. In fact, from

$$L_c(u^\lambda) = \int\left(\frac{\lambda^2 u_x^2 + \lambda^2 v^2 + \lambda\beta u_x\mathscr{H}u + cu^2}{2} - \frac{\lambda^{3p/2} u^{p+2}}{(p+2)(p+1)}\right),$$

and $Q(u)=0$, we obtain that

$$L_c(u) - L_c(u^\lambda) = \int (g_3(\lambda)u_x^2 + g_4(\lambda)\beta u_x \mathscr{H} u + g_3(\lambda)v^2), \tag{3.6}$$

where $g_3(\lambda)=\frac{1}{2}(1-\lambda^2)-\frac{2}{3p}(1-\lambda^{3p/2})$ and $g_4(\lambda)=\frac{1}{2}(1-\lambda)-\frac{1}{3p}(1-\lambda^{3p/2})$. Using an argument similar to those in the proofs of $g_i(\lambda)\geqslant 0$ $(i=1,2)$ for any $\lambda\in(0,1)$, we have that $g_j(\lambda)\geqslant 0$ $(j=3,4)$ for any $\lambda\in(0,1)$. Thus $L_c(u)\geqslant L_c(u^\lambda)$ for all $\lambda\in(0,1)$. In particular, $L_c(u)\geqslant L_c(u^{\lambda_2})$ and we complete the proof. □

Now we are in a position to define several invariant sets.

$$\mathscr{R}_+ = \{u\in X_s; L_c(u)<d, N(u)<0, R(u)>0\};$$
$$\mathscr{R}_- = \{u\in X_s; L_c(u)<d, N(u)<0, R(u)<0\};$$
$$\mathscr{Q}_+ = \{u\in X_s; L_c(u)<d, N(u)<0, Q(u)>0\};$$
$$\mathscr{Q}_- = \{u\in X_s; L_c(u)<d, N(u)<0, Q(u)<0\}.$$

**Lemma 3.4** For any $c>0$ and $2\leqslant p<4$, $\mathscr{R}_\pm$ and $\mathscr{Q}_\pm$ are invariant under the flow generated by the Cauchy problem (1.1) and (1.2).

**Proof** We only prove that $\mathscr{Q}_-$ is invariant under the flow generated by the Cauchy problem (1.1) and (1.2) since the proofs of the others are similar. Let $u(x,y,t)$ $(t\in[0,T))$ be the solution of (1.1) and (1.2) with initial data $u_0$. We simply denote the function $t\mapsto u(x,y,t)$ by $u(t)$. First, by Lemma 3.1, we know that if $L_c(u_0)<d$, then $L_c(u(t))\equiv L_c(u_0)<d$.

Second, we show that $N(u(t))<0$ for $t\in[0,T)$. If this is not true, from the continuity, $t_0\in(0,T)$ such that $N(u(t_0))=0$. Then since $u(t_0)\neq 0$, we know that $L_c(u(t_0))\geqslant d$. This contradicts with $L_c(u(t))<d$ for all $t\in(0,T)$. Therefore $N(u(t))<0$ for $t\in[0,T)$.

Finally, we show that $Q(u(t))<0$ for $t\in[0,T)$. If this is not true, from the continuity, $t_1\in(0,T)$ such that $Q(u(t_1))=0$. Since we have proved that $N(u(t))<0$, we can have that $u(t_1)\in\mathscr{Q}$. So that $L_c(u(t_1))\geqslant d_{\mathscr{Q}}\geqslant d$, which contradicts with $L_c(u(t))<d$ for all $t\in(0,T)$. Therefore $Q(u(t))<0$ for $t\in[0,T)$. □

## 4 Improved Blow-up

Recalling that, in a different setting, if $S(t)$ is a solution of the nonlinear Schrödinger equation

$$\mathrm{i}S_t + \Delta S + |S|^\sigma S = 0, \quad S(0) = S_0, \quad x\in\mathbf{R}^m,$$

then it is shown that $S(t)$ either exists globally in the energy space $H^1(\mathbf{R}^m)$ or blows up in a finite time in the energy space, i.e. $\exists\, T_1>0$ such that

$$\lim_{t\to T_1^-}\|S(t)\|_{H^1(\mathbf{R}^m)} = \infty$$

(e.g. [15-17]). For the problem considered here, we know from Lemma 3.4, (3.1) and

(3.2) that if $u_0 \in \mathscr{R}_+$, then the solution $u(t)$ with $u(0)=u_0$ satisfies $\|u(t)\|_Y \leqslant C$. Consequently, $u(t)$ exists globally in $Y$ for $t \geqslant 0$ and hence blow up cannot occur in a finite time in $Y$. But as we will see below, we do have a blow-up result that is only due to the transverse dispersion.

Guo & Han[4] proved that the solution of (1.1) and (1.2) blows up in a finite time if the initial data $u_0$ satisfies $E_\beta(u_0)<0$ and $p \geqslant 4$. However, using the invariant sets constructed in Section 3, we are able to extend this blow-up result to allow the energy $E_\beta$ to be positive and $2 \leqslant p < 4$.

**Theorem 4.1** If $2 \leqslant p < 4$, $u_0 \in \mathscr{R}_+ \cap \mathscr{Q}_-$, and $yu_0 \in L^2(\mathbf{R}^2)$, then the solution $u(x,y,t)$ of (1.1) and (1.2) with initial data $u_0$ blows up at finite time.

**Proof** For any $u_0 \in \mathscr{R}_+ \cap \mathscr{Q}_-$, we have that $u(x,y,t) \in \mathscr{R}_+ \cap \mathscr{Q}_-$. Now, for any fixed $t$, we simply denote $u(x,y,\cdot)$ by $u$. Thus $L_c(u)<d$, $N(u)<0$, $Q(u)<0$ and $R(u)>0$. For any $\lambda>0$, we denote that $u^\lambda(x,y)=\lambda^{3/2}u(\lambda x, \lambda^2 y)$, then similar to the computations used in the proof of Lemma 3.3, we have that $N(u^\lambda) \to c\int u^2 > 0$ as $\lambda \to 0$ and $N(u^\lambda) \to N(u)<0$ as $\lambda \to 1$. Hence there is $\lambda^* \in (0,1)$ such that $N(u^{\lambda^*})=0$ and $N(u^\lambda)<0$ for $\lambda \in (\lambda^*, 1]$. For $\lambda \in [\lambda^*, 1]$, $Q(u^\lambda)$ has the following three possibilities:

(i) $Q(u^\lambda)<0$ for $\lambda \in [\lambda^*, 1]$;

(ii) $Q(u^{\lambda^*})=0$;

(iii) there is $\mu$ such that $\lambda^* < \mu < 1$ and $Q(u^\mu)=0$.

For cases (i) and (ii), we all have that $N(u^{\lambda^*})=0$ and $Q(u^{\lambda^*}) \leqslant 0$. It follows that $L_c(u^{\lambda^*}) \geqslant d$. Moreover we have that

$$
\begin{aligned}
L_c(u) - L_c(u^{\lambda^*}) &= \int\left(\frac{(1-\lambda^{*2})u_x^2 + (1-\lambda^*)\beta u_x \mathscr{H}u + (1-\lambda^{*2})v^2}{2} - \frac{(1-\lambda^{*3p/2})}{(p+2)(p+1)}u^{p+2}\right) \\
&\geqslant \int\left(\frac{2}{3p}((1-\lambda^{*2})u_x^2 + (1-\lambda^*)\beta u_x \mathscr{H}u + (1-\lambda^{*2})v^2) - \right. \\
&\qquad \left. \frac{(1-\lambda^{*3p/2})}{(p+2)(p+1)}u^{p+2}\right) \\
&= \frac{2}{3p}(Q(u) - Q(u^{\lambda^*})) \geqslant \frac{2}{3p}Q(u).
\end{aligned} \tag{4.1}
$$

For case (iii), we have $N(u^\mu)<0$ and $Q(u^\mu)=0$, i.e. $u^\mu \in \mathscr{Q}$. Thus $L_c(u^\mu) \geqslant d_{\mathscr{Q}} \geqslant d$ and we can also obtain from a similar computation that

$$
L_c(u) - L_c(u^\mu) \geqslant \frac{2}{3p}(Q(u) - Q(u^\mu)) \geqslant \frac{2}{3p}Q(u).
$$

In all cases, we have that

$$
Q(u) \leqslant \frac{3p}{2}(L_c(u_0) - d) = -\delta_0 < 0.
$$

Note that from Guo & Han ([4], Theorem 2.4), $yu \in L^\infty(0,T; L^2(\mathbf{R}^2))$ as long as $yu_0 \in$

$L^2(\mathbf{R}^2)$. Moreover by Guo & Han ([4], Theorem 3.1), for $I(t) = \int y^2 u^2(x,y,t)\mathrm{d}x\mathrm{d}y$, there holds

$$I''(t) = 8\int\left(v^2 - \frac{pu^{p+2}}{2(p+2)(p+1)}\right) = 8(Q(u) - R(u)).$$

On the other hand, $R(u)>0$ since $u_0 \in \mathscr{R}_+$. Thus we find that there exists $t_0>0$ such that $I(t_0)=0$. The conserved momentum $\frac{1}{2}\int |u|^2\mathrm{d}x\mathrm{d}y$ and the classical inequality $|u|_2^2 \leqslant 2|yu|_2 \cdot |u_y|_2$ imply that there exists a blow-up time $T_1 \leqslant t_0$ such that

$$\lim_{t\to T_1^-} |\partial_y u|_2 = +\infty. \qquad \square$$

**Remark 4.2** If $2\leqslant p<4$ and $u_0 \in \mathscr{R}_+$, then

$$E_\beta(u_0) = \int\left(\frac{u_{0x}^2 + v_0^2 + \beta u_{0x}Hu_0}{2} - \frac{u_0^{p+2}}{(p+2)(p+1)}\right)$$
$$> \int\left(\left(\frac{1}{2} - \frac{1}{p}\right)u_{0x}^2 + \frac{1}{2}v^2 + \left(\frac{1}{2} - \frac{1}{2p}\right)\beta u_{0x}Hu_0\right) > 0.$$

# 5 Instability of Solitary Waves

In this section, we will prove that the solitary wave of (1.1) is strongly unstable. We assume throughout this section that $c > \beta/2$ and if $\varphi$ is a solitary wave of (1.1), then $y\varphi \in L^2(\mathbf{R}^2)$.

**Definition 5.1** Suppose for $c>0, \varphi \in Y$ is a groundstate solution of (2.1) (of course a solitary wave of (1.1) with wave speed $c$ in the $x$-direction). We say that $\varphi$ is strongly unstable if for any $\delta>0$ there is $u_0 \in X_s (s\geqslant 3)$, with $\| u_0 - \varphi \|_Y < \delta$ such that the solution $u$ of (1.1) with initial data $u_0$ blows up in a finite time. More precisely, there is $0< T_2 < \infty$ such that

$$\lim_{t\to T_2^-} \|\partial_y u\|_2 = \infty. \tag{5.1}$$

**Theorem 5.2** Suppose that $\varphi$ is a solitary wave solution of (1.1) corresponding to the groundstate solution of (2.1) and $2\leqslant p<4$. If $c>\beta/2$, then $\varphi$ is strongly unstable in the sense of Definition 5.1.

Before proving Theorem 5.2, we need the following lemmas.

**Lemma 5.3** Suppose that $\varphi$ is the groundstate solution of (2.1) obtained in Section 2. Then $Q(\varphi)=0$ and $R(\varphi)=0$.

**Proof** Since $\varphi$ is a groundstate solution of (2.1), $\partial_\lambda E_\beta(\varphi^\lambda)|_{\lambda=1}=0$ with $\varphi^\lambda(x,y) = \lambda^{3/2}\varphi(\lambda x, \lambda^2 y)$ and $\partial_\lambda E_\beta(\varphi_\lambda)|_{\lambda=1}=0$ with $\varphi_\lambda(x,y) = \lambda\varphi(\lambda x, \lambda y)$. Direct computations arrive at $Q(\varphi) = \partial_\lambda E_\beta(\varphi^\lambda)|_{\lambda=1}=0$ and $R(\varphi) = \partial_\lambda E_\beta(\varphi_\lambda)|_{\lambda=1}=0$. The proof is complete. $\square$

**Lemma 5.4** There exists $\lambda>0$ and $\eta>0$ such that for $w_0(x,y) = \lambda\varphi(x, \eta y)$,

$$L_c(w_0) < d, \quad N(w_0) < 0, \quad Q(w_0) < 0 \quad \text{and} \quad R(w_0) > 0.$$

**Proof** Let $\varphi$ be the groundstate solution of (2.1). We denote that $\varphi_y = \psi_x$. Then

$$N(\varphi) = \int \left( \varphi_x^2 + \psi^2 + \beta \varphi_x \mathscr{H} \varphi + c\varphi^2 - \frac{\varphi^{p+2}}{p+1} \right) = 0. \tag{5.2}$$

Lemma 5.3 implies that

$$R(\varphi) = \int \left( \varphi_x^2 + \frac{\beta}{2} \varphi_x \mathscr{H} \varphi - \frac{p}{(p+2)(p+1)} \varphi^{p+2} \right) = 0, \tag{5.3}$$

$$Q(\varphi) = \int \left( \varphi_x^2 + \psi^2 + \frac{\beta}{2} \varphi_x \mathscr{H} \varphi - \frac{3p}{2(p+2)(p+1)} \varphi^{p+2} \right) = 0. \tag{5.4}$$

From (5.2) and (5.4), we have that

$$\int c\varphi^2 = \int \left( \frac{4-p}{2(p+2)(p+1)} \varphi^{p+2} - \frac{\beta}{2} \varphi_x \mathscr{H} \varphi \right). \tag{5.5}$$

By (5.3) and (5.4), we get that

$$\int \psi^2 = \frac{p}{2(p+2)(p+1)} \int \varphi^{p+2}. \tag{5.6}$$

Let $w_0(x, y) = \lambda \varphi(x, \eta y)$. We want to find $\lambda$ and $\eta$ such that

$$L_c(w_0) < d, \quad N(w_0) < 0, \quad Q(w_0) < 0 \quad \text{and} \quad R(w_0) > 0. \tag{5.7}$$

Since

$$L_c(w_0) = \int \left( \frac{\lambda^2}{2\eta} \varphi_x^2 + \frac{\lambda^2 \eta}{2} \psi^2 + \frac{\lambda^2}{2\eta} \beta \varphi_x \mathscr{H} \varphi + \frac{\lambda^2}{2\eta} c\varphi^2 - \frac{\lambda^{p+2} \varphi^{p+2}}{\eta(p+2)(p+1)} \right), \tag{5.8}$$

$$N(w_0) = \int \left( \frac{\lambda^2}{\eta} \varphi_x^2 + \lambda^2 \eta \psi^2 + \frac{\lambda^2}{\eta} \beta \varphi_x \mathscr{H} \varphi + \frac{\lambda^2}{\eta} c\varphi^2 - \frac{\lambda^{p+2} \varphi^{p+2}}{\eta(p+1)} \right), \tag{5.9}$$

$$Q(w_0) = \int \left( \frac{\lambda^2}{\eta} \varphi_x^2 + \lambda^2 \eta \psi^2 + \frac{\lambda^2}{2\eta} \beta \varphi_x \mathscr{H} \varphi - \frac{3p\lambda^{p+2} \varphi^{p+2}}{2\eta(p+2)(p+1)} \right), \tag{5.10}$$

$$R(w_0) = \int \left( \frac{\lambda^2}{\eta} \varphi_x^2 + \frac{\lambda^2}{2\eta} \beta \varphi_x \mathscr{H} \varphi - \frac{p\lambda^{p+2} \varphi^{p+2}}{\eta(p+2)(p+1)} \right), \tag{5.11}$$

we have from (5.3), (5.5), (5.6) and delicate computations that condition (5.7) is equivalent to conditions

$$R(w_0) = \left( \frac{\lambda^2}{\eta} - \frac{\lambda^{p+2}}{\eta} \right) \int \frac{p\varphi^{p+2}}{(p+2)(p+1)} > 0, \tag{5.12}$$

$$Q(w_0) = \left( \frac{\lambda^2}{\eta} + \frac{\lambda^2 \eta}{2} - \frac{3\lambda^{p+2}}{2\eta} \right) \int \left( \varphi_x^2 + \frac{\beta}{2} \varphi_x \mathscr{H} \varphi \right) < 0, \tag{5.13}$$

$$N(w_0) = \left( \frac{\lambda^2 \eta}{2} + \frac{(4+p)\lambda^2}{2p\eta} - \frac{(p+2)\lambda^{p+2}}{p\eta} \right) \int \left( \varphi_x^2 + \frac{\beta}{2} \varphi_x \mathscr{H} \varphi \right) < 0, \tag{5.14}$$

$$\begin{aligned} L_c(w_0) &= \left( \frac{\lambda^2 \eta}{2} + \frac{(4+p)\lambda^2}{2p\eta} - \frac{2\lambda^{p+2}}{p\eta} \right) \int \left( \frac{\varphi_x^2}{2} + \frac{\beta}{4} \varphi_x \mathscr{H} \varphi \right) \\ &< \int \left( \frac{\varphi_x^2}{2} + \frac{\beta}{4} \varphi_x \mathscr{H} \varphi \right). \end{aligned} \tag{5.15}$$

Conditions (5.12) and (5.15) are equivalent to

$$\left.\begin{aligned} \lambda^p &< 1, \\ 3\lambda^p &> \eta^2 + 2, \\ \frac{2(p+2)}{p}\lambda^p &> \eta^2 + \frac{4+p}{p}, \\ \frac{4}{p}\lambda^p &> \eta^2 + \frac{4}{p} - 1. \end{aligned}\right\} \tag{5.16}$$

Taking $\eta_0^2 = 1 - \varepsilon_0$ (where $\varepsilon_0 > 0$ but is still small enough) and $\lambda_0$ with

$$1 > \lambda_0^p > \max\left\{1 - \frac{\varepsilon_0}{3}, 1 - \frac{p}{2(p+2)}\varepsilon_0\right\},$$

we know that (5.7) holds for $w_0(x, y) = \lambda_0 \varphi(x, \eta_0 y)$. The proof is complete. □

**Lemma 5.5** For any fixed $u, w \in Y$ with $\|w\|_Y \leqslant K$, there exist positive constants $K_j$ ($j = 1,2,3,4$) independent of $w$, such that

$$L_c(u + w) < L_c(u) + K_1 \|w\|_Y, \tag{5.17}$$

$$N(u + w) < N(u) + K_2 \|w\|_Y, \tag{5.18}$$

$$Q(u + w) < Q(u) + K_3 \|w\|_Y, \tag{5.19}$$

$$R(u + w) > R(u) - K_4 \|w\|_Y. \tag{5.20}$$

**Proof** We prove (5.19) in detail and indicate the differences in the proofs of the others. First using the elementary inequality

$$|u + w|^{p+2} \geqslant |u|^{p+2} + |w|^{p+2} - C(p)(|u||w|^{p+1} + |u|^{p+1}|w|),$$

and the anisotropic Sobolev imbedding theorem[8], we have that

$$\begin{aligned} \int(|u|^{p+2} - |u + w|^{p+2}) &\leqslant \|w\|_Y^{p+2} + C(\|u\|_Y \|w\|_Y^{p+1} + \|u\|_Y^{p+1}\|w\|_Y) \\ &\leqslant C\|w\|_Y. \end{aligned} \tag{5.21}$$

Second using the properties (Ⅰ)~(Ⅴ) of the Hilbert operator $\mathscr{H}$, we have that

$$\int(u_x + w_x)\mathscr{H}(u + w) \leqslant \int u_x \mathscr{H} u + C\|w\|_Y. \tag{5.22}$$

Finally, we have that

$$\begin{aligned} Q(u + w) &\leqslant Q(u) + C_1\|w\|_Y + C_2\|w\|_Y^2 + C_3\int(|u|^{p+2} - |u + w|^{p+2}) \\ &\leqslant Q(u) + K_3\|w\|_Y, \end{aligned}$$

proving (5.19). The proofs of (5.17) and (5.18) are almost the same and omitted here. For (5.20), firstly using the properties (Ⅰ)~(Ⅴ) of the Hilbert operator $\mathscr{H}$, we have that

$$\int(u_x + w_x)\mathscr{H}(u + w) \geqslant \int u_x \mathscr{H} u - C_4\|w\|_Y.$$

Secondly, using the anisotropic Sobolev imbedding theorem and

$$|u + w|^{p+2} \leqslant |u|^{p+2} + |w|^{p+2} + C_5(|u||w|^{p+1} + |u|^{p+1}|w|),$$

we have that

$$\int(u + w)^{p+2} \leqslant \int u^{p+2} + C_6\|w\|_Y.$$

Thus

$$R(u+w) \geqslant R(u) + \int (2u_x w_x + w_x^2) - C_4 \| w \|_Y - C_6 \| w \|_Y \geqslant R(u) - K_4 \| w \|_Y.$$

We complete the proof of Lemma 5.5. □

Now we are in a position to prove Theorem 5.2.

**Proof of Theorem 5.2** For any $\delta > 0$, we choose sufficiently small $\varepsilon_0$,

$$\eta_0^2 = 1 - \varepsilon_0 \quad \text{and} \quad 1 > \lambda_0^p > \max\left\{1 - \frac{\varepsilon_0}{3}, 1 - \frac{p}{2(p+2)}\varepsilon_0\right\},$$

such that

$$\| w_0 - \varphi \|_Y < \frac{\delta}{2},$$

where $w_0$ is defined as in Lemma 5.4. Since $X_s$ is dense in $Y$, we find $u_0 \in X_s$ such that

$$\| u_0 - w_0 \|_Y < \min\left\{\frac{\delta}{2}, \frac{\delta_1}{K_1 + K_2 + K_3 + K_4}\right\},$$

where $K_j (j = 1,2,3,4)$ are chosen as in Lemma 5.5 and $\delta_1$ is defined as

$$\delta_1 = \min\left\{d - L_c(w_0), -N(w_0), -Q(w_0), R(w_0), \frac{\delta}{2}\right\}.$$

Therefore

$$\| u_0 - \varphi \|_Y \leqslant \| u_0 - w_0 \|_Y + \| w_0 - \varphi \|_Y < \delta,$$

and we obtain from Lemmas 5.4 and 5.5 that

$$u_0 \in \mathscr{R}_+ \cap \mathscr{Q}_-.$$

Since $\delta$ is arbitrary and $y\varphi \in L^2(\mathbf{R}^2)$, we have that $yu_0 \in L^2(\mathbf{R}^2)$. Theorem 4.1 implies that the solution $u(x, y, t)$ of (1.1) with initial data $u(x, y, 0) = u_0$ blows up in a finite time. Hence for any $c > \beta/2$, the solitary wave $\varphi(x - ct, y)$ is strongly unstable in the sense of Definition 5.1. □

## References


[1] Liu Y. Blow up and instability of solitary-wave solutions to a generalized Kadomtsev Petviashvili equation. Trans AMS, 2001, 353: 191-208.

[2] Kadomtsev B B, Petviashvili V I. On the stability of solitary waves in weakly dispersing media. Soviet Phys Dokl, 1970, 15: 539-543.

[3] Ablowitz M J, Clarkson P A. Solitons, nonlinear evolution equations and inverse scattering. London Math Soc Lecture Note Series No. 149. Cambridge: Cambridge University Press, 1991.

[4] Guo B, Han Y. Remarks on the generalized Kadomtsev-Petviashvili equations and twodimensional Benjamin-Ono equations. Proc R Soc A, 1996, 452: 1585-1595.

[5] Liu Y, Wang X P. Nonlinear stability of solitary waves of a generalized Kadomtsev-Petviashvili equation. Commun Math Phys, 1997, 183: 253-266.

[6] De Bouard A, Saut J C. Solitary waves of generalized Kadomtsev-Petviashvili equations. Ann Inst H Poincare Anal Non Lineaire, 1997, 14: 211-236.

[7] Guo B, Tan S. Cauchy problem for a generalized nonlinear dispersive equation. J Part Differ Eqns, 1992, 5: 37-50.

[8] Besov O V, Il'in V P, Nikolskii S M. Integral representations of functions and imbedding theorems: I. New York: Wiley, 1978.

[9] Lions P L. The concentration-compactness principle in the calculus of variations. The locally compact case, part Ⅰ. Ann Inst H Poincaré Anal Non Linéaire, 1984, 1: 109－145.

[10] Lions P L. The concentration-compactness principle in the calculus of variations. The locally compact case, part Ⅱ. Ann Inst H Poincaré Anal Non Linéaire, 1984, 1: 223－283.

[11] Ambrosetti A, Wang Z Q. Positive solutions to a class of quasilinear elliptic equations on **R**. Discrete Contin Dyn Syst, 2003, 9: 55－68.

[12] Willem M. Minimax theorems. Boston: Birkhäuser, 1996.

[13] Brezis H, Lieb E. A relation between pointwise convergence of functions and convergence of functionals. Proc AMS, 1983, 88: 486－490.

[14] Chen J, Guo B. On "Blow up" and instability result for a quasilinear Schrödinger equation. Applied Mathematical Modelling, 2009, 33(11): 4192－4200.

[15] Berestycki H, Cazenave T. Instabilite des etats stationnaires dans les equations de Schrödinger et de Klein-Gordon nonlinearires. C R Acad Sci Paris Seire Ⅰ. 1981, 293: 489－492.

[16] Weinstein M I. Nonlinear Schrödinger equations and sharp interpolations estimates. Commun Math Phys, 1983, 87: 567－576.

[17] Zhang J. Cross-constrained variational problem and nonlinear Schrödinger equations // Proc Int Conf on Foundations of Computational Mathematics in Honor of Professor S. Smale's 70th Birthday, Hong Kong, 2001.

# Global Smooth Solutions of the Spin Polarized Transport Equation *

Guo Boling(郭柏灵) Pu Xueke(蒲学科)

**Abstract**

In this paper, we prove the existence of global smooth solutions of the spin-polarized transport equation using energy estimates method in dimension two.

**Keywords** spin-polarized transport equations; global smooth solutions; Landau-Lifshitz equations.

## 1 Introduction

It is well known that the Landau-Lifshitz equation has a fundamental importance in understanding the ferromagnetism in materials, see [11]. It reads as

$$m_t = m \times \Delta m - \mu m \times (m \times \Delta m),$$

where $m \in S^2$ is the magnetization field and $\mu > 0$ is the Gilbert damping coefficients. There is a long list of work contributed to this equation regarding its existence, uniqueness, self-similar solutions, blowup and partially regularities, the interested reader can refer to [1,3,8,7,11, 12] for more details. Mathematically, it is closely related to harmonic maps to the sphere, see [8,10]. There are also several equations closely related to Landau-Lifshitz equation, such as Landau-Lifshitz-Maxwell equation[4], and the equations in ferrimagnetic materials[9] and so on. Many authors also take the spin polarize into account, to consider the spin polarized current-driven magnetization in materials, see [13, 14]. Recently, C. J. Garcia-Cervera and X. P. Wang[6] considered the weak solutions of the following spin-polarized transport equations in materials

$$\left.\begin{aligned} \frac{\partial s}{\partial t} &= -\operatorname{div} J_s - D_0(x)s - D_0(x)s \times m, \\ \frac{\partial m}{\partial t} &= -m \times (h+s) + \alpha m \times \frac{\partial m}{\partial t}, \\ s(x,0) &= s_0(x), \quad m(x,0) = m_0(x). \end{aligned}\right\} \tag{1.1}$$

In this equation, $(s, m)$ is the unknown, $s = (s_1, s_2, s_3): \Omega \to \mathbf{R}^3$ denotes the spin accumulation and $m = (m_1, m_2, m_3): \Omega \to S^2$ is the magnetization field, where $S^2$ is the unit sphere in $\mathbf{R}^3$. $J_s$ is the spin current

* Electronic Journal of Differential Equations, 2008 (63): 1-15.

$$J_s = m \otimes J_e - D_0(x)[\nabla s - \beta m \otimes (\nabla s \cdot m)], \tag{1.2}$$

where $J_e$ is the applied electronic current, $0<\beta<1$ is the spin polarized parameter and $D_0(x)$ is the diffusion parameter depending on the material. In the second equation of (1.1), $h = -\nabla_m \Phi + \Delta m + h_d$ denotes the anisotropy, exchange and self-induced energy respectively.

As can be seen, this equation is closely related to the Landau-Lifshitz equation in the magnetization field $m$ and to quasilinear parabolic equations in the spin accumulation $s$. It has great importance physically and interesting mathematical structures as well. In [6], the authors considered the global weak solutions of this equation and obtained interesting results by Galerkin's approximating method. A natural question is whether or not this system has global smooth solutions. Taking into account the results in [8], in particular Theorem 2.6, it is not right in general by technical reasons. However, we can show that under some additional conditions on the initial data, this system admits global smooth solutions. For simplicity, we consider the periodic boundary conditions and assume that the diffusion material $D_0(x)$ is constant, which is set to be $D_0(x)\equiv 1$ in this paper. Furthermore, we neglect to consider the anisotropic energy $\nabla_m \Phi$ and self-induced energy $h_d$. Under these assumptions and simplifications, we show that there exists a global smooth solution for this system under smallness condition (1.3) in space dimension two. Our main result is the following theorem.

**Theorem 1.1** Let $k\geqslant 2$, $J_e\in H^k(\mathbf{R}_+\times\Omega)$, $0<\beta<1$ and $s_0\in H^k(\Omega)$, $m_0\in H^k(\Omega)$. Then there exists a positive constan $\lambda_0>0$, such that if the smallness condition

$$\|s_0\|_{L^2}^2 + \|J_e\|_{L^2(\mathbf{R}_+\times\Omega)}^2 + \|\nabla m_0\|_{L^2}^2 < \lambda_0 \tag{1.3}$$

holds, then there exists a global solution $(s, m)$ of (1.1) in $H^k(\Omega)$ satisfying

$$\partial_t^j \partial_x^\alpha \in L^\infty([0,T];L^2\Omega), \quad \partial_t^k \partial_x^\beta \in L^2([0,T];L^2\Omega),$$

with $2j+|\alpha|\leqslant k$ and $2k+|\beta|\leqslant k+1$. In particular, if the initial data are smooth, the solution is globally smooth.

In the sequel, we denote by $\|\cdot\|_X$ the norm of functions on space $X$. In particular, if $X = L^2(\Omega)$, we simply denote $\|\cdot\|$ instead of $\|\cdot\|_{L^2(\Omega)}$.

This paper is organized as follows. In the next Section, we show that there exists a local solution in Sobolev spaces $W^{2,p}$ for $p>2$. By standard bootstrap method, this solution is indeed smooth. In Section 3, we give some *a priori* global estimates of the solution in $H^2$. Together with the local theory, we show that the system has a global $H^2$ solution provided that the initial data are in $H^2$, and small in the sense of (1.3). Finally in Section 4, we give some *a priori* estimates in $H^k$ to deduce that the solution remains as smooth as the initial data for all time if the initial data are small in the sense of (1.3). In particular, if the solution is smooth, we get a global smooth solution for the system under consideration.

## 2 Local Smooth Solutions

In this section, we show that there exists a local smooth solution for the system we considered in this paper. For this purpose, we rewrite

$$\operatorname{div}(m\otimes(\nabla s\cdot m))=(\Delta s\cdot m)m+\widetilde{\operatorname{div}}(m\otimes(\nabla s\cdot m)),$$

where $\widetilde{\operatorname{div}}(m\otimes(\nabla s\cdot m))=\nabla m\cdot(\nabla s\cdot m)+(\nabla s\cdot\nabla m)m$. For the rest of this paper, we set

$$A=A(m)=\begin{pmatrix}1-\beta m_1^2 & -\beta m_1m_2 & -\beta m_1m_3\\ -\beta m_2m_1 & 1-\beta m_2^2 & -\beta m_2m_3\\ -\beta m_3m_1 & -\beta m_3m_2 & 1-\beta m_3^2\end{pmatrix}.$$

Since $0<\beta<1$, we know there two exists positive numbers $\lambda,\Lambda>0$, such that

$$\lambda|\xi|^2\leqslant\xi A(m)\xi^{\mathrm{T}}\leqslant\Lambda|\xi|^2,\quad\forall\xi\in\mathbf{R}^3,\ \xi\neq0.\tag{2.1}$$

On the other hand, since $m\in S^2$, we can write the equation for $m$ in its equivalent form

$$(1+\alpha^2)\frac{\partial m}{\partial t}=-m\times(\Delta m+s)-\alpha m\times(m\times(\Delta m+s)).\tag{2.2}$$

Using this notation, system (1.1) can be rewritten as

$$\left.\begin{aligned}&\frac{\partial s}{\partial t}-A(m)\Delta s+s=-\operatorname{div}(m\otimes J_e)-\beta\widetilde{\operatorname{div}}(m\otimes(\nabla s\cdot m))-s\times m,\\&(1+\alpha^2)\frac{\partial m}{\partial t}-\alpha\Delta m=-m\times\Delta m-m\times s+\alpha|\nabla m|^2m-\alpha m\times(m\times s).\end{aligned}\right\}\tag{2.3}$$

Obviously, this system is strongly parabolic[5,8].

In the following, we use Hamilton's idea in [10] to derive the local existence of the solution of (1.1) with smooth initial data $(s_0,m_0)$. As in [8], we prove the following result.

**Lemma 2.1** Let $1<p<\infty$. For every function $g_1,g_2\in L^p(\Omega\times[\gamma,\omega])$, there exists a unique solution $(l,k)$ of

$$\begin{pmatrix}\partial_t l\\(1+\alpha^2)\partial_t k\end{pmatrix}=\begin{pmatrix}A(m)\Delta l+d(s,m)k+e(s,m,J_e)\nabla k+f(m)l+h(m)\nabla l+g_1\\ \alpha\Delta k+m\times\Delta k+a(s,m)\nabla k+b(s,m)k+c(m)l+g_2\end{pmatrix}$$

on $\Omega\times[\gamma,\omega]$, where $a,b,c,d,e,f,s,m$ are given smooth functions.

**Proof** Denote

$$H\begin{pmatrix}l\\k\end{pmatrix}=\begin{pmatrix}\partial_t l-A(m)\Delta l\\(1+\alpha^2)\partial_t k-\alpha\Delta k-m\times\Delta k\end{pmatrix},$$

$$K\begin{pmatrix}l\\k\end{pmatrix}=\begin{pmatrix}d(s)k+e(s,m,J_e)\nabla k+f(m)l+h(m)\nabla l\\a(s,m)\nabla k+b(s,m)k+c(m)l\end{pmatrix}.$$

Since $H$ is a strong parabolic linear operator, the map $(l,k)\to H(l,k)$ is an isomorphism from $W^{2,p}(\Omega\times[\gamma,\omega])$ onto $L^p(\Omega\times[\gamma,\omega])$. On the other hand, $K:W^{2,p}(\Omega\times[\gamma,\omega])\to L^p(\Omega\times[\gamma,\omega])$ is compact. By the Fredholm theory, the indices of the map $(H-K)$ is 0 by definition of the indices of a Fredholm operator. Thus it suffices to show that its kernel is zero to deduce that it is an isomorphism.

Let $(l,k)\in W^{2,p}(\Omega\times[\gamma,\omega])$, and $f|_{\Omega\times\gamma}=0$. Since $(H-K)((l,k)^{\mathrm{T}})=0$, taking the scalar product of this equation with $(l,k)$, we have respectively

$$\frac{1}{2}\frac{\mathrm{d}}{\mathrm{d}t}\int_\Omega |l|^2 + \lambda\int_\Omega |\nabla l|^2 \leqslant \int_\Omega |\nabla A(m)||\nabla l||l| + C_1\int_\Omega |\nabla l||l| + C_2\int_\Omega |\nabla k||l| + C_3\int_\Omega |k||l| + C_4\int_\Omega |l|^2$$
$$\leqslant \frac{\lambda}{4}\int_\Omega |\nabla l|^2 + \frac{\alpha}{4}\int_\Omega |\nabla k|^2 + C_5\int_\Omega |l|^2 + |k|^2$$

and similarly,

$$(1+\alpha^2)\frac{1}{2}\frac{\mathrm{d}}{\mathrm{d}t}\int_\Omega |k|^2 + \alpha\int_\Omega |\nabla k|^2 \leqslant \int_\Omega k\cdot(m\times\Delta k) + C_6\int_\Omega |\nabla k||k| + C_7\int_\Omega |k||l| + C_8\int_\Omega |k|^2$$
$$\leqslant \frac{\lambda}{4}\int_\Omega |\nabla l|^2 + \frac{\alpha}{4}\int_\Omega |\nabla k|^2 + C_9\int_\Omega |l|^2 + |k|^2.$$

In the last inequality, we used the fact that

$$\int_\Omega k\cdot(m\times\Delta k) = -\int_\Omega k\cdot\nabla m\times\nabla k.$$

Adding the two inequalities together we obtain a Gronwall type inequality

$$\frac{\mathrm{d}}{\mathrm{d}t}\int |l|^2 + (1+\alpha^2)\int |k|^2 + \lambda\int |\nabla l|^2 + \alpha\int |\nabla k|^2 \leqslant C\int |l|^2 + |k|^2.$$

Hence by Gronwall inequality, we have $(l,k)=0$, which shows that the kernel of the map $(H-K)$ is zero and finishes the proof the lemma. □

**Lemma 2.2** Let $(s_0, m_0)$ be a smooth map. Then there exists a positive number $\varepsilon>0$ and a unique local solution $(s,m)\in W^{2,p}(\Omega\times[0,\varepsilon])$ for $p>2$ strictly.

**Proof** Define a nonlinear map $\mathscr{L}: W^{2,p}(\Omega\times[0,\omega])\to L^p(\Omega\times[0,\omega])$ as follows:

$$\mathscr{L}\begin{pmatrix} s\\ m\end{pmatrix} = \begin{bmatrix} A(m)\Delta s - \mathrm{div}(m\otimes J_e) - \beta\,\widetilde{\mathrm{div}}(m\otimes(\nabla s\cdot m)) - s - s\times m\\ \alpha\Delta m + \alpha|\nabla m|^2 m - am\times(m\times s) - m\times(\Delta m + s)\end{bmatrix}.$$

Then for a smooth map $(s,m)$, the derivative of $L$ at $(s,m)$ is given by

$$D\mathscr{L}(s,m)\begin{pmatrix} l\\ k\end{pmatrix} = \begin{bmatrix} A(m)\Delta l + d(s,m)k + e(s,m,J_e)\nabla k + f(m)l + h(m)\nabla l\\ \alpha\Delta k + m\times\Delta k + a(s,m)\nabla k + b(s,m)k + c(m)l\end{bmatrix},$$

where $a,b,c,d,e,f$ are smooth matrix-valued functions. Denote

$$\mathscr{R}\begin{pmatrix} s\\ m\end{pmatrix} = \begin{bmatrix} \partial_t s\\ (1+\alpha^2)\partial_t m\end{bmatrix} - \mathscr{L}\begin{pmatrix} s\\ m\end{pmatrix}.$$

Let $(s_\omega, m_\omega):\Omega\times[0,\ \omega]\to\mathbf{R}^3$ be smooth maps with $(s_\omega, m_\omega)=(s_0,m_0)$ on $\Omega\times 0$, we denote $(s,m)=(s_\omega,m_\omega)+(s_*,\ m_*)$. The derivative of $\mathscr{R}$ at $(s,m)$ is given by

$$D\mathscr{R}(s,m)\begin{pmatrix} l\\ k\end{pmatrix} = \begin{bmatrix} \partial_t l\\ (1+\alpha^2)\partial_t k\end{bmatrix} - D\mathscr{L}(s,m)\begin{pmatrix} l\\ k\end{pmatrix}. \tag{2.4}$$

Consider $(s_*,\ m_*)$ as a variable function, then $(s_*,\ m_*)\to\mathscr{R}\begin{bmatrix} s_\omega + s_*\\ m_\omega + m_*\end{bmatrix}$ defines a continuously differential map of $W^{2,p}(\Omega\times[0,\omega])\to L^p(\Omega\times[0,\omega])$. Its derivative at $[(s_*,$

$m_*)=0]$ is given by (2.4) with $(s,m)$ replaced by $(s_\omega,m_\omega)$ and from the above lemma we know it is an isomorphism from $W^{2,p}(\Omega\times[0,\omega])$ onto $L^p(\Omega\times[0,\omega])$.

Therefore by the inverse function theorem, the set of all $\mathscr{R}\begin{pmatrix} s_\omega+s_* \\ m_\omega+m_* \end{pmatrix}$ for $(s_*,m_*)$ in a neighborhood $\mathscr{N}\subset W^{2,p}(\Omega\times[0,\omega])$ of 0 covers a neighborhood $\mathscr{O}$ of $\mathscr{R}(s_\omega,m_\omega)$ in $L^p(\Omega\times[0,\omega])$. If we choose $\varepsilon>0$ small enough, the function

$$(\tilde{l},\tilde{k})=\begin{cases}(0,0), & 0\leqslant t\leqslant\varepsilon\\ \mathscr{R}(s_\omega,m_\omega), & \varepsilon<t\leqslant\omega\end{cases}$$

will be in this neighborhood $\mathscr{O}$. So $(s,m)|_{\Omega\times[0,\varepsilon]}$ solves the equation (1.1). On the other hand, from our choice $p>2$ and $m\in S^2$, we can show the smoothness of the solution by standard bootstrap method. We stress that $p>2$ here plays an important role in the bootstrap method, because for $p>2$, we have $W^{k,p}(\Omega)\hookrightarrow W^{k-1,\infty}(\Omega)$ in space dimension 2.

## 3 *A Priori* Estimates

In this section, we show some *a priori* estimates for the solution. The following lemma can be found in [6] and we omit its proof here.

**Lemma 3.1** Let $J_e\in(H^1(\mathbf{R}_+\times\Omega))^3$ and $s(x,t)\in L^\infty(\mathbf{R}_+,L^2(\Omega))$ and $m(x,t)\in L^\infty(\mathbf{R}_+,H^1(\Omega))$ be a weak solution to the problem (1.1). Then the solution $(s,m)$ satisfies the following relations:

$$m\in L^\infty(\mathbf{R}_+,H^1(\Omega)),\quad \frac{\partial m}{\partial t}\in L^2(\mathbf{R}_+,L^2(\Omega)),$$

$$s\in L^\infty(\mathbf{R}_+,L^2(\Omega))\cap L^2(\mathbf{R}_+,H^1(\Omega)),\quad \frac{\partial s}{\partial t}\in L^2(\mathbf{R}_+,H^{-1}(\Omega))$$

with the following estimates: for any $T>0$,

$$\sup_{0\leqslant t\leqslant T}\int_\Omega|s(x,t)|^2+\int_0^T\int_\Omega|s|^2+\int_0^T\int_\Omega|\nabla s|^2\leqslant C\int_\Omega|s_0|^2+C\int_0^T\int_\Omega|J_e|^2,\tag{3.1}$$

and

$$\int_0^T\int_\Omega\left|\frac{\partial m}{\partial t}\right|^2+\sup_{0\leqslant t\leqslant T}\int_\Omega|\nabla m(x,t)|^2\leqslant C\int_\Omega|s_0|^2+C\int_0^T\int_\Omega|J_e|^2+C\int_\Omega|\nabla m_0|^2,\tag{3.2}$$

where $C$ only depends on the coefficients $(\alpha,\beta)$ of the equation (1.1), in particular, $C$ is independent of $T>0$.

**Remark 3.2** Because of the highly nonlinear terms and that the embedding $H^1(\Omega)\hookrightarrow L^\infty(\Omega)$ fails for dimension greater than one, we cannot expect more regularity than that stated in Lemma 3.1. One can also see from (3.2) that if the initial data are sufficiently small

$$C\|s_0\|^2+C\|J_e\|^2_{L^2(\mathbf{R}_+\times\Omega)}+C\|\nabla m_0\|^2<\lambda_0,\tag{1.3}$$

one can keep $\|\nabla m(\cdot,t)\|^2<\lambda_0$ for all time $t>0$.

Below we will give the $H^2$ *a priori* estimates. We will use the following equivalent form for $m$,

$$(1+\alpha^2)\frac{\partial m}{\partial t}=-m\times(\Delta m+s)-\alpha m\times(m\times(\Delta m+s)).$$

**Lemma 3.3** Assume that $J_e\in H^2(\mathbf{R}_+\times\Omega)$, $s_0\in H^2(\Omega)$ and $m_0\in H^2(\Omega)$. Then the smooth solution $(s,m)$ satisfies

$$\nabla^2 m\in L^\infty(\mathbf{R}_+,L^2)\text{ and }\nabla^3 m\in L^2(\mathbf{R}_+,L^2),\tag{3.3}$$

provided the smallness condition (1.3) holds. Further more, one can get $\nabla m\in L^2(\mathbf{R}_+,L^\infty)$ by sobolev embedding $H^2\hookrightarrow L^\infty$ in dimension 2.

**Proof** Denote $D^\alpha=\partial_{x_1}^{\alpha_1}\partial_{x_2}^{\alpha_2}$, where $\alpha$ is a multi-index, $\alpha_1,\alpha_2$ are nonnegative integers and $|\alpha|=\alpha_1+\alpha_2$. For simplicity, we will use $D^k$ to denote any kind of the differential operator $D^\alpha$ with $|\alpha|=k$.

Differentiating equation (2.2) with $D^2$, taking inner product with $D^2u$ and integrating over $\Omega$, we have

$$\begin{aligned}\frac{1+\alpha^2}{2}\frac{\mathrm{d}}{\mathrm{d}t}\|D^2m\|^2=&-\int_\Omega\langle D^2(m\times(\Delta m+s)),D^2m\rangle+\\&\alpha\int_\Omega\langle\Delta D^2m,D^2m\rangle+\alpha\int_\Omega\langle D^2(|\nabla m|^2m),D^2m\rangle-\\&\alpha\int_\Omega\langle D^2((m\cdot s)m),D^2m\rangle+\alpha\int_\Omega\langle D^2s,D^2m\rangle\\=&\ I+II+III+IV+V.\end{aligned}\tag{3.4}$$

- Estimates of term $I=:VI+VII$.

$$\begin{aligned}VI&=\left|\int_\Omega\langle D^2(m\times\nabla m),\nabla D^2m\rangle\right|\\&=\left|\int_\Omega\langle D^2m\times\nabla m+2Dm\times D\nabla m+m\times\nabla D^2m,\nabla D^2m\rangle\right|\\&\leqslant C\|\nabla m\|_{L^4(\Omega)}\|\nabla^2m\|_{L^4(\Omega)}\|\nabla^3m\|.\end{aligned}$$

Using Gagliardo-Nirenberg inequality, we have (in dimension $d=2$),

$$\|\nabla m\|_{L^4}\leqslant C\|\nabla m\|_{L^2}^{\frac34}\|\nabla m\|_{H^2}^{\frac14},\tag{3.5}$$

$$\|D^2m\|_{L^4}\leqslant C\|\nabla m\|_{L^2}^{\frac14}\|\nabla m\|_{H^2}^{\frac34}.\tag{3.6}$$

Then writing the $H^2$-norm explicitly and using the $\varepsilon$-Young's inequality, we can bound the term $VI$ by

$$\begin{aligned}|VI|\leqslant&(C\|\nabla m\|_{L^2}+\varepsilon)\|\nabla\Delta m\|_{L^2}^2+\\&C(\|\nabla m\|_{L^2}+\|\nabla m\|_{L^2}^2)\|\Delta m\|_{L^2}^2+C\|\nabla m\|_{L^2}^4,\end{aligned}\tag{3.7}$$

where the constant $C$ can be different from line to line.

$$\begin{aligned}|VII|&=\left|\int_\Omega\langle D(m\times s),D^3m\rangle\right|\\&\leqslant\varepsilon\|D^3m\|^2+C(\varepsilon)\|Ds\|^2+C(\varepsilon)(\|Dm\|^4+\|s\|^4)\end{aligned}$$

$$\leqslant \varepsilon\|D^3 m\|^2 + C(\varepsilon)(\|\nabla m\|^2\|\Delta m\|^2 + \|s\|^2\|\nabla s\|^2 + \|\nabla s\|^2 + \|s\|_{L^2}^4 + \|\nabla m\|_{L^2}^4). \tag{3.8}$$

Since $s, \nabla m \in L^\infty(0, T; L^2)$ and $\nabla s, \Delta m \in L^2(0, T; L^2)$, the last term on the right is integrable on $[0, T]$ for any $T > 0$.

• Estimates of term $II$. Since we consider the periodic case, no boundary terms appears, thus

$$\alpha\int_\Omega \langle \Delta D^2 m, D^2 m\rangle = -\alpha\|\nabla D^2 m\|^2. \tag{3.9}$$

• Estimates of term $III$. Since we have

$$\|D(|\nabla m|^2 m)\| \leqslant \|\nabla m\|_{L^6(\Omega)}^3 + 2\|m\|_{L^\infty(\Omega)}\|\nabla m\|_{L^4(\Omega)}\|D\nabla m\|_{L^4(\Omega)},$$

the term $III$ can be estimated by

$$|III| = \left|\int_\Omega \langle D(|\nabla m|^2 m), D^3 m\rangle\right|$$
$$\leqslant (C\|\nabla m\| + \varepsilon)\|\nabla\Delta m\|^2 + C(\|\nabla m\|_{L^2} + \|\nabla m\|_{L^2}^2)\|\Delta m\|_{L^2}^2 + C\|\nabla m\|_{L^2}^4, \tag{3.10}$$

thanks to the Gagliardo-Nirenberg inequality (GN).

• Estimates of the last two terms $IV$ and $V$.

$$|IV| + |V| \leqslant \varepsilon\|D^3 m\|^2 + C(\varepsilon)(\|D((m\cdot s)m)\|_{L^2}^2 + \|Ds\|_{L^2}^2)$$
$$\leqslant \varepsilon\|D^3 m\|^2 + C(\varepsilon)(\|\nabla s\|_{L^2}^2 + \|s\|_{L^2}^4 + \|\nabla m\|_{L^2}^4)$$
$$\leqslant \varepsilon\|D^3 m\|^2 + C(\varepsilon)(\|\nabla s\|_{L^2}^2 + \|s\|_{L^2}^4 + \|s\|_{L^2}^2\|\nabla s\|_{L^2}^2 + \|\nabla m\|_{L^2}^4 + \|\nabla m\|_{L^2}^2\|\nabla^2 m\|_{L^2}^2), \tag{3.11}$$

where in the last inequality, we used the Gagliardo-Nirenberg inequality (GN). Summarizing (3.4)~(3.11), we have

$$\frac{1+\alpha^2}{2}\frac{\mathrm{d}}{\mathrm{d}t}\|\nabla^2 m\|_{L^2(\Omega)}^2 + \alpha\|\nabla^3 m\|_{L^2(\Omega)}^2$$
$$\leqslant (C\|\nabla m\|_{L^2} + 4\varepsilon)\|\nabla^3 m\|_{L^2}^2 + C(\|\nabla m\|_{L^2} + \|\nabla m\|_{L^2}^2)\|\nabla^2 m\|_{L^2}^2 + C(\|\nabla m\|_{L^2}^4 + \|\nabla s\|_{L^2}^2 + \|s\|_{L^2}^4 + \|s\|_{L^2}^2\|\nabla s\|_{L^2}^2). \tag{3.12}$$

By Remark 3.2, one can keep $\|\nabla m(t)\|_{L^2(\Omega)}^2 \leqslant \lambda_0$ for all time $t > 0$ provided (1.3) holds. Thus there exists a $\lambda_0 > 0$ such that if (1.3) holds, we have

$$C\|\nabla m\|_{L^2} < \frac{\alpha}{4}.$$

Furthermore, set $4\varepsilon < \frac{\alpha}{4}$, we then have

$$C\|\nabla m\|_{L^2} + 4\varepsilon < \frac{\alpha}{2}.$$

This together with (3.12) insures us to deduce the Gronwall-type inequality for $\|\nabla^2 m\|_{L^2(\Omega)}^2$:

$$\frac{\mathrm{d}}{\mathrm{d}t}\|\nabla^2 m\|_{L^2(\Omega)}^2 + \frac{\alpha}{2}\|\nabla^3 m\|_{L^2(\Omega)}^2$$
$$\leqslant C(\|\nabla m\|_{L^2} + \|\nabla m\|_{L^2}^2)\|\nabla^2 m\|_{L^2}^2 +$$

$$C(\|\nabla m\|_{L^2}^4 + \|\nabla s\|_{L^2}^2 + \|s\|_{L^2}^4 + \|s\|_{L^2}^2\|\nabla s\|_{L^2}^2). \tag{3.13}$$

Since both the coefficient of $\|\nabla^2 m\|_{L^2(\Omega)}^2$ and the second term on the right hand side are integrable by Lemma 3.1, we deduce by Gronwall inequality that for any $T>0$,

$$\sup_{0\leqslant t\leqslant T}\|\nabla^2 m\|_{L^2(\Omega)}^2 \leqslant C, \quad \forall T > 0. \tag{3.14}$$

Then integrating (3.13), we have

$$\|\nabla^3 m\|_{L^2(0,T;L^2(\Omega))}^2 \leqslant C, \quad \forall T > 0, \tag{3.15}$$

and by Sobolev embedding $H^2(\Omega)\hookrightarrow L^\infty(\Omega)$, we get $\nabla m\in L^2(0,T;L^\infty(\Omega))$. □

**Remark 3.4** It can be easily seen from this lemma and the expression (2.2), that $\partial_t m\in L^\infty(0,\mathbf{R}_+;L^2(\Omega))$ for any positive times $T>0$. Furthermore, differentiating equation (2.2) with respect to spatial variable $x$, we have $\partial_t\nabla m \in L^2(0,\mathbf{R}_+;L^2(\Omega))$. This observation is rather simple, thus we omit the details here.

Below, we will focus ourselves on the estimates for the $s$ variable.

**Lemma 3.5** Under the conditions of Lemma 3.3, we have

$$\frac{\partial s}{\partial t} \in L^2(\mathbf{R}_+,L^2(\Omega)), \quad s \in L^\infty(\mathbf{R}_+,H^1(\Omega))$$

with the estimation: for any $T>0$,

$$\begin{aligned}&\int_0^T\left\|\frac{\partial s}{\partial t}\right\|_{L^2(\Omega)}^2 + \sup_{0\leqslant t\leqslant T}\|s\|_{H^1(\Omega)}^2\\ &\leqslant Ce^{C\|\nabla m\|_{L^2(0,T;L^\infty)}^2}\Big(\|s_0\|_{H^1(\Omega)}^2 + \sup_{0\leqslant t\leqslant T}\|\nabla m\|_{L^2(\Omega)}^2\|\nabla m\|_{L^2(0,T;L^\infty(\Omega))}^2 +\\ &\|\nabla J_e\|_{L^2([0,T]\times\Omega)}^2\Big).\end{aligned} \tag{3.16}$$

**Proof** Taking the inner product of $s$-equation with $\frac{\partial s}{\partial t}$, and then integrating over $\Omega$, we get

$$\int_\Omega\left|\frac{\partial s}{\partial t}\right|^2 + \frac{1}{2}\frac{\mathrm{d}}{\mathrm{d}t}\int_\Omega|s|^2 = -\int_\Omega\langle \operatorname{div} J_s,\frac{\partial s}{\partial t}\rangle - \int_\Omega\langle s\times m,\frac{\partial s}{\partial t}\rangle =: I + II . \tag{3.17}$$

Since the second term on the right can be easily controlled by

$$|II| \leqslant \varepsilon\int_\Omega\left|\frac{\partial s}{\partial t}\right|^2 + C(\varepsilon)\int_\Omega|s|^2, \tag{3.18}$$

we need only to handle the first term $I$ on the right hand side carefully. Using the notations introduced in Section 2, we can rewrite $I$ as follow:

$$I = -\int_\Omega\langle\operatorname{div}(m\otimes J_e),\frac{\partial s}{\partial t}\rangle + \int_\Omega\langle A(m)\Delta s,\frac{\partial s}{\partial t}\rangle - \beta\int_\Omega\langle\widetilde{\operatorname{div}}(m\otimes(\nabla s\cdot m)),\frac{\partial s}{\partial t}\rangle. \tag{3.19}$$

The first and the last term on the right side of (3.19) can be estimated by

$$\left|\int_\Omega\langle\operatorname{div}(m\otimes J_e),\frac{\partial s}{\partial t}\rangle\right| \leqslant \varepsilon\int_\Omega\left|\frac{\partial s}{\partial t}\right|^2 + C(\varepsilon)\left(|J_e|_{L^\infty(\Omega)}^2\int_\Omega|\nabla m|^2 + \int_\Omega|\nabla J_e|^2\right), \tag{3.20}$$

and

$$\left|\int_\Omega \langle \widetilde{\operatorname{div}}(m\otimes(\nabla s\cdot m)),\frac{\partial s}{\partial t}\rangle\right|\leqslant \varepsilon\int_\Omega\left|\frac{\partial s}{\partial t}\right|^2+C(\varepsilon)|\nabla m|_{L^\infty}^2\int_\Omega|\nabla s|^2 \tag{3.21}$$

respectively, where $\nabla m\in L^\infty(0,T;L^2(\Omega))$ and $J_e\in H^2(\mathbf{R}_+\times\Omega)$.

In the following, we focus on estimates of the second term of (3.19). For this purpose, write

$$A\Delta s = A\,\nabla\cdot\nabla s = \nabla\cdot(A\,\nabla s)-(\nabla\cdot A)\cdot\nabla s.$$

Then the second term on the right of (3.19) can be rewritten as

$$\begin{aligned}\int_\Omega\langle A\Delta s,\frac{\partial s}{\partial t}\rangle &= -\frac{1}{2}\frac{\mathrm{d}}{\mathrm{d}t}\int_\Omega\langle A\,\nabla s,\nabla s\rangle-\frac{1}{2}\int_\Omega\langle(\partial_t A)\,\nabla s,\nabla s\rangle-\int_\Omega\langle\nabla\cdot A\cdot\nabla s,\partial_t s\rangle\\ &=: \mathrm{III}+\mathrm{IV}+\mathrm{V}.\end{aligned}$$

• Since $A$ is strictly positively definite, we can write III as

$$\mathrm{III} = -\frac{1}{2}\frac{\mathrm{d}}{\mathrm{d}t}\int_\Omega\langle A^{1/2}\,\nabla S,A^{1/2}\,\nabla S\rangle. \tag{3.22}$$

• Estimates of term IV. From the positivity of $A$ in (2.1), we know that for any vector $\xi\in\mathbf{R}^3$, we have $\lambda^2|\xi|^2\leqslant|A\xi|^2$. This together with the $s$-equation in (2.3),

$$A\Delta s = \frac{\partial s}{\partial t}+\operatorname{div}(m\otimes J_e)+\beta\,\widetilde{\operatorname{div}}(m\otimes(\nabla s\cdot m))+s+s\times m \tag{3.23}$$

implies

$$\begin{aligned}\lambda^2\|\Delta s\|_{L^2}^2&\leqslant\|A\Delta s\|_{L^2}^2\\ &\leqslant\left\|\frac{\partial s}{\partial t}\right\|_{L^2}^2+2\|s\|_{L^2}^2+C\|\nabla m\|_{L^2}^2+\|\nabla J_e\|_{L^2}^2+C\|\nabla m\|_{L^\infty}^2\|\nabla s\|_{L^2}^2.\end{aligned}$$

On the other hand, by Lemma 3.1, Lemma 3.3 and (2.2), we can choose $\varepsilon_1$ small enough to satisfy $\varepsilon_1\|m_t\|_{L^2(\Omega)}<\frac{1}{8}\lambda^2$ for all $t\geqslant0$. Therefore, we can estimate IV by

$$\begin{aligned}|\mathrm{IV}|&\leqslant\int_\Omega|m_t|\,|\nabla s|^2\leqslant\|m_t\|_{L^2}\|\nabla s\|_{L^4}^2\\ &\leqslant C\|m_t\|_{L^2}\|\nabla s\|_{H^1}\|\nabla s\|_{L^2}\\ &\leqslant\varepsilon_1\|m_t\|_{L^2}\|\Delta s\|_{L^2}^2+C(\varepsilon_1)\|m_t\|_{L^2}\|\nabla s\|_{L^2}^2\\ &\leqslant\frac{1}{8}\left\|\frac{\partial s}{\partial t}\right\|_{L^2}^2+(\|s\|_{L^2}^2+C\|\nabla m\|_{L^2}^2+\|\nabla J_e\|_{L^2}^2)+\\ &\quad C(\|\nabla m\|_{L^\infty}^2+\|m_t\|_{L^2})\|\nabla s\|_{L^2}^2.\end{aligned}$$

• Estimates of term V.

$$\|\mathrm{V}\|\leqslant\varepsilon\left\|\frac{\partial s}{\partial t}\right\|_{L^2}^2+C(\varepsilon)\|\nabla m\|_{L^\infty}^2\|\nabla s\|_{L^2}^2. \tag{3.24}$$

Finally, since $A$ is positively definite and can be bounded from above and below by

$$\sqrt{\lambda}\,\|\nabla s\|_X\leqslant\|A^{1/2}\,\nabla s\|_X\leqslant\sqrt{\Lambda}\,\|\nabla s\|_X, \tag{3.25}$$

we can regard $\|A^{1/2}\nabla s\|_X$ and $\|\nabla s\|_X$ as the same in the sequel. From (3.17) and the above estimates, if we set $\varepsilon$ small enough, say $\varepsilon<\frac{1}{8}$, we can get the Gronwall inequality:

$$\begin{aligned}&\frac{1}{2}\int_{\Omega}\left|\frac{\partial s}{\partial t}\right|^2+\frac{1}{2}\frac{\mathrm{d}}{\mathrm{d}t}\int_{\Omega}|s|^2+\frac{\mathrm{d}}{\mathrm{d}t}\int_{\Omega}|\nabla s|^2\\&\leqslant C(\|\nabla m\|_{L^\infty}^2+\|m_t\|_{L^2}+\|m_t\|_{L^2}\|\nabla m\|_{L^\infty}^2)\|\nabla s\|_{L^2}^2+C\|s\|_{L^2}^2+\\&\quad C(\|J_e\|_{L^\infty}^2\|\nabla m\|_{L^2}^2+\|\nabla J_e\|_{L^2}^2+\|m_t\|_{L^2}\|s\|_{L^2}^2+\|m_t\|_{L^2}\|\nabla m\|_{L^2}^2).\end{aligned}\tag{3.26}$$

By the previous lemmas, $|\nabla m|\in L^2(0,T;L^\infty),\nabla m\in L^\infty(0,T;L^2)$ and $|m_t|\in L_t^\infty L_x^2\cap L_t^2L_x^2$, thus both the coefficients before $\|s\|_{L^2}^2$ and $\|\nabla s\|_{L^2}^2$ and the last term are integrable in time. Then the Gronwall inequality immediately implies

$$\frac{\partial s}{\partial t}\in L^2(\mathbf{R}_+,L^2),\quad s\in L^\infty(\mathbf{R}_+,H^1(\Omega)),\tag{3.27}$$

with the estimates (3.16). This concludes the proof. □

**Lemma 3.6** Under the conditions of Lemma 3.3, we have $s\in L^2(\mathbf{R}_+,H^2(\Omega))$. Indeed, by Lemma 3.1 and Lemma 3.5, we get that the right hand side of (3.23) belongs to $L^2(\Omega)$ for a.e. $0\leqslant t\leqslant T$, which implies that $\Delta s\in L^2(\Omega)$ for a.e. $0\leqslant t\leqslant T$. On the other hand

$$\begin{aligned}\|\widetilde{\mathrm{div}}(m\otimes(\nabla s\cdot m))\|_{L^2(0,T;L^2)}^2&\leqslant C\int_0^T\|\nabla m\|_{L^\infty}^2\|\nabla s\|_{L^2}^2\mathrm{d}t\\&\leqslant C\|\nabla m\|_{L^2(0,T;L^\infty)}^2\sup_{0\leqslant t\leqslant T}\|\nabla s\|_{L^2}^2.\end{aligned}$$

By equation (3.23), we immediately have $\|\Delta s\|_{L^2(\Omega)}\in L^2(0,T)$ for any $T>0$. We can also get $s\in L^2(\mathbf{R}_+,L^\infty(\Omega))$ by embedding $H^2\hookrightarrow L^\infty$ in dimension 2.

However we can expect more about the regularity of the $s$-variable. In the following lemma, we improve the regularity of the $s$-variable.

**Lemma 3.7** Suppose that $J_e$ is smooth, and $(s_0,m_0)\in H^2(\Omega)\times H^2(\Omega)$ is small in the sense of (1.3), then for any $T>0$, the solution of (1.1) satisfy

$$\partial_t s\in L^\infty(0,T;L^2),\quad \partial_t\nabla s\in L^2(0,T;L^2),\tag{3.28}$$

$$s\in L^\infty(0,T;H^2),\quad s\in L^2(0,T;H^3).\tag{3.29}$$

Also, for $s''$, we have

$$s''\in L^2(0,T;H^{-1}(\Omega)).\tag{3.30}$$

**Proof** (1) Differentiating the $s$-equation with respect to $t$, taking inner product with $\partial_t s$ and then integrating over $\Omega$, we obtain

$$\frac{1}{2}\frac{\mathrm{d}}{\mathrm{d}t}\|\partial_t s\|^2=\int_{\Omega}\langle\partial_t J_s,\partial\nabla s\rangle-\|\partial_t s\|^2-\int_{\Omega}\langle s\times\partial_t m,\partial_t s\rangle.\tag{3.31}$$

The last term can be estimated as

$$\left|\int_{\Omega}\langle s\times\partial_t m,\partial_t s\rangle\right|\leqslant C\|\partial_t s\|^2+C\|\partial_t m\|^2\|s\|_{L^\infty(\Omega)}^2,\tag{3.32}$$

where the second term on the right is integrable on $[0,T]$ since $s\in L^2(\mathbf{R}_+,L^\infty(\Omega))$ by Lemma 3.6 and $\partial_t m\in L^\infty(0,T;L^2(\Omega))$ by Remark 3.2. The first term on the right of

(3.31) can be estimated as

$$\begin{aligned}\int_\Omega \langle \partial_t J_s, \partial_t \nabla s\rangle &= \int_\Omega \langle \partial_t (m \otimes J_e), \partial_t \nabla s\rangle - \|\partial_t \nabla s\|^2 + \\ &\quad \beta \int_\Omega \langle \partial_t (m \otimes (\nabla s \cdot m)), \partial_t \nabla s\rangle \\ &\leqslant -(1-\beta-\varepsilon)\|\partial_t \nabla_s\|^2 + C(\varepsilon)\|\partial_t (m \otimes J_e)\|^2 + \\ &\quad C(\varepsilon)\|\partial_t m\|_{L^2}^2 \|\partial_t m\|_{H^1}^2 + C(\varepsilon)\|\nabla s\|^2 \|\nabla s\|_{H^1}^2,\end{aligned} \tag{3.33}$$

where in the last step, we used the Gagliardo-Nirenberg inequality, and $\|\partial_t (m \otimes J_e)\|^2$ can be bounded by

$$\|\partial_t (m \otimes J_e)\|^2 \leqslant \|\partial_t J_e\|^2 + C\|J_e\|_{L^\infty(\Omega)}^2 \|\partial_t m\|^2. \tag{3.34}$$

Combining (3.31)~(3.34), and setting $\varepsilon$ small enough, say $\varepsilon = \frac{1-\beta}{2}$, we have

$$\begin{aligned}&\frac{1}{2}\frac{\mathrm{d}}{\mathrm{d}t}\|\partial_t s\|^2 + \delta\|\partial_t \nabla s\|^2 + \frac{1}{2}\|\partial_t s\|^2 \\ &\leqslant C\|\partial_t s\|^2 + C(\|\partial_t m\|^2 \|s\|_{L^\infty(\Omega)}^2 + \|\partial_t J_e\|^2 + \|J_e\|_{L^\infty(\Omega)}^2 \|\partial_t m\|^2 + \\ &\quad \|\partial_t m\|^2 \|\partial_t m\|_{H^1}^2 + \|\nabla s\|^2 \|\nabla s\|_{H^1}^2),\end{aligned} \tag{3.35}$$

where $\delta = (1-\beta) - \varepsilon > 0$ and the terms in the parentheses $(\cdots)$ is integrable due to the above lemmas. Applying the Gronwall lemma and integrating on $[0, T]$, we have (3.28).

(2) Differentiating the $s$-equation in (1.1) with $\Delta$, taking inner product with $\Delta s$ and integrating over $\Omega$, we have

$$\frac{1}{2}\frac{\mathrm{d}}{\mathrm{d}t}\|\Delta s\|^2 = \int \langle \Delta J_s, \nabla \Delta s\rangle - \|\Delta s\|^2 - \int \langle \nabla(s \times m), \nabla \Delta s\rangle. \tag{3.36}$$

The last term of the right hand side can be controlled by

$$\left|\int \langle \nabla(s \times m), \nabla \Delta s\rangle\right| \leqslant \varepsilon\|\nabla \Delta s\|^2 + C(\varepsilon)(\|\nabla s\|^2 + \|s\|_{L^\infty}^2 \|\nabla m\|^2), \tag{3.37}$$

where the second term on the right of which is integrable on $[0, T]$ thanks to Lemma 3.6. For the first term on the right of (3.36), we have

$$\begin{aligned}&\int \langle \Delta J_s, \nabla \Delta s\rangle \\ &= \int \langle \Delta(m \otimes J_e, \nabla \Delta s)\rangle - \|\nabla \Delta s\|^2 + \beta \int \langle \Delta(m \otimes (\nabla s \cdot m)), \nabla \Delta s\rangle \\ &=: I + II + III.\end{aligned} \tag{3.38}$$

For $I$, we have

$$\begin{aligned}|I| &\leqslant \varepsilon\|\nabla \Delta s\|^2 + C(\varepsilon)(\|J_e\|_{L^\infty}^2 \|\Delta m\|^2 + \|\Delta J_e\|^2) + C(\varepsilon)(\|\nabla m\|_{L^4}^4 + \|\nabla J_e\|_{L^4}^4) \\ &\leqslant \varepsilon\|\nabla \Delta s\|^2 + C(\varepsilon)(\|J_e\|_{L^\infty}^2 \|\Delta m\|^2 + \|\Delta J_e\|^2) + \\ &\quad C(\varepsilon)(\|\nabla m\|^2 \|\nabla m\|_{H^1}^2 + \|\nabla J_e\|^2 \|\nabla J_e\|_{H^1}^2),\end{aligned} \tag{3.39}$$

where the last two terms are integrable on $[0, T]$.

For $III$, we have

$$|III| \leqslant (\beta+\varepsilon)\|\nabla\Delta s\|^2 + C(\varepsilon)(\|\nabla m\|_{L^\infty}^2\|\Delta s\|^2 + \|\nabla s\|^2\|\nabla s\|_{H^1}^2 +$$
$$\|\Delta m\|^2\|\Delta m\|_{H^1}^2 + \|\nabla m\|_{L^\infty}^2(\|\nabla s\|^2\|\nabla s\|_{H^1}^2 + \|\nabla m\|^2\|\nabla m\|_{H^1}^2)). \quad (3.40)$$

Setting $\varepsilon$ small enough, say $\varepsilon = \frac{1-\beta}{4}$, from (3.36) ~ (3.40), we have

$$\frac{1}{2}\frac{\mathrm{d}}{\mathrm{d}t}\|\Delta s\|^2 + \delta\|\nabla\Delta s\|^2$$
$$\leqslant C(\|\nabla m\|_{L^\infty}^2 + \|\nabla s\|^2 + \|\nabla m\|_{L^\infty}^2\|\nabla s\|^2)\|\Delta s\|_{L^2}^2 + R, \quad (3.41)$$

where $\delta = (1-\beta) - 3\varepsilon > 0$ and the remainder term $R$ is integrable on $[0, T]$. Thus the Gronwall inequality implies our result (3.29).

(3) Prove (3.30). Differentiating the $s$-equation with respect to $t$, we get

$$\frac{\partial^2 s}{\partial t^2} = -\operatorname{div}\partial_t J_s - \partial_t s - \partial_t(s \times m). \quad (3.42)$$

Then for any $v \in H_0^1(\Omega)$, we have

$$\left|\int\langle\partial_t s, v\rangle\right| \leqslant \|\partial_t s\|\|v\|, \quad (3.43)$$

$$\left|\int\langle\partial_t(s\times m), v\rangle\right| \leqslant (\|\partial_t s\| + \|s\|_{L^\infty(\Omega)}\|\partial_t m\|)\|v\|. \quad (3.44)$$

For the first term on the right of (3.42), we have

$$-\int\langle\operatorname{div}\partial_t J_s, v\rangle = \int\langle\partial_t J_s, \nabla v\rangle, \quad (3.45)$$

which can be bounded by

$$\left|\int\langle\partial_t J_s, \nabla v\rangle\right| \leqslant C\|\nabla v\|_{L^2} \quad (3.46)$$

thanks to the above Lemma 3.5, Remark 3.4 and the first two parts of this lemma. Thus $s'' \in L^2(0, T; H^{-1})$ for all $T > 0$. This completes the proof of the lemma. □

Finally, These *a priori* estimates imply well-posedness of the problem in $H^2$. We can do this by approximating the initial data $(s_0, m_0)$ by smooth data $(s_0^l, m_0^l) \in C^\infty$ such that

$$(s_0^l, m_0^l) \to (s_0, m_0) \quad \text{in } H^2 \times H^2$$

as $l \to \infty$. Thus the *a priori* estimates imply a uniform (in $l$) $H^2$ bound on the approximating solutions $(s^l, m^l)$. Therefore we can extract a subsequence (if necessary) $\{(s^l, m^l)\}$ which converges weakly in $H^2$ to an $H^2$ solution $(s, m)$ of the problem. The *a priori* estimates also implies that the solution is global.

For the reader's convenience, we summarize the above lemmas in the following theorem.

**Theorem 3.8** Suppose that $J_e \in H^2(\mathbf{R}_+ \times \Omega)$. Let $(s_0, m_0) \in H^2(\Omega)$ satisfy the smallness condition (1.3). Then there exists a global solution in $H^2$ such that for any $T > 0$, we have

$$m \in L^\infty(0, T; H^2), \quad m \in L^2(0, T; H^3),$$
$$\partial_t m \in L^\infty(0, T; L^2), \quad \partial_t \nabla m \in L^2(0, T; L^2);$$

for $s$,

$$s \in L^\infty(0, T; H^2), \quad s \in L^2(0, T; H^3),$$

$$\partial_t s \in L^\infty(0,T;L^2),\quad \partial_t \nabla s \in L^2(0,T;L^2);$$

and for $s''$, we have $s'' \in L^2(0,T;H^{-1}(\Omega))$.

# 4 Conclusions

In the previous section, we obtained some *a priori* estimates of the solution to ensure that the solution is in $L^\infty(0,T;H^2(\Omega))$ for all positive times $T>0$ for given initial data in $H^2(\Omega)$. In this section, we provide the higher order *a priori* estimates of the solution of (1.1) to show that the solution is indeed smooth. In the sequel, we will abuse the notation of $D$ with $\nabla$ and we get our result by induction.

**Lemma 4.1** Let $k \geqslant 3$. Suppose $J_e \in H^k(\mathbf{R}_+ \times \Omega)$, and $(s,m)$ is a smooth solution of the equation (1.1) with initial data $(s_0, m_0)$. Furthermore, if the smallness condition (1.3) holds, then

$$m \in L^\infty(0,T;H^k(\Omega)),\quad m \in L^2(0,T;H^{k+1}(\Omega))$$
$$s \in L^\infty(0,T;H^k(\Omega)),\quad s \in L^2(0,T;H^{k+1}(\Omega)).$$

**Proof** We prove this lemma by induction. Let us first assume that for the initial data $s_0, m_0 \in H^{k-1}$, we have proved our result, i.e.,

$$m \in L^\infty(0,T;H^{k-1}(\Omega)),\quad m \in L^2(0,T;H^k(\Omega))$$
$$s \in L^\infty(0,T;H^{k-1}(\Omega)),\quad s \in L^2(0,T;H^k(\Omega)).$$

We divide this proof into two parts and we omit the constant $\alpha$, which will not influence the result.

(1) In this part, we prove the regularity for $m$. Differentiating (2.2) with $D^k$, and taking inner product with $D^k m$, then integrating over $\Omega$ leads us to

$$\begin{aligned}
&\frac{1}{2}\frac{\mathrm{d}}{\mathrm{d}t}\|D^k m\|^2 + \|\nabla D^k m\|^2 \\
=&\langle D^k(m\times\nabla m),\nabla D^k m\rangle - \langle D^k(m\times s),D^k m\rangle + \\
&\langle D^{k-1}(|\nabla m|^2 m),D^{k+1}m\rangle - \langle D^{k-1}((m\cdot s)m),D^{k+1}m\rangle + \langle D^k s,D^k m\rangle,
\end{aligned}\tag{4.1}$$

where $\langle\cdot,\cdot\rangle$ denotes the inner product over $\Omega$. Below we will bound the terms one by one. The last term can be estimated

$$|\langle D^k s, D^k m\rangle| \leqslant \|D^k s\|^2 + \|D^k m\|^2,$$

where the first term on the right is integrable on $[0,T]$. Recalling Kato's inequality, we have

$$\|D^l(fg) - fD^l g\|_{L^p} \leqslant C\|Df\|_{L^\infty}\|D^{l-1}g\|_{L^p} + C\|D^l f\|_{L^p}\|g\|_{L^\infty}$$

for any two functions $f$ and $g$, see [8]. Then the second term can be estimated by

$$\begin{aligned}
&|\langle D^k(m\times s),D^k m\rangle| \\
\leqslant&\|D^k m\|^2 + \|D^k(m\times s)\|^2 \\
\leqslant&C\|Ds\|_{L^\infty}^2\|D^{k-1}m\|^2 + C\|D^k s\|^2 + C(\|s\|_{L^\infty}^2+1)\|D^k m\|^2,
\end{aligned}\tag{4.2}$$

where both the coefficients of $\|D^k m\|^2$ and the first two terms are integrable on $[0,T]$ by the induction assumption. The fourth term can be estimated the same. Since

$$D^k(m\times\nabla m)=D^km\times\nabla m+m\times D^k\nabla m+\sum_{h=1}^{k-1}c_hD^hm\times D^{k-h}\nabla m,\qquad(4.3)$$

the first term can be estimated by

$$\begin{aligned}&|\langle D^k(m\times\nabla m),\nabla D^km\rangle|\\&\leqslant|\langle D^km\times\nabla m,\nabla D^km\rangle|+\sum_{h=1}^{k-1}c_h|\langle D^hm\times D^{k-h}\nabla m,\nabla D^km\rangle|\\&\leqslant\frac{1}{4}\|\nabla D^km\|^2+C\|\nabla^km\|^2+R,\end{aligned}$$

where the remainder term $R$ depends on the lower order derivatives of $m$ and is integrable on $[0,T]$.

Finally from (4.2), we have

$$\begin{aligned}&\|D^{k-1}(|\nabla m|^2m)\|_{L^2}\\&\leqslant\|mD^{k-1}(|\nabla m|^2)\|+C\|\nabla m\|_{L^\infty}\|D^{k-2}|\nabla m|^2\|+C\|D^{k-1}m\|\,\||\nabla m|^2\|_{L^\infty}\\&\leqslant C_1\|\nabla^km\|+C_2.\end{aligned}$$

Then the third term of (4.1) can be estimated by

$$|\langle D^{k-1}(|\nabla m|^2m),D^{k+1}m\rangle|\leqslant\frac{1}{4}\|\nabla D^km\|^2+C\|\nabla^km\|^2+C.\qquad(4.4)$$

Summarizing (4.1)~(4.4), we arrive at the Gronwall type inequality

$$\frac{\mathrm{d}}{\mathrm{d}t}\|D^km\|^2+\|D^{k+1}m\|^2\leqslant C\|\nabla^km\|^2+C,$$

from which we deduce that for all $T>0$,

$$\sup_{0\leqslant t\leqslant T}\|\nabla^km(\cdot,t)\|_{L^2(\Omega)}\leqslant C,\quad|\nabla^{k+1}m|_{L^2(0,T;L^2(\Omega))}\leqslant C.$$

Furthermore, we have $\|\nabla m\|_{L^\infty(\Omega)}\leqslant C$.

(2) In this part, we prove the regularity for $s$. This is similar with that in the first part. Differentiating the $s$ equation with $D^k$, and then taking inner product with $D^ks$ over $\Omega$, we have

$$\frac{1}{2}\frac{\mathrm{d}}{\mathrm{d}t}\|D^ks\|^2=\langle D^kJ_s,\nabla D^ks\rangle-\|D^ks\|^2-\langle D^k(s\times m),D^ks\rangle=\mathrm{I}+\mathrm{II}+\mathrm{III}.\qquad(4.5)$$

• Estimates of $\mathrm{I}$. We expand $\mathrm{I}$ as

$$\langle D^kJ_s,\nabla D^ks\rangle+\|\nabla D^ks\|^2=\langle D^k(m\times J_e),\nabla D^ks\rangle+\beta\langle D^k(m\otimes(\nabla s\cdot m)),\nabla D^ks\rangle,$$

where the terms on right hand side can be estimated by

$$|\langle D^k(m\times J_e),\nabla D^ks\rangle|\leqslant\varepsilon\|\nabla D^ks\|^2+C\|D^k(m\times J_e)\|^2\leqslant\varepsilon\|\nabla D^ks\|^2+C$$

and

$$|\beta\langle D^k(m\otimes(\nabla s\cdot m)),\nabla D^ks\rangle|\leqslant(\beta+\varepsilon)\|\nabla D^ks\|^2+C\|D^ks\|^2+C.$$

• Estimates of $\mathrm{III}$.

$$|\mathrm{III}|\leqslant\|D^ks\|^2+C,\qquad(4.6)$$

where $C$ is integrable on $[0,T]$.

Letting $\varepsilon$ sufficiently small, we can deduce from (4.5) ~ (4.6) the Gronwall type inequality

$$\frac{\mathrm{d}}{\mathrm{d}t}\|D^k s\|^2 + \|\nabla D^k s\|^2 \leqslant C\|D^k s\|^2 + C,$$

from which we can deduce

$$\nabla^k s \in L^\infty(0,T;L^2(\Omega)), \quad \nabla^{k+1} s \in L^2(0,T;L^2(\Omega)),$$

which concludes the proof. □

One can also give the regularity in time as what we did in Section 3, which is standard and we omit the details here. The local existence theorem in Section 2 and the *a priori* estimates then imply global existence of solutions for system (1.1) in $H^k$. In particular, the above estimates hold for any $k \geqslant 2$. Thus if the initial data is smooth, we have global smooth solutions as stated in Theorem 1.1.

## References

[1] Alouges F, Soyeur A. On global weak solutions for Landau-Lifshitz equations: existence and nonuniqueness. Nonlinear Analysis TMA, 1992, 18(11): 1071 - 1084.

[2] Chen Y. The weak solutions to the evolution problems of harmonic maps. Math Z, 1989, 201: 69 - 74.

[3] Ding S, Guo B. Hausdorff measure of the singular set of Landau-Lifshitz equations with a nonlocal term. Commun Math Phys, 2004, 250: 95 - 117.

[4] Ding S, Guo B, Lin J, et al. Global existence of weak solutions for Landau-Lifshitz-Maxwell equations. Discre Contin Dynam Syst, 2007, 17(4): 867 - 890.

[5] Friedman A. Partial differential equations of parabolic type. [S.l.]: Prentice-Hall, 1964.

[6] Garcia-Cervera C J, Wang X P. Spin-polarized transport: existence of weak solutions. Discre Contin Dynam Syst Series B, 2007, 7: 87 - 100.

[7] Guo B, Han Y. Global smooth solution of hydrodynamical equation for the Heisenberg paramagnet. Math Meth Appl Sci, 2004, 27: 181 - 191.

[8] Guo B, Hong M. The Landau-Lifshitz equation of the ferromagnetic spin chain and harmonic maps. Calc Var, 1993, 1: 311 - 334.

[9] Guo B, Pu X. Well-posedness of the ferrimagnetic equations. J Math Anal Appl, 2008, 339: 312 - 323.

[10] Hamilton R. Harmonic maps of manifold with boundary. Lect Notes Math, Vol 471. Berlin: Springer, 1975.

[11] Landau L, Lifshitz E M. On the theory of the dispersion of magnetic permeability in ferroagnetic bodies. Phsy Z Sowj, 1935, 8: 153.

[12] Moser R. Partial regularity for the Landau-Lifshitz equation in small dimensions. MPI preprint 26, 2002.

[13] Shpiro A, Levy P M, Zhang S. Self-consistent treatment of nonequlibrium spin torques in magnetic multilayers. Phys Rev B, 2003, 67: 104430.

[14] Zhang S, Levy P M, Fert A. Mechanism of spin-polarized current-driver magnetization switching. Phys Rev Lett, 2002, 88: 236601.

# Existence of Partially Regular Weak Solutions to Landau-Lifshitz-Maxwell Equations *

Ding Shijin(丁时进) Guo Boling(郭柏灵)

**Abstract**

In this paper, the authors establish the existence of partially regular weak solutions to the Landau-Lifshitz equations coupling with static Maxwell systems in 3 dimensions by Ginzburg-Landau approximation. It is proved that the Hausdorff measure of the singular set is locally finite. This extends the similar results of Ding and Guo[14] from the stationary solutions to weak solutions and the results of Wang[36] from Landau-Lifshitz equations to Landau-Lifshitz-Maxwell equations.

**Keywords** Landau-Lifshitz equations; maxwell equations; weak solutions; partial regularity

## 1 Introduction

Minimizers of the total micromagnetic energy

$$E = d^2\int_\Omega |\nabla u|^2 + Q\int_\Omega \phi(u) - \int_\Omega u \cdot H - 2\int_\Omega u \cdot H_0 \tag{1.1}$$

capture most features of the microstructure of ferromagnetic materials such as domain patterns, domain wall structures. In functional (1.1), $u = (u_1, u_2, u_3): \Omega \to S^2$ denotes the magnetic moment, $H$ is the induced field, $H_0$ is the applied field, $d$ is called exchange length, $Q$ is the anisotropy constant, $\Omega \subset \mathbf{R}^3$, a smooth bounded domain, stands for the ferromagnetic material. The component $d^2\int_\Omega |\nabla u|^2$ is called exchange energy, $Q\int_\Omega \phi(u)$ the anisotropy energy with polynomial $\phi(u)$, $\int_\Omega u \cdot H$ the magnetostatic energy induced by the demagnetization field, $2\int_\Omega u \cdot H_0$ is the Zeem energy from the applied field $H_0$. The magnetic moment $u$ links the magnetic field by Maxwell equation and Faraday's law as follows

$$\operatorname{curl} H = 0, \quad \operatorname{div}(H + u) = 0 \quad \text{in } D'(\mathbf{R}^3), \tag{1.2}$$

where $u = 0$ outside $\Omega$.

Without the applied field and the anisotropy field and for $d^2 = \frac{1}{2}$, the energy reads as

$$E = \frac{1}{2}\int_\Omega |\nabla u|^2 - \int_\Omega u \cdot H. \tag{1.3}$$

---

* J Differential Equations, 2008, 244: 2448–2472.

Let $\Phi$ be the induced magnetic potential such that $H=\nabla\Phi$, then (1.2) becomes

$$-\Delta\Phi = \operatorname{div} u \quad \text{in } \mathbf{R}^3 \tag{1.4}$$

where $u$ is zero outside $\Omega$. The energy can be expressed as

$$E = \frac{1}{2}\int_{\Omega}|\nabla u|^2 + \int_{\mathbf{R}^3}|\nabla\Phi|^2. \tag{1.5}$$

The Euler-Lagrange equation for the energy under the constraint $|u|=1$ is

$$-\Delta u - u|\nabla u|^2 = H - \langle H,u\rangle u. \tag{1.6}$$

In the last decade, the minimization problem of the functional (1.1) has got extensive study. In variety of multiscale regimes, the authors of [3, 11 – 13, 19, 23 – 25, 31 – 33] investigated the domain structures.

In this paper, we shall consider the following Landau-Lifshitz flow which describes the precession of the magnetic moment subject to the Gilbert damping

$$u_t = -\lambda_1 u\times\left(u\times\frac{\delta E(u)}{\delta u}\right) + \lambda_2 u\times\frac{\delta E(u)}{\delta u}, \tag{1.7}$$

where $\lambda_1>0$ is the Gilbert damping constant, $\lambda_2$ is a constant too.

In the classical sense and for $\lambda_1=\lambda_2=1$, Eq. (1.7) can be equivalently rewritten as

$$\frac{1}{2}u_t - \frac{1}{2}(u\times u_t) = \Delta u + u|\nabla u|^2 + H - \langle H,u\rangle u \quad \text{in } \Omega\times\mathbf{R}_+, \tag{1.8}$$

where $H=\nabla\Phi$ is the induced field with $\Phi$ determined by (1.4) subject to the conditions crossing $\partial\Omega$: $[\Phi]=0$, $[\partial\Phi/\partial\nu]=u\cdot\nu$ where $\nu$ is the outward unit normal to $\partial\Omega$.

We impose to Eqs. (1.8) the initial condition

$$u(x,0) = u_0(x) \tag{1.9}$$

and the Dirichlet boundary condition

$$u = u_0(x), \quad \forall (x,t)\in\partial\Omega\times[0,\infty). \tag{1.10}$$

In the following, we always assume that $u_0(x)$ is a smooth map with the constraint $|u_0(x)|\equiv 1$ and denote by $\bar{u}$ the zero extension of $u$ from $\Omega$ to $\mathbf{R}^3$. We should notice that this extension guarantees $\bar{u}\in L^\infty(\mathbf{R}^3\times\mathbf{R}_+)\cap L^\infty(0,\infty;W^{-1,\infty}(\mathbf{R}^3))$.

The latest developments in the studies of Eq. (1.8) coupling with static Maxwell equations (1.2) are about the stability of static solution which matches $u(-\infty)=-e_1$ and $u(+\infty)=e_1$[7], the result on the evolution of boundary vortices[33], and boundary layers[6] in some special limit regimes. Other dynamic behaviors of the domains and domain walls can be found in [22, 28] and references therein.

For Eq. (1.8) without the nonlocal term, that is

$$\frac{1}{2}u_t - \frac{1}{2}(u\times u_t) = \Delta u + u|\nabla u|^2 \quad \text{in } \Omega\times\mathbf{R}_+, \tag{1.11}$$

there have been many works concerning the existence and regularities of weak solutions. In 1984 – 1987, Zhou and Guo[37,38] proved the global existence of weak solutions. The unique smooth solution in one dimension was given in [39]. In 1992, F. Alouges and A. Soyeur[1], using penalty method, proved that if $\lambda_2 = 1$, and the initial data $u_0 \in H^1_{loc}(\mathbf{R}^3), \nabla u_0 \in L^2(\mathbf{R}^3), |u_0| = 1$ a.e., then there exists a global weak solution. If $u_0 \in H^1(\Omega), \lambda_1 > 0$, then the Neumann boundary problem admits infinitely many weak solutions. If $\lambda_1 \to 0$, then the equation tends to $u_t - u \times \Delta u$; but if $\lambda_1 \to \infty$, the equation tends to $u_t = \Delta u + u|\nabla u|^2$, harmonic map heat flow.

For the problem with a nonlocal term ((1.7) or (1.8)), Carbou and Fabrie in [4] have got the global smooth solution for small initial data, and the local existence of smooth solutions by Galerkin method. In [4], the full Maxwell equations are contained, see also [21] for the similar results.

In this paper our main concern is the regularity of the weak solutions. This can be compared with the regularity problem of harmonic map heat flow (see [20]).

As we know, for the high dimensional heat flow of harmonic maps, Chen and Struwe in [9] established the existence of partially smooth weak solutions of harmonic map heat flow by Ginzburg-Landau approximation. In their proof, the key point is the parabolic energy monotonicity formula.

However, Coron[10] observed that there are infinitely many weak solutions to the flow different from those constructed in [9]. On the other hand, Riviére's example[34] showed that a weakly harmonic map from $B^3$ into $S^2$ may be everywhere singular on $B^3$. Therefore, one should ask, under what conditions are the weak solutions partially smooth? To answer such a question, Feldman[17] introduced, motivated by the studies on stationary harmonic maps by Evans[16], a notion of "stationary weak solution" for the flow and proved that a stationary solution must be partially regular since under such stationary conditions, the parabolic energy monotonicity inequality holds. He also pointed out that the solution constructed by Chen and Struwe[9] is stationary.

We should notice that the "stationary" conditions are unnatural, or at least, it is not easy to be verified.

The regularity problem concerning Landau-Lifshitz equation is of importance in physics. As we know, since $|u| = 1$, the singular point of Landau-Lifshitz equations at which a sudden change of the direction of the magnetization $u$ appears stands for the defects by vortices or phase transition in domain walls, see [22 – 24, 30, 31] and references therein.

The first progress on the existence of partially regular solutions to Landau-Lifshitz-Gilbert equations was made by Guo and Hong[20] in 1993 in which they revealed the links between 2-d system (1.11) and the harmonic maps heat flow, and established the existence of Chen-Struwe solution (referred to [9]). The uniqueness of weak solution with finite energy for 2-dimensional problem was obtained by Chen, Ding and Guo[8] in 1998.

Moser[30] observed, for Landau-Lifshitz equation (1.11) in dimension $n \leqslant 4$, that under "stationary" conditions, there holds the parabolic energy monotonicity inequality. He used this monotonicity to prove a partial regularity for weak solutions of the Landau-Lifshitz equations satisfying a "stationary" condition similar to [17].

Recently, there have also been more works on the partial regularity for suitable weak solutions to Landau-Lifshitz equation (1.11) by Liu[27] and Landau-Lifshitz-Maxwell equation (1.8) by Ding and Guo[14] under certain "stationary" conditions. For example, the stationary condition for (1.8) was derived in [14] which states:

A weak solution $u$ of (1.1) is said a stationary solution if for any $\xi(x,t)\in C_0^1(\Omega\times\mathbf{R}_+,\mathbf{R}^3)$, $\theta(x,t)\in C_0^1(\Omega\times\mathbf{R}_+,\mathbf{R})$ with $\xi(x,t)$, $\theta(x,t)$, $\nabla_{(x,t)}\xi$, $\nabla_{(x,t)}\theta$ bounded on $\Omega\times\mathbf{R}_+$ and $\xi,\theta\equiv 0$ for $t=0$ and $t\geqslant t^*>0$ such that $x+\tau\xi|_{\partial\Omega}=\mathrm{Id}$, $t+\tau\theta|_{\partial\Omega}=\mathrm{Id}$, there holds

$$\int_0^{+\infty}\int_\Omega\left(\frac{1}{2}u_t-\frac{1}{2}u\times u_t\right)\left(\frac{\partial u^\tau}{\partial\tau}\right)_{\tau=0}+\partial_\tau^+\int_0^{+\infty}\int_\Omega e(u^\tau)+|H(u^\tau)|^2\mathrm{d}x\mathrm{d}t\leqslant 0,$$

where $u^\tau(x,t)=u(x+\tau\xi(x,t),t+\tau\theta(x,t))$, $e(u)=\frac{1}{2}|\nabla u(x,t)|^2$.

But, it is unknown whether these problems admits solutions satisfying such "stationary" conditions. Or, in other words, one should ask about the regularity of weak solutions instead of "stationary" weak solutions. The definition of weak solution is standard, that is:

A function $u\in L^\infty(0,T;H^1(\Omega))$ with $u_t\in L^2(0,T;L^2(\Omega))$ is said a weak solution if (1.8) holds in the sense of distribution.

Very recently, Melcher[29] discussed this issue and proved the existence of partially smooth weak solutions to the Landau-Lifshitz equations for $\Omega=\mathbf{R}^3$. Melcher pointed out that his argument does not work if $n\geqslant 4$.

Wang[36], using the idea of [35] and Ginzburg-Landau approximation approach, proved the existence of partially smooth weak solutions to the Landau-Lifshitz equation (1.11) in bounded domain $\Omega$ of dimension $\leqslant 4$ which is the first result of the existence of partially regular solution to Landau-Lifshitz equations in dimension 4. His method is different from that in [29].

This paper is concerned with the existence of partially regular solution to the Landau-Lifshitz-Maxwell equations (1.8) by using the main idea of [36]. We deal with, in this paper, the Dirichlet problem which is more delicate to obtain the boundary estimates than Neumann problem.

Our main results are the following.

**Main Theorem** For any bounded smooth domain $\Omega\subset\mathbf{R}^3$ and $g=u_0(x)\in C^\infty(\bar{\Omega},S^2)$, there exist a global weak solution $u:\Omega\times\mathbf{R}_+\to S^2$ of problem (1.8)~(1.10) and a closed subset $\Sigma\subset\bar{\Omega}\times\mathbf{R}_+$ with $\mathscr{H}^3(\Sigma\cap K)<\infty$ for any compact subset $K\subset\bar{\Omega}\times\mathbf{R}_+$ such that $u\in C^\infty(\bar{\Omega}\times\mathbf{R}_+\setminus\Sigma;S^2)$ where $\mathscr{H}^3$ is the 3-dimensional parabolic Hausdorff measure with respect to the parabolic metric $d((x,t),(y,s))=\max\{|x-y|,|t-s|^{\frac{1}{2}}\}$.

For the Cauchy problem, as a consequence, we also prove

**Corollary** If $g=u_0(x)\in C^\infty(\mathbf{R}^3,S^2)$ with $\nabla g\in L^2(\mathbf{R}^3)$, then there exist a global weak solution $u:\mathbf{R}^3\times\mathbf{R}_+\to S^2$ of problem (1.8)~(1.9) and a closed subset $\Sigma\subset\mathbf{R}^3\times\mathbf{R}_+$ with $\mathscr{H}^3(\Sigma\cap K)<\infty$ for any compact subset $K\subset\mathbf{R}^3\times\mathbf{R}_+$ such that $u\in C^\infty(\mathbf{R}^3\times\mathbf{R}_+\setminus\Sigma;S^2)$.

In 2007, Ding and Wang[15] proved the existence of weak solutions with finite time singularity to Landau-Lifshitz-Gilbert equations. But, for the technical reasons, we do not know the types of the singularities.

## 2 Some Preliminaries

We shall prove our main theorem by considering the following Ginzburg-Landau approximations

$$\frac{1}{2}u_{\varepsilon t}-\frac{1}{2}u_{\varepsilon}\times u_{\varepsilon t}=\Delta u_{\varepsilon}+\frac{(1-|u_{\varepsilon}|^2)}{\varepsilon^2}u_{\varepsilon}-u_{\varepsilon}\times(u_{\varepsilon}\times H_{\varepsilon}),\tag{2.1}$$

where $H_{\varepsilon}=\nabla\Phi_{\varepsilon}$ and

$$\Delta\Phi_{\varepsilon}=-\operatorname{div}\bar{u}_{\varepsilon}\quad\text{in }\mathscr{D}'(\mathbf{R}^3).\tag{2.2}$$

We first prove that (2.1) with conditions (1.9) and (1.10) admits global smooth solutions $u_{\varepsilon}$. Then we prove that these solutions subsequently converge to the global weak solution of the problem (1.8)~(1.10) weakly. We also prove that such a convergence is also true in $C^{\infty}$ outside a set of $\Omega\times\mathbf{R}_+$ with locally finite Hausdorff measure. Hence, such a weak solution of the problem (1.8)~(1.10) obtained above is partially regular.

Now let us recall some results on the quasi-static Maxwell equations.

**Lemma 2.1** (See [2]) Let $u\in H^1(\Omega,S^2)$. Let $H=\nabla\Phi\in L^2(\mathbf{R}^3,\mathbf{R}^3)$ be the solution of

$$\operatorname{curl}H=0,\quad\operatorname{div}(H+\tilde{u})=0\quad\text{in }\mathscr{D}'(\mathbf{R}^3),\tag{2.3}$$

where $\tilde{u}$ is equal to $u$ in $\Omega$ and zero outside $\Omega$. Then

$$H\in\bigcap_{1\leqslant p<\infty}L^p(\mathbf{R}^3)\tag{2.4}$$

and for all $p\in(1,\infty)$ there exists a constant $K_p>0$ such that

$$\|H\|_{L^p(\mathbf{R}^3)}\leqslant K_p\|u\|_{L^p(\Omega)}.\tag{2.5}$$

**Remark 2.1** Lemma 2.1 implies (see [14])

$$\|\Phi\|_{L^{\infty}}\leqslant C\|\nabla u\|_{L^2}\tag{2.6}$$

and

$$\|\Phi\|_{L^p}\leqslant K_p\|u\|_{L^p(\Omega)}\leqslant C\|\nabla u\|_{L^2},\quad\forall\,1\leqslant p<\infty.\tag{2.7}$$

**Remark 2.2** We have

$$-\int_{\Omega}u\cdot H=-\int_{\mathbf{R}^3}u\cdot\nabla\Phi=\int_{\mathbf{R}^3}\Phi\operatorname{div}u=-\int_{\mathbf{R}^3}\Phi\Delta\Phi$$

$$=\int_{\mathbf{R}^3}|\nabla\Phi|^2=\int_{\mathbf{R}^3}|H|^2.\tag{2.8}$$

**Remark 2.3** Let $G(r)=\dfrac{C}{r}$ be the Laplace kernel in $\mathbf{R}^3$. Then the solution of the Poisson equation with parameter $t$

$$\Delta\Phi(x,t)=-\operatorname{div}u\quad\text{in }\mathscr{D}'(\mathbf{R}^3)$$

can be expressed by[2,14]

$$\Phi(x,t) = -\int_{\Omega} G(|x-y|)\mathrm{div}u(y,t)\mathrm{d}y + \int_{\partial\Omega} G(|x-y|)\langle u(y,t),\mathbf{n}(\mathbf{y})\rangle \mathrm{d}\sigma(y). \tag{2.9}$$

**Remark 2.4** It follows from (2.9) that, if $u\in C^{1,\alpha}(\bar{\Omega})$, then $\Phi\in C^{2,\alpha}(\bar{\Omega})$. In fact, since $u\in C^{1,\alpha}(\bar{\Omega})$, the conclusion that the first integral

$$-\int_{\Omega} G(|x-y|)\mathrm{div}\, u(y,t)\mathrm{d}y$$

belongs to $C^{2,\alpha}(\bar{\Omega})$ follows from [18]. On the other hand, the potential estimate for the second integral $w = \int_{\partial\Omega} G(|x-y|)\langle u(y,t),\mathbf{n}(\mathbf{y})\rangle \mathrm{d}\sigma(y)$ can be done in the similar manner as in [18] if noticing that

$$\begin{aligned}\nabla_x w &= \int_{\partial\Omega} \nabla_x G(|x-y|)\langle u(y,t),\mathbf{n}(\mathbf{y})\rangle \mathrm{d}\sigma(y)\\ &= -\int_{\partial\Omega} (\nabla_y G(|x-y|))\langle u(y,t),\mathbf{n}(\mathbf{y})\rangle \mathrm{d}\sigma(y)\\ &= \int_{\partial\Omega} G(|x-y|)(\nabla_y\langle u(y,t),\mathbf{n}(\mathbf{y})\rangle)\mathrm{d}\sigma(y).\end{aligned}$$

In the following, we turn to investigate the properties for the solutions of problem (2.1) ~(2.2) with conditions (1.9)~(1.10).

First of all, the existence of global weak solutions to the Ginzburg-Landau penalty problem (2.1) for fixed $\varepsilon>0$ follows from [5].

Next, we claim that the weak solutions to the penalty problem (2.1)~(2.2) and (1.9) ~(1.10) are in fact smooth solutions. We also derive some basic uniform in $\varepsilon$ estimates for the solution $u_\varepsilon$.

**Lemma 2.2** Let $u_\varepsilon$ be a solution of (2.1)~(2.2) and (1.9)~(1.10) and $|u_0(x)|=1$. Then there holds

$$|u_\varepsilon|\leqslant 1, \quad |\nabla u_\varepsilon|\leqslant C\varepsilon^{-1}. \tag{2.10}$$

**Proof** The first inequality of (2.10) follows from the standard maximum principle if noticing that $(u_t\wedge u)\cdot u=(u\times(u\times H))\cdot u=0$ (see [1] for details).

Now we prove the second inequality of (2.10). Let $\Omega_\varepsilon=\varepsilon^{-1}\Omega$, $w_\varepsilon(x,t)=u_\varepsilon(\varepsilon x, \varepsilon^2 t)$, $\psi_\varepsilon(x,t)=\Phi_\varepsilon(\varepsilon x,\varepsilon^2 t)$ and $h_\varepsilon=\nabla\psi_\varepsilon$. We get

$$\frac{1}{2}w_{\varepsilon t}-\frac{1}{2}w_\varepsilon\times w_{\varepsilon t} = \Delta w_\varepsilon+(1-|w_\varepsilon|^2)w_\varepsilon-\varepsilon w_\varepsilon\times(w_\varepsilon\times h_\varepsilon) \quad \text{in } \Omega_\varepsilon\times\mathbf{R}_+, \tag{2.11}$$

where

$$\Delta\psi_\varepsilon = -\varepsilon\mathrm{div}\,\bar{w}_\varepsilon \quad \text{in } \mathscr{D}'(\mathbf{R}^3)$$

with initial condition and boundary condition

$$w_\varepsilon(x,t) = u_0(\varepsilon x), \quad (x,t)\in\Omega_\varepsilon\times\{t=0\}\cup\partial\Omega_\varepsilon\times\{t>0\}. \tag{2.12}$$

Therefore the standard regularity theory of parabolic systems yields that

$$|\nabla w_\varepsilon|\leqslant C$$

with $C$ independent of $\varepsilon$. Rescaling back to variable $x$, we get the conclusion of the lemma. □

**Lemma 2.3** (Global Energy Estimates) For any given $T>0$, there exists a constant $C(T)>0$ independent of $\varepsilon$ such that for any solution of (2.1)~(2.2) and (1.9)~(1.10), there holds

$$\frac{1}{4}\int_0^T\int_\Omega |u_{\varepsilon t}|^2+\int_\Omega\left[\frac{1}{2}|\nabla u_\varepsilon|^2+\frac{(1-|u_\varepsilon|^2)^2}{4\varepsilon^2}\right]\leqslant C(T). \tag{2.13}$$

**Proof** Multiplying Eq. (2.1) by $u_{\varepsilon t}$ and integrating by parts, one gets

$$\frac{1}{2}\int_\Omega |u_{\varepsilon t}|^2=\int_\Omega u_{\varepsilon t}\Delta u_\varepsilon+\int_\Omega\frac{(1-|u_\varepsilon|^2)}{\varepsilon^2}u_\varepsilon u_{\varepsilon t}-\int_\Omega u_\varepsilon\times(u_\varepsilon\times H_\varepsilon)\cdot u_{\varepsilon t}. \tag{2.14}$$

This gives, from the initial boundary conditions that

$$\frac{1}{2}\int_\Omega |u_{\varepsilon t}|^2+\frac{\mathrm{d}}{\mathrm{d}t}\int_\Omega\left[\frac{1}{2}|\nabla u_\varepsilon|^2+\frac{(1-|u_\varepsilon|^2)^2}{4\varepsilon^2}\right]$$
$$=-\int_\Omega u_\varepsilon\times(u_\varepsilon\times H_\varepsilon)\cdot u_{\varepsilon t}. \tag{2.15}$$

It follows from Hölder inequality and $|u_\varepsilon|\leqslant 1$ that

$$\frac{1}{4}\int_\Omega |u_{\varepsilon t}|^2+\frac{\mathrm{d}}{\mathrm{d}t}\int_\Omega\left[\frac{1}{2}|\nabla u_\varepsilon|^2+\frac{(1-|u_\varepsilon|^2)^2}{4\varepsilon^2}\right]\leqslant\int_\Omega |H_\varepsilon|^2. \tag{2.16}$$

Since $H_\varepsilon=\nabla\Phi_\varepsilon$, it follows from Lemma 2.1 and Remarks 2.1~2.4 that

$$\int_\Omega |H_\varepsilon|^2=\int_{\mathbf{R}^3}|\nabla\Phi_\varepsilon|^2=-\int_{\mathbf{R}^3}\Phi_\varepsilon\Delta\Phi_\varepsilon=\int_{\mathbf{R}^3}\Phi_\varepsilon\operatorname{div}\bar{u}=-\int_{\mathbf{R}^3}\bar{u}\,\nabla\Phi_\varepsilon\leqslant\frac{1}{2}\int_{\mathbf{R}^3}|\nabla\Phi_\varepsilon|^2+\frac{1}{2}\int_{\mathbf{R}^3}|\bar{u}|^2.$$

However, there holds

$$\int_{\mathbf{R}^3}|\bar{u}|^2\leqslant\int_\Omega |u_\varepsilon|^2\leqslant|\Omega|$$

and

$$\int_{\mathbf{R}^3}|\nabla\Phi_\varepsilon|^2\leqslant\int_{\mathbf{R}^3}|\bar{u}|^2\leqslant\int_\Omega |u_\varepsilon|^2\leqslant|\Omega|,$$

we finally get

$$\frac{1}{4}\int_\Omega |u_{\varepsilon t}|^2+\frac{\mathrm{d}}{\mathrm{d}t}\int_\Omega\left[\frac{1}{2}|\nabla u_\varepsilon|^2+\frac{(1-|u_\varepsilon|^2)^2}{4\varepsilon^2}\right]\leqslant C_0. \tag{2.17}$$

Hence we have the global energy estimate

$$\frac{1}{4}\int_0^T\int_\Omega |u_{\varepsilon t}|^2+\int_\Omega\left[\frac{1}{2}|\nabla u_\varepsilon|^2+\frac{(1-|u_\varepsilon|^2)^2}{4\varepsilon^2}\right]\leqslant C(T), \tag{2.18}$$

where $C(T)$ is determined by $\int_\Omega|\nabla u_0(x)|^2$. □

**Lemma 2.4** Suppose that $u_0(x)\in C^\infty(\bar{\Omega})$. Then for the solution of (2.1)~(2.2) and (1.9)~(1.10), we have

$$u_\varepsilon\in C^\infty(\bar{\Omega}\times\mathbf{R}_+;\mathbf{R}^3). \tag{2.19}$$

**Proof** Since $|u_\varepsilon|\leqslant 1$, we have from Lemma 2.1 that $\{H_\varepsilon\}$ is bounded in $L^p(\Omega)$ for any $p>1$. It follows from the theory of strongly parabolic systems (see [26]) that the solution

$\{u_\varepsilon\}$ belongs to $W_p^{2,1}(\Omega\times\mathbf{R}_+)$ for any $p>1$. For $p$ large enough, space $W_p^{2,1}(\Omega\times[0,T])$ is continuously imbedded into $C^{1+\alpha,(1+\alpha)/2}(\bar{\Omega}\times[0,T])$, therefore, $u_\varepsilon\in C^{1+\alpha,(1+\alpha)/2}(\bar{\Omega}\times[0,T])$. Substituting this result into

$$\Phi(x,t)=-\int_\Omega G(|x-y|)\mathrm{div}u(y,t)\mathrm{d}y+\int_{\partial\Omega}G(|x-y|)\langle u(y,t),\mathbf{n}(\mathbf{y})\rangle\mathrm{d}\sigma(y),$$

and using Remark 2.4, we know that $\Phi_\varepsilon\in C^{2+\alpha,1+\alpha/2}(\bar{\Omega}\times[0,T])$, i. e. $H_\varepsilon\in C^{1+\alpha,(1+\alpha)/2}(\bar{\Omega}\times[0,T])$. This result in turn yields that the solution of problem (2.1) belongs to $C^{3+\alpha,(3+\alpha)/2}(\bar{\Omega}\times[0,T])$. Repeating this iteration, we finish the proof of the lemma. □

The following lemma is needed in the following section. Denote $P_r(z_0)=B_r(x_0)\times(t_0-r^2,t_0)$ for $z_0=(x_0,t_0)$.

**Lemma 2.5** (Local Energy Estimates) For any $p>2$, there exists a constant $C>0$ independent of $\varepsilon$ such that

$$\begin{aligned}&r^{-1}\int_{(\Omega\cap\mathbf{R}_+)\cap P_{r/2}(z_0)}|u_{\varepsilon t}|^2+r^{-1}\int_{\Omega\cap B_{r/2}(x_0)}e_\varepsilon(u_\varepsilon)\\&\leqslant Cr^{-3}\int_{(\Omega\cap\mathbf{R}_+)\cap P_r(z_0)}|\nabla u_\varepsilon|^2+Cr^{4-6/p}\|H_\varepsilon\|^2_{L^\infty(0,t_0;L^p(B_r(x_0)))}.\end{aligned}\tag{2.20}$$

**Proof** Let $x_0\in\bar{\Omega}$, $0<r<\sqrt{t_0}$. By Fubini's theorem, one may choose $\alpha\in\left(\frac{1}{2},\frac{7}{8}\right)$ such that

$$\int_{\Omega\cap B_r(x_0)}|\nabla u|^2(x,t-\alpha^2r^2)\mathrm{d}x\leqslant Cr^{-2}\int_{(\Omega\times\mathbf{R}_+)\cap P_r(z_0)}|\nabla u|^2\mathrm{d}x\mathrm{d}t.\tag{2.21}$$

Since $\frac{1}{2}<\alpha<\frac{7}{8}$, there holds $P_{r/2}(z_0)\subset P_{\alpha r}(z_0)\subset P_r(z_0)$.

Take $\phi(x)\in C_0^\infty(B_r(x_0))$. Multiplying Eq. (2.1) by $\Phi^2\frac{\partial u_\varepsilon}{\partial t}$ and integrating by parts, we get

$$\begin{aligned}&\frac{1}{2}\int_{\Omega\cap B_r(x_0)}|u_{\varepsilon t}|^2\phi^2+\frac{1}{2}\frac{\mathrm{d}}{\mathrm{d}t}\int_{\Omega\cap B_r(x_0)}\phi^2|\nabla u_\varepsilon|^2+\frac{1}{4\varepsilon^2}\frac{\mathrm{d}}{\mathrm{d}t}\int_{\Omega\cap B_r(x_0)}\phi^2(1-|u_\varepsilon|^2)^2\\&=-2\int_{\Omega\cap B_r(x_0)}\phi\nabla\phi\cdot\nabla u_\varepsilon\cdot u_{\varepsilon t}-\int_{\Omega\cap B_r(x_0)}\phi^2u_\varepsilon\times(u_\varepsilon\times H_\varepsilon)u_{\varepsilon t}.\end{aligned}\tag{2.22}$$

Equality (2.22) combined with Hölder inequality yields

$$\begin{aligned}&\frac{1}{4}\int_{\Omega\cap B_r(x_0)}|u_{\varepsilon t}|^2\phi^2+\frac{\mathrm{d}}{\mathrm{d}t}\int_{\Omega\cap B_r(x_0)}\phi^2e_\varepsilon(u_\varepsilon)\\&\leqslant 8\int_{\Omega\cap B_r(x_0)}\phi^2|\nabla\phi|^2|\nabla u_\varepsilon|^2+2\int_{\Omega\cap B_r(x_0)}\phi^2|H_\varepsilon|^2.\end{aligned}\tag{2.23}$$

In (2.23), $e_\varepsilon(u_\varepsilon)=\frac{1}{2}|\nabla u_\varepsilon|^2+\frac{1}{4\varepsilon^2}(1-|u_\varepsilon|^2)^2$. Integrating (2.23) over $[t_0-\alpha^2r^2,t_0]$, one obtains

$$\frac{1}{4}\int_{(\Omega\cap\mathbf{R}_+)\cap P_{\alpha r}(z_0)}|u_{\varepsilon t}|^2\phi^2+\int_{\Omega\cap B_r(x_0)}\phi^2e_\varepsilon(u_\varepsilon)(t_0)-\int_{\Omega\cap B_r(x_0)}\phi^2e_\varepsilon(u_\varepsilon)(t_0-\alpha^2r^2)$$

$$\leqslant 8\int_{(\Omega\cap\mathbf{R}_+)\cap P_r(z_0)}\phi^2|\nabla\phi|^2|\nabla u_\varepsilon|^2+2\int_{(\Omega\cap\mathbf{R}_+)\cap P_r(z_0)}\phi^2|H_\varepsilon|^2. \tag{2.24}$$

This combined with (2.21) yields

$$\int_{(\Omega\cap\mathbf{R}_+)\cap P_{r/2}(z_0)}|u_{\varepsilon t}|^2+\int_{\Omega\cap B_{r/2}(x_0)}e_\varepsilon(u_\varepsilon)$$
$$\leqslant Cr^{-2}\int_{(\Omega\cap\mathbf{R}_+)\cap P_r(z_0)}|\nabla u_\varepsilon|^2+2\int_{(\Omega\cap\mathbf{R}_+)\cap P_r(z_0)}|H_\varepsilon|^2. \tag{2.25}$$

Using Lemma 2.1, one has

$$\int_{(\Omega\cap\mathbf{R}_+)\cap P_r(z_0)}|H_\varepsilon|^2=\int_{t_0-r^2}^{t_0}\int_{B_r(x_0)}|H_\varepsilon|^2\leqslant\int_{t_0-r^2}^{t_0}\left(\int_{B_r(x_0)}|H_\varepsilon|^p\right)^{2/p}\left(\int_{B_r(x_0)}\mathrm{d}x\right)^{1-2/p}$$
$$\leqslant Cr^2(r^3)^{1-2/p}\sup_{0<t<t_0}\left(\int_{B_r(x_0)}|H_\varepsilon|^p\right)^{2/p}$$
$$\leqslant Cr^{5-6/p}\|H_\varepsilon\|^2_{L^\infty(0,t_0;L^p(B_r(x_0)))} \tag{2.26}$$

and gets the lemma. □

## 3 Generalized Monotonicity at Time Slices

In the previous section, we have given some basic estimates for the approximate solution $u_\varepsilon$. Now we further give some fine estimates for the smooth solution $u_\varepsilon$ of the approximate equation. In order to derive the partial regularity, the most important inequality is the parabolic monotonicity inequality which is generally untrue for Landau-Lifshitz equations. Nevertheless, we may derive the inequalities called generalized monotonicity inequalities at time slices which finally yield the desired energy decay estimates. In Liu[27] and Ding and Guo[14], the parabolic generalized monotonicity inequalities come from the stationary conditions. However we do not have such conditions now. As it was done in [36], we derive these inequalities at time slices by Pohozaev method.

**Lemma 3.1** (Interior Generalized Monotonicity) For the solution of the approximation problem (2.1)~(2.2) and $x_0\in\Omega, t>0, 0<r<R<1, p>2$, there hold

$$r^{-1}E_\varepsilon(u_\varepsilon,B_r(x_0))$$
$$\leqslant 4R^{-1}E_\varepsilon(u_\varepsilon,B_R(x_0))+C_0R\int_{B_R(x_0)}\left|\frac{\partial u_\varepsilon}{\partial t}\right|^2+$$
$$C_0R^{4-\frac{6}{p}}\|H_\varepsilon\|^2_{L^\infty(t-R^2,t;L^p(B_R(x_0)))} \tag{3.1}$$

and

$$\int_{B_R}|x-x_0|^{-1}\frac{(1-|u_\varepsilon|^2)^2}{\varepsilon^2}$$
$$\leqslant 4R^{-1}E_\varepsilon(u_\varepsilon,B_R(x_0))+C_0R\int_{B_R(x_0)}\left|\frac{\partial u_\varepsilon}{\partial t}\right|^2+$$
$$C_0R^{4-\frac{6}{p}}\|H_\varepsilon\|^2_{L^\infty(t-R^2,t;L^p(B_R(x_0)))}, \tag{3.2}$$

where $B_R(x_0)$ denotes the circle centered at $x_0$ with radius $R$ and $E_\varepsilon(A) = \int_A \left(\frac{1}{2}|\nabla u_\varepsilon|^2 + \frac{3(1-|u_\varepsilon|^2)^2}{4\varepsilon^2}\right)$.

**Proof** Denote $R(u_\varepsilon)u_{\varepsilon t} = \frac{1}{2}u_{\varepsilon t} - \frac{1}{2}u_\varepsilon \times u_{\varepsilon t}$, let $x_0 \in \Omega$, $t>0, 0<r\leqslant R<\min\{1, \operatorname{dist}(x_0,\partial\Omega)\}$. For simplicity, we assume $x_0=0$.

Multiplying Eq. (2.1) by $x\cdot\nabla u_\varepsilon$ and integrating over $B_r(0)$ by parts, we have

$$\begin{aligned}\int_{B_r(0)} R(u_\varepsilon)u_{\varepsilon t}(x\cdot\nabla u_\varepsilon) &= \int_{B_r(0)}\left[\Delta u_\varepsilon + \frac{1}{\varepsilon^2}(1-|u_\varepsilon|^2)u_\varepsilon - u_\varepsilon\times(u_\varepsilon\times H_\varepsilon)\right](x\cdot\nabla u_\varepsilon)\\ &= \int_{\partial B_r(0)}(x\cdot\nabla u_\varepsilon)\frac{\partial u_\varepsilon}{\partial r} - \int_{B_r(0)}\partial_k(x_i\partial_i u_\varepsilon)\partial_k u_\varepsilon - \\ &\quad \frac{1}{4\varepsilon^2}\int_{B_r(0)} x\cdot\nabla(1-|u_\varepsilon|^2)^2 - \int_{B_r(0)} u_\varepsilon\times(u_\varepsilon\times H_\varepsilon)(x\cdot\nabla u_\varepsilon).\end{aligned}$$

Hence we have

$$\begin{aligned}\int_{B_r(0)} R(u_\varepsilon)u_{\varepsilon t}(x\cdot\nabla u_\varepsilon) &= r\int_{\partial B_r(0)}\left[\left|\frac{\partial u_\varepsilon}{\partial r}\right|^2 - \frac{1}{2}|\nabla u_\varepsilon|^2 - \frac{(1-|u_\varepsilon|^2)^2}{4\varepsilon^2}\right]\mathrm{d}\sigma + \\ &\quad \int_{B^r(0)}\left[\frac{1}{2}|\nabla u_\varepsilon|^2 + \frac{3(1-|u_\varepsilon|^2)^2}{4\varepsilon^2}\right] - \int_{B_r(0)} u_\varepsilon\times(u_\varepsilon\times H_\varepsilon)(x\cdot\nabla u_\varepsilon).\end{aligned}\tag{3.3}$$

A simple computation yields

$$\begin{aligned}&\frac{\mathrm{d}}{\mathrm{d}t}(r^{-1}E_\varepsilon(u_\varepsilon, B_r(0)))\\ &= -r^{-2}E_\varepsilon(u_\varepsilon, B_r(0)) + r^{-1}\int_{\partial B_r(0)}\left[\frac{1}{2}|\nabla u_\varepsilon|^2 + \frac{3(1-|u_\varepsilon|^2)^2}{4\varepsilon^2}\right].\end{aligned}\tag{3.4}$$

We get from (3.3)~(3.4) that

$$\begin{aligned}\frac{\mathrm{d}}{\mathrm{d}r}(r^{-1}E_\varepsilon(u_\varepsilon, B_r(0))) &= -r^{-2}\int_{B_r(0)}\left[R(u_\varepsilon)\frac{\partial u_\varepsilon}{\partial t} + u_\varepsilon\times(u_\varepsilon\times H_\varepsilon)\right](x\cdot\nabla u_\varepsilon) + \\ &\quad r^{-1}\int_{\partial B_r(0)}\left[\left|\frac{\partial u_\varepsilon}{\partial r}\right|^2 + \frac{(1-|u_\varepsilon|^2)^2}{2\varepsilon^2}\right],\end{aligned}\tag{3.5}$$

which yields

$$\begin{aligned}&\frac{\mathrm{d}}{\mathrm{d}r}\left(r^{-1}E_\varepsilon(u_\varepsilon, B_r(0)) - r^{-1}\int_{B_r(0)} R(u_\varepsilon)\frac{\partial u_\varepsilon}{\partial t}(x\cdot\nabla u_\varepsilon)\right)\\ &= -r^{-2}\int_{B_r(0)} u_\varepsilon\times(u_\varepsilon\times H_\varepsilon)(x\cdot\nabla u_\varepsilon) + \\ &\quad r^{-1}\int_{\partial B_r(0)}\left[\left|\frac{\partial u_\varepsilon}{\partial r}\right|^2 + \frac{(1-|u_\varepsilon|^2)^2}{2\varepsilon^2} - R(u_\varepsilon)\frac{\partial u_\varepsilon}{\partial t}(x\cdot\nabla u_\varepsilon)\right].\end{aligned}\tag{3.6}$$

The first term on the right-hand side of (3.6) can be estimated as follows:

$$-r^{-2}\int_{B_r(0)} u_\varepsilon\times(u_\varepsilon\times H_\varepsilon)(x\cdot\nabla u_\varepsilon)$$

$$\geqslant -r^{-2}\left(\int_{B_r(0)}|H_\varepsilon|^2\right)^{1/2}\left(\int_{B_r(0)}|x\cdot\nabla u_\varepsilon|^2\right)^{1/2}$$

$$\geqslant -r^{-2}r^{\frac{3}{2}-\frac{3}{p}}\|H_\varepsilon\|_{L^\infty(t-R^2,t;L^p(B_R(0)))}\left(\int_{B_r(0)}\left|r\frac{\partial u_\varepsilon}{\partial r}\right|^2\right)^{1/2}$$

$$\geqslant -C_0r^{\frac{1}{2}-\frac{3}{p}}\left(\int_{B_r(0)}\left|\frac{\partial u_\varepsilon}{\partial r}\right|^2\right)^{1/2}\|H_\varepsilon\|_{L^\infty(t-R^2,t;L^p(B_R(0)))}. \tag{3.7}$$

Substituting (3.7) into (3.6), one gets that for any $p>2$ there holds

$$\frac{\mathrm{d}}{\mathrm{d}r}\left(r^{-1}E_\varepsilon(u_\varepsilon,B_r(0))-r^{-1}\int_{B_r(0)}R(u_\varepsilon)\frac{\partial u_\varepsilon}{\partial t}(x\cdot\nabla u_\varepsilon)\right)$$

$$\geqslant -C_0r^{\frac{1}{2}-\frac{3}{p}}\left(\int_{B_r(0)}\left|\frac{\partial u_\varepsilon}{\partial r}\right|^2\right)^{1/2}\|H_\varepsilon\|_{L^\infty(t-R^2,t;L^p(B_R(0)))}+$$

$$r^{-1}\int_{\partial B_r(0)}\left[\left|\frac{\partial u_\varepsilon}{\partial r}\right|^2+\frac{(1-|u_\varepsilon|^2)^2}{2\varepsilon^2}-R(u_\varepsilon)\frac{\partial u_\varepsilon}{\partial t}(x\cdot\nabla u_\varepsilon)\right]. \tag{3.8}$$

Inequality (3.8) is just similar to the one in the proof of (8) in Lemma 2.2 of [30], see also (4.12) in the proof of Lemma 4.1 of [14].

Integrating (3.8) on $[r,R]$ with respect to $r$ and by the same argument from (4.15)~(4.20) of Lemma 4.1 in [14], one gets

$$r^{-1}E_\varepsilon(u_\varepsilon,B_r(0))+\int_r^R s^{-1}\int_{\partial B_s(0)}\left[\left|\frac{\partial u_\varepsilon}{\partial r}\right|^2+\frac{(1-|u_\varepsilon|^2)^2}{2\varepsilon^2}\right]$$

$$\leqslant 4R^{-1}E_\varepsilon(u_\varepsilon,B_R(0))+C_0R\int_{B_R}\left|\frac{\partial u_\varepsilon}{\partial t}\right|^2+$$

$$C_0\|H_\varepsilon\|_{L^\infty(t-R^2,t;L^p(B_R(0)))}\int_r^R s^{\frac{1}{2}-\frac{3}{p}}\left(\int_{B_s(0)}\left|\frac{\partial u_\varepsilon}{\partial r}\right|^2\right)^{1/2}. \tag{3.9}$$

Noticing that

$$\|H_\varepsilon\|_{L^\infty(t-R^2,t;L^p(B_R(0)))}\int_r^R s^{\frac{1}{2}-\frac{3}{p}}\left(\int_{B_s(0)}\left|\frac{\partial u_\varepsilon}{\partial r}\right|^2\right)^{1/2}$$

$$\leqslant\|H_\varepsilon\|_{L^\infty(t-R^2,t;L^p(B_R(0)))}\left(\int_{B_R(0)}\left|\frac{\partial u_\varepsilon}{\partial r}\right|^2\right)^{1/2}\int_r^R s^{\frac{1}{2}-\frac{3}{p}}\mathrm{d}s$$

$$\leqslant\|H_\varepsilon\|_{L^\infty(t-R^2,t;L^p(B_R(0)))}R^{\frac{3}{2}-\frac{3}{p}}\left(\int_{B_R(0)}\left|\frac{\partial u_\varepsilon}{\partial r}\right|^2\right)^{1/2}$$

$$\leqslant R^{-1}\int_{B_R(0)}\frac{1}{2}\left\|\frac{\partial u_\varepsilon}{\partial r}\right\|^2+C_0R^{4-\frac{6}{p}}\|H_\varepsilon\|^2_{L^\infty(t-R^2,t;L^p(B_R(0)))}$$

$$\leqslant R^{-1}E(u_\varepsilon,B_R(0))+C_0R^{4-\frac{6}{p}}\|H_\varepsilon\|^2_{L^\infty(t-R^2,t;L^p(B_R(0)))} \tag{3.10}$$

and

$$\int_r^R s^{-1}\int_{\partial B_s(0)}\left[\left|\frac{\partial u_\varepsilon}{\partial r}\right|^2+\frac{(1-|u_\varepsilon|^2)^2}{2\varepsilon^2}\right]$$

$$=\int_{B_R\setminus B_r(0)}|x|^{-1}\left[\left|\frac{\partial u_\varepsilon}{\partial r}\right|^2+\frac{(1-|u_\varepsilon|^2)^2}{\varepsilon^2}\right], \tag{3.11}$$

we finally get from (3.9)~(3.11) that

$$r^{-1}E_\varepsilon(u_\varepsilon,B_r(0))$$
$$\leqslant 4R^{-1}E_\varepsilon(u_\varepsilon,B_R(0))+C_0R\int_{B_R}\left|\frac{\partial u_\varepsilon}{\partial t}\right|^2+C_0R^{4-\frac{6}{p}}\|H_\varepsilon\|^2_{L^\infty(t-R^2,t;L^p(B_R(0)))} \tag{3.12}$$

and

$$\int_{B_R\setminus B_r(0)}|x|^{-1}\left[\left|\frac{\partial u_\varepsilon}{\partial r}\right|^2+\frac{(1-|u_\varepsilon|^2)^2}{\varepsilon^2}\right]$$
$$\leqslant 4R^{-1}E_\varepsilon(u_\varepsilon,B_R(0))+C_0R\int_{B_R}\left|\frac{\partial u_\varepsilon}{\partial t}\right|^2+C_0R^{4-\frac{6}{p}}\|H_\varepsilon\|^2_{L^\infty(t-R^2,t;L^p(B_R(0)))}. \tag{3.13}$$

The lemma is proved if we send $r$ to zero in (3.13). □

Now we give the boundary generalized monotonicity as follows.

**Lemma 3.2** For the solution of the approximation problem (2.1)~(2.2), there exist $C_g>0, 0<R_0=R_0(\Omega)<1$ such that for any $x_0\in\partial\Omega, t>0, 0<r<R\leqslant R_0, p>2$, there hold

$$C_gr+e^{C_gr}r^{-1}E(u,B_r^+)\leqslant C_gR+e^{C_gR}R^{-1}E(u,B_R^+)+C_gR\int_{B_R^+}|u_t|^2+$$
$$C_gR^{4-\frac{6}{p}}\|H_\varepsilon\|^2_{L^\infty(t-R^2,t;L^p(B_R^+(x_0)))} \tag{3.14}$$

and

$$\int_{B_R^+}|x-x_0|^{-1}\frac{(1-|u|^2)^2}{\varepsilon^2}\leqslant C_gR+e^{C_gR}R^{-1}E(u,B_R^+)+C_gR\int_{B_R^+}|u_t|^2+$$
$$C_gR^{4-\frac{6}{p}}\|H_\varepsilon\|^2_{L^\infty(t-R^2,t;L^p(B_R^+(x_0)))}, \tag{3.15}$$

where $B_R^+(x_0)=B_R(x_0)\cap\Omega$.

**Proof** Denote $C_1=C_0\Lambda_g, C_2=C_0\Lambda_g^2, \Lambda_g=\|\nabla g\|_{C^1(B_{R_0}(x_0))}$. For simplicity, we assume that $x_0=0$, $\Omega=\{x\in\mathbf{R}^3: x_3>0\}$. For $0<r<1$, denote $B_r^+=B_r\cap\mathbf{R}^3_+$, $T_r=\{x\in\mathbf{R}^3: |x|<r, x_3=0\}, (\partial B_r)^+=\{x\in\partial B_r, x_3>0\}, \partial(B_r^+)=(\partial B_r)^+\cup T_r$. Denote $u_\varepsilon$ by $u$.

Multiplying Eq. (2.1) by $x\cdot\nabla(u-g)$ and integrating over $B_r^+$, we obtain

$$\int_{B_r^+}R(u)u_t(x\cdot\nabla(u-g))=\int_{B_r^+}\Delta u(x\cdot\nabla(u-g))+\int_{B_r^+}\frac{(1-|u|^2)u}{\varepsilon^2}(x\cdot\nabla(u-g))-$$
$$\int_{B_r^+}(u\times(u\times H))(x\cdot\nabla(u-g))$$
$$=I+II+III. \tag{3.16}$$

The estimates for $I$, $II$, $III$ and then the conclusions of the lemma can be obtained just as the argument in [36] combining with the argument in Lemma 3.1. We omit the details. □

# 4 Lower Bound of $|u_\varepsilon|$

Let $z_0=(x_0,t_0)\in\Omega\times\mathbf{R}^+$, $0<r<\sqrt{t_0}$, $\lambda>0$. We shall show in this section that if $r^{-3}\int_{P_r(z_0)}e_\varepsilon(u_\varepsilon)$ is small enough, then $|u_\varepsilon|\geqslant\frac{1}{2}$ on $\left[t_0-\frac{r^2}{4},t_0\right)$.

**Definition** For any $\varepsilon\in\left(0,\frac{1}{2}\right)$, we define good slice

$$G^\Lambda_{z_0,r}=\left\{t\in[t_0-r^2,t_0):r\int_{\Omega\cap B_r(x_0)}|u_t|^2\leqslant\frac{\Lambda}{r}\int_{(\Omega\times\mathbf{R}_+)\cap P_r(z_0)}|u_t|^2\right\}\tag{4.1}$$

and bad slice

$$B^\Lambda_{z_0,r}=[t_0-r^2,t_0)\setminus G^\Lambda_{z_0,r}.\tag{4.2}$$

It follows from Fubini's theorem that

$$|B^\Lambda_{z_0,r}|\leqslant\frac{r^2}{\Lambda}.\tag{4.3}$$

Similarly to [36], we can prove

**Lemma 4.1** Let $u_\varepsilon$ be smooth solutions of the approximate problem. For any given $\Lambda>0$, there exist $\varepsilon_0>0$ and $r_0>0$ such that for any $z_0=(x_0,t_0)$, if $x_0\in\Omega$, $0<r<\operatorname{dist}(x_0,\partial\Omega)$ or $x_0\in\partial\Omega$, $0<r\leqslant\min\{\sqrt{t_0},r_0\}$ and there holds

$$r^{-3}\int_{(\Omega\times\mathbf{R}_+)\cap P_r(z_0)}e_\varepsilon(u_\varepsilon)\leqslant\varepsilon_0^2,\tag{4.4}$$

then

$$|u_\varepsilon(x,t)|\geqslant\frac{1}{2},\quad\forall x\in\Omega\cap B_{r/4}(x_0),\quad\forall t\in G^\Lambda_{z_0,r/2}.\tag{4.5}$$

**Proof** Let $x_0\in\partial\Omega$. If there exists $(x_1,t_1)\in\bar\Omega\cap B_{r/4}(x_0)\times G^\Lambda_{z_0,r/2}$ such that $|u_\varepsilon|<\frac{1}{2}$, since $|u_\varepsilon|=1$ on $\Omega$ and $|\nabla u_\varepsilon|\leqslant\frac{C}{\varepsilon}$ then $\operatorname{dist}(x_1\partial\Omega)>\frac{\varepsilon}{2C}$. Therefore, for any $0<\theta<\frac{1}{4C}$, we have $B_{\theta\varepsilon}(x_1)\subset\Omega$ and

$$|u_\varepsilon(x,t_1)|\leqslant\frac{1}{2}+C\theta,\quad\forall x\in B_{\theta\varepsilon}(x_1),\tag{4.6}$$

so that

$$\int_{B_{\frac{\varepsilon}{4C}}}|x-x_1|^{-1}\frac{(1-|u_\varepsilon|^2)^2(x,t_1)}{\varepsilon^2}\geqslant C_0.\tag{4.7}$$

The hypothesis in this lemma implies

$$\sup_{x\in\bar\Omega\cap B_{r/2}(x_0)}(r/2)^{-3}\int_{P_{r/2}(z_0)}e_\varepsilon(u_\varepsilon)\leqslant 8\varepsilon_0^2.\tag{4.8}$$

Local energy estimate in Lemma 2.5 implies

$$\sup_{t\in\left[t_0-\frac{r^2}{16},t_0\right)}\sup_{x\in\bar\Omega\cap B_{r/4}(x_0)}(r/4)^{-1}\int_{\Omega\cap B_{r/4}(x_0)}e_\varepsilon(u_\varepsilon)$$
$$\leqslant C\varepsilon_0^2+Cr^{4-6/p}\|H_\varepsilon\|^2_{L^\infty(L^p(B_R(x_0)))}.\tag{4.9}$$

The definition of $G^{\Lambda}_{z_0,\frac{r}{4}}$ yields

$$\sup_{t\in G^{\Lambda}_{z_0,\frac{r}{4}}}\sup_{x\in\bar{\Omega}\cap B_{r/2}(x_0)} r\int_{\Omega\cap B_{r/4}(z_0)}|u_t|^2\leqslant\sup_{x\in\bar{\Omega}\cap B_{r/2}(x_0)}\frac{\Lambda}{r}\int_{(\Omega\times\mathbf{R}_+)\cap P_{r/4}(z_0)}|u_t|^2. \quad (4.10)$$

It follows from (2.20) that

$$\begin{aligned}&\sup_{x\in\bar{\Omega}\cap B_{r/2}(x_0)}\frac{\Lambda}{r}\int_{(\Omega\times\mathbf{R}_+)\cap P_{r/4}(z_0)}|u_t|^2\\ &\leqslant C\Lambda r^{-3}\int_{(\Omega\times\mathbf{R}_+)\cap P_{r/2}(z_0)}|\nabla u_\varepsilon|^2+Cr^{4-6/p}\|H_\varepsilon\|^2_{L^\infty(L^p(B_R(x_0)))}.\end{aligned} \quad (4.11)$$

We deduce from (4.4), (4.10) and (4.11) that

$$\sup_{t\in G^{\Lambda}_{z_0,\frac{r}{4}}}\sup_{x\in\bar{\Omega}\cap B_{r/2}(x_0)} r\int_{\Omega\cap B_{r/4}(z_0)}|u_t|^2\leqslant C\Lambda\varepsilon_0^2+Cr^{4-6/p}\|H_\varepsilon\|^2_{L^\infty(L^p(B_R(x_0)))}. \quad (4.12)$$

Since $x_1\in\Omega\cap B_{r/4}(x_0)$ and $r_1<r/4$, (4.9) and (4.12) imply

$$\begin{aligned}&\sup_{t\in G_{z_0,r/4}}\left\{r_1^{-1}\int_{B_{r_1}}e_\varepsilon(u_\varepsilon)+r_1\int_{B_{r_1}(x_0)}|u_t|^2\right\}\\ &\leqslant C(1+\Lambda)\varepsilon_0^2+Cr^{4-6/p}\|H_\varepsilon\|^2_{L^\infty(L^p(B_R(x_0)))}.\end{aligned} \quad (4.13)$$

Substituting (4.13) into (3.2) or (3.15), one obtains

$$\begin{aligned}&\int_{B_{\frac{\varepsilon}{4C}}(x_1)}|x-x_1|^{-1}\frac{(1-|u_\varepsilon|^2)^2}{\varepsilon_0^2}\\ &\leqslant C(1+\Lambda)\varepsilon_0^2+Cr^{4-6/p}\|H_\varepsilon\|^2_{L^\infty(L^p(B_R(x_0)))}+Cr.\end{aligned} \quad (4.14)$$

This contradicts (4.7) if one chooses $r_0$ and $\varepsilon_0$ small enough and $p>2$. □

## 5 Energy Decay

In this section, we derive the energy decay for the solutions of the approximate problem (2.1)~(2.2). The aim of this section is to prove the following two lemmas. As above, denote $\|H_\varepsilon\|^2_{L^\infty(L^p(B_r(x_0)))}=\|H_\varepsilon\|^2_{L^\infty(t_0-r^2,t_0;L^p(B_r(x_0)))}$.

**Lemma 5.1** There exist $\varepsilon_0>0$ and $0<\theta_0<\frac{1}{2}$ such that for any smooth solution $u_\varepsilon$ of (2.1)~(2.2) and (1.9)~(1.10), $z_0\in\Omega\times\mathbf{R}_+$, $0<r<\min\{\mathrm{dist}(x_0,\partial\Omega),\sqrt{t_0}\}$, $0<\varepsilon\leqslant\varepsilon_0$, the inequality

$$r^{-3}\int_{P_r(z_0)}e_\varepsilon(u_\varepsilon)\leqslant\varepsilon_0^2 \quad (5.1)$$

implies

$$(\theta_0 r)^{-3}\int_{P_{\theta_0 r}(z_0)}e_\varepsilon(u_\varepsilon)\leqslant\frac{1}{2}\max\left\{r^{-3}\int_{P_r(z_0)}e_\varepsilon(u_\varepsilon),C_0r^{4-\frac{6}{p}}\|H_\varepsilon\|^2_{L^\infty(L^p(B_r(x_0)))}\right\}. \quad (5.2)$$

**Lemma 5.2** There exist $\varepsilon_0>0, C>0, r_0>0$ and $0<\theta_0<\frac{1}{2}$ such that for any smooth solution $u_\varepsilon$ of (2.1)~(2.2) and (1.9)~(1.10), $x_0\in\partial\Omega, 0<r<r_0, 0<\varepsilon\leqslant\varepsilon_0$, the inequality

$$r^{-3}\int_{(\Omega\times\mathbf{R}_+)\cap P_r(z_0)} e_\varepsilon(u_\varepsilon)\leqslant\varepsilon_0^2 \tag{5.3}$$

implies

$$(\theta_0 r)^{-3}\int_{(\Omega\times\mathbf{R}_+)\cap P_{\theta_0 r}(z_0)} e_\varepsilon(u_\varepsilon)$$
$$\leqslant\frac{1}{2}\max\left\{r^{-3}\int_{(\Omega\times\mathbf{R}_+)\cap P_r(z_0)} e_\varepsilon(u_\varepsilon), Cr^2\|\nabla g\|^2_{C^1(\Omega\cap B_r(x_0))}, Cr^{4-\frac{6}{p}}\|H_\varepsilon\|^2_{L^\infty(L^p(B_r(x_0)))}\right\}. \tag{5.4}$$

We only prove the second lemma. Before the proof, we give a remark.

**Remark 5.1** This energy decay is different from that in [36] since we have the term $r^{4-\frac{6}{p}}\|H_\varepsilon\|^2_{L^\infty(L^p(B_r(x_0)))}$ which comes from the local energy estimate (2.20) and the generalized monotonicity inequalities (3.1)~(3.2) and (3.14)~(3.15). Although this term is not scaling invariant, we can also handle it, see the proof below.

**Proof of (5.4)** For simplicity, we assume $r=1$, otherwise, we may directly proof the conclusions without letting $r=1$. We also let $\Omega=\mathbf{R}^3_+, x_0=0, t_0=1, u=u_\varepsilon$ and $P_r^+(0,1)=\mathbf{R}^3_+\cap P_r(0,1), r>0$. Denote $\Lambda_g=\|\nabla g\|^2_{C^1(\Omega\cap B_{r_0}(x_0))}$. If $\Lambda_g>\varepsilon_0$ or $\|H_\varepsilon\|_{L^\infty(L^p(B_1(0)))}>\varepsilon_0$, then (5.4) holds. So, we may assume

$$\Lambda_g\leqslant\varepsilon_0 \quad\text{and}\quad \|H_\varepsilon\|_{L^p(B_1(0))}\leqslant\varepsilon_0. \tag{5.5}$$

In order to estimate $\int_{P^+_{\theta_0}(0,1)} e_\varepsilon(u_\varepsilon)=\int_{1-\theta_0^2}^1\int_{B^+_{\theta_0}(0)} e_\varepsilon(u_\varepsilon)$, we divided it into

$$\int_{P^+_{\theta_0}(0,1)} e_\varepsilon(u_\varepsilon)=\int_{(1-\theta_0^2,1)\cap G^\Lambda_{(0,1),1/2}}\int_{B^+_{\theta_0}(0)} e_\varepsilon(u_\varepsilon)+\int_{(1-\theta_0^2,1)\cap B^\Lambda_{(0,1),1/2}}\int_{B^+_{\theta_0}(0)} e_\varepsilon(u_\varepsilon).$$

**Case 1** $t\in G^\Lambda_{(0,1),1/2}$.

Since $4|u|^2|\nabla u|^2=4|\nabla u\times u|^2+|\nabla|u|^2|^2\leqslant 4|\nabla u\times u|^2+2|\nabla|u||^2$ and $|u|>1/2$ if $t\in G^\Lambda_{(0,1),1/2}$, we have

$$\int_{B^+_{\theta_0}(0)} e_\varepsilon(u)=\int_{B^+_{\theta_0}(0)}\frac{1}{2}|\nabla u|^2+\frac{3(1-|u|^2)^2}{4\varepsilon^2}$$
$$\leqslant 2\int_{B^+_{\theta_0}(0)}|\nabla u\times u|^2+\int_{B^+_{\theta_0}(0)}|\nabla|u||^2+\frac{3(1-|u|^2)^2}{4\varepsilon^2}=2A+B. \tag{5.6}$$

We estimate the terms on the right-hand side of (5.6) as follows. We first estimate $A$.

Using the definition of $G^\Lambda_{(0,1),1/2}$ and the local energy estimate near the boundary (2.20), we have

$$\int_{B_{1/2}^+} e(u) + \int_{B_{1/2}^+} |u_t|^2 \leqslant \int_{B_{1/2}^+} e(u) + \Lambda \int_{P_{1/2}^+(0,1)} |u_t|^2$$
$$\leqslant C\Lambda\left( \| H_\varepsilon \|^2_{L^\infty(0,1;L^p(B_1^+))} + \int_{P_1^+(0,1)} e(u) \right)$$
$$\leqslant C\Lambda\varepsilon_0^2. \tag{5.7}$$

We should notice that, if we do not assume $r = 1$, then it follows from the definition of $G^{\Lambda}_{(0,1),r/2}$ and (2.20) that the right-hand side of (5.7) should be

$$C\Lambda\left( r^{4-\frac{6}{p}} \| H_\varepsilon \|^2_{L^\infty(0,1;L^p(B_r^+))} + r^{-3}\int_{P_r^+(0,1)} e(u) \right),$$

which just yields what we have claimed in the lemma. With this and (3.14)~(3.15) at hand, the following estimates can be done in the similar manner in [36]. Hence we only need to sketch them.

Therefore, for $t \in G^{\Lambda}_{(0,1),1/2}$, there holds

$$\sup_{x\in B_{1/4}^+} \left\{ \int_{B_{1/4}^+(x)} e(u) + \int_{B_{1/4}^+(x)} |u_t|^2 \right\}$$
$$\leqslant C\Lambda\left( \| H_\varepsilon \|^2_{L^\infty(0,1;L^p(B_1^+))} + \int_{P_1^+(0,1)} e(u) \right) \leqslant C\Lambda\varepsilon_0^2. \tag{5.8}$$

Now it follows from (5.8) and the monotonicity inequality (3.14) that

$$\sup\left\{ s^{-1}\int_{B_s^+(x)} |\nabla u|^2 : x \in B_{1/4}^+, 0 < s < 1/4 \right\}$$
$$\leqslant C\Lambda_g^2 + C\Lambda\left( \| H_\varepsilon \|^2_{L^\infty(0,1;L^p(B_1^+))} + \int_{P_1^+(0,1)} e(u) \right) \leqslant C(\Lambda\varepsilon_0^2 + \Lambda_g^2). \tag{5.9}$$

Let $\theta_0 \in (0, 1/8)$. Taking $\phi \in C_0^\infty(B_1)$ be a cut-off function even with respect to $x_3$, $0 \leqslant \phi \leqslant 1$, $\phi \equiv 1$ in $B_{\theta_0}$ and $\phi \equiv 0$ outside $B_{2\theta_0}$, $|\nabla\phi| \leqslant c_0\theta_0^{-1}$, extending $u$ from $B_1^+$ to $B_1$ evenly with respect to $x_3$ and defining

$$u_g(x) = (u-g)(x), \quad \text{for } x_3 \geqslant 0, \quad u_g(x) = -(u-g)(x', -x_3), \quad \text{for } x_3 < 0, \tag{5.10}$$

we have

$$\int_{B_{\theta_0}^+} |\nabla u \times u|^2 \leqslant \int_{\mathbf{R}^3} \phi^2 |\nabla u_g \times \bar{u}|^2 + C\Lambda_g^2\theta_0^3. \tag{5.11}$$

The first term on the right-hand side of (5.11) can be rewritten as

$$\int_{\mathbf{R}^3} \phi^2 |\nabla u_g \times \bar{u}|^2 = \int_{\mathbf{R}^3} \phi^2 \nabla u_g \times \bar{u} \cdot \nabla u_g \times \bar{u}$$
$$= \int_{\mathbf{R}^3} \nabla\cdot(\phi^2 \nabla u_g \times \bar{u}) \cdot u_g \times \bar{u} + \int_{\mathbf{R}^3} \phi^2 (\nabla u_g \times \bar{u}) \cdot ((\nabla\bar{u} - \lambda) \times u_g) + \lambda\int_{\mathbf{R}^3} \phi u_g, \tag{5.12}$$

where

$$\lambda = \frac{\int_{\mathbf{R}^3} \phi^2 (\nabla u_g \times \bar{u}) \times \nabla\bar{u}}{\int_{\mathbf{R}^3} \phi},$$

which can be estimated in the following way

$$|\lambda| \leqslant C\theta_0^{-3}\int_{B_{2\theta_0}} (|\nabla u_g|^2 + |\nabla u|^2) \leqslant C\left(\int_{B_{1/2}^+} e(u) + \int_{B_{1/2}^+} |u_t|^2 + \Lambda_g^2\right) \leqslant C\left(\Lambda\int_{P_1^+} e(u) + \Lambda_g^2\right).$$

This combined with the following

$$\| u_g \|_{L^2(B_{2\theta_0})} \leqslant C\Lambda\theta_0^{3/2}\varepsilon_0$$

yields

$$\left|\lambda\int_{\mathbf{R}^3}\phi u_g\right| \leqslant C\theta_0^{3/2}\varepsilon_0\left(\Lambda\int_{P_1^+} e(u) + \Lambda_g^2\right).$$

Similarly to [36] again, we obtain

$$\begin{aligned}\int_{\mathbf{R}_+^3} |\nabla\cdot(\phi\nabla u_g\times\bar{u})|^2 &\leqslant 8\int_{\mathbf{R}_+^3}[|\nabla\phi|^2(|\nabla u|^2+|\nabla g|^2)+\phi^2(|u_t|^2+|\nabla^2 g|^2+|\nabla g|^2|\nabla u|^2)]\\ &\leqslant C\theta_0^2\Lambda_g^2 + C(\Lambda_g^2+\theta_0^2)\int_{B_{2\theta_0}^+}|\nabla u|^2 + C\int_{B_{1/2}^+}|u_t|^2\\ &\leqslant C\Lambda_g^2 + C\Lambda\left(\| H_\varepsilon\|^2_{L^\infty(0,1;L^p(B_1^+))} + \int_{P_1^+(0,1)} e(u)\right)\end{aligned} \tag{5.13}$$

and then, by Hölder inequality, Poincaré inequality we get

$$\begin{aligned}&\int_{\mathbf{R}_+^3} |\nabla\cdot(\phi^2\nabla u_g\times\bar{u})\cdot u_g\times\bar{u}|\\ &\leqslant C\Lambda\theta_0^{3/2}\varepsilon_0\left(C\Lambda_g^2 + C\Lambda\| H_\varepsilon\|^2_{L^\infty(0,1;L^p(B_1^+))} + C\Lambda\int_{P_1^+(0,1)} e(u)\right)^{1/2}.\end{aligned} \tag{5.14}$$

On the other hand, by the definition of BMO space and the monotonicity inequality and the Poincaré inequality as in [36], we have

$$\begin{aligned}\left|\int_{\mathbf{R}^3}\phi^2(\nabla u_g\times\bar{u})\cdot((\nabla\bar{u}-\lambda)\times u_g)\right| &\leqslant C\|\phi^2(\nabla u_g\times\bar{u})\cdot(\nabla\bar{u}-\lambda)\|_{\mathscr{H}^1(\mathbf{R}^3)}[u_g]_{\mathrm{BMO}(B_{1/4})}\\ &\leqslant C\Lambda\varepsilon_0\left(\Lambda_g^2 + \Lambda\int_{P_1^+(0,1)} e(u)\right).\end{aligned} \tag{5.15}$$

Substituting these estimates into (5.12) and then into (5.11), we get

$$\int_{B_{\theta_0}^+} |\nabla u\times u|^2 \leqslant C\Lambda^2\varepsilon_0\left(C\| H_\varepsilon\|^2_{L^\infty(0,1;L^p(B_1^+))} + C\int_{P_1^+(0,1)} e(u)\right) + C\Lambda_g^2\theta_0^3\varepsilon_0. \tag{5.16}$$

We finally get

$$A \leqslant C(\Lambda,\delta,\theta_0)\Lambda_g^2 + (\delta\theta_0^2 + C\Lambda_2\varepsilon_0)\int_{P_1^+(0,1)} e(u) + \frac{C\Lambda}{\delta\theta_0^2}\int_{B_{2\theta_0}^+} |u-g|^2. \tag{5.17}$$

Just as in [36], we may estimate $B = \int_{B_{\theta_0}^+(0)} |\nabla|u||^2 + \frac{3(1-|u|^2)^2}{4\varepsilon^2}$ to get

$$B = \int_{B_{\theta_0}^+(0)} |\nabla|u||^2 + \frac{3(1-|u|^2)^2}{4\varepsilon^2}$$

$$\leqslant C\Lambda\varepsilon\int_{P_1^+(0,1)} e_\varepsilon(u) + C\int_{B_{2\theta_0}^+} |\nabla u \times u|^2 + \frac{\varepsilon}{2}\int_{B_{2\theta_0}^+} |H|^2. \tag{5.18}$$

It follows from (5.6), (5.17) and (5.18) that

$$\int_{B_{\theta_0}^+(0)} e(u) \leqslant C(\Lambda, \delta, \theta_0)\Lambda_g^2 + (\delta\theta_0^2 + C\Lambda^2\varepsilon_0 + C\Lambda\varepsilon)\int_{P_1^+(0,1)} e(u) + \frac{C\Lambda}{\delta\theta_0^2}\int_{B_{2\theta_0}^+} |u-g|^2 + \frac{\varepsilon}{2}\int_{B_{2\theta_0}^+} |H|^2. \tag{5.19}$$

**Case 2** $t \in B^\Lambda_{(0,1),1/2}$.

$$\theta_0^{-3}\int_{B_{\theta_0}^+(0)\times(B^\Lambda_{(0,1),1/2}\cap[1-\theta_0^2,1))} e(u) \leqslant \theta_0^{-3}|B^\Lambda_{(0,1),1/2}|\int_{P_1^+(0,1)} e(u)$$

$$\leqslant \frac{1}{\theta_0^3\Lambda}\int_{P_1^+(0,1)} e(u). \tag{5.20}$$

Combining Case 1 with Case 2, we have

$$\theta_0^{-3}\int_{P_{\theta_0}^+(0,1)} e_\varepsilon(u_\varepsilon) \leqslant C(\Lambda, \delta, \theta_0)\Lambda_g^2 + \left(\delta\theta_0^2 + C\Lambda^2\varepsilon_0 + C\Lambda\varepsilon + \frac{1}{\theta_0^3\Lambda}\right)\int_{P_1^+(0,1)} e(u) + \frac{C\Lambda}{\delta\theta_0^5}\int_{P_{2\theta_0}^+} |u-g|^2 + \frac{\varepsilon}{2}\int_{B_{2\theta_0}^+} |H|^2. \tag{5.21}$$

In the following, we finally estimate $\frac{C\Lambda}{\delta\theta_0^5}\int_{P_{2\theta_0}^+} |u-g|^2$ by compactness method.

**Lemma 5.3** Under the same assumptions of Lemma 5.2, we have

$$\theta_0^{-5}\int_{P_{\theta_0}^+(0,1)} |u_\varepsilon - g|^2 \leqslant \max\left\{\theta_0^2\int_{P_1^+(0,1)} e_\varepsilon(u_\varepsilon), C_0\|\nabla g\|^2_{C^1(B_1^+)}, C_0\|H_\varepsilon\|^2_{L^2(B_1^+)}\right\}. \tag{5.22}$$

**Proof** Suppose the conclusion of Lemma 5.3 is false, then for any $\theta \in (0, 1/2)$, there exists $\varepsilon_k \to 0$, $C_k \to \infty$, $\delta_k \to 0$, $g_k \in C^2(B_1^+)$ and $H_k \in L^2(B_1^+)$ such that

$$\int_{P_1^+(0,1)} e_{\varepsilon_k}(u_{\varepsilon_k}) = \delta_k^2, \quad \|\nabla g_k\|^2_{C^1(B_1^+)} \leqslant C_k^{-1}\delta_k, \quad \|H_k\|^2_{L^2(B_1^+)} \leqslant C_k^{-1}\delta_k \tag{5.23}$$

but

$$\theta^{-5}\int_{P_\theta^+(0,1)} |u_k - g|^2 \geqslant \theta^2\delta_k^2. \tag{5.24}$$

Define $v_k = \delta_k^{-1}(u_k - g_k)$. We know that $\{v_k\}$ is uniformly bounded in $H^1(P_1^+(0,1))$ and $v_k = 0$ on $T_1$. From the above we may assume as $\varepsilon_k \to 0$, there holds $v_k \to v$ weakly in $H^1(P_1^+(0,1))$ and strongly in $L^2(P_1^+(0,1))$, $u_k \to p$ for some $p \in S^2$, $H_k \to 0$ in $L^2$. Now we can claim that $v$ solves

$$R(p)v_t - \Delta v = 0. \tag{5.25}$$

The proof left over is just like that in [36] if we notice that $H_k \to 0$ in $L^2$. We omit the details. □

Putting the above discussions together, we finally get the decay estimates. The proof of Lemma 5.2 (decay near the boundary) is finished. The decay estimate in the interior can be done in the similar manner. That is, the proof of Lemma 5.1 can be omitted. □

# 6 Partial Regularity

In this section, we prove the main theorem on the partial regularity.

**Proposition 6.1** Under the same assumptions of the main theorem, there exist $\varepsilon_0>0$, $\alpha\in(0,1)$, $C_0>0$ such that if either

(1) for $x_0\in\Omega$, $t_0>0$, $0<r<\min\{d(x_0,\partial\Omega),\sqrt{t_0}\}$, and $\varepsilon<(\rho/r)^3 r$

or

(2) for $x_0\in\partial\Omega$, $t_0>0$, $r_0=r_0(\Omega)>0$, $0<r<\min\{r_0,\sqrt{t_0}\}$, and $\varepsilon<(\rho/r)^3 r$ if $u_\varepsilon$ satisfies

$$r^{-3}\int_{(\Omega\times\mathbf{R}_+)\cap P_r(x_0,t_0)} e_\varepsilon(u_\varepsilon)\leqslant\varepsilon_0^2, \tag{6.1}$$

then either

$$(\text{i})\quad \rho^{-3}\int_{P_\rho(x_0,t_0)} e_\varepsilon(u_\varepsilon)+\rho^2\left|\frac{\partial u_\varepsilon}{\partial t}\right|^2\leqslant C_0\varepsilon_0^2(\rho/r)^{2\alpha},\quad \forall 0<\rho\leqslant r/4, \tag{6.2}$$

or

$$(\text{ii})\quad \rho^{-3}\int_{(\Omega\times\mathbf{R}_+)\cap P_\rho(x_0,t_0)} e_\varepsilon(u_\varepsilon)+\rho^2\left|\frac{\partial u_\varepsilon}{\partial t}\right|^2$$
$$\leqslant C_0\max\{\varepsilon_0^2,r^2\}(\rho/r)^{2\alpha},\quad \forall 0<\rho<r/4. \tag{6.3}$$

**Proof** We only consider the interior case, for the near boundary case can be handled in the similar manner. Let $k\geqslant 1$ be such that $\theta_0^{k+1}r\leqslant\rho\leqslant\theta_0^k r$. Then by iterating Lemma 5.1 $k$ times, we get

$$\theta_0^k r^{-3}\int_{P_{\theta_0^k r}(x_0,t_0)} e_\varepsilon(u_\varepsilon)+\theta_0^k r^2\left|\frac{\partial u_\varepsilon}{\partial t}\right|^2\leqslant 2^{-k}r^{-3}\int_{P_r(x_0,t_0)} e_\varepsilon(u_\varepsilon). \tag{6.4}$$

Since $\dfrac{\ln(\rho/r)}{\ln(\theta_0)}-1\leqslant k\leqslant\dfrac{\ln(\rho/r)}{\ln(\theta_0)}$, one sees from (6.4) that there exists $\alpha\in(0,1)$ such that (i) holds. The proposition follows. □

**Proof of the Main Theorem** Let $\varepsilon_0$ be as in Proposition 6.1, and define the concentrate set of the sequence $\{u_\varepsilon\}$ by

$$\Sigma=\bigcap_{r>0}\left\{z\in\bar{\Omega}\times\mathbf{R}_+:\lim\inf_{k\to\infty} r^{-3}\int_{(\Omega\times\mathbf{R}_+)\cap P_r(z)} e_{\varepsilon_k}(u_{\varepsilon_k})\geqslant\varepsilon_0^2\right\}. \tag{6.5}$$

The standard covering argument (see [9]) shows that $\mathcal{H}^3(\Sigma\cap K)<\infty$ for any compact subset of $\bar{\Omega}\times\mathbf{R}_+$. Let $u$ be a weak limits of $u_{\varepsilon_k}$ in $H^1_{\text{loc}}(\Omega\times\mathbf{R}_+;\mathbf{R}^3)$. Then for any $z_0\in\bar{\Omega}\times\mathbf{R}_+\setminus\Sigma$, Proposition 6.1 and the definition of $\Sigma$ imply that there exists $r_0>0$ such that for any $z\in\bar{\Omega}\times\mathbf{R}_+\cap P_{r/2}(z_0)$ and $0<\rho\leqslant r/4$ we have by sending $k\to\infty$

$$\rho^{-3}\int_{(\Omega\times\mathbf{R}_+)\cap P_\rho(z)}|\nabla u|^2+\rho^2\left|\frac{\partial u}{\partial t}\right|^2\leqslant C(\rho/r)^{2\alpha}. \tag{6.6}$$

It follows from Morrey's Lemma that $u\in C^\alpha((\bar{\Omega}\times\mathbf{R}_+)\cap P_{r_0/4}(z_0);S^2)$. The higher smoothness follows from standard bootstrap argument (see [17,30], or [36]). The theorem follows. □

## References


[1] Alouges F, Soyeur A. On global weak solutions for Landau-Lifshitz equations: existence and nonuniqueness. Nonlinear Anal, 1992, 18 (11): 1071－1084.

[2] Carbou G. Regularity for critical points of a nonlocal energy. Calc Var Partial Differential Equations, 1997, 5: 409－433.

[3] Carbou G. Thin layers in micromagnetism. Math Models Methods Appl Sci, 2001, 11 (9): 1529－1546.

[4] Carbou G, Fabrie P. Regular solutions for Landau-Lifshitz equation in a bounded domain. Differential Integral Equations, 2001, 14: 213－229.

[5] Carbou G, Fabrie P. Time average in micromagnetism. J Differential Equations, 1998, 147: 383－409.

[6] Carbou G, Fabrie P, Gués O. Boundary layers for Landau-Lifshitz equations. Phys B, 2004, 343: 331－336.

[7] Carbou G, Labbe S. Stability for static walls in ferromagnetic nanowires. Preprint, 2006.

[8] Chen Y, Ding S, Guo B. Partial regularity for two-dimensional Landau-Lifshitz equations. Acta Math Sinica, 1998, 14 (3): 423－432.

[9] Chen Y, Struwe M. Existence and partial regularity results for the heat flow of harmonic maps. Math Z, 1989, 201: 83－103.

[10] Coron J. Nonuniqueness for the heat flow of harmonic maps. Ann Inst H Poincaré Anal Non Linéaire, 1990, 7: 335－344.

[11] DeSimone A, Kohn R V, Müller S, et al. A reduced theory for thin film micromagnetics. Comm Pure Appl Math, 2002, 55: 1408－1460.

[12] DeSimone A, Kohn R V, Müller S, et al. Magnetic microstructure: a paradigm of multiscale problem// Ball J, Hunt J C R. Proc ICIAM 99, Edinburgh, 1999. Oxford: Oxford Univ Press, 2000: 175－199.

[13] DeSimone A, Kohn R V, Müller S, el al. A compactness result in the gradient theory of phase transitions. Proc Roy Soc Edinburgh Sect A, 2001, 131: 833－844.

[14] Ding S, Guo B. Hausdorff measure of the singular set of Landau-Lifshitz equations with a nonlocal term. Comm Math Phys, 2004, 250 (1): 95－117.

[15] Ding S, Wang C. Finite time singularity of the Landau-Lifshitz-Gilbert equation. Int Math Res Not, 2007, Article ID rnm012.

[16] Evans L C. Partial regularity for harmonic maps into spheres. Arch Ration Mech Anal, 1991, 116: 101－113.

[17] Feldman M. Partial regularity for harmonic maps of evolution into spheres. Comm Partial Differential Equations, 1994, 19: 761－790.

[18] Gilbarg D, Trudinger N S. Elliptic partial differential equations of second order. Berlin: Springer-Verlag, 2001.

[19] Gioia G, James R D. Micromagnetics of very thin films. Proc R Soc Lond Ser A, 1997, 453: 213－223.

[20] Guo B, Hong M. The Landau-Lifshitz equations of the ferromagnetic spin chain and harmonic maps. Calc Var Partial Differential Equations, 1993, 1: 311 - 334.

[21] Guo B, Su F. Global weak solution for the Landau-Lifshitz-Maxwell equation in three space dimensions. J Math Anal Appl, 1997, 211 (1): 326 - 346.

[22] Gues O, Sueur F. On 3D domain walls for the Landau-Lifshitz equations. Dyn Partial Differ Equ, 2007, 4 (2): 143 - 165.

[23] Kurzke M, Melcher C, Moser R. Domain walls and vortices in the thin ferromagnetic films. Phys B, 2004, 343: 331 - 336.

[24] Kurzke M, Melcher C, Moser R. Domain walls and vortices in thin ferromagnetic films. MPI Leipzig, preprint No.45, 2006.

[25] Kohn R V, Slastikov V V. Another thin film limit of micromagnetics. Arch Ration Mech Anal, 2005, 178 (2):227 - 245.

[26] Ladyzhenskaya O A, Uraltzeva N N, Solonnikov V A. Linear and quasilinear equations of parabolic type. Providence, RI: Amer Math Soc, 1967.

[27] Liu X. Partial regularity for Landau-Lifshitz system of ferromagnetic spin chain. Calc Var Partial Differential Equations, 2004, 20 (2): 153 - 173.

[28] Melcher C. Domain wall motion in ferromagnetic films. Phys D, 2004, 192: 249 - 264.

[29] Melcher C. Existence of partially regular solutions for Landau-Lifshitz equations in $\mathbf{R}^3$. Comm Partial Differential Equations, 2005, 30: 567 - 587.

[30] Moser R. Partial regularity for the Landau-Lifshitz equation in small dimensions. MPI Leipzig, preprint, 2002.

[31] Moser R. Ginzburg-Landau vortices in thin ferromagnetic films. ARMX Appl Math Res Ex, 2003, 1: 1 - 32.

[32] Moser R. Boundary vortices for thin ferromagnetic films. Arch Ration Mech Anal, 2004, 174 (2): 267 - 300.

[33] Moser R. Moving boundary vortices for a thin film limit in micromagnetics. Comm Pure Appl Math, 2005, 58: 701 - 721.

[34] Riviére T. Everywhere discontinuous harmonic maps from $B^3$ into $S^2$. C R Acad Sci Paris, 1992, 314: 719 - 723.

[35] Wang C. On moving Ginzburg-Landau vortices. Comm Anal Geom, 2004, 12 (5): 1185 - 1199.

[36] Wang C. On Landau-Lifshitz equations in dimensions at most four. Indiana Univ Math J, 2006, 55 (5): 1615 - 1644.

[37] Zhou Y, Guo B. Existence of weak solution for boundary problems of systems of ferromagnetic chain. Sci China Ser A, 1984, 27: 799 - 811.

[38] Zhou Y, Guo B. Weak solution systems of ferromagnetic chain with several variables. Sci China Ser A, 1987, 30: 1251 - 1266.

[39] Zhou Y, Guo B, Tan S. Existence and uniqueness of smooth solution of system of ferromagnetic chain. Sci China Ser A, 1991, 34: 257 - 266.

# The Limit Behavior of the Solutions to a Class of Nonlinear Dispersive Wave Equations *

Tian Lixin(田立新) Gui Guilong(桂贵龙) Guo Boling(郭柏灵)

**Abstract**

In this paper, we study the limit behavior of the solutions to a class of nonlinear dispersive wave equations. We also demonstrate that the solutions to Eq. (1.1) converge to the solution to the corresponding BBM equation as the parameter $\gamma$ converges to zero. And the convergence of solutions to Eq. (1.1) as $\alpha^2\to 0$ is studied in $H^s(\mathbf{R})$, $s>\frac{3}{2}$. Given the discussion of the parameters in the nonlinear dispersive wave equation (1.1), one will obtain some conditions in which compacton and peakon occur.

**Keywords** nonlinear dispersive wave equation; Bi-Hamiltonian structure; well-posedness; weak and strong limit; peakon; compacton

## 1 Introduction

The nonlinear dispersive wave equation

$$\begin{cases} u_t - \alpha^2 u_{xxt} + ku_x + 3uu_x + \beta u_{xxx} = \gamma\alpha^2(2u_xu_{xx} + uu_{xxx}) \\ u(0,x) = u_0(x), \quad t > 0, x \in \mathbf{R} \end{cases} \tag{1.1}$$

has recently been investigated with regard to well-posedness and singularity formation by Yin[1]. He also showed the existence of smooth solitary waves for certain values of the parameters $k$ and $\gamma$. For $k=0$, Eq. (1.1) was found by Dai and Huo[2] as a model for nonlinear waves in cylindrical hyperelastic rods with $u(t,x)$ representing the radial stretch related to a pre-stressed state. The study of special compressible materials leads to values of $\gamma$ ranging from $-29.476$ to $3.417$. In this case, the equation is not a long-wave model, but rather a narrow-banded spectrum approximation. In [3], the dynamic stability of solitary wave solutions of (1.1) was considered by H. Kalisch. It shows that for all $\gamma$, the equation admits smooth stable solitary waves. When $\gamma<1$, all solitary waves are stable. For $\gamma=1$, $\beta=0$ the equation is known as the Camassa-Holm equation

$$u_t - \alpha^2 u_{xxt} + ku_x + 3uu_x = \alpha^2(2u_xu_{xx} + uu_{xxx}). \tag{1.2}$$

It was first found by Fokas and Fuchssteiner[4], and later recovered as a water wave model by Camassa, Holm and Hyman in [5,6]. They show that for all $k$, (1.2) has a Lax pair

* J Math Anal Appl, 2008, 341: 1311-1333.

formulation, and for $k=0$, (1.2) has traveling wave solutions to the form $ce^{-|x-ct|}$, which is called peakon because they have a discontinuous first derivative at the wave peak. For every $k$, Eq. (1.2) is bi-Hamiltonian and thus possesses an infinite number of conservation law. Moreovcr, Eq. (1.2) has a simple multi-peakon which shows many fantastic properties. Recently, Johnson has given an account of the role of the Camassa-Holm equation as a long-wave model for water waves in a long uniform channel[7]. Besides being a model equation for water waves, the Camassa-Holm equation is also interesting due to its integrable bi-Hamiltonian structure. This property has led to many interesting developments, a sample of which can be found in [4−6,8−11]. One aspect of the integrability of the equation in case $\gamma=1$ is that the solitary wave solutions are soliton (see [5,12]), and it is similar to the solitary wave solutions to the Korteweg-de Vries equation. In [22], the existence of global weak solutions to the Camassa-Holm equation is proved by using the method of compensated compactness. G. Rodriguez-Blanco demonstrated the Cauchy problem for the Camassa-Holm equation by the method of T. Kato (see [28]). In [24], Tang and Yang obtained the general explicit expressions of the two wave solutions to Eq. (1.2) by using the bifurcation phase portraits of the traveling wave system. Yet another application of the Camassa-Holm equation arises in the context of differential geometry, where it can be seen as a re-expression for geodesic flow on an infinite-dimensional Lie group (see [13,14]). In [34 − 36], the authors study the solution property of peakon in generalized Dullin-Gottwald-Holm equation and generalized Degasperis-Procesi equation. When $\gamma\neq0$, the diffeomorphism transformation

$$u(t,x)\mapsto\frac{1}{\gamma}u(t,x)$$

will formally simplify Eq. (1.1) to the following equation

$$u_t-\alpha^2u_{xxt}+ku_x+\frac{3}{\gamma}uu_x+\beta u_{xxx}=\alpha^2(2u_xu_{xx}+uu_{xxx}).$$

Considering the above transformation, one can see that the solutions to Eq. (1.1) are related in form to the solutions to the corresponding Camassa-Holm equation (1.2) as $\gamma\to1$. When $\gamma\to0$, however, the transformation can not be used to Eq. (1.1). But, for $\gamma=\beta=0$, Eq. (1.1) is related in form to become the BBM equation as follows:

$$u_t-\alpha^2u_{xxt}+2\omega u_x+3uu_x=0,$$

where $k=2\omega$. It is a well-known model for surface waves in a channel[15], and the solitary waves are smooth. Despite having a Hamiltonian structure, the equation is not integrable and its solitary waves are not solitons[16].

With $\alpha=0$ in Eq. (1.1) we find the well-known Korteweg-de Vries equation, which describes the unidirectional propagation of waves at the free surface of shallow water under the influence of gravity. $u(x,t)$ represents the wave height above a flat bottom. $x$ is proportional to distance in the direction of propagation and $t$ is proportional to the time elapsed. The Cauchy problem of the KdV equation has been studied extensively (see [21,25 − 30]). As soon as $u_0\in(H^1\mathbf{R})$, the solution to the KdV equation is global (see [30]). The equation is integrable (see [31]) and its solitary waves are solitons (see [32]). The traveling wave of the equation is linearly unstable (see [33]). And in [25], J. L. Bona and R. Scott demonstrated KdV equation as the limit of BBM equation. On the basis of the above researches, this paper studies the Cauchy problem, the behavior of solutions to the Cauchy problem for the nonlinear dispersive wave equation (1.1) as the dispersive parameter $\gamma$ tends to zero. Given some *a priori* estimates, the locally strong or weak limit of the solution to the nonlinear dispersive wave equation (1.1) as the dispersive parameter $\gamma$ tends to zero is obtained. By using the method in [17], we can demonstrate that the solutions to the nonlinear dispersive wave equation (1.1) as $\alpha^2$ tends to zero converges to the solution of corresponding KdV equation. Finally, we can obtain some condition in which compactons and peakons do occur after the discussion of the parameters in the nonlinear dispersive wave equation (1.1).

In the paper, main theorems are Theorems 3.2, 4.6, 5.3 and 5.4.

**Notations** We shall use the standard notation $|\cdot|_p$ for the norm of the space $L^p(\mathbf{R})$, $1\leqslant p<\infty$, i.e., $|f|_p=\left(\int_{\mathbf{R}}|f|^p\mathrm{d}x\right)^{1/p}$. The space $L^\infty=L^\infty(\mathbf{R})$ consists of all essentially bounded, Lebesgue measurable functions with the standard norm

$$|f|_\infty=|f|_{L^\infty}=\inf_{m(e)=0}\sup_{x\in\mathbf{R}\setminus e}|f(x)|.$$

And we denote the norm in the Sobolev space $H^s=H^s(\mathbf{R})$ by

$$\|f\|_s=\|f\|_{H^s}=\left(\int_{\mathbf{R}}(1+|\xi|^2)^s|\hat{f}(\xi)|^2\mathrm{d}\xi\right)^{1/2}$$

for $s\in\mathbf{R}$. Here $\hat{f}(\xi)$ is the Fourier transform of $f(x)$. We also define the operator $\Lambda_\alpha^s$ for any $\alpha, s\in\mathbf{R}$ by the formula

$$\Lambda_\alpha^s=(1-\alpha^2\partial_x^2)^{\frac{s}{2}}\ (\text{respectively } \Lambda^s:=\Lambda_1^s=(1-\partial_x^2)^{\frac{s}{2}},\ \Lambda=\Lambda_1^1)$$

and denote $\langle\cdot,\cdot\rangle_s$ as the inner product on $H^s$.

In the paper, Section 1 gives the introduction. Section 2 deals with some *a priori* estimates. In Section 3, we research on the strong and weak limit behavior of solutions to (1.1) as $\gamma\to0$ and get the result that the solution converges to the solution of BBM equation under $H^s, s>3/2$. In Section 4, we study the behavior of solution of (1.1) as parameter $\alpha$ tends to zero and have that solution of (1.1) converges to the solution of KdV equation as $\varepsilon\to0$ under $H^s$, $s>3$, where $\varepsilon$ denotes $\alpha^2$. In Section 5, we obtain some conditions in which compacton and peakon occur.

## 2 *A Priori* Estimates

We define $p(x)=\frac{1}{2\alpha}\mathrm{e}^{-\frac{|x|}{\alpha}}, x\in\mathbf{R}$, then $(1-\alpha^2\partial_x^2)^{-1}f=p*f$ for all $f\in L^2(\mathbf{R})$, where $*$ denotes convolution. Using this identity, we can rewrite Eq. (1.1) as the following nonlocal form:

$$u_t + \gamma u u_x - \frac{\beta}{\alpha^2} u_x + \partial_x p * \left[\frac{3-\gamma}{2} u^2 + \frac{\gamma \alpha^2}{2} u_x^2 + \left(k + \frac{\beta}{\alpha^2}\right) u\right] = 0. \qquad (2.1)$$

Eq. (1.1) can be written in Hamiltonian form and has the following invariants:

$$E(u) = \frac{1}{2}\int_{\mathbf{R}} (u^2 + \alpha^2 u_x^2) \mathrm{d}x, \quad E(u) = \frac{1}{2}\int_{\mathbf{R}} (u^3 + \gamma u u_x^2 + k u^2 - \beta u_x^2) \mathrm{d}x.$$

Similar to the discussion of Proposition 2.1 in [17], one can see that Eq. (1.1) is suitable for applying Kato's theory (see [18]).

**Lemma 2.1** Given $u_0 \in H^s(\mathbf{R}), s > \frac{3}{2}$, there exists a maximal time $T(\alpha, \omega, \gamma, u_0) > 0$, and a unique solution $u$ to Eq. (1.1), such that

$$u = u(\cdot, u_0) \in C([0,T); H^s(\mathbf{R})) \cap C^1([0,T); H^{s-1}(\mathbf{R})).$$

Moreover, the solution $u$ to (1.1) depends continuously on the initial data $u_0$.

According to the inequalities

$$\int_{\mathbf{R}} (u^2 + u_x^2) \mathrm{d}x \leqslant \int_{\mathbf{R}} (u^2 + \alpha^2 u_x^2) \mathrm{d}x = 2E(u) \quad (\alpha \geqslant 1),$$

$$\int_{\mathbf{R}} (u^2 + u_x^2) \mathrm{d}x \leqslant \frac{1}{\alpha^2}\int_{\mathbf{R}} (u^2 + \alpha^2 u_x^2) \mathrm{d}x = \frac{2}{\alpha^2} E(u) \quad (\alpha \leqslant 1),$$

the invariance of $E(u)$ guarantees that all solutions to Eq. (1.1) are uniformly bounded as long as they exist. Considering the above, we can obtain the following results by applying the method of Theorem 2.2's proof in [17].

**Theorem 2.2** Given $u_0 \in H^s(\mathbf{R}), s > \frac{3}{2}$, the solution $u = u(\cdot, u_0)$ to Eq. (1.1) is uniformly bounded on $[0, T)$. Moreover, $T < \infty$ if and only if

$$\liminf_{t \uparrow T} \{ \inf_{x \in \mathbf{R}} [\gamma u_x(t, x)] \} = -\infty.$$

For later estimation of Sobolev norms of solutions, we will require the following basic inequalities, due to Y. A. Li and P. J. Olver[19].

**Lemma 2.3** Given $q \geqslant 0$, let $u = u(x) \in H^q$ be any function so $\| u_x \|_{L^\infty} < \infty$. Then there is a constant $c_q$ depending only on $q$, such that the following inequalities hold:

$$\left| \int_{\mathbf{R}} \Lambda^q u \Lambda^q (u u_x) \mathrm{d}x \right| \leqslant c_q \| u_x \|_{L^\infty} \| u \|_{H^q}^2, \qquad (2.2)$$

$$\left| \int_{\mathbf{R}} \Lambda^q u \Lambda^q (u^2) \mathrm{d}x \right| \leqslant c_q \| u \|_{L^\infty} \| u \|_{H^q}^2. \qquad (2.3)$$

Moreover, if $u$ and $f$ are functions in $H^{q+1} \cap \{ \| u_x \|_{L^\infty} < \infty \}$, then

$$\left| \int_{\mathbf{R}} \Lambda^q u \Lambda^q (uf)_x \mathrm{d}x \right| \leqslant \begin{cases} c_q \| f \|_{H^{q+1}} \| u \|_{H^q}^2, & q \in \left(\frac{1}{2}, 1\right] \\ c_q (\| f \|_{H^{q+1}} \| u \|_{L^\infty} \| u \|_{H^q} + \| f \|_{H^q} \| u_x \|_{L^\infty} \| u \|_{H^q} + \| f_x \|_{L^\infty} \| u \|_{H^q}^2), & q \in (1, \infty) \end{cases}. \qquad (2.4)$$

Now we study norms of solution to (1.1) using energy estimates. Without loss of generality, we always assume the parameter $\alpha=1$ in the remainder of this section. The proof of the following theorem is motivated for Theorem 4.1 in [19].

**Theorem 2.4** Suppose that for some $s>\frac{3}{2}$, the function $u(x,t)$ is a solution of Eq. (1.1) corresponding to the initial data $u_0\in H^s$. Then for any real number $q\in(0,s-1]$, there exists a constant $c$ depending only on $q$, such that

$$\begin{aligned}&\|u\|_{H^q}^2+\|u_x\|_{H^q}^2\\ \leqslant&\|u_0\|_{H^q}^2+\|u_{0x}\|_{H^q}^2+c\int_0^t\|u_x\|_{L^\infty}\left(\int_{\mathbf{R}}(\Lambda^q u)^2+(\Lambda^q u_x)^2\mathrm{d}x\right)\mathrm{d}t.\end{aligned}\tag{2.5}$$

For any $q\in(1/2,s-1]$ and any $r\in(1/2,q]$, there exists a constant $c$ depending only on $r$ and $q$, such that

$$\begin{aligned}&\|u\|_{H^q}^2+\|u_x\|_{H^q}^2\\ \leqslant&\|u_0\|_{H^q}^2+\|u_{0x}\|_{H^q}^2+c\int_0^t\left(\int_{\mathbf{R}}(\Lambda^{r+1}u)^2\mathrm{d}x\right)^{1/2}\left(\int_{\mathbf{R}}(\Lambda^q u)^2+(\Lambda^q u_x)^2\mathrm{d}x\right)\mathrm{d}t.\end{aligned}\tag{2.6}$$

Moreover, for any $q\in(0,s-1]$, there exists a constant $c$ depending only on $q$, such that

$$\|u_t\|_{H^q}\leqslant c(\|u\|_{H^1}+1)\|u\|_{H^{q+1}}.\tag{2.7}$$

**Proof** For any $q\in(0,s-1]$, applying $(\Lambda^q u)\Lambda^q$ to both sides of Eq. (1.1) and integrating on $\mathbf{R}$ with respect to $x$ again, one obtains the equation

$$\begin{aligned}&\frac{1}{2}\frac{\mathrm{d}}{\mathrm{d}t}\int_{\mathbf{R}}(\Lambda^q u)^2+(\Lambda^q u_x)^2\mathrm{d}x\\ =&-3\int_{\mathbf{R}}\Lambda^q u\Lambda^q(uu_x)\mathrm{d}x+\frac{\gamma}{2}\int_{\mathbf{R}}\Lambda^q u\Lambda^q(\partial_x^3(u^2))+\Lambda^q u_x\Lambda^q(u_x^2)\mathrm{d}x\\ =&(-3+\gamma)\int_{\mathbf{R}}\Lambda^q u\Lambda^q(uu_x)\mathrm{d}x-\gamma\int_{\mathbf{R}}\Lambda^{q+1}u\Lambda^{q+1}(uu_x)\mathrm{d}x+\frac{\gamma}{2}\int_{\mathbf{R}}\Lambda^q u_x\Lambda^q(u_x^2)\mathrm{d}x\end{aligned}$$

using integration by parts. It follows from the inequalities (2.2) and (2.3) that there is a constant $c_q$ so that

$$\begin{aligned}\frac{1}{2}\frac{\mathrm{d}}{\mathrm{d}t}(\|u\|_{H^q}^2+\|u_x\|_{H^q}^2)&\leqslant 4c_q\|u_x\|_{L^\infty}\|u\|_{H^q}^2+\frac{3}{2}\gamma c_q\|u_x\|_{L^\infty}\|u_x\|_{H^q}^2\\ &\leqslant 4c_q\|u_x\|_{L^\infty}(\|u\|_{H^q}^2+\gamma\|u_x\|_{H^q}^2)\leqslant 4c_q\|u_x\|_{L^\infty}(\|u\|_{H^q}^2+\|u_x\|_{H^q}^2).\end{aligned}$$

Integrating with respect to $t$ on both sides of the above inequality leads to inequality (2.5). Applying the inequality

$$\|u_x\|_{L^\infty}\leqslant c_{r+1}\|u\|_{H^{r+1}}\leqslant c_{r+1}\left(\int_{\mathbf{R}}(\Lambda^{r+1}u)^2\mathrm{d}x\right)^{1/2}$$

for $r>\frac{1}{2}$ to the right-hand side of (2.5) yields the estimate (2.6). For any $q\in(0,s-1]$, applying $(\Lambda^q u_t)\Lambda^q$ to both sides of Eq. (2.1)

$$u_t=-\gamma uu_x+\beta u_x-\partial_x p*\left(\frac{3-\gamma}{2}u^2+\frac{\gamma}{2}u_x^2+(k+\beta)u\right)$$

and integrating on $\mathbf{R}$ with respect to $x$ again, one obtains the equation

$$\frac{1}{2}\int_{\mathbf{R}}(\Lambda^q u_t)^2\mathrm{d}x = -\gamma\int_{\mathbf{R}}\Lambda^q u_t\Lambda^q(uu_x)\mathrm{d}x - 2\omega\int_{\mathbf{R}}\Lambda^q u_t\Lambda^{q-2}(u_x)\mathrm{d}x - \frac{3-\gamma}{2}\int_{\mathbf{R}}\Lambda^q u_t\Lambda^{q-2}(\partial_x(u^2))\mathrm{d}x - \frac{\gamma}{2}\int_{\mathbf{R}}\Lambda^q u_t\Lambda^{q-2}(u_x^2)_x\mathrm{d}x. \quad (2.8)$$

By using the inequalities (2.2), (2.3), one may obtain the following inequalities

$$\left|\int_{\mathbf{R}}\Lambda^q u_t\Lambda^q(uu_x)\mathrm{d}x\right| \leqslant \|\Lambda^q u_t\|_{L^2}\left(\int_{\mathbf{R}}(1+\xi^2)^{q+1}\mathrm{d}\xi\left(\int_{\mathbf{R}}\hat{u}(\xi-\eta)\hat{u}(\eta)\mathrm{d}\eta\right)^2\right)^{1/2}$$
$$\leqslant c_q\|u_t\|_{H^q}\|\hat{u}\|_{L^1}\|u\|_{H^{q+1}} \leqslant c_q\|u_t\|_{H^q}\|u\|_{H^1}\|u\|_{H^{q+1}}.$$

Using the above inequalities and integration by parts, we have

$$\left|\int_{\mathbf{R}}\Lambda^q u_t\Lambda^{q-2}(\partial_x(u^2))\mathrm{d}x\right| = \left|\int_{\mathbf{R}}\Lambda^{q-1}u_t\Lambda^{q-1}(\partial_x(u^2))\mathrm{d}x\right|$$
$$\leqslant \|\Lambda^{q-1}u_t\|_{L^2}\left(\int_{\mathbf{R}}(1+\xi^2)^{q-1}\mathrm{d}\xi\left(\int_{\mathbf{R}}\hat{u}(\xi-\eta)\hat{u}(\eta)\mathrm{d}\eta\right)^2\right)^{1/2}$$
$$\leqslant c_q\|u_t\|_{H^{q-1}}\|\hat{u}\|_{L^1}\|u\|_{H^q} \leqslant c_q\|u_t\|_{H^{q-1}}\|u\|_{H^1}\|u\|_{H^q}$$

and

$$\left|\gamma\int_{\mathbf{R}}\Lambda^q u_t\Lambda^{q-2}(u_x^2)_x\mathrm{d}x\right| \leqslant c_q\gamma\|u_t\|_{H^q}\left(\int_{\mathbf{R}}(1+\xi^2)^{q-1}\mathrm{d}\xi\left(\int_{\mathbf{R}}\hat{u}_x(\xi-\eta)\hat{u}_x(\eta)\mathrm{d}\eta\right)^2\right)^{1/2}$$
$$\leqslant c_q\gamma\left(\int_{\mathbf{R}}(1+\xi^2)^{-1}\|u_x\|_{L^2}^2\|u_x\|_{H^q}^2\mathrm{d}\xi\right)^{1/2}$$
$$\leqslant c_q\gamma\|u_t\|_{H^q}\|u\|_{H^1}\|u\|_{H^{q+1}}.$$

And we also obtain the following inequalities by using integration by parts

$$\left|\int_{\mathbf{R}}\Lambda^q u_t\Lambda^{q-2}(u_x)\mathrm{d}x\right| \leqslant c_q\|u_t\|_{H^q}\|u_x\|_{H^{q-2}}.$$

Applying the above six inequalities to (1.1) yields the inequality

$$\|u_t\|_{H^q}^2 \leqslant c_q(\|u\|_{H^1}+1)\|u\|_{H^{q+1}}\|u_t\|_{H^q},$$

so

$$\|u_t\|_{H^q} \leqslant c_q(\|u\|_{H^1}+1)\|u\|_{H^{q+1}}.$$

□

**Lemma 2.5** Let $u_{\varepsilon 0}$ be the convolution $u_{\varepsilon 0}=\phi_\varepsilon * u_0$ of the functions $\phi_\varepsilon(x) = \varepsilon^{-1/4}\phi(\varepsilon^{-1/4}x)$ and $u_0$ such that the Fourier transform $\hat{\phi}$ of $\phi$ satisfies $\hat{\phi}\in C_0^\infty([0,\infty);H^\infty)$, $\widehat{\phi(\xi)}\geqslant 0$, and $\widehat{\phi(\xi)}=1$ for any $\xi\in(-1,1)$, the following estimates hold for any $\varepsilon$ with $0<\varepsilon<1/4$:

$$\|u_{\varepsilon 0}\|_{H^q} \leqslant c \quad (q\leqslant s), \quad (2.9)$$

$$\|u_{\varepsilon 0}\|_{H^q} \leqslant c\varepsilon^{\frac{s-q}{4}} \quad (q>s), \quad (2.10)$$

$$\|u_{\varepsilon 0}-u_0\|_{H^q} \leqslant c\varepsilon^{\frac{s-q}{4}} \quad (q\leqslant s), \quad (2.11)$$

$$\|u_{\varepsilon 0}-u_0\|_{H^s} = o(1). \quad (2.12)$$

Combining the estimates in Lemma 2.3, we shall evaluate norms of the function $u_\varepsilon$ in the following theorem, which will be used to show the convergence of $\{u_\varepsilon\}$. For convenience, we shall always omit $\varepsilon$ in the solution $u_\varepsilon$ of Eq. (1.1) in the remainder of the paper.

**Lemma 2.6** There exist constants $c_1$ and $c_2$ independent of $\varepsilon$ such that the following inequalities hold

$$\|u\|_{H^r}^2 \leqslant \frac{4M_r}{(2-cM_r^{1/2}t)^2} \leqslant \frac{M}{(2-Mt)^2}, \quad \frac{3}{2} < r \leqslant s, \tag{2.13}$$

where $M_r = \|u_0\|_{H^r}^2, M = \max\{4M_r, cM_r^{1/2}\}$,

$$\|u\|_{H^s} \leqslant \frac{c_1}{(2-Mt)^{c_2}}, \tag{2.14}$$

$$\|u\|_{H^{s+p}} \leqslant \frac{c_1 \varepsilon^{-\frac{p}{4}}}{(2-Mt)^{c_2}}, \quad p > 0, \tag{2.15}$$

and

$$\|u_t\|_{H^{s+p}} \leqslant \frac{c_1 \varepsilon^{-\frac{p+1}{4}}}{(2-Mt)^{c_2}}, \quad p > -1, \tag{2.16}$$

hold for any $\varepsilon$ sufficiently small and $t < \frac{2}{M}$.

**Proof** Choose a fixed number $r$ with $3/2 < r \leqslant s$. It follows from (2.6) that

$$\int_{\mathbf{R}} (\Lambda^r u)^2 \mathrm{d}x \leqslant \int_{\mathbf{R}} (\Lambda^r u_0)^2 \mathrm{d}x + c\int_0^t \left(\int_{\mathbf{R}} (\Lambda^r u)^2 \mathrm{d}x\right)^{3/2} \mathrm{d}t. \tag{2.17}$$

Multiplying (2.17) with $\left(\int_{\mathbf{R}} (\Lambda^r u)^2 \mathrm{d}x\right)^{1/2}$, we obtain the inequality

$$\begin{aligned} &\left(\int_{\mathbf{R}} (\Lambda^r u)^2 \mathrm{d}x\right)^{3/2} \\ &\leqslant \left(\int_{\mathbf{R}} (\Lambda^r u)^2 \mathrm{d}x\right)^{1/2} \left[\int_{\mathbf{R}} (\Lambda^r u_0)^2 \mathrm{d}x + c\int_0^t \left(\int_{\mathbf{R}} (\Lambda^r u)^2 \mathrm{d}x\right)^{3/2} \mathrm{d}t\right]. \end{aligned} \tag{2.18}$$

Denote that

$$V(t) := \int_{\mathbf{R}} (\Lambda^r u_0)^2 \mathrm{d}x + c\int_0^t \left(\int_{\mathbf{R}} (\Lambda^r u)^2 \mathrm{d}x\right)^{3/2} \mathrm{d}t,$$

and we have

$$\frac{\mathrm{d}}{\mathrm{d}t} V(t) \leqslant \left(\int_{\mathbf{R}} (\Lambda^r u)^2 \mathrm{d}x\right)^{1/2} V(t) \leqslant V(t) \tag{2.19}$$

from the inequality (2.17).

It follows from Gronwall's inequality and (2.17) that

$$\|u\|_{H^r}^2 \leqslant \frac{4M_r}{(2-cM_r^{1/2}t)^2} \leqslant \frac{M}{(2-Mt)^2}, \quad 3/2 < r \leqslant s,$$

holds for any $t \in [0, 2/M)$. Substituting the inequality (2.13) into (2.6) with $q = s$ and $u = u_\varepsilon$, one obtains the estimate

$$\|u\|_{H^s}^2 \leqslant \|u_0\|_{H^s}^2 + c\int_0^t \frac{\sqrt{M}}{(2-Mt)} \left(\int_{\mathbf{R}} (\Lambda^s u)^2 \mathrm{d}x\right) \mathrm{d}t$$

for any $t \in [0, 2/M)$. It follows from Gronwall's inequality and (2.9), (2.10) that there are constants $c_1$, $c_2$ depending on $M$ such that

$$\| u \|_{H^s} \leqslant \frac{c_1}{(2 - Mt)^{c_2}}.$$

According to the inequalities (2.9), (2.10), one can see that

$$M_s = \| u_0 \|_{H^s}^2 \leqslant c_0^2 \leqslant c_1,$$

where $c_0$, $c$ are independent of $\varepsilon$. So $M$, $c_1$, $c_2$ are independent of $\varepsilon$.

In a similar way, one may obtain $\| u \|_{H^{s+p}} \leqslant \frac{c_1 \varepsilon^{-\frac{p}{4}}}{(2 - Mt)^{c_2}}$, $p > 0$. And by using the inequalities (2.7), (2.15), we obtain the inequality

$$\| u_t \|_{H^{s+p}} \leqslant \frac{c_1 \varepsilon^{-\frac{p+1}{4}}}{(2 - Mt)^{c_2}}, \quad p > -1. \qquad \square$$

**Remark** According to the inequality (2.14) in Lemma 2.4, one can see that $\| u \|_{H^s}$ is uniformly bounded and the boundedness is independent of $\varepsilon$.

## 3 The Strong and Weak Limit Behavior of Solutions as $\gamma \to 0$

In this section, we shall demonstrate the behavior of solutions of the Cauchy problem (1.1) as the parameter $\gamma$ tends to zero. Firstly, we shall prove that $\{u_\varepsilon\}$ is a Cauchy sequence in the space $C([0, T); H^s(\mathbf{R}))$. We consider Eq. (1.1) as the following formula:

$$u_t + \gamma u u_x - \beta u_x + \partial_x p * \left( \frac{3 - \gamma}{2} u^2 + \frac{\gamma}{2} u_x^2 + (k + \beta) u \right) = 0,$$

where we assumed $\alpha = 1$.

Let $u = u_\varepsilon$ and $v = u_\gamma$ be solutions of (2.1), corresponding to the parameters $\gamma = \varepsilon$ and $\gamma$, respectively, and let $w = u - v$ and $f = u + v$. Then $w$ satisfies the problem

$$w_t + \frac{(\gamma - \varepsilon)}{2} (u^2)_x + \frac{\varepsilon}{2} (wf)_x - \beta w_x$$
$$= - p * \left( \frac{\varepsilon - \gamma}{2} (u^2)_x + \frac{3 - \varepsilon}{2} (wf)_x + \frac{\gamma - \varepsilon}{2} (u_x^2)_x + \frac{\varepsilon}{2} (w_x f_x)_x + (k + \beta) w_x \right). \tag{3.1}$$

Without loss of generality, we always assume $0 < \varepsilon \leqslant \gamma \leqslant 2\varepsilon \leqslant 1/8$.

**Theorem 3.1** Given the initial data $u_0 \in H^s(\mathbf{R})$ $\left( s > \frac{3}{2} \right)$. There exists $T = T(\| u_0 \|_{H^s}) > 0$, independent of $\varepsilon$, such that $\{u_\varepsilon\}$ and $\{u_{\varepsilon t}\}$ are Cauchy sequences in the space $C([0, T); H^s(\mathbf{R}))$ and $C([0, T); H^{s-1}(\mathbf{R}))$, respectively.

**Proof** First, for a constant $q$ with $1/2 < q \leqslant \min\{1, s - 1\}$, multiplying to $\Lambda^{2q} w$ both sides of Eq. (3.1) and integrating with respect to $x$ again, we may obtain the equation

$$\frac{1}{2} \frac{\mathrm{d}}{\mathrm{d}t} \int_{\mathbf{R}} (\Lambda^q w)^2 \mathrm{d}x = (\varepsilon - \gamma) \int_{\mathbf{R}} (\Lambda^q w)(\Lambda^q (u u_x)) \mathrm{d}x - \frac{\varepsilon}{2} \int_{\mathbf{R}} (\Lambda^q w)(\Lambda^q (wf)_x) \mathrm{d}x +$$

$$\int_{\mathbf{R}}(\Lambda^q w)\Big(\Lambda^{q-2}\Big(\frac{\varepsilon-\gamma}{2}(u^2)_x+\frac{3-\varepsilon}{2}(wf)_x+\frac{\gamma-\varepsilon}{2}(u_x^2)_x+$$
$$\frac{3}{2}(w_xf_x)_x+(k+\beta)w_x\Big)\Big)\mathrm{d}x.$$

Consider the following inequalities:

$$\Big|(\varepsilon-\gamma)\int_{\mathbf{R}}(\Lambda^q w)(\Lambda^q(uu_x))\mathrm{d}x\Big|\leqslant\varepsilon\|\Lambda^q w\|_{L^2}\Big(\int_{\mathbf{R}}(1+\xi^2)^{q+1}\mathrm{d}\xi\Big(\int_{\mathbf{R}}\hat{u}(\xi-\eta)\hat{u}(\eta)\mathrm{d}\eta\Big)^2\Big)$$
$$\leqslant\varepsilon\|u\|_{H^q}\|u\|_{H^{q+1}}\|w\|_{H^q}\leqslant c_1\varepsilon\|w\|_{H^q},$$

$$\Big|\frac{\varepsilon}{2}\int_{\mathbf{R}}(\Lambda^q w)(\Lambda^q(wf)_x)\mathrm{d}x\Big|\leqslant\frac{\varepsilon}{2}c_3(\|w_x\|_{L^\infty}+\|v\|_{H^q})\|w\|_{H^q}^2+\frac{\varepsilon}{2}c_3(\|v\|_{H^{q+1}}+\|w\|_{H^q})\|w\|_{H^q}$$
$$\leqslant\frac{\varepsilon}{2}c_3(\|w\|_{H^{\frac{3}{2}+\beta}}+\|v\|_{H^q})\|w\|_{H^q}^2+\frac{\varepsilon}{2}c_3(\|v\|_{H^{q+1}}+\|w\|_{H^q})\|w\|_{H^q},$$

$$\Big|\int_{\mathbf{R}}(\Lambda^q w)\Big(\Lambda^{q-2}\Big(\frac{\varepsilon-\gamma}{2}(u^2)_x\Big)\Big)\mathrm{d}x\Big|\leqslant\varepsilon\|u\|_{H^q}\|u\|_{H^{q-1}}\|w\|_{H^{q-1}},$$

$$\Big|\int_{\mathbf{R}}(\Lambda^q w)\Big(\Lambda^{q-2}\Big(\frac{3-\varepsilon}{2}(wf)_x\Big)\Big)\mathrm{d}x\Big|\leqslant c\|f\|_{H^q}\|w\|_{H^q}^2,$$

$$\Big|\int_{\mathbf{R}}(\Lambda^q w)\Big(\Lambda^{q-2}\Big(\frac{\gamma-\varepsilon}{2}(u_x^2)_x\Big)\Big)\mathrm{d}x\Big|\leqslant\varepsilon\|u\|_{H^q}\|u\|_{H^{q+1}}\|w\|_{H^q},$$

$$\Big|\int_{\mathbf{R}}(\Lambda^q w)\Big(\Lambda^{q-2}\Big(\frac{\varepsilon}{2}(w_xf_x)_x\Big)\Big)\mathrm{d}x\Big|\leqslant\varepsilon c_6\|f\|_{H^q}\|w\|_{H^q}^2,$$

$$\int_{\mathbf{R}}(\Lambda^q w)(\Lambda^{q-2}(w_x))\mathrm{d}x=0,$$

and

$$\frac{1}{2}\frac{\mathrm{d}}{\mathrm{d}t}\int_{\mathbf{R}}(\Lambda^q w)^2\mathrm{d}x\leqslant\varepsilon(\|u\|_{H^q}\|u\|_{H^{q+1}})\|w\|_{H^q}+\frac{\varepsilon}{2}c_3(\|w_x\|_{L^\infty}+\|v\|_{H^q})\|w\|_{H^q}^2+$$
$$\frac{\varepsilon}{2}c_3(\|v\|_{H^{q+1}}+\|w\|_{H^q})\|w\|_{H^q}+\varepsilon(\|u\|_{H^q}\|u\|_{H^{q-1}})\|w\|_{H^{q-1}}+$$
$$\varepsilon(\|u\|_{H^q}\|u\|_{H^{q+1}})\|w\|_{H^q}+\varepsilon c_6\|f\|_{H^q}\|w\|_{H^q}^2$$
$$\leqslant c(\varepsilon\|u\|_{H^q}\|u\|_{H^{q+1}}\|w\|_{H^q}+\|u\|_{H^q}\|w\|_{H^q}^2).$$

By using the inequalities (2.14), (2.15), one can see that

$$\frac{\mathrm{d}}{\mathrm{d}t}\int_{\mathbf{R}}(\Lambda^q w)^2\mathrm{d}x\leqslant c(\varepsilon\|w\|_{H^q}+\|w\|_{H^q}^2).$$

Combining with the inequality (2.10), we obtain the following inequality

$$\|w\|_{H^q}\leqslant c_1\varepsilon^{\frac{s-q}{4}}\mathrm{e}^{ct}+\varepsilon(\mathrm{e}^{ct}-1).$$

In a similar way, combining with the inequalities (2.10) and (2.9), one can obtain

$$\|w\|_{H^s}\leqslant c_1\|w_0\|_{H^s}\mathrm{e}^{ct}+\varepsilon(\mathrm{e}^{ct}-1)$$

and

$$\|w\|_{H^{s+p}}\leqslant c_1\varepsilon^{\frac{-p}{4}}\mathrm{e}^{ct}+\varepsilon^{\frac{3-p}{4}}(\mathrm{e}^{ct}-1),\tag{3.2}$$

for all $p>0$. So $\|w\|_{H^s}\to 0$, as $\varepsilon,\gamma\to 0$.

Next, we consider convergence of the sequence $\{u_{\varepsilon t}\}$. Multiplying to $\Lambda^{2s-2}w_t$ both sides of Eq. (3.1) and integrating with respect to $x$ again, we may obtain the equation

$$\begin{aligned}\frac{1}{2}\frac{\mathrm{d}}{\mathrm{d}t}\int_{\mathbf{R}}(\Lambda^{s-1}w_t)^2\mathrm{d}x &= -2\omega\int_{\mathbf{R}}(\Lambda^{s-1}w_t)(\Lambda^{s-3}w_x)\mathrm{d}x-(\gamma-\varepsilon)\int_{\mathbf{R}}(\Lambda^{s-1}w_t)(\Lambda^{s-1}(uu_x))\mathrm{d}x-\\&\frac{\varepsilon}{2}\int_{\mathbf{R}}(\Lambda^{s-1}w_t)(\Lambda^{s-1}(wf)_x)\mathrm{d}x-(\gamma-\varepsilon)\int_{\mathbf{R}}(\Lambda^{s-1}w_t)(\Lambda^{s-3}(uu_x))\mathrm{d}x-\\&\frac{\varepsilon}{2}\int_{\mathbf{R}}(\Lambda^{s-1}w_t)(\Lambda^{s-3}(w_xf_x)_x)\mathrm{d}x-\frac{\varepsilon-\gamma}{2}\int_{\mathbf{R}}(\Lambda^{s-1}w_t)(\Lambda^{s-3}(u_x^2)_x)\mathrm{d}x-\\&\frac{3-\varepsilon}{2}\int_{\mathbf{R}}(\Lambda^{s-1}w_t)(\Lambda^{s-3}(wf)_x)\mathrm{d}x.\end{aligned}$$

By using the following inequalities

$$\left|2\omega\int_{\mathbf{R}}\Lambda^{s-1}w_t(\Lambda^{s-3}w_x)\mathrm{d}x\right|\leqslant 2\omega\|w_x\|_{H^{s-3}}\|w_t\|_{H^{s-1}},$$

$$\left|(\gamma-\varepsilon)\int_{\mathbf{R}}(\Lambda^{s-1}w_t)(\Lambda^{s-1}(uu_x))\mathrm{d}x\right|\leqslant\varepsilon\|v\|_{H^s}^2\|w_t\|_{H^{s-1}},$$

$$\left|\frac{\varepsilon}{2}\int_{\mathbf{R}}(\Lambda^{s-1}w_t)(\Lambda^{s-1}(wf)_x)\mathrm{d}x\right|\leqslant c\varepsilon(\|w\|_{H^s}^2+\|w\|_{H^s}\|u\|_{H^s})\|w_t\|_{H^{s-1}},$$

$$\left|(\gamma-\varepsilon)\int_{\mathbf{R}}(\Lambda^{s-1}w_t)(\Lambda^{s-3}(uu_x))\mathrm{d}x\right|\leqslant\varepsilon\|u\|_{H^{s-2}}^2\|w_t\|_{H^{s-1}},$$

$$\left|\frac{3-\varepsilon}{2}\int_{\mathbf{R}}(\Lambda^{s-1}w_t)(\Lambda^{s-3}(wf)_x)\mathrm{d}x\right|\leqslant c(\|w\|_{H^{s-2}}^2+\|w\|_{H^{s-2}}\|u\|_{H^{s-2}})\|w_t\|_{H^{s-1}},$$

$$\left|\frac{\varepsilon}{2}\int_{\mathbf{R}}(\Lambda^{s-1}w_t)(\Lambda^{s-3}(w_xf_x)_x)\mathrm{d}x\right|\leqslant c\varepsilon(\|w\|_{H^{s-1}}^2+\|w\|_{H^{s-1}}\|u\|_{H^{s-1}})\|w_t\|_{H^{s-1}},$$

and

$$\left|\frac{\varepsilon-\gamma}{2}\int_{\mathbf{R}}(\Lambda^{s-1}w_t)(\Lambda^{s-3}(u_x^2)_x)\mathrm{d}x\right|\leqslant c\varepsilon\|u\|_{H^s}^2\|w_t\|_{H^{s-1}},$$

and combining with (2.9), (2.10), (2.14), (2.15), (2.21), we have

$$\|w_t\|_{H^{s-1}}\leqslant c(\varepsilon\|w\|_{H^s}+\varepsilon)\to 0,$$

as $\varepsilon,\gamma\to 0$. □

From Theorem 3.1, one can see that

**Theorem 3.2** Under the assumption $\varepsilon>0$, let $u_\gamma$ be the solution of (1.1) with the initial data $u_{\gamma 0}$ in $H^s(\mathbf{R})$, $s>\frac{3}{2}$. Then $u_\gamma$ converges to the solution to the following BBM equation (3.3) with the initial data $u_0$ in $H^s(\mathbf{R})$, $s>\frac{3}{2}$, as $\gamma\to 0$ and $k=2\omega,\beta=O(\gamma)$:

$$\begin{cases}u_t-u_{xxt}+2\omega u_x+3uu_x=0\\u(0,x)=u_0(x)\end{cases}.\tag{3.3}$$

## 4 The Strong and Weak Limit Behavior of Solutions as $\alpha\to 0$

Now let us study the behavior of solutions to the Cauchy problem (1.1) as the parameter $\alpha$ tends to zero. For the convenience, we denote $\varepsilon$ as $\alpha^2$, and we shall demonstrate that $\{u_\varepsilon\}$

is a Cauchy sequence as $\varepsilon\to 0$.

### 4.1 *A Priori* Estimates

Firstly, we shall give the following estimates about norm in $H^s(\mathbf{R})$, $s\geqslant 0$.

**Theorem 4.1** Suppose that for some $s>\frac{3}{2}$, the function $u(x,t)$ is a solution to Eq. (1.1) corresponding to the initial data $u_0\in H^s(\mathbf{R})$. Then for any real number $q\in(0,s]$, there exists a constant $c$ depending only on $q$, so that

$$\|u\|_{H^q}^2+\varepsilon\|u_x\|_{H^q}^2$$
$$\leqslant\|u_0\|_{H^q}^2+\varepsilon\|u_{0x}\|_{H^q}^2+c\int_0^t\|u_x\|_{L^\infty}\left(\int_{\mathbf{R}}(\Lambda^q u)^2+\varepsilon(\Lambda^q u_x)^2\mathrm{d}x\right)\mathrm{d}t. \tag{4.1}$$

For any $q\in(1/2,s]$ and any $r\in(1/2,q]$, there exists a constant $c$ depending only on $r$ and $q$, such that

$$\|u\|_{H^q}^2+\varepsilon\|u_x\|_{H^q}^2\leqslant\|u_0\|_{H^q}^2+\varepsilon\|u_{0x}\|_{H^q}^2+c\int_0^t\left(\int_{\mathbf{R}}(\Lambda^{r+1}u)^2+\varepsilon(\Lambda^{r+1}u_x)^2\mathrm{d}x\right)^{1/2}\times$$
$$\left(\int_{\mathbf{R}}(\Lambda^q u)^2+\varepsilon(\Lambda^q u_x)^2\mathrm{d}x\right)\mathrm{d}t. \tag{4.2}$$

Moreover, for any $q\in(0,s)$ there exists a constant $c$ depending only on $q$, such that

$$\|u_t\|_{H^q}+\varepsilon\|u_t\|_{H^{q+1}}\leqslant c\|u_t\|_{H^{q+3}}(c(k,\gamma)+\|u\|_{H^{1+\theta}}) \tag{4.3}$$

for all $\theta>1/2$. And under the assumption $\beta=O(\varepsilon)$, one can see that

$$\|u_t\|_{H^q}+\varepsilon\|u_t\|_{H^{q+1}}\leqslant c\|u\|_{H^{q+2}}(c(k)+\|u\|_{H^{1+\theta}}).$$

**Proof** For any $q\in(0,s]$, applying $(\Lambda^q u)\Lambda^q$ to both sides of Eq. (4.1) and integrating on $\mathbf{R}$ with respect to $x$ again, one obtains the equation

$$\frac{1}{2}\frac{\mathrm{d}}{\mathrm{d}t}\int_{\mathbf{R}}(\Lambda^q u)^2+\varepsilon(\Lambda^q u_x)^2\mathrm{d}x$$
$$=-3\int_{\mathbf{R}}\Lambda^q u\Lambda^q(uu_x)\mathrm{d}x+\frac{\varepsilon}{2}\int_{\mathbf{R}}\Lambda^q u\Lambda^q(\partial_x^3(u^2))+\Lambda^q u_x\Lambda^q(u_x^2)\mathrm{d}x$$
$$=(-3+\varepsilon)\int_{\mathbf{R}}\Lambda^q u\Lambda^q(uu_x)\mathrm{d}x-\varepsilon\int_{\mathbf{R}}\Lambda^{q+1}u\Lambda^{q+1}(uu_x)\mathrm{d}x+\frac{\varepsilon}{2}\int_{\mathbf{R}}\Lambda^q u_x\Lambda^q(u_x^2)\mathrm{d}x$$

using integration by parts. It follows from the inequalities (4.2) and (4.3) that there is a constant $c_q$ such that

$$\frac{1}{2}\frac{\mathrm{d}}{\mathrm{d}t}(\|u\|_{H^q}^2+\varepsilon\|u_x\|_{H^q}^2)\leqslant 4c_q\|u_x\|_{L^\infty}\|u\|_{H^q}^2+\frac{3}{2}\varepsilon c_q\|u_x\|_{L^\infty}\|u_x\|_{H^q}^2$$
$$\leqslant 4c_q\|u_x\|_{L^\infty}(\|u\|_{H^q}^2+\varepsilon\|u_x\|_{H^q}^2).$$

Integrating with respect to $t$ on both sides of the above inequality leads to inequality (4.1). Applying the inequality

$$\|u_x\|_{L^\infty}\leqslant c_{r+1}\|u\|_{H^{r+1}}\leqslant c_{r+1}\left(\int_{\mathbf{R}}(\Lambda^{r+1}u)^2+\varepsilon(\Lambda^{r+1}u_x)^2\mathrm{d}x\right)^{1/2}$$

for $r>\frac{1}{2}$ to the right-hand side of (4.1) yields the estimate (4.2). For any $q\in(0,s]$, applying $(\Lambda^q u_t)\Lambda^q$ to both sides of Eq. (4.1) and integrating with respect to $x$ again, one obtains the equation

$$\frac{1}{2}\frac{\mathrm{d}}{\mathrm{d}t}\int_{\mathbf{R}}(\Lambda^q u_t)^2+\varepsilon(\Lambda^q u_{xt})^2\mathrm{d}x$$

$$=-3\int_{\mathbf{R}}\Lambda^q u_t\Lambda^q(uu_x)\mathrm{d}x-k\int_{\mathbf{R}}\Lambda^q u_t\Lambda^q(u_x)\mathrm{d}x-\gamma\int_{\mathbf{R}}\Lambda^q u_t\Lambda^q(u_{xxx})\mathrm{d}x+$$

$$\frac{\varepsilon}{2}\int_{\mathbf{R}}\Lambda^q u_t\Lambda^q(\partial_x^3(u^2))-\Lambda^q u_t\Lambda^q(u_x^2)_x\mathrm{d}x$$

$$=(-3+\varepsilon)\int_{\mathbf{R}}\Lambda^q u_t\Lambda^q(uu_x)\mathrm{d}x-\varepsilon\int_{\mathbf{R}}\Lambda^{q+1}u_t\Lambda^{q+1}(uu_x)\mathrm{d}x+$$

$$\frac{\varepsilon}{2}\int_{\mathbf{R}}\Lambda^q u_{xt}\Lambda^q(u_x^2)\mathrm{d}x-k\int_{\mathbf{R}}\Lambda^q u_t\Lambda^q(u_x)\mathrm{d}x-\gamma\int_{\mathbf{R}}\Lambda^q u_t\Lambda^q(u_{xxx})\mathrm{d}x. \tag{4.4}$$

By using the inequalities (4.2), (4.3), one may obtain the following inequalities:

$$\left|\int_{\mathbf{R}}\Lambda^q u_t\Lambda^q(uu_x)\mathrm{d}x\right|\leqslant c_q\|u_t\|_{H^q}\|\hat{u}\|_{L^1}\|u\|_{H^{q+1}}\leqslant c_q\|u_t\|_{H^q}\|u\|_{H^1}\|u\|_{H^{q+1}},$$

$$\left|\varepsilon\int_{\mathbf{R}}\Lambda^{q+1}u_t\Lambda^{q+1}(uu_x)\mathrm{d}x\right|\leqslant c_q\varepsilon\|u_t\|_{H^{q+1}}\|\hat{u}\|_{L^1}\|u\|_{H^{q+2}}\leqslant c_q\varepsilon\|u_t\|_{H^{q+1}}\|u\|_{H^1}\|u\|_{H^{q+2}},$$

$$\left|\varepsilon\int_{\mathbf{R}}\Lambda^q u_t\Lambda^q(u_x^2)_x\mathrm{d}x\right|\leqslant c_q\varepsilon\|u_{xt}\|_{H^q}\|\hat{u}_x\|_{L^1}\|u_x\|_{H^q}\leqslant c_q\varepsilon\|u_{xt}\|_{H^q}\|u\|_{H^{1+\theta}}\|u_x\|_{H^q}$$

$$\leqslant c_q\varepsilon\|u_t\|_{H^{q+1}}\|u\|_{H^{1+\theta}}\|u\|_{H^{q+1}}\quad\left(\forall\theta>\frac{1}{2}\right),$$

$$\left|\frac{\varepsilon}{2}\int_{\mathbf{R}}\Lambda^q u_{xt}\Lambda^q(u_x^2)\mathrm{d}x\right|\leqslant c_q\varepsilon\|u_{xt}\|_{H^q}\|u\|_{H^{1+\theta}}\|u_x\|_{H^q}\quad\left(\forall\theta>\frac{1}{2}\right).$$

$$\left|\int_{\mathbf{R}}\Lambda^q u_t\Lambda^q(u_x)\mathrm{d}x\right|\leqslant c_q\|u_t\|_{H^q}\|u_x\|_{H^q},$$

and

$$\left|\gamma\int_{\mathbf{R}}\Lambda^q u_t\Lambda^q(u_{xxx})\mathrm{d}x\right|\leqslant c_q\gamma\|u_t\|_{H^q}\|u_{xxx}\|_{H^q}.$$

Applying the above six inequalities to (4.4) yields the inequality

$$\|u_t\|_{H^q}^2+\varepsilon\|u_{xt}\|_{H^q}^2\leqslant c(\|u_t\|_{H^q}+\varepsilon\|u_{xt}\|_{H^q})\|u\|_{H^{q+3}}(k+\beta+\|u\|_{H^{1+\theta}}),$$

where $c=c(q)$ is constant independent of $\varepsilon$, i.e.,

$$\|u_t\|_{H^q}+\varepsilon\|u_t\|_{H^{q+1}}\leqslant c\|u\|_{H^{q+3}}(k+\beta+\|u\|_{H^{1+\theta}})$$

or

$$\|u_t\|_{H^q}+\varepsilon\|u_t\|_{H^{q+1}}\leqslant c\|u\|_{H^{q+2}}(k+\|u\|_{H^{1+\theta}})\ (\text{with }\beta=O(\varepsilon)).$$

So we have

$$\|u_t\|_{H^q}+\varepsilon\|u_t\|_{H^{q+1}}\leqslant c\|u\|_{H^{q+3}}(k+\beta+\|u\|_{H^{1+\theta}})$$

for all $\theta>\frac{1}{2}$. And under the assumption $\beta=O(\varepsilon)$, one can see that

$$\|u_t\|_{H^q}+\varepsilon\|u_t\|_{H^{q+1}}\leqslant c\|u\|_{H^{q+2}}(k+\|u\|_{H^{1+\theta}})$$

for some constant $c$ independent of $\varepsilon$. The proof of Theorem 4.1 is complete. □

Now let us apply some lemma as follows:

**Lemma 4.2** (See [17]) Under the above assumptions, the following estimates hold for any $\varepsilon$ with $0<\varepsilon<1/8$

$$\|u_{\varepsilon 0}\|_{H^q}\leqslant c\quad (q\leqslant s),\tag{4.5}$$

$$\|u_{\varepsilon 0}\|_{H^q}\leqslant c\varepsilon^{\frac{s-q}{8}}\quad (q>s),\tag{4.6}$$

$$\|u_{\varepsilon 0}-u_0\|_{H^q}\leqslant c\varepsilon^{\frac{s-q}{8}}\quad (q\leqslant s),\tag{4.7}$$

$$\|u_{\varepsilon 0}-u_0\|_{H^s}=o(1),\tag{4.8}$$

here $u_{\varepsilon 0}$ is the convolution $u_{\varepsilon 0}=\phi_\varepsilon * u_0$ of the functions $\phi_\varepsilon(x)=\varepsilon^{-1/8}\phi(\varepsilon^{-1/8}x)$ and $u_0$ such that the Fourier transform $\hat{\phi}$ of $\phi$ satisfies $\hat{\phi}\in C_c^\infty([0,\infty);H^\infty)$, $\hat{\phi}(\xi)\geqslant 0$, and $\hat{\phi}(\xi)=1$ for any $\xi\in(-1,1)$.

**Lemma 4.3** There exist constants $c_1$ and $c_2$ independent of $\varepsilon$ such that the following inequalities hold

$$\|u\|_{H^r}^2+\varepsilon\|u_x\|_{H^r}^2\leqslant\frac{4M_r}{(2-cM_r^{1/2}t)^2}\leqslant\frac{M}{(2-Mt)^2},\quad \frac{3}{2}<r\leqslant s,\tag{4.9}$$

where $M_r=\|u_0\|_{H^r}^2+\varepsilon\|u_{0x}\|_{H^r}^2$, $M=\max\{4Mr, cM_r^{1/2}\}$,

$$(\|u\|_{H^s}^2+\varepsilon\|u_x\|_{H^s}^2)^{1/2}\leqslant\frac{c_1}{(2-Mt)^{c_2}},\tag{4.10}$$

$$(\|u\|_{H^{s+p}}^2+\varepsilon\|u_x\|_{H^{s+p}}^2)^{1/2}\leqslant\frac{c_1\varepsilon^{-\frac{p}{8}}}{(2-Mt)^{c_2}},\quad p>0,\tag{4.11}$$

and

$$(\|u_t\|_{H^{s+p}}^2+\varepsilon\|u_{xt}\|_{H^{s+p}}^2)^{1/2}\leqslant\frac{c_1\varepsilon^{-\frac{p+1}{8}}}{(2-Mt)^{c_2}},\quad p>-1,\tag{4.12}$$

for any $\varepsilon$ sufficiently small.

**Proof** Choose a fixed number $r$ with $3/2<r\leqslant s$. It follows from (4.2) that

$$\begin{aligned}&\int_{\mathbf{R}}(\Lambda^r u)^2+\varepsilon(\Lambda^r u_x)^2\mathrm{d}x\\ \leqslant&\int_{\mathbf{R}}(\Lambda^r u_0)^2+\varepsilon(\Lambda^r u_{0x})^2\mathrm{d}x+c\int_0^t\left(\int_{\mathbf{R}}(\Lambda^r u)^2+\varepsilon(\Lambda^r u_x)^2\mathrm{d}x\right)^{3/2}\mathrm{d}t.\end{aligned}\tag{4.13}$$

Multiplying (4.13) with $\left(\int_{\mathbf{R}}(\Lambda^r u)^2+\varepsilon(\Lambda^r u_x)^2\mathrm{d}x\right)^{1/2}$, we obtain the inequality

$$\begin{aligned}&\left(\int_{\mathbf{R}}(\Lambda^r u)^2+\varepsilon(\Lambda^r u_x)^2\mathrm{d}x\right)^{3/2}\\ \leqslant&\left(\int_{\mathbf{R}}(\Lambda^r u)^2+\varepsilon(\Lambda^r u_x)^2\mathrm{d}x\right)^{1/2}\times\\ &\left[\int_{\mathbf{R}}(\Lambda^r u_0)^2+\varepsilon(\Lambda^r u_{0x})^2\mathrm{d}x+c\int_0^t\left(\int_{\mathbf{R}}(\Lambda^r u)^2+\varepsilon(\Lambda^r u_x)^2\mathrm{d}x\right)^{3/2}\mathrm{d}t\right].\end{aligned}\tag{4.14}$$

Denote that

$$V(t) := \int_{\mathbf{R}} (\Lambda^r u_0)^2 + \varepsilon(\Lambda^r u_{0x})^2 \mathrm{d}x + c\int_0^t \left(\int_{\mathbf{R}} (\Lambda^r u)^2 + \varepsilon(\Lambda^r u_x)^2 \mathrm{d}x\right)^{3/2} \mathrm{d}t,$$

and we have

$$\frac{\mathrm{d}}{\mathrm{d}t} V(t) \leqslant \left(\int_{\mathbf{R}} (\Lambda^r u)^2 + \varepsilon(\Lambda^r u_x)^2 \mathrm{d}x\right)^{1/2} V(t) \qquad (4.15)$$

from the inequality (4.13). It follows from Gronwall's inequality and (4.13) that

$$\| u \|_{H^r}^2 + \varepsilon \| u_x \|_{H^r}^2 \leqslant \frac{4M_r}{(2 - cM_r^{1/2} t)^2} \leqslant \frac{M}{(2 - Mt)^2}, \quad 3/2 < r \leqslant s,$$

holds for any $t \in [0, 2/M)$. Substituting the inequality (4.9) into (4.2) with $q = s$ and $u = u_\varepsilon$, one obtains the estimate

$$\| u \|_{H^s}^2 + \varepsilon \| u_x \|_{H^s}^2 \leqslant \| u_0 \|_{H^s}^2 + \varepsilon \| u_{0x} \|_{H^s}^2 + c\int_0^t \frac{\sqrt{M}}{(2 - Mt)} \left(\int_{\mathbf{R}} (\Lambda^s u)^2 + \varepsilon(\Lambda^s u_x)^2 \mathrm{d}x\right) \mathrm{d}t$$

for any $t \in [0, 2/M)$. It follows from Gronwall's inequality and (4.5), (4.6) that there are constants $c_1$, $c_2$ depending on $M$ such that

$$(\| u \|_{H^s}^2 + \varepsilon \| u_x \|_{H^s}^2)^{1/2} \leqslant \frac{c_1}{(2 - Mt)^{c_2}}.$$

According to the inequalities (4.5), (4.6), one can see that

$$M_s = \| u \|_{H^s}^2 + \varepsilon \| u_{0x} \|_{H^s}^2 \leqslant c_0(1 + \varepsilon^{3/4}) \leqslant c,$$

where $c_0$, $c$ are independent of $\varepsilon$. So $M$, $c_1$, $c_2$ are independent of $\varepsilon$. In a similar way, one may obtain

$$(\| u \|_{H^{s+p}}^2 + \varepsilon \| u_x \|_{H^{s+p}}^2)^{1/2} \leqslant \frac{c_1 \varepsilon^{-\frac{p}{8}}}{(2 - Mt)^{c_2}}, \quad p > 0.$$

And by using the inequalities (4.3), (4.11), we obtain the inequality

$$(\| u_t \|_{H^{s+p}}^2 + \varepsilon \| u_{xt} \|_{H^{s+p}}^2)^{1/2} \leqslant \frac{c_1 \varepsilon^{-\frac{p+1}{8}}}{(2 - Mt)^{c_2}}, \quad p > -1. \qquad \square$$

**Remark** According to the inequality (4.10) in Lemma 4.3, one can see that $\| u \|_{H^s}$ is uniformly bounded and the boundedness is independent of $\varepsilon$.

### 4.2 The Convergence of Solutions as $\varepsilon \to 0$

We shall demonstrate that $\{u_\varepsilon\}$ is a Cauchy sequence. We rewrite Eq. (1.1) as

$$u_t - \varepsilon u_{xxt} + k u_x + (3 - 3\varepsilon\gamma) u u_x + \beta u_{xxx} = \varepsilon\gamma(-3 u u_x + 2 u_x u_{xx} + u u_{xxx}),$$

i.e.,

$$u_t - \varepsilon u_{xxt} + k u_x + (3 - 3\gamma\varepsilon) u u_x + \beta u_{xxx} = -\gamma\varepsilon(1 - \partial_x^2)(u u_x) - \gamma\varepsilon\partial_x\left(u^2 + \frac{1}{2} u_x^2\right).$$

Let $u_\varepsilon$ and $u_\delta$ be solutions of (1.1), corresponding to the parameters $\varepsilon$ and $\delta$, respectively, and let $w = u^\varepsilon - v^\delta$ and $f = u^\varepsilon + v^\delta$. Then $w$ satisfies the problem for $\beta_\varepsilon = \beta = O(\varepsilon)$ and $\beta_\delta = \beta = O(\delta)$:

$$w_t - \varepsilon w_{xxt} + (\delta - \varepsilon) v_{xxt} + k w_x + \frac{1}{2}(3 - 3\varepsilon)(w f)_x + (\delta - \varepsilon) v v_x + \beta_\varepsilon w_{xxx} + (\beta_\varepsilon - \beta_\delta) v_{xxx}$$

$$= -\frac{\varepsilon}{2}\gamma(1-\partial_x^2)(wf)_x + \gamma(\delta-\varepsilon)(1-\partial_x^2)(vv_x) - \varepsilon\gamma\partial_x\left(wf + \frac{1}{2}w_xf_x\right) + (\delta-\varepsilon)\gamma\partial_x\left(v^2 + \frac{1}{2}v_x^2\right). \tag{4.16}$$

Without loss of generality, we always assume that $0<\varepsilon\leqslant\delta\leqslant 2\varepsilon\leqslant 1/8$. And considering the different properties under the assumption $\beta=O(\varepsilon)$ or without the assumption $\beta=0(\varepsilon)$, one can see

**Theorem 4.4** Under the assumption $\beta=O(\varepsilon)$, there exists $T>0$, independent of $\varepsilon$, such that $\{u_\varepsilon\}$ and $\{u_{\varepsilon t}\}$ are Cauchy sequences in the space $C([0,T);H^s(\mathbf{R}))$ and $C([0,T);H^{s-1}(\mathbf{R}))$, respectively.

**Proof** First, for a constant $q$ with $1/2<q\leqslant\min\{1,s-1\}$, multiplying to $\Lambda^{2q}w$ both sides of Eq. (4.16) and integrating with respect to $x$ again, we may obtain the equation

$$\begin{aligned}
&\frac{1}{2}\frac{\mathrm{d}}{\mathrm{d}t}\int_{\mathbf{R}}(\Lambda^q w)^2 + \varepsilon(\Lambda^q w_x)^2\mathrm{d}x \\
&= -(\delta-\varepsilon)\int_{\mathbf{R}}(\Lambda^q w)(\Lambda^q v_{xxt})\mathrm{d}x - k\int_{\mathbf{R}}(\Lambda^q w)(\Lambda^q w_x)\mathrm{d}x - \\
&\quad \frac{1}{2}(3-3\varepsilon)\int_{\mathbf{R}}(\Lambda^q w)(\Lambda^q(wf)_x)\mathrm{d}x - (\delta-\varepsilon)\int_{\mathbf{R}}(\Lambda^q w)(\Lambda^q(vv_x))\mathrm{d}x - \\
&\quad \gamma\int_{\mathbf{R}}(\Lambda^q w)(\Lambda^q(w_{xxx}))\mathrm{d}x - (\gamma_\varepsilon-\gamma_\delta)\int_{\mathbf{R}}(\Lambda^q w)(\Lambda^q(v_{xxx}))\mathrm{d}x - \\
&\quad \frac{\varepsilon}{2}\int_{\mathbf{R}}(\Lambda^q w)(\Lambda^{q+2}(wf)_x)\mathrm{d}x + (\delta-\varepsilon)\int_{\mathbf{R}}(\Lambda^q w)(\Lambda^{q+2}(vv_x))\mathrm{d}x - \\
&\quad \varepsilon\int_{\mathbf{R}}(\Lambda^q w)(\Lambda^q(wf)_x)\mathrm{d}x - \frac{1}{2}\varepsilon\int_{\mathbf{R}}(\Lambda^q w)(\Lambda^q(w_xf_x)_x)\mathrm{d}x + \\
&\quad (\delta-\varepsilon)\int_{\mathbf{R}}(\Lambda^q w)(\Lambda^q(v^2)_x)\mathrm{d}x + (\delta-\varepsilon)\int_{\mathbf{R}}(\Lambda^q w)\left(\Lambda^q\left(\frac{1}{2}v_x^2\right)_x\right)\mathrm{d}x \\
&= -(\delta-\varepsilon)\int_{\mathbf{R}}(\Lambda^q w)(\Lambda^q v_{xxt})\mathrm{d}x - \frac{1}{2}(3-3\varepsilon)\int_{\mathbf{R}}(\Lambda^q w)(\Lambda^q(wf)_x)\mathrm{d}x - \\
&\quad (\delta-\varepsilon)\int_{\mathbf{R}}(\Lambda^q w)(\Lambda^q(vv_x))\mathrm{d}x - (\gamma_\varepsilon-\gamma_\delta)\int_{\mathbf{R}}(\Lambda^q w)(\Lambda^q(v_{xxx}))\mathrm{d}x - \\
&\quad \frac{\varepsilon}{2}\int_{\mathbf{R}}(\Lambda^q w)(\Lambda^{q+2}(wf)_x)\mathrm{d}x + (\delta-\varepsilon)\int_{\mathbf{R}}(\Lambda^q w)(\Lambda^{q+2}(vv_x))\mathrm{d}x - \\
&\quad \varepsilon\int_{\mathbf{R}}(\Lambda^q w)(\Lambda^q(wf)_x)\mathrm{d}x - \frac{1}{2}\varepsilon\int_{\mathbf{R}}(\Lambda^q w)(\Lambda^q(w_xf_x)_x)\mathrm{d}x + \\
&\quad (\delta-\varepsilon)\int_{\mathbf{R}}(\Lambda^q w)(\Lambda^q(v^2)_x)\mathrm{d}x + (\delta-\varepsilon)\int_{\mathbf{R}}(\Lambda^q w)\left(\Lambda^q\left(\frac{1}{2}v_x^2\right)_x\right)\mathrm{d}x.
\end{aligned}$$

Consider the following inequalities:

$$\left|(\delta-\varepsilon)\int_{\mathbf{R}}(\Lambda^q w)(\Lambda^q v_{xxt})\mathrm{d}x\right| \leqslant \delta\|v_t\|_{H^{q+2}}\|w\|_{H^q},$$

$$\left|\int_{\mathbf{R}}(\Lambda^q w)(\Lambda^q(wf)_x)\mathrm{d}x\right| \leqslant c_3(\|w_x\|_{L^\infty} + \|v\|_{H^q})\|w\|_{H^q}^2 + c_3(\|v\|_{H^{q+1}} + \|w\|_{H^q})\|w\|_{H^q},$$

$$\left|(\delta-\varepsilon)\int_{\mathbf{R}}(\Lambda^q w)(\Lambda^q(vv_x))\mathrm{d}x\right| \leqslant \delta(\|v\|_{H^q}\|v\|_{H^{q+1}})\|w\|_{H^q},$$

$$\left|(\gamma_\varepsilon-\gamma_\delta)\int_{\mathbf{R}}(\Lambda^q w)(\Lambda^q(v_{xxx}))\mathrm{d}x\right|\leqslant c\delta\|v\|_{H^{q+3}}\|w\|_{H^q}\left|\frac{\varepsilon}{2}\int_{\mathbf{R}}(\Lambda^q w)(\Lambda^{q+2}(wf)_x)\mathrm{d}x\right|$$

$$\leqslant c_3\|w\|_{H^q}[\varepsilon\|f\|_{H^{q+3}}(\|f\|_{H^{q+3}}+\|v\|_{H^{q+3}})]$$

$$\leqslant c_3\varepsilon^{1/4}\|w\|_{H^q}\left|(\delta-\varepsilon)\int_{\mathbf{R}}(\Lambda^q w)(\Lambda^{q+2}(vv_x))\mathrm{d}x\right|$$

$$\leqslant \delta(\|v\|_{H^{q+2}}\|v\|_{H^{q+3}})\|w\|_{H^q},$$

$$\left|\varepsilon\int_{\mathbf{R}}(\Lambda^q w)(\Lambda^q(wf)_x)\mathrm{d}x\right|\leqslant\varepsilon(2\|v\|_{H^{q+1}}\|f\|_{H^{q+1}})\|w\|_{H^q},$$

$$\left|\frac{\varepsilon}{2}\int_{\mathbf{R}}(\Lambda^q w)(\Lambda^q(w_xf_x)_x)\mathrm{d}x\right|\leqslant c_4\varepsilon\|f\|_{H^{q+1}}\|w\|^2_{H^{q+1}},$$

$$\left|(\delta-\varepsilon)\int_{\mathbf{R}}(\Lambda^q w)(\Lambda^q(v^2)_x)\mathrm{d}x\right|\leqslant\delta\|v\|^2_{H^{q+1}}\|w\|_{H^q},$$

and

$$\left|(\delta-\varepsilon)\int_{\mathbf{R}}(\Lambda^q w)\left(\Lambda^q\left(\frac{1}{2}v_x^2\right)_x\right)\mathrm{d}x\right|\leqslant\delta\|v\|^2_{H^{q+2}}\|w\|_{H^q}.$$

By using the inequalities (4.10), (4.11), one can see that

$$\frac{\mathrm{d}}{\mathrm{d}t}\int_{\mathbf{R}}(\Lambda^q w)^2+\varepsilon(\Lambda^q w_x)^2\mathrm{d}x\leqslant c(\delta^\rho\|w\|_{H^q}+\delta^{1/4}\|w\|_{H^q}+\|w\|^2_{H^q}),$$

where

$$\rho=\begin{cases}1, & s\geqslant 3+q\\ \dfrac{1+s-q}{4}, & s<3+q\end{cases}.$$

Combining with the inequality (4.6), we obtain the following inequality:

$$\|w\|_{H^q}\leqslant c_1\delta^{\frac{s-q}{4}}\mathrm{e}^{ct}+\delta^\rho(\mathrm{e}^{ct}-1)+\delta^{1/4}(\mathrm{e}^{ct}-1).$$

In a similar way, combining with the inequalities (4.6) and (4.5), one can obtain

$$\|w\|_{H^s}\leqslant c_1(\|w_0\|_{H^s}+\delta^{\frac{3}{4}})\mathrm{e}^{ct}+\delta^m(\mathrm{e}^{ct}-1),$$

where $m=\min\left\{\frac{1}{4},\frac{s-q-1}{4}\right\}$ and

$$\|w\|_{H^{s+p}}\leqslant c_1\delta^{\frac{-p}{4}}\mathrm{e}^{ct}+\delta^{\frac{1-p}{4}}(\mathrm{e}^{ct}-1),\tag{4.17}$$

for all $p>0$. So $\|w\|_{H^s}\to 0$, as $\varepsilon,\delta\to 0$.

Next, we consider convergence of the sequence $\{u_{\varepsilon t}\}$. Multiplying both sides of Eq. (4.16) by $\Lambda^{2s-2}w_t$ and integrating on $\mathbf{R}$ with respect to $x$ again, we may obtain the equation

$$\frac{1}{2}\frac{\mathrm{d}}{\mathrm{d}t}\int_{\mathbf{R}}(\Lambda^{s-1}w_t)^2+\varepsilon(\Lambda^{s-1}w_{xt})^2\mathrm{d}x$$

$$=-(\delta-\varepsilon)\int_{\mathbf{R}}(\Lambda^{s-1}w_t)(\Lambda^{s-1}v_{xxt})\mathrm{d}x-k\int_{\mathbf{R}}(\Lambda^{s-1}w_t)(\Lambda^{s-1}w_x)\mathrm{d}x-$$

$$\frac{1}{2}(3-3\varepsilon)\int_{\mathbf{R}}(\Lambda^{s-1}w_t)(\Lambda^{s-1}(wf)_x)\mathrm{d}x-(\delta-\varepsilon)\int_{\mathbf{R}}(\Lambda^{s-1}w_t)(\Lambda^{s-1}(vv_x))\mathrm{d}x-$$

$$\int_{\mathbf{R}}(\Lambda^{s-1}w_t)(\Lambda^{s-1}(w_{xxx}))\mathrm{d}x-(\gamma_\varepsilon-\gamma_\delta)\int_{\mathbf{R}}(\Lambda^{s-1}w_t)(\Lambda^{s-1}(v_{xxx}))\mathrm{d}x-$$

$$\frac{\varepsilon}{2}\int_{\mathbf{R}}(\Lambda^{s-1}w_t)(\Lambda^{s+1}(wf)_x)\mathrm{d}x+(\delta-\varepsilon)\int_{\mathbf{R}}(\Lambda^{s-1}w_t)(\Lambda^{s+1}(vv_x))\mathrm{d}x-$$

$$\varepsilon\int_{\mathbf{R}}(\Lambda^{s-1}w_t)(\Lambda^{s-1}(wf)_x)\mathrm{d}x-\frac{1}{2}\varepsilon\int_{\mathbf{R}}(\Lambda^{s-1}w_t)(\Lambda^{s-1}(w_xf_x)_x)\mathrm{d}x+$$

$$(\delta-\varepsilon)\left(\int_{\mathbf{R}}(\Lambda^{s-1}w_t)(\Lambda^{s-1}(v^2)_x)\mathrm{d}x+\int_{\mathbf{R}}(\Lambda^{s-1}w_t)\left(\Lambda^{s-1}\left(\frac{v_x^2}{2}\right)_x\right)\mathrm{d}x\right).$$

By using the following inequalities:

$$\left|(\delta-\varepsilon)\int_{\mathbf{R}}(\Lambda^{s-1}w_t)(\Lambda^{s-1}v_{xxt})\mathrm{d}x\right|\leqslant\delta(\|v_{xxt}\|_{H^{s-1}})\|w_t\|_{H^{s-1}},$$

$$\left|k\int_{\mathbf{R}}(\Lambda^{s-1}w_t)(\Lambda^{s-1}w_x)\mathrm{d}x\right|\leqslant k(\|w_x\|_{H^{s-1}})\|w_t\|_{H^{s-1}},$$

$$\left|\int_{\mathbf{R}}(\Lambda^{s-1}w_t)(\Lambda^{s-1}(wf)_x)\mathrm{d}x\right|\leqslant c(\|w\|_{H^s}^2+\|w\|_{H^s}\|v\|_{H^s})\|w_t\|_{H^{s-1}},$$

$$\left|(\delta-\varepsilon)\int_{\mathbf{R}}(\Lambda^{s-1}w_t)(\Lambda^{s-1}(vv_x))\mathrm{d}x\right|\leqslant\delta(\|v\|_{H^s}^2)\|w_t\|_{H^{s-1}},$$

$$\left|\gamma_\varepsilon\int_{\mathbf{R}}(\Lambda^{s-1}w_t)(\Lambda^{s-1}(w_{xxx}))\mathrm{d}x\right|\leqslant(\gamma_\varepsilon\|w\|_{H^{s+2}})\|w_t\|_{H^{s-1}},$$

$$\left|(\gamma_\varepsilon-\gamma_\delta)\int_{\mathbf{R}}(\Lambda^{s-1}w_t)(\Lambda^{s-1}(v_{xxx}))\mathrm{d}x\right|\leqslant(\gamma_\delta\|v\|_{H^{s+2}})\|w_t\|_{H^{s-1}},$$

and

$$\left|\frac{\varepsilon}{2}\int_{\mathbf{R}}(\Lambda^{s-1}w_t)(\Lambda^{s+1}(wf)_x)\mathrm{d}x\right|\leqslant c(\varepsilon^{1/2}\|w\|_{H^{s+2}}\varepsilon^{1/2}\|f\|_{H^{s+2}})\|w_t\|_{H^{s-1}},$$

and combining with (4.5), (4.6), (4.10), (4.11), (4.17), we have

$$\|w_t\|_{H^{s-1}}\leqslant c(\delta^{1/2}\|w\|_{H^{s+2}}+\delta^{1/2})\to 0$$

as $\varepsilon,\delta\to 0$. □

From Theorem 4.4, one can see that

**Corollary 4.5** Under the assumption $\varepsilon>0$, let $u_\varepsilon$ be the solution to (1.1) with the initial data $u_{\varepsilon 0}\in H^s$, $s>\frac{3}{2}$. Then $u_\varepsilon$ converges to the solution to the following equation (4.18) with the initial data $u_0\in H^s$, $s>\frac{3}{2}$, as $\varepsilon\to 0$,

$$\begin{cases}u_t+ku_x+3uu_x=0\\u(0,x)=u_0(x)\end{cases}.\tag{4.18}$$

From the proof of Theorem 4.2 and *a priori* estimates in the proof of Theorem 4.1, one can see that, without the assumption $\beta=O(\varepsilon)$, we have

**Theorem 4.6** If $u=u^\varepsilon$ and $v=u^\delta$ are the solutions to the problem (1.1) in $C([0,T),H^s)$, $s>\frac{3}{2}$, with $\varepsilon$ and $\varepsilon=\delta$, respectively, then $\|u-v\|_{H^{s-3}}$ converges to zero as $\varepsilon,\delta\to 0$. Moreover, $u_\varepsilon$ converges to the solution to the following KdV equation (4.19) with the initial data $u_0$ in $H^s$, $s>3$, as $\varepsilon\to 0$,

$$\begin{cases}u_t+ku_x+3uu_x+\beta u_{xxx}=0\\u(0,x)=u_0(x)\end{cases}\tag{4.19}$$

**Remark** According to some *a priori* estimates in the proof of Theorem 4.2, and analogue with the proof of Theorem 3.1 in Section 3 of [17], we can obtain that the solutions to Eq. (1.1)

$$u_t - \alpha^2 u_{xxt} + ku_x + 3uu_x + \beta u_{xxx} = \gamma\alpha^2(2u_xu_{xx} + uu_{xxx})$$

converge to the solution of the following equation

$$u_t - \alpha^2 u_{xxt} + ku_x + 3uu_x = \gamma\alpha^2(2u_xu_{xx} + uu_{xxx})$$

as $\beta \to 0, \gamma \neq 0$, particularly, to the solution of the Camassa-Holm equation as $\beta \to 0, \gamma = 1$. And for $\gamma = 0$, the above convergence is also evident.

# 5 Convergence of Solitary Wave Solutions

Substituting the traveling wave solution $u = \varphi(x - Vt)$, for constant wave speed $V$, into Eq. (1.1), one can obtain the ordinary differential equation

$$-V\varphi' + \alpha^2 V\varphi''' + 3\varphi\varphi' + k\varphi' + \beta\varphi''' = \gamma\alpha^2(2\varphi'\varphi'' + \varphi\varphi'''),$$

i.e.,

$$(\alpha^2 V + \beta - \gamma\alpha^2\varphi)\varphi''' = -3\varphi\varphi' + (V - k)\varphi' - 2\gamma\alpha^2\varphi'\varphi''. \tag{5.1}$$

Considering the transformation $\varphi = \psi + a$, we have the following ordinary differential equation

$$(\alpha^2 V + \beta - \gamma\alpha^2 a - \gamma\alpha^2\psi)\psi''' = -3\psi\psi' + (V - k - 3a)\psi' - 2\gamma\alpha^2\psi\psi'. \tag{5.2}$$

**Definition 5.1** A solitary wave $\varphi(x)$ with undisturbed depth $a = \lim_{|x| \to \infty} \phi(x)$ is a weak solution of the ordinary differential equation (5.1) if and only if $\psi = \phi - a \in H^1$, and

$$\left\langle \frac{3}{2}\phi^2 - \frac{\gamma\alpha^2}{2}(\phi')^2 + (k - V)\phi, g' \right\rangle + \left\langle (\alpha^2 V + \beta)\phi - \frac{\gamma\alpha^2}{2}\phi^2, g''' \right\rangle = 0, \tag{5.3}$$

for any $g \in C_c^\infty(\mathbf{R})$.

For (5.1), integrating on $\mathbf{R}$ with respect to $\eta$, and letting integrate constants be zero, and integrating by parts, we have

$$(\alpha^2 V + \beta - \gamma\alpha^2\varphi)\varphi'' = -\frac{3}{2}\varphi^2 + (V - k)\varphi + \frac{\gamma\alpha^2}{2}(\varphi')^2. \tag{5.4}$$

Let $z = \varphi'$, and we can reduce (5.4) to the dynamical system

$$\begin{cases} \dfrac{d\varphi}{d\eta} = z \\ \dfrac{dz}{d\eta} = \dfrac{2(V - k)\varphi - 3\varphi^2 + \gamma\alpha^2 z^2}{2(\alpha^2 V + \beta - \gamma\alpha^2\varphi)} \end{cases}, \tag{5.5}$$

where $\varphi = \dfrac{1}{\gamma}\left(V + \dfrac{\beta}{\alpha^2}\right)$ is a singular line.

The transformation $d\eta = (\alpha^2 V + \beta - \gamma\alpha^2\varphi)d\tau$ reduce (5.5) to the Hamilton system

$$\begin{cases} \dfrac{d\varphi}{d\tau} = z(\alpha^2 V + \beta - \gamma\alpha^2\varphi) \\ \dfrac{dz}{d\tau} = (V - k)\varphi - \dfrac{3}{2}\varphi^2 + \dfrac{\gamma\alpha^2}{2}z^2 \end{cases}. \tag{5.6}$$

The first integral is

$$H(\varphi,z)=\frac{z^2(\alpha^2V+\beta-\gamma\alpha^2\varphi)}{2}+\frac{\varphi^2(\varphi+k-V)}{2}=h,$$

where $h$ is a constant.

Noting

$$\chi=\gamma(V\alpha^2+\beta)[(3-2\gamma)V\alpha^2+3\beta+2k\gamma\alpha^2],\quad z_{\pm}^0=\pm\frac{1}{|\gamma^2\alpha^3|}\sqrt{\chi},$$

$$\varphi_1=\frac{2}{3}(V-k),\quad \varphi_2=\frac{1}{\gamma}\left(V+\frac{\beta}{\alpha^2}\right),$$

one can see that

(ⅰ) when $V=k$, there is only a singular point $O(0,0)$ on $\varphi$-axis of the system (5.5);

(ⅱ) when $V\neq k$, there are two singular points (i.e., $O(0,0)$ and $(\varphi_1,0)$) on $\varphi$-axis of the system (5.5);

(ⅲ) when $\chi>0$, there are two singular points (i.e., $(\varphi_2,z_+^0)$ and $(\varphi_2,z_-^0)$) on the line $\varphi=\varphi_2$ of the system (5.5);

(ⅳ) when $\chi=0$, there is only a singular point $(\varphi_2,0)$ on the line $\varphi=\varphi_2$ of the system (5.5);

(ⅴ) when $\chi<0$, the singular point does not exist on the line $\varphi=\varphi_2$.

Considering Hamilton function $H(\varphi, z)$, we have

$$h_0=H(0,0)=0,\quad h_1=H(\varphi_1,0)=\frac{2}{27}(k-V)^3,$$

$$h_2=H(\varphi_2,z_\pm^0)-\frac{1}{2\gamma^3\alpha^6}(V\alpha^2+\beta)^2[(1-\gamma)V\alpha^2+\beta+k\gamma\alpha^2].$$

Letting $\lambda(\varphi, z)$ be an eigenvalue of the point $(\varphi, z)$, and $f_\lambda$ be its corresponding eigenfunction, we have

$$f_\lambda=\begin{vmatrix}\lambda+\gamma\alpha^2z & \gamma\alpha^2\varphi-(V\alpha^2+\beta)\\ 3\varphi-(V-k) & \lambda-\gamma\alpha^2z\end{vmatrix},$$

i.e.,

$$\lambda^2-[\gamma^2\alpha^4z^2+(\gamma\alpha^2\varphi-(V\alpha^2+\beta))(3\varphi-(V-k))]=0.$$

So

$$\lambda^2(0,0)-(V\alpha^2+\beta)(V-k)=0,$$

i.e.,

$$\lambda(0,0)=\pm\sqrt{(V\alpha^2+\beta)(V-k)},$$

$$\lambda^2(\varphi_1,0)-(\gamma\alpha^2\varphi_1-(V\alpha^2+\beta))(3\varphi_1-(V-k))=0,$$

i.e.,

$$\lambda(\varphi_1,0)=\pm\sqrt{(V-k)\left[\left(\frac{2}{3}\gamma-1\right)V\alpha^2-\frac{2}{3}\gamma k\alpha^2-\beta\right]}$$

and

$$\lambda(\varphi_2,z_\pm^0)=|\gamma\alpha^2|z_\pm^0=\pm\frac{1}{|\gamma\alpha|}\sqrt{\chi}.$$

According to the proof of Proposition 1 in [20], we obtain

**Lemma 5.2** If $(\phi(\eta), z(\eta))$ is a homoclinic orbit representation of system (5.6), then $u=\varphi(x-Vt)$ is a solitary solution to Eq. (1.1).

The transformation

$$\boldsymbol{\xi}=\begin{pmatrix}\xi_1\\ \xi_2\\ \xi_3\end{pmatrix}=\begin{pmatrix}\psi\\ \psi'\\ \psi''\end{pmatrix}$$

reduces (5.2) to the dynamical system

$$\boldsymbol{\xi}'=\begin{pmatrix}\psi'\\ \psi''\\ \psi'''\end{pmatrix}=\begin{pmatrix}\xi_2\\ \xi_3\\ \dfrac{-3\xi_1\xi_2+(V-k-3a)\xi_2-2\gamma\alpha^2\xi_2\xi_3}{\alpha^2V+\beta-\gamma\alpha^2a-\gamma\alpha^2\xi_1}\end{pmatrix}.$$

Let

$$D=\nabla f(0)=\begin{pmatrix}0&1&0\\0&0&1\\0&\dfrac{V-k-3a}{\alpha^2V+\beta-\gamma\alpha^2a}&0\end{pmatrix},$$

using the transformation

$$\boldsymbol{y}=\boldsymbol{\xi}+\boldsymbol{a}=\begin{pmatrix}\xi_1+a\\ \xi_2\\ \xi_3\end{pmatrix}$$

as before yields the system of equations

$$\boldsymbol{\xi}'=\begin{pmatrix}\psi'\\ \psi''\\ \psi'''\end{pmatrix}=\begin{pmatrix}\xi_2\\ \xi_3\\ \dfrac{-3\xi_1\xi_2+(V-k-3a)\xi_2-2\gamma\alpha^2\xi_2\xi_3}{\alpha^2V+\beta-\gamma\alpha^2a-\gamma\alpha^2\xi_1}\end{pmatrix}=D\boldsymbol{\xi}+g(\boldsymbol{\xi}),\qquad(5.7)$$

where

$$\phi=\psi+a,\quad D=\nabla f(0)=\begin{pmatrix}0&1&0\\0&0&1\\0&\dfrac{V-k-3a}{\alpha^2V+\beta-\gamma\alpha^2a}&0\end{pmatrix}.$$

The set of fixed points and singularities of system (5.3) consists of all points of the $y_1$-axis. Next, we discuss properties of each fixed point or singularity $(a,0,0)$ of system (5.1) in different cases.

Considering

$$\lambda I-D=\begin{pmatrix}\lambda&-1&0\\0&\lambda&-1\\0&-\dfrac{V-k-3a}{\alpha^2V+\beta-\gamma\alpha^2a}&\lambda\end{pmatrix},$$

one can see that

$$f_A(\lambda)=\lambda\left(\lambda^2+\frac{3a+k-V}{\alpha^2 V+\beta-\gamma\alpha^2 a}\right)=\lambda\left[\lambda^2-\frac{3}{\gamma\alpha^2}\frac{a-\frac{V-k}{3}}{a-\frac{1}{\gamma}\left(V+\frac{\beta}{\alpha^2}\right)}\right].$$

The constants $\frac{1}{\gamma}\left(V+\frac{\beta}{\alpha^2}\right)$ and $\frac{V-k}{3}$ divide the $y_1$-axis into three intervals.

**Case Ⅰ** When $\gamma<0$ and $\frac{V-k}{3}>\frac{1}{\gamma}\left(V+\frac{\beta}{\alpha^2}\right)$, the intervals are

$$\left(-\infty,\frac{1}{\gamma}\left(V+\frac{\beta}{\alpha^2}\right)\right),\quad \left(\frac{1}{\gamma}\left(V+\frac{\beta}{\alpha^2}\right),\frac{V-k}{3}\right)\quad\text{and}\quad\left(\frac{V-k}{3},\infty\right).$$

If $a\in\left(-\infty,\frac{1}{\gamma}\left(V+\frac{\beta}{\alpha^2}\right)\right)\cup\left(\frac{V-k}{3},\infty\right)$, one can see that

$$\frac{3a+k-V}{\alpha^2 V+\beta-\gamma\alpha^2 a}=\frac{1}{-\gamma\alpha^2}\frac{a-\frac{V-k}{3}}{a-\frac{1}{\gamma}\left(V+\frac{\beta}{\alpha^2}\right)}>0,$$

the system (5.3) shows that (5.1) has a three-dimensional center manifold near the point $(a,0,0)$. On the other hand, for any $a\in\left(\frac{1}{\gamma}\left(V+\frac{\beta}{\alpha^2}\right),\frac{V-k}{3}\right)$, one can see that

$$\frac{3a+k-V}{\alpha^2 V+\beta-\gamma\alpha^2 a}=\frac{1}{-\gamma\alpha^2}\frac{a-\frac{V-k}{3}}{a-\frac{1}{\gamma}\left(V+\frac{\beta}{\alpha^2}\right)}<0.$$

Then there is a one-dimensional center manifold, a one-dimensional stable manifold, and a one-dimensional unstable manifold at the point $(a,0,0)$, near which there is a unique, analytic, homoclinic orbit represented by the solution

$$\phi(x-Vt)=\xi(x-Vt)+a.$$

We consider the periodic solutions $\phi_\delta[x-(V+\delta)t]=a+\frac{\delta}{\gamma}+\zeta_\delta[x-(V+\delta)t]$ to Eq. (5.2), the function $\zeta_\delta$ satisfy the differential equations

$$(\zeta_\delta')^2=\frac{(\zeta_\delta-H(\delta))^2\left(\zeta_\delta+2H(\delta)-B+\frac{3-\gamma}{\gamma}\delta\right)}{\gamma\alpha^2\zeta_\delta-A},\tag{5.8}$$

where $H=H(\delta)\to 0$ (as $\delta\to 0$), $A:=\alpha^2 V+\beta-\gamma\alpha^2 a$ and $B:=V-k-3a$. In particular, letting $H(\delta):=\frac{\delta}{-\gamma}$, we can see that $\zeta_\delta\in H^1$.

The solution $\phi(x-Vt)=\xi(x-Vt)+a$ is obtained as the limit, as $\delta\to 0$, of periodic solutions

$$\phi_\delta[x-(V+\delta)t]=a+\frac{\delta}{\gamma}+\zeta_\delta[x-(V+\delta)t],\tag{5.9}$$

where the function $\xi$ satisfies the respective differential equations

$$(\zeta')^2 = \frac{\zeta^2(\zeta - B)}{\gamma\alpha^2\zeta - A}. \tag{5.10}$$

The point $\left(\frac{1}{\gamma}\left(V+\frac{\beta}{\alpha^2}\right),0,0\right)$ forms a singular point of system, providing the compacton solution

$$\phi_0(x - Vt) = \begin{cases} \frac{1}{\gamma}\left(V+\frac{\beta}{\alpha^2}\right)+\left[V-k-\frac{3}{\gamma}\left(V+\frac{\beta}{\alpha^2}\right)\right]\cos^2\frac{x}{2|\alpha|\sqrt{-\gamma}}, & |x|\leqslant|\alpha|\sqrt{-\gamma}\pi \\ \frac{1}{\gamma}\left(V+\frac{\beta}{\alpha^2}\right), & \text{otherwise} \end{cases} \tag{5.11}$$

For each $a\in\left(\frac{1}{\gamma}\left(V+\frac{\beta}{\alpha^2}\right),\frac{V-k}{3}\right)$, we substitute $\phi = \psi + a$ into (5.1), integrating the resulting equation once and setting the integral constant to zero, we obtain

$$(\alpha^2 V + \beta - \gamma\alpha^2 a - \gamma\alpha^2\psi)\psi'' = -\frac{3}{2}\psi^2 + (V - k - 3a)\psi + \frac{\gamma\alpha^2}{2}(\psi')^2. \tag{5.12}$$

The system (5.12) has two fixed points—the origin $O(0,0)$ and $\left(\frac{2(V-k)}{3}-2a,0\right)$. The origin is a saddle point whose unique homoclinic orbit represents an analytic solitary wave solution. Near the point $\left(\frac{2}{3}(V-k)-2a,0\right)$, then exists a two-dimensional center manifold having periodic orbits converging to the homoclinic orbit at the origin. Substitute $\psi = \xi + \frac{2}{3}(V-k)-2a$ into (5.12) and comparing (5.1) with the resulting equation

$$\left(\alpha^2 V + \beta - \frac{2}{3}\gamma\alpha^2(V-k) + \gamma\alpha^2 a - \gamma\alpha^2\xi\right)\xi''$$
$$= -\frac{3}{2}\xi^2 + (V - k - 3a)\xi + \frac{\gamma\alpha^2}{2}(\xi')^2, \tag{5.13}$$

one may recognize that periodic orbits near the fixed point $\left(\frac{2(V-k)}{3}-2a,0\right)$ of (5.2) come from the center manifold of the fixed point $\left(\frac{2(V-k)}{3}-a,0,0\right)$ in system (5.1). Therefore the homeomorphism

$$\Psi(a) = \frac{2}{3}(V-k) - a \tag{5.14}$$

of $\left(\frac{1}{\gamma}\left(V+\frac{\beta}{\alpha^2}\right),\frac{V-k}{3}\right)$ onto $\left(\frac{V-k}{3},\frac{2}{3}(V-k)-\frac{1}{\gamma}\left(V+\frac{\beta}{\alpha^2}\right)\right)$ determines a one-to-one mapping from the set $\left\{(a,0,0);a\in\left(\frac{1}{\gamma}\left(V+\frac{\beta}{\alpha^2}\right),\frac{V-k}{3}\right)\right\}$ of quasi-hyperbolic points to the set of points $\left\{(\Psi(a),0,0);\ a\in\left(\frac{1}{\gamma}\left(V+\frac{\beta}{\alpha^2}\right),\frac{V-k}{3}\right)\right\}$ whose center manifolds contain periodic orbits converging to homoclinic orbits at the corresponding quasi-hyperbolic fixed points. One may also notice that the mapping $\Psi$ is defined in such a way that the points $O(0,0)$ and $(\Psi(a)-a,\ 0)$ always appear as a pair of fixed points in (5.12).

**Case Ⅱ** $\gamma>0$.

(Ⅱ)$_1$ When $\gamma>3$ and $\frac{V-k}{3}>\frac{1}{\gamma}\left(V+\frac{\beta}{\alpha^2}\right)$, the intervals are

$$\left(-\infty,\frac{1}{\gamma}\left(V+\frac{\beta}{\alpha^2}\right)\right),\quad \left(\frac{1}{\gamma}\left(V+\frac{\beta}{\alpha^2}\right),\frac{V-k}{3}\right)\quad \text{and}\quad \left(\frac{V-k}{3},\infty\right).$$

If $a\in\left(\frac{1}{\gamma}\left(V+\frac{\beta}{\alpha^2}\right),\frac{V-k}{3}\right)$, one can see that

$$\frac{3a+k-V}{\alpha^2 V+\beta-\gamma\alpha^2 a}=\frac{1}{-\gamma\alpha^2}\frac{a-\frac{V-k}{3}}{a-\frac{1}{\gamma}\left(V+\frac{\beta}{\alpha^2}\right)}>0,$$

then there is three-dimensional center manifold near the point $(a,0,0)$. On the other hand, for any $a\in\left(-\infty,\frac{1}{\gamma}\left(V+\frac{\beta}{\alpha^2}\right)\right)\cup\left(\frac{V-k}{3},\infty\right)$, one can see that

$$\frac{3a+k-V}{\alpha^2 V+\beta-\gamma\alpha^2 a}=\frac{1}{-\gamma\alpha^2}\frac{a-\frac{V-k}{3}}{a-\frac{1}{\gamma}\left(V+\frac{\beta}{\alpha^2}\right)}<0,$$

the system (5.3) shows that (5.1) has a one-dimensional center manifold, a one-dimensional stable manifold, and a one-dimensional unstable manifold at the point $(a,0,0)$, near which there is a unique, analytic, homoclinic orbit represented by the solution

$$\phi(x-Vt)=\xi(x-Vt)+a.$$

(Ⅱ)$_2$ When $0<\gamma<3$ and $\frac{V-k}{3}<\frac{1}{\gamma}\left(V+\frac{\beta}{\alpha^2}\right)$, the intervals are

$$\left(-\infty,\frac{V-k}{3}\right),\quad \left(\frac{V-k}{3},\frac{1}{\gamma}\left(V+\frac{\beta}{\alpha^2}\right)\right)\quad \text{and}\quad \left(\frac{1}{\gamma}\left(V+\frac{\beta}{\alpha^2}\right),\infty\right).$$

When $\gamma>0$, we have

**Theorem 5.3** If the coefficients of Eq. (1.1) satisfy the inequalities

$$\alpha\neq 0,\quad \gamma>0,\quad \text{and}\quad \frac{V-k}{3}>\frac{1}{\gamma}\left(V+\frac{\beta}{\alpha^2}\right),$$

then there exists an orbitally unique and analytic solitary wave solution $a+\zeta_a(x-Vt)$ for each $a\in\left(\frac{1}{\gamma}\left(V+\frac{\beta}{\alpha^2}\right),\frac{V-k}{3}\right)$. Moreover, as an approaches $\frac{V-k}{2}-\frac{\alpha^2 V+\beta}{2\gamma\alpha^2}$, the sequence of the solitary wave solutions $\{a+\zeta_a(x-Vt)\}$ converges to the peakon solution

$$\phi(x)=\left(\frac{V-k}{2}-\frac{\alpha^2 V+\beta}{2\gamma\alpha^2}\right)-Ae^{-\gamma^{1/2}|\alpha x|},\tag{5.15}$$

where $A=\alpha^2 V+\beta-\gamma\alpha^2 a$.

**Theorem 5.4** When $\beta=0,\alpha\to 0,\phi_p(x)\equiv-\frac{k}{2}-A$ is the unique stationary solution to the following KdV equation

$$u_t+ku_x+3uu_x+\beta u_{xxx}=0,\tag{5.16}$$

and compactons and peakons do not occur.

And when $\gamma<0$, one can see that

**Theorem 5.5** If the coefficients of Eq. (1.1) satisfy the inequalities

$$\alpha \neq 0, \quad \gamma < 0, \quad \text{and} \quad \frac{V-k}{3} > \frac{1}{\gamma}\left(V+\frac{\beta}{\alpha^2}\right),$$

then there exists an orbitally unique and analytic solitary wave solution $a+\zeta_a(x-Vt)$ for each $a \in \left(\frac{1}{\gamma}\left(V+\frac{\beta}{\alpha^2}\right), \frac{V-k}{3}\right)$. Moreover, as an approaches $\frac{1}{\gamma}\left(V+\frac{\beta}{\alpha^2}\right)$, the sequence of the solitary wave solutions $\{a+\zeta_a(x-Vt)\}$ converges to the compacton solution given in (5.16):

$$\phi_0(x-Vt)=\begin{cases}\frac{1}{\gamma}\left(V+\frac{\beta}{\alpha^2}\right)+\left[V-k-\frac{3}{\gamma}\left(V+\frac{\beta}{\alpha^2}\right)\right]\cos^2\frac{x}{2|\alpha|\sqrt{-\gamma}}, & |x|\leqslant|\alpha|\sqrt{-\gamma}\pi \\ \frac{1}{\gamma}\left(V+\frac{\beta}{\alpha^2}\right), & \text{otherwise}\end{cases}. \tag{5.17}$$

**Proof** We consider the periodic solutions

$$\phi_\delta[x-(V+\delta)t]=a+\frac{\delta}{\gamma}+\zeta_\delta[x-(V+\delta)t]$$

to Eq. (5.2). Let $A=\alpha^2V+\beta-\gamma\alpha^2a$, $B=V-k-3a$ and $B_0=V-k-\frac{3}{\gamma}\left(V+\frac{\beta}{\alpha^2}\right)$. Then Eq. (5.2) for the solitary wave solution reduces to

$$(\zeta_a')^2=\frac{\zeta_a^2(\zeta_a-B)}{\gamma\alpha^2\zeta_a-A}. \tag{5.18}$$

Using the inequality $0\leqslant\zeta_\delta\leqslant B-2H(\delta)-\frac{3-\gamma}{\gamma}\delta$ valid for all $x\in\mathbf{R}$, one may show that sequence of functions $\{\zeta_a'\}$ and $\{\zeta_a\}$ are uniformly bounded on the real axis. Therefore, the Ascoli-Arzela Theorem shows that, as $a\to\frac{1}{\gamma}\left(V+\frac{\beta}{\alpha^2}\right)$, there exist subsequences of the families $\{\zeta_a\}$ and $\{\zeta_a'\}$, without loss of generality still denoted by $\{\zeta_a\}$ and $\{\zeta_a'\}$, which are uniformly convergent to a function and its derivative, respectively, on any compact set of $\mathbf{R}$. Here we are relying on the fact that each is an even function, since it is symmetric with respect to its elevation and translation invariant. Taking the limit on both sides of (5.9) as $a\to\frac{1}{\gamma}\left(V+\frac{\beta}{\alpha^2}\right)$, or as $A\to0$ leads to the equation

$$\gamma\alpha^2\zeta(\zeta')^2=\zeta^2(\zeta-B_0) \tag{5.19}$$

satisfied by the function $\zeta$. Since $\lim\limits_{A\to0}\max\limits_{x\in\mathbf{R}}\zeta_a(x)=\lim\limits_{A\to0}B=B_0>0$ and each $\zeta_a$ is even, monotone on each side of the origin and exponentially decaying to zero at infinity, the limiting function $\zeta$ is a nontrivial solution to (5.11). Thus, function $\zeta$ satisfies the equation $\gamma\alpha^2(\zeta')^2=\zeta(\zeta-B_0)$. Therefore, as an even and monotone decreasing function on the positive real axis, $\phi = \zeta + \frac{1}{\gamma}\left(V + \frac{\beta}{\alpha^2}\right)$ is the compacton solution (5.11). This completes the proof of Theorem 5.5. □

**References**


[1] Yin Z. On the Cauchy problem for a nonlinearly dispersive wave equation. J Nonlinear Math Phys, 2003, 10: 10-15.

[2] Dai H H, Huo Y. Solitary shock waves and other travelling waves in a general compressible hyperelastic rod. Proc R Soc Lond Ser A, 2000, 456: 331-363.

[3] Kalisch Henrik, Stability of solitary waves for a nonlinearly dispersive equation. Discrete Contin Dyn Syst, 2004, 10 (3): 709-717.

[4] Fuchssteiner B, Fokas A S. Symplectic structures, their Backlund transformation and hereditary symmetries. Phys D, 1981, 4: 47-66.

[5] Camassa R, Holm D. An integrable shallow water equation with peaked solitons. Phys Rev Lett, 1993, 71: 1661-1664.

[6] Camassa R, Holm D, Hyman J. A new integrable shallow water equation. Adv Appl Mech, 1994, 31: 1-33.

[7] Johnson R S. Camassa-Holm, Korteweg-de Vries and related models for water waves. J Fluid Mech, 2002, 455: 63-82.

[8] Camassa R. Characteristics and the initial value problem of a completely integrable shallow water equation. Discrete Contin Dyn Syst Ser B, 2003, 3: 115-139.

[9] Constantin A. On the scattering problem for the Camassa-Holm equation. Proc R Soc Lond Ser A, 2001, 457: 953-970.

[10] Fokas A S. On a class of physically important integrable equations. Phys D, 1995, 87: 145-150.

[11] Fuchssteiner B. Some tricks from the symmetry-toolbox for nonlinear equations: generalizations of the Camassa-Holm equation. Phys D, 1996, 95: 229-243.

[12] Constantin A, Strauss W A. Stability of the Camassa-Holm solitons. J Nonlinear Sci, 2002, 12: 415-422.

[13] Constantin A, Kolev B. On the geometric approach to the motion of inertial mechanical systems. J Phys A, 2002, 35: R51-R79.

[14] Misiolek G. A shallow water equation as a geodesic flow on the Bott-Virasoro group. J Geom Phys, 1998, 24: 203-208.

[15] Benjamin B, Bona J, Mahony J. Model equations for long waves in nonlinear dispersive systems. Philos Trans R Soc Lond, 1972, 272: 47.

[16] Dodd R K, Eilbeck J C, Gibbon J D, et al. Solitons and nonlinear wave equations. New York: Academic Press, 1984.

[17] Tian Lixin, Gui Guilong, Liu Yue. On the well-posedness problem and the scattering problem for the Dullin-Gottwald-Holm equation. Comm Math Phys, 2005, 257: 671-704.

[18] Kato T. On the Cauchy problem for the (generalized) KdV equation// Studies in applied mathematics. Advances in Mathematics Supplementary, Vol 8. New York: Academic Press, 1983: 93-128.

[19] Li Y A, Olver P J. Well-posedness and blow-up solutions for an integrable nonlinearly dispersive model

wave equation. J Differential Equations, 2000, 162: 27 – 63.

[20] Guo B L, Liu Z R. Peaked wave solutions of $CH-\gamma$ equation. Sci China Ser A, 2003, 33 (4): 325 – 337.

[21] Temam R. Navier-Stokes equations and nonlinear functional analysis. Philadelphia: SIAM, 1983.

[22] Rodriguez-Blanco G. On the Cauchy problem for the Camassa-Holm equation. Nonlinear Anal, 2001, 46: 309 – 327.

[23] Constantin A, Escher J. Global weak solutions for a shallow water equation. Indiana Univ Math J, 1998, 47 (4): 1527 – 1545.

[24] Tang Minying, Yang Chengxi. Extension on peaked wave solutions of $CH-\gamma$ equation. Chaos Solitons Fractals, 2004, 20: 815 – 825.

[25] Bona J L, Scott R. Solutions of the Korteweg-de Vries equation in fractional order Sobolev spaces. Duke Math J, 1976, 43: 87 – 99.

[26] Bona J L, Smith R. The initial-value problem for the Korteweg-de Vries equation. Philos Trans R Soc Lond, 1975, 278:555 – 601.

[27] Kato T. Quasi-linear equations of evolution, with applications to partial differential equations// Everitt W N. Spectral theory and differential equations. Lecture Notes in Math, Vol 448. Berlin: Springer-Verlag, 1975: 25 – 70.

[28] Kato T. On the Korteweg-de Vries equation. Manuscripta Math, 1979, 28: 89 – 99.

[29] Carlos E Kenig, Gustavo Ponce, Luis Vega. The Cauchy problem for the Korteweg-de Vries equation in Sobolev spaces of negative indices. Duke Math, 1993, 71: 1 – 21.

[30] Kenig C, Ponce G, Vega L. Well-posedness and scattering results for the generalized Korteweg-de Vries equation via the contraction principle. Comm Pure Appl Math, 1993, 46: 527 – 620.

[31] Mckean H P. Integrable systems and algebraic curves// Grmela M, Marsden J E. Global analysis. Lecture Notes in Math, Vol 755. Berlin: Springer-Verlag, 1979: 83 – 200.

[32] Drazin P G, Johnson R S. Solitons: an introduction. Cambridge: Cambridge Univ Press, 1989.

[33] Lopes O. A linearized instability result for solitary waves. Discrete Contin Dyn Syst, 2002, 8: 115 – 119.

[34] Tian Lixin, Fan Guochang, Gui Guilong. Well-posedness and blow-up for an integrable shallow water equation with strong dispersive term. Int J Nonlinear Sci, 2006, 1 (1): 3 – 13.

[35] Tian Lixin, Li Xiuming. On the well-posedness problem for the generalized Degasperis-Procesi equation. Int J Nonlinear Sci, 2006, 2 (2): 67 – 76.

[36] Lu Dianchen, Peng Dejun, Tian Lixin. On the well-posedness problem for the generalized Dullin-Gottwald-Holm equation. Int J Nonlinear Sci, 2006, 3 (1): 178 – 186.

# Cauchy Problem for the Multi-dimensional Boussinesq Type Equation *

Yang Zhijian(杨志坚) Guo Boling(郭柏灵)

**Abstract**

The paper studies the existence and non-existence of global weak solutions to the Cauchy problem for the multi-dimensional Boussinesq type equation $u_{tt}-\Delta u+\Delta^2 u=\Delta\sigma(\mu)$. It proves that the Cauchy problem admits a global weak solution under the assumptions that $\sigma\in C(\mathbf{R})$, $\sigma(s)$ is of polynomial growth order, say $p(>1)$, either $0\leqslant\sigma(s)s\leqslant\beta\int_0^s\sigma(\tau)\mathrm{d}\tau$, $s\in\mathbf{R}$, where $\beta>0$ is a constant, or the initial data belong to a potential well. And the weak solution is regularized and the strong solution is unique when the space dimension $N=1$. In contrast, any weak solution of the Cauchy problem blows up in finite time under certain conditions. And two examples are shown.

**Keywords** Cauchy problem; Boussinesq type equation; global solutions; blowup of solutions

## 1 Introduction

In this paper, we are concerned with the existence and non-existence of global weak solutions to the Cauchy problem for the multi-dimensional Boussinesq type equation

$$u_{tt}-\Delta u+\mu\Delta^2 u=\Delta\sigma(u)\quad \text{in } \mathbf{R}^N\times(0,\infty), \tag{1.1}$$

$$u(x,0)=u_0(x),\quad u_t(x,0)=u_1(x),\quad x\in\mathbf{R}^N, \tag{1.2}$$

where $\mu>0$ is a constant and $\sigma(s)$ is a nonlinear function specified later.

When the space dimension $N=1$, Eq. (1.1) becomes

$$u_{tt}-u_{xx}+\mu u_{xxxx}=\sigma(u)_{xx}, \tag{1.3}$$

and Eq. (1.3) is a generalization of the well-known Boussinesq equation

$$u_{tt}-u_{xx}+\mu u_{xxxx}=a(u^2)_{xx}, \tag{1.4}$$

where $\mu>0$ denotes the dispersive parameter depending on the compression and rigidity characteristics of the materials, $a=\text{const}\in\mathbf{R}$ is a constant coefficient controlling nonlinearity, $u$ is the vertical deflection, and the quadratic nonlinearity accounts for the curvature of the bending beam. Equation (1.4) was derived by Boussinesq[4] in 1872 and is called the "good" Boussinesq equation because of its linear stability, it models small oscillations of nonlinear beams (see [24]), and it governs the oscillation of the nonlinear strings and the two-dimensional irrotational flows of a inviscid liquid in a uniform rectangular channel as well (see [2,3,14]).

* J Math Anal Appl, 2008, 340: 64-80.

Equation (1.4) with $\mu<0$ is known as the "bad" Boussinesq equation because of its linear instability, it describes the propagation of long surface waves on shallow water, and it also arises in a large range of physical phenomena including the propagation of ion-sound waves in a uniform isotropic plasma and nonlinear lattice waves (see [17]). For the "bad" Boussinesq equation and its generalizations, there have been many researches from various points of view (cf. [1,6,7,10,13,26-28]).

With regard to the Cauchy problem of the "good" Boussinesq type equation, Bona and Sachs[3] established the global existence of smooth solutions under the assumptions that: "$1<p<5, (p-1)/4<c^2<1$, where $c>0$ is the wave speed, $\sigma(u)$ being either $u^p$ with $p$ integer or $|u|^{p-1}u$ with $p$ real and initial data lie relatively close to stable solitary waves." Tsutsumi and Matahashi[20] proved the local and global well posedness by means of transforming the Cauchy problem of Eq. (1.3) into the system of nonlinear Schrödinger equations. Linares[14,15] improved the results in [20] and further discussed the asymptotic behavior of the solutions provided that $\sigma(u)=|u|^{p-1}u$ with $p>1$ and the initial data are small. Sachs[19] proved that for a large set of initial data, the Cauchy problem of Eq. (1.4), with $\mu=\alpha=1$, has no smooth solution for all time, which was claimed by Kalantarov and Ladyzhenskaya[11]. Liu[16] studied the instability of solitary waves for Eq. (1.3). And some related works can be seen in [9,18]. Recently, Chen and Yang[5] proved the existence and uniqueness of global solution to the IBVP of Eq. (1.3) under the assumptions that "$\sigma\in C^3(\mathbf{R})$, $\sigma'(s)$ is bounded below, $\sigma''(0)=0$ and $\sigma'''(s)$ satisfies local Lipschitz condition", and some blowup results were also given.

But to the authors' best knowledge, there are very few works on the multi-dimensional Boussinesq type equation. Recently, taking into account the role of viscosity in real process, Varlamov[21-25] studied in detail the existence, uniqueness and long-time asymptotics of solutions to the Cauchy problem and the IBVP of the damped Boussinesq equation

$$u_{tt}-2b\Delta u_t-\Delta u+\alpha\Delta^2 u=\beta\Delta(u^2) \tag{1.5}$$

in the cases of space dimension $N=1, 2, 3$, respectively. When $N=1$, the long-time asymptotic expansion of the classical solution was computed for a Cauchy problem with periodic small initial data (see [22]), and the blowup of solutions (even for small initial data) was also studied (see [23]). When $N=2$, Eq. (1.5) models the small nonlinear oscillations of elastic membranes, and the author constructed the global solution by using eigenfunction expansion method and computed the longtime asymptotics on the basis of obtained the serious representation for the IBVP of Eq. (1.5) in both a disk and a ball, respectively (see [24, 25]).

When the space dimension $N=2$, Akmel[1] established the global existence of small solutions for the "bad" Boussinesq equation provided that the potential $v(x)$ verifying $\hat{v}(\xi)\in C_0^\infty(\mathbf{R}^2)$, and he showed further that the small solution decays to zero when the nonlinear term possesses an appropriate lower bound.

But for general space dimension, does Cauchy problem (1.1), (1.2) still admit a global solution? In the case of space dimension $N=1$, does the corresponding Cauchy problem still admit a global solution provided that the conditions in [3] are not valid (say $p\geqslant 5$)? Can the theoretically qualitative condition: "smallness of initial data" in [14, 15] be replaced by a quantitative one? All these questions are still open.

Various versions of the Boussinesq equation discussed in literature all possess one obvious characteristic: they are perturbations of the linear wave equation that takes into account the effects of small nonlinearity and dispersion. From the physical points of view, in real process, dispersion plays an important spreading role for the energy gather arising from nonlinearity, and the interaction of it with the nonlinearity accompanies accumulation, balance and dissipation of the energy in the configurations. And hence, it is interesting to consider the interaction between the two factors. In fact, despite the status of Boussinesq type equations as the first model for nonlinear dispersive wave propagation, the mathematical theory for such equations is not nearly so complete as is the case for the Korteweg-de Vries-type equations (see [3,24]).

In the present paper, first, by using a Galerkin approximation scheme combined with a limiting process of the solutions of a series of periodic boundary value problems (PBVP), which is different from those used in [1, 21 - 25], the authors establish the existence of global solutions to Cauchy problem (1.1), (1.2) under the assumptions that $\sigma\in C(\mathbf{R})$, $\sigma(s)$ is of polynomial growth order, say $p(>1)$, either $0\leqslant\sigma(s)s\leqslant\beta\int_0^s\sigma(\tau)\mathrm{d}\tau$, $s\in\mathbf{R}$, where $\beta>0$ is a constant, or the initial data belong to a potential well $W_0$.

Second, the above-mentioned weak solution is regularized and the strong solution is proved to be unique when the space dimension $N=1$. And where the restrictions for both $p$, i.e. "$1<p<5$ and $(p-1)/4<c^2<1$" and the initial data, i.e. "the initial data lie relatively close to stable solitary waves" (see [3]) are removed.

Third, any weak solution of the Cauchy problem blows up in finite time under certain conditions.

The plan of the paper is as follows. The main results and some notations are stated in Section 2. The global existence of weak solutions to the PBVP and Cauchy problem (1.1), (1.2) is discussed in Sections 3 and 4, respectively. In Section 5, the weak solutions are regularized and so the strong solution is proved to be unique in the case of space dimension $N=1$. And in Section 6, a blowup theorem is proved and two examples are shown.

## 2 Statement of Main Results

We first introduce the following abbreviations:

$$\Omega=(-L,L)^N,\quad Q_T=\Omega\times(0,T),\quad L_p=L_p(\Omega),\quad W^{m,p}=W^{m,p}(\Omega),$$
$$C_0^\infty=C_0^\infty(\Omega),\quad H^m=W^{m,2},\quad \|\cdot\|_p=\|\cdot\|_{L^p},\quad \|\cdot\|_{k,p}=\|\cdot\|_{W^{k,p}},$$
$$\|\cdot\|_{H^k}=\|\cdot\|_{H^k(\Omega)},\quad \|\cdot\|=\|\cdot\|_{L^2},\quad \dot{H}^s(\mathbf{R}^N)=(-\Delta)^{-\frac{s}{2}}L_2(\mathbf{R}^N).\tag{2.1}$$

We denote the operators $G=(-\Delta)^{-1}, G^{1/2}=(-\Delta)^{-1/2}, p'=\frac{p}{p-1}$, with $p\in(1,\infty)$, and we denote the measure of $\Omega$ by $|\Omega|$. And the notation $(\cdot,\cdot)$ for the $L_2$-inner product will also be used for the notation of duality paring between dual spaces.

Without loss of generality, we take $\mu=1$ in Eqs. (1.1) and (1.3).

Applying $G$ to both sides of Eq. (1.1), we have

$$Gu_{tt}+u-\Delta u+\sigma(u)=0 \quad \text{in } \mathbf{R}^N\times(0,\infty). \tag{2.2}$$

**Definition** The function $u\in L_\infty([0,T];H^1(\mathbf{R}^N))$, with $G^{1/2}u_t\in L_\infty([0,T];L_2(\mathbf{R}^N))$, is called a weak solution of Cauchy problem (1.1), (1.2), if for any $\chi\in H^1(\mathbf{R}^N)$, supp $\chi$ is abounded set in $\mathbf{R}^N$, and a.e. $t\in[0,T]$,

$$(G^{1/2}u_{tt},G^{1/2}\chi)+(u,\chi)+(\nabla u,\nabla\chi)+(\sigma(u),\chi)=0, \tag{2.3}$$

$$u(\cdot,0)=u_0 \quad \text{in } H^1(\mathbf{R}^N), \quad G^{1/2}u_t(\cdot,0)=G^{1/2}u_1 \quad \text{in } L_2(\mathbf{R}^N). \tag{2.4}$$

Let $v(x,t)=e^{-\lambda t}u(x,t)$, where $\lambda>0$ is a constant. Then problem (1.1), (1.2) is equivalent to the problem

$$G(v_{tt}+2\lambda v_t+\lambda^2 v)+v-\Delta v+\tilde{\sigma}(t,v)=0 \quad \text{in } \mathbf{R}^N\times(0,\infty), \tag{2.5}$$

$$v(x,0)=v_0(x), \quad v_t(x,0)=v_1(x), \quad x\in\mathbf{R}^N, \tag{2.6}$$

where $v_0(x)=u_0(x)$, $v_1(x)=u_1(x)-\lambda u_0(x)$, $\tilde{\sigma}(t,v)=e^{-\lambda t}\sigma(e^{\lambda t}v)$.

For any $u\in H_0^1(\Omega)$, let

$$\tilde{u}(x)=\begin{cases}u(x), & x\in\Omega\\ 0, & x\in\mathbf{R}^N-\Omega\end{cases},$$

and define

$$Gu=G\tilde{u}, \quad G^{1/2}u=G^{1/2}\tilde{u}. \tag{2.7}$$

Then $Gu\in\dot{H}^3(\mathbf{R}^N)$ and $G^{1/2}u\in\dot{H}^2(\mathbf{R}^N)$, $Gu$ and $G^{1/2}u$ are meaningful.

We first consider the PBVP of Eq. (2.5)

$$v|_{\partial\Omega}=0, \quad v(x,t)=v(x+2Le_i,t), \quad x\in\mathbf{R}^N,\ t>0, \tag{2.8}$$

$$v(x,0)=\varphi(x), \quad v_t(x,0)=\psi(x), \quad x\in\mathbf{R}^N, \tag{2.9}$$

where $x+2Le_i=(x_1,\cdots,x_{i-1},x_i+2L,x_{i+1},\cdots,x_N)$, $L>0$ is a real number, $\varphi\in H_0^1(\Omega)$, $\psi\in L_2(\Omega)$, with $\varphi|_{\partial\Omega}=\psi|_{\partial\Omega}=0$, $\varphi(x)=\varphi(x+2Le_i)$, $\psi(x)=\psi(x+2Le_i)$, $x\in\mathbf{R}^N$, $i=1,\cdots,N$.

**Definition** The function $v\in L_\infty([0,T];H_0^1)$, with $G^{1/2}v_t\in L_\infty([0,T];L_2)$, is called a weak solution of problem (2.5), (2.8), (2.9), if for any $\chi\in H_0^1$, and a. e. $t\in[0,T]$,

$$(G^{1/2}v_{tt},G^{1/2}\chi)+2\lambda(G^{1/2}v_t,G^{1/2}\chi)+\lambda^2(G^{1/2}v,G^{1/2}\chi)+$$
$$(v,\chi)+(\nabla v,\nabla\chi)+(\tilde{\sigma}(t,v),\chi)=0, \tag{2.10}$$

$$v(\cdot,0)=\varphi \quad \text{in } H^1, \quad G^{1/2}v_t(\cdot,0)=G^{1/2}\psi \quad \text{in } L_2. \tag{2.11}$$

Define the potential wells with parameter $t$,

$$W_t=\{v(t)\in H_0^1 \mid I(v(t))=\|v(t)\|_{H^1}^2-e^{\lambda(p-1)t}\|v(t)\|_{p+1}^{p+1}>0\}\cup\{0\}, \tag{2.12}$$

$$\widetilde{W}_t = \{v(t) \in H^1(\mathbf{R}^N) \mid I(v(t)) = \|v(t)\|^2_{H^1(\mathbf{R}^N)} - e^{\lambda(p-1)t}\|v(t)\|^{p+1}_{L_{p+1}(\mathbf{R}^N)} > 0\} \cup \{0\}, \quad t \in [0,T]. \tag{2.13}$$

For later purposes we introduce the functional $J$ defined by

$$J(v(t)) := \frac{1}{2}\|v(t)\|^2_{H^1} - \frac{b}{p+1}e^{\lambda(p-1)t}\|v(t)\|^{p+1}_{p+1}, \quad t \in [0,T], \tag{2.14}$$

for suitable $v(t)$. Obviously,

$$J(v(t)) = \frac{1}{2}I(v(t)) + b_1 e^{\lambda(p-1)t}\|v(t)\|^{p+1}_{p+1}$$
$$= \frac{b}{p+1}I(v(t)) + b_1\|v(t)\|^2_{H^1} \tag{2.15}$$

for all such $v(t)$, where and in the sequel $b_1 = \dfrac{p+1-2b}{2(p+1)}$.

**Lemma 2.1** For any $T>0$ and each $t \in [0, T]$, $W_t$ is a neighborhood of 0 in $H_0^1$ provided $1<p\leqslant\dfrac{N+2}{(N-2)^+}$ $(p<+\infty)$, where $a^+=\max\{a,0\}$.

**Proof** By the Gagliardo-Nirenberg inequality,

$$e^{\lambda(p-1)t}\|v(t)\|^{p+1}_{p+1} \leqslant C_* e^{\lambda(p-1)t}\|v(t)\|^{(p+1)(1-\theta)}\|\nabla v(t)\|^{(p+1)\theta}$$
$$\leqslant C_* e^{\lambda(p-1)t}\|v(t)\|^{p-1}_{H^1}\|v(t)\|^2_{H^1}$$
$$< \|v(t)\|^2_{H^1}, \quad t \in [0,T], \tag{2.16}$$

as long as $\|v(t)\|_{H^1}<(1/C_*)^{1/(p-1)}e^{-\lambda T}, t\in[0,T]$, where $\theta=\dfrac{N(p-1)}{2(p+1)}$ $(\leqslant 1)$. Lemma 2.1 is proved. □

**Remark 1** From (2.16) we know that $H_0^1 \hookrightarrow L_{p+1}$, with the embedding constant $C_*$ independent of $\Omega$.

Now we state the main results of the paper.

**Theorem 2.1** Assume that

$(\mathrm{H}_1)$ $\sigma \in C(\mathbf{R}), |\sigma(s)| \leqslant b|s|^p, s \in \mathbf{R}$, with $1 < p \leqslant \dfrac{N+2}{(N-2)^+}$ $(p<+\infty)$.

$(\mathrm{H}_2)$ $G^{1/2}\psi \in L_2, \varphi \in H_0^1$, and one of the following conditions holds:

(i)

$$0 \leqslant \sigma(s)s \leqslant \beta\int_0^s \sigma(\tau)\mathrm{d}\tau, \quad s \in \mathbf{R}, \tag{2.17}$$

where $\beta>0$ is a constant;

(ii) $b<\dfrac{p+1}{2}, \varphi\in W_0$, i.e. $\varphi=0$ or

$$I(\varphi) = \|\varphi\|^2_{H^1} - \|\varphi\|^{p+1}_{p+1} > 0, \tag{2.18}$$

and

$$E^*(0) < \frac{b_1}{2e} C_*^{\frac{-2}{p+1}}, \tag{2.19}$$

where $b_1 = \dfrac{p+1-2b}{2(p+1)}$ $(>0)$, $C_*$ as shown in (2.16) and

$$E^*(0) = \frac{1}{2}(\| G^{1/2}\psi\|^2 + \lambda^2 \| G^{1/2}\varphi\|^2 + \|\varphi\|_{H^1}^2) + \int_\Omega \int_0^\varphi \sigma(s)\mathrm{d}s\mathrm{d}x + \lambda(G^{1/2}\psi, G^{1/2}\varphi) + \lambda^2\| G^{1/2}\varphi\|^2, \tag{2.20}$$

with $\lambda = 2/(3T)$.

Then PBVP (2.5), (2.8), (2.9) admits a weak solution on $[0,T]$.

**Theorem 2.2** In addition to assumption ($H_1$), we assume that

($H_3$) $G^{1/2}u_1 \in L_2(\mathbf{R}^N)$, $u_0 \in H^1(\mathbf{R}^N)$, and one of the following conditions holds:

(i) inequality (2.17);

(ii) $b < \dfrac{p+1}{2}$, $u_0 \in \widetilde{W}_0$, i.e. $u_0 = 0$ or

$$I(u_0) = \|u_0\|_{H^1(\mathbf{R}^N)}^2 - \|u_0\|_{L_{p+1}(\mathbf{R}^N)}^{p+1} > 0, \tag{2.21}$$

and

$$\widetilde{E}^*(0) < \frac{b_1}{2e} C_*^{\frac{-2}{p+1}}, \tag{2.22}$$

where $C_*$ as shown in (2.16), and

$$\widetilde{E}^*(0) = \frac{1}{2}(\| G^{1/2}(u_1 - \lambda u_0)\|_{L_2(\mathbf{R}^N)}^2 + \lambda^2\| G^{1/2}u_0\|_{L_2(\mathbf{R}^N)}^2 + \|u_0\|_{H^1(\mathbf{R}^N)}^2) + \int_{\mathbf{R}^N}\int_0^{u_0}\sigma(s)\mathrm{d}s\mathrm{d}x + \lambda(G^{1/2}u_1, G^{1/2}u_0), \tag{2.23}$$

with $\lambda = 2/(3T)$.

Then Cauchy problem (1.1), (1.2) admits a weak solution on $[0,T]$.

**Remark 2** Since for $T>1$,

$$\widetilde{E}^*(0) \leqslant 2(\| G^{1/2}u_1\|_{L_2(\mathbf{R}^N)}^2 + \| G^{1/2}u_0\|_{L_2(\mathbf{R}^N)}^2) + \|u_0\|_{H^1(\mathbf{R}^N)}^2 + \int_{\mathbf{R}^N}\int_0^{u_0}\sigma(s)\mathrm{d}s\mathrm{d}x \equiv \overline{E}^*(0),$$

(2.22) holds true for all $T>1$ as long as $\overline{E}^*(0) < b_1 C_*^{\frac{-2}{p+1}}/2e$.

In the case of space dimension $N=1$, the weak solutions can be regularized.

**Theorem 2.3** Under the assumptions of Theorem 2.2 (with $N=1$), if also $\sigma \in C^4(\mathbf{R})$, $u_0 \in H^4(\mathbf{R})$ and $u_1 \in H^2(\mathbf{R})$, then for any $T>0$, Cauchy problem (1.3), (1.2), with $N=1$, admits a unique strong solution

$$u \in C([0,T];H^4(\mathbf{R})) \cap C^1([0,T];H^2(\mathbf{R})) \cap C^2([0,T];L_2(\mathbf{R})).$$

**Theorem 2.4** Assume that

($H_4$) there exists a constant $\alpha > 0$ such that

$$\sigma(s)s \leqslant 2(2\alpha+1)\int_0^s \sigma(\tau)\mathrm{d}\tau, \quad s \in \mathbf{R}. \tag{2.24}$$

($H_5$) $u_0 \in H^1(\mathbf{R}^n)$, $G^{1/2}u_1 \in L_2(\mathbf{R}^n)$ such that either

$$E(0) = \| G^{1/2}u_1\|^2_{L_2(\mathbf{R}^N)} + \| u_0\|^2_{H^1(\mathbf{R}^N)} + 2\int_{\mathbf{R}^N}\int_0^{u_0}\sigma(s)\mathrm{d}s\mathrm{d}x < 0 \tag{2.25}$$

or

$$0 \leqslant E(0) < \frac{(G^{1/2}u_0, G^{1/2}u_1)^2}{\| G^{1/2}u_0\|^2_{L_2(\mathbf{R}^N)}} \text{ with } (G^{1/2}u_0, G^{1/2}u_1) > 0. \tag{2.26}$$

Then any weak solution $u$ of Cauchy problem (1.1), (1.2) blows up in finite time $\widetilde{T}$, i.e.

$$\| G^{1/2}u(t)\|_{L_2(\mathbf{R}^N)} \to +\infty \text{ as } t \to \widetilde{T}^-. \tag{2.27}$$

**Remark 3** We set $\sigma(s) = a|s|^{p-1}s$ and $\sigma(s) = as^p$, with $a \neq 0, p > 1$, in Eq. (1.1), respectively. A simple verification shows that (2.24) is valid, with $\alpha = \frac{p-1}{4}$ ($>0$), and then we have by Theorem 2.4 that any weak solution of the corresponding Cauchy problem blows up in finite time as long as the initial data satisfy condition ($H_5$).

# 3 Global Existence of Weak Solutions to the PBVP

**Lemma 3.1** (See [8]) Let $\Omega \subset \mathbf{R}^N$ be a bounded domain, and $\{w_j\}_{j=1}^\infty$ a complete orthogonal system in $L_2(\Omega)$. Then for every $\varepsilon > 0$ there is a positive integer $N_\varepsilon$ such that for all $u \in W_0^{1,p}(\Omega)$,

$$\| u\| \leqslant \left[\sum_{j=1}^{N_\varepsilon}(u, w_j)^2\right]^{1/2} + \varepsilon\| u\|_{1,p},$$

where $2 \leqslant p < +\infty$.

**Proof of Theorem 2.1** We look for approximate solutions $v^n$ of problem (2.5), (2.8), (2.9) of the form

$$v^n(t) := \sum_{j=1}^n T_{jn}(t)w_j, \quad t \geqslant 0, \tag{3.1}$$

where $\{w_j\}_{j=1}^\infty$ is a Schauder basis in $H_0^1$ and at the same time an orthonormal basis in $L_2$, with $w_j(x) = w_j(x + 2Le_i)$, $x \in \mathbf{R}^N$, $i = 1, \cdots, N$, and the coefficients $\{T_{jn}\}_{j=1}^n$ satisfy $T_{jn}(t) = (v^n(t), w_j)$ with

$$(G^{1/2}v_{tt}^n, G^{1/2}w_j) + 2\lambda(G^{1/2}v_t^n, G^{1/2}w_j) + \lambda^2(G^{1/2}v^n, G^{1/2}w_j) +$$
$$(v^n, w_j) + (\nabla v^n, \nabla w_j) + (\tilde{\sigma}(t, v^n), w_j) = 0, \quad t > 0, \tag{3.2}$$
$$v_n(0) = \varphi^n, \quad G^{1/2}v_t^n(0) = G^{1/2}\psi^n, \tag{3.3}$$

for $j = 1, \cdots, n$, $n \in \mathbf{N}$, and where $\varphi^n$ and $\psi^n$ are in $C_0^\infty$ such that

$$G^{1/2}\varphi^n \to G^{1/2}\varphi \text{ in } H^3, \quad G^{1/2}\psi^n \to G^{1/2}\psi \text{ in } L_2 \text{ as } n \to \infty. \tag{3.4}$$

Substituting $w_j$ in (3.2) by $v_t^n$ and $v^n$, respectively, we have

$$\frac{\mathrm{d}}{\mathrm{d}t}\left[\frac{1}{2}(\| G^{1/2}v_t^n(t)\|^2 + \lambda^2\| G^{1/2}v^n(t)\|^2 + \| v^n(t)\|^2_{H^1}) + \mathrm{e}^{-2\lambda t}\int_\Omega\int_0^{\mathrm{e}^{\lambda t}v^n}\sigma(s)\mathrm{d}s\mathrm{d}x\right] +$$
$$2\lambda\left(\| G^{1/2}v_t^n(t)\|^2 + \mathrm{e}^{-2\lambda t}\int_\Omega\int_0^{\mathrm{e}^{\lambda t}v^n}\sigma(s)\mathrm{d}s\mathrm{d}x\right)$$
$$= \lambda\mathrm{e}^{-2\lambda t}(\sigma(\mathrm{e}^{\lambda t}v^n), \mathrm{e}^{\lambda t}v^n), \quad t > 0, \tag{3.5}$$

$$\frac{\mathrm{d}}{\mathrm{d}t}[(G^{1/2}v_t^n, G^{1/2}v^n) + \lambda\|G^{1/2}v^n(t)\|^2] + \lambda^2\|G^{1/2}v^n(t)\|^2 + \|v^n(t)\|_{H^1}^2 + \mathrm{e}^{-2\lambda t}(\sigma(\mathrm{e}^{\lambda t}v^n), \mathrm{e}^{\lambda t}v^n)$$

$$= \|G^{1/2}v_t^n(t)\|^2, \quad t > 0. \tag{3.6}$$

The combination of (3.5) with (3.6) yields

$$\frac{\mathrm{d}}{\mathrm{d}t}E_n^*(t) + \overline{E}_n(t) = 0, \quad t > 0, \tag{3.7}$$

where

$$E_n^*(t) = \frac{1}{2}(\|G^{1/2}v_t^n(t)\|^2 + 3\lambda^2\|G^{1/2}v^n(t)\|^2 + \|v^n(t)\|_{H^1}^2) +$$

$$\mathrm{e}^{-2\lambda t}\int_\Omega\int_0^{\mathrm{e}^{\lambda t}v^n}\sigma(s)\mathrm{d}s\mathrm{d}x + \lambda(G^{1/2}v_t^n, G^{1/2}v^n), \tag{3.8}$$

$$\overline{E}_n(t) = \lambda\|G^{1/2}v_t^n(t)\|^2 + \lambda^3\|G^{1/2}v^n(t)\|^2 + \lambda\|v^n(t)\|_{H^1}^2 + 2\lambda\mathrm{e}^{-2\lambda t}\int_\Omega\int_0^{\mathrm{e}^{\lambda t}v^n}\sigma(s)\mathrm{d}s\mathrm{d}x, \quad t \geqslant 0. \tag{3.9}$$

(1) If $(H_2)$(ⅰ) holds true, by noticing $(G^{1/2}v_t^n, G^{1/2}v^n) \leqslant \frac{\varepsilon}{2}\|G^{1/2}v^n\|^2 + \frac{1}{2\varepsilon}\|G^{1/2}v_t^n\|^2$, and by taking $\varepsilon = 2\lambda$ and $\varepsilon = \lambda/2$, respectively, we obtain

$$\frac{1}{4}\|G^{1/2}v_t^n(t)\|^2 + \frac{1}{2}\lambda^2\|G^{1/2}v^n(t)\|^2 + \frac{1}{2}\|v^n(t)\|_{H^1}^2 + \mathrm{e}^{-2\lambda t}\int_\Omega\int_0^{\mathrm{e}^{\lambda t}v^n}\sigma(s)\mathrm{d}s\mathrm{d}x$$

$$\leqslant E_n^*(t) \leqslant \frac{2}{\lambda}\overline{E}_n(t), \quad t > 0. \tag{3.10}$$

It follows from $(H_1)$, Remark 1 and (3.4) that

$$\left|\int_\Omega\int_\varphi^{\varphi^n}\sigma(s)\mathrm{d}s\mathrm{d}x\right| \leqslant \left|\int_\Omega\sigma(k^n)(\varphi^n - \varphi)\mathrm{d}x\right| \leqslant b\|k^n\|_{p+1}^p\|\varphi^n - \varphi\|_{p+1} \to 0$$

as $n \to \infty$, where $k^n = \theta\varphi^n + (1-\theta)\varphi, 0<\theta<1$. As a result,

$$E_n^*(0) \to E^*(0)\ (>0) \text{ as } n \to \infty, \tag{3.11}$$

where $E^*(0)$ as shown in (2.20). Without loss of generality we assume that

$$E_n^*(0) < 2E^*(0) \text{ for all } n. \tag{3.12}$$

The combination of (3.7) with (3.10) and (3.12) leads to

$$\frac{\mathrm{d}}{\mathrm{d}t}E_n^*(t) + \frac{\lambda}{2}E_n^*(t) \leqslant 0,$$

$$\frac{1}{4}\|G^{1/2}v_t^n(t)\|^2 + \frac{1}{2}\lambda^2\|G^{1/2}v^n(t)\|^2 + \frac{1}{2}\|v^n(t)\|_{H^1}^2 +$$

$$\mathrm{e}^{-2\lambda t}\int_\Omega\int_0^{\mathrm{e}^{\lambda t}v^n}\sigma(s)\mathrm{d}s\mathrm{d}x \leqslant 2E^*(0)\mathrm{e}^{\frac{-\lambda t}{2}}, \quad t > 0. \tag{3.13}$$

By $(H_2)$(ⅰ) and (3.13),

$$\int_\Omega|\tilde{\sigma}(t, v^n)|^{(p+1)'}\mathrm{d}x = \mathrm{e}^{-(p+1)'\lambda t}\int_\Omega|\sigma(\mathrm{e}^{\lambda t}v^n)|^{(p+1)'}\mathrm{d}x$$

$$\leqslant b^{1/p}\mathrm{e}^{-(p+1)'\lambda t}\int_\Omega\sigma(\mathrm{e}^{\lambda t}v^n)\mathrm{e}^{\lambda t}v^n\mathrm{d}x$$

$$\leqslant \beta b^{1/p} e^{-(p+1)'\lambda t} \int_{\Omega} \int_0^{e^{\lambda t} v^n} \sigma(s) \mathrm{d}s \mathrm{d}x$$

$$\leqslant 2\beta b^{1/p} e^{\frac{p-2}{2p}\lambda t} E^*(0), \quad t > 0. \tag{3.14}$$

(2) If $(H_2)$(ⅱ) holds true, by (3.4) and Remark 1,

$$I(\varphi^n) = \|\varphi^n\|_{H^1}^2 - \|\varphi^n\|_{p+1}^{p+1} \to I(\varphi) > 0 \quad \text{as} \quad n \to \infty. \tag{3.15}$$

Without loss of generality we assume that $I(\varphi^n) > 0$, i.e. $\varphi^n \in W_0$ for all $n$. By the continuity of $I(v^n(t))$, there exists a right neighborhood $(0, \delta)$ such that $I(v^n(t)) > 0$, $t \in (0, \delta)$. We claim that for any $T > 0$, $v^n(t) \in W_t$, $t \in [0, T]$, i.e.

$$I(v^n(t)) > 0, \quad t \in [0, T]. \tag{3.16}$$

In fact, if there exists $t_0: 0 < t_0 \leqslant T$ such that $v^n(t) \in W_t$, $t \in [0, t_0)$, and $v^n(t_0) \in \partial W_{t_0}$, i.e. $I(v^n(t_0)) = 0$. Then by $(H_1)$, $(H_2)$(ⅱ), (2.14) and (2.15),

$$\frac{1}{2}\|v^n(t)\|_{H^1}^2 + e^{-2\lambda t} \int_{\Omega} \int_0^{e^{\lambda t} v^n} \sigma(s) \mathrm{d}s \mathrm{d}x$$

$$\geqslant J(v^n(t)) \geqslant \begin{cases} b_1 \|v^n(t)\|_{H^1}^2, \\ b_1 e^{\lambda(p-1)t} \|v^n(t)\|_{p+1}^{p+1}, \quad t \in [0, t_0]. \end{cases} \tag{3.17}$$

Hence (3.10) and (3.13) hold true on $[0, t_0]$. The combination of (3.17) with (3.13) yields

$$b_1 \|v^n(t)\|_{H^1}^2 \leqslant 2E^*(0) e^{\frac{-\lambda t}{2}}, \quad t \in [0, t_0]. \tag{3.18}$$

By taking $\lambda = 2/(3T)$, we get from (2.16) and (2.19) that

$$e^{\lambda(p-1)t} \|v^n(t)\|_{p+1}^{p+1} \leqslant C_* \left(\frac{2E^*(0)}{b_1}\right)^{\frac{p-1}{2}} e^{\frac{3}{4}\lambda(p-1)T} \|v^n(t)\|_{H^1}^2$$

$$\leqslant C_* \left(\frac{2eE^*(0)}{b_1}\right)^{\frac{p-1}{2}} \|v^n(t)\|_{H^1}^2 < \|v^n(t)\|_{H^1}^2, \quad t \in [0, t_0]. \tag{3.19}$$

In particular, $I(v^n(t_0)) > 0$, which violates the assumption. So (3.16) is valid. And hence (3.17)~(3.19) hold true on $[0, T]$.

The combination of (3.13) with (3.17) leads to

$$\frac{1}{4}\|G^{1/2} v_t^n(t)\|^2 + \frac{1}{2}\lambda^2 \|G^{1/2} v^n(t)\|^2 + \frac{b_1}{2}(\|v^n(t)\|_{H^1}^2 + e^{\lambda(p-1)t}\|v^n(t)\|_{p+1}^{p+1})$$

$$\leqslant 2E^*(0) e^{\frac{-\lambda t}{2}}, \quad t \in [0, T]. \tag{3.20}$$

By $(H_2)$(ⅱ) and (3.20),

$$\int_{\Omega} |\tilde{\sigma}(t, v^n)|^{(p+1)'} \mathrm{d}x \leqslant b^{(p+1)'} e^{-(p+1)'\lambda t} \int_{\Omega} |e^{\lambda t} v^n(t)|^{(p+1)} \mathrm{d}x$$

$$\leqslant \frac{4}{b_1} b^{(p+1)'} e^{\frac{p-2}{2p}\lambda t} E^*(0), \quad t \in [0, T], \tag{3.21}$$

where $\lambda = 2/(3T)$. Integrating (3.2) over $(0, t)$, we obtain

$$(G^{1/2} v_t^n(t), G^{1/2} w_j) + 2\lambda (G^{1/2} v^n(t), G^{1/2} w_j) + \int_0^t (\lambda^2 (G^{1/2} v^n, G^{1/2} w_j) +$$

$$(v^n, w_j) + (\nabla v^n, \nabla w_j) + (\tilde{\sigma}(t, v^n), w_j))\mathrm{d}\tau$$
$$= (G^{1/2}\psi^n, G^{1/2}w_j) + 2\lambda(G^{1/2}\varphi^n, G^{1/2}w_j), \quad t > 0, \tag{3.22}$$
$$v^n(0) = \varphi^n. \tag{3.23}$$

For any $T>0$, it follows from (3.13), (3.14) (if $(H_2)$( i ) holds true) or (3.20), (3.21) (if $(H_2)$( ii ) holds true) that the nonlinear terms appearing in (3.22) are uniformly bounded on $[0, T]$, so the solution of problem (3.22), (3.23) exists on $[0, T]$ for each $n$.

From (3.13), (3.14) (if $(H_2)$( i ) holds true) or (3.20), (3.21) (if $(H_2)$( ii ) holds true) we can extract a subsequence from $\{v^n\}$, still denoted by $\{v^n\}$, such that

$$G^{1/2}v^n \to G^{1/2}v \quad \text{in } L_\infty([0, T]; L_2) \text{ weak}^*, \tag{3.24}$$
$$v^n \to v \quad \text{in } L_\infty([0, T]; H_0^1) \text{ weak}^*, \tag{3.25}$$
$$G^{1/2}v_t^n \to G^{1/2}v_t \quad \text{in } L_\infty([0, T]; L_2) \text{ weak}^*, \tag{3.26}$$
$$\tilde{\sigma}(t, v^n) \to \gamma \quad \text{in } L_\infty([0, T]; L_{(p+1)'}) \text{ weak}^*, \tag{3.27}$$

as $n \to \infty$. It follows from Lemma 3.1, (3.25), (3.13) (if $(H_2)$( i ) holds true) or (3.20) (if $(H_2)$( ii ) holds true) that for every $\varepsilon > 0$, there exists a positive integer $N_\varepsilon$ such that for all $v^n$, when $n \to \infty$,

$$\| v^n(t) - v(t) \| \leqslant \left[ \sum_{j=1}^{N_\varepsilon} (v^n - v, w_j)^2 \right]^{1/2} + \varepsilon \| v^n(t) - v(t) \|_{H^1}$$
$$\leqslant CE^*(0)\varepsilon, \quad t \in [0, T]. \tag{3.28}$$

By the arbitrariness of $\varepsilon$,

$$v^n \to v \quad \text{in } L_\infty([0, T]; L_2) \text{ and a.e. on } Q_T, \tag{3.29}$$

as $n \to \infty$. By (3.29) and the continuity of $\sigma$, for a.e. $t \in [0, T]$,

$$\tilde{\sigma}(t, v^n) \to \tilde{\sigma}(t, v), \quad \text{a.e. on } \Omega \tag{3.30}$$

as $n \to \infty$. It follows from the Egoroff theorem that for any $\delta > 0$, there exists a measurable set $\Omega_\delta \subset \Omega$, $|\Omega_\delta| < \delta$ such that for a.e. $t \in [0, T]$, $\tilde{\sigma}(t, v^n) \to \tilde{\sigma}(t, v)$ uniformly on $\tilde{\Omega} = \Omega - \Omega_\delta$ as $n \to \infty$. Hence, for any $p_1 : p_1 > p$, by the embedding theorem, (3.14) (if $(H_2)$( i ) holds true) or (3.21) (if $(H_2)$( ii ) holds true), we have

$$\| \tilde{\sigma}(t, v^n) - \tilde{\sigma}(t, v) \|_{L_{(p_1+1)'}(\Omega_\delta)} \leqslant \| \tilde{\sigma}(t, v^n) - \tilde{\sigma}(t, v) \|_{L_{(p+1)'}(\Omega_\delta)} \delta^{\frac{p_1 - p}{(p+1)(p_1+1)}}$$
$$\leqslant M(T) \delta^{\frac{p_1 - p}{(p+1)(p_1+1)}}, \quad t \in [0, T], \tag{3.31}$$

where and in the sequel $M(T)$ denotes positive constants depending only on $T$. Therefore, when $n \to \infty$,

$$\| \tilde{\sigma}(t, v^n) - \tilde{\sigma}(t, v) \|_{L_{(p_1+1)'}(\Omega)} \leqslant \| \tilde{\sigma}(t, v^n) - \tilde{\sigma}(t, v) \|_{L_{(p_1+1)'}(\tilde{\Omega})} + \| \tilde{\sigma}(t, v^n) - \tilde{\sigma}(t, v) \|_{L_{(p_1+1)'}(\Omega_\delta)}$$
$$\leqslant M(T) \delta^{\frac{p_1 - p}{(p+1)(p_1+1)}}, \quad t \in [0, T]. \tag{3.32}$$

By the arbitrariness of $\delta$,

$$\tilde{\sigma}(t, v^n) \to \tilde{\sigma}(t, v) \quad \text{in } L_\infty([0, T]; L_{(p_1+1)'}) \text{ as } n \to \infty. \tag{3.33}$$

It follows from (3.33) and the uniqueness of weak* limit that $\gamma(t)=\tilde{\sigma}(t,v)$ in $L_{(p_1+1)'}$ for a.e. $t\in[0,T]$. Since $C_0^\infty$ is dense in $L_{p+1}$, for any $\kappa\in L_{p+1}$ there exists a sequence $\{\kappa_n\}$, $\kappa_n\in C_0^\infty$, such that $\kappa_n\to\kappa$ in $L_{p+1}$ as $n\to\infty$. Hence,

$$|(\tilde{\sigma}(t,v)-\gamma(t),\kappa_n-\kappa)|$$
$$\leqslant\|\tilde{\sigma}(t,v)-\gamma(t)\|_{(p+1)'}\|\kappa_n-\kappa\|_{p+1}\to 0 \quad \text{as } n\to\infty, t\in[0,T], \tag{3.34}$$
$$(\tilde{\sigma}(t,v)-\gamma(t),\kappa)=\lim_{n\to\infty}(\tilde{\sigma}(t,v)-\gamma(t),\kappa_n)=0, \tag{3.35}$$
$$\tilde{\sigma}(t,v)=\gamma(t) \quad \text{in } L_{(p+1)'} \text{ for a.e. } t\in[0,T]. \tag{3.36}$$

For any $\chi\in H_0^1$, it follows from (2.7) that $G^{1/2}\chi\in H^2$. Hence, letting $n\to\infty$ in (3.22) and making use of (3.24) ~ (3.27), (3.36) and (3.4) we obtain, for any $\chi\in H_0^1$ and a. e. $t\in[0,T]$,

$$(G^{1/2}v_t(t),G^{1/2}\chi)+2\lambda(G^{1/2}v(t),G^{1/2}\chi)+$$
$$\int_0^t[\lambda^2(G^{1/2}v,G^{1/2}\chi)+(v,\chi)+(\nabla v,\nabla\chi)+(\tilde{\sigma}(\tau,v),\chi)]d\tau$$
$$=(G^{1/2}\psi,G^{1/2}\chi)+2\lambda(G^{1/2}\varphi,G^{1/2}\chi). \tag{3.37}$$

Differentiating (3.37) we get that for any $\chi\in H_0^1$ and a.e. $t\in[0,T]$, (2.10) is valid. By (3.24), (3.26) and (3.4),

$$(G^{1/2}v^n,w_j)\to(G^{1/2}v,w_j) \quad \text{in } H^1[0,T] \text{ and in } C[0,T],$$
$$G^{1/2}v(0)=G^{1/2}\varphi \quad \text{in } L_2. \tag{3.38}$$

Since $\varphi\in H_0^1$, applying the operators $G^{-1/2}=(-\Delta)^{1/2}$ and $G^{-1}=-\Delta$ to both sides of (3.38), we get

$$v(0)=\varphi \quad \text{in } H^1. \tag{3.39}$$

Letting $t=0$ in (3.37) and using (3.38) and the density of $H_0^1$ in $L_2$, we have

$$Gv_t(0)=G\psi \quad \text{in } L_2. \tag{3.40}$$

Applying $G^{-1/2}=(-\Delta)^{1/2}$ to both sides of (3.40) leads to

$$G^{1/2}v_t(0)=G^{1/2}\psi. \tag{3.41}$$

The combination of (2.10) with (3.39) and (3.41) indicates that $v$ is a weak solution of PBVP (2.5), (2.8), (2.9). Theorem 2.1 is proved. □

## 4 Global Weak Solutions to the Cauchy Problem

**Proof of Theorem 2.2** We take a sequence $\{L_s\}$, where $L_s(>1)$ are real numbers, and $L_s\to+\infty$ as $s\to\infty$. Let $\Omega_s=(-L_s,L_s)^N$. For each $s$ we construct the potential wells with parameter $t$

$$W_t^s=\{v(t)\in H_0^1(\Omega_s)\mid I_s(v(t))=\|v(t)\|_{H^1(\Omega_s)}^2-e^{\lambda(p-1)t}\|v(t)\|_{L_{p+1}(\Omega_s)}^{p+1}>0\}\cup\{0\}, \tag{4.1}$$

and the periodic functions $\varphi_s\in H_0^1(\Omega_s)$, $\psi_s\in L_2(\Omega_s)$, with $\varphi_s|_{\partial\Omega_s}=\psi_s|_{\partial\Omega_s}=0$, and

(i) $\varphi_s(x)=\varphi_s(x+2L_se_i), \psi_s(x)=\psi_s(x+2L_se_i), x\in\mathbf{R}^N, i=1,\cdots,N.$

(ii) $\varphi_s(x) = v_0(x), \psi_s(x) = v_1(x)$ on $\Omega_s^* = (-L_s+1, L_s-1)^N$, and

$$\|G^{1/2}\varphi_s\|_{L_2(\Omega_s)} \leqslant \|G^{1/2}v_0\|_{L_2(\mathbf{R}^N)}, \quad \|\varphi_s\|_{H^1(\Omega_s)} \leqslant \|v_0\|_{H^1(\mathbf{R}^N)},$$
$$\|\varphi_s\|_{L_{p+1}(\Omega_s)} \leqslant \|v_0\|_{L_{p+1}(\mathbf{R}^N)}, \quad \|G^{1/2}\psi_s\|_{L_2(\Omega_s)} \leqslant \|G^{1/2}v_1\|_{L_2(\mathbf{R}^N)}. \tag{4.2}$$

Let

$$\tilde{\varphi}_s(x) = \begin{cases} \varphi_s(x), & x \in \Omega_s \\ 0, & x \in \mathbf{R}^N - \Omega_s \end{cases}, \quad \tilde{\psi}_s(x) = \begin{cases} \psi_s(x), & x \in \Omega_s \\ 0, & x \in \mathbf{R}^N - \Omega_s \end{cases}. \tag{4.3}$$

By (4.2),

$$\|G^{1/2}\tilde{\varphi}_s - G^{1/2}v_0\|_{L_2(\mathbf{R}^N)} \leqslant 2\|G^{1/2}v_0\|_{L_2(\mathbf{R}^N)}.$$

Hence,

$$\|G^{1/2}\tilde{\varphi}_s - G^{1/2}v_0\|_{L_2(\mathbf{R}^N)} = \|G^{1/2}\tilde{\varphi}_s - G^{1/2}v_0\|_{L_2(\mathbf{R}^N - \Omega_s^*)} \to 0 \tag{4.5}$$

as $s \to \infty$. Similarly,

$$\tilde{\varphi}_s \to v_0 \quad \text{in } H^1(\mathbf{R}^N) \cap L_{p+1}(\mathbf{R}^N), \quad G^{1/2}\tilde{\psi}_s \to G^{1/2}v_1 \quad \text{in } L_2(\mathbf{R}^N) \tag{4.6}$$

as $s \to \infty$. It follows from (4.6) that

$$I_s(\varphi_s) = \|\varphi_s\|_{H^1(\Omega_s)}^2 - \|\varphi_s\|_{L_{p+1}(\Omega_s)}^{p+1} \to I(v_0) = \|v_0\|_{H^1(\mathbf{R}^N)}^2 - \|v_0\|_{L_{p+1}(\mathbf{R}^N)}^{p+1} \tag{4.7}$$

as $s \to \infty$. By the integral mean value theorem, the Hölder inequality and (4.6),

$$\left|\int_{\Omega_s}\int_0^{\varphi_s}\sigma(\tau)\mathrm{d}\tau\mathrm{d}x - \int_{\mathbf{R}^N}\int_0^{v_0}\sigma(\tau)\mathrm{d}\tau\mathrm{d}x\right| \leqslant \|\sigma(\xi_s)\|_{L_{(p+1)'}(\mathbf{R}^N)}\|\tilde{\varphi}_s - v_0\|_{L_{p+1}(\mathbf{R}^N)}$$
$$\leqslant C\|\xi_s\|_{L_{p+1}(\mathbf{R}^N)}^p\|\tilde{\varphi}_s - v_0\|_{L_{p+1}(\mathbf{R}^N)} \to 0 \tag{4.8}$$

as $s \to \infty$, where $\xi_s = v_0 + \theta_s\tilde{\varphi}_s, 0 < \theta_s < 1$. By (4.5), (4.6) and (4.8),

$$E_s^*(0) \to \widetilde{E}^*(0) \quad \text{as } s \to \infty, \tag{4.9}$$

where $E_s^*(0)$ as shown in (2.20) (substituting $\varphi$, $\psi$ and $\Omega$ there by $\varphi_s$, $\psi_s$ and $\Omega_s$ respectively), and $\widetilde{E}^*(0)$ as shown in (2.23), where the facts: $v_0 = u_0$ and $v_1 = u_1 - \lambda u_0$ have been used. By (2.21), (2.22), (4.7) and (4.9), without loss of generality we assume that

$$I(\varphi_s) > 0, \quad E_s^*(0) < \min\left\{\frac{b_1}{2\mathrm{e}}C_*^{\frac{-2}{p-1}}, 2\widetilde{E}^*(0)\right\} \quad \text{for all } s. \tag{4.10}$$

Repeating the arguments of Theorem 2.1, for each $s$ we get a weak solution $v^s$ of corresponding PBVP (2.5), (2.8), (2.9) (substituting $L$, $\Omega$, $\varphi$ and $\psi$ there by $L_s$, $Q_s$, $\varphi_s$ and $\psi_s$, respectively), and $v^s$ is a weak* limit of a sequence $\{v^{n_s}\}$, where $v^{n_s}$ are the solutions of problem (3.2), (3.3) (substituting $\Omega$, $v^n$, $\varphi^n$ and $\psi^n$ there by $\Omega_s$, $v^{n_s}$, $\varphi^{n_s}$ and $\psi^{n_s}$, respectively). And inequalities (3.13), (3.14) (if $(H_2)$(i) holds true) or (3.20), (3.21) (if $(H_2)$(ii) holds true) are valid for $v^{n_s}$ (substituting $v^n$ and $\Omega$ there by $v^{n_s}$ and $\Omega_s$, respectively).

(1) If $(H_2)$(i) holds true, then $F(\mathrm{e}^{\lambda t}v^{n_s}) = \int_0^{\mathrm{e}^{\lambda t}v^{n_s}}\sigma(\tau)\mathrm{d}\tau \geqslant 0$ are measurable functions

defined on $\Omega_s$ for a.e. $t\in[0,T]$. Since $u^{n_s}\to v^s$ a.e. on $\Omega_s$ for a.e. $t\in[0,T]$ as $n_s\to\infty$ (see (3.29)), and by the continuity of $F(e^{\lambda t}v^{n_s})$,

$$F(e^{\lambda t}v^{n_s})\to F(e^{\lambda t}v^s)\quad \text{a.e. on } \Omega_s \text{ as } n_s\to\infty,$$

i.e. there exists a set $\widetilde{\Omega}_s\subset\Omega_s$, $|\widetilde{\Omega}_s|=0$, such that

$$F(e^{\lambda t}v^{n_s})\to F(e^{\lambda t}v^s)\quad \text{on } \Omega_s-\widetilde{\Omega}_s \text{ as } n_s\to\infty.$$

Then, by the Fatou theorem, for a.e. $t\in[0,T]$,

$$\begin{aligned}\int_{\Omega_s}F(e^{\lambda t}v^s)dx&=\int_{\Omega_s-\widetilde{\Omega}_s}F(e^{\lambda t}v^s)dx\leqslant \liminf_{n_s\to\infty}\int_{\Omega_s-\widetilde{\Omega}_s}F(e^{\lambda t}v^{n_s})dx\\&=\liminf_{n_s\to\infty}\int_{\Omega_s}F(e^{\lambda t}v^{n_s})dx.\end{aligned}\tag{4.11}$$

From the sequential lower semi-continuity of the norm and (4.11) we conclude that inequalities (3.13) and (3.14) are still valid for $v^s$ (substituting $v^n$, $\Omega$ and $E^*(0)$ there by $v^s$, $\Omega_s$ and $\widetilde{E}^*(0)$, respectively).

(2) If $(H_2)$(ii) holds true, from the sequential lower semi-continuity of the norm we deduce that inequalities (3.20) and (3.21) are still valid for $v^s$ (substituting $v^n$, $\Omega$ and $E^*(0)$ there by $v^s$, $\Omega_s$ and $\widetilde{E}^*(0)$, respectively).

Hence, by the same arguments used in the proof of Theorem 2.1 we can extract a subsequence from $\{v^s\}$, still denoted by $\{v^s\}$, such that, for any $\Omega_L=(-L,L)^N$ and $T>0$,

$$\begin{aligned}&G^{1/2}v^s\to G^{1/2}v\quad \text{in } L_\infty([0,T];L_2(\Omega_L)) \text{ weak}^*,\\&v^s\to v\quad \text{in } L_\infty([0,T];H_0^1(\Omega_L)) \text{ weak}^*,\\&G^{1/2}v_t^s\to G^{1/2}v_t\quad \text{in } L_\infty([0,T];L_2(\Omega_L)) \text{ weak}^*,\\&\tilde{\sigma}(t,v^s)\to\tilde{\sigma}(t,v)\quad \text{in } L_\infty([0,T];L_{(p+1)'}(\Omega_L)) \text{ weak}^*,\end{aligned}\tag{4.12}$$

as $s\to\infty$. And by the same method used above we see that inequalities (3.13) and (3.20) still hold true for the limit function $v$ (substituting $v^n$, $\Omega$ and $E^*(0)$ there by $v$, $\Omega_L$ and $\overline{E}^*(0)$, respectively).

From the arbitrariness of $L$ and the bounded-ness of the norm of $v$ (see (3.13) and (3.20)) we know that the limiting function

$$v\in L_\infty([0,T];H^1(\mathbf{R}^N)),\quad \text{with } G^{1/2}v_t\in L_\infty([0,T];L_2(\mathbf{R}^N)).\tag{4.13}$$

For any $\chi\in H^1(\mathbf{R}^N)$, supp $\chi$ is a bounded set in $\mathbf{R}^N$, there must be an $L>0$ such that supp $\chi\subset\Omega_L$ and $\chi\in H_0^1(\Omega_L)$. Substituting $v^n$, $w_j$, $\varphi^n$ and $\psi^n$ in (3.22) by $v^s$, $\chi$, $\varphi_s$ and $\psi_s$, respectively, and letting $s\to\infty$, we obtain

$$\begin{aligned}&(G^{1/2}v_t(t),G^{1/2}\chi)+2\lambda(G^{1/2}v(t),G^{1/2}\chi)+\\&\int_0^t(\lambda^2(G^{1/2}v(\tau),G^{1/2}\chi)+(v(\tau),\chi)+(\nabla v(\tau),\nabla\chi)+(\tilde{\sigma}(\tau,v(\tau)),\chi))d\tau\\&=(G^{1/2}v_1,G^{1/2}\chi)+2\lambda(G^{1/2}v_0,G^{1/2}\chi),\quad t\in[0,T].\end{aligned}\tag{4.14}$$

By (4.12), for any $\chi\in C_0^\infty(\mathbf{R}^N)$,

$$(G^{1/2}v^s,\chi)\to(G^{1/2}v,\chi)\quad \text{in } H^1[0,T] \text{ and in } C[0,T]. \tag{4.15}$$

Then by (4.5),

$$(G^{1/2}v(0),\chi)=(G^{1/2}v_0,\chi). \tag{4.16}$$

Since $C_0^\infty(\mathbf{R}^N)$ is dense in $L_2(\mathbf{R}^N)$, from the continuity of the inner-product in $L_2$ we get that (4.16) holds true for arbitrary $\chi\in L_2(\mathbf{R}^N)$, and hence

$$G^{1/2}v(0)=G^{1/2}v_0\quad \text{in } L_2(\mathbf{R}^N). \tag{4.17}$$

Since $G^{1/2}v_0\in H^2(\mathbf{R}^N)$, applying $G^{-1/2}$ and $G^{-1}$ to both sides of (4.17), we have

$$v(0)=v_0\quad \text{in } H^1(\mathbf{R}^N). \tag{4.18}$$

For any $\chi\in C_0^\infty(\mathbf{R}^N)$, letting $t=0$ in (4.14), exploiting (4.17) and using the same method as above, we get

$$(Gv_t(0),\chi)=(Gv_1,\chi),\quad Gv_t(0)=Gv_1\quad \text{in } L_2(\mathbf{R}^N). \tag{4.19}$$

Since $G^{1/2}v_1\in L_2(\mathbf{R}^N)$, applying $G^{-1/2}$ to both sides of (4.19), we obtain

$$G^{1/2}v_t(0)=G^{1/2}v_1\quad \text{in } L_2(\mathbf{R}^N). \tag{4.20}$$

Differentiating (4.14) and combining the result with (4.18) and (4.20) we see that $v$ is a weak solution of Cauchy problem (2.5), (2.6) on $[0,T]$. Therefore, $u=e^{\lambda t}v$ is a weak solution of Cauchy problem (1.1), (1.2) on $[0,T]$. Theorem 2.2 is proved. □

# 5 The Case in One Dimension

**Lemma 5.1** (See [29]) Assume that $G(z_1,\cdots,z_h)$ is a $k$-times continuously differentiable function with respect to variables $z_1,\cdots,z_h$ and $z_i\in L_\infty([0,T];H^k(\Omega))(i=1,\cdots,h)$. Then

$$\left\|\frac{\partial^k}{\partial x^k}G(z_1(\cdot,t),\cdots,z_h(\cdot,t))\right\|^2\leqslant C(\overline{M},k,h)\sum_{i=1}^{h}\|z_i(t)\|_{H^k}^2$$

where $\overline{M}=\max_{1\leqslant i\leqslant h}\max_{(x,t)\in\overline{Q}_T}|z_i(x,t)|$, $C(\overline{M},k,h)$ is a positive constant depending only on $\overline{M}$, $k$ and $h$.

In this section, we still use the notations in (2.1), with $\Omega=(-L,L)$. Moreover, $\|\cdot\|_C=\|\cdot\|_{C(\Omega)}$.

**Proof of Theorem 2.3** We first consider the PBVP of Eq. (1.3)

$$u|_{\partial\Omega}=0,\quad u(x,t)=u(x+2L,t),\quad x\in\mathbf{R},\quad t>0, \tag{5.1}$$

$$u(x,0)=\xi(x),\quad u_t(x,0)=\eta(x),\quad x\in\mathbf{R}, \tag{5.2}$$

where $\xi\in H^4\cap H_0^1$, $\eta\in H^2\cap H_0^1$, with $\xi(x)=\xi(x+2L)$, $\eta(x)=\eta(x+2L)$, $x\in\mathbf{R}$. And it follows from (2.9) (with $N=1$) that $\xi(x)=\varphi(x)$, $\eta(x)=\psi(x)+\lambda\varphi(x)$.

We still start with the approximate solutions of PBVP (1.3), (5.1), (5.2) of the form

$$u^n(t):=\sum_{j=1}^{m}\widetilde{T}_{jn}(t)w_j,\quad t\geqslant 0, \tag{5.3}$$

where $\{w_j\}_{j=1}^\infty$ is a Schauder basis in $H^4\cap H_0^1$ and at the same time an orthonormal basis in $L_2$, with $w_j(x)=w_j(x+2L)$, $x\in\mathbf{R}$, and the coefficients $\{\widetilde{T}_{jn}\}_{j=1}^n$ satisfy $\widetilde{T}_{jn}(t)=$

$(u^n(t), w_j)$ with

$$(u_{tt}^n, w_j) - (u_{xx}^n, w_j) + (u_{x^4}^n, w_j) = (\sigma(u^n)_{xx}, w_j), \quad t > 0, \tag{5.4}$$

$$u^n(0) = \xi^n, \quad u_t^n(0) = \eta^n, \tag{5.5}$$

for $j = 1, \cdots, n, n \in \mathbf{N}$, where and in the sequel $u_{x^k} = \dfrac{\partial^k u}{\partial x^k}$, $\xi^n, \eta^n \in C_0^\infty$ and

$$\xi^n \to u_0 \quad \text{in } H^4, \quad \eta^n \to u_1 \quad \text{in } H^2 \quad \text{as } n \to \infty. \tag{5.6}$$

By (3.13) (if $(H_2)$( i ) holds true) or (3.20) (if $(H_2)$( ii ) holds true), and the Gagliardo-Nirenberg inequality,

$$\|u^n(t)\|_{H^1} = e^{\lambda t}\|v^n(t)\|_{H^1} \leqslant C(T),$$

$$\|u^n(t)\|_C \leqslant C\|u^n(t)\|^{\frac{1}{2}}\|u_x^n(t)\|^{\frac{1}{2}} \leqslant M(T), \quad t \in [0, T]. \tag{5.7}$$

Substituting $w_j$ in (5.4) by $u_t^n + u_{x^4 t}^n$, integrating by parts and exploiting Lemma 5.1, (5.7) and the Gronwall inequality, we get

$$\frac{1}{2}\frac{d}{dt}(\|u_t^n(t)\|^2 + \|u_{x^2 t}^n(t)\|^2 + \|u^n(t)\|_{H^4}^2)$$

$$= (\sigma(u^n)_{xx}, u_t^n) + (\sigma(u^n)_{x^4}, u_{x^2 t}^n) + (u^n, u_t^n) \leqslant M(T)(\|u_t^n(t)\|^2 + \|u_{x^2 t}^n(t)\|^2 + \|u^n(t)\|_{H^4}^2),$$

$$\|u_t^n(t)\|_{H^2}^2 + \|u^n(t)\|_{H^4}^2 \leqslant (\|\eta^n\|_{H^2}^2 + \|\xi^n\|_{H^4}^2)e^{2M(T)T}$$

$$\leqslant M(T), \quad t \in [0, T]. \tag{5.8}$$

Substituting $w_j$ in (5.4) by $u_{tt}^n$ and using the Hölder inequality, Lemma 5.1 and (5.8), we obtain

$$\|u_{tt}^n(t)\| \leqslant \|u_{x^2}^n(t)\| + \|u_{x^4}^n(t)\| + \|\sigma(u^n(t))_{xx}\| \leqslant M(T), \quad t \in [0, T]. \tag{5.9}$$

Let $w^n = u^n - u^{n-1}$, then $w^n(t)$ satisfy

$$(w_{tt}^n, w_j) - (w_{xx}^n, w_j) + (w_{x^4}^n, w_j)$$

$$= (\sigma(u^n)_{xx} - \sigma(u^{n-1})_{xx}, w_j), \quad t > 0, \tag{5.10}$$

$$w^n(0) = \xi^n - \xi^{n-1}, \quad w_t^n(0) = \eta^n - \eta^{n-1}, \tag{5.11}$$

for $j = 1, \cdots, n, n \in \mathbf{N}$. Substituting $w_j$ in (5.10) by $w^n + w_{x^4 t}^n$ and exploiting the Lagrange mean value theorem, (5.8), Lemma 5.1 and the Gronwall inequality, we arrive at

$$\frac{1}{2}\frac{d}{dt}(\|w_t^n(t)\|^2 + \|w_{x^2 t}^n(t)\|^2 + \|w^n(t)\|_{H^4}^2)$$

$$= ((\sigma(u^n) - \sigma(u^{n-1}))_{xx}, w_t^n) + ((\sigma(u^n) - \sigma(u^{n-1}))_{x^4}, w_{x^2 t}^n) + (w^n, w_t^n)$$

$$\leqslant M(T)(\|w_t^n(t)\|^2 + \|w_{x^2 t}^n(t)\|^2 + \|w^n(t)\|_{H^4}^2), \quad t \in [0, T], \tag{5.12}$$

$$\|w_t^n(t)\|_{H^2}^2 + \|w^n(t)\|_{H^4}^2 \leqslant (\|\eta^n - \eta^{n-1}\|_{H^2}^2 + \|\xi^n - \xi^{n-1}\|_{H^4}^2)e^{2M(T)T} \to 0 \tag{5.13}$$

uniformly on $[0, T]$ as $n \to \infty$. Substituting $w_j$ in (5.10) by $w_{tt}^n$ and exploiting Lemma 5.1 and (5.13), we have

$$\|w_{tt}^n(t)\| \leqslant \|w_{x^2}^n(t)\| + \|w_{x^4}^n(t)\| + \|(\sigma(u^n(t)) - \sigma(u^{n-1}(t)))_{xx}\|$$

$$\leqslant M(T)\|w^n(t)\|_{H^4} \to 0 \tag{5.14}$$

uniformly on $[0,T]$ as $n\to\infty$. Then,

$$\|u_t^n(t)-u_t^m(t)\|_{H^2}^2+\|u^n(t)-u^m(t)\|_{H^4}^2+\|u_{tt}^n(t)-u_{tt}^m(t)\|^2\to 0 \quad (5.15)$$

uniformly on $[0,T]$ as $m,n\to\infty$, i.e. $\{u^n\}$ is a Cauchy sequence in $C([0,T];H^4)\cap C^1([0,T];H^2)\cap([0,T];L_2)$. Hence

$$u^n\to u \quad \text{in } C([0,T];H^4)\cap C^1([0,T];H^2)\cap C^2([0,T];L_2) \quad (5.16)$$

as $n\to\infty$. Letting $n\to\infty$ in (5.4) and (5.5) we get that $u$ is a strong solution of PBVP (1.3), (5.1), (5.2).

Now, we restrict our attention to Cauchy problem (1.3), (1.2), with $N=1$.

We still take a sequence $\{L_s\}$ as usual and construct the periodic functions $\xi_s(x)=\varphi_s(x)$, $\eta_s(x)=\psi_s(x)+\lambda\varphi_s(x)$, where $\varphi_s\in H^4(\Omega_s)\cap H_0^1(\Omega_s)$, $\psi_s\in H^2(\Omega_s)\cap H_0^1(\Omega_s)$, and $\varphi_s$ and $\psi_s$ satisfy conditions (ⅰ) and (ⅱ) in the proof of Theorem 2.2, with $N=1$ and $\Omega_s=(-L_s,L_s)$. Obviously, $\xi_s|_{\partial\Omega_s}=\eta_s/\partial\Omega_s=0$, $\xi_s(x)=\xi_x(x+2L)$, $\eta_s(x)=\eta_s(x+2L)$, $x\in\mathbf{R}$, and $\xi_s(x)=u_0(x)$, $\eta_s(x)=u_1(x)$ in $\Omega_s^*=(-L_s+1,L_s-1)$. Moreover, let

$$\|\xi_s\|_{H^4(\Omega_s)}\leqslant\|u_0\|_{H^4(\mathbf{R})},\quad \|\eta_s\|_{H^2(\Omega_s)}\leqslant\|u_1\|_{H^2(\mathbf{R})}, \quad (5.17)$$

and $\tilde\xi_s(x)=\tilde\varphi_s(x)$, $\tilde\eta_s(x)=\tilde\psi_s(x)+\lambda\tilde\varphi_s(x)$, where $\tilde\varphi_s(x)$, $\tilde\psi_s(x)$ as shown in (4.3), with $N=1$. As a result, (4.5)~(4.10) hold true. Similarly,

$$\tilde\xi_s\to u_0 \quad \text{in } H^4(\mathbf{R}),\quad \tilde\eta_s\to u_1 \quad \text{in } H^2(\mathbf{R}) \quad \text{as } s\to\infty. \quad (5.18)$$

As shown in the proof of Theorem 2.2, for each $s$ we get a strong solution $u^s\in C([0,T];H^4(\Omega_s))\cap C^1([0,T];H^2(\Omega_s))\cap C^2([0,T];L_2(\Omega_s))$ of corresponding problem (1.3), (5.1), (5.2) (substituting $\xi$, $\eta$ and $L$ there by $\xi_s$, $\eta_s$ and $L_s$, respectively), and $u^s$ is a weak* limit of a sequence $\{u^{n_s}\}$, where $u^{n_s}$ are the solutions of problem (5.4), (5.5) (substituting $u^n$, $\xi^n$, $\eta^n$, $\xi$ and $\eta$ there by $u^{n_s}$, $\xi^{n_s}$, $\eta^{n_s}$, $\xi_s$ and $\eta_s$, respectively). From the sequential lower semi-continuity of the norm we know that inequalities (5.7)~(5.9) still hold true for $u^s$ (substituting $u^n$, $\xi^n$ and $\eta^n$ there by $u^s$, $\xi_s$ and $\eta_s$, respectively). And inequalities (5.13), (5.14) still hold true for $w^s=u^s-u^{s-1}$ (substituting $w^n$, $\xi^n$ and $\eta^n$ there by $w^s$, $\xi_s$ and $\eta_s$, respectively). Then $\{u^s\}$ is a Cauchy sequence in $C([0,T];H^4(\Omega_L))\cap C^1([0,T];H^2(\Omega_L))\cap C^2([0,T];L_2(\Omega_L))$ and

$$u^s\to u \quad \text{in } C([0,T];H^4(\Omega_L))\cap C^1([0,T];H^2(\Omega_L))\cap C^2([0,T];L_2(\Omega_L)) \quad (5.19)$$

as $s\to\infty$. And inequalities (5.7)~(5.9) are still valid for the limit function $u$. From the arbitrariness of $L$ and the bounded-ness of the norm of $u$ (see (5.8)~(5.9)) we know that the limiting function

$$u\in C([0,T];H^4(\mathbf{R}))\cap C^1([0,T];H^2(\mathbf{R}))\cap C^2([0,T];L_2(\mathbf{R})), \quad (5.20)$$

and $u$ is a strong solution of Cauchy problem (1.3), (1.2), with $N=1$.

By using the standard method we easily get the uniqueness of the strong solution, so we omit the process. Theorem 2.3 is proved. □

# 6 Blowup of Solutions

**Lemma 6.1** (See [12]) Assume that $\phi \in C^2, \phi(t) \geqslant 0$ satisfies

$$\phi(t)\phi''(t) - (1+\delta)\phi'^2(t) \geqslant 0$$

for certain real number $\delta > 0, \phi(0) > 0$ and $\phi'(0) > 0$. Then there exists a real number $\widetilde{T}$: $0 < \widetilde{T} \leqslant \dfrac{\phi(0)}{\delta\phi'(0)}$ such that

$$\phi(t) \to \infty \text{ as } t \to \widetilde{T}^-.$$

**Proof of Theorem 2.4** Taking $L_2$-inner product of Eq. (2.2) with $u$ and $u_t$, respectively, we get

$$(G^{1/2}u_{tt}, G^{1/2}u) + \|u(t)\|^2 + \|\nabla u(t)\|^2 + \int_{\mathbf{R}^N} \sigma(u)u\,dx = 0, \quad t > 0, \tag{6.1}$$

$$E'(t) = 0, \quad E(t) = E(0), \quad t > 0, \tag{6.2}$$

where

$$E(t) = \|G^{1/2}u_t\|^2 + \|u(t)\|^2 + \|\nabla u(t)\|^2 + 2\int_{\mathbf{R}^N}\int_0^u \sigma(s)\,ds\,dx. \tag{6.3}$$

(1) If (2.25) holds true, we set

$$\phi(t) = \|G^{1/2}u(t)\|^2 + \beta(t+t_0)^2. \tag{6.4}$$

Obviously,

$$\phi'(t) = 2(G^{1/2}u, G^{1/2}u_t) + 2\beta(t+t_0), \tag{6.5}$$

$$\phi'^2(t) \leqslant 4\phi(t)(\|G^{1/2}u_t(t)\|^2 + \beta), \tag{6.6}$$

$$\begin{aligned}\phi''(t) = {} & 2\Big[2\alpha(\|u(t)\|^2 + \|\nabla u(t)\|^2) + 2(2\alpha+1)\int_{\mathbf{R}^N}\int_0^u \sigma(s)\,ds\,dx - \\ & \int_{\mathbf{R}^N}\sigma(u)u\,dx - (2\alpha+1)(E(0)+\beta)\Big] + \\ & 4(1+\alpha)(\|G^{1/2}u_t(t)\|^2 + \beta), \quad t > 0,\end{aligned} \tag{6.7}$$

where (6.1)~(6.3) have been used. Then

$$\phi(t)\phi''(t) - (1+\alpha)\phi'^2(t) \geqslant \phi(t)Q(t), \quad t \geqslant 0, \tag{6.8}$$

where

$$\begin{aligned}Q(t) = {} & 2\Big[2\alpha(\|u(t)\|^2 + \|\nabla u(t)\|^2) + 2(2\alpha+1)\int_{\mathbf{R}^N}\int_0^u \sigma(s)\,ds\,dx - \\ & \int_{\mathbf{R}^N}\sigma(u)u\,dx - (2\alpha+1)(E(0)+\beta)\Big].\end{aligned} \tag{6.9}$$

By taking $\beta = -E(0)$ ($>0$), we get from (2.24) and (6.9) that $Q(t) \geqslant 0$. And hence

$$\phi(t)\phi''(t) - (1+\alpha)\phi'^2(t) \geqslant 0, \quad t \geqslant 0. \tag{6.10}$$

Obviously $\phi(0) = \|G^{1/2}u_0\|^2 + \beta t_0^2 > 0$. By taking $t_0$ large enough such that $\phi'(0) = 2(G^{1/2}u_0, G^{1/2}u_1) + 2\beta t_0 > 0$, we deduce from Lemma 6.1 that there must be $\widetilde{T} \leqslant \phi(0)/\alpha\phi'(0)$ such that $\phi(t) \to +\infty$ as $t \to \widetilde{T}^-$, i.e.

$$\|G^{1/2}u(t)\|^2 \to +\infty \quad \text{as } t \to \widetilde{T}^-. \tag{6.11}$$

(2) If (2.26) holds true, we set

$$\phi(t) = \|G^{1/2}u(t)\|^2, \quad t \geqslant 0. \tag{6.12}$$

As a result, (6.5)~(6.9) still hold true, with $\beta=0$. And it follows from (6.8) and (6.9) that

$$\phi(t)\phi''(t) - (1+\alpha)\phi'^2(t) \geqslant -2(2\alpha+1)E(0)\phi(t), \quad t \geqslant 0. \tag{6.13}$$

①If $E(0)=0$, it follows from the facts: $\phi(0)=\|G^{1/2}u_0\|^2>0, \phi'(0)=2(G^{1/2}u_0, G^{1/2}u_1)>0$ and Lemma 6.1 that (6.11) holds true.

②If $E(0)>0$, we set

$$P(t) = \phi^{-\alpha}(t), \quad t \geqslant 0. \tag{6.14}$$

Obviously,

$$\begin{aligned} P'(t) &= -\alpha\phi^{-\alpha-1}(t)\phi'(t), \\ P''(t) &= -\alpha\phi^{-\alpha-2}(t)(\phi(t)\phi''(t) - (1+\alpha)\phi'^2(t)) \\ &\leqslant 2\alpha(2\alpha+1)E(0)\phi^{-\alpha-1}(t), \quad t \geqslant 0. \end{aligned} \tag{6.15}$$

By assumption, $P'(0) = -\alpha\phi^{-\alpha-1}(0)\phi'(0)<0$. We claim

$$P'(t) < 0, \quad t > 0. \tag{6.16}$$

In fact, if there exists $t_0>0$ such that $P'(t)<0$, $t\in[0, t_0)$ and $P'(t_0)=0$, multiplying (6.15) by $2P'(t)$ we obtain

$$\frac{\mathrm{d}}{\mathrm{d}t}P'^2(t) \geqslant 4\alpha^2E(0)(\phi^{-2\alpha-1}(t))', \tag{6.17}$$

$$P'^2(t) \geqslant 4\alpha^2E(0)\phi^{-2\alpha-1}(t) + P'^2(0) - 4\alpha^2E(0)\phi^{-2\alpha-1}(0), \tag{6.18}$$

$$P'(t) \leqslant -(P'^2(0) - 4\alpha^2E(0)\phi^{-2\alpha-1}(0))^{1/2} < 0, \quad t\in[0,t_0]. \tag{6.19}$$

In particular, $P'(t_0)<0$, which violates the assumption. So (6.16) is valid.

Therefore, (6.19) holds true for $t>0$. Integrating (6.19) over $(0,t)$, we have

$$P(t) \leqslant P(0) - (P'^2(0) - 4\alpha^2E(0)\phi^{-2\alpha-1}(0))^{1/2}t, \quad t>0. \tag{6.20}$$

It follows from (6.20) that there exists a number $\widetilde{T}$: $0<\widetilde{T}\leqslant P(0)/(P'^2(0) - 4\alpha^2E(0)\phi^{-2\alpha-1}(0))^{1/2}$ such that $P(\widetilde{T})=0$, i. e. $\phi(t)\to+\infty$ as $t\to\widetilde{T}^-$. Theorem 2.4 is proved. □

**Example 1** We set $\sigma(s) = a|s|^{p-1}s$ in Eq. (1.1), where $a\neq0$, $p>1$ are real numbers.

(1) When $a>0$, $1<p\leqslant\frac{N+2}{(N-2)^+}$ ($p<+\infty$), obviously, (2.17) holds true, with $\beta=p+1$. So Cauchy problem (1.1), (1.2) admits a weak solution on $[0,T]$ (for any $T>0$) as long as $G^{1/2}u_1\in L_2(\mathbf{R}^N)$, $u_0\in H^1(\mathbf{R}^n)$ (see Theorem 2.2).

(2) When $a<0$, $|a|<\frac{p+1}{2}$, $1<p\leqslant\frac{N+2}{(N-2)^+}$ ($p<+\infty$), Cauchy problem (1.1), (1.2) admits a weak solution on $[0,T]$ (for any $T>0$) as long as $u_0\in\widetilde{W}_0$ (see (2.21)), $G^{1/2}u_1\in L_2(\mathbf{R}^N)$ such that (2.22) holds true (see Theorem 2.2).

(3) When the space dimension $N=1$ and $p>4$, obviously $\sigma\in C^4(\mathbf{R})$. Then the weak solution of Cauchy problem (1.3), (1.2) in both cases 1 and 2 can be regularized, the strong solution $u$ is unique and

$$u\in C([0,T];H^4(\mathbf{R}))\cap C^1([0,T];H^2(\mathbf{R}))\cap C([0,T];L_2(\mathbf{R})) \quad (6.21)$$

as long as $u_0\in H^4(\mathbf{R}),u_1\in H^2(\mathbf{R})$ (see Theorem 2.3).

(4) A simple verification shows that (2.24) is valid, with $\alpha=\frac{p-1}{4}$ ($>0$). As a result, any weak solution of Cauchy problem (1.1), (1.2) blows up in finite time, i.e. (2.27) holds true as long as the initial data $u_0$ and $u_1$ satisfy condition ($H_5$) of Theorem 2.4 (see Theorem 2.4).

**Example 2** We set $\sigma(s)=as^p$ in Eq. (1.1), where $a\neq0$, $p>1$ are real numbers.

(1) When $a>0$, $p$ is an odd integer and $1<p\leqslant\frac{N+2}{(N-2)^+}$ ($p<+\infty$), a simple computation shows that (2.17) holds true, with $\beta=p+1$. So the same conclusion as in the case 1 of Example 1 is valid.

(2) When $|a|<\frac{p+1}{2}$ and $1<p\leqslant\frac{N+2}{(N-2)^+}$ ($p<+\infty$), we have the same conclusion as in the case 2 of Example 1.

(3) When the space dimension $N=1,p>4$ or $p$ is an integer, the same conclusion as in the case 3 of Example 1 holds true.

(4) Obviously (2.24) holds, with $\alpha=\frac{p-1}{4}$ ($>0$). And hence, we have the same conclusion as in the case 4 of Example 1.

### References

[1] Akmel D G. Global existence and decay for solutions to the "bad" Boussinesq equation in two space dimensions. Appl Anal 2004, 83: 17-36.

[2] Bona J L, Smith R A. A model for the two-way propagation of water waves in a channel. Math Proc Cambridge Philos Soc, 1976, 79:167-182.

[3] Bona J L, Sachs R L. Global existence of smooth solutions and stability of solitary waves for a generalized Boussinesq equation. Comm Math Phys, 1988, 118: 15-29.

[4] Boussinesq J. Theorie des ondes et des remous qui se propagent le long d'un canal rectangulaire horizontal, en comuniquant au liquide contene dans ce canal des vitesses sunsiblement parielles de la surface au fond. J Math Pures Appl Ser, 1872, 2 (17): 55-108.

[5] Chen G W, Yang Z J. Existence and non-existence of global solutions for a class of non-linear wave equation. Math Methods Appl Sci, 2000, 23: 615-631.

[6] Clarkson P A. New exact solutions of the Boussinesq equations. European J Appl Math, 1990, 1: 279-300.

[7] Clarkson P A, Kruskal M D. New similarity reductions of the Boussinesq equation. J Math Phys, 1989, 30: 2201-2213.

[8] Clements J. Existence theorems for a quasi-linear evolution equation. SIAM J Appl Math, 1974, 26: 745 – 752.

[9] Defrutos J, Ortega T, Sanz-Serna J M. Pseudo-spectral method for the "good" Boussinesq equation. Math Comp, 1991, 57: 109 – 122.

[10] Deift P, Tomei C, Trubowitz E. Inverse scattering and the Boussinesq equation. Comm Pure Appl Math, 1982, 35: 567 – 628.

[11] Kalantarov V K, Ladyzhenskaya O A. The occurrence of collapse for quasi-linear equations of parabolic and hyperbolic type. J Sov Math, 1978, 10: 53 – 70.

[12] Levine H A. Instability and nonexistence of global solutions to nonlinear wave equations of the form $Pu_{tt} = -Au + F(u)$. Trans Amer Math Soc, 1974, 192: 1 – 21.

[13] Levine H A, Sleeman B D. A note on the non-existence of global solutions of initial boundary value problems for the Boussinesq equation $u_{tt} = 3u_{xxxx} + u_{xx} - 12(u^2)_{xx}$. J Math Anal Appl, 1985, 107: 206 – 210.

[14] Linares F. Global existence of small solutions for a generalized Boussinesq equation. J Differential Equations, 1993, 106: 257 – 293.

[15] Linares F. Asymptotic behavior of solutions of a generalized Boussinesq type equation. Nonlinear Anal, 1995, 25: 1147 – 1158.

[16] Liu Y. Instability of solitary waves for generalized Boussinesq equation. J Dynam Differential Equations, 1993, 5: 537 – 558.

[17] Makhankov V G. Dynamics of classical solutions. Phys Rev Lett, 1978, 35 (c): 1 – 128.

[18] Pani A K, Saranga H. Finite element Galerkin method for the "good" Boussinesq equation. Nonlinear Anal, 1997, 29: 937 – 956.

[19] Sachs R L. On the blow-up of certain solutions of the "good" Boussinesq equation. Appl Anal, 1990, 36: 145 – 152.

[20] Tsutsumi M, Matahashi T. On the Cauchy problem for the Boussinesq-type equation. Math Japan, 1991, 36: 371 – 379.

[21] Varlamov V. Existence and uniqueness of a solution to the Cauchy problem for the damped Boussinesq equation. Math Methods Appl Sci, 1996, 19: 639 – 649.

[22] Varlamov V. On spatially periodic solutions of the damped Boussinesq equation. Differential Integral Equations, 1997, 10: 1197 – 1211.

[23] Varlamov V. Long-time asymptotics of solutions of the second initial-boundary value problem for the damped Boussinesq equation. Abstr Appl Anal, 1998, 2: 281 – 289.

[24] Varlamov V. Eigenfunction expansion method and the long-time asymptotics for the damped Boussinesq equation. Discrete Contin Dyn Syst, 2001, 7: 675 – 702.

[25] Varlamov V, Balogh A. Forced nonlinear oscillations of elastic membranes. Nonlinear Anal Real World Appl, 2006, 7: 1005 – 1028.

[26] Yang Z J, Chen G W. Blowup of solutions for a class of generalized Boussinesq equations. Acta Math Sci, 1996, 16: 31 – 40.

[27] Yang Z J. On local existence of initial boundary value problems for the "bad" Boussinesq-type equation. Nonlinear Anal, 2002, 51: 1259 – 1271.

[28] Yang Z J, Wang X. Blowup of solutions for the "bad" Boussinesq-type equation. J Math Anal Appl, 2003, 285: 282 – 293.

[29] Zhou Y L, Fu H Y. Nonlinear hyperbolic systems of higher order generalized Sine-Gordon type. Acta Math Sinica, 1983, 26: 234 – 249.

# Well-posedness of the Ferrimagnetic Equations *

Guo Boling(郭柏灵) Pu Xueke(蒲学科)

**Abstract**

In this paper, we show the existence of global weak solutions of the ferrimagnetic equations on compact Riemannian manifold using the penalty method. We also show the uniqueness of the solution and its well-posedness by the energy estimates method in lower dimensions. In particular, when the space dimension is one, we can prove that the problem is globally well-posed.

**Keywords** ferrimagnetic equation; penalty method; well-posedness; weak solutions; contraction mapping principle

## 1 Introduction

In this paper, we consider the following ferrimagnetic equation

$$\begin{cases} u_t = u \times \Box u, & \text{in } M \times \mathbf{R}_+ \\ u(x,0) = u_0, & u_t(x,0) = u_1 \end{cases}, \tag{1.1}$$

where $u = (u^1, u^2, u^3): M \to S^2$ is the unknown, $|u_0| = 1$, $M$ denotes the $d$-dimensional compact Riemannian manifold without boundary, "$\times$" denotes the vector cross product in $\mathbf{R}^3$ and $\Box = \partial_{tt} - \Delta$ denotes the wave operator. This equation describes the ferrimagnetic phenomena in ferrimagnetic materials which was first derived by A. B. Borisov, V. V. Kiseliev and G. G. Talutz in [1]. It plays a fundamental role in studying ferrimagnetic materials or ferrits as the Landau-Lifshitz equation does in studying the evolution of spin fields in continuum ferromagnetism. The Landau-Lifshitz equation has aroused the interest of many authors and has been broadly studied, one can refer to [4,5,7,8] and references therein. Although aroused the interests of both mathematicians and physicists greatly, the only available rigorous mathematical theory of Eq. (1.1) is the existence of weak solutions of the equation in case of Neumann boundary condition by [6] recently. They obtained their result by using the penalized method which was broadly used by many authors to get the existence of weak solutions of many other PDEs, in particular, the Landau-Lifshitz equation and harmonic maps, see for example F. Alouges and A. Soyeur[2], and Y. Chen[3]. Let us recall that in [6], they first prove that there exists a weak solution of Eq. (1.1), and in the case of dimension one they

* J Math Anal Appl, 2008, 339: 312-323.

use difference method to obtain the unique smooth solution for this equation.

In this paper, we first show the existence of global weak solutions of Eq. (1.1) by using the penalty method, then we prove the existence and uniqueness of solutions in some Sobolev spaces by contraction mapping principle. Finally we show that we can improve the well-posedness of the problem in some lower regularity Sobolev space by the *a priori* energy estimates method. To put forward our problem precisely, we give some notations and definitions in the sequel.

Throughout this paper, $M$ always denotes the $d$-dimensional compact Riemannian manifold without boundary, $M_T$ denotes $M\times[0,T]$, i.e. the product of $M$ with $[0,T]$, $T_uS^2$ denotes the tangent space of the 2-sphere $S^2$ at $u$, $Du=(u_t,\nabla u)$ denotes the space-time derivative of $u$ and $C$ denotes the universal constant which can vary from line to line. We also use $\|\cdot\|$ to denote the $L^2$ norm and $\|\cdot\|_X$ to denote the $X$-norm.

**Definition 1.1** we call $u(x,t)\in L^2(0,T;\ H^1(M))$, $u_t(x,t)\in L^2(0,T;L^2(M))$ (i.e. $u\in H^1(M_T)$) a weak solution of Eq. (1.1) with the prescribed initial conditions, provided that the following identity holds:

$$\int_{M_T} u_t\cdot\Phi \mathrm{d}x\mathrm{d}t+\int_{M_T}(u\times u_t)\cdot\Phi_t \mathrm{d}x\mathrm{d}t+$$
$$\int_M (u_0\times u_1)\cdot\Phi(x,0)\mathrm{d}x-\sum_{i=1}^{m}\int_{M_T} u\times\frac{\partial u}{\partial x_i}\cdot\frac{\partial\Phi}{\partial x_i}=0$$

for any $\Phi\in H^1(M_T)$, with $\Phi(x,T)=0$.

Then our main results read as follows.

**Theorem 1.2** Let $M$ be the $d$-dimensional compact Riemannian manifold without boundary. Suppose also that $u_0\in H^1(M,S^2)$ and $u_1\in L^2(M,T_uS^2)$. Then there exists a global weak solution of the problem (1.1) with the initial value $(u_0,u_1)$.

**Theorem 1.3** Suppose $d\leqslant 3$, then for any data $(u_0,u_1)\in H^2\times H^1$, there exists a unique local solution $u$ of class $H^2$. Further more, if $d=1$, the solution extends uniquely for all time. If $(u_0,u_1)\in H^s$ with $s>2$, then so does $u$.

We organize this paper as follows. In Section 2, we show the existence of global weak solutions of the equation. In this section, we first derive an equivalent form of Eq. (1.1) then we use the penalty method to show the existence of global weak solutions. We first show the existence of weak solutions of the penalized problem, and by passing to the limit we show the existence of global weak solutions of the original problem. Section 3 is devoted to the local well-posedness of the problem in some Sobolev space $H^k(M)\left(k>\frac{d}{2}+2\right)$ by the contraction mapping principle. In Section 4 we derive some energy estimates of the solution which are necessary for our proof of Theorem 1.3. With these *a priori* estimates, we can show the well-posedness of the problem in $H^2(M)$ in the last section. Further more in the case of dimension one ($d=1$), we find that the solution is global.

## 2 Existence of Global Weak Solutions

This section is devoted to the existence of the global weak solutions of the equation. For this purpose, we exploit the equivalent form

$$\Box u + u \times u_t + (|u_t|^2 - |\nabla u|^2) u = 0 \tag{2.1}$$

of Eq. (1.1). Equation (2.1) is very similar to wave maps, for which there is a long list of results and one can refer to [9 – 11] and references therein. We first prove the following:

**Lemma 2.1** In the classical sense, (1.1) and (2.1) are equivalent in the time interval $[0, T]$ on which the solution exists.

**Proof** In deed, suppose $u$ solves (1.1). From the identity $a \times (b \times c) = (a \cdot c) b - (a \cdot b) c$, where $a$, $b$, $c$ are vectors in $\mathbf{R}^3$, we have

$$u \times u_t = u \times (u \times \Box u) = -\Box u + (u \cdot \Box u) u.$$

Taking inner product of (1.1) with $u$, we have

$$u \cdot u_t = 0 \quad \forall (x, t) \in M \times [0, T),$$

which together with $|u_0| = 1$ implies

$$|u(x,t)| = 1 \quad \text{and} \quad u \cdot u_{tt} = -|u_t|^2, \quad \forall (x,t) \in M \times [0, T).$$

From these facts, we obtain

$$u \cdot \nabla u = 0 \quad \text{and} \quad (u \cdot \Box u) = -|u_t|^2 + |\nabla u|^2.$$

Thus

$$\Box u = -(|u_t|^2 - |\nabla u|^2) u - u \times u_t.$$

To prove the contrary, we need only to prove $|u| = 1$, if $u$ solves (2.1). Taking the inner product with $u_t$ of (2.1), and integrating over $M$, we can get

$$\frac{1}{2} \frac{\mathrm{d}}{\mathrm{d}t} \int_M (|u_t|^2 + |\nabla u|^2) \leqslant \sup_{x,t} |u| |u_t| \int_M (|u_t|^2 + |\nabla u|^2).$$

Setting $Z = |u|^2$, we then have

$$\partial_t Z = 2u \cdot u_t, \quad \nabla Z = 2u \cdot \nabla u, \quad \Delta Z = 2u \cdot \Delta u + 2|\nabla u|^2,$$

and

$$\begin{aligned} \partial_{tt} Z &= 2|u_t|^2 + 2u \cdot u_{tt} \\ &= 2|u_t|^2 + 2u \cdot (\Delta u - (|u_t|^2 - |\nabla u|^2) u - u \cdot u_t) \\ &= 2|u_t|^2 + \Delta Z - 2|\nabla u|^2 - 2(|u_t|^2 - |\nabla u|^2)|u|^2. \end{aligned}$$

Thus we have

$$\partial_{tt} Z - \Delta Z = -2(|u_t|^2 - |\nabla u|^2)(|u|^2 - 1), \tag{2.2}$$

with the initial condition

$$Z(x, 0) = 1, \quad Z_t(x, 0) = 2u \cdot u_t|_{t=0} = 0, \tag{2.3}$$

with the latter one follows from the compatible conditions.

Then we set $W = Z - 1$, we can deduce that

$$\partial_{tt}W - \Delta W = -2(|u_t|^2 - |\nabla u|^2)W, \tag{2.4}$$

$$W(x,0) = 0, \quad W_t(x,0) = 0. \tag{2.5}$$

Taking inner product of (2.4) with $W_t$, and integrating over $M$, one can get

$$\frac{1}{2}\frac{\mathrm{d}}{\mathrm{d}t}\int_M |W_t|^2 + |\nabla W|^2 \leqslant 2\max_{x,t}(|u_t|^2 + |\nabla u|^2)\int_M |W||W_t|$$

$$\leqslant 2\max_{x,t}(|u_t|^2 + |\nabla u|^2)(|u|^2 + 1)|M|^{\frac{1}{2}}\left(\int_M |W_t|^2\right)^{\frac{1}{2}}.$$

Then it follows from Gronwall's inequality and (2.5) that $W(x,t)=0$ for any $(x,t)\in M\times[0,T]$, i.e. $|u|^2=1$, which concludes our proof. □

Now we construct the penalized problem of Eq. (1.1)

$$\begin{cases} u_{tt}^k - \Delta u^k + u^k \times u_t^k + k(|u^k|^2 - 1)u^k = 0 \\ u^k(x,0) = u_0, \quad u_t^k(x,0) = u_1 \end{cases}. \tag{2.6}$$

In the following, we prove that for every $k\geqslant 1$, the penalized problem (2.6) has a global weak solution $u^k$. Then we show that the limit $u$ of $u^k$ is a weak solution of (1.1) in the distribution sense. We divide this into the following two lemmas.

**Lemma 2.2** The penalized problem (2.6) has a global weak solution $u^k$ in the distribution sense.

**Proof** Fix an integer $k$, we apply Galerkin's method to solve the above penalized problem. More precisely, let $w_i = w_i(x)$ be a smooth orthonormal basis of $L^2(M)$ consisting of eigenvectors for the operator $L = -\Delta$ in $H^1$.

Fix now a positive integer $m$. We will look for a function $u_m:[0,T]\to H^1(M)$ of the form

$$u_m = \sum_{i=1}^{m}\varphi_i(t)w_i,$$

where $\varphi_i(t)$ are $\mathbf{R}^3$-valued vectors, and we hope to select the coefficients $\varphi_i(t)(0\leqslant t\leqslant T$, $i=1,\cdots,m)$ so that for all $1\leqslant i\leqslant m$

$$\left\langle \frac{\partial^2 u_m}{\partial t^2} - \Delta u_m + u_m \times \frac{\partial u_m}{\partial_t} + k(|u_m|^2 - 1)u_m, w_i \right\rangle = 0, \tag{2.7}$$

where $\langle\cdot,\cdot\rangle$ denotes the inner product in $L^2(M)$ with the initial value condition

$$u_{m0} := u_m(x,0) = \sum_{i=1}^{m}\varphi_i(0)w_i(x) \to u_0 \quad \text{in } H^1(M),$$

$$\frac{\partial u_{m0}}{\partial t} := \frac{\partial u_m}{\partial t}(x,0) = \sum_{i=0}^{m}\varphi_{it}(0)w_i(x) \to u_1 \quad \text{in } L^2(M).$$

Inserting the expression of $u_m$ into (2.7), we get an ODE system for $\varphi_i(t)$, and from standard existence theory of ordinary differential equations, we have the local existence of approximate solutions $\{u_m\}$ for (2.7).

Next, we establish estimates of $\{u_m\}$ on $[0, T]$ to get a global solution of the penalized equation (2.6). Multiplying (2.7) with $\dfrac{\partial\varphi_i}{\partial t}$ and summing for $1\leqslant i\leqslant m$, we have

$$\frac{1}{2}\frac{\mathrm{d}}{\mathrm{d}t}\int_M\left|\frac{\partial u_m}{\partial t}\right|^2\mathrm{d}x+\frac{1}{2}\frac{\mathrm{d}}{\mathrm{d}t}\int_M|\nabla u_m|^2\mathrm{d}x+\frac{1}{4}\frac{\mathrm{d}}{\mathrm{d}t}\int_M k(|u_m|^2-1)^2\mathrm{d}x=0. \quad (2.8)$$

Integrating (2.8) over $[0,t)$, where $0<t\leqslant T$, we have

$$\int_M\left|\frac{\partial u_m}{\partial t}\right|^2\mathrm{d}x+\int_M|\nabla u_m|^2\mathrm{d}x+\frac{1}{2}\int_M k(|u_m|^2-1)^2\mathrm{d}x$$
$$\leqslant C+\frac{1}{2}\int_M k(|u_{m0}|^2-1)^2\mathrm{d}x, \quad (2.9)$$

where $C=\int_M\left|\frac{\partial u_{m0}}{\partial t}\right|^2+\int_M|\nabla u_{m0}|^2$ is independent of $k$. From this inequality, we know that $\{|u_m|^2-1\}$, $\left\{\frac{\partial u_m}{\partial t}\right\}$, and $\{\nabla u_m\}$ are bounded in $L^\infty(0,T;L^2(M))$. Further more, we get by Young's inequality that

$$\int_M|u_m|^2\leqslant\frac{3}{2}|M|+\frac{1}{2}\int_M(|u_m|^2-1)^2,$$

where $|M|$ is the measure of $M$, thus $u_m$ are bounded in $L^\infty(0,T;\ H^1(M))$.

Thus we can select from $\{u_m\}$ a subsequence (still denoted by $\{u_m\}$) such that

$$\begin{cases} u_m\to u^k & \text{weakly star in } L^\infty(0,T;H^1(M)) \text{ and weakly in } H^1(M_T)\\ u_m\to u^k & \text{strongly in } L^2(M_T) \text{ and a.e.}\\ |u_m|^2-1\to|u^k|^2-1 & \text{weakly in } L^2(M_T)\end{cases}. \quad (2.10)$$

Taking limit $m\to\infty$ in (2.7), we deduce that for all $\psi$ in $\mathrm{span}\{w_i\}_{i=1}^m$ and $\varphi\in C^\infty[0,T]$ with $\varphi(T)=0$, there holds

$$-\int_{M_T}u_t^k\cdot(\varphi'(t)\psi(x))\mathrm{d}x\mathrm{d}t-\int_M\varphi(0)u_1\cdot\psi\mathrm{d}x+\sum_{i=1}^m\int_{M_T}\varphi(t)\frac{\partial u^k}{\partial x_i}\cdot\frac{\partial\psi}{\partial x_i}\mathrm{d}x\mathrm{d}t+$$
$$\int_{M_T}u^k\times u_t^k\cdot(\varphi(t)\psi(x))\mathrm{d}x\mathrm{d}t+\int_{M_T}k(|u^k|^2-1)u^k\cdot(\varphi(t)\psi(x))\mathrm{d}x\mathrm{d}t=0.$$

By a density argument, we can show from the above equation that

$$-\int_{M_T}u_t^k\cdot\phi_t\mathrm{d}x\mathrm{d}t-\int_M u_1\cdot\phi(x,0)\mathrm{d}x+\sum_{i=1}^m\int_{M_T}\frac{\partial u^k}{\partial x_i}\cdot\frac{\partial\phi}{\partial x_i}\mathrm{d}x\mathrm{d}t+$$
$$\int_M u^k\times u_t^k\cdot\phi(x,t)\mathrm{d}x\mathrm{d}t+\int_{M_T}k(|u^k|^2-1)u^k\cdot\phi(x,t)\mathrm{d}x\mathrm{d}t=0 \quad (2.11)$$

for all $\phi(x,t)\in H^1(M_T)$ with $\phi(x,T)=0$. □

In the above lemma, we obtained the existence of global weak solutions for the penalized method in the distribution sense. Next, we make uniform estimates to let $k\to\infty$ to get a weak solution of the problem (1.1).

**Lemma 2.3** The $u^k$ we obtained of the penalized problem (2.6) can be taken to limit to get a weak solution of (1.1).

**Proof** Passing to the limit $m\to\infty$ in (2.9), we know that

$$\int_M\left|\frac{\partial u^k}{\partial t}\right|^2\mathrm{d}x+\int_M|\nabla u^k|^2\mathrm{d}x+\frac{1}{2}\int_M k(|u^k|^2-1)^2\mathrm{d}x\leqslant C, \quad (2.12)$$

thus

$$\left\|\frac{\partial u^k}{\partial t}\right\|_{L^2(M)}, \quad \|\nabla u^k\|_{L^2(M)} \leqslant C, \tag{2.13}$$

moreover

$$\int_M (|u^k|^2 - 1)^2 dx \leqslant \frac{C}{k}. \tag{2.14}$$

Thus we can extract from $u^k$ a subsequence (still denoted by $\{u^k\}$), such that

$$\begin{cases} u^k \to u & \text{weakly star in } L^\infty(0,T;H^1(M)) \text{ and weakly in } H^1(M_T) \\ |u^k|^2 - 1 \to 0 & \text{strongly in } L^2(M_T) \text{ and a.e.} \end{cases} . \tag{2.15}$$

which also shows that for all $t \in [0,T)$, $|u| = 1$ a.e. in $M$.

Now we will show that the limit $u$ is a weak solution of our problem in the distribution sense. Let $\Phi \in C^\infty(\overline{M_T})$ with $\Phi(T) = 0$. Set $\phi = u^k \times \Phi$ in (2.11). As $\phi \in (H^1(M_T))^3$ with $\phi(x, T) = 0$, there holds

$$-\int_{M_T} u_t^k \cdot (u^k \times \Phi_t) dx dt - \int_M u_1 \cdot (u_0 \times \Phi(0)) dx + \sum_{i=1}^m \int_{M_T} \frac{\partial u^k}{\partial x_i} \cdot \left(u^k \times \frac{\partial \Phi}{\partial x_i}\right) dx dt +$$

$$\int_{M_T} |u^k|^2 \Phi \cdot u_t^k dx dt - \int_{M_T} (u^k \cdot \Phi) u^k \cdot u_t^k dx dt = 0. \tag{2.16}$$

On the other hand, we can show that

$$\int_{M_T} (u^k \cdot \Phi) u^k \cdot u_t^k dx dt \to 0.$$

In fact, we can write

$$\int_{M_T} (u^k \cdot \Phi) u^k \cdot u_t^k dx dt = \int_{M_T} (u^k \cdot \Phi) \frac{1}{2} \frac{\partial}{\partial t} (|u^k|^2 - 1) dx dt.$$

Since we can rewrite the integrand on the right-hand side as

$$\frac{1}{2} \frac{\partial}{\partial t} [(u^k \cdot \Phi)(|u^k|^2 - 1)] - \frac{1}{2} (|u^k|^2 - 1) \frac{\partial}{\partial t} (u^k \cdot \Phi),$$

we have

$$\int_{M_T} (u^k \cdot \Phi) u^k \cdot u_t^k dx dt = -\frac{1}{2} \int_{M_T} (|u^k|^2 - 1)(u_t^k \Phi + u^k \Phi_t),$$

where we used the fact that $\Phi(x, T) = 0$ and $|u^k(x, 0)|^2 - 1 = 0$. Thus passing to the limit in (2.16) ($k \to \infty$), we are lead to

$$\int_{M_T} u_t \cdot \Phi dx dt + \int_M (u_0 \times u_1) \cdot \Phi(x, 0) dx + \int_{M_T} (u \times u_t) \cdot \Phi_t dx -$$

$$\sum_{i=1}^m \int_{M_T} \left(u \times \frac{\partial u}{\partial x_i} \cdot \frac{\partial \Phi}{\partial x_i}\right) dx dt = 0. \tag{2.17}$$

By density argument, the above equality also holds for all $\Phi \in H^1(M_T)$ with $\Phi(x, T) = 0$.

□

Now, we in fact finished our proof of our Theorem 1.2, i.e. the problem (1.1) has a global weak solution in the distribution sense.

# 3 Local Well-posedness

In this section, we prove the local well-posedness by the contraction mapping principle. Before we do this, we need some energy estimates of the equation.

**Lemma 3.1** Let

$$E_k(t) = \|u(t)\|_{H^k} + \|\partial_t u(t)\|_{H^{k-1}}.$$

Then for $k > \frac{d}{2} + 2$, solutions of Eq. (2.1) satisfy the *a priori* estimates

$$E_k(t) \leqslant C(E_k(0)) + \int_0^t G(E_k(\tau))\mathrm{d}\tau \tag{3.1}$$

where $G$ is a function depending on $k$.

**Proof** Since $s > \frac{d}{2}$, $H^k$ is an algebra. Thus for any $l$ and any smooth function $g$, we can estimates $\|g(u, Du, \cdots, D^l u)\|_{H^s}$ in terms of $\|u\|_{H^{l+s}}$. Furthermore, we mention that in the expression of $E_k(t)$, $\|u(t)\|_{L^2}$ and $\|\partial_t u(t)\|_{L^2}$ are both bounded, the latter one can be seen from Lemma 4.1.

Differentiating the equation $k-1$ times, then multiplying by $D^{k-1}\partial_t u$ and integrating on $M$ leading to the following:

$$\begin{aligned}&\frac{\mathrm{d}}{\mathrm{d}t}\int_M |D^{k-1}\nabla u|^2 + |D^{k-1}u_t|^2 \mathrm{d}x \\ \leqslant{}& C\|g(u, Du, D^2 u)\|_{H^{k-2}}\|\partial_t u\|_{H^{k-1}} \leqslant G(E_k(t))\end{aligned} \tag{3.2}$$

and then integrating over $[0, t]$ leads to our energy estimates. □

**Theorem 3.2** (Local well-posedness) Given $(u_0, u_1) \in H^k \times H^{k-1}$ for some $k > \frac{d}{2} + 2$, there exists a $T > 0$ such that the Cauchy problem (2.1) is locally well-posed in $(u, u_t) \in C([0, T]; H^k) \times C([0, T]; H^{k-1})$.

**Proof** We prove this lemma by the contraction mapping principle. Let

$$X = \{u \in L^2([0, T] \times M) : Du \in L^\infty([0, T] \times H^{k-1}) \cap C([0, T]; L^2)\} \tag{3.3}$$

with $\|u\|_X := \sup\limits_{0 \leqslant t \leqslant T} E_k(t)$. For $\delta > 0$, we define

$$X_\delta = \{u \in X, u(0) = u_0, u_t(0) = u_1, \|u\|_X \leqslant \delta\}$$

which is a complete metric space under the metric

$$\rho(u, v) = \sup_{0 \leqslant t \leqslant T}(\|u(t) - v(t)\|_{H^k} + \|\partial_t u(t) - \partial_t v(t)\|_{H^{k-1}})$$

by lower semicontinuity of weak limits.

Define the map $L : v \in X_\delta \to u$, where $u$ solves the linear equation

$$\partial_t^2 u - \Delta u = -v \times v_t - (|v_t|^2 - |\nabla v|^2)v. \tag{3.4}$$

Now applying energy estimates to this equation we established in the above lemma, we obtain that

$$\| u \|_X \leqslant C_0(E_k(0)) + TC(\| v \|_X)(\| u \|_X + 1).$$

Choosing $\delta = 2C_0(E_k(0))$ and $T$ small enough, we arrive at

$$\| u \|_X \leqslant 2C_0(E_k(0)) = \delta$$

which implies $L$ maps $X_\delta$ into itself.

For $v_1, v_2 \in X_\delta$, let $u_1, u_2$ be the corresponding solutions of (3.4). Then

$$\begin{aligned} &\partial_t^2(u_1 - u_2) - \Delta(u_1 - u_2) \\ &= - v_1 \times \partial_t v_1 - (|\partial_t v_1|^2 - |\nabla v_1|^2) v_1 + v_2 \times \partial_t v_2 + (|\partial_t v_2|^2 - |\nabla v_2|^2) v_2. \end{aligned}$$

From the energy estimates, we have

$$\rho(u_1, u_2) \leqslant TC(\delta)\rho(v_1, v_2),$$

which implies

$$\rho(u_1, u_2) \leqslant \frac{1}{2}\rho(v_1, v_2),$$

for sufficiently small $T$.

Therefore by the contraction mapping principle, $L$ has a unique fixed point and we complete our proof. □

## 4 *A Priori* Estimates

In this section, we focus ourselves on *a priori* estimates of the solution as we promised before. We write our results in the following lemmas.

**Lemma 4.1** Assume $u_0 \in H^1$, $u_1 \in L^2$, if $u$ is a solution for Eq. (1.1), then the following identity holds for all time

$$E(u(t)) = \frac{1}{2}\int_M |u_t|^2 + |\nabla u|^2 dx = \text{Const}. \tag{4.1}$$

That is to say, $u \in L^\infty(0, T; H^1(M))$ and $u_t \in L^\infty(0, T; L^2(M))$, for any $T > 0$.

**Proof** Taking inner product of (1.1) with $u$, and then integrating in the variable $x$ in $M$, we get

$$\int_M \partial_t u \cdot u dx = 0,$$

which follows directly $\| u(t) \|^2 = \| u_0 \|^2$. Then taking inner product of (1.1) with $\Box u$, and then integrating in $M$, we get

$$\int_M \partial_t u \cdot \Box u dx = 0$$

from which we get

$$\frac{d}{dt}\int_M |u_t(t)|^2 + |\nabla u(t)|^2 dx = 0.$$

This directly implies our result

$$E(u(t)) = \frac{1}{2}\int_M |u_t|^2 + |\nabla u|^2 dx = \frac{1}{2}\int_M |u_1|^2 + |\nabla u_0|^2 dx = \text{Const}. \qquad \Box$$

In the next lemma, we show the $H^2$-estimates of the solution.

**Lemma 4.2** Assume $u_0\in H^2, u_1\in H^1$, and suppose that $u$ is a solution for Eq. (1.1), then in fact the following inequality holds locally:

$$\|D^2 u\|_{L^2(M)}^2 \leqslant C,$$

where $C$ depends on the initial data $u_0$ and $u_1$ of the equation.

**Proof** We use the equivalent form (2.1) to deduce our result. Denote $\partial$ any first order spatial derivative operator. Applying $\partial$ to Eq. (2.1), we then get

$$\begin{aligned}\Box(\partial u) &= -\partial\{(|u_t|^2-|\nabla u|^2)u\} - \partial\{u\times u_t\}\\ &= -(|u_t|^2-|\nabla u|^2)\partial u - \partial(|u_t|^2-|\nabla u|^2)u - \partial u\times u_t - u\times\partial u_t. \quad (4.2)\end{aligned}$$

If we take the inner product of this resulting equation with $\partial_t\partial u$, we get

$$\mathrm{LHS} = \int_M \langle\Box(\partial u),\partial_t\partial u\rangle \mathrm{d}x = \frac{1}{2}\frac{\mathrm{d}}{\mathrm{d}t}\int_M |\partial_t\partial u|^2 + |\partial\nabla u|^2 \mathrm{d}x, \quad (4.3)$$

$$\begin{aligned}\mathrm{RHS} &= -\int_M\langle(|\partial_t u|^2-|\nabla u|^2)\partial u,\partial_t\partial u\rangle - \int_M\langle\partial(|\partial_t u|^2-|\nabla u|^2)u,\partial_t\partial u\rangle\\ &\quad -\int_M\langle\partial u\times\partial_t u,\partial_t\partial u\rangle - \int_M\langle u\times\partial_t\partial u,\partial_t\partial u\rangle\\ &= I + II + III + IV. \quad (4.4)\end{aligned}$$

Where LHS and RHS denote "Left-Hand Side" and "Right-Hand Side" respectively. We immediately have

$$|I| \leqslant \int_M |Du|^3|D^2u| \leqslant \|Du\|_{L^6}^3\|D^2u\|_{L^2},$$

then using the embedding theorem,

$$\|Du\|_{L^6} \leqslant \|D^2u\|^{\frac{d}{3}}\|Du\|_{L^2}^{1-\frac{d}{3}},$$

we get

$$|I| \leqslant \|D^2u\|_{L^2}^{1+d}\|Du\|_{L^2}^{3-d}.$$

For part $II$, using integration by parts and noticing that there is no boundary of $M$, we have

$$\begin{aligned}II &= -\int_M\langle\partial(|u_t|^2-|\nabla u|^2)u,\partial_t\partial u\rangle\\ &= \int_M\langle\partial^2(|u_t|^2-|\nabla u|^2)u,\partial_t u\rangle + \int_M\langle\partial(|u_t|^2-|\nabla u|^2)\partial u,\partial_t u\rangle\\ &= \int_M\langle\partial(|u_t|^2-|\nabla u|^2)\partial u,\partial_t u\rangle \quad (4.5)\end{aligned}$$

where the last equality is due to the fact that $u\cdot\partial_t u=0$ from (1.1).

Then from (4.5), we get

$$|II| \leqslant \int_M |Du|^3|D^2u| \leqslant \|D^2u\|_{L^2}^{1+d}\|Du\|_{L^2}^{3-d}.$$

For part $III$, we have

$$|III| \leqslant \int_M |Du|^2|D^2u| \leqslant \|Du\|_{L^4}^2\|D^2u\|_{L^2},$$

then using the interpolation inequalities

$$\| Du \|_{L^4}^2 \leqslant \| D^2 u \|_{L^2}^{\frac{d}{2}} \| Du \|_{L^2}^{2-\frac{d}{2}},$$

we get

$$| Ⅲ | \leqslant \| D^2 u \|_{L^2}^{1+\frac{d}{2}} \| Du \|_{L^2}^{2-\frac{d}{2}}.$$

For part Ⅳ, we have Ⅳ $=0$.

Summing up the inequalities for Ⅰ, Ⅱ, Ⅲ, Ⅳ, we get

$$\frac{\mathrm{d}}{\mathrm{d}t}\int_M |\partial_t \partial u|^2 + |\partial \nabla u|^2 \leqslant \| D^2 u \|^{1+d} \| Du \|^{3-d} + \| D^2 u \|^{1+\frac{d}{2}} \| Du \|^{2-\frac{d}{2}}. \tag{4.6}$$

We need also the estimates of $\partial_{tt} u$. For this purpose, applying the operator $\partial_t$ to $u$, we then get

$$\square(\partial_t u) = -\partial_t(|u_t|^2 - |\nabla u|^2)u - (|u_t|^2 - |\nabla u|^2)\partial_t u - u \times u_{tt}. \tag{4.7}$$

Taking inner product of this equation with $\partial_{tt} u$, and then integrating over $\Omega$, we get

$$\text{LHS} = \frac{1}{2}\frac{\mathrm{d}}{\mathrm{d}t}\int_M |u_{tt}|^2 + |\nabla u_t|^2,$$

$$\begin{aligned}\text{RHS} &= -\int_M \langle \partial_t(|u_t|^2 - |\nabla u|^2)u, \partial_{tt} u\rangle - \int_M \langle (|u_t|^2 - |\nabla u|^2)\partial_t u, \partial_{tt} u\rangle \\ &= Ⅰ + Ⅱ.\end{aligned} \tag{4.8}$$

From (1.1), we get $\partial_{tt} u = \partial_t u \times \square u + u \times \partial_t \square u$. Then taking the inner product with $u$, we have

$$u \cdot \partial_{tt} u = u \cdot (\partial_t u \times \square u) = \partial_t u \cdot (\square u \times u) = -|\partial_t u|^2. \tag{4.9}$$

Then from (4.9), we get

$$Ⅰ = -\int_M \partial_t(|u_t|^2 - |\nabla u|^2)(-|\partial_t u|^2) = \int_M \partial_t(|u_t|^2 - |\nabla u|^2)|\partial_t u|^2, \tag{4.10}$$

from which we get the estimates

$$|Ⅰ| \leqslant \int_M |Du|^3 |D^2 u| \leqslant \| D^2 u \|_{L^2}^{1+d} \| Du \|_{L^2}^{3-d}. \tag{4.11}$$

As before, for part Ⅱ, we get

$$|Ⅱ| \leqslant \int_M |Du|^3 |D^2 u| \leqslant \| D^2 u \|_{L^2}^{1+d} \| Du \|_{L^2}^{3-d}. \tag{4.12}$$

From (4.7), and the inequality (4.11)~(4.12), we get

$$\frac{\mathrm{d}}{\mathrm{d}t}\int_M |u_{tt}|^2 + |\nabla u_t|^2 \leqslant \| D^2 u \|^{1+d} \| Du \|^{3-d}. \tag{4.13}$$

Adding (4.6) and (4.13), we get

$$\frac{\mathrm{d}}{\mathrm{d}t}\| D^2 u \|^2 \leqslant \| D^2 u \|^{1+d} \| Du \|^{3-d} + \| D^2 u \|^{1+\frac{d}{2}} \| Du \|^{2-\frac{d}{2}}. \tag{4.14}$$

Since $\| Du \|$ is constant, we arrive at a Gronwall-type inequality

$$\frac{\mathrm{d}}{\mathrm{d}t}\| D^2 u \|^2 \leqslant C(\| D^2 u \|^{1+d} + \| D^2 u \|^{1+\frac{d}{2}}), \tag{4.15}$$

where $C$ depends on the $\| Du \|$ and $d$. This inequality implies a local in time $H^2$ *a priori* bound. □

From the proof of the lemma, we see that when $d=1$, we can get a global bound for the second derivative of the solution, that is we have

**Corollary 4.3** If $d=1$, we can bound the second derivative of the solution globally, and the solution extends for all time.

With the second order derivative estimates, it is enough for us to get our Theorem 1.3, but we postpone it to Section 5. For completeness, we write out the $d$th-order estimates of our solution with a brief proof.

**Lemma 4.4** Suppose $u_0 \in H^k$, $u_1 \in H^{k-1}$ for any $k \geqslant 2$, if $u$ is a solution for Eq. (1.1), then there exists $T>0$ such that the following estimates hold locally:

$$\sup_{0 \leqslant t \leqslant T} \| D^k u(t) \|_{L^2(M)}^2 \leqslant C. \tag{4.16}$$

**Proof** We prove this lemma by induction. For this purpose, we differentiate Eq. (2.1) $(k-1)$ times and take inner product with some $k$-order derivative of $u$. Precisely, if we differentiate the equation with the operator $\partial_t^l \partial^s$, we then make the inner product with $\partial_t^{l+1} \partial^s u$ with $l+s=k-1$. In the following we will sketch our proof only in the case of differentiating with the operator $\partial^{k-1}$, and the other cases can be handled similarly.

Differentiating Eq. (2.1) with $\partial^{k-1}$, we have

$$\Box(\partial^{k-1} u) = -\partial^{k-1}((|u_t|^2 - |\nabla u|^2) u) - \partial^{k-1}(u \times u_t).$$

Multiplying this by $\partial_t \partial^{k-1} u$, and then integrating over $M$, we get

$$\text{LHS} = \frac{1}{2} \frac{\mathrm{d}}{\mathrm{d}t} \int_M |\partial_t \partial^{k-1} u|^2 + |\partial^k u|^2,$$

$$\text{RHS} = -\int_M \langle \partial^{k-1}((|u_t|^2 - |\nabla u|^2) u), \partial_t \partial^{k-1} u \rangle - \int_M \langle \partial^{k-1}(u \times u_t), \partial_t \partial^{k-1} u \rangle.$$

Denoting this by $\text{RHS} = \sum_{h=0}^{k-1} I_{(k-1-h,h)} + \sum_{h=0}^{k-1} II_{(k-1-h,h)}$, the meaning of $I_{(k-1-h,h)}$ is obvious: acting the operator $\partial^{k-1}$ $h$ times on $u$ and $k-1-h$ times on $(|u_t|^2 - |\nabla u|^2)$.

From the fact $\partial^{k-1}(u \cdot u_t) = 0$, we have

$$u \cdot \partial^{k-1} \partial_t u = -\sum_{|\alpha|=1}^{k-1} \binom{k-1}{|\alpha|} (\partial^{|\alpha|} u \cdot \partial^{k-1-|\alpha|} \partial_t u), \quad \text{where } \alpha = (\alpha_1, \cdots, \alpha_m).$$

That is we can use the lower order derivatives to replace the highest order derivative, which can make our estimates much easier. Inserting this into the term $I_{(k-1,0)}$, we get

$$|I_{(k-1,0)}| \leqslant C \int_M |\partial^{k-1}(|u_t|^2 - |\nabla u|^2)| \sum_{|\alpha|=1}^{k-1} |D^{|\alpha|} u| |D^{k-1-|\alpha|} u|.$$

Using Hölder inequalities and interpolation inequalities we can bound this term by $C \| D^k u \|^\alpha$, where $\alpha$ is positive depending on $d$. By the same reasoning we can get all the estimates of RHS and all the estimates of the $k$th derivatives. Adding all the estimates of the $k$th-order derivatives, we arrive at the Gronwall-type inequality

$$\frac{\mathrm{d}}{\mathrm{d}t} \| D^k u \|^2 \leqslant \| D^k u \|^\alpha. \tag{4.17}$$

Then we can conclude our proof by using Gronwall lemma. □

As in Corollary 4.3, we can state

**Corollary 4.5** For any data $(u_0, u_1) \in H^k$, then so does $u$. Particularly, if $d = 1$, the problem (1.1) has a global smooth solution $u(x, t)$ provided $k$ large enough.

# 5 Conclusions

In this section, we will prove our Theorem 1.3. Uniqueness of $H^2$ solutions follows from the energy estimates. Again recall that $u \cdot u_t = 0$, and hence for $H^2$ solutions $u$ and $v$, we have

$$\Box w + (|u_t|^2 - |\nabla u|^2)u - (|v_t|^2 - |\nabla v|^2)v + u \times \frac{\partial w}{\partial t} + w \times \frac{\partial v}{\partial t} = 0.$$

Taking inner product with $\frac{\partial w}{\partial t}$, and then integrating on $M$, we have

$$\text{LHS} = \frac{1}{2}\frac{\mathrm{d}}{\mathrm{d}t}\|D(u - v)\|^2, \quad \text{RHS} = I + II,$$

where

$$I = \int_M \left\langle \frac{\partial(u - v)}{\partial t}, (|u_t|^2 - |\nabla u|^2)u - (|v_t|^2 - |\nabla v|^2)v \right\rangle,$$

$$II = \int_M \left\langle w \times \frac{\partial v}{\partial t}, \frac{\partial w}{\partial t} \right\rangle.$$

For part $I$, using the fact that $u \cdot u_t = 0$ we have

$$\begin{aligned} I &= \int_M (\langle u_t, (|v_t|^2 - |\nabla v|^2)u - (|v_t|^2 - |\nabla v|^2)v\rangle - \\ &\quad \langle v_t, (|u_t|^2 - |\nabla u|^2 v)u - (|u_t|^2 - |\nabla u|^2)v\rangle) \\ &\leqslant C\int_M |u - v|\,|Du - Dv|\,(|Du|^2 + |Dv|^2) \end{aligned} \tag{5.1}$$

which implies

$$I \leqslant C\|D(u - v)\|_{L^2}\| |u - v|(|Du|^2 + |Dv|^2)\|_{L^2}.$$

Then by Sobolev's embedding theorem, if $m \leqslant 3$ we have

$$I \leqslant C\|D(u - v)\|_{L^2}\|u - v\|_{L^6}(\|Du\|_{L^6}^2 + \|Dv\|_{L^6}^2) \leqslant C\|D(u - v)\|_{L^2}^2. \tag{5.2}$$

For part $II$, if $d \leqslant 3$ we have

$$II \leqslant \|u - v\|_{L^4}\left\|\frac{\partial v}{\partial t}\right\|_{L^4}\|D(u - v)\|_{L^2} \leqslant C\|D(u - v)\|_{L^2}^2. \tag{5.3}$$

Thus we arrive at a Gronwall-type inequality, which implies the uniqueness for $H^2$ solutions.

*A priori* estimates also imply well-posedness of the problem (1.1) in $H^2$ for $d \leqslant 3$. In fact, we can reason as follows. Let $(u_0^l, u_1^l) \in C^\infty$, and

$$(u_0^l, u_1^l) \to (u_0, u_1) \quad \text{in } H^2 \times H^1 \text{ as } l \to \infty.$$

Then inequality (4.15) implies a uniform $l$ local $H^2$ bound on the approximating solutions $u^l$, and hence the solution $\{u^l\}$ extend as smooth solutions to (1.1) on a time interval whose length depends on the norm of the data in $H^2 \times H^1$. Therefore we can extract a subsequence (still denoted $\{u^l\}$) which converges weakly in $H^2$ to an $H^2$ solution $u$ of the problem (1.1). That is we have proved the local well-posedness of the problem (1.1).

In particular, if $d = 1$, from Theorem 3.2, Corollary 4.5 and the uniqueness we just proved, it is easy to see that there exists a unique global smooth solution of the problem (1.1) provided the initial data are smooth enough. Thus we finished the proof of Theorem 1.3.

**References**


[1] Borisov A B, Kiseliev V V, Talutz G G. Solitons in a ferrimagnet. Solid State Commun, 1982, 44 (3): 411-412.

[2] Alouges F, Soyeur A. On global weak solutions for Landau-Lifshitz equations: existence and nonuniqueness. Nonlinear Anal, 1992, 18(11): 1071-1084.

[3] Chen Y. The weak solutions to the evolution problems of harmonic maps. Math Z, 1989, 201: 69-74.

[4] Guo Boling, Han Yongqian. Global regular solutions for Landau-Lifshitz equation. Front Math China, 2006, 4 : 538-568.

[5] Guo B, Hong M. The Landau-Lifshitz equation of the ferromagnetic spin chain and harmonic maps. Calc Var, 1993, 1: 311-334.

[6] Guo Boling, Zeng Ming. Existence of solutions for a dynamics equation of ferrimagnets. Progress in Natural Science, 2007, 17(11): 1279-1283.

[7] Stephen Gustafson, Jalal Shatah. The stability of localized solutions of Landau-Lifshitz equations. Comm Pure Appl Math, 2002, 55: 1136-1159.

[8] Porsezian K, Lakshmanan M. On the dynamics of the radially symmetric Heisenberg ferromagnetic spin system. J Math Phys, 1991, 32 (10): 2923-2928.

[9] Shatah J. Weak solutions and development of singularities in the $SU(2)\sigma$-model. Comm Pure Appl Math, 1988, 41: 459-469.

[10] Shatah J, Tahvlidar-Zadeh A S. On the Cauchy problem for equivariant wave maps. Comm Pure Appl Math, 1994, 47: 719-754.

[11] Shatah J, Struwe M. Geometric wave equations. New York: Amer Math Soc, 2000.

# The Asymptotic Behavior of the Stochastic Ginzburg-Landau Equation with Additive Noise *

Wang Guolian(王国联)  Guo Boling(郭柏灵)  Li Yangrong(李扬荣)

**Abstract**

We show that the stochastic Ginzburg-Landau equation with additive noise can be solved pathwise and the unique solution generates a random dynamic system. Then we prove that the system possesses a compact random attractor in $L^2(D)$ when the spatial dimension of $D$ is one and two, respectively.

**Keywords** Ginzburg-Landau equation; random dynamical systems; random attractor

## 1 Introduction

The Ginzburg-Landau equation arises in various areas of physics and chemistry. It has been studied extensively as a deterministic model[1-3]. However, some perturbations may neglect in the derivation of this ideal model (such as molecular collisions in gases and liquids and electric fluctuations in resistors[8]). When considering the perturbations of each microscopic units to the models, which will lead to a very large complex system, people usually represent the micro effects by random perturbations in the dynamics of the macro observables. Thus, it is interesting to study the following stochastic Ginzburg-Landau equation with additive white noise

$$du = (\lambda + i\mu)\Delta u dt - (\kappa + i\beta)|u|^2 u dt + \gamma u dt + \sum_{j=1}^{m} \varphi_j dw_j(t), \tag{1.1}$$

with boundary-initial conditions

$$u|_{\partial D} = 0, \tag{1.2}$$

$$u(t_0) = u_0, \tag{1.3}$$

where $\lambda$, $\mu$, $\kappa$, $\beta$, $\gamma$ are real coefficients, $D \subset \mathbf{R}^n (n = 1, 2)$ is an open bounded set with boundary $\partial D$ sufficiently regular, the functions $\varphi_j, j = 1, \cdots, m$ are time independent, and $w_j(t), j = 1, \cdots, m$ are independent two-sided real-valued standard wiener processes on a complete probability space $(\Omega, \mathscr{F}, P)$.

Our main aim is to study the long time behavior of the stochastic system (1.1)~(1.3). It is well known that attractors are quite well investigated to describe the long time behavior for deterministic systems (see [1]). But all analysis breaks down as soon as one wants to take

* Applied Mathematics and Computation, 2008, 198: 849-857.

random influences on the system under investigation into account. In particular, when subjecting the system to additive white noise, there is no chance that bounded subsets of the state space remain invariant. White noise pushes the system out of every bounded set with probability one. Recently, the notion of random attractors for a stochastic dynamical system has been introduced in [4,5]. Random attractors are in fact compact invariant sets – however, they are not fixed, but they depend on chance, and they move with time (in a coherent, "stationary" manner). It seems to be a good generalization of the classical concept of global attractors for deterministic dynamical systems, and it has been successfully applied to many infinite dimensional stochastic dynamical systems (see [4–6,9]).

In [6], the author studied the Ginzburg-Landau equation perturbed by multiplicative white noise

$$du = (\lambda + i\mu)\Delta u dt - (\kappa + i\beta)|u|^2 u dt + \gamma u dt + \sigma u dw(t).$$

Under the conditions that $w(t)$ is a two-sided standard wiener process and $\sigma$ is a positive constant, he obtained the existence of the random attractor in spacial dimension one and two.

Although the main conceptual line in proving the existence of global attractors in the additive noise case (in this paper) and the multiplicative case (see [6]) is similar, due to the difference of the translating variants in the two cases, we have to deal with spacial dimension one and two respectively at technical level in the present paper, and we also obtain the existence of the random attractors.

The paper is arranged as follows. In Section 2, we present some preliminaries of random dynamical system (RDS) and the solution of (1.1)~(1.3), and get the corresponding RDS. In Section 3, we prove the existence of the random attractors in dimension one and two, respectively.

# 2 Preliminaries

## 2.1 Preliminaries on Random Dynamical System

For the details of the definitions of this part, one can refer to [4,5,7].

Let $(\Omega, \mathscr{F}, P)$ be a probability space and $\{\theta_t : \Omega \to \Omega,\ t \in \mathbf{R}\}$ a family of measure preserving transformations such that $(t, \omega) \mapsto \theta_t\omega$ is measurable, $\theta_0 = \mathrm{id}$, and $\theta_{t+s} = \theta_t\theta_s$ for all $s, t \in \mathbf{R}$. The flow $\theta_t$ together with the corresponding probability space $(\Omega, \mathscr{F}, P)$ is called a metric dynamical system.

A continuous random dynamical system (RDS) on a polish space $(X, d)$ with Borel sigma-algebra $\mathscr{F}$ over $\theta$ on $(\Omega, \mathscr{F}, P)$ is by definition a measurable map

$$S: \mathbf{R}_+ \times \Omega \times X \to X, \quad (t, w, x) \mapsto S(t, \omega)x,$$

such that $P$-a.s.

(i) $S(0, \omega) = \mathrm{id}$ on $X$;

(ii) $S(t+s, \omega) = S(t, \theta_s\omega)S(s, \omega)$ for all $s, t \in \mathbf{R}_+$; (cocycle property)

(iii) $S(t,\omega):X\to X$ is continuous.

A random compact set $\{K(\omega)\}_{\omega\in\Omega}$ is a family of compact sets indexed by $\omega$ such that for every $x\in X$ the map $\omega\mapsto d(x, K(\omega))$ is measurable with respect to $\mathscr{F}$.

Let $\mathscr{A}(\omega)$ be a random set and $B\subset X$, one says $\mathscr{A}(\omega)$ attracts $B$ if

$$\lim_{t\to\infty}\operatorname{dist}(S(t,\theta_{-t}\omega)B,\mathscr{A}) = 0,\quad P\text{-a.s.}$$

A random set $\mathscr{A}(\omega)$ is said to be a random attractor for the RDS $S$ if $P$-a.s.

(i) $\mathscr{A}(\omega)$ is a random compact set;

(ii) $\mathscr{A}(\omega)$ is invariant, that is, $S(t,\omega)\mathscr{A}(\omega)=\mathscr{A}(\theta_t\omega)$, for all $t>0$;

(iii) $\mathscr{A}(\omega)$ attracts all deterministic bounded sets $B\subset X$.

**Theorem 2.1** If there exists a random compact set absorbing every bounded set $B\subset X$, then the RDS $S$ possesses a random attractor $\mathscr{A}(\omega)$,

$$\mathscr{A}(\omega) = \overline{\bigcup_{B\subset X}\Lambda_B(\omega)},$$

where $\Lambda_B(\omega):\bigcap_{s\geqslant 0}\overline{\bigcup_{t\geqslant s}S(t,\theta_{-t\omega})B}$ is the omega-limit set of $B$.

For Wiener process $w_j$ in Eq. (1.1), we take

$$\Omega = \{\omega\in C(\mathbf{R},\mathbf{R}^m)\mid\omega(0) = 0\},$$

with $P$ being a product measure of two Wiener measures on the negative and the positive time of $\Omega$. Then we have

$$(w_1(t,\omega),w_2(t,\omega),\cdots,w_m(t,\omega)) = \omega(t),\quad t\in\mathbf{R}.$$

The time shift is simply defined by

$$\theta_s\omega(t) = \omega(t+s)-\omega(s),\quad t,s\in\mathbf{R},$$

then $(\Omega,\mathscr{F},P,\theta_t)$ is an ergodic metric dynamical system.

## 2.2 Solve the Equations and Generate a RDS

For the mathematical setting of the problem, we introduce complex Sobolev spaces. Denote $(\cdot,\cdot)$ and $|\cdot|$ the inner product and the norm in $L^2(D)$, respectively, and $\|\cdot\|$ the norm in $H_0^1(D)$, where

$$(u,v) = \operatorname{Re}\int_D u(x)\bar{v}(x)\mathrm{d}x,$$

for $u,v\in L^2(D)$.

Without confusion, we also use $|\cdot|$ denoting the norm of the complex number below.

We always write $H=L^2(D)$, $V=H_0^1(D)$, $Au=-\Delta u$ and $f(u)=|u|^2u$. The operator $A$ is an isomorphism form $D(A)=V\cap H^2(D)$ onto $H$. Let $\{e_n\}$ be the orthonormal basis in $H$ of its eigenvectors with the corresponding eigenvalues $\lambda_n$, $\lambda_n>0$, $\lambda\nearrow\infty(n\to\infty)$. For the first eigenvalue, we have the inequality $\lambda_1|u|^2\leqslant\|u\|^2$.

We now write (1.1) as follows in the abstract from:

$$\mathrm{d}u = -(\lambda+\mathrm{i}\mu)Au\mathrm{d}t-(\kappa+\mathrm{i}\beta)f(u)\mathrm{d}t+\gamma u\mathrm{d}t+\sum_{j=1}^m\varphi_j\mathrm{d}w_j(t). \tag{2.1}$$

It is supplemented with the initial condition

$$u(t_0) = u_0, \tag{2.2}$$

where $\lambda, \mu, \kappa, \beta, \gamma \in \mathbf{R}, \lambda > 0, k > 0$.

From now on we always assume that the functions $\varphi_j \in D(A), j = 1, \cdots, m$ are time independent.

In the study of (2.1), it is usual to translate the unknown. Following the idea of [4,5], we introduce the unknown

$$v = u - z,$$

where

$$z = \sum_{j=1}^{m} \varphi_j z_j,$$

and $z$ is the Ornstein-Uhlenbeck process[10]

$$z_j = \int_{-\infty}^{t} e^{-\alpha(t-s)} dw_j(s),$$

which depends on a parameter $\alpha > 0$. It is well-known that $z$ is a stationary process and its trajectories are $P$-a.s. continuous. We also need the following fact

$$(E|z_j(0)|)^2 \leqslant E|z_j(0)|^2 = \int_{-\infty}^{0} e^{2\alpha\tau} d\tau = \frac{1}{2\alpha} \to 0,$$

as $\alpha \to \infty$. We also have (using, e.g., the law of the iterated logarithm)

$$\lim_{t \to -\infty} \frac{z_j(t)}{t} = 0, \quad P\text{-a.s.}$$

Then the process $v$ follows the random differential equation

$$\frac{dv}{dt} = -(\lambda + i\mu)Av - (\kappa + i\beta)f(v + z) + \gamma v + (\gamma + \alpha)z - (\lambda + i\mu)Az, \quad (2.3)$$

$$v(t_0, \omega) = v_0(\omega) = u_0 - z(t_0, \omega), \quad (2.4)$$

which can be solved pathwise.

By the same proof as that of Theorem Ⅳ of [1], using Galërkin approximation and some estimates, one can show that for $P$-a.s. $\omega \in \Omega$, the following holds:

(ⅰ) For $s \in \mathbf{R}$, $\xi \in H$, there exists a unique solution $v \in C([s, T); H) \cap L^2(s, T; V)$ $\forall T < \infty$ of (2.3) with $v(s) = \xi$;

(ⅱ) If $\xi \in V$, then the solution $v \in C([s, T); V) \cap L^2(s, T; D(A)), \forall T < \infty$;

(ⅲ) The mapping $\xi = v(s) \mapsto v(t)$ from $H$ to $H$ is continuous for all $t \geqslant s$.

Denote $v(t, \omega; s, \xi)$ the above solution of Eq. (2.3) with $v(s) = \xi$, then

$$u(t, \omega; t_0, u_0) = v(t, \omega; t_0, u_0 - z(t, \omega)) + z(t, \omega),$$

is the solution of (2.1) and (2.2), i.e.

$$S(t, \omega)u_0 = u(t, \omega; t_0, u_0) = v(t, \omega; 0, u_0 - z(0, \omega)) + z(t, \omega),$$

is a continuous random dynamical system. This will be called the stochastic flow associated with the Ginzburg-Landau equation with additive noise.

It is also useful to note that

$$S(s, \theta_{-s}\omega)u_0 = u(0, \omega; -s, u_0).$$

# 3 The Existence of Random Attractors

In the following paper, we only consider the case $\gamma > 0$. For the case $\gamma \leqslant 0$, the existence of the random attractor is obvious. Let $v$ be the solution of (2.3) and (2.4) for $\omega \in \Omega$. We now prove the existence of a compact absorbing set at time $t = -1$ in $H$.

## 3.1 Absorption in $H$ at Time $t = -1$

Multiplying (2.3) in $H$ by $\bar{v}$, integrating over $D$, and taking the real part, we obtain

$$\frac{1}{2}\frac{\mathrm{d}}{\mathrm{d}t}|v|^2 + \lambda\|v\|^2 = \gamma|v|^2 - \mathrm{Re}(\kappa + \mathrm{i}\beta)(f(v+z), v) + \mathrm{Re}(\gamma + \alpha)(z, v) - \mathrm{Re}(\lambda + \mathrm{i}\mu)(Az, v). \tag{3.1}$$

Noticing the following inequalities:

$$(f(v+z), v) = (|v+z|^2(v+z), v) = (v^2\bar{v} + z^2\bar{z} + z^2\bar{v} + v^2\bar{z} + 2|v|^2 z + 2|z|^2 v, v),$$

$$\begin{aligned} &- \mathrm{Re}(\kappa + \mathrm{i}\beta)(f(v+z), v) \\ &= -\kappa\int_D |v|^4 \mathrm{d}x - \mathrm{Re}(\kappa + \mathrm{i}\beta)(z^2\bar{z} + z^2\bar{v} + v^2\bar{z} + 2|v|^2 z + 2|z|^2 v, v) \\ &\leqslant -\kappa\int_D |v|^4 \mathrm{d}x + |\kappa + \mathrm{i}\beta|\int_D (|z|^3|v| + 3|z|^2|v|^2 + 3|v|^3|z|)\mathrm{d}x \\ &\leqslant -\kappa\int_D |v|^4 \mathrm{d}x + \frac{|\kappa + \mathrm{i}\beta|^2}{\gamma}|z|_{L^6(D)}^6 + \frac{\gamma}{4}|v|^2 + \frac{9|\kappa + \mathrm{i}\beta|^2}{\kappa}|z|_{L^4(D)}^4 + \\ &\quad \frac{\kappa}{4}|v|_{L^4(D)}^4 + \frac{3^7|\kappa + \mathrm{i}\beta|^4}{4\kappa^3}|z|_{L^4(D)}^4 + \frac{\kappa}{4}|v|_{L^4(D)}^4, \end{aligned} \tag{3.2}$$

$$\mathrm{Re}(\gamma + \alpha)(z, v) \leqslant |\gamma + \alpha|\int_D |z||v|\mathrm{d}x \leqslant \frac{|\gamma + \alpha|^2}{\gamma}|z|^2 + \frac{\gamma}{4}|v|^2, \tag{3.3}$$

$$-\mathrm{Re}(\lambda + \mathrm{i}\mu)(Az, v) \leqslant \frac{|\lambda + \mathrm{i}\mu|}{2\lambda}\|z\|^2 + \frac{\lambda}{2}\|v\|^2, \tag{3.4}$$

then, we have

$$\frac{\mathrm{d}}{\mathrm{d}t}|v|^2 + \lambda\|v\|^2 \leqslant -\kappa\int_D |v|^4 \mathrm{d}x + 3\gamma|v|^2 + g_1(t),$$

where

$$g_1(t) = \frac{|\kappa + \mathrm{i}\beta|^2}{\gamma}|z|_{L^6(D)}^6 + \left(\frac{9|\kappa + \mathrm{i}\beta|^2}{\kappa} + \frac{3^7|\kappa + \mathrm{i}\beta|^4}{4\kappa^3}\right)|z|_{L^4(D)}^4 + |\gamma + \alpha|\frac{|\gamma + \alpha|^2}{\gamma}|z|^2 + \frac{|\lambda + \mathrm{i}\mu|}{2\lambda}\|z\|^2,$$

and $g_1(t)$ has at most polynomial growth as $t \to \infty$, for $P$-a. s. $\omega \in \Omega$ (by the law of large numbers), and $g_1(t) \geqslant 0$. We will see that $g_i(i = 2,3,4,5,6)$ below have the same properties as $g_1$.

$$\frac{\mathrm{d}}{\mathrm{d}t}|v|^2 + \lambda\|v\|^2 + \gamma|v|^2 + \frac{\kappa}{2}|v|_{L^4(D)}^4 \leqslant -\int_D \frac{\kappa}{2}\left(|v|^4 - \frac{4\gamma}{\kappa}|v|^2\right)\mathrm{d}x + g_1(t)$$

$$\leqslant -\int_D \frac{\kappa}{2}\left(|v|^2-\frac{2\gamma}{\kappa}\right)^2 dx+\frac{2\gamma^2}{\kappa}|D|+g_1(t)\leqslant g_2(t), \tag{3.5}$$

where $g_2(t)=g_1(t)+\frac{2\gamma^2}{\kappa}|D|$.

By the Gronwall's lemma

$$|v(t)|^2\leqslant |v(s)|^2 e^{-\gamma(t-s)}+\int_s^t g_2(\sigma)e^{-\gamma(t-\sigma)}d\sigma,$$

for all $t_0\leqslant s\leqslant t$.

For $t=-1, s=t_0$, we get

$$|v(-1)|^2\leqslant |v(t_0)|^2 e^{\gamma(1+t_0)}+\int_{t_0}^{-1} g_2(\sigma)e^{\gamma(1+\sigma)}d\sigma$$

$$\leqslant |u_0|^2 e^{\gamma(1+t_0)}+|z(t_0)|^2 e^{\gamma(1+t_0)}+\int_{-\infty}^{-1} g_2(\sigma)e^{\gamma(1+\sigma)}d\sigma.$$

**Lemma 3.1** There exists a random radius $r_1(\omega)$, such that for all $\rho>0$ there exists (a deterministic) $\bar{t}\leqslant -1$ such that the following holds $P$-a.s. For all $t_0\leqslant \bar{t}$ and for all $u_0\in H$ with $|u_0|\leqslant\rho$, the solution $v(t,\omega;u_0-z(t_0,\omega))$ of (2.3) and (2.4) over $[t_0,\infty)$, satisfies the inequality

$$|v(t,\omega;u_0-z(t_0,\omega))|^2\leqslant r_1^2(\omega).$$

Of course one can deduce a similar absorption result for $u(-1)$ instead of $v(-1)$ from this lemma, but we will not need it in the following.

**Proof** Put

$$r_1^2(\omega)=2+2\sup_{t_0\leqslant -1}|z(t_0)|^2 e^{\gamma(1+t_0)}+\int_{-\infty}^{-1} g_2(\sigma)e^{\gamma(1+\sigma)}d\sigma,$$

which is finite $P$-a.s. since $|z(t_0)|^2$ and $g_2(\sigma,\omega)$ have at most polynomial growth for $t_0$ and $\sigma$, respectively, tending to $-\infty$. Given $\rho>0$, choose $\bar{t}$ such that

$$e^{\gamma(1+t_0)}\rho^2\leqslant 1.$$

Integrating (3.5) over $[-1,0]$, we have

$$\int_{-1}^0 \lambda|v(s)|^2 ds+\int_{-1}^0 \frac{\kappa}{2}|v|_{L^4(\Omega)}^4 ds\leqslant |v(-1)|^2+\int_{-1}^0 g_2(\sigma)d\sigma. \tag{3.6}$$

Consequently, from Lemma 3.1 we immediately get □

**Lemma 3.2** There exist two random variables $c_1(\omega)$ and $c_2(\omega)$, such that for all $\rho>0$ there exists (a deterministic) $\bar{t}\leqslant -1$ such that the following holds $P$-a.s. For all $t_0\leqslant\bar{t}$ and far all $u_0\in H$ with $|u_0|\leqslant\rho$, the solution $v(t,\omega;u_0-z(t_0,\omega))$ of (2.3) and (2.4) over $[t_0,\infty)$, satisfies

$$\int_{-1}^0 \|v(s)\|^2\leqslant c_1(\omega),\quad\text{and}\quad \int_{-1}^0 |v(s)|_{L^4(\Omega)}^4 ds\leqslant c_2(\omega),$$

where $u(s)=v(s)+z(s)$.

### 3.2 Absorption in $V$ at Time $t=0$ when $n=1$

Multiplying Eq. (2.3) in $H$ by $\Delta\bar{v}$, integrating over $D$, and taking the real part, we obtain

$$\frac{1}{2}\frac{\mathrm{d}}{\mathrm{d}t}\|v\|^2+\lambda|\Delta v|^2$$
$$=\gamma\|v\|^2+\mathrm{Re}((\kappa+\mathrm{i}\beta)(f(u),\Delta v)-(\gamma+\alpha)(z,\Delta v)+(\lambda+\mathrm{i}\mu)(Az,\Delta v)),$$
$$(f(v+z),\Delta v)=(v^2\bar{v},\Delta v)+(z^2\bar{z}+z^2\bar{v}+v^2\bar{z}+2|v|^2z+2|z|^2v,\Delta v),$$
$$(v^2\bar{v},\Delta v)=(|v|^2v,\Delta v)=-\int_D(|v|^2|\nabla|^2+v\nabla\bar{v}\,\nabla|v|^2)\mathrm{d}x,$$
$$\mathrm{Re}(\kappa+\mathrm{i}\beta)(v^2\bar{v},\Delta v)$$
$$=-\kappa\int_D|v|^2|\nabla v|^2\mathrm{d}x-\kappa\int_D\mathrm{Re}(v\nabla\bar{v})\nabla|v|^2\mathrm{d}x+\beta\int_D\mathrm{Im}(v\nabla\bar{v})\nabla|v|^2\mathrm{d}x$$
$$=-\kappa\int_D|v|^2|\nabla v|^2\mathrm{d}x-\frac{\kappa}{2}\int_D(\nabla|v|^2)^2\mathrm{d}x+\frac{\beta}{2}\int_D\mathrm{i}(v\nabla\bar{v}-\bar{v}\nabla v)\nabla|v|^2\mathrm{d}x$$
$$=-\frac{1}{4}\int_D(3\kappa(\nabla|v|^2)^2-2\beta\mathrm{i}(v\nabla\bar{v}-\bar{v}\nabla v)\nabla|v|^2+\kappa|v\nabla\bar{v}-\bar{v}\nabla v|^2)\mathrm{d}x$$
$$\leqslant 0,$$

where in the last inequality we use the condition $\sqrt{3}\kappa\geqslant|\beta|$.

$$\mathrm{Re}(\kappa+\mathrm{i}\beta)(z^2\bar{z}+z^2\bar{v}+v^2\bar{z}+2|v|^2z+2|z|^2v,\Delta v)$$
$$\leqslant|\kappa+\mathrm{i}\beta|\int_D(3|z|^2|v|+3|v|^2|z|+|z|^3)|\Delta v|\mathrm{d}x$$
$$\leqslant\frac{1}{2}|v|^4_{L^4(D)}+\frac{3\lambda}{8}|\Delta v|^2+\frac{\gamma}{2}\|v\|^2+\frac{162|\kappa+\mathrm{i}\beta|^4}{\lambda^2}|z|^8_{L^8(D)}+$$
$$\frac{2|\kappa+\mathrm{i}\beta|^2}{\lambda}|z|^6_{L^6(D)}+\frac{3^{11}|\kappa+\mathrm{i}\beta|^8}{2\lambda^4\gamma^3}|z|^8_4|v|^{10},$$
$$\mathrm{Re}(\gamma+\alpha)(z,\Delta v)\leqslant\frac{2|\gamma+\alpha|^2}{\lambda}|z|^2+\frac{\lambda}{8}|\Delta v|^2,$$
$$-\mathrm{Re}(\lambda+\mathrm{i}\mu)(Az,\Delta\bar{v})\leqslant\frac{|\lambda+\mathrm{i}\mu|^2}{2\gamma}\|z\|^2+\frac{\gamma}{2}\|v\|^2.$$

From the above inequalities, we get

$$\frac{\mathrm{d}}{\mathrm{d}t}\|v\|^2+\lambda|\Delta v|^2\leqslant 3\gamma\|v\|^2+|v|^4_{L^4(D)}+g_3(t),\tag{3.7}$$

where

$$g_3(t)=\frac{162|\kappa+\mathrm{i}\beta|^4}{\lambda^2}|z|^8_{L^8(D)}+\frac{2|\kappa+\mathrm{i}\beta|^2}{\lambda}|z|^6_{L^6(D)}+\frac{3^{11}|\kappa+\mathrm{i}\beta|^8}{2\lambda^4\gamma^3}|z|^8_4|v|^{10}+$$
$$\frac{2|\gamma+\alpha|^2}{\lambda}|z|^2+\frac{|\lambda+\mathrm{i}\mu|^2}{2\gamma}\|z\|^2.$$

Integrating over an arbitrary interval $[s,0]$, we get

$$\|v(0)\|^2\leqslant\|v(s)\|^2+\int_s^0(3\gamma\|v(\sigma)\|^2+|v(\sigma)|^4_{L^4(D)}+g_3(\sigma))\mathrm{d}\sigma.$$

Integrating again in $s$ over $[-1,0]$, we finally have

$$\|v(0)\|^2\leqslant\int_{-1}^0\|v(s)\|^2\mathrm{d}s+\int_s^0(3\gamma\|v(\sigma)\|^2+|v(\sigma)|^4_{L^4(D)}+g_3(\sigma))\mathrm{d}\sigma.$$

From Lemma 3.2 we readily have

**Lemma 3.3** Let $\sqrt{3}\kappa \geqslant |\beta|$, then there exists a random radius $r_2(\omega)$ such that for all $\rho>0$ there exists $\bar{t} \leqslant -1$ in such a way that the following holds $P$-a.s. For all $t_0 \leqslant \bar{t}$ and all $u_0 \in H$ with $|u_0| \leqslant \rho$ denote by $v(t,\omega;t_0,u_0-z(t_0,\omega))$ the solution of (2.3) over $[t_0,\infty)$ with $v(t_0)=u_0-z(t_0)$, and put $u(t,\omega;t_0,u_0)=z(t_0)+v(t,\omega;t_0,u_0-z(t_0,\omega))$. Then

$$\|u(0,\omega;t_0,u_0)\|^2 \leqslant r_2^2(\omega).$$

Using Theorem 2.1, Lemmas 3.1 and 3.3, we obtain

**Theorem 3.4** Under the assumption $\sqrt{3}\kappa \geqslant \beta$, and $\varphi_j \in D(A), j=1,\cdots,m$, the RDS associated the stochastic Ginzburg-Landau equation possesses a global random attractor $\mathscr{A}(\omega)$ when the dimension is one.

### 3.3 Absorption in $V$ at Time $t=0$ when $n=2$

**Lemma 3.5** We have the following inequality:

$$\frac{1}{4}\frac{\mathrm{d}}{\mathrm{d}t}|v|_{L^4(D)}^4 \leqslant -\frac{\kappa}{2}|v|_{L^6(D)}^6+\gamma|v|_{L^4(D)}^4-\frac{1}{4}\int_D(3\lambda(\nabla|v|^2)^2- 2\mu\mathrm{i}(v\nabla\bar{v}-\bar{v}\nabla v)\nabla|v|^2+\lambda|v\nabla\bar{v}-\bar{v}\nabla v|^2)\mathrm{d}x+g_4(t), \quad (3.8)$$

where

$$g_4(t)=\frac{9|\kappa+\mathrm{i}\beta|^3}{\varepsilon_1^3}|z|_{L^6(D)}^6+\frac{3^6|\kappa+\mathrm{i}\beta|^6}{6\varepsilon_2^6}|z|_{L^6(D)}^6+ \frac{|\kappa+\mathrm{i}\beta|^2}{2\varepsilon_3}|z|_{L^6(D)}^6+\frac{|\gamma+\alpha|}{2\varepsilon_4}|z|^2+\frac{|\gamma+\mathrm{i}\mu|}{2\varepsilon_5}|Az|^2.$$

**Proof** Taking real part of the inner product of (2.3) with $|v|^2v$ in $H$,

$$\mathrm{Re}\left(\frac{\mathrm{d}v}{\mathrm{d}t},|v|^2v\right)=\int_D|v|^2\mathrm{Re}(v_t\bar{v})\mathrm{d}x=\frac{1}{2}\int_D|v|^2\mathrm{Re}(v_t\bar{v}+\bar{v}_tv)\mathrm{d}x =\frac{1}{2}\int_D|v|^2\frac{\mathrm{d}}{\mathrm{d}t}|v|^2\mathrm{d}x=\frac{1}{4}\frac{\mathrm{d}}{\mathrm{d}t}|v|_{L^4(D)}^4,$$

$$(Av,|v|^2v)=-\int_D\Delta v|v|^2\bar{v}\mathrm{d}x=\int_D|\nabla v|^2|v|^2\mathrm{d}x+\int_D\bar{v}\nabla v\nabla|v|^2\mathrm{d}x,$$

$$|\nabla v|^2|v|^2=\frac{1}{4}|\nabla|v|^2|+\frac{1}{4}|v\nabla\bar{v}-\bar{v}\nabla v|^2,$$

$$(\lambda+\mathrm{i}\mu)(Av,|v|^2v)=\frac{1}{4}\int_D(3\lambda(\nabla|v|^2)^2-2\mu\mathrm{i}(v\nabla\bar{v}-\bar{v}\nabla v)\nabla|v|^2-\lambda|v\nabla\bar{v}-\bar{v}\nabla v|^2)\mathrm{d}x,$$

$$\begin{aligned}&-\mathrm{Re}(\kappa+\mathrm{i}\mu)(f(v+z),|v|^2v)\\&=-\mathrm{Re}(\kappa+\mathrm{i}\mu)(v^2\bar{v},|v|^2v)-\mathrm{Re}(\kappa+\mathrm{i}\mu)(z^2\bar{z}+z^2\bar{v}+v^2\bar{z}+2|v|^2z+2|z|^2v,|v|^2v)\\&\leqslant-\kappa|v|_{L^6(D)}^6+\frac{9|\kappa+\mathrm{i}\beta|^3}{\varepsilon_1^3}|z|_{L^6(D)}^6+\frac{2\varepsilon_1^{3/2}}{3}|v|_{L^6(D)}^6+\frac{3^6|\kappa+\mathrm{i}\beta|^6}{6\varepsilon_2^6}|z|_{L^6(D)}^6+\\&\quad\frac{5\varepsilon_2^{6/5}}{6}|v|_{L^6(D)}^6+\frac{|\kappa+\mathrm{i}\beta|^2}{3\varepsilon_3}|z|_{L^6(D)}^6+\frac{\varepsilon_3}{2}|v|_{L^6(D)}^6,\end{aligned}$$

$$(\gamma+\alpha)(z,|v|^2v)\leqslant\frac{|\gamma+\alpha|}{2\varepsilon_4}|z|^2+\frac{\varepsilon_4}{2}|v|_{L^6(D)}^6,$$

$$(\lambda + i\mu)(Az, |v|^2 v) \leqslant \frac{|\lambda + i\mu|}{2\varepsilon_5}|Az|^2 + \frac{\varepsilon_5}{2}|v|^6_{L^6(D)},$$

we can choose $\varepsilon_i (i = 1, \cdots, 5)$ suitably such that $\frac{2}{3}\varepsilon_1 + \frac{5}{6}\varepsilon_2 + \frac{1}{2}\varepsilon_3 + \frac{1}{2}\varepsilon_4 + \frac{1}{2}\varepsilon_5 = \frac{1}{2}\kappa$. Then we get the result of Lemma 3.5. □

Noticing that

$$\begin{aligned}
&\mathrm{Re}(\kappa + i\beta)(z^2\bar{z} + z^2\bar{v} + v^2\bar{z} + 2|v|^2 z + 2|z|^2 v, \Delta v) \\
&\leqslant |\kappa + i\beta| \int_D (3|z|^2|v| + 3|v|^2|z| + |z|^3)|\Delta v| \mathrm{d}x \\
&\leqslant \frac{1}{2}|v|^4_{L^4(D)} + \frac{3\lambda}{8}|\Delta v|^2 + \frac{\kappa}{2}|v|^6_{L^6(D)} + \frac{162|\kappa + i\beta|^4}{\lambda^2}|z|^8_{L^8(D)} + \\
&\quad \frac{2|\kappa + i\beta|^2}{\lambda}|z|^6_{L^6(D)} + \frac{3^3 2^5 |\kappa + i\beta|^8}{\lambda^3 \kappa}|z|^6_{L^6(D)},
\end{aligned}$$

we have

$$\begin{aligned}
&\frac{1}{2}\frac{\mathrm{d}}{\mathrm{d}t}\|v\|^2 + \frac{1}{2}\lambda|\Delta v|^2 \\
&\leqslant \gamma\|v\|^2 + \frac{1}{2}|v|^4_{L^4(D)} + \frac{\kappa}{2}|v|^6_{L^6(D)} - \frac{1}{4}\int_D (3\kappa(\nabla|v|^2)^2 - \\
&\quad 2\beta i(v\nabla\bar{v} - \bar{v}\nabla v)\nabla|v|^2 + \kappa|v\nabla\bar{v} - \bar{v}\nabla v|^2)\mathrm{d}x + g_5(t),
\end{aligned} \tag{3.9}$$

where

$$\begin{aligned}
g_5(t) = {} & \frac{162|\kappa + i\beta|^4}{\lambda^2}|z|^8_{L^8(\Omega)} + \frac{2|\kappa + i\beta|^2}{\lambda}|z|^6_{L^6(\Omega)} + \frac{3^3 2^5|\kappa + i\beta|^8}{\lambda^3\delta^2\kappa}|z|^6_{L^6(\Omega)} + \\
& \frac{2|\lambda + \alpha|^2}{\lambda}|z|^2 + \frac{|\lambda + i\mu|^2}{2\gamma}\|z\|^2.
\end{aligned}$$

From (3.8) and (3.9), we obtain

$$\begin{aligned}
&\frac{1}{2}\frac{\mathrm{d}}{\mathrm{d}t}\left(\|v\|^2 + \frac{\delta^2}{2}|v|^4_{L^4(\Omega)}\right) + \frac{\lambda}{2}|\Delta v|^2 \\
&\leqslant \frac{3\gamma}{2}\|v\|^2 + \frac{1 + 2\gamma}{2}|v|^4_{L^4(\Omega)} + \delta^2 g_4(t) + g_5(t) - \\
&\quad \frac{1}{4}\int_\Omega (3(\delta^2\lambda + \kappa)(\nabla|v|^2)^2 - 2(\delta^2\mu + \beta)i(v\nabla\bar{v} - \bar{v}\nabla v)\nabla|v|^2 + \\
&\quad (\delta^2\lambda + \kappa)|v\nabla\bar{v} - \bar{v}\nabla v|^2)\mathrm{d}x,
\end{aligned} \tag{3.10}$$

we can choose $\delta$ such that the matrix below is nonnegative definite

$$\boldsymbol{M} = \begin{pmatrix} 3(\delta^2\lambda + \kappa) & -(\delta^2\mu + \beta) \\ -(\delta^2\mu + \beta) & \delta^2\lambda + \kappa \end{pmatrix}.$$

Then we have

$$\begin{aligned}
&\frac{1}{2}\frac{\mathrm{d}}{\mathrm{d}t}\left(\|v\|^2 + \frac{\delta^2}{2}|v|^4_{L^4(D)}\right) + \frac{\lambda}{2}|\Delta v|^2 \\
&\leqslant \frac{3\gamma}{2}\|v\|^2 + \frac{1 + 2\gamma}{2}|v|^4_{L^4(D)} + g_6(t),
\end{aligned} \tag{3.11}$$

where $g_6(t) = \delta^2 g_4(t) + g_5(t)$.

Integrating over an arbitrary interval $[s,0]$ we get

$$\|v(0)\|^2 \leqslant \|v(s)\|^2 + \frac{\delta^2}{2}|v(s)|_{L^4(D)}^4 + \int_s^0 \left(\frac{3\gamma}{2}\|v(\sigma)\|^2 + \frac{1+2\gamma}{2}|v(\sigma)|_{L^4(D)}^4 + g_6(\sigma)\right)\mathrm{d}\sigma.$$

Integrating again in $s$ over $[-1,0]$ we finally have

$$\|v(0)\|^2 \leqslant \int_{-1}^0 \|v(s)\|^2 \mathrm{d}x + \int_{-1}^0 \frac{\delta^2}{2}|v(s)|_{L^4(D)}^4 \mathrm{d}x + \int_{-1}^0 \left(\frac{3\gamma}{2}\|v(\sigma)\|^2 + \frac{1+2\gamma}{2}|v(\sigma)|_{L^4(D)}^4 + g_6(\sigma)\right)\mathrm{d}\sigma.$$

**Lemma 3.6** If the above matrix $\boldsymbol{M}$ is nonnegative definite, then there exists a random radius $r_3(\omega)$ such that for all $\rho>0$ there exists $\bar{t}\leqslant -1$ in such a way that the following holds $P$-a.s. for all $t_0\leqslant \bar{t}$ and all $u_0\in H$ with $|u_0|\leqslant\rho$ denote by $v(t,\omega;t_0,u_0-z(t_0,\omega))$ the solution of (2.3) over $[t_0,\infty)$ with $v(t_0)=u_0-z(t_0)$, and put $u(t,\omega;t_0,u_0)=z(t_0)+v(t,\omega;t_0,u_0-z(t_0,\omega))$. Then

$$\|u(0,\omega;t_0,u_0)\|^2 \leqslant r_3^2(\omega).$$

Using Theorem 2.1, Lemmas 3.5 and 3.6, we obtain

**Theorem 3.7** Under the assumption that the above matrix $\boldsymbol{M}$ is nonnegative definite, and $\varphi_j\in D(A), j=1,\cdots,m$, the RDS associated the stochastic Ginzburg-Landau equation possesses a global random attractor $\mathscr{A}(\omega)$ when the dimension is two.

## References


[1] Temam R. Infinite-dimensional systems in mechanics and physics. New York: Springer-Verlag, 1988.

[2] Ginibre J, Velo G. The Cauchy problem in local spaces for the complex Ginzburg-Landau equation, I: compactness, methods. Physica D, 1996, 95: 191-228.

[3] Ghidaglia J M, Héron B. Dimension of the attractor associated to the Ginzburg-Landau equation. Physica D, 1987, 28: 282-304.

[4] Crauel H, Flaudoli F. Attractors for random dynamical systems. Probab Theory Relat Fields, 1994, 100: 365-393.

[5] Crauel H, Debussche A, Franco F, et al. J Dyn Diff Eq, 1997, 9 (2): 307-341.

[6] Yang D. The asymptotic behavior of the stochastic Ginzburg-Landau equation with multiplicative noise. J Math Phys, 2004, 45 (11): 4064-4076.

[7] Arnold L. Random dynamical system. Springer Monographs in Mathematics. Berlin: Springer-Verlag, 1998.

[8] Gardiner C W. Handbooks of stochastic methods for physics, chemistry and natural sciences. Berlin: Springer-Verlag, 1983.

[9] Debussche A. Hausdorff dimension of random invariant set. J Math Pure Appl, 1998, 77: 967-988.

[10] Da Prato G, Zabczyk J. Stochastic equations in infinite dimensions. Cambridge: Cambridge University Press, 1992.

# Blow-up and Global Smooth Solutions for Incompressible Three-dimensional Navier-Stokes Equations *

Guo Boling(郭柏灵) Yang Ganshan(杨干山) Pu Xueke(蒲学科)

**Abstract**

We present some explicit self-similar blow-up solutions and some other solutions of the incompressible three-dimensional Navier-Stokes equations. These solutions indicate that in $C^{\infty}$ the solution of Navier-Stokes equations does not always tend to a solution of Euler equations.

On the blow-up problem of compressible Navier-Stokes equations, Xin proved that any smooth solution to the multidimensional Navier-Stokes equations for polytropic fluids in the absence of heat conduction will blow-up in a finite time interval if the initial density is compactly supported[1]. On the blow-up problem of incompressible Navier-Stokes equations, Tsai[2] proved that the Leray self-similar solutions to the Navier-Stokes equations must be zero if they satisfy local energy estimates. The key idea of the proof is to obtain a Liouville-type lemma for the "total" pressure function by applying certain results on singular integrals on some weighted spaces. The nonexistence of self-similar solutions with small local energy has also been implied due to the partial regularity obtained by Caffarelli et al[3]. Later in 1998, Tian and Xin[4] constructed a one-parameter family of explicit smooth solutions with one-point singularity for the incompressible Navier Stokes equations in three spatial dimensions. These solutions are axisymmetrically homogeneous of degree $-1$ and are the only possible one-point singular solutions for viscous axisymmetric flows. However, these solutions are steady solutions, i.e. the solutions are independent of time variable $t$. In this letter, we construct some blow-up solutions of the three-dimensional (3D) Navier Stokes equations which depend on $t$ explicitly.

Precisely, we consider the following three-dimensional Navier-Stokes equation:

$$\frac{\partial \boldsymbol{u}}{\partial t} - \sigma\Delta \boldsymbol{u} + (\boldsymbol{u} \cdot \nabla)\boldsymbol{u} + \nabla p = 0, \tag{1}$$

$$\nabla \cdot \boldsymbol{u} = 0, \tag{2}$$

where $\boldsymbol{u} = (u_1, u_2, u_3)$ is the speed vector in $\mathbf{R}^3$ and $p$ is a scalar function denoting the pressure. We construct some explicit solutions to this equation which depends on time explicitly and will blow up at any given $T$. In the following Theorem 1, we set the viscous parameter

* Supported by Institute of Mathematical Sciences, The Chinese University of Hong Kong.
Chin Phys Lett, 2008, 25(6):2115-2117.

$\sigma=1$ for simplicity. For a general viscous parameter, one can construct other blow-up solutions by the same method. We use $\boldsymbol{x}=(x_1,x_2,x_3)$ to denote the coordinate in space $\mathbf{R}^3$.

**Theorem 1** The 3D Navier-Stokes equations (1) admit the following blow-up solutions

$$\boldsymbol{u}=\begin{pmatrix}\dfrac{(2\alpha+\beta)x_1}{(2\alpha+\beta)C-(4\alpha-2\beta)t}\\ \dfrac{(2\alpha+\beta)x_2}{(2\alpha+\beta)C-(4\alpha-2\beta)t}\\ \dfrac{-2(2\alpha+\beta)x_3}{(2\alpha+\beta)C-(4\alpha-2\beta)t}\end{pmatrix},$$

$$p=-\frac{3(2\alpha+\beta)(\alpha x_1^2+\alpha x_2^2+\beta x_3^2)}{((2\alpha+\beta)C-(4\alpha-\beta)t)^2},$$

where $\alpha,\beta$ and $C$ are some constants.

**Proof** One can find this type of solutions by the method of separation of variables as follows. Set

$$u_1(\boldsymbol{x},t)=u_1(\boldsymbol{x})T(t), \tag{3}$$

$$u_2(\boldsymbol{x},t)=u_2(\boldsymbol{x})T(t), \tag{4}$$

$$u_3(\boldsymbol{x},t)=u_3(\boldsymbol{x})T(t), \tag{5}$$

$$p(\boldsymbol{x},t)=p(\boldsymbol{x})T_1(t). \tag{6}$$

To let the solution satisfy the divergence free condition, we set

$$u_1(\boldsymbol{x})=x_1,\quad u_2(\boldsymbol{x})=x_2,\quad u_3(\boldsymbol{x})=-2x_3.$$

Substituting ansatz (3)~(6) into Eq. (1) and (2) yields

$$x_1T'+x_1T^2+\frac{\partial p}{\partial x_1}T_1=0,$$

$$x_2T'+x_2T^2+\frac{\partial p}{\partial x_2}T_1=0,$$

$$2x_3T'-4x_3T^2-\frac{\partial p}{\partial x_3}T_1=0.$$

Then by separation of variables one can obtain

$$\frac{T'+T^2}{T_1}=-\frac{1}{x_1}\frac{\partial p}{\partial x_1}=a, \tag{7}$$

$$\frac{T'+T^2}{T_1}=-\frac{1}{x_2}\frac{\partial p}{\partial x_2}=b, \tag{8}$$

$$\frac{T'+T^2}{T_1}=\frac{1}{2x_3}\frac{\partial p}{\partial x_3}=c, \tag{9}$$

from which we necessarily have $a=b$ and $p(\boldsymbol{x})=\alpha(x_1^2+x_2^2)+\beta x_3^2$. Equations (7)~(9) can be reduced to

$$T^2/T_1=-(2\alpha+\beta)/3, \tag{10}$$

$$T'/T_1 = -(4\alpha - \beta)/3, \tag{11}$$

thus $T' = \left(\dfrac{4\alpha - \beta}{2\alpha + \beta}\right)T^2$, which can be solved by

$$T = \frac{2\alpha + \beta}{(2\alpha + \beta)C - (4\alpha - \beta)t}.$$

Substituting this solution into Eqs. (10) and (11), we obtain

$$T_1 = \frac{-3(2\alpha + \beta)}{((2\alpha + \beta)C - (4\alpha - \beta)t)^2}.$$

The proof is completed.

In the above theorem, we find some solutions of the 3D incompressible Navier-Stokes equations which will surely blow up at some finite time $T_0$. Moreover, for the solution we have constructed, one have $\Delta \boldsymbol{u} = 0$, i.e. they are also the blow-up solutions of the Euler equations ($\sigma = 0$). In what follows, we construct some explicit solutions of the 3D incompressible Navier-Stokes equations with $\Delta \boldsymbol{u} \neq 0$, thus drastically different from the above ones.

**Theorem 2** Let $0 < T \leqslant \infty$. The 3D Navier-Stokes equations (1) and (2) admit a class of smooth blow-up solutions which are smooth in $(0, T)$ and blow-up exactly at arbitrary time $T$:

$$u_1 = \frac{1}{\sqrt{T-t}}\left(-1 + c_1\exp\left(\frac{1}{12\sigma(T-t)}s^2 - \frac{1}{\sigma\sqrt{T-t}}s + c\right)\right),$$

$$u_2 = \frac{1}{\sqrt{T-t}}\left(-1 + c_2\exp\left(\frac{1}{12\sigma(T-t)}s^2 - \frac{1}{\sigma\sqrt{T-t}}s + c\right)\right),$$

$$u_3 = -\frac{1}{\sqrt{T-t}}\left(1 + (c_1 + c_2)\exp\left(\frac{1}{12\sigma(T-t)}s^2 - \frac{1}{\sigma\sqrt{T-t}}s + c\right)\right),$$

$$p = \frac{1}{T-t}\left(\frac{1}{2\sqrt{T-t}}s + c_3\right),$$

with the initial data given by $\boldsymbol{u}_0 = \boldsymbol{u}(\boldsymbol{x}, 0)$ and $p_0 = p(\boldsymbol{x}, 0)$, where $s = \sum_{i=1}^{3}(x_i - x_{i0})$.

Here a solution of Eq. (1) is called a smooth blow-up solution at time $T$, if $u_i, p \in C^\infty([0, T_1] \times B_M) \cap W^{m_1, p_1}(0, T_1; W^{w_2, p_2}(B_M))$ $(i = 1, 2, 3)$ for any $T_1 \in [0, T)$, any non-negative integers $m_1$ and $m_2$, any positive real numbers $p_1$ and $p_2$, and any $M > 0$, but

$$\lim_{t \to T^-} \| u \|_{W^{m,p}(B_M)} = \infty,$$

$$\lim_{t \to T^-} \| p \|_{W^{m,p}(B_M)} = \infty,$$

for any non-negative integers $m$ and real numbers $q \geqslant 1$ and for all $M > 0$, where $B_M$ denotes the ball of radius $M$ centred at the origin.

**Proof** We compute directly as follows. For simplicity, we set

$$f(\boldsymbol{x}, t) = \exp\left(\frac{1}{12\sigma(T-t)}s^2 - \frac{1}{\sigma\sqrt{T-t}}s + c\right).$$

From the expression, we have

$$u_{1t}=\frac{1}{2\sqrt{(T-t)^3}}(-1+c_1f)+\frac{c_1f}{\sqrt{T-t}}\left(\frac{s^2}{12\sigma(T-t)^2}-\frac{s}{2\sigma\sqrt{(T-t)^3}}\right),$$

$$u_{2t}=\frac{1}{2\sqrt{(T-t)^3}}(-1+c_2f)+\frac{c_2f}{\sqrt{T-t}}\left(\frac{s^2}{12\sigma(T-t)^2}-\frac{s}{2\sigma\sqrt{(T-t)^3}}\right),$$

$$u_{3t}=-\frac{1}{2\sqrt{(T-t)^3}}(1+(c_1+c_2)f)-\frac{(c_1+c_2)f}{\sqrt{T-t}}\left(\frac{s^2}{12\sigma(T-t)^2}-\frac{s}{2\sigma\sqrt{(T-t)^3}}\right),$$

$$\Delta u_1=\frac{3c_1f}{\sqrt{T-t}}\left(\frac{s-6\sqrt{T-t}}{6\sigma(T-t)}\right)^2+\frac{c_1f}{2\sigma\sqrt{(T-t)^3}},$$

$$\Delta u_2=\frac{3c_2f}{\sqrt{T-t}}\left(\frac{s-6\sqrt{T-t}}{6\sigma(T-t)}\right)^2+\frac{c_2f}{2\sigma\sqrt{(T-t)^3}},$$

$$\Delta u_3=-\frac{3(c_1+c_2)f}{\sqrt{T-t}}\left(\frac{s-6\sqrt{T-t}}{6\sigma(T-t)}\right)^2-\frac{(c_1+c_2)f}{2\sigma\sqrt{(T-t)^3}},$$

$$(\boldsymbol{u}\cdot\nabla)\boldsymbol{u}=\begin{pmatrix}-\dfrac{3c_1f}{\sqrt{(T-t)^3}}\left(\dfrac{s}{6\sigma(T-t)}-\dfrac{1}{\sigma\sqrt{T-t}}\right)\\ -\dfrac{3c_2f}{\sqrt{(T-t)^3}}\left(\dfrac{s}{6\sigma(T-t)}-\dfrac{1}{\sigma\sqrt{T-t}}\right)\\ \dfrac{3(c_1+c_2)f}{\sqrt{(T-t)^3}}\left(\dfrac{s}{6\sigma(T-t)}-\dfrac{1}{\sigma\sqrt{T-t}}\right)\end{pmatrix}$$

and finally

$$\nabla p=\left(\frac{1}{2\sqrt{(T-t)^3}},\frac{1}{2\sqrt{(T-t)^3}},\frac{1}{2\sqrt{(T-t)^3}}\right).$$

From these computations, we have

$$\frac{\partial\boldsymbol{u}}{\partial t}-\sigma\Delta\boldsymbol{u}+(\boldsymbol{u}\cdot\nabla)\boldsymbol{u}+\nabla p=0,$$

$$\nabla\cdot\boldsymbol{u}=0,$$

by which we finish our proof.

A function $f(t)$ is piecewise smooth on $[0,\infty)$, if there exists $0=t_0<t_1<t_2<\cdots<t_k<+\infty$ such that $f(t)$ is smooth on each interval $(t_i,t_{i+1})$ $(i=0,1,\cdots,k-1)$ and $(t_k,\infty)$.

**Theorem 3** Let $f(t)$ be a piecewise smooth function, then Eqs. (1) and (2) have a class of solutions which satisfy the equation almost everywhere

$$\boldsymbol{u}=f(t)(c_{11}x_1+c_{21}x_2+c_{31}x_3,\ c_{12}x_1+c_{22}x_2+c_{32}x_3,\ c_{13}x_1+c_{23}x_2+c_{33}x_3),$$

$$p=-\frac{1}{2}f^2(t)\sum_{i,j,k=1}^{3}c_{ik}c_{kj}x_ix_j-\frac{1}{2}f'(t)\sum_{i,j=1}^{3}c_{ij}x_ix_j,$$

for any constants $c_{ij}$ satisfying $c_{ij}=c_{ji}$, $c_{33}+c_{11}+c_{22}=0$. Moreover the sufficient and necessary condition for smooth $\boldsymbol{u}$ is that $f$ is smooth and the blow-up time of $\boldsymbol{u}$ is just the blow-up time of $f$.

**Proof** We compute term by term as follows:

$$\nabla \cdot \boldsymbol{u} = f(t)(c_{11} + c_{22} + c_{33}) = 0,$$

$$\boldsymbol{u}_t = f'(t)(c_{11}x_1 + c_{21}x_2 + c_{31}x_3,\ c_{12}x_1 + c_{22}x_2 + c_{32}x_3,\ c_{13}x_1 + c_{23}x_2 + c_{33}x_3),$$

$$\Delta \boldsymbol{u} = 0,$$

$$(\boldsymbol{u} \cdot \nabla)\boldsymbol{u} = f^2(t)\sum_{j=1}^{3}(c_{1j}x_1 + c_{2j}x_2 + c_{3j}x_3)(c_{j1}, c_{j2}, c_{j3}),$$

$$\nabla p = -\frac{1}{2}f^2(t)\sum_{j=1}^{3}(c_{1j}x_1 + c_{2j}x_2 + c_{3j}x_3)(c_{j1}, c_{j2}, c_{j3}) - \frac{1}{2}f'(t)(c_{11}x_1 + c_{21}x_2 + c_{31}x_3, c_{12}x_1 + c_{22}x_2 + c_{32}x_3, c_{13}x_1 + c_{23}x_2 + c_{33}x_3).$$

Thus

$$\frac{\partial \boldsymbol{u}}{\partial t} - \sigma\Delta \boldsymbol{u} + (\boldsymbol{u} \cdot \nabla)\boldsymbol{u} + \nabla p = 0,$$

which concludes the proof.

**Example 1** Let $c_{11} = c_{22} = 1, c_{33} = -2$, and all other $c_{ij} = 0$, and $f(t) = 1/[(T_1 - t)\cdot(T_2 - t)\cdots(T_k - t)]$, then

$$\boldsymbol{u} = (f(t)x_1, f(t)x_2, -2f(t)x_3),$$

$$p = -\frac{1}{2}(f^2(t) + f'(t))(x_1^2 + x_2^2 + 4x_3^2),$$

is a solution to Eqs. (1) and (2), and $u$ blows up at $t = T_1, T_2, \cdots, T_k$.

**Example 2** In example 1, for any smooth function $f(t)$,

$$\boldsymbol{u} = (f(t)x_1, f(t)x_2, -2f(t)x_3),$$

$$p = -\frac{1}{2}(f^2(t) + f'(t))(x_1^2 + x_2^2 + 4x_3^2)$$

is a smooth solution to Eqs. (1) and (2).

When the pressure term $p$ in Eqs. (1) and (2) does not depend on spatial variables, i.e. $\nabla p = 0$, then this equation can be rewritten as

$$\frac{\partial \boldsymbol{u}}{\partial t} - \sigma\Delta \boldsymbol{u} + (\boldsymbol{u} \cdot \nabla)\boldsymbol{u} = 0,$$

$$\nabla \cdot \boldsymbol{u} = 0,$$

where the pressure $p$ may well depends on time, which appears when one studies the molecular gas dynamics, one can refer to the newly published book of Soné and Yoshio[5]. Here we find a particular kind of nontrivial smooth solutions of this type of equations.

**Theorem 4** The 3D Navier-Stokes equations (1) and (2) admits a class of smooth solutions

$$\boldsymbol{u} = \mathrm{e}^{\sigma(c_1^2 + c_2^2 + (c_1 + c_2)^2)t + c_1x_1 + c_2x_2 - (c_1 + c_2)x_3 + c}(1, 1, 1)^{\mathrm{T}},$$

$$p = g(t),$$

where the pressure term $p$ depends only on time.

**Proof** For simplicity, taking

$$f(t) = e^{\sigma(c_1^2+c_2^2+(c_1+c_2)^2 t+c)},$$

we can compute as follows:

$$\nabla \cdot \boldsymbol{u} = f(t)e^{c_1x_1+c_2x_2-(c_1+c_2)x_3}(c_1 + c_2 - c_1 - c_2) = 0,$$

$$\boldsymbol{u}_t = \sigma(c_1^2 + c_2^2 + (c_1 + c_2)^2)f(t)e^{c_1x_1+c_2x_2-(c_1+c_2)x_3}(1,1,1)^{\mathrm{T}},$$

$$\Delta\boldsymbol{u} = (c_1^2 + c_2^2 + (c_1 + c_2)^2)f(t)e^{c_1x_1+c_2x_2-(c_1+c_2)x_3}(1,1,1)^{\mathrm{T}},$$

$$(\boldsymbol{u} \cdot \nabla)\boldsymbol{u} = f(t)e^{c_1x_1+c_2x_2-(c_1+c_2)x_3}(c_1 + c_2 - c_1 - c_2)(1,1,1)^{\mathrm{T}} = 0.$$

All these computations lead to

$$\frac{\partial \boldsymbol{u}}{\partial t} - \sigma\Delta\boldsymbol{u} + (\boldsymbol{u} \cdot \nabla)\boldsymbol{u} + \nabla p = 0,$$

by which we finish our proof.

Authors are grateful to Professor Zhouping Xin for his enthusiastic support.

**References**

[1] Xin Z P. Blowup of smooth solutions to the compressible Navier-Stokesequation with compact density. Comm Pure Appl Math, 1998, 51(3): 229-240.

[2] Tsai T P. On Leray's self-similar solutions of the Navier-Stokes equations satisfying local energy estimates. Arch Rational Mech Anal, 1998, 143(1): 29-51.

[3] Caffarelli L, Kohn R, Nirenberg L. Partial regularity of suitable weak solutions of the Navier-Stokesequations. Comm Pure Appl Math, 1982, 35(6): 771-831.

[4] Tian G, Xin Z. One-point singular solutions to the Navier-Stokes equations. Topol Methods NonlinearAnal, 1998, 11(1):135-145.

[5] Yoshio S. Molecular gas dynamics: theory, techniques and applications. Boston: Birkhauser, 2006.

# Blow up and Strong Instability Result for a Quasilinear Schrödinger Equation *

Chen Jianqing(陈建清) Guo Boling(郭柏灵)

**Abstract**

Variational methods are used to prove that the solution of the quasilinear Schrödinger equation

$$iu_t + u_{xx} + |u|^{p-2}u + (|u|^2)_{xx}u = 0, \quad u|_{t=0} = u_0(x), \quad x \in \mathbf{R}$$

must blow up in a finite time for suitable initial data with positive initial energy and some restrictions on $p$. Then using this we prove that the standing wave is $H^1(\mathbf{R})$ strongly unstable with respect to the quasilinear Schrödinger equation.

**Keywords** variational methods; blow up; instability of standing wave; quasilinear Schrödinger equation

## 1 Introduction

In many physical situations, the following quasilinear Schrödinger equation:

$$iu_t + u_{xx} + \beta|u|^{p-2}u + \theta(|u|^2)_{xx}u = 0, \quad t \geqslant 0, \quad x \in \mathbf{R} \tag{1.1}$$

has been derived, where $u := u(t,x): \mathbf{R}_+ \times \mathbf{R} \to \mathbf{C}$ is a complex-valued function, $i^2 = -1$, $p>2$ and $\theta, \beta \in \mathbf{R}$. $u_{xx}$ denotes the second order derivatives of $u$ with respect to $x$. For example, it models the time evolution of condensate wave function in superfluid film[1]. Eq. (1.1) has also appeared in plasma physics and fluid mechanics, in the theory of Heisenberg ferromagnet and magnons and in dissipative quantum mechanics[2-5]. Comparing (1.1) with the usual semilinear Schrödinger equation $iu_t + u_{xx} + |u|^{p-2}u = 0$, there is an additional quasilinear term $(|u|^2)_{xx}u$. Physically this quasilinear term is used to stabilize the standing waves, but it will creates great difficulties in mathematical studies. The mathematical interests of (1.1) have at least three aspects.

Firstly, study the existence and multiplicity of standing waves. By a standing wave, we mean a special periodic solution of (1.1) of the form $e^{i\omega t}v(x)$, where $v(x)$ is the minimal action solution of

$$-\omega v + v_{xx} + \beta|v|^{p-2}v + \theta(|v|^2)_{xx}v = 0.$$

When $\theta = 0$, the existence of standing waves of (1.1) has been obtained in [6,7]. When $\theta \neq 0$, Ambrosetti and Wang[8] and Poppenberg et al[9] got the existence of standing waves

* Supported in part by Youth Foundation of NSFC (10501006) and the Program of NCETFJ.
Applied Mathematical Modelling, 2009, 33: 4192-4200.

under different conditions.

Secondly, consider the Cauchy problem of (1.1) with initial value

$$u(t,x)|_{t=0}=u_0(x),\quad x\in \mathbf{R}. \tag{1.2}$$

When $\theta=0$, local and global well posedness of (1.1) and (1.2) with $u_0\in H^1(\mathbf{R})$ has been studied extensively, see e.g. [10]. When $\theta\neq 0$, local well posedness of (1.1) and (1.2) and almost global existence of (1.1) with small initial value $u_0$ have been obtained in [11,12]. They need $u_0$ contained in higher order Sobolev space $H^\infty(\mathbf{R})$ (see Section 2). However, in the case of $\theta\neq 0$ the general global existence result of (1.1) is still *open* mainly due to the quasilinear term.

Thirdly, study the stability and instability of standing waves of (1.1), which was an important and difficult issue. For $\theta=0$, it is proved that when $2<p<6$, the standing wave is stable and when $p\geqslant 6$, the standing wave is unstable, see e.g. [13-16] and the references therein. But for $\theta\neq 0$, we do not see the stability results in the literature. The main difficulty is due to the quasilinear term $(|u|^2)_{xx}u$.

In the present paper, we are going to study the blowing up and instability of standing waves of (1.1) and (1.2) in the case of $\theta\neq 0$. For simplicity, we assume throughout this paper that $\beta=1$ and $\theta=1$ even though our arguments still work for $\beta>0$ and $\theta\geqslant 0$. Recall that the instability of standing waves by blowing up in the case of $\theta=0$ has been given in [13]. Our arguments will extend those results to the case of $\theta\neq 0$. We will prove that when $p\geqslant 8$ and $u_0\in K$ (see Lemma 2.6), then the standing waves of (1.1) is strongly unstable by blowing up, see Theorem 4.5. Our result implies that the presence of quasilinear term makes the standing wave more stable in the sense that we need larger $p$ to get the unstable standing wave, which is consistent with the physical phenomena. We overcome the difficulties created by the quasilinear term by introducing two new minimization problem $d_\infty$ and $d_{M_\infty}$ and comparing $d_M$ with $d_1$, see Section 2 for details.

This paper is organized as follows. The forthcoming section is some preliminaries. In Section 3, we prove a blowing up result for one class of initial value which allows the initial energy positive, see Theorem 3.1. Section 4 is due to the instability result of standing wave of (1.1), which is contained in Theorem 4.5. The final section is a concluding remark.

## 2 Preliminaries

Throughout this paper, $\beta=1$ and $\theta=1$, $\int_{\mathbf{R}}\cdot\,\mathrm{d}x$ is simply denoted by $\int\cdot$, unless stated otherwise. $C_j, B_j$ denote various positive constants whose exact value are not important. For any $t$, the function $x\mapsto u(t,x)$ is also denoted by $u(t)$ if no confusion occurs. Since we are studying the stability of standing wave of (1.1), we need the local well posedness result of Poppenberg[12], which was established for the initial value in a higher order Sobolev space. Let $H^k(\mathbf{R})=W^{k,2}(\mathbf{R})$ $(k\geqslant 1)$ be the standard Sobolev space with the norm $\|u\|^2_{H^k(\mathbf{R})}=\int(\omega|u|^2$

$+\sum_{j=1}^{k}|u^{(j)}|^2)\mathrm{d}x$ for positive constant $\omega$ and $u^{(j)}$ the $j$th derivative of $u$. The norm in $H^1(\mathbf{R})$ is simply denoted by $\|\cdot\|$. Define $H^\infty(\mathbf{R})=\bigcap_{k=1}^{\infty}H^k(\mathbf{R})$. The following local well posedness result of Poppenberg[12] will play an important role in what follows.

**Proposition 2.1**[12] For any $u_0\in H^\infty(\mathbf{R})$, there is $T>0$ and a unique solution $u(t):=u(t,x)\in C^1([0,T);H^\infty(\mathbf{R}))$ of

$$\begin{cases} \mathrm{i}u_t+u_{xx}+|u|^{p-2}u+(|u|^2)_{xx}u=0, & x\in\mathbf{R} \\ u(0,x)=u_0(x), & x\in\mathbf{R} \end{cases}. \tag{2.1}$$

Moreover, on the time interval of existence, there are two conserved identities:

$$E_1(u)=\int|u(t)|^2\mathrm{d}x\equiv\int|u_0|^2\mathrm{d}x\equiv\text{const} \tag{2.2}$$

and

$$E_2(u)=\int\left(|u_x|^2-\frac{2}{p}|u|^p+\frac{1}{2}|(|u|^2)_x|^2\right)\equiv E_2(u_0)=\text{const}. \tag{2.3}$$

Next, in order to study the blowing up result, we need the following virial type identity. The proof of Proposition 2.2 is based on an idea of Glassey[17], see e.g. [18].

**Proposition 2.2**[18] Let $u(t,x)$ be the solution of (2.1). Denote $I(t)=\frac{1}{2}\int|x|^2|u|^2$. If $|x|u_0\in L^2(\mathbf{R})$, then

$$I''(t)=4\int\left(|u_x|^2+\frac{3}{4}|(|u|^2)_x|^2-\frac{p-2}{2p}|u|^p\right).$$

Now for $v\in H^1(\mathbf{R})$ and $\omega>0$, we define the following functional:

$$J(v)=\int\left(\frac{1}{2}(|v_x|^2+\omega|v|^2)+\frac{1}{4}|(|v|^2)_x|^2-\frac{1}{p}|v|^p\right)\mathrm{d}x. \tag{2.4}$$

Since $H^1(\mathbf{R})\hookrightarrow L^\infty(\mathbf{R})$, we know that $J$ is well defined and $C^1$ on $H^1(\mathbf{R})$, for the detailed proofs see [9] (please note that this is usually not the case on $H^1(\mathbf{R}^N)$ due to the quasilinear term, see [19]). We also need two other functionals on $H^1(\mathbf{R})$:

$$\mathcal{N}(v)=\int(|v_x|^2+\omega|v|^2+|(|v|^2)_x|^2-|v|^p)\mathrm{d}x, \tag{2.5}$$

$$Q(|v|)=\int\left(|v_x|^2+\frac{3}{4}|(|v|^2)_x|^2-\frac{p-2}{2p}|v|^p\right)\mathrm{d}x. \tag{2.6}$$

Define the following set:

$$S_1=\{v\in H^1(\mathbf{R});\mathcal{N}(v)=0,v\neq0\}$$

and the related minimization problem

$$d_1=\inf_{v\in S_1}J(v). \tag{2.7}$$

Then the following lemma is due to Ambrosetti and Wang[8].

**Lemma 2.3**[8] If $p\geqslant4$, then $d_1>0$ and $d_1$ is achieved by $v\in H^1(\mathbf{R})$.

In order to study the stability of standing waves, we need to define another set:

$$S_\infty=\{v\in H^\infty(\mathbf{R}):\mathcal{N}(v)=0,v\neq0\}$$

and define the related minimization problem

$$d_\infty = \inf_{v \in S_\infty} J(v).$$

Then we have that

**Lemma 2.4** If $p \geqslant 4$, then $d_1 = d_\infty$.

**Proof** We first prove that $d_1 = d_\infty$. It is easy to see that $d_\infty \geqslant d_1$. To prove that $d_\infty \leqslant d_1$, we only need to show that for any $v^* \in S_1$ and any $\varepsilon > 0$, there holds

$$J(v^*) \geqslant \inf_{v \in S_\infty} J(v) - \varepsilon. \tag{2.8}$$

Since $H^\infty(\mathbf{R})$ is dense in $H^1(\mathbf{R})$, we can find $v_n \in H^\infty(\mathbf{R})$ such that

$$v^* = v_n + w_n \quad \text{and} \quad w_n \to 0 \quad \text{strongly in} \quad H^1(\mathbf{R}) \quad \text{as} \quad n \to \infty.$$

From $v^* \neq 0$, $\mathcal{N}(v^*) = 0$ and $w_n \to 0$ in $H^1(\mathbf{R})$, we know that $\lim\limits_{n\to\infty}\int |v_n|^p \neq 0$ and $\lim\limits_{n\to\infty}\mathcal{N}(v_n) = 0$. [19, Page 883, Lemma 2.3] implies that there is a sequence $\{\rho_n\}$ such that $\rho_n v_n \in S_\infty$ and $\lim\limits_{n\to\infty}\rho_n = 1$.

Next, denoting $v^* = \rho_n v_n + (1-\rho_n)v_n + w_n$, we compare $J(v^*)$ with $J(\rho_n v_n)$. Note that $1-\rho_n \to 0$ and $w_n \to 0$ strongly in $H^1(\mathbf{R})$, one immediately have that

$$\frac{1}{2}\int (|(v^*)_x|^2 + \omega |v^*|^2)\mathrm{d}x \geqslant \frac{1}{2}\int (|(\rho_n v_n)_x|^2 + \omega |\rho_n v_n|^2)\mathrm{d}x - \frac{\varepsilon}{3}$$

and

$$-\frac{1}{p}\int |v^*|^p \mathrm{d}x \geqslant -\frac{1}{p}\int |\rho_n v_n|^p \mathrm{d}x - \frac{\varepsilon}{3}.$$

We now expand the quasilinear term $\frac{1}{4}\int |(|\rho_n v_n + (1-\rho_n)v_n + w_n|^2)_x|^2 \mathrm{d}x$. Using the fact that $(w_n)_x \to 0$ strongly in $L^2(\mathbf{R})$, $1-\rho_n \to 0$ and $w_n \to 0$ strongly in $L^q(\mathbf{R})$ for any $2 \leqslant q \leqslant \infty$, we know that all other terms go to zero as $n \to \infty$ but $\int |(|\rho_n v_n|^2)_x|^2 \mathrm{d}x$. Hence we get that for $n$ large

$$\frac{1}{4}\int |(|v^*|^2)_x|^2 \mathrm{d}x \geqslant \frac{1}{4}\int |(|\rho_n v_n|^2)_x|^2 \mathrm{d}x - \frac{\varepsilon}{3}.$$

Therefore

$$J(v^*) \geqslant J(\rho_n v_n) - \varepsilon \geqslant \inf_{v \in S_\infty} J(v) - \varepsilon.$$

(2.8) holds. The proof is complete. □

To continue, we define two other minimization problem

$$d_M = \inf_{v \in M} J(v), \quad d_{M_\infty} = \inf_{v \in M_\infty} J(v), \tag{2.9}$$

where

$$M = \{v \in H^1(\mathbf{R}); Q(v) = 0, \mathcal{N}(v) < 0\} \tag{2.10}$$

and

$$M_\infty = \{v \in H^\infty(\mathbf{R}); Q(v) = 0, \mathcal{N}(v) < 0\}.$$

Then using an argument similar to those used in the proof of Lemma 2.4, one has that

$$d_M = d_{M_\infty}. \tag{2.11}$$

Moreover, we have that

**Lemma 2.5** If $p\geqslant 8$ and $\omega>0$, then $d_M\geqslant d_1$.

**Proof** It suffices to show that for any $v\in M$, we can find a $w\in S_1$ such that $J(v)\geqslant J(w)$. To see this, for any $v\in M(Q(v)=0$ and $\mathcal{N}(v)<0)$, we set $v_\lambda(x)=\lambda^{\frac{1}{2}}v(\lambda x)$ with $\lambda>0$. By simple calculations we have that

$$\mathcal{N}(v_\lambda)=\int[\lambda^2|v_x|^2+w|v|^2+\lambda^3|(|v|^2)_x|^2-\lambda^{\frac{p-2}{2}}|v|^p]. \tag{2.12}$$

Note that $\mathcal{N}(v_\lambda)\to\mathcal{N}(v)<0$ as $\lambda\to 1$ and $\mathcal{N}(v_\lambda)\to\omega\int|v|^2>0$ as $\lambda\to 0$, we can find a $0<\lambda^*<1$ such that $v_{\lambda^*}\neq 0$ and $\mathcal{N}(v_{\lambda^*})=0$. Hence $v_{\lambda^*}\in S_1$ and we have that $J(v_{\lambda^*})\geqslant d_1$.

Next, we claim that $J(v)\geqslant J(v_\lambda)$ for all $0<\lambda\leqslant 1$. Indeed, keeping the expression of $J$ in mind, we have that

$$\begin{aligned}&J(v)-J(v_\lambda)\\&=\int\left[\frac{1}{2}(1-\lambda^2)|v_x|^2+\frac{1}{4}(1-\lambda^3)|(|v|^2)_x|^2-\frac{1}{p}(1-\lambda^{\frac{p-2}{2}})|v|^p\right]dx.\end{aligned} \tag{2.13}$$

It follows from $Q(v)=0$ that:

$$\begin{aligned}&J(v)-J(v_\lambda)\\&=\int\left[\frac{1}{2}(1-\lambda^2)|v_x|^2+\frac{1}{4}(1-\lambda^3)|(|v|^2)_x|^2-\frac{2}{p-2}(1-\lambda^{\frac{p-2}{2}})(|v_x|^2+\frac{3}{4}|(|v|^2)_x|^2)\right]dx\\&=\int\left\{\left[\frac{1-\lambda^2}{2}-\frac{2}{p-2}(1-\lambda^{\frac{p-2}{2}})\right]|v_x|^2+\frac{1}{4}\left[(1-\lambda^3)-\frac{6}{p-2}(1-\lambda^{\frac{p-2}{2}})\right]|(|v|^2)_x|^2\right\}dx.\end{aligned} \tag{2.14}$$

Denoting $g_1(\lambda)=\frac{1-\lambda^2}{2}-\frac{2}{p-2}(1-\lambda^{\frac{p-2}{2}})$, we have that

$$g_1(\lambda)\to\frac{p-6}{2(p-2)}\quad\text{as}\quad\lambda\to 0\quad\text{and}\quad g_1(\lambda)\to 0\quad\text{as}\quad\lambda\to 1.$$

Since $g_1'(\lambda)=\lambda(\lambda^{\frac{p-6}{2}}-1)<0$ for $0<\lambda\leqslant 1$, we obtain that $g_1(\lambda)\geqslant g_1(1)=0$ for all $0<\lambda\leqslant 1$.

Similarly, we obtain that $g_2(\lambda)=(1-\lambda^3)-\frac{6}{p-2}(1-\lambda^{\frac{p-2}{2}})\geqslant 0$ for all $0<\lambda\leqslant 1$, where we use $p\geqslant 8$. Thus the claim is proved.

In particular, $J(v)\geqslant J(v_{\lambda^*})$ and therefore $d_M\geqslant d_1$. □

In the following, we will construct a set which is invariant under the flow generated by the Cauchy problem (1.1) and (1.2).

**Lemma 2.6** Let $p\geqslant 8$ and $d_1$ be the minimum defined in (2.7). Denote

$$K=\{u\in H^\infty(\mathbf{R});J(u)<d_1,\mathcal{N}(u)<0,Q(u)<0\}.$$

Then $K$ is invariant under the flow of (1.1) and (1.2). That is to say, if $u(t,x)$ is a solution of (1.1) and (1.2) and $u_0\in K$, then $u(t,\cdot)\in K$ for all $t\in[0,T)$.

**Proof** Let $u(t)$ be the solution of (1.1) with initial value $u_0$. In the first place, from the conserved identities

$$E_1(u) = \int |u_0|^2 dx \equiv \text{const and}$$

$$E_2(u) = \int \left( |u_x|^2 - \frac{2}{p}|u|^p + \frac{1}{2}|(|u|^2)_x|^2 \right) \equiv E_2(u_0) \equiv \text{const},$$

we get immediately that $J(u) \equiv J(u_0)$. Thus if $u_0 \in K$, then $J(u) = J(u_0) < d_1$.

In the second place, we show $\mathcal{N}(u(t)) < 0$ for $t \in [0, T)$. If otherwise, from the continuity, there were a $t_0 \in (0, T)$ such that $\mathcal{N}(u(t_0)) = 0$. Then since $u(t_0) \neq 0$, we know that $J(u(t_0)) \geqslant d_1$. This contradicts to $J(u(t)) < d_1$ for all $t \in (0, T)$. Therefore $\mathcal{N}(u(t)) < 0$ for $t \in [0, T)$.

Finally, we show $Q(u(t)) < 0$ for $t \in [0, T)$. If otherwise, from the continuity, there were a $t_1 \in (0, T)$ such that $Q(u(t_1)) = 0$. Since we have proved that $\mathcal{N}(u(t)) < 0$, we can have that $u(t_1) \in M_\infty$. So that $J(u(t_1)) \geqslant d_{M_\infty} = d_M \geqslant d_1$, which contradicts to $J(u(t)) < d_1$ for all $t \in (0, T)$. Therefore we have $Q(u(t)) < 0$ for $t \in [0, T)$. □

## 3 Blowing up Result

In this section, we will use the invariant set constructed in Lemma 2.6 to prove a blowing up result of the solutions of (1.1) and (1.2) for suitable $u_0$. The idea has been originated from [13,15] and later used in [16].

**Theorem 3.1** Let $p \geqslant 8$, $u_0 \in K$ and $|\cdot| u_0(\cdot) \in L^2(\mathbf{R})$. Then the solutions of (1.1) and (1.2) must blow up in a finite time, i.e. there is $T > 0$ such that

$$\lim_{t \to T^-} \| u_x \|_{L^2(\mathbf{R})}^2 = +\infty. \tag{3.1}$$

**Proof** Note that from the local well posedness of (1.1) and (1.2) (see Proposition 2.1), we always have $u(t) \in H^\infty(\mathbf{R})$ as long as $u_0 \in H^\infty(\mathbf{R})$. Moreover since $u_0 \in K$, we have $u(t) \in K$. Denote $u_\lambda(x) = \lambda^{\frac{1}{2}} u(t, \lambda x)$ with $\lambda > 0$. Then by direct calculations we have that

$$\mathcal{N}(u_\lambda) = \int (\lambda^2 |u_x|^2 + \omega |u|^2 + \lambda^3 |(|u|^2)_x|^2 - \lambda^{(p-2)/2} |u|^p) dx, \tag{3.2}$$

$$Q(u_\lambda) = \int \left( \lambda^2 |u_x|^2 + \frac{3\lambda^3}{4} |(|u|^2)_x|^2 - \frac{p-2}{2p} \lambda^{(p-2)/2} |u|^p \right) dx. \tag{3.3}$$

Note that $\mathcal{N}(u) < 0$, we have $\mathcal{N}(u_\lambda) \to \mathcal{N}(u) < 0$ as $\lambda \to 1$ while $\mathcal{N}(u_\lambda) \to \int |u|^2 > 0$ as $\lambda \to 0$, we find that there is $\lambda_*$ such that $\mathcal{N}(u_{\lambda_*}) = 0$ and when $\lambda \in (\lambda_*, 1]$, $\mathcal{N}(u_\lambda) < 0$. For $\lambda \in (\lambda_*, 1]$, $Q(u_\lambda)$ has the following three possibilities:

(i) $Q(u_\lambda) < 0$ for all $\lambda \in [\lambda_*, 1]$;

(ii) $Q(u_{\lambda_*}) = 0$;

(iii) $\exists \lambda_* < \mu < 1$ such that $Q(u_\mu) = 0$.

For the cases (i) and (ii), we all have $\mathcal{N}(u_{\lambda_*}) = 0$ and $Q(u_{\lambda_*}) \leqslant 0$. Of course, $u_{\lambda_*} \in S_\infty$ at this time. Therefore

$$J(u_{\lambda_*}) \geqslant d_\infty = d_1$$

and we have that

$$\begin{aligned}
& J(u) - J(u_{\lambda_*}) \\
&= \int \Big( \frac{1}{2}(1-\lambda_*^2)|u_x|^2 + \frac{1}{4}(1-\lambda_*^3)|(|u|^2)_x|^2 - \frac{1}{p}(1-\lambda_*^{(p-2)/2})|u|^p \Big) dx \\
&\geqslant \frac{1}{3}\int \Big( (1-\lambda_*^2)|u_x|^2 + \frac{3}{4}(1-\lambda_*^3)|(|u|^2)_x|^2 - \frac{p-2}{2p}(1-\lambda_*^{(p-2)/2})|u|^p \Big) dx \\
&= \frac{1}{3}(Q(u) - Q(u_{\lambda_*})) \geqslant \frac{1}{3}Q(u).
\end{aligned} \tag{3.4}$$

For the case (iii), we have $\mathcal{N}(u_\mu) < 0$ and $Q(u_\mu) = 0$. Thus $u_\mu \in M_\infty$ and we have that $J(u_\mu) \geqslant d_M \geqslant d_1$. It is now deduced from a similar computation that

$$J(u) - J(u_\mu) \geqslant \frac{1}{3}(Q(u) - Q(u_\mu)) \geqslant \frac{1}{3}Q(u). \tag{3.5}$$

In all cases, we have that

$$Q(u) < 3(J(u) - d_1).$$

Noticing from Proposition 2.2, we have that for $I(t) = \frac{1}{2}\int |x|^2 |u|^2 dx$, there holds

$$I''(t) = 4Q(u).$$

Now using $J(u) = J(u_0)$ and $u_0 \in K$, we obtain that

$$I''(t) < 12(J(u) - d_1). \tag{3.6}$$

Denote $\delta_0 := 12(d_1 - J(u))$. Then $I''(t) < -\delta_0 < 0$. Thus combining this with $I(0) = \frac{1}{2}\int |x|^2 |u_0|^2 dx$, we can find a $T_0 > 0$ such that

$$\lim_{t \to T_0^-} I(t) = 0. \tag{3.7}$$

Observing that

$$\int |u|^2 \leqslant C\Big(\int |x|^2 |u|^2 dx\Big)^{\frac{1}{2}} \Big(\int |u_x|^2 dx\Big)^{\frac{1}{2}}$$

and $C$ is independent of $u$, we obtain from the conservation of mass, i.e. $\int |u|^2 = \int |u_0|^2 \equiv$ const $> 0$ that there is $T > 0$ such that (3.1) holds. □

## 4 Instability of the Standing Waves

In this section, we will prove the instability of standing waves of (1.1) by blowing up. Recalling that the existence of standing waves has been obtained in [8,9], we have that the minimization problem

$$d_1 = \inf_{v \in S_1} J(v) \tag{4.1}$$

was solved by $v \in H^1(\mathbf{R})$ and $v$ is a solution of the following elliptic equation

$$-\omega v + v_{xx} + |v|^{p-2}v + (|v|^2)_{xx} v = 0. \tag{4.2}$$

Thus $u(t,x)=e^{i\omega t}v(x)$ is a standing wave of (1.1) and $G=\{e^{i\omega t}v(x);\ t\geqslant 0\}$ is an orbit of $v$. Before proving the instability, we need further several lemmas.

**Lemma 4.1** Let $v\in H^1(\mathbf{R})\setminus\{0\}$ and $p\geqslant 4$. Then there is unique $\mu>0$ such that $N(\mu v)=0$ and $J(\mu v)>J(\lambda v)$ for any $\lambda>0$ and $\lambda\neq\mu$. In particular, if $v\in S_1$ then $J(v)>J(\lambda v)$ for any $\lambda>0$ and $\lambda\neq 1$.

**Proof** For the proof, see Liu et al [19, Lemma 2.1]. □

**Lemma 4.2** Let $v$ be a minimizer of $d_1$. Then $Q(v)=0$.

**Proof** The idea is based on Pohozaev type identities. Define $\chi_0\in C_0^\infty(\mathbf{R})$ such that $0\leqslant\chi_0\leqslant 1$ and $\chi_0(s)=1$ if $0\leqslant|s|\leqslant 1$, $\chi_0(s)=0$ if $0\leqslant|s|\leqslant 2$. We set $\chi_n(x)=\chi_0\left(\frac{|x|^2}{n^2}\right)$, $n=1,2,\cdots$.

First noting that $v$ is a minimizer of $d_1$, we have that

$$\int(|v_x|^2+\omega|v|^2+|(|v|^2)_x|^2-|v|^p)\mathrm{d}x=0 \tag{4.3}$$

and $v$ is a solution of the following equation:

$$-\omega v+v_{xx}+(|v|^2)_{xx}v+|v|^{p-2}v=0. \tag{4.4}$$

Multiplying (4.4) by $x(\chi_n(x)v)_x$ and integrating over $\mathbf{R}$, we get that

$$\int x(\chi_n(x)v)_x(-\omega v+v_{xx}+(|v|^2)_{xx}v+|v|^{p-2})v=0.$$

Note that $x(\chi_n(x)v)_x=\frac{2}{n^2}|x|^2v\chi_n'\left(\frac{|x|^2}{n^2}\right)+\chi_n(x)xv_x$. We obtain from integrating by parts that

$$\int x(\chi_n(x)v)_x(-\omega v)=-\omega\int\frac{2|x|^2}{n^2}|v|^2\chi_n'\left(\frac{|x|^2}{n^2}\right)+\frac{1}{2}\omega\int|v|^2.$$

By Lebesgue dominated convergence theorem, we get that

$$\int x(\chi_n(x)v)_x(-\omega v)\to\frac{1}{2}\omega\int|v|^2\quad(n\to\infty). \tag{4.5}$$

Since

$$\int x(\chi_n(x)v)_xv_{xx}=\int xvv_{xx}(\chi_n(x))_x+\int x\chi_n(x)v_{xx}v_x,$$

$$\int x\chi_n(x)v_{xx}v_x=-\int v_x(\chi_n(x))_x(xv_x)-\int\chi_n(x)v_x(xv_x)_x$$

and

$$\begin{aligned}-\int\chi_n(x)v_x(xv_x)_x&=-\int\chi_n(x)|v_x|^2-\frac{1}{2}\int\chi_n(x)x(|v_x|^2)_x\\&=-\int\chi_n(x)|v_x|^2+\int x|v_x|^2(\chi_n(x))_x+\frac{1}{2}\int\chi_n(x)|v_x|^2\\&=-\frac{1}{2}\int\chi_n(x)|v_x|^2+\int x|v_x|^2(\chi_n(x))_x.\end{aligned}$$

Hence by Lebesgue dominated convergence theorem, we get that

$$\int x(\chi_n(x)v)_xv_{xx}\to-\frac{1}{2}\int|v_x|^2\quad(n\to\infty). \tag{4.6}$$

Similarly we can obtain that as $n \to \infty$

$$\int |v|^{p-2} v x(\chi_n(x)v)_x \to -\frac{1}{p}\int |v|^p, \tag{4.7}$$

$$\int (|v|^2)_{xx} v x(\chi_n(x)v)_x \to -\frac{1}{4}\int |(|v|^2)_x|^2. \tag{4.8}$$

Therefore

$$\int\left(\frac{1}{2}\omega|v|^2 - \frac{1}{2}|v_x|^2 - \frac{1}{p}|v|^p - \frac{1}{4}|(|v|^2)_x|^2\right) = 0. \tag{4.9}$$

Combining (4.3) and (4.9) we get that

$$Q(v) = \int\left(|v_x|^2 + \frac{3}{4}|(|v|^2)_x|^2 - \frac{p-2}{2p}|v|^p\right)dx = 0.$$

The proof is complete. □

**Lemma 4.3** Let $1<\lambda<2$, $p>2$ and $v, w_1 \in H^1(\mathbf{R})$ with $\|w_1\|<1$. Then there are positive constants $D_j$, $j=1,2,3$, which are independent of $w_1$ such that

$$J(\lambda(v+w_1)) \leqslant J(\lambda v) + D_1\|w_1\|, \tag{4.10}$$

$$\mathcal{N}(\lambda(v+w_1)) \leqslant \mathcal{N}(\lambda v) + D_2\|w_1\|, \tag{4.11}$$

$$Q(\lambda(v+w_1)) \leqslant Q(\lambda v) + D_3\|w_1\|. \tag{4.12}$$

**Proof** Using the embedding $H^1(\mathbf{R}) \hookrightarrow L^\infty(\mathbf{R})$, Hölder and Sobolev inequalities and elementary computations, we obtain that

$$\begin{aligned} J(\lambda(v+w_1)) \leqslant{} & J(\lambda v) + \int\Big\{\lambda^2\Big(|v_x||w_{1x}| + \frac{1}{2}|w_{1x}|^2 + \frac{\omega}{2}|w_1|^2 + \omega|v||w_1|\Big) + \\ & \lambda^4(2|v||w_1||v_x|^2 + |w_1|^2|v_x|^2 + 2|v_x||w_{1x}||v|^2 + \\ & 4|v||w_1||v_x||w_{1x}| + 2|v_x||w_{1x}||w_1|^2 + |v|^2|w_{1x}|^2 + \\ & 2|v||w_1||w_{1x}|^2 + |w_1|^2|w_{1x}|^2) + \frac{1}{p}\lambda^p(|v|^p - |v+w_1|^p)\Big\}dx \\ \leqslant{} & J(\lambda v) + \lambda^2(B_1\|w_1\| + B_2\|w_1\|^2) + \lambda^4(B_3\|w_1\| + B_4\|w_1\|^2 + \\ & B_5\|w_1\|^3 + B_6\|w_1\|^4) + \lambda^p(B_7\|w_1\| + B_8\|w_1\|^{p-1} + B_9\|w_1\|^p) \\ \leqslant{} & J(\lambda v) + D_1\|w_1\| \quad (\text{since } \|w_1\|<1). \end{aligned} \tag{4.13}$$

Similarly, we can find $D_2$ and $D_3$ such that

$$\mathcal{N}(\lambda(v+w_1)) \leqslant \mathcal{N}(\lambda v) + D_2\|w_1\|$$

and

$$Q(\lambda(v+w_1)) \leqslant Q(\lambda v) + D_3\|w_1\|.$$

This completes the proof. □

**Definition 4.4** A standing wave $u(t,x) = e^{i\omega t}v(x)$ is called $H^1(\mathbf{R})$ strongly unstable with respect to (1.1) if for any $\delta>0$, there is $u_0 \in H^\infty(\mathbf{R})$ satisfying

$$\|u_0 - v\| < \delta,$$

but the solution $u(t)$ of (1.1) with initial value $u(0) = u_0$ must blow up in a finite time, i.e. there is $T>0$ such that

$$\lim_{t\to T^-}\|u_x\|_{L^2(\mathbf{R})}=+\infty. \tag{4.14}$$

**Theorem 4.5** Let $p\geqslant 8$, $\omega>0$ and $v$ be the minimizer $d_1$. Then the standing wave $u(t,x)=e^{i\omega t}v(x)$ is $H^1(\mathbf{R})$ strongly unstable.

**Proof** Since $v$ is the minimizer of $d_1$, (4.1), (4.2) and Lemma 4.2 imply that $\mathcal{N}(v)=0$, $Q(v)=0$ and $J(v)=d_1$. Lemma 4.1 implies that

$$J(\lambda v)<J(v)=d_1 \quad \text{for any} \quad \lambda>1.$$

Moreover, using the fact that $\mathcal{N}(v)=0$ and $Q(v)=0$ we have that for any $\lambda>1$

$$\mathcal{N}(\lambda v)=(\lambda^2-\lambda^4)\int(|v_x|^2+\omega|v|^2)+(\lambda^4-\lambda^p)\int|v|^p<0$$

and

$$Q(\lambda v)=(\lambda^2-\lambda^4)\int|v_x|^2dx+\frac{p-2}{2p}(\lambda^4-\lambda^p)\int|v|^pdx<0.$$

Now for any $\delta>0$ and any fixed $1<\lambda<2$ satisfies $\lambda-1<\frac{\delta}{4}\|v\|$. Since $H^\infty(\mathbf{R})$ is dense in $H^1(\mathbf{R})$, we can find $\phi_0$ such that

$$\|\phi_0-v\|<\min\left\{\frac{\delta_1}{D_1+D_2+D_3},\frac{\delta}{2},\|v\|\right\}, \tag{4.15}$$

where $D_1$, $D_2$ and $D_3$ are chosen as in Lemma 4.3 and

$$\delta_1=\min\left\{d_1-J(\lambda v),-\mathcal{N}(\lambda v),-Q(\lambda v),\frac{\delta}{2}\right\}.$$

Write $\phi_0=v+w_1$ with $w_1$ satisfying

$$\|w_1\|<\min\left\{\frac{\delta_1}{D_1+D_2+D_3},\frac{\delta}{2},\|v\|\right\},$$

we obtain from Lemma 4.3 that

$$J(\lambda\phi_0)\leqslant J(\lambda v)+D_1\|w_1\|<d_1, \tag{4.16}$$

$$\mathcal{N}(\lambda\phi_0)\leqslant\mathcal{N}(\lambda v)+D_2\|w_1\|<0, \tag{4.17}$$

$$Q(\lambda\phi_0)\leqslant Q(\lambda v)+D_3\|w_1\|<0. \tag{4.18}$$

In other word, $\lambda\phi_0\in K$. On the other hand, the choice of $\lambda$ implies that

$$\|\lambda\phi_0-v\|\leqslant(\lambda-1)\|\phi_0\|+\|\phi_0-v\|<\delta.$$

Because $v$ is exponentially decay at infinity (see [19, Lemma 5.10]), we get from (4.15) that $\lambda|\cdot|\phi_0(\cdot)\in L^2(\mathbf{R})$. Denote $u_0=\lambda\phi_0$. We obtain from Theorem 3.1 that the solution $u(t)$ of (1.1) with initial value $u(0)=u_0$ must blow up at finite time. In view of Definition 4.4, the standing wave $u(t,x)=e^{i\omega t}v(x)$ is $H^1(\mathbf{R})$ strongly unstable. □

## 5 Concluding Remark

In this section, we give some remarks on the initial datum $u_0$ such that the solutions of (1.1) and (1.2) blow up in a finite time. We want to emphasize that if the initial energy $E(u_0)$ is nonpositive, we can get a blowing up result for Cauchy problem (1.1) and (1.2) in any spatial dimensions, see [18, Theorem 1.1].

Theorem 3.1 indicates that we can get the blowing up results in the case of negative and positive initial energy. The reasons are as follows. Let $v$ be the minimizer of $d_1$. Then firstly from Lemma 4.2, we have that $Q(v)=0$, which reads

$$\frac{1}{p}\int |v|^p = \frac{2}{p-2}\int |v_x|^2 + \frac{3}{2(p-2)}\int |(|v|^2)_x|^2.$$

Hence for any $\lambda > 1$, we have that

$$\begin{aligned} E_2(\lambda v) &= \lambda^2\int |v_x|^2 + \frac{1}{2}\lambda^4\int |(|v|^2)_x|^2 - 2\lambda^p\left(\frac{2}{p-2}\int |v_x|^2 + \frac{3}{2(p-2)}\int |(|v|^2)_x|^2\right) \\ &= \left(\lambda^2 - \frac{4}{p-2}\lambda^p\right)\int |v_x|^2 + \left(\frac{1}{2}\lambda^4 - \frac{3}{p-2}\lambda^p\right)\int |(|v|^2)_x|^2 > 0 \end{aligned}$$

since $p \geqslant 8$.

Secondly from the proof of Theorem 4.5, we have for any $\lambda > 1, J(\lambda v) < d_1, \mathcal{N}(\lambda v) < 0$ and $Q(\lambda v) < 0$. Since $H^\infty(\mathbf{R})$ is dense in $H^1(\mathbf{R})$, we find a $\psi \in H^\infty(\mathbf{R})$ such that $E_2(\psi) > 0$ and $\psi \in K$.

Thirdly, for any $\phi \in H^\infty(\mathbf{R})$, we have from $p \geqslant 8$ that $J(\mu\phi) < d_1, \mathcal{N}(\mu\phi) < 0$ and $Q(\lambda\phi) < 0$ for $\mu$ large enough. Hence $\mu\phi \in K$ for $\mu$ large. Moreover $E(\mu\phi) > 0$ for $\mu$ large.

However, due to the quasilinear term, the blow up result for Cauchy problem (1.1) and (1.2) in higher spatial dimensions ($N \geqslant 2$) and positive initial energy $E_2(u_0)$ is still open, which will be a problem for further study.

## References


[1] Kurihura S. Large amplitude quasi-solitons in superfluid film. J Phys Soc Jpn, 1981, 50: 3262 - 3267.

[2] Hasse R W. A general method for the solution of nonlinear soliton and kink Schrödinger equations. Z Phys B, 1980, 37: 83 - 87.

[3] Porkolab M, Goddman M V. Upper hybrid solitons and oscillating two-stream instabilities. Phys Fluids, 1976, 19: 872 - 881.

[4] Spatschek K H, Tagary S G. Nonlinear propagation of ion-cyclotron modes. Phys Fluids, 1977, 20: 1505 - 1509.

[5] Yu M Y, Shunla P K. On the formation of upper-hybrid solitons. Plasma Phys, 1977, 19: 889 - 893.

[6] Rabinowitz P. On a class of nonlinear Schrödinger equations. Z Angew Math Phys, 1992, 43: 270 - 291.

[7] Strauss W. Existence of solitary waves in higher dimensions. Commun Math Phys, 1977, 55: 149 - 162.

[8] Ambrosetti A, Wang Z Q. Positive solutions to a class of quasilinear elliptic equations on **R**. Discrete Cont Dyn Syst, 2003, 9: 55 - 68.

[9] Poppenberg M, Schmitt K, Wang Z Q. On the existence of soliton solutions to quasilinear Schrödinger equations. Cal Var PDEs, 2002, 14: 329 - 344.

[10] Ginibre J, Velo G. On the class of nonlinear Schrödinger equations Ⅰ, Ⅱ. J Funct Anal, 1979, 32: 1 - 32, 33 - 37.

[11] Lange H, Poppenperg M, Teismann H. Nash-Moser methods for the solutions of quasilinear Schrödinger equations. Commun PDE, 1999, 24: 1399 - 1418.

[12] Poppenberg M. On the local well posedness of quasilinear Schrödinger equations in arbitrary space dimension. J Differ Eq, 2001, 172: 83 - 115.

[13] Berestycki H, Cazenave T. Instabilite des etats stationnaires dans les equations de Schrodinger et de Klein-Gordon non linearires. C R Acad Sci Paris, Seire Ⅰ, 1981, 293: 489 - 492.

[14] Cazenave T, Lions P L. Orbital stability of standing waves for some nonlinear Schrödinger equations. Commun Math Phys, 1982, 85: 549 - 561.

[15] Shatah J, Strauss W. Instability of nonlinear bound states. Commun Math Phys, 1985, 100: 173 - 190.

[16] Zhang J. Cross-constrained variational problem and nonlinear Schrödinger equations // Proceedings of the International Conference on Foundations of Computational Mathematics in Honor of Professor S. Smale's 70th Birthday, Hong Kong, 2001.

[17] Glassey R T. On the blowing-up of solutions to the Cauchy problem for the nonlinear Schrödinger equation. J Math Phys, 1977, 18: 1794 - 1797.

[18] Guo B, Chen J, Su F. The "Blow up" problem for a quasilinear Schrödinger equation. J Math Phys, 2005, 46: 073510. 1 - 10.

[19] Liu J Q, Wang Y, Wang Z Q. Solutions for quasilinear Schrödinger equations via Nehari method. Commun PDE, 2004, 29: 879 - 901.

# Orbital Stability of Solitary Waves for Kundu Equation *

Zhang Weiguo(张卫国) Qin Yinghao(秦英豪) Zhao Yan(赵岩) Guo Boling(郭柏灵)

**Abstract**

In this paper, we consider the Kundu equation which is not a standard Hamiltonian system. The abstract orbital stability theory proposed by Grillakis et al[17,18] cannot be applied directly to study orbital stability of solitary waves for this equation. Motivated by the idea of Guo and Wu[10], we construct three invariants of motion and use detailed spectral analysis to obtain orbital stability of solitary waves for Kundu equation. Since Kundu equation is more complex than the derivative Schrödinger equation, we utilize some techniques to overcome some difficulties in this paper. It should be pointed out that the results obtained in this paper are more general than those obtained by Guo and Wu[10]. We present a sufficient condition under which solitary waves are orbitally stable for $2c_3+s_2v<0$, while Guo and Wu[10] only considered the case $2c_3+s_2v>0$. We obtain the results on orbital stability of solitary waves for the derivative Schrödinger equation given by Colin and Ohta[11] as a corollary in this paper. Furthermore, we obtain orbital stability of solitary waves for Chen-Lee-Lin equation and Gerdjikov-Ivanov equation, respectively.

**Keywords** Kundu cquation; solitary wave; orbital stability; spectral analysis

## 1 Introduction

In this paper, we consider orbital stability of solitary waves for Kundu equation[1] with fifth order nonlinear term

$$\mathrm{i}u_t+u_{xx}+c_3|u|^2u+c_5|u|^4u-\mathrm{i}s_2(|u|^2u)_x-\mathrm{i}r(|u|^2)_xu=0,\quad x\in\mathbf{R},\tag{1a}$$

where $c_3$, $c_5$, $s_2$ and $r$ are real constants. Eq. (1a) was derived by Kundu[1] in the study of integrability and it is an important special case of the generalized complex Ginzburg-Landau equation[2]. Meanwhile, Eq. (1a) and its special cases arise in various physical and mechanical applications, such as plasma physics, nonlinear fluid mechanics, nonlinear optics and quantum physics (see [2-7]). For convenience, we denote the Kundu equation as follow:

$$u_t=\mathrm{i}u_{xx}+\mathrm{i}(c_3|u|^2+c_5|u|^4)u+\alpha|u|^2u_x+\beta u^2\bar{u}_x,\quad x\in\mathbf{R},\tag{1b}$$

where $\alpha=(2s_2+r)$, $\beta=(s_2+r)$ .

Obviously, if $r=0$, then Eq. (1) reduces to the derivative Schrödinger equation

$$u_t=\mathrm{i}u_{xx}+\mathrm{i}(c_3|u|^2+c_5|u|^4)u+s_2(|u|^2u)_x;\tag{2}$$

* J Differential Equations, 2009, 247: 1591-1615.

if $c_3=0, c_5=0, s_2=-\delta, r=-s_2$, then Eq. (1) reduces to Chen-Lee-Lin equation[8]

$$\mathrm{i}u_t+u_{xx}+\mathrm{i}\delta|u|^2u_x=0; \tag{3}$$

if $c_5=2\delta^2, s_2=2\delta, r=-2s_2$, then Eq. (1) reduces to Gerdjikov-Ivanov equation[9]

$$\mathrm{i}u_t+u_{xx}+c_3|u|^2u+2\delta^2|u|^4u+2\mathrm{i}\delta u^2\bar{u}_x=0. \tag{4}$$

In [2], by using primary integral method, Saarloos and Hohenberg obtained the exact solitary waves of the form

$$u(x,t)=\mathrm{e}^{-\mathrm{i}\omega t}\mathrm{e}^{\mathrm{i}\psi(x-vt)}a(x-vt)=\mathrm{e}^{-\mathrm{i}\omega t}\mathrm{e}^{\mathrm{i}\psi(\xi)}a(\xi),\quad \xi=x-vt, \tag{5}$$

for Eq. (2). Furthermore, Guo and Wu proved that this kind of solitary waves for Eq. (2) are orbitally stable in [10]. For more results on Eq. (2), we refer the readers to [12 – 15].

The existence of solutions for initial value problem of Eq. (1) has been proved in [16]. However, to our best knowledge, orbital stability of solitary waves in the form (5) for Eq. (1) have not been considered. We will focus on studying this problem in this paper. For clarity, a solution in the form (5) is called a solitary wave throughout this paper. As mentioned above, Eqs. (2) ~ (4) are special cases of Eq. (1). In other words, Eq. (1) is more general than Eq. (2), moreover, Eq. (2) does not include Eqs. (3) ~ (4).

It is worth pointing out that it is difficult for us to transform Eq. (1) into a standard Hamiltonian system. Therefore, the results about orbital stability obtained in [17] and [18] cannot be applied directly to prove orbital stability of solitary waves for Eq. (1). Motivated by the approach in [10], we study this problem by constructing three appropriate invariants of motion and using detailed spectral analysis. Due to Eq. (1) being more complex than Eq. (2), we overcome some difficulties by using some techniques. The results obtained in this paper are more general than those in [10]: (a) Using Theorem 3.4 in this paper, we obtain not only the results shown in [10], but also the results which have not been obtained in [10]. The conclusions in [10] were given in the case of $2c_3+s_2v>0$. When $s_2<0, c_3>0$, [10] obtained the orbital stability of solitary waves traveling left. In this paper, we give not only the whole results obtained in [10], but also a sufficient condition under which solitary waves are orbitally stable as $2c_3+s_2v<0$. Corollary 3.5 obtained in this paper can be used to discriminate orbital stability for both solitary waves traveling left and solitary waves traveling right of Eq. (2). (b) Corollary 3.6 obtained in this paper enables us to derive the same results on orbital stability of solitary waves for the following derivative Schrödinger equation:

$$\mathrm{i}u_t+u_{xx}+\mathrm{i}(|u|^2u)_x=0 \tag{6}$$

as those presented by Colin and Ohta[11]. By using variational method, they proved that solitary waves of Eq. (6) are orbitally stable if the wave speed $v$ satisfies $v^2<-4\omega$. Since Eq. (2) becomes Eq. (6) when $s_2=-1$, $c_3=c_5=0$, from the results obtained in [10], it is easy to see that solitary waves of Eq. (6) are orbitally stable as $v<0$ and $v^2<-4\omega$. However, it is unable to derive the conclusion that solitary waves of Eq. (6) are orbitally stable as $v>0$ and $v^2<-4\omega$. By Theorem 3.4 obtained in this paper, solitary waves of Eq. (6) are orbitally stable as $v^2<-4\omega$ and $v\neq 0$. (c) By means of the results obtained in this paper, we have

Corollary 3.7 on orbital stability of solitary waves for Chen-Lee-Lin equation, and Corollary 3.8 on orbital stability of solitary waves for Gerdjikov-Ivanov equation. Eqs. (3) ~ (4) are special cases of Eq. (1), but Eq. (2) does not include them.

## 2 Exact Solitary Waves of Kundu Equation

Due to the complexity of transforming Eq. (1) to primary integral, Saarloos and Hohenberg[2] obtained the exact solitary waves in the form (5) for Eq. (2), but they did not give any exact solution to Eq. (1). In this section, we give exact solitary waves of Eq. (1) by using proper transformations and undetermined coefficient method.

Assume that Eq. (1) has solutions of the form (5). Let

$$\hat{a}(\xi) = \hat{a}(x - vt) = e^{i\psi(x-vt)}a(x - vt), \tag{7}$$

and substitute $u(x,t) = e^{-i\omega t}\hat{a}(x - vt)$ into Eq. (1), then $\hat{a}(\xi)$ satisfies

$$-\hat{a}_{xx} - g(a^2)\hat{a} + i\beta(a^2\hat{a})_x + i(\alpha - 2\beta)a^2\hat{a}_x - \omega\hat{a} + iv\hat{a}_x = 0, \tag{8}$$

where $g(a^2) = c_3a^2 + c_5a^4$. Substituting (7) into Eq. (8) and making the real part and imaginary part of Eq. (8) equal to zero, we have

$$\psi''a + 2\psi'a' - \alpha a^2a' - \beta a^2a' - va' = 0, \tag{9}$$

$$a'' + [g(a^2) + \alpha\psi'a^2 - \beta\psi'a^2 - (\psi')^2 + \omega + v\psi']a = 0. \tag{10}$$

Let

$$\psi'(\xi) = E + Da^2(\xi). \tag{11}$$

Substituting (11) into Eq. (9) and equating the coefficients of these terms $a, a', a''$ to zero, we get $E = \frac{v}{2}, D = \frac{(\alpha+\beta)}{4}$. Therefore, when

$$\psi'(\xi) = \frac{v}{2} + \frac{(\alpha+\beta)}{4}a^2(\xi), \tag{12}$$

Eq. (9) is identical to zero. Combining (12) and (10), we know that $a(\xi)$ satisfies the following equation:

$$a'' - d_1a - 2d_2a^3 - 3d_4a^5 = 0, \tag{13}$$

where $d_1 = -\omega - \frac{v^2}{4}, d_2 = -\frac{c_3}{2} - \frac{(\alpha-\beta)v}{4}, d_4 = -\frac{1}{3}\left(c_5 + \frac{(\alpha+\beta)(3\alpha-5\beta)}{16}\right)$.

In order to solve Eq. (13), we make the following transformation

$$a(\xi) = \sqrt{\varphi(\xi)}. \tag{14}$$

Then, $\varphi(\xi)$ satisfies

$$2\varphi\varphi'' - \varphi'^2 - 4d_1\varphi^2 - 8d_2\varphi^3 - 12d_4\varphi^4 = 0. \tag{15}$$

Now, we assume that Eq. (15) has solutions in the following form:

$$\varphi(\xi) = \frac{Ae^{C(\xi+\xi_0)}}{(1 + e^{C(\xi+\xi_0)})^2 + Be^{C(\xi+\xi_0)}} = \frac{A\operatorname{sech}^2\frac{C}{2}(\xi+\xi_0)}{4 + B\operatorname{sech}^2\frac{C}{2}(\xi+\xi_0)}, \tag{16}$$

where $A$, $B$, $C$ are undetermined constants, and $\xi_0$ is an arbitrary constant.

By substituting (16) into Eq. (15), we have

$$\begin{cases} C^2 - 4d_1 = 0 \\ -4d_2A - 4d_1(2+B) - C^2(2+B) = 0 \\ -5C - 2d_1(2+(2+B)^2) - 4d_2A(2+B) - 6d_4A^2 = 0 \end{cases} \tag{17}$$

Further, we obtain

$$A = \pm \frac{4d_1}{\sqrt{d_2^2 - 4d_1d_4}}, \quad B = -2 \pm \frac{-2d_2}{\sqrt{d_2^2 - 4d_1d_4}}, \quad C = \pm\sqrt{4d_1}, \quad d_1 > 0. \tag{18}$$

From (18) and (16), we get two solutions for Eq. (15) as follows:

$$\varphi_1(\xi) = \frac{\dfrac{2d_1}{\sqrt{d_2^2 - 4d_1d_4}}\operatorname{sech}^2\sqrt{d_1}(\xi+\xi_0)}{2 - \left(1 + \dfrac{d_2}{\sqrt{d_2^2 - 4d_1d_4}}\right)\operatorname{sech}^2\sqrt{d_1}(\xi+\xi_0)}, \tag{19}$$

$$\varphi_2(\xi) = \frac{\dfrac{-2d_1}{\sqrt{d_2^2 - 4d_1d_4}}\operatorname{sech}^2\sqrt{d_1}(\xi+\xi_0)}{2 + \left(-1 + \dfrac{d_2}{\sqrt{d_2^2 - 4d_1d_4}}\right)\operatorname{sech}^2\sqrt{d_1}(\xi+\xi_0)}. \tag{20}$$

Under the condition $d_1>0$, it is easy to verify that

(1) if $d_4<0$ or if $d_4\geqslant 0, d_2<0$ and $d_2^2-4d_1d_4>0$, then for any $\xi\in\mathbf{R}$, $\varphi_1(\xi)>0$;

(2) if $d_4\geqslant 0$, $d_2>0$ and $d_2^2-4d_1d_4>0$, then $\varphi_1(\xi)$ is an unbounded function;

(3) if $d_4<0$ or if $d_4\geqslant 0$, $d_2>0$ and $d_2^2-4d_1d_4>0$, then for any $\xi\in\mathbf{R}, \varphi_2(\xi)<0$;

(4) if $d_4\geqslant 0, d_2<0$ and $d_2^2-4d_1d_4>0$, then $\varphi_2(\xi)$ is an unbounded function.

Note that if we let

$$d_3 = -\frac{d_2}{2d_1}, \quad d_5^2 = \frac{d_2^2 - 4d_1d_4}{4d_1^2}, \quad d_6^2 = 4d_1, \tag{21}$$

then (19) can be rewritten as

$$\varphi_1(\xi) = \frac{1}{d_3 + d_5\cosh d_6(\xi+\xi_0)}, \quad \xi\in\mathbf{R}. \tag{22}$$

Substituting (22) into (14), and noticing that if $a(\xi)$ is a solution of Eq. (13), then $-a(\xi)$ is also a solution of Eq. (13), we have the following lemma.

**Lemma 2.1** Suppose that $d_1>0$. If $d_4<0$ or if $d_4\geqslant 0$, $d_2<0$ and $d_2^2-4d_1d_4>0$, then Eq. (13) has bounded analytic solutions of the form

$$a(\xi) = \pm\left[\frac{1}{d_3 + d_5\cosh d_6(\xi+\xi_0)}\right]^{\frac{1}{2}}. \tag{23}$$

Furthermore, we get the following theorem.

**Theorem 2.2** Suppose that $d_1>0$. If $d_4<0$ or if $d_4\geqslant 0$, $d_2<0$ and $d_2^2-4d_1d_4>0$, then Eq. (1) has solitary waves $u(x,t)=e^{-i\omega t}e^{i\psi(x-vt)}a(x-vt)$, where $a(\xi)$ and $\psi(\xi)$ are given by (23) and (12), respectively.

Theorem 2.2 implies that when $d_1>0$, $d_4<0$, Eq. (1) has solitary waves of the form

(5) no matter whether $d_2<0$ or $d_2\geqslant 0$. It is also true for Eq. (2) in this case. Theorem 2.2 obtained in [10] does not include the result that Eq. (2) has solitary waves of the form (5) in the case of $d_2\geqslant 0$.

## 3 Orbital Stability of Solitary Waves

In this section, we only consider the positive solutions in (23) (the negative solutions in (23) can be discussed similarly). Therefore, we assume that $a(\xi)>0$ in the remainder of this paper.

Now, we consider the following initial value problem of Kundu equation:

$$u_t = \mathrm{i}u_{xx} + \mathrm{i}(c_3|u|^2 + c_5|u|^4)u + \alpha|u|^2u_x + \beta u^2\bar{u}_x, \tag{24}$$

$$u(0,x) = u_0(x), \quad x \in \mathbf{R}. \tag{25}$$

The complex space $X=H^1(\mathbf{R})$ with real inner product

$$(u,v) = \mathrm{Re}\int_{\mathbf{R}}(u_x\bar{v}_x + u\bar{v})\mathrm{d}x, \quad \forall\, u,v \in X, \tag{26}$$

is chosen as the function space. The dual space of $X$ is denoted by $X^* = H^{-1}(\mathbf{R})$ and the natural isomorphism $I: X\to X^*$ is defined by

$$\langle Iu, v\rangle = (u,v), \tag{27}$$

where $\langle\cdot,\cdot\rangle$ denotes the pairing between $X$ and $X^*$,

$$\langle f, u\rangle = \mathrm{Re}\int_{\mathbf{R}} f\bar{u}\,\mathrm{d}x. \tag{28}$$

From (26)~(28), we have

$$I = -\frac{\partial^2}{\partial x^2} + 1. \tag{29}$$

Let $T_1, T_2$ be one-parameter groups of unitary operator on $X$ defined by

$$T_1(s_1)\phi(\cdot) = \mathrm{e}^{-s_1\mathrm{i}}\phi(\cdot), \quad \forall\,\phi(\cdot)\in X,\ s_1\in\mathbf{R}, \tag{30}$$

$$T_2(s_2)\phi(\cdot) = \phi(\cdot - s_2), \quad \forall\,\phi(\cdot)\in X,\ s_2\in\mathbf{R}. \tag{31}$$

Obviously, $T_1'(0) = -\mathrm{i}$, $T_2'(0) = -\frac{\partial}{\partial x}$.

According to [16] (or [19-21]), we know that for any $u_0\in H^1(\mathbf{R})$, Eq. (24) has a unique solution $u\in C([0,T_{\max}], H^1(\mathbf{R}))$ satisfying $u(0)=u_0$.

From the above definitions, we write the solitary waves of Eq. (24) in Theorem 2.2 as $T_1(\omega t)T_2(\upsilon t)\hat{a}_{\omega,\upsilon}(x)$, where $\hat{a}_{\omega,\upsilon}$ is defined by (7) and (23). In what follows, we mainly study the orbital stability of solitary waves $T_1(\omega t)T_2(\upsilon t)\hat{a}_{\omega,\upsilon}(x)$. Note that Eq. (24) has the symmetries of phase and translation, we define the orbital stability as follows (see also [22]).

**Definition 3.1** The solitary wave $T_1(\omega t)T_2(\upsilon t)\hat{a}_{\omega,\upsilon}(x)$ is called orbitally stable if for any $\varepsilon>0$, there exists $\delta>0$ with the following property. If $\|u_0-\hat{a}_{w,\upsilon}\|_X<\delta$ and $u(t)$ is a solution of Eq. (24) in the interval $[0, t_0)$ with $u(0)=u_0$, then $u(t)$ can be continued to a solution in $0\leqslant t<\infty$ and

$$\sup_{0<t<\infty}\inf_{s_1\in\mathbf{R}}\inf_{s_2\in\mathbf{R}}\|u(t)-T_1(s_1)T_2(s_2)\hat{a}_{w,v}\|_X<\varepsilon.$$

Otherwise, $T_1(\omega t)T_2(vt)\hat{a}_{\omega,v}(x)$ is called orbitally unstable.

It is difficult for us to transform Eq. (24) into a standard form of Hamiltonian system when $\alpha$ and $\beta$ are equal to zero asynchronously. (In [10], Eq. (2) cannot be changed into a standard form of Hamiltonian system as well.) Therefore, the abstract theory on orbital stability of solitary waves of nonlinear Hamiltonian system in [17] and [18] can not be applied directly to study Eq. (24). According to the "stability theorem" in the introduction of [18], the assumptions 1~3 which make that theorem hold, and the deductions in Sections 3~4 of [18], we know that when a local solution of initial value problem of an equation exists, we only need to find $E(u)$, $Q_\sigma(u)$ such that assumptions 2~3 hold. The results on orbital stability can be obtained even the equation cannot been changed into a standard form of Hamiltonian system. According to the above analysis, we construct three new invariants of motion as follows:

$$E(u)=\frac{1}{2}\int_{\mathbf{R}}\left\{|u_x|^2+\frac{\beta(\alpha+\beta)}{6}|u|^6-G(|u|^2)+\frac{\alpha+\beta}{2}\mathrm{Im}(|u|^2u\bar{u}_x)\right\}\mathrm{d}x,\tag{32}$$

where $G(u)=\int_0^u g(s)\mathrm{d}s$, $g(s)=c_3s^2+c_5s^4$,

$$Q_1(u)=\frac{1}{2}\int_{\mathbf{R}}|u|^2\mathrm{d}x,\tag{33}$$

$$Q_2(u)=\int_{\mathbf{R}}\left[\frac{1}{2}\mathrm{Im}(\bar{u}u_x)-\frac{\beta}{4}|u|^4\right]\mathrm{d}x.\tag{34}$$

It is easy to check that $E(u)$, $Q_1(u)$ and $Q_2(u)$ are $C^2$ functionals defined on the complex space $X$. Let their derivatives be $\langle E'(u),v\rangle$, $\langle Q_1'(u),v\rangle$ and $\langle Q_2'(u),v\rangle$, respectively, where $E',Q_1',Q_2':X\to X^*$, and their second derivatives be $\langle E''(u)\omega,v)$, $\langle Q_1''(u)\omega,v\rangle$ and $\langle Q_2''(u)\omega,v\rangle$, respectively.

By computation, we have

$$E'(u)=-u_{xx}+\frac{\beta(\alpha+\beta)}{2}|u|^4u-g(|u|^2)u+\mathrm{i}(\alpha+\beta)|u|^2u_x,$$

$$Q_1'(u)=u,$$

$$Q_2'(u)=-\mathrm{i}u_x-\beta|u|^2u.$$

We can prove that $E(u)$, $Q_1(u)$ and $Q_2(u)$ are invariant under $T_1$ and $T_2$, which means for any $s_1,s_2\in\mathbf{R}$, we have

$$\begin{cases}E(T_1(s_1)T_2(s_2)u)=E(u),\\Q_1(T_1(s_1)T_2(s_2)u)=Q_1(u),\\Q_2(T_1(s_1)T_2(s_2)u)=Q_2(u),\end{cases}\tag{35}$$

and for any $t\in\mathbf{R}$, $u(t)$ is a flow of (24),

$$E(u(t))=E(u(0)),\quad Q_1(u(t))=Q_1(u(0)),\quad Q_2(u(t))=Q_2(u(0)).\tag{36}$$

(35) can be obtained by direct substitution. We will verify that (36) holds next. First, we

prove

$$\mathrm{Re}\int_{\mathbf{R}} |u|^2 u\bar{u}_x \mathrm{d}x = \mathrm{Re}\int_{\mathbf{R}} |u|^4 u\bar{u}_x \mathrm{d}x = \mathrm{Re}\int_{\mathbf{R}} |u|^6 u\bar{u}_x \mathrm{d}x = 0.$$

Let $u = u_1 + \mathrm{i}u_2$, where $u_1, u_2$ are real functions and $u_1, u_2 \in H^1(\mathbf{R})$, then

$$\begin{aligned}\mathrm{Re}\int_{\mathbf{R}} |u|^2 u\bar{u}_x \mathrm{d}x &= \int_{\mathbf{R}} (u_1^2 + u_2^2)(u_1 u_{1x} + u_2 u_{2x})\mathrm{d}x \\ &= \int_{\mathbf{R}} (u_1^2 u_2 u_{2x} + u_1 u_2^2 u_{1x})\mathrm{d}x \\ &= \int_{\mathbf{R}} u_1^2 u_2 u_{2x}\mathrm{d}x - \int_{\mathbf{R}} u_1^2 u_2 u_{2x}\mathrm{d}x = 0,\end{aligned}$$

and

$$\begin{aligned}\mathrm{Re}\int_{\mathbf{R}} |u|^4 u\bar{u}_x \mathrm{d}x &= \int_{\mathbf{R}} (u_1^4 + 2u_1^2 u_2^2 + u_2^4)(u_1 u_{1x} + u_2 u_{2x})\mathrm{d}x \\ &= \int_{\mathbf{R}} (u_1^4 u_2 u_{2x} + 2u_1^3 u_2^2 u_{1x} + 2u_1^2 u_2^3 u_{2x} + u_1 u_2^4 u_{1x})\mathrm{d}x.\end{aligned}$$

Due to

$$\begin{aligned}\int_{\mathbf{R}} (2u_1^3 u_2^2 u_{1x} + 2u_1^2 u_2^3 u_{2x})\mathrm{d}x &= \frac{1}{2}\int_{\mathbf{R}} u_2^2 \mathrm{d}u_1^4 + \frac{1}{2}\int_{\mathbf{R}} u_1^2 \mathrm{d}u_2^4 \\ &= -\int_{\mathbf{R}} (u_1^4 u_2 u_{2x} + u_1 u_2^4 u_{1x})\mathrm{d}x,\end{aligned}$$

we have $\mathrm{Re}\int_{\mathbf{R}} |u|^4 u\bar{u}_x \mathrm{d}x = 0$ and

$$\begin{aligned}\mathrm{Re}\int_{\mathbf{R}} |u|^6 u\bar{u}_x \mathrm{d}x &= \int_{\mathbf{R}} (u_1^6 + 3u_1^4 u_2^2 + 3u_1^2 u_2^4 + u_2^6)(u_1 u_{1x} + u_2 u_{2x})\mathrm{d}x \\ &= \int_{\mathbf{R}} (u_1^6 u_2 u_{2x} + 3u_1^5 u_2^2 u_{1x} + 3u_1^2 u_2^5 u_{2x} + u_1 u_2^6 u_{1x})\mathrm{d}x + \\ &\quad \int_{\mathbf{R}} (3u_1^4 u_2^3 u_{2x} + 3u_1^3 u_2^4 u_{1x})\mathrm{d}x.\end{aligned}$$

Let

$$I = \int_{\mathbf{R}} (u_1^6 u_2 u_{2x} + 3u_1^5 u_2^2 u_{1x} + 3u_1^2 u_2^5 u_{2x} + u_1 u_2^6 u_{1x})\mathrm{d}x,$$

$$II = \int_{\mathbf{R}} (3u_1^4 u_2^3 u_{2x} + 3u_1^3 u_2^4 u_{1x})\mathrm{d}x,$$

then

$$\begin{aligned}I &= \int_{\mathbf{R}} (u_1^6 u_2 u_{2x} + 3u_1^5 u_2^2 u_{1x} + 3u_1^2 u_2^5 u_{2x} + u_1 u_2^6 u_{1x})\mathrm{d}x \\ &= \int_{\mathbf{R}} u_1^6 u_2 \mathrm{d}u_2 + \int_{\mathbf{R}} (3u_1^5 u_2^2 u_{1x} + 3u_1^2 u_2^5 u_{2x})\mathrm{d}x + \int_{\mathbf{R}} u_1 u_2^6 \mathrm{d}u_1 \\ &= -\int_{\mathbf{R}} (u_1^6 u_2 u_{2x} + 3u_1^5 u_2^2 u_{1x} + 3u_1^2 u_2^5 u_{2x} + u_1 u_2^6 u_{1x})\mathrm{d}x \\ &= -I,\end{aligned}$$

and

$$II = \int_{\mathbf{R}} (3u_1^4 u_2^3 u_{2x} + 3u_1^3 u_2^4 u_{1x})\mathrm{d}x$$

$$= \int_{\mathbf{R}} \frac{3}{4} u_1^4 \mathrm{d} u_2^4 + \int_{\mathbf{R}} 3 u_1^3 u_2^4 u_{1x} \mathrm{d} x$$
$$= - \int_{\mathbf{R}} 3 u_1^2 u_2^4 u_{1x} \mathrm{d} x + \int_{\mathbf{R}} 3 u_1^3 u_2^4 u_{1x} \mathrm{d} x = 0.$$

Further, we obtain $I = 0$ and $\operatorname{Re}\int_{\mathbf{R}} |u|^6 u \bar{u}_x \mathrm{d} x = I + II = 0$.

Now, we prove that $E(u)$ given by (32) is an invariant of motion. By computations, we have

$$\frac{\mathrm{d}E(u)}{\mathrm{d}t} = \langle E'(u), u_t \rangle$$
$$= \left\langle \left( - u_{xx} + \frac{\beta(\alpha+\beta)}{2} |u|^4 u - g(|u|^2) u + \mathrm{i}(\alpha+\beta) |u|^2 u_x \right), \right.$$
$$\left. (\mathrm{i} u_{xx} + \mathrm{i} g(|u|^2) u + \alpha |u|^2 u_x + \beta u^2 \bar{u}_x) \right\rangle.$$

Moreover,

$$\langle -u_{xx}, \mathrm{i} u_{xx} \rangle = \operatorname{Re}\int_{\mathbf{R}} (-u_{xx} \bar{\mathrm{i}} \cdot \bar{u}_{xx}) \mathrm{d} x = \operatorname{Re}\int_{\mathbf{R}} \mathrm{i}(u_{xx} \bar{u}_{xx}) \mathrm{d} x = 0,$$
$$\langle -u_{xx}, \mathrm{i} g(|u|^2) u \rangle + \langle -g(|u|^2) u, \mathrm{i} u_{xx} \rangle$$
$$= \operatorname{Re}\int_{\mathbf{R}} (-u_{xx} \bar{\mathrm{i}} \cdot g(|u|^2) \bar{u}) \mathrm{d} x + \operatorname{Re}\int_{\mathbf{R}} (-g(|u|^2) u \bar{\mathrm{i}} \cdot \bar{u}_{xx}) \mathrm{d} x$$
$$= \operatorname{Re}\int_{\mathbf{R}} \mathrm{i}(u_{xx} g(|u|^2) \bar{u}) \mathrm{d} x - \operatorname{Re}\int_{\mathbf{R}} \mathrm{i}(g(|u|^2) \bar{u} \cdot u_{xx}) \mathrm{d} x$$
$$= 0,$$
$$\langle -u_{xx}, \alpha |u|^2 u_x \rangle + \langle -u_{xx}, \beta u^2 u_x \rangle + \langle \mathrm{i}(\alpha+\beta) |u|^2 u_x, \mathrm{i} u_{xx} \rangle$$
$$= -\operatorname{Re}\int_{\mathbf{R}} \alpha |u|^2 u_{xx} \bar{u}_x \mathrm{d} x - \operatorname{Re}\int_{\mathbf{R}} \beta u_{xx} \bar{u}^2 u_x \mathrm{d} x + \operatorname{Re}\int_{\mathbf{R}} (\alpha+\beta) |u|^2 u_x \bar{u}_{xx} \mathrm{d} x$$
$$= -\operatorname{Re}\int_{\mathbf{R}} \alpha |u|^2 u_x \bar{u}_{xx} \mathrm{d} x - \operatorname{Re}\int_{\mathbf{R}} \beta u^2 \bar{u}_x \bar{u}_{xx} \mathrm{d} x + \operatorname{Re}\int_{\mathbf{R}} (\alpha+\beta) |u|^2 u_x \bar{u}_{xx} \mathrm{d} x$$
$$= -\operatorname{Re}\int_{\mathbf{R}} \beta u \bar{u}_{xx} (u \bar{u}_x - \bar{u} u_x) \mathrm{d} x.$$

Let $u = u_1 + \mathrm{i} u_2$, and $III = \operatorname{Re}\int_{\mathbf{R}} \beta u \bar{u}_{xx} (u \bar{u}_x - \bar{u} u_x) \mathrm{d} x$, then

$$III = \operatorname{Re}\int_{\mathbf{R}} \beta (u_1 + \mathrm{i} u_2)(u_{1xx} - \mathrm{i} u_{2xx})((u_1 + \mathrm{i} u_2)(u_{1x} - \mathrm{i} u_{2x}) - (u_1 - \mathrm{i} u_2)(u_{1x} + \mathrm{i} u_{2x})) \mathrm{d} x$$
$$= -2\beta \int_{\mathbf{R}} (u_2^2 u_{1x} u_{1xx} - u_1 u_2 u_{1xx} u_{2x} - u_1 u_2 u_{1x} u_{2xx} + u_1^2 u_{2x} u_{2xx}) \mathrm{d} x$$
$$= -2\beta \left( \int_{\mathbf{R}} u_2^2 u_{1x} \mathrm{d} u_{1x} - \int_{\mathbf{R}} u_1 u_2 u_{2x} \mathrm{d} u_{1x} - \int_{\mathbf{R}} u_1 u_2 u_{1x} \mathrm{d} u_{2x} + \int_{\mathbf{R}} u_1^2 u_{2x} \mathrm{d} u_{2x} \right)$$
$$= -2\beta \left( -\int_{\mathbf{R}} u_2^2 u_{1x} u_{1xx} \mathrm{d} x + \int_{\mathbf{R}} u_1 u_2 u_{1x} u_{2xx} \mathrm{d} x + \int_{\mathbf{R}} u_1 u_2 u_{1xx} u_{2x} \mathrm{d} x - \int_{\mathbf{R}} u_1^2 u_{2x} u_{2xx} \mathrm{d} x \right)$$
$$= -III.$$

Therefore, $III = 0$. Furthermore, we obtain

$$\langle -u_{xx},\alpha|u|^2u_x\rangle+\langle -u_{xx},\beta u^2\bar{u}_x\rangle+\langle \mathrm{i}(\alpha+\beta)|u|^2u_x,\mathrm{i}u_{xx}\rangle=-\mathit{III}=0,$$

$$\left\langle\frac{\beta(\alpha+\beta)}{2}|u|^4u,\mathrm{i}u_{xx}\right\rangle+\langle \mathrm{i}(\alpha+\beta)|u|^2u_x,\beta u^2\bar{u}_x\rangle$$

$$=-\frac{\beta(\alpha+\beta)}{2}\mathrm{Re}\int_{\mathbf{R}}\mathrm{i}(u^3\bar{u}^2)\mathrm{d}\bar{u}_x+\beta(\alpha+\beta)\mathrm{Re}\int_{\mathbf{R}}\mathrm{i}|u|^2u_x^2\bar{u}^2\mathrm{d}x$$

$$=\frac{3\beta(\alpha+\beta)}{2}\mathrm{Re}\int_{\mathbf{R}}\mathrm{i}|u|^4|u_x|^2\mathrm{d}x+\beta(\alpha+\beta)\mathrm{Re}\int_{\mathbf{R}}\mathrm{i}|u|^2u^2(\bar{u}_x)^2\mathrm{d}x+$$

$$\beta(\alpha+\beta)\mathrm{Re}\int_{\mathbf{R}}\mathrm{i}|u|^2u_x^2\bar{u}^2\mathrm{d}x$$

$$=-\beta(\alpha+\beta)\mathrm{Re}\int_{\mathbf{R}}\mathrm{i}|u|^2u_x^2\bar{u}^2\mathrm{d}x+\beta(\alpha+\beta)\mathrm{Re}\int_{\mathbf{R}}\mathrm{i}|u|^2u_x^2\bar{u}^2\mathrm{d}x=0,$$

$$\left\langle\frac{\beta(\alpha+\beta)}{2}|u|^4u,\mathrm{i}g(|u|^2)u\right\rangle=-\frac{\beta(\alpha+\beta)}{2}\mathrm{Re}\int_{\mathbf{R}}\mathrm{i}g(|u|^2)|u|^6\mathrm{d}x=0,$$

$$\left\langle\frac{\beta(\alpha+\beta)}{2}|u|^4u,\alpha(|u|^2)u_x\right\rangle=\frac{\alpha\beta(\alpha+\beta)}{2}\mathrm{Re}\int_{\mathbf{R}}|u|^6u\bar{u}_x\mathrm{d}x=0,$$

$$\left\langle\frac{\beta(\alpha+\beta)}{2}|u|^4u,\beta u^2\bar{u}_x\right\rangle=\frac{\beta^2(\alpha+\beta)}{2}\mathrm{Re}\int_{\mathbf{R}}|u|^6u\bar{u}_x\mathrm{d}x=0,$$

$$\langle -g(|u|^2)u,\mathrm{i}g(|u|^2)u\rangle=\mathrm{Re}\int_{\mathbf{R}}\mathrm{i}g^2(|u|^2)|u|^2\mathrm{d}x=0,$$

$$\langle -g(|u|^2)u,\alpha|u|^2u_x\rangle=-\alpha\mathrm{Re}\int_{\mathbf{R}}g(|u|^2)|u|^2u\bar{u}_x\mathrm{d}x$$

$$=-c_3\alpha\mathrm{Re}\int_{\mathbf{R}}|u|^4u\bar{u}_x\mathrm{d}x-c_5\alpha\mathrm{Re}\int_{\mathbf{R}}|u|^6u\bar{u}_x\mathrm{d}x$$

$$=0,$$

$$\langle -g(|u|^2)u,\beta u^2\bar{u}_x\rangle=-\beta\mathrm{Re}\int_{\mathbf{R}}g(|u|^2)|u|^2u\bar{u}_x\mathrm{d}x=0,$$

$$\langle \mathrm{i}(\alpha+\beta)|u|^2u_x,\mathrm{i}g(|u|^2)u\rangle=(\alpha+\beta)\mathrm{Re}\int_{\mathbf{R}}g(|u|^2)|u|^2u\bar{u}_x\mathrm{d}x=0,$$

$$\langle \mathrm{i}(\alpha+\beta)|u|^2u_x,\alpha|u|^2u_x\rangle=\alpha(\alpha+\beta)\mathrm{Re}\int_{\mathbf{R}}\mathrm{i}|u|^4|u_x|^2\mathrm{d}x=0.$$

It follows from the above computations that $\frac{\mathrm{d}E(u)}{\mathrm{d}t}=\langle E'(u),u_t\rangle=0$, i.e., for any $t\in\mathbf{R}$, $E(u(t))=E(u(0))$. Similarly, we can verify that $Q_1(u(t))=Q_1(u(0))$ and $Q_2(u(t))=Q_2(u(0))$ for any $t\in\mathbf{R}$.

We can also prove that $\hat{a}$ given by (7) and (23) satisfies

$$E'(\hat{a}(x))-\omega Q_1'(\hat{a}(x))-\upsilon Q_2'(\hat{a}(x))=0. \tag{37}$$

In fact,

$$E'(\hat{a}(x))-\omega Q_1'(\hat{a}(x))-\upsilon Q_2'(\hat{a}(x))$$

$$=-\hat{a}_{xx}+\frac{\beta(\alpha+\beta)}{2}|u|^4\hat{a}-g(|a|^2)\hat{a}+\mathrm{i}(\alpha+\beta)|a|^2\hat{a}_x-\omega\hat{a}+\mathrm{i}\upsilon\hat{a}_x+\beta\upsilon|a|^2\hat{a}$$

$$\xlongequal{(8)} \frac{\beta(\alpha+\beta)}{2} a^4 \hat{a} + \beta v a^2 \hat{a} - 2\beta i a a_x \hat{a} + 2\beta i a^2 \hat{a}_x$$

$$\xlongequal{(7)(12)} e^{i\psi}\left[\frac{\beta(\alpha+\beta)}{2} a^5 + \beta v a^3 - 2\beta i a^2 a_x + 2\beta i a^2 (i\psi' a + a')\right] = 0.$$

For any $u, \phi \in X$, by computation, we have

$$E''(u)\phi = -\phi_{xx} + \frac{\beta(\alpha+\beta)}{2}(3|u|^4\phi + 2|u|^2 u^2 \bar{\phi}) - g(|u|^2)\phi -$$
$$g'(|u|^2)(\bar{u}\phi + u\bar{\phi})u + i(\alpha+\beta)(|u|^2\phi_x + \phi\bar{u}u_x + \bar{\phi}uu_x),$$
$$Q_1''(u)\phi = \phi,$$
$$Q_2''(u)\phi = -i\phi_x - 2\beta|u|^2\phi - \beta u^2\bar{\phi}.$$

Defining an operator from $X$ to $X^*$ by

$$H_{\omega,v} = E''(\hat{a}) - \omega Q_1''(\hat{a}) - v Q_2''(\hat{a}), \tag{38}$$

then for any $\phi \in X$, we have

$$H_{\omega,v}\phi = -\phi_{xx} + \frac{\beta(\alpha+\beta)}{2}(3|a|^4\phi + 2|a|^2\hat{a}^2\hat{\bar{\phi}}) - g(|a|^2)\phi -$$
$$g'(|a|^2)(\bar{\hat{a}}\phi + \hat{a}\,\hat{\bar{\phi}})\hat{a} + i(\alpha+\beta)(|a|^2\phi_x + \phi\,\bar{\hat{a}}\,\hat{a}_x + \bar{\phi}\hat{a}\,\hat{a}_x) -$$
$$\omega\phi + iv\phi_x + 2v\beta|a|^2\phi + \beta v\hat{a}^2\bar{\phi}. \tag{39}$$

It is easy to see that $H_{\omega,v}$ is self-adjoint, which shows that $I^{-1}H_{\omega,v}$ is a bounded self-conjugate operator on $X$. The spectrum of $H_{\omega,v}$ consists of the real numbers $\lambda$ such that $H_{\omega,v} - \lambda I$ is not invertible. $\lambda = 0$ belongs to the spectrum of $H_{\omega,v}$.

From $T_1'(0) = -i$ and $T_2'(0) = -\frac{\partial}{\partial x}$, we obtain

$$T_1'(0)\hat{a}(x) = -ia(x)e^{i\psi(x)} = -i\hat{a}(x),$$
$$T_2'(0)\hat{a}(x) = -(a'(x) + i\psi'(x)a)e^{i\psi(x)} = -\hat{a}_x(x).$$

Consequently, it is easy to obtain that

$$H_{\omega,v}T_1'(0)\hat{a}(x) = 0, \tag{40}$$
$$H_{\omega,v}T_2'(0)\hat{a}(x) = 0. \tag{41}$$

In fact,

$$H_{\omega,v}T_1'(0)\hat{a}(x) = H_{\omega,v}(-i\hat{a}(x))$$
$$= i\hat{a}_{xx} + \frac{\beta(\alpha+\beta)}{2}(-3i|a|^4\hat{a} + 2i|a|^2\hat{a}^2\bar{\hat{a}}) + ig(|a|^2)\hat{a} -$$
$$g'(|a|^2)(-i\bar{\hat{a}}\hat{a} + i\hat{a}\bar{\hat{a}})\hat{a} + i(\alpha+\beta)(-i|a|^2\hat{a}_x - i\hat{a}\,\bar{\hat{a}}\,\hat{a}_x + i\hat{a}\,\bar{\hat{a}}\,\hat{a}_x) +$$
$$i\omega\hat{a} + v\hat{a}_x - i2v\beta|a|^2\hat{a} + i\beta v\hat{a}^2\bar{\hat{a}}$$
$$= -i[E'(\hat{a}) - \omega Q_1'(\hat{a}) - vQ_2'(\hat{a})] = 0,$$

and

$$H_{\omega,v}T_2'(0)\hat{a}(x) = H_{\omega,v}(-\hat{a}_x(x))$$
$$= \hat{a}_{xxx} - \frac{\beta(\alpha+\beta)}{2}(3|a|^4\hat{a}_x + 2|a|^2\hat{a}^2\bar{\hat{a}}_x) + g(|a|^2)\hat{a}_x +$$

$$g'(|a|^2)(\overline{\hat{a}}\hat{a}_x+\hat{a}\overline{\hat{a}}_x)\hat{a}-\mathrm{i}(\alpha+\beta)(|a|^2\hat{a}_{xx}+\hat{a}_x\overline{\hat{a}}\hat{a}_x+\hat{a}\overline{\hat{a}}_x\hat{a}_x)+$$
$$\omega\hat{a}_x-\mathrm{i}\upsilon\hat{a}_{xx}-2\upsilon\beta|a|^2\hat{a}_x-\beta\upsilon\hat{a}^2\overline{\hat{a}}_x$$
$$=-\frac{\partial}{\partial x}[E'(\hat{a}(x))-\omega Q_1'(\hat{a}(x))-\upsilon Q_2'(\hat{a}(x))]=0.$$

Let $Z=\{k_1T_1'(0)\hat{a}(x)+k_2T_2'(0)\hat{a}(x)\mid k_1,k_2\in\mathbf{R}\}$, from (40) and (41), we know that $Z$ is contained in the kernel of $H_{\omega,\upsilon}$.

**Assumption 3.2** Spectral decomposition of operator $H_{\omega,\upsilon}$: the space $X$ can be decomposed as a direct sum

$$X=N+Z+P, \tag{42}$$

where $Z$ is the space defined above, $N$ is a finite-dimensional subspace such that

$$\langle H_{\omega,\upsilon}u,u\rangle<0,\quad \forall 0\neq u\in N, \tag{43}$$

$P$ is a closed subspace, and for any $u\in P$, there exists $\delta>0$ independent of $u$, such that

$$\langle H_{\omega,\upsilon}u,u\rangle\geqslant\delta\|u\|_X^2,\quad \forall u\in P. \tag{44}$$

We define $d(\omega,\upsilon):\mathbf{R}\times\mathbf{R}\to\mathbf{R}$ as

$$d(\omega,\upsilon)=E(\hat{a}_{\omega,\upsilon})-\omega Q_1(\hat{a}_{\omega,\upsilon})-\upsilon Q_2(\hat{a}_{\omega,\upsilon}), \tag{45}$$

and $d''(\omega,\upsilon)$ as the Hessian matrix of function $d(\omega,\upsilon)$, which has a symmetric bilinear form. In addition, we use $p(d'')$ to express the numbers of positive eigenvalue of $d''$ and $n(H_{\omega,\upsilon})$ to express the numbers of negative eigenvalue of $H_{\omega,\upsilon}$.

[16] indicates the existence of the local solution of initial value problem of Kundu equation. Based on the above discussions, we know that Eq. (24) has three invariants of motion satisfying (35) and (36). Moreover, we prove that the solitary waves of Eq. (24) satisfy (37). Furthermore, we give the definition of the operator $H_{\omega,\upsilon}$. According to the "stability theorem" in the introduction of [18] or Theorem 4.1 in [18], we can obtain the following abstract orbital stability theorem for the solitary waves of Eq. (24).

**Theorem 3.3** Suppose that there exist three functionals $E(u)$, $Q_1(u)$, $Q_2(u)$ satisfying (35) and (36), and solitary waves $T_1(\omega t)T_2(\upsilon t)\hat{a}_{\omega,\upsilon}(x)$ satisfying (37). Moreover, suppose that the operator $H_{\omega,\upsilon}$ given by (38) satisfies Assumption 3.2. If $d(\omega,\upsilon)$ is non-degenerative and $p(d'')=n(H_{\omega,\upsilon})$, then solitary waves $T_1(\omega t)T_2(\upsilon t)\hat{a}_{\omega,\upsilon}(x)$ are orbitally stable.

From Theorem 3.3, the results about orbital stability of solitary waves for Eq. (24) can be obtained.

**Theorem 3.4** For any fixed real constants $c_3,c_5,\alpha=2s_2+r,\beta=s_2+r$, if $\omega,\upsilon$ satisfy $4\omega+\upsilon<0$ and one of the following conditions holds:

(a) $16c_5+(\alpha+\beta)(3\alpha-5\beta)=0$, $2c_3+(\alpha-\beta)\upsilon<0$,

(b) $16c_5+(\alpha+\beta)(3\alpha-5\beta)>0$, $2c_3+(\alpha-\beta)\upsilon\geqslant 0$,

(c) $16c_5+(\alpha+\beta)(3\alpha-5\beta)<0$, $2c_3+(\alpha-\beta)\upsilon>0$, and $3(2c_3+(\alpha-\beta)\upsilon)^2-(16c_5+(\alpha+\beta)(3\alpha-5\beta))(4\omega+\upsilon^2)>0$,

(d) $16c_5+(\alpha+\beta)(3\alpha-5\beta)>0$, $2c_3+(\alpha-\beta)\upsilon<0$, and $3(\alpha-\beta)^2-(16c_5+(\alpha+$

$\beta)(3\alpha-5\beta))\geqslant\frac{\sqrt{3}(\alpha-\beta)^2\sqrt{-(4\omega+\upsilon^2)(16c_5+(\alpha+\beta)(3\alpha-5\beta))}}{\pi(2c_3+(\alpha-\beta)\upsilon)}$,

then the solitary waves $e^{-i\omega t}\hat{a}(x-\upsilon t)$ of Eq. (24) are orbitally stable.

Since Eq. (24) reduces to Eq. (2) as $r=0$, we have the following corollary for Eq. (2).

**Corollary 3.5** For any fixed real constants $c_3, c_5, s_2$, if $\omega, \upsilon$ satisfy $4\omega+\upsilon<0$ and one of the following conditions holds:

(a) $16c_5+2s_2^2=0$, $2c_3+s_2\upsilon>0$,

(b) $16c_5+2s_2^2>0$, $2c_3+s_2\upsilon\geqslant 0$,

(c) $16c_5+2s_2^2<0$, $2c_3+s_2\upsilon>0$, and $3(2c_3+s_2\upsilon)^2-(16c_5+2s_2^2)(4\omega+\upsilon^2)>0$,

(d) $16c_5+2s_2^2>0$, $2c_3+s_2\upsilon<0$, and $16c_5\leqslant-\frac{\sqrt{3}s_2^2\sqrt{-(4\omega+\upsilon^2)(16c_5+2s_2^2)}}{\pi(2c_3+s_2\upsilon)}$,

then the solitary waves of Eq. (2) are orbitally stable.

When $c_3=c_5=0, s_2=-1$, $r=0$, Eq. (24) reduces to Eq. (6), we have the following corollary for Eq. (6).

**Corollary 3.6** For any wave speed $\upsilon\neq 0$, if $\upsilon^2<-4\omega$, then the solitary waves of Eq. (6) are orbitally stable.

**Proof** When $\upsilon^2<-4\omega$, from Theorem 3.4(b), it is easy to prove that the solitary waves of Eq. (6) are orbitally stable as $\upsilon<0$, while from Theorem 3.4(d), it is easy to prove that the solitary waves of Eq. (6) are orbitally stable as $\upsilon>0$.

Similarly, we can obtain the orbital stability of solitary waves for Eqs. (3)~(4) from Theorem 3.4. □

**Corollary 3.7** For any wave speed $\upsilon\neq 0$, if $\upsilon^2<-4\omega$, then solitary waves of Eq. (3) are orbitally stable.

**Corollary 3.8** Suppose that $\delta\neq 0$, for any wave speed $\upsilon\neq 0$, if $\upsilon^2<-4\omega$, then solitary waves of Eq. (4) are orbitally stable.

## 4 Proof of Theorem 3.4

By the above discussions, we know that if the conditions in Theorem 2.2 hold, then Eq. (24) has solitary waves of the form $e^{-i\omega t}\hat{a}(x-\upsilon t)$. In this section, we mainly prove that these solitary waves are orbitally stable. According to (32)~(38), we only need to prove that Assumption 3.2 and $n(H_{\omega,\upsilon})=p(d'')$ hold. Consequently, solitary waves of Eq. (24) are orbitally stable from Theorem 3.3.

### 4.1 Proof of Assumption 3.2

First of all, we study the decomposition of $H_{\omega,\upsilon}$ and the spectral properties of each part.

For any $\phi(x)\in X$, let

$$\phi(x)=e^{i\varphi(x)}z(x),\quad z(x)=z_1(x)+iz_2(x),\quad z_1(x)=\mathrm{Re}\, z(x),\qquad(46)$$

then

$$H_{\omega,v}\phi = [L_{11}z_1 + L_{12}z_2 + \mathrm{i}(L_{21}z_1 + L_{22}z_2)]\mathrm{e}^{\mathrm{i}\psi},$$

where

$$L_{11} = -\frac{\partial}{\partial x^2} + (\psi')^2 - g(a^2) - 2g'(a^2)a^2 - \omega - v\psi' + \frac{5\beta(\alpha+\beta)}{2}a^4 - 3(\alpha+\beta)a^2\psi' + 3v\beta a^2,$$

$$L_{12} = \psi'' - \frac{\alpha+\beta}{2}a^2\frac{\partial}{\partial x},$$

$$L_{21} = -\psi'' + \frac{\alpha+\beta}{2}a^2\frac{\partial}{\partial x} + 2(\alpha+\beta)aa',$$

$$L_{22} = -\frac{\partial}{\partial x^2} + (\psi')^2 - g(a^2) - \omega - v\psi' - \alpha a^2\psi' + \beta a^2\psi'.$$

Furthermore, we have

$$\langle H_{\omega,v}\phi, \phi\rangle = \langle \overline{L}_{11}z_1, z_1\rangle + \langle L_{22}z_2, z_2\rangle - \langle(\alpha+\beta)a^2z_2', z_1\rangle + \left\langle \frac{(\alpha+\beta)^2}{4}a^4z_1 + (\alpha+\beta)aa'z_2, z_1\right\rangle, \tag{47}$$

where

$$\overline{L}_{11} = -\frac{\partial}{\partial x^2} + (\psi')^2 - g(a^2) - 2g'(a^2)a^2 - \omega - v\psi' - 3\alpha a^2\psi' + 3\beta a^2\psi' - \frac{\alpha+\beta}{2}a^2(v - 2\psi' + \alpha a^2 - \beta a^2), \tag{48}$$

due to

$$\langle H_{\omega,v}\phi, \phi\rangle = \langle L_{11}z_1, z_1\rangle + \langle L_{12}z_2, z_1\rangle + \langle L_{21}z_1, z_2\rangle + \langle L_{22}z_2, z_2\rangle,$$

$$\langle L_{11}z_1, z_1\rangle + \langle L_{12}z_2, z_1\rangle + \langle L_{21}z_1, z_2\rangle$$

$$= \langle L_{11}z_1, z_1\rangle + \left\langle -\frac{\alpha+\beta}{2}a^2z_2', z_1\right\rangle + \left\langle \frac{\alpha+\beta}{2}a^2z_1', z_2\right\rangle + \langle 2(\alpha+\beta)aa'z_1, z_2\rangle,$$

$$\left\langle \frac{\alpha+\beta}{2}a^2z_1', z_2\right\rangle = \frac{\alpha+\beta}{2}\int_{\mathbf{R}} a^2z_2\mathrm{d}z_1 = -\frac{\alpha+\beta}{2}\int_{\mathbf{R}} z_1(2aa'z_2 + 2a^2z_2')\mathrm{d}x$$

$$= \langle -(\alpha+\beta)aa'z_2, z_1\rangle + \left\langle -\frac{\alpha+\beta}{2}a^2z_2', z_1\right\rangle.$$

It follows from Eq.(10) that

$$L_{22}a(x) = 0. \tag{49}$$

Differentiating Eq.(10) with respect to $x$, and combining (12) and (48), we have

$$\overline{L}_{11}a'(x) = 0. \tag{50}$$

In view of (23), we know that $a'(x)$ only has a simple zero point $x = 0$ and it changes the sign only once. Therefore, by Sturm-Liouville theorem, it is easy to see that zero is the second eigenvalue of $\overline{L}_{11}$ and $\overline{L}_{11}$ only has one negative eigenvalue $-\lambda_{11}^2$, whose corresponding eigenfunction is denoted by $\chi_{11}$, namely,

$$L_{11}\chi_{11} = -\lambda_{11}^2\chi_{11}. \tag{51}$$

In addition, $\overline{L}_{11}$ can be rewritten as

$$\overline{L}_{11} = -\frac{\partial}{\partial x^2} + d_1 + M_1(x), \tag{52}$$

where

$$M_1(x) = \frac{(\alpha+\beta)^2}{16}a^4 - g(a^2) - 2g'(a^2)a^2 - 3\alpha a^2\psi' + 3\beta a^2\psi' - \frac{\alpha+\beta}{2}a^2(\upsilon - 2\psi' + \alpha a^2 - \beta a^2).$$

From (23), it can be seen that $a^2 \to 0$ as $|x| \to \infty$, therefore, we have

$$M_1(x) \to 0 \quad (|x| \to \infty). \tag{53}$$

Based on (52)~(53) and Weyl's essential spectral theorem obtained in [23], we have

$$\sigma_{\text{ess}}(\overline{L}_{11}) = [d_1, +\infty), \quad d_1 > 0. \tag{54}$$

Thus, we have the spectral properties of $\overline{L}_{11}$ as follows.

**Proposition 4.1** $\overline{L}_{11}$ only has a simple negative eigenvalue and its kernel is spanned by $a'(x)$. Moreover, the rest of its spectrums are positive and bounded away from zero.

It follows from Proposition 4.1 and [17] that for any real function $z_1 \in H^1(\mathbf{R})$, if it satisfies

$$\langle z_1, a'\rangle = \langle z_1, \chi_{11}\rangle = 0, \tag{55}$$

then there exists a positive constant $\overline{\delta}_1 > 0$ independent of $z_1$, such that

$$\langle \overline{L}_{11}z_1, z_1\rangle \geqslant \overline{\delta}_1 \| z_1 \|_{L^2}^2. \tag{56}$$

Furthermore, according to (52)~(56), we obtain the following lemma (see Appendix in [24]).

**Lemma 4.2** For any real function $z_1 \in H^1(\mathbf{R})$ satisfying (55), there exists a positive $\delta_1 > 0$ which is independent of $z_1$, such that

$$\langle \overline{L}_{11}z_1, z_1\rangle \geqslant \delta_1 \| z_1 \|_{H^1}^2. \tag{57}$$

Next, we discuss $L_{22}$.

From (23) and (49), it is easy to see that $a(x)$ has a fixed sign and zero is the first eigenvalue of $L_{22}$. Note that

$$L_{22} = -\frac{\partial}{\partial x^2} + d_1 + M_2(x), \tag{58}$$

where $M_2(x) = \dfrac{(\alpha+\beta)^2}{16}a^4 - g(a^2) - \alpha a^2\psi' + \beta a^2\psi'$.

Since $a^2 \to 0$ as $|x| \to \infty$, we have that

$$M_2(x) \to 0 \quad (|x| \to \infty). \tag{59}$$

Moreover,

$$\sigma_{\text{ess}}(L_{22}) = [d_1, +\infty), \quad d_1 > 0. \tag{60}$$

Therefore, we have the following spectral properties of $L_{22}$.

**Proposition 4.3** The kernel of $L_{22}$ is spanned by $a(x)$. Moreover, the rest of its spectrum is positive and bounded away from zero.

Similarly, by Proposition 4.3 and (58)~(60), we have

**Lemma 4.4** For any real function $z_2 \in H^1(\mathbf{R})$ satisfying

$$\langle z_2, a \rangle = 0, \tag{61}$$

there exists a positive number $\delta_2 > 0$ such that

$$\langle L_{22} z_2, z_2 \rangle \geqslant \delta_2 \| z_2 \|_{H^1}^2, \tag{62}$$

where $\delta_2$ is independent of $z_2$.

In the following, we verify that Assumption 3.2 holds and $n(H_{\omega,v}) = 1$.

For any $\phi(x) \in X$, let

$$\phi(x) = e^{i\psi(x)}(z_1(x) + iz_2(x)), \quad z_2(x) = a(x) z_3(x), \tag{63}$$

where $z_1, z_2, z_3$ are real functions and $z_1, z_2 \in H^1(\mathbf{R})$. Note that

$$\begin{aligned}\langle L_{22} z_2(x), z_2(x) \rangle &= \langle -z_2'', z_2 \rangle + \langle d_1 z_2, z_2 \rangle + \langle M_2 z_2, z_2 \rangle \\ &= \langle -a'' z_3 + d_1 a z_3 + M_2 a z_3, a z_3 \rangle + \langle -2a' z_3' - a z_3'', a z_3 \rangle \\ &= \langle L_{22} a, a z_3^2 \rangle - \langle (a^2 z_3')', z_3 \rangle \\ &= -\int_{\mathbf{R}} z_3 \mathrm{d}(a^2 z_3') \\ &= \int_{\mathbf{R}} (a z_3')(a z_3') \mathrm{d}x \\ &= \langle a z_3', a z_3' \rangle.\end{aligned} \tag{64}$$

Then, it follows from (47) and (63)~(64) that

$$\begin{aligned}\langle H_{\omega,v}\phi, \phi \rangle &= \langle \overline{L}_{11} z_1, z_1 \rangle + \langle a z_3', a z_3' \rangle + \\ &\quad \left\langle \frac{(\alpha+\beta)^2}{4} a^4 z_1 + (\alpha+\beta) a a' z_2, z_1 \right\rangle - \langle (\alpha+\beta) a^2 z_2', z_1 \rangle \\ &= \langle \overline{L}_{11} z_1, z_1 \rangle + \int_{\mathbf{R}} \left( (a z_3')^2 + \left( \frac{\alpha+\beta}{2} a^2 z_1 \right)^2 - 2(a z_3') \left( \frac{\alpha+\beta}{2} a^2 z_1 \right) \right) \mathrm{d}x \\ &= \langle \overline{L}_{11} z_1, z_1 \rangle + \int_{\mathbf{R}} \left( \frac{\alpha+\beta}{2} a^2 z_1 - a z_3' \right)^2 \mathrm{d}x.\end{aligned} \tag{65}$$

Choose

$$\chi_- = (\chi_{11} + i\chi_{12}) e^{i\psi(x)}, \tag{66}$$

and

$$\chi_{12} = a\chi_{13} = a\left( \frac{\alpha+\beta}{2} \int_{-\infty}^{x} a(s) \chi_{11}(s) \mathrm{d}s + k_1 \right), \tag{67}$$

where $k_1$ is an arbitrary real constant, then from (63) and (65)~(67), we have

$$\langle H_{\omega,v} \chi_-, \chi_- \rangle = \langle \overline{L}_{11} \chi_{11}, \chi_{11} \rangle = -\lambda_{11}^2 < 0. \tag{68}$$

Choose $k_1$ such that

$$\langle \chi_{12}, a \rangle = 0, \tag{69}$$

and set

$$N = \{ k\chi_- \mid k \in \mathbf{R} \}, \tag{70}$$

then we have (43) holds from (68) and (70).

Let

$$\chi_1 = \left(a'(x) + \mathrm{i}a(x)\left(k_2 + \frac{\alpha+\beta}{4}a^2(x)\right)\right)\mathrm{e}^{\mathrm{i}\psi(x)}, \tag{71}$$

$$\chi_{12} = \mathrm{i}a(x)\mathrm{e}^{\mathrm{i}\psi(x)}, \tag{72}$$

and choose $k_2$ such that

$$\left\langle \left(k_2 + \frac{\alpha+\beta}{4}a^2\right)a, a\right\rangle = 0, \tag{73}$$

then subspace $Z$ can be rewritten as

$$Z = \{k_3\chi_1 + k_4\chi_2 \mid k_3, k_4 \in \mathbf{R}\}. \tag{74}$$

Denoted subspace $P$ by

$$P = \{p \in X \mid p = (p_1 + \mathrm{i}p_2)\mathrm{e}^{\mathrm{i}\psi(x)},\quad \langle p_1, \chi_{11}\rangle = \langle p_1, a'\rangle = \langle p_2, a\rangle = 0\}, \tag{75}$$

then by the method in Appendix 1 of [10], we have the following lemma.

**Lemma 4.5** For any $p \in P$, there exists a constant $\delta > 0$ independent of $p$ such that

$$\langle H_{\omega,v}p, p\rangle \geqslant \delta\|p\|_{H^1}^2. \tag{76}$$

For any $\phi(x) = \mathrm{e}^{\mathrm{i}\psi(x)}(z_1 + \mathrm{i}z_2) \in X$, we choose

$$a_1 = \langle z_1, \chi_{11}\rangle,\quad b_1 = \frac{\langle z_1, a'\rangle}{\|a'\|_{L^2}^2},\quad b_2 = \frac{\langle z_2, a\rangle}{\|a\|_{L^2}^2}, \tag{77}$$

then $\phi(x)$ can be rewritten as

$$\phi(x) = a_1\chi_- + b_1\chi_1 + b_2\chi_2 + p. \tag{78}$$

From (50)～(51) and (69)～(73), we obtain that $\phi(x)$ is expressed uniquely by (78). Then, it follows from (70) and (74)～(78) that Assumption 3.2 holds and $n(H_{\omega,v}) = 1$.

### 4.2 Proof of $p(d'') = n(H_{\omega,v}) = 1$

In this subsection, we focus on proving that $p(d'') = 1$ under the conditions in Theorem 3.4. Consequently, $p(d'') = n(H_{\omega,v}) = 1$.

From

$$d(\omega, v) = E(\hat{a}) - \omega Q_1(\hat{a}) - vQ_2(\hat{a}),$$

we have

$$d_\omega = -Q_1(\hat{a}),\quad d_v = -Q_2(\hat{a}),$$

$$d_{\omega\omega} = -\left\langle Q_1'(\hat{a}), \frac{\partial\hat{a}}{\partial\omega}\right\rangle = -\left\langle \hat{a}, \frac{\partial\hat{a}}{\partial\omega}\right\rangle = -\frac{1}{2}\frac{\partial}{\partial\omega}\langle a, a\rangle,$$

$$d_{\omega v} = d_{v\omega} = -\left\langle Q_1'(\hat{a}), \frac{\partial\hat{a}}{\partial v}\right\rangle = -\left\langle \hat{a}, \frac{\partial\hat{a}}{\partial v}\right\rangle = -\frac{1}{2}\frac{\partial}{\partial v}\langle a, a\rangle,$$

$$\begin{aligned} d_{vv} &= -\left\langle Q_2'(\hat{a}), \frac{\partial\hat{a}}{\partial v}\right\rangle = \left\langle \mathrm{i}\hat{a}_x + \beta a^2\hat{a}, \frac{\partial\hat{a}}{\partial v}\right\rangle \\ &= \mathrm{Re}\int_{\mathbf{R}}(-\psi'a + \mathrm{i}a' + \beta a^3)\left(\frac{\partial a}{\partial v} - \mathrm{i}\frac{\partial\psi}{\partial v}\right)\mathrm{d}x \end{aligned}$$

$$= \int_{\mathbf{R}} \left( (-\psi' a + \beta a^3) \frac{\partial a}{\partial \upsilon} + a' a \frac{\partial \psi}{\partial \upsilon} \right) \mathrm{d}x$$

$$= \int_{\mathbf{R}} \left( -\left( \frac{\upsilon}{2} + \frac{\alpha + \beta}{4} a^2 \right) a + \beta a^3 \right) \frac{\partial a}{\partial \upsilon} \mathrm{d}x + \frac{1}{2} \int_{\mathbf{R}} \frac{\partial \psi}{\partial \upsilon} \mathrm{d}a^2$$

$$= \int_{\mathbf{R}} \left( -\frac{\upsilon}{2} a \frac{\partial a}{\partial \upsilon} \right) \mathrm{d}x - \int_{\mathbf{R}} \frac{1}{4} a^2 \mathrm{d}x - \frac{\alpha - \beta}{2} \int_{\mathbf{R}} a^3 \frac{\partial a}{\partial \upsilon} \mathrm{d}x$$

$$= -\frac{\upsilon}{4} \frac{\partial}{\partial \upsilon} \langle a, a \rangle - \frac{1}{4} \langle a, a \rangle - \frac{\alpha - \beta}{8} \frac{\partial}{\partial \upsilon} \int_{\mathbf{R}} a^4 \mathrm{d}x.$$

Therefore, we obtain

$$d'' = \begin{pmatrix} d_{\omega\omega} & d_{\omega\upsilon} \\ d_{\omega\upsilon} & d_{\upsilon\upsilon} \end{pmatrix} = \begin{pmatrix} -\frac{1}{2} \frac{\partial}{\partial \omega} \langle a, a \rangle & -\frac{1}{2} \frac{\partial}{\partial \upsilon} \langle a, a \rangle \\ -\frac{1}{2} \frac{\partial}{\partial \upsilon} \langle a, a \rangle & d_{\upsilon\upsilon} \end{pmatrix}.$$

Moreover,

$$\det(d'') = \frac{\upsilon}{8} \frac{\partial}{\partial \omega} \langle a, a \rangle \frac{\partial}{\partial \upsilon} \langle a, a \rangle + \frac{1}{8} \langle a, a \rangle \frac{\partial}{\partial \omega} \langle a, a \rangle + \frac{\alpha - \beta}{16} \frac{\partial}{\partial \omega} \langle a, a \rangle \frac{\partial}{\partial \upsilon} \int_{\mathbf{R}} a^4 \mathrm{d}x - \frac{1}{4} \left( \frac{\partial}{\partial \upsilon} \langle a, a \rangle \right)^2. \tag{79}$$

Note that

$$a^2(x) = \frac{1}{d_3 + d_5 \cosh d_6 x},$$

$$d_3 = -\frac{d_2}{2d_1}, \quad d_5^2 = \frac{d_2^2 - 4d_1 d_4}{4d_1^2}, \quad d_6^2 = 4d_1,$$

$$d_1 = -\omega - \frac{\upsilon^2}{4}, \quad d_2 = -\frac{1}{2} c_3 - \frac{\alpha - \beta}{4} \upsilon,$$

$$d_4 = -\frac{1}{3} \left( c_5 + \frac{(\alpha + \beta)(3\alpha - 5\beta)}{16} \right),$$

it is obviously that

$$\frac{\partial d_1}{\partial \omega} = -1, \quad \frac{\partial d_1}{\partial \upsilon} = -\frac{\upsilon}{2},$$

$$\frac{\partial d_2}{\partial \upsilon} = -\frac{1}{4}(\alpha - \beta), \quad \frac{\partial d_2}{\partial \omega} = 0,$$

$$\frac{\partial d_4}{\partial \omega} = \frac{\partial d_4}{\partial \upsilon} = 0.$$

We will prove that $p(d'') = 1$ from the following four cases. That is, we will prove that $\det(d'') < 0$.

**Case 1** $d_4 = 0, d_2 < 0$.

In this case, we have

$$a^2(x) = \frac{1}{d_3 + d_5 \cosh d_6 x} = -\frac{2d_1}{d_2} \frac{1}{(1 + \cosh d_6 x)},$$

$$\langle a, a \rangle = -\frac{2d_1}{d_2} \int_{\mathbf{R}} \frac{1}{(1 + \cosh d_6 x)} \mathrm{d}x = -\frac{2\sqrt{d_1}}{d_2},$$

$$\frac{\partial}{\partial\omega}\langle a,a\rangle=-\frac{2}{2d_2}\cdot\frac{1}{2\sqrt{d_1}}\cdot\frac{\partial d_1}{\partial\omega}=\frac{1}{d_2\sqrt{d_1}}<0,$$

$$\frac{\partial}{\partial\upsilon}\langle a,a\rangle=\frac{-\sqrt{d_1}}{2d_2^2}\left((\alpha-\beta)-\frac{\upsilon d_2}{d_1}\right),$$

$$\int_{\mathbf{R}}a^4(x)\mathrm{d}x=\frac{4d_1^2}{d_2^2}\int_{\mathbf{R}}\left(\frac{1}{1+\cosh d_6x}\right)^2\mathrm{d}x=\frac{4d_1^2}{d_2^2}\int_{\mathbf{R}}\frac{4\mathrm{e}^{2d_6x}}{(\mathrm{e}^{2d_6x}+2\mathrm{e}^{d_6x}+1)^2}\mathrm{d}x$$

$$=\frac{16d_1^2}{d_2^2d_6}\int_0^{+\infty}\frac{y}{(y+1)^4}\mathrm{d}y=\frac{4d_1\sqrt{d_1}}{3d_2^2},$$

$$\frac{\partial}{\partial\upsilon}\int_{\mathbf{R}}a^4(x)\mathrm{d}x=\frac{d_1^{\frac{3}{2}}}{d_2^3}\left((\alpha-\beta)-\frac{\upsilon d_2}{d_1}\right)-\frac{1}{3}(\alpha-\beta)\frac{d_1^{\frac{3}{2}}}{d_2^3}.$$

Furthermore, from (79), we have

$$\det(d'')=\frac{1}{8}\langle a,a\rangle\frac{\partial}{\partial\omega}\langle a,a\rangle+\frac{1}{8}\frac{\partial}{\partial\upsilon}\langle a,a\rangle\left(\upsilon\frac{\partial}{\partial\omega}\langle a,a\rangle-2\frac{\partial}{\partial\upsilon}\langle a,a\rangle\right)+$$
$$\frac{\alpha-\beta}{16}\frac{\partial}{\partial\omega}\langle a,a\rangle\frac{\partial}{\partial\upsilon}\int_{\mathbf{R}}a^4\mathrm{d}x$$
$$=-\frac{1}{4d_2^2}-\frac{(\alpha-\beta)^2d_1}{48d_2^4},$$

that it is negative because $d_1>0$ and $d_2\neq0$.

Consequently, $d''$ has exactly one negative and one positive eigenvalues. Namely, $p(d'')=1$ holds for Case 1.

**Case 2** $d_4<0$, $d_2\leqslant0$.

In this case, we have

$$d_5^2-d_3^2=\frac{d_2^2-4d_1d_4}{4d_1^2}-\frac{d_2^2}{4d_1^2}=\frac{-d_4}{d_1}>0,\quad d_5^2>d_3^2,$$

$$\langle a,a\rangle=\int_{\mathbf{R}}\frac{\mathrm{d}x}{d_3+d_5\cosh d_6x}=\frac{1}{\sqrt{-d_4}}\left(\frac{\pi}{2}-\arctan\frac{-d_2}{2\sqrt{-d_1d_4}}\right)>0,$$

$$\frac{\partial}{\partial\omega}\langle a,a\rangle=\frac{d_2}{(d_2^2-4d_1d_4)\sqrt{d_1}}\leqslant0,$$

$$\frac{\partial}{\partial\upsilon}\langle a,a\rangle=\frac{-\sqrt{d_1}}{2(d_2^2-4d_1d_4)}\left((\alpha-\beta)-\frac{\upsilon d_2}{d_1}\right),$$

$$\int_{\mathbf{R}}a^4(x)\mathrm{d}x=\int_{\mathbf{R}}\left(\frac{1}{d_3+d_5\cosh d_6x}\right)^2\mathrm{d}x$$
$$=\frac{-\sqrt{d_1}}{d_4}-\frac{d_2}{2d_4\sqrt{-d_4}}\left(\frac{\pi}{2}-\arctan\frac{-d_2}{2\sqrt{-d_1d_4}}\right),$$

$$\frac{\partial}{\partial\upsilon}\int_{\mathbf{R}}a^4(x)\mathrm{d}x=\frac{\upsilon}{4d_4\sqrt{d_1}}+\frac{\alpha-\beta}{8d_4\sqrt{-d_4}}\left(\frac{\pi}{2}-\arctan\frac{-d_2}{2\sqrt{-d_1d_4}}\right)+$$
$$\frac{d_2\sqrt{d_1}}{4d_4(d_2^2-4d_1d_4)}\left(\alpha-\beta-\frac{\upsilon d_2}{d_1}\right).$$

Then, we have

$$\det(d'') = \frac{1}{8}\langle a,a\rangle \frac{\partial}{\partial\omega}\langle a,a\rangle + \frac{1}{8}\frac{\partial}{\partial v}\langle a,a\rangle\left(v\frac{\partial}{\partial\omega}\langle a,a\rangle - 2\frac{\partial}{\partial v}\langle a,a\rangle\right)+$$

$$\frac{\alpha-\beta}{16}\frac{\partial}{\partial\omega}\langle a,a\rangle\frac{\partial}{\partial v}\int_{\mathbf{R}} a^4 \mathrm{d}x$$

$$= \frac{1}{8}\langle a,a\rangle\frac{\partial}{\partial\omega}\langle a,a\rangle + \frac{(\alpha-\beta)^2}{64d_4(d_2^2-4d_1d_4)}\cdot$$

$$\left[1+\frac{d_2}{2\sqrt{-d_1d_4}}\left(\frac{\pi}{2} - \arctan\frac{-d_2}{2\sqrt{-d_1d_4}}\right)\right]$$

$$= I + II .$$

Evidently, $I \leqslant 0$. Let $y = \frac{-d_2}{2\sqrt{-d_1d_4}}$, then we have $y \geqslant 0$. Furthermore, we have $II = \frac{(\alpha-\beta)^2}{64d_4(d_2^2-4d_1d_4)}Y_1(y)$, where $Y_1(y) = 1 - y\left(\frac{\pi}{2} - \arctan y\right)$. It is clearly that $Y_1(0)=1$. By L'Hospital's rule of limit, we have

$$Y_1(+\infty) = 1 - \lim_{y\to\infty}\frac{\frac{\pi}{2}-\arctan y}{1/y} = 0. \tag{80}$$

Further, we have

$$Y'_1(y) = -\frac{\pi}{2} + \arctan y + \frac{y}{1+y^2},$$

$$Y'_1(0) = -\frac{\pi}{2},\quad Y'_1(+\infty) = 0, \tag{81}$$

$$Y''_1(y) = \frac{2}{1+y^2} - \frac{2y^2}{(1+y^2)^2} = \frac{2}{(1+y^2)^2} > 0. \tag{82}$$

From (81) and (82), it is easy to see that $Y'_1(y)<0$ for any $y\in\mathbf{R}$. It follows from (80) that $Y_1(y)>0$ for any $y\in\mathbf{R}$. Since $d_4<0, II<0$. Accordingly, $p(d'')=1$.

**Case 3** $d_4>0, d_2<0, d_2^2-4d_1d_4>0$.

In this case, we have

$$d_5^2 - d_3^2 = \frac{d_2^2-4d_1d_4}{4d_1^2} - \frac{d_2^2}{4d_1^2} = -\frac{d_4}{d_1},\quad d_5^2 < d_3^2,$$

$$\langle a,a\rangle = \int_{\mathbf{R}}\frac{\mathrm{d}x}{d_3+d_5\cosh d_6x} = \int_{\mathbf{R}}\frac{2\mathrm{e}^{d_6x}}{d_5\mathrm{e}^{2d_6x}+2d_3\mathrm{e}^{d_6x}+d_5}\mathrm{d}x = \frac{1}{2\sqrt{d_4}}\ln\left|\frac{-d_2+2\sqrt{d_1d_4}}{-d_2-2\sqrt{d_1d_4}}\right|,$$

$$\frac{\partial}{\partial\omega}\langle a,a\rangle = \frac{d_2}{\sqrt{d_1}(d_2^2-4d_1d_4)},$$

$$\frac{\partial}{\partial v}\langle a,a\rangle = -\frac{\sqrt{d_1}}{2(d_2^2-4d_1d_4)}\left((\alpha-\beta) - \frac{vd_2}{d_1}\right).$$

From

$$\int_{\mathbf{R}} a^4(x)\mathrm{d}x = \int_{\mathbf{R}}\left(\frac{1}{d_3+d_5\cosh d_6x}\right)^2\mathrm{d}x = -\frac{\sqrt{d_1}}{d_4} - \frac{d_2}{4d_4\sqrt{d_4}}\ln\left|\frac{-d_2+2\sqrt{d_1d_4}}{-d_2-2\sqrt{d_1d_4}}\right|,$$

we have

$$\frac{\partial}{\partial \upsilon}\int_{\mathbf{R}} a^4(x)\mathrm{d}x = \frac{\upsilon}{4d_4\sqrt{d_1}} + \frac{\alpha-\beta}{16d_4\sqrt{d_4}}\ln\left|\frac{-d_2+2\sqrt{d_1d_4}}{-d_2-2\sqrt{d_1d_4}}\right| + \frac{\sqrt{d_1}d_2}{4d_4(d_2^2-4d_1d_4)}\left((\alpha-\beta)-\frac{\upsilon d_2}{d_1}\right).$$

Furthermore, we obtain

$$\begin{aligned}\det(d'') &= \frac{1}{8}\langle a,a\rangle\frac{\partial}{\partial\omega}\langle a,a\rangle + \frac{1}{8}\frac{\partial}{\partial\upsilon}\langle a,a\rangle\left(\upsilon\frac{\partial}{\partial\omega}\langle a,a\rangle - 2\frac{\partial}{\partial\upsilon}\langle a,a\rangle\right)+\\ &\quad \frac{\alpha-\beta}{16}\frac{\partial}{\partial\omega}\langle a,a\rangle\frac{\partial}{\partial\upsilon}\int_{\mathbf{R}} a^4\mathrm{d}x\\ &= \frac{1}{8}\langle a,a\rangle\frac{\partial}{\partial\omega}\langle a,a\rangle + \frac{(\alpha-\beta)^2}{64d_4(d_2^2-4d_1d_4)}\left[1+\frac{d_2}{4\sqrt{d_1d_4}}\ln\left|\frac{-d_2+2\sqrt{d_1d_4}}{-d_2-2\sqrt{d_1d_4}}\right|\right]\\ &= I + II.\end{aligned}$$

It is obviously that $I<0$. In what follows, we will prove $II<0$.

Assume that $y=\dfrac{-d_2}{2\sqrt{d_1d_4}}>0$, then we have $y>1$ and $II=\dfrac{(\alpha-\beta)^2}{64d_4(d_2^2-4d_1d_4)}Y_2(y)$, where $Y_2(y)=1-\dfrac{y}{2}\ln\left(\dfrac{y+1}{y-1}\right)$. Evidently, $Y_2(1)=-\infty$. By L'Hospital's rule of limit, we have

$$\lim_{y\to\infty}Y_2(y) = 1-\lim_{y\to\infty}\left(\frac{y-1}{y+1}\right)\left(\frac{y^2}{(y-1)^2}\right)=0. \tag{83}$$

Notice that

$$Y_2'(y) = -\frac{1}{2}\ln\left(\frac{y+1}{y-1}\right)+\frac{y}{y^2-1},\quad Y_2'(+\infty)=0, \tag{84}$$

$$Y_2''(y) = -\frac{2}{(y^2-1)^2}<0,\quad \forall\ 1<y<+\infty. \tag{85}$$

From (84) and (85), we have $Y_2'(y)>0$, $\forall\ 1<y<+\infty$, $\forall\ y\in\mathbf{R}$. From (83), it is easy to obtain that $Y_2(y)<0$, $\forall\ 1<y<+\infty$. Since $d_4>0, d_2^2-4d_1d_4>0, II<0$. Furthermore, we have $p(d'')=1$.

**Case 4** $d_4<0, d_2>0$.

Based on the above results, we have

$$\frac{1}{8}\langle a,a\rangle\frac{\partial}{\partial\omega}\langle a,a\rangle = \frac{d_2}{8(d_2^2-4d_1d_4)\sqrt{-d_1d_4}}\left[\frac{\pi}{2}-\arctan\frac{-d_2}{2\sqrt{-d_1d_4}}\right],$$

$$\begin{aligned}\det(d'') &= \frac{1}{8}\langle a,a\rangle\frac{\partial}{\partial\omega}\langle a,a\rangle + \frac{(\alpha-\beta)^2}{64d_4(d_2^2-4d_1d_4)}\cdot\\ &\quad\left[1+\frac{d_2}{2\sqrt{-d_1d_4}}\left[\frac{\pi}{2}-\arctan\frac{-d_2}{2\sqrt{-d_1d_4}}\right]\right]\\ &= \frac{1}{64d_4(d_2^2-4d_1d_4)}\left\{(\alpha-\beta)^2+[(\alpha-\beta)^2+16d_4]\cdot\right.\end{aligned}$$

$$\frac{d_2}{2\sqrt{-d_1d_4}}\left(\frac{\pi}{2}+\arctan\frac{d_2}{2\sqrt{-d_1d_4}}\right)\Bigg\}.$$

To prove that $\det(d'')<0$, we only need to prove that

$$(\alpha-\beta)^2+16d_4\geqslant\frac{-(\alpha-\beta)^2}{\frac{d_2}{2\sqrt{-d_1d_4}}\left(\frac{\pi}{2}+\arctan\frac{d_2}{2\sqrt{-d_1d_4}}\right)}.$$

Since

$$\frac{-(\alpha-\beta)^2}{\frac{d_2}{2\sqrt{-d_1d_4}}\left(\frac{\pi}{2}+\arctan\frac{d_2}{2\sqrt{-d_1d_4}}\right)}<-\frac{(\alpha-\beta)^2}{\frac{d_2}{2\sqrt{-d_1d_4}}\left(\frac{\pi}{2}+\frac{\pi}{2}\right)}=-\frac{(\alpha-\beta)^2 2\sqrt{-d_1d_4}}{\pi d_2},$$

the parameters only need to satisfy

$$(\alpha-\beta)^2+16d_4\geqslant-\frac{(\alpha-\beta)^2\sqrt{-d_1d_4}}{\pi d_2},\tag{86}$$

which is equal to

$$-2(4c_5-s_2r-r^2)\geqslant\frac{s_2^2\sqrt{-3(4\omega+\upsilon^2)(16c_5+3s_2^2-4s_2r-4r^2)}}{2\pi(2c_3+s_2\upsilon)}.\tag{87}$$

Under the condition (86) or (87), we have $\det(d'')<0$, which implies $p(d'')=1$.

Summarized the above results, we have $p(d'')=1$ under the conditions in Theorem 3.4. Furthermore, $n(H_{\omega,\upsilon})=p(d'')$. Thus, Theorem 3.4 is proved completely.

## References


[1] Kundu A. Landau-Lifshitz and higher-order nonlinear systems gauge generated from nonlinear Schrödinger-type equation. J Math Phys, 1984, 25: 3433-3436.

[2] Saarloos W V, Hohenberg P C. Fronts, pulses, sources and sinks in generalized complex Ginzburg-Landau equations. Phys D, 1992, 56: 303-367.

[3] Mjolhus E. On the modulational instability of hydromagnetic waves parallel to the magnetic field. J Plasma Phys, 1976, 16:321-334.

[4] Mio K, Ogino T, Minami K, et al. Modified nonlinear Schrödinger equation for Alfvén waves propagating along the magnetic fields in cold plasma. J Phys Soc Jpn, 1976, 41: 265-271.

[5] Deissler R J, Brand H R. Generation of counter propagating nonlinear interacting travelling waves by localized noise. Phys Lett A, 1977, 130: 293-296.

[6] Tsutsumi M, Fukuda I. On solution of the derivative nonlinear Schrödinger equation, Ⅱ. Funkcial Ekvac, 1981, 24: 85-94.

[7] Haken H. Synergetics: an Introduction. New York: Springer, 1987.

[8] Chen H H, Lee Y C, Lin C S. Integrability of nonlinear Hamiltonian system by inverse scattering method. Phys Scr, 1979, 20: 490-492.

[9] Gerdjikov V S, Ivanov I. The quadratic bundle of general form and the nonlinear evolution equation, Ⅱ. Bulg J Phys, 1983, 10(2): 130-143.

[10] Guo B L, Wu Y P. Orbital stability of solitary waves for the nonlinear derivative Schrödinger equation. J

Differential Equations, 1995, 123: 35 – 55.

[11] Colin M, Ohta M. Stability of solitary waves for derivative Schrödinger equation. Ann Inst H Poincaré Anal Non Linéaire, 2006, 23: 753 – 764.

[12] Clarckson P A, Cosgrove C M. Painlevé analysis of the nonlinear Schrödinger family of equations. J Phys A, 1987, 20: 2003 – 2024.

[13] Clarckson P A. Dimensional reductions and exact solutions of a generalized Schrödinger equation. Nonlinearity, 1992, 5: 453 – 472.

[14] Florjanczyk F, Gagnon L. Exact solutions for higher-order nonlinear Schrödinger equation. Phys Rev A, 1990, 41 (8): 4478 – 4485.

[15] Nakkeeran K. On the integrability of the extended nonlinear Schrödinger equation and the coupled extended nonlinear Schrödinger equations. J Phys A, 2000, 33: 3947 – 3949.

[16] Ozawa T. On the nonlinear Schrödinger equation of derivative type. Indiana Univ Math J, 1996, 45: 137 – 163.

[17] Grillakis M, Shatah J, Strauss W. Stability theory of solitary waves in the presence of symmetry I. J Funct Anal, 1987, 74: 160 – 197.

[18] Grillakis M, Shatah J, Strauss W. Stability theory of solitary waves in the presence of symmetry, Ⅱ. J Funct Anal, 1990, 94: 308 – 348.

[19] Hayashi N, Ozawa T. The initial value problem for the derivative nonlinear Schrödinger equation in the energy space. Nonlinear Anal, 1993, 20: 823 – 833.

[20] Hayashi N, Ozawa T. On the derivative nonlinear Schrödinger equation. Phys D, 1992, 55: 14 – 36.

[21] Hayashi N, Ozawa T. Finite energy solutions of the nonlinear Schrödinger equation of derivative type. SIAM J Math Anal, 1994, 25: 1488 – 1503.

[22] Weinstein M I. Lyapunov stability of ground states of nonlinear dispersive evolution equations. Comm Pure Appl Math, 1986, 39: 51 – 68.

[23] Reed M, Simon B. Methods of modern mathematical physics, Vol Ⅳ. New York: Academic Press, 1978.

[24] Albert J, Bona J L. Total positivity and the stability of internal waves in stratified fluids of finite depth. IMA J Appl Math, 1991, 46: 1 – 19.

# Some Exact Solutions to Multidimensional Landau-Lifshitz Equation with Uprush External Field and Anisotropy Field *

Yang Ganshan(杨干山) Guo Boling(郭柏灵)

**Abstract**

In this paper we give a method that can construct some exact solutions to a multidimensional Landau-Lifshitz equation with uprush external field and anisotropy field.

**Keywords** multidimensional Landau-Lifshitz equation; uprush external field; anisotropy field; exact transform; exact solution

## 1 Introduction

In 1935, the equation proposed by Landau and Lifshitz in [1]

$$\begin{cases} \dfrac{\partial u}{\partial t} = \lambda_1 u \times H^e - \lambda_2 u \times (u \times H^e) & \text{in } \Omega \times (0,T) \\ H^e := -H(u) + \sum\limits_{i,j=1}^{3} \dfrac{\partial}{\partial x_j}\left(a_{ij}\dfrac{\partial u}{\partial x_i}\right) + H & \text{in } \Omega \times (0,T) \end{cases}, \tag{1.1}$$

where $\lambda_1$, $\lambda_2$ are constants, $\lambda_2 > 0$, describes an evolution of spin fields in continuum ferromagnets, it bears a fundamental role in the understanding of non-equilibrium magnetism, just as the Navier-Stokes equation does in that of fluid dynamics.

In 1974, Nakamura and Sasada studied the following one-dimensional equation

$$\frac{\partial u}{\partial t} = u \times (u_{xx} + H) \tag{1.2}$$

where the spin vector $u = u(x,t) = (u_1(x,t), u_2(x,t), u_3(x,t)), x \in \mathbf{R}$, $H = (0,0,h)$. They found some solitary wave solutions (see [2]). Tjon and Wright also found some solitons under the case with non-vanishing external magnetic field in 1977 (see [3]). In 1976, Lakshmanan, Ruijgrok and Thompson constructed a class of solutions of (1.2) under a vanishing external magnetic field (i.e. $H=0$) in [4]. They found solitary waves having total energy localized in a finite region, with the velocity of propagation inversely proportional to the width of this region.

In 1975, V. M. Eleonskiĭ, N. N. Kirova, V. M. Petrov found soliton solutions of the Landau-Lifshitz equation $\frac{\partial u}{\partial t} = u \times H_{\text{eff}}$, where $u$ is the magnetic-moment density and $H_{\text{eff}} =$

* Nonlinear Analysis, 2009, 71: 3999–4006.

$H + A\Delta u + Knu_z$ is the effective field with the external magnetic field $H$ in the $z$-direction and the unit vector $n$ parallel to the $z$-axis. The equation is supposed to describe spin-dynamics in ferromagnets with uniaxial anisotropy (see [5]).

In 1998, Yue, Chen, Huang[6] gave a method of inverse scattering transform to solve the one-dimensional Landau-Lifshitz equation for a spin chain with an easy plane in the case of spin non-flip $\partial_t u(x, t) = u(x,t) \times u(x,t)_{xx} - u(x,t) \times Ju(x,t)$ with $|u| = 1$ and the diagonal matrix $J = \mathrm{diag}(0, -16\rho^2, 0)$.

In 2001, T. Sadakane considers the one-dimensional Landau-Lifshitz (LL) equation

$$\partial_t u(x,t) = u(x,t) \times u(x,t)_{xx} - u(x,t) \times Ju(x,t),$$

where the vector $u(x,t)$ describes the spin in a ferromagnetic material and $J$ is a diagonal matrix whose entries $J_1$, $J_2$, $J_3$ are coupling constants. Sadakane takes into account two partially isotropic cases: the easy-axis case (corresponding to $J_1 < J_2 = J_3$) and the easy-plane case (corresponding to $J_1 < J_2 = J_3$) (see [7]).

One may see by [5,8] that the multidimensional Landau-Lifshitz equation is hard to solve exactly. In two space dimensions, (1.2) admits a large family of explicit, static solutions that are known (see [18]). However explicit dynamic solutions have rarely been seen. In 2000 – 2001 by using many techniques[9] constructed several exact blow up solutions for the following cylindrical symmetric $n$-dimension Landau-Lifshitz equation

$$z_t = z \times z_{rr} + \frac{n-1}{r} z \times z_r, \quad n \geqslant 2 \tag{1.3}$$

in [10,11]. Here is a solution $u \in C^1([0,T); C^2(\mathbf{R}^n \setminus B(\delta))) \cap C^1([0,T); H^2(\mathbf{R}^n \setminus B(\delta)))$ of (1.2)which is called as blow up solution, if there exists $x \in \mathbf{R}^n \setminus B(\delta)$ such that

$$\lim_{t \to T-0} \sqrt{|\nabla u_1|^2 + |\nabla u_2|^2 + |\nabla u_3|^2} = \infty$$

for every small enough $\delta > 0$, where

$$B(\delta) = \left\{(x_1, x_2, \cdots, x_n) \in \mathbf{R}^n \mid \sqrt{x_1^2 + x_2^2 + \cdots + x_n^2} < (\delta)\right\}$$

(see [10,11]). In 2001, Guo, Yang constructed a family of exact disc solutions for the two-dimension Landau-Lifshitz equation[12]. In 2003, Yang and Chang obtained an exact transform between the case with external field and the case without an external field. By the transform, some sufficient and necessary conditions so that the solution with an external field converges to a certain solution without an external field are given (see [13]).

In particular, when both external field and anisotropy field are present there are no papers discussing the explicit solution of the multidimensional Landau-Lifshitz equation until present. Indeed only when there is a lack of an external field or an anisotropy field, many authors had attempted Hirota's famous method[14], Lie algebraic structure[15] and Bäcklund transform[16], Double transform[17], etc, but exact solutions were still not constructed. In this paper we discuss the following multidimensional Landau-Lifshitz equation with an uprush external field and anisotropy field

$$\begin{cases}\dfrac{\partial u}{\partial t}=u\times(\Delta u+\mu(t)H+H(u)), & \text{in } \mathbf{R}^n\times(0,\infty)\\ u\in S^2, \quad \text{in } \mathbf{R}^n\times[0,\infty)\end{cases}, \tag{1.4}$$

where $\mu(t)\in L^2(0,\infty)$ and $\int_0^t \mu(s)\mathrm{d}s\in C[0,\infty)$, $H(u)=\sum_{i,j=1}^{3} b_{ij}u_i(b_1,b_2,b_3)$ satisfies

$$b_{ij}=\begin{cases}\alpha+(\beta-\alpha)b_ib_j, & i=j\\ (\beta-\alpha)b_ib_j, & i\neq j\end{cases}. \tag{1.5}$$

The spin vector $u=u(x,t)=u(x_1,x_2,\cdots,x_n;t)=(u_1,u_2,u_3)$ is a 3-dimensional vector valued unknown function with respect to space variables $x=(x_1,x_2,\cdots,x_n)$ and time $t$. $\mu(t)$ depends on $t$. The external magnetic field $\mu(t)H=\mu(h_1,h_2,h_3)$ is a 3-dimensional vector-valued function connect with time. " $\times$ " denotes the cross-product of two 3-dimensional vectors.

Notice that here the external field $\mu(t)H$ does not need to be continuous and bounded. For example, we take

$$\mu(t)=\begin{cases}\dfrac{1}{\sqrt[4]{T-t}}, & 0\leqslant t\leqslant\dfrac{T}{2}\\ \dfrac{2}{\sqrt[4]{T-t}}, & \dfrac{T}{2}<t<T\\ C, & t=T\\ \dfrac{1}{\sqrt[4]{t-T+(t-T)^2}}, & t>T\end{cases}, \tag{1.6}$$

then $\mu(t)$ is discontinuous and unbounded. Therefore the external field $\mu(t)H$ is also discontinuous, even unbounded, i. e. like an uprush vector function. Although $\mu(t)$ is discontinuous and unbounded, still $\int_0^t \mu(s)\mathrm{d}s\in C[0,\infty)$ and $\mu(t)\in L^2(0,\infty)$.

In order to get explicit solution of (1.4), we first show a theorem in the Section 2. Then we give some explicit solutions of (1.4).

## 2 Main Results

Let $0<T\leqslant\infty$, $\Omega$ is $\mathbf{R}^n$ or a bounded domain in $\mathbf{R}^n$. For the sake of convenience, we let $h_2^2+h_3^2\neq 0$ in this paper. We first generalize the Theorem 1 in [13] as follows.

**Lemma 2.1** (Ref. [13]) Let

$$v=uA, \tag{2.1}$$

be a transform between the initial boundary problem of the Landau-Lifshitz equation with an external magnetic field

$$
\text{(IBPH)}\quad \begin{cases} \dfrac{\partial u}{\partial t} = u \times (\Delta u + \mu(t) H), & (x,t) \in \Omega \times (0,T) \\ u(x,0) = \phi(x), & x \in \Omega \\ u(x,t) = \psi(x,t), & (x,t) \in \partial\Omega \times (0,T) \\ u \in S^2, & \text{in } \mathbf{R}^n \times [0,\infty) \end{cases}, \tag{2.2}
$$

and the initial boundary problem of the Landau-Lifshitz equation without an external magnetic field

$$
\text{(IBP)}\quad \begin{cases} \dfrac{\partial v}{\partial t} = v \times \Delta v, & (x,t) \in \Omega \times (0,T) \\ v(x,0) = \phi(x) Q, & x \in \Omega \\ v(x,t) = \psi(x,t) A, & (x,t) \in \partial\Omega \times (0,T) \\ v \in S^2, & \text{in } \mathbf{R}^n \times [0,\infty) \end{cases}, \tag{2.3}
$$

where

$$
A = \left\{ \begin{array}{ccc} \dfrac{\sqrt{h_2^2 + h_3^2}\cos\rho}{|H|} & \dfrac{\sqrt{h_2^2 + h_3^2}\sin\rho}{|H|} & \dfrac{h_1}{|H|} \\ -\dfrac{h_1 h_2 \cos\rho}{|H|\sqrt{h_2^2 + h_3^2}} - \dfrac{h_3 \sin\rho}{\sqrt{h_2^2 + h_3^2}} & -\dfrac{h_1 h_2 \sin\rho}{|H|\sqrt{h_2^2 + h_3^2}} + \dfrac{h_3 \cos\rho}{\sqrt{h_2^2 + h_3^2}} & \dfrac{h_2}{|H|} \\ -\dfrac{h_1 h_3 \cos\rho}{|H|\sqrt{h_2^2 + h_3^2}} + \dfrac{h_2 \sin\rho}{\sqrt{h_2^2 + h_3^2}} & -\dfrac{h_1 h_3 \sin\rho}{|H|\sqrt{h_2^2 + h_3^2}} - \dfrac{h_2 \cos\rho}{\sqrt{h_2^2 + h_3^2}} & \dfrac{h_3}{|H|} \end{array} \right\}, \tag{2.4}
$$

$$
Q = \left\{ \begin{array}{ccc} \dfrac{\sqrt{h_2^2 + h_3^2}}{|H|} & 0 & \dfrac{h_1}{|II|} \\ -\dfrac{h_1 h_2}{|H|\sqrt{h_2^2 + h_3^2}} & \dfrac{h_3}{\sqrt{h_2^2 + h_3^3}} & \dfrac{h_2}{|H|} \\ -\dfrac{h_1 h_2}{|H|\sqrt{h_2^2 + h_3^2}} & -\dfrac{h_2}{\sqrt{h_2^2 + h_3^2}} & \dfrac{h_3}{|H|} \end{array} \right\} \tag{2.5}
$$

and $|H| = \sqrt{h_1^2 + h_2^2 + h_3^2}, \rho = |H| \int_0^t \mu(s) ds$. If

$$
\phi \in C^2(\Omega), \quad \psi \in C^1(0,T; C^2(\partial\Omega; \mathbf{R}^3)),
$$

then $u \in C^1(0,T; C^2(\mathbf{R}^n; \mathbf{R}^3))$ is a solution of (IBPH) if and only if $v$ is a solution of (IBP), provided that $\mu(t)$ is continuous on $(0, T)$. $u \in C^1([0,T]; C^2(\mathbf{R}^n \setminus B(\delta))) \cap C^1([0,T]; H^2(\mathbf{R}^n \setminus B(\delta)))$ if and only if $v \in C^1([0,T]; C^2(\mathbf{R}^n \setminus B(\delta))) \cap C^1([0,T]; H^2(\mathbf{R}^n \setminus B(\delta)))$. Simply put that it is equivalent in this paper.

**Proof** Let $u$ be a solution of (IBPH), $B$ a three square matrix which does not depend on $x$, suppose

$$
\frac{\partial}{\partial t}(uB) = (uB) \times \Delta(uB). \tag{2.6}
$$

Then by using the method of undetermined coefficients we found that $B$ can be taken as $A$ in (2.2). It can be checked directly that $v = uA$ is a solution of (IBP) when $u$ is a solution of

(IBPH). In fact, let

$$H^* = \begin{pmatrix} 0 & -h_3 & h_2 \\ h_3 & 0 & -h_1 \\ -h_2 & h_1 & 0 \end{pmatrix},$$

then

$$\frac{\partial}{\partial t}A = \mu(t)\begin{pmatrix} -\sqrt{h_2^2+h_3^2}\sin\rho & \sqrt{h_2^2+h_3^2}\cos\rho & 0 \\ \frac{h_1h_2\sin\rho - h_3|H|\cos\rho}{\sqrt{h_2^2+h_3^2}} & -\frac{h_1h_2\cos\rho + h_3|H|\sin\rho}{\sqrt{h_2^2+h_3^2}} & 0 \\ \frac{h_1h_3\sin\rho + h_2|H|\cos\rho}{\sqrt{h_2^2+h_3^2}} & \frac{-h_1h_3\cos\rho + h_2|H|\sin\rho}{\sqrt{h_2^2+h_3^2}} & 0 \end{pmatrix}$$

$$= -\mu(t)H^*A. \tag{2.7}$$

similarly we also have $u\times H = uH^*$, $(uA)\times\Delta(uA) = (u\times\Delta u)A$. Since $u$ is a solution of (IBPH), we obtain

$$\begin{aligned}\frac{\partial}{\partial t}(uA) - (uA)\times\Delta(uA) &= u\frac{\partial}{\partial t}A + \mu(t)(u\times H)A \\ &= u\frac{\partial}{\partial t}A + \mu(t)u(H^*A) \\ &= 0.\end{aligned}$$

Thus $v = uA$ is also a solution of (IBP).

One may also verify that $v = uA^{\mathrm{T}}$ is a solution of (IBPH) when $u$ is a solution of (IBP), and those conclusions on the initial-boundary value condition. Thus we have completed the proof of lemma. □

We introduce the following second transformation in $\mathbf{R}^3$.

**Lemma 2.2** Let $b_3\neq 0$,

$$\begin{cases} u_1 = \frac{\sqrt{b_2^2+b_3^2}}{\sqrt{b_1^2+b_2^2+b_3^2}}v_1 + \frac{b_1}{\sqrt{b_1^2+b_2^2+b_3^2}}v_3 \\ u_2 = -\frac{b_1b_2}{\sqrt{b_1^2+b_2^2+b_3^2}\sqrt{b_2^2+b_3^2}}v_1 + \frac{b_3}{\sqrt{b_2^2+b_3^2}}v_2 + \frac{b_2}{\sqrt{b_1^2+b_2^2+b_3^2}}v_3, \\ u_3 = -\frac{b_1b_3}{\sqrt{b_1^2+b_2^2+b_3^2}\sqrt{b_2^2+b_3^2}}v_1 - \frac{b_2}{\sqrt{b_2^2+b_3^2}}v_2 + \frac{b_3}{\sqrt{b_1^2+b_2^2+b_3^2}}v_3 \end{cases} \tag{2.8}$$

where

$$v = (v_1, v_2, v_3) \in C^2(\mathbf{R}^n;\mathbf{R}^3)\cap S^2.$$

By the transformation (2.8), the equation

$$\begin{cases} u_t = u\times(\Delta u + \mu(t)H + H(u)), & \text{in } \mathbf{R}^n\times(0,\infty) \\ u|_{t=0} = u_0, & \text{in } \mathbf{R}^n \\ u\in S^2, & \text{in } \mathbf{R}^n\times[0,\infty) \end{cases}, \tag{2.9}$$

can be changed into

$$\begin{cases} v_t = v \times (\Delta v + \mu(t)\widetilde{H} + \widetilde{H}(v)), & \text{in } \mathbf{R}^n \times (0,\infty) \\ v|_{t=0} = v_0, & \text{in } \mathbf{R}^n \\ v \in S^2, & \text{in } \mathbf{R}^n \times [0,\infty) \end{cases}, \tag{2.10}$$

where

$$\widetilde{H} = (\tilde{h}_1, \tilde{h}_2, \tilde{h}_3),$$

$$\begin{cases} \tilde{h}_1 = \dfrac{\sqrt{b_2^2+b_3^2}}{\sqrt{b_1^2+b_2^2+b_3^2}} h_1 - \dfrac{b_1 b_2}{\sqrt{b_1^2+b_2^2+b_3^2}\sqrt{b_2^2+b_3^2}} h_2 - \dfrac{b_1 b_3}{\sqrt{b_1^2+b_2^2+b_3^2}\sqrt{b_2^2+b_3^2}} h_3 \\ \tilde{h}_2 = \dfrac{b_3}{\sqrt{b_2^2+b_3^2}} h_2 - \dfrac{b_2}{\sqrt{b_2^2+b_3^2}} h_3 \\ \tilde{h}_3 = \dfrac{b_1}{\sqrt{b_1^2+b_2^2+b_3^2}} h_1 + \dfrac{b_2}{\sqrt{b_1^2+b_2^2+b_3^2}} h_2 + \dfrac{b_3}{\sqrt{b_1^2+b_2^2+b_3^2}} h_3 \end{cases}. \tag{2.11}$$

$\widetilde{H}(v)$ and $v_0$ will be given below.

Now we give third transform. Let us consider the following Landau-Lifshitz equation,

$$\begin{cases} \dfrac{\partial u}{\partial t} = u \times (\Delta u + \mu(t)H + H(u)), & \text{in } \mathbf{R}^n \times (0,\infty) \\ u|_{t=0} = u_0, & \text{in } \mathbf{R}^n \\ u \in S^2 & \text{in } \mathbf{R}^n \times [0,\infty) \end{cases}, \tag{2.12}$$

where $\int_0^t \mu(s)\mathrm{d}s \in C[0,\infty)$ and $\mu(t) \in L^2(0,\infty)$, $H(u) = \sum_{i,j=1}^{3} b_{ij}u_i(b_1,b_2,b_3)$,

$$b_{ij} = \begin{cases} \alpha + (\beta-\alpha)b_ib_j, & i = j, \\ (\beta-\alpha)b_ib_j, & i \neq j, \end{cases} \quad b_3 \neq 0. \tag{2.13}$$

We denote the transform (2.8) by $u = vB^{\mathrm{T}}$, where

$$B = \begin{pmatrix} \dfrac{\sqrt{b_2^2+b_3^2}}{\sqrt{b_1^2+b_2^2+b_3^2}} & 0 & \dfrac{b_1}{\sqrt{b_1^2+b_2^2+b_3^2}} \\ -\dfrac{b_1b_2}{\sqrt{b_1^2+b_2^2+b_3^2}\sqrt{b_2^2+b_3^2}} & \dfrac{b_3}{\sqrt{b_2^2+b_3^2}} & \dfrac{b_2}{\sqrt{b_1^2+b_2^2+b_3^2}} \\ -\dfrac{b_1b_3}{\sqrt{b_1^2+b_2^2+b_3^2}\sqrt{b_2^2+b_3^2}} & -\dfrac{b_2}{\sqrt{b_2^2+b_3^2}} & \dfrac{b_3}{\sqrt{b_1^2+b_2^2+b_3^2}} \end{pmatrix}. \tag{2.14}$$

By Lemma 2.2, the (2.12) becomes

$$\begin{cases} v_t = v \times (\Delta v + \mu(t)\widetilde{H} + \widetilde{H}(v)), & \text{in } \mathbf{R}^n \times (0,\infty) \\ v|_{t=0} = v_0, & \text{in } \mathbf{R}^n \\ v \in S^2, & \text{in } \mathbf{R}^n \times [0,\infty) \end{cases}, \tag{2.15}$$

where

$$
\begin{cases}
v_0 = u_0 B \\
\widetilde{H} = (\tilde{h}_1, \tilde{h}_2, \tilde{h}_3) \triangleq (h_1, h_2, h_3) B \quad (\text{see } (2.11)) \\
\widetilde{H}(v) = (\alpha' v_1, \alpha' v_2, \beta' v_3) \\
\alpha' = \alpha, \quad \beta' = \alpha + (\beta - \alpha)(b_1^2 + b_2^2 + b_2^2)
\end{cases}, \tag{2.16}
$$

under the transform $u = vB^{\mathrm{T}}$.

Let

$$
\widetilde{A} = \begin{pmatrix}
\dfrac{\sqrt{\tilde{h}_2^2 + \tilde{h}_3^2}\cos\tilde{\rho}}{|\widetilde{H}|} & \dfrac{\sqrt{\tilde{h}_2^2 + \tilde{h}_3^2}\sin\tilde{\rho}}{|\widetilde{H}|} & \dfrac{\tilde{h}_1}{|\widetilde{H}|} \\
-\dfrac{\tilde{h}_1 \tilde{h}_2 \cos\tilde{\rho}}{|\widetilde{H}|\sqrt{\tilde{h}_2^2 + \tilde{h}_3^2}} - \dfrac{\tilde{h}_3 \sin\tilde{\rho}}{\sqrt{\tilde{h}_2^2 + \tilde{h}_3^2}} & -\dfrac{\tilde{h}_1 \tilde{h}_2 \sin\tilde{\rho}}{|\tilde{h}|\sqrt{\tilde{h}_2^2 + \tilde{h}_3^2}} + \dfrac{\tilde{h}_3 \cos\tilde{\rho}}{\sqrt{\tilde{h}_2^2 + \tilde{h}_3^2}} & \dfrac{\tilde{h}_2}{|\widetilde{H}|} \\
-\dfrac{\tilde{h}_1 \tilde{h}_3 \cos\tilde{\rho}}{|\widetilde{H}|\sqrt{\tilde{h}_2^2 + \tilde{h}_3^2}} + \dfrac{\tilde{h}_2 \sin\tilde{\rho}}{\sqrt{\tilde{h}_2^2 + \tilde{h}_3^2}} & -\dfrac{\tilde{h}_1 \tilde{h}_3 \sin\tilde{\rho}}{|\widetilde{H}|\sqrt{\tilde{h}_2^2 + \tilde{h}_3^2}} - \dfrac{\tilde{h}_2 \cos\tilde{\rho}}{\sqrt{\tilde{h}_2^2 + \tilde{h}_3^2}} & \dfrac{\tilde{h}_3}{|\widetilde{H}|}
\end{pmatrix}, \tag{2.17}
$$

where $|\widetilde{H}| = \sqrt{\tilde{h}_1^2 + \tilde{h}_2^2 + \tilde{h}_3^2}$, $\tilde{\rho} = |\widetilde{H}| \int_0^t \mu(s)\mathrm{d}s$.

Under the transform $v = z\widetilde{A}^{\mathrm{T}}$, the (2.15) becomes the following equation by the Lemma 2.1

$$
\begin{cases}
z_t = z \times (\Delta z + \overline{H}(z)), & \text{in } \mathbf{R}^n \times (0, \infty) \\
z|_{t=0} = z_0, & \text{in } \mathbf{R}^n \\
z \in S^2, & \text{in } \mathbf{R}^n \times [0, \infty)
\end{cases}, \tag{2.18}
$$

where

$$
\begin{cases}
z_0 = u_0 B\widetilde{Q} \\
\overline{H}(z) = \sum\limits_{i,j=1}^{3} c_{ij} z_i (c_1, c_2, c_3) \\
c_{ij} = \begin{cases} \alpha' + (\beta' - \alpha') c_i c_j, & i = j \\ (\beta' - \alpha') c_i c_j, & i \neq j \end{cases} \\
\alpha' = \alpha, \quad \beta' = \alpha + (\beta - \alpha)(b_1^2 + b_2^2 + b_3^2)
\end{cases}, \tag{2.19}
$$

and

$$
\boldsymbol{Q} = \begin{pmatrix}
\dfrac{\sqrt{\tilde{h}_2^2 + \tilde{h}_3^2}}{|\widetilde{H}|} & 0 & \dfrac{\tilde{h}_1}{|\widetilde{H}|} \\
-\dfrac{\tilde{h}_1 \tilde{h}_2}{|\widetilde{H}|\sqrt{\tilde{h}_2^2 + \tilde{h}_3^2}} & \dfrac{\tilde{h}_3}{\sqrt{\tilde{h}_2^2 + \tilde{h}_3^2}} & \dfrac{\tilde{h}_2}{|\widetilde{H}|} \\
-\dfrac{\tilde{h}_1 \tilde{h}_3}{|\widetilde{H}|\sqrt{\tilde{h}_2^2 + \tilde{h}_3^2}} & -\dfrac{\tilde{h}_2}{\sqrt{\tilde{h}_2^2 + \tilde{h}_3^2}} & \dfrac{\tilde{h}_3}{|\widetilde{H}|}
\end{pmatrix}, \tag{2.20}
$$

$$(c_1, c_2, c_3) = \left(-\frac{\tilde{h}_1\tilde{h}_3}{|\widetilde{H}|\sqrt{\tilde{h}_2^2+\tilde{h}_3^2}}, -\frac{\tilde{h}_2}{\tilde{h}_2^2+\tilde{h}_3^2}, \frac{\tilde{h}_3}{|\widetilde{H}|}\right) \quad (\text{see } (2.16) \text{ for } \tilde{h}_i).$$

Under the transform $z = \omega\widetilde{B}^{\mathrm{T}}$, the (2.18) becomes the following equation by the Lemma 2.1

$$\begin{cases} w_t = w \times (\Delta w + H_1(w)), & \text{in } \mathbf{R}^n \times (0,\infty) \\ w|_{t=0} = w_0, & \text{in } \mathbf{R}^n \\ w \in S^2, & \text{in } \mathbf{R}^n \times [0,\infty) \end{cases}, \tag{2.21}$$

where

$$\begin{cases} w_0 = u_0 B\widetilde{Q}\widetilde{B} \\ H_1(w) = (0,0,(\beta-\alpha)(b_1^2+b_2^2+b_3^2)w_3) \end{cases}, \tag{2.22}$$

and

$$\widetilde{B} = \begin{pmatrix} \sqrt{b_2^2+b_3^2} & 0 & b_1 \\ -\dfrac{b_1b_2}{\sqrt{b_2^2+b_3^2}} & \dfrac{b_3}{\sqrt{b_2^2+b_3^2}} & b_2 \\ -\dfrac{b_1b_3}{\sqrt{b_2^2+b_3^2}} & -\dfrac{b_2}{\sqrt{b_2^2+b_3^2}} & b_3 \end{pmatrix}. \tag{2.23}$$

Thus we has the following Theorem.

**Theorem 2.3** By the transform $u = w\widetilde{B}_{\mathrm{T}}\widetilde{A}^{\mathrm{T}}B^{\mathrm{T}}$, we have that under the condition (2.13), (2.12) is equivalent to (2.21), where $B, \widetilde{A}, \widetilde{B}, H_1(w)$ see (2.14), (2.17), (2.23), (2.22), respectively.

By [8], we may see if $n \geqslant 2$ and the anisotropy field are present, it is very important and difficult to get the explicit solution of the following Landau-Lifshitz equation

$$\begin{cases} \dfrac{\partial w}{\partial t} = w \times (\Delta w + \lambda(0,0,w_3)), & \text{in } \mathbf{R}^n \times (0,\infty) \\ w \in S^2, & \text{in } \mathbf{R}^n \times [0,\infty) \end{cases}. \tag{2.24}$$

Though (2.24) admits a large family of explicit, static solutions that are known, explicit dynamic solutions have be not seen. Next we give two types of examples of an explicit solution to (2.24) and (1.4) with initial data, or (2.12).

**Example 2.4** Let $f(x) = b + c_1x_1 + \cdots + c_nx_n$, one may verify that for constants $a$ ($|a| \leqslant 1$), $b$,

$$w = \left(\sqrt{1-a^2}\cos\left(a\left(\lambda + \sum_{i=1}^n c_i^2\right)t + f(x)\right), \sqrt{1-a^2}\sin\left(a\left(\lambda + \sum_{i=1}^n c_i^2\right)t + f(x)\right), a\right) \tag{2.25}$$

is a explicit multidimensional plane solution of the following equation

$$\begin{cases} \dfrac{\partial w}{\partial t} = w \times (\Delta w + \lambda(0,0,w_3)), & \text{in } \mathbf{R}^n \times (0,\infty) \\ w|_{t=0} = w_0 = (\sqrt{1-a^2}\cos f(x), \sqrt{1-a^2}\sin f(x), a), & \text{in } \mathbf{R}^n. \\ w \in S^2, \quad \text{in } \mathbf{R}^n \times [0,\infty) \end{cases} \tag{2.26}$$

**Example 2.5** When $n = 2$, the multidimensional Landau-Lifshitz equation with anisotropy field

$$\begin{cases} \dfrac{\partial w}{\partial t} = w \times (\Delta w + \lambda(0,0,w_3)), & \text{in } \mathbf{R}^2 \times (0,\infty) \\ w \in S^2, \quad \text{in } \mathbf{R}^2 \times [0,\infty) \end{cases} \tag{2.27}$$

has a explicit cylindrical symmetric solution $w = (w_1, w_2, w_3)$, where

$$\begin{cases} w_1 = \dfrac{1}{\sqrt{1+(4c_1t+c_2)^2}}\cos\left(\dfrac{c_1(x_1^2+x_2^2)}{\sqrt{1+(4c_1t+c_2)^2}} - \dfrac{\lambda\sqrt{1+(4c_1t+c_2)^2}}{4c_1}\right) \\ w_2 = \dfrac{1}{\sqrt{1+(4c_1t+c_2)^2}}\sin\left(\dfrac{c_1(x_1^2+x_2^2)}{\sqrt{1+(4c_1t+c_2)^2}} - \dfrac{\lambda\sqrt{1+(4c_1t+c_2)^2}}{4c_1}\right) \\ w_3 = \dfrac{4c_1t+c_2}{\sqrt{1+(4c_1t+c_2)^2}} \end{cases} \tag{2.28}$$

satisfying the initial data

$$\begin{cases} w_1 = \dfrac{1}{\sqrt{1+c_2^2}}\cos\left(\dfrac{c_1(x_1^2+x_2^2)}{\sqrt{1+c_2^2}} - \dfrac{\lambda\sqrt{1+c_2^2}}{4c_1}\right) \\ w_2 = \dfrac{1}{\sqrt{1+c_2^2}}\sin\left(\dfrac{c_1(x_1^2+x_2^2)}{\sqrt{1+c_2^2}} - \dfrac{\lambda\sqrt{1+c_2^2}}{4c_1}\right). \\ w_3 = \dfrac{c_2}{\sqrt{1+c_2^2}} \end{cases} \tag{2.29}$$

According to Theorem 2.3, we have the following theorem.

**Theorem 2.6** For any constants $h_1, h_2, h_3, b_1, b_2, b_3, \alpha$, if $h_2^2 + h_3^2 \neq 0$, $b_3 \neq 0$, let $H = (h_1, h_2, h_3)$, $\mu(t)$ be uprush function,

$$b_{ij} = \begin{cases} \alpha + \lambda b_i b_j, & i = j \\ \lambda b_i b_j, & i \neq j \end{cases}, \tag{2.30}$$

$H(u) = \sum\limits_{i,j=1}^{3} b_{ij}u_i(b_1, b_2, b_3)$. Then $u = w\widetilde{B}^{\mathrm{T}}\widetilde{A}^{\mathrm{T}}B^{\mathrm{T}}$ all are solutions of the following initial data problem for multidimensional Landau-Lifshitz equation with uprush external field and anisotropy field

$$\begin{cases} \dfrac{\partial u}{\partial t} = u \times (\Delta u + \mu(t)H + H(u)), & \text{in } \mathbf{R}^n \times (0,\infty) \\ u|_{t=0} = u_0, \quad \text{in } \mathbf{R}^n \\ u \in S^2 \quad \text{in } \mathbf{R}^n \times [0,\infty) \end{cases}, \tag{2.31}$$

where $u_0 = w_0\widetilde{B}^{\mathrm{T}}\widetilde{Q}^{\mathrm{T}}B^{\mathrm{T}}$, $B, \widetilde{A}, \widetilde{B}$ see (2.14), (2,17), (2,23), respectively.

## References

[1] Landau L D. Lifshitz E M. On the theory of the dispersion of magnetic permeability in ferromagnetic bodies //Collected papers of L. D. Landu. New York: Pergamon, 1965: 101 - 114.

[2] Nakamura K, Sasada T. Soliton and wave trains in ferromagnets. Phys Lett A, 1974 (48): 321 - 322.

[3] Tjon J, Wright J. Soliton in the Hesenberg chain. Phys Rev B, 1977 (15): 3470 - 3476.

[4] Lakshmanan M, Ruijgrok T W, Thompson C J. On the dynamics of a continuous spin systems. Physica A, 1976 (84): 577 - 590.

[5] Eleonskiĭ V M, Kirova N N, Petrov V M. General type of solutions of the Landau-Lifshitz equations. Sov Phys JETP, 1975, 41 (5): 966 - 970.

[6] Yue Hong, Chen Xiang jun, Huang Nianning. An inverse scattering transform for the Landau-Lifshitz equation for a spin chain with an easy plane. J Phys A, 1998, 31 (10): 2491 - 2501.

[7] Sadakane T. Soliton solutions of *XXZ* lattice Landau-Lifshitz equation. J Math Phys, 2001, 42(11): 5457 - 5471.

[8] Gustafson S, Shatah J. The stability of localized solutions of Landau-Lifshitz equations. Commun Pure Appl Math, 2002, 55 (5): 1136 - 1159.

[9] Yang Ganshan. $\delta$ -viscosity solution, blow up solution and global solution of multidimensional Landau-Lifshitz equations. Beijing: Graduate School of China Academy of Engineering Physics, 2002.

[10] Guo Boling, Han Yongqian, Yang Ganshan. Blow up problem for Landau-Lifshitz equations in two dimensions. Commun Nonlinear Sci Numer Simul, 2000, 5 (1): 43 - 44.

[11] Guo Boling, Han Yongqian, Yang Ganshan. Exact blow-up solutions for multidimensional Landau-Lifshitz equations. Adv Math (China), 2001, 30 (1): 91 - 93.

[12] Guo Boling, Yang Ganshan. Some exact nontrivial global solutions with values in unit sphere for two-dimensional Landau-Lifshitz equations. J Math Phys, 2001, 42: 5223 - 5227.

[13] Yang Ganshan, Chang Qianshun. Limit behavior of solution for multidimensional Landau-Lifshitz equations with external magnetic field. Phys Lett A, 2003, 318 (3): 270 - 280.

[14] Michael Svendsen, Hans C Fogedby. Phase shift analysis of the Landau-Lifshitz equation. J Phys A, 1993, 26 (7): 1717 - 1730.

[15] Shibani Sen, Roy Chowdhury A. On the invariance properties and conservation laws for Landau-Lifshitz equation in three dimensions with Tijon-Wright ansatz. J Math Phys, 1991, 32 (9): 2343 - 2346.

[16] Roelofs G H M, Martini R. Prolongation structure of the Landau-Lifshitz equation. J Math Phys, 1993, 34 (6): 2394 - 2399.

[17] Skrypnyk T. "Doubled" generalized Landau-Lifshitz hierarchies and special quasigraded Lie algebras. J Phys A, 2004, 37 (31): 7755 - 7768.

[18] Kosevich A M, Ivanov B A, Kovalev A S. Magnetic solitons. Phys Rep, 1990, 194 (3/4): 117 - 238.

# Dynamics of Stochastic Zakharov Equations *

Guo Boling(郭柏灵) Lv Yan(吕艳) Yang Xiaoping(杨孝平)

**Abstract**

The asymptotic behavior of Zakharov equations driven by random force is studied. The force, which is smooth enough and homogeneous in space and white noise in time, acts on both equations of the Zakharov system independently. By some *a priori* estimates we prove the existence of a solution in energy spaces $\varepsilon_1$ and $\varepsilon_2$ via the Galerkin approximation. This solution is defined on the given probability space rather than a martingale solution. Then a global random attractor is constructed in energy space $\varepsilon_2$ equipped with weak topology. Further the existence of a stationary measure is proved in energy space $\varepsilon_2$ with usual topology.

## 1 Introduction

In the theory of laser plasma interaction, the most general useful model for the Langmuir turbulence is described by Zakharov equations, which consist of one equation for the envelope of the Langmuir wave of the high-frequency electric field and another for the ion-acoustic wave density perturbation[16,27].

$$\frac{1}{\lambda^2}n_{tt} - \Delta(n + |E|^2) = 0, \tag{1.1}$$

$$\mathrm{i}E_t + \Delta E - nE = 0. \tag{1.2}$$

where $E: D \times \mathbf{R}_+ \to \mathbf{C}$ and $n: D \times \mathbf{R}_+ \to \mathbf{R}$. The complex number $E$ represents the envelope of the electric field and $n$ the fluctuation of the ion density about its equilibrium value. The parameter $\lambda$ is proportional to the ion-acoustic speed[26]. Zakharov equations play important roles in the turbulence theory for plasma waves and resemble closely nonlinear Schrödinger equations. An important effect is the so-called Langmuir collapse, which is related to the modulational instability. Zakharov equations have one-dimensional (1D) soliton solutions. In the quasistatic state, $n_{tt}$ may be neglected and (1.1) and (1.2) reduce to a cubic nonlinear Schrödinger equation. Because of its importance, Zakharov equations draw great attention of many mathematicians and physicists and have been quite extensively studied theoretically and numerically. The existence of solutions can be seen in Refs. [9,10,13 and 24]. Some numerical experiments suggest that the solutions of two-dimensional (2D) and three-dimensional (3D)

* Journal of Mathematical Physics, 2009, 50: 052703

Zakharov equations may develop singularly in finite time[15,18,19,22]. These are consistent with the physical phenomenon of Langmuir collapse. Later, Glangetas and Merle[7] proved theoretically that 2D Zakharov equations have exact self-similar singular solutions. Masselin[17] obtained a blow up rate of the blow up solutions to 3D Hamiltonian Zakharov equations (if such solutions exist). However, smooth global solutions to 1D problems always exist[9,24].

Compared with the actual experimental results, solutions of the conserved Zakharov equations (1.1) and (1.2) possess good qualitative agreement. But in order to get good quantitative agreement it is necessary to include damping effects, or effects of the loss of energy. Then we have dissipative Zakharov equations of the form

$$\frac{1}{\lambda^2}n_{tt} + \alpha n_t - \Delta(n + |E|^2) = f,$$

$$\mathrm{i}E_t + \Delta E - nE + \mathrm{i}\gamma E = g,$$

where positive constants $\alpha$ and $\gamma$ are damping coefficients and $f$ and $g$ are external forces. Flahaut[5] studied the long time behavior of the dissipative Zakharov equations with Dirichlet boundary conditions on a bounded interval $D = (0, L)$ and proved the existence of a finite-dimensional compact global attractor in the weak topology of energy space $\varepsilon_2$ (see Sec.2) by a series of *a priori* estimates. Later, Goubet and Moise[8] improved the result by a decomposition of a semigroup. They showed that the weakly compact global attractor is actually strongly compact.

Recently the need to take random effects into account in modeling, analyzing, simulating, and predicting complex phenomena has been widely recognized in geophysical and climate dynamics, materials science, chemistry, biology, and other areas (Refs. [4,11], for example). Stochastic partial differential equations (SPDEs) are appropriate mathematical models for complex systems under random influences or noise. Stochastic force is also considered in Zakharov equations. As an example we mention that stochastic force is used to enhance the Zakharov equations to estimate by how much over the thermal level the spectrum seen by an incoherent scatter radar will be enhanced[6]. On the other hand noise is also a simple approximation of turbulence in fluids.

In this paper we consider the following random forced dissipative Zakharov equations on a regular domain $D$ in the Itô sense:

$$\frac{1}{\lambda^2}\mathrm{d}n_t + [\alpha n_t - \Delta(n + |E|^2)]\mathrm{d}t = f\mathrm{d}t + \mathrm{d}W_1, \tag{1.3}$$

$$\mathrm{id}E + [\Delta E - nE + \mathrm{i}\gamma E]\mathrm{d}t = g\mathrm{d}t + \mathrm{d}W_2, \tag{1.4}$$

where $W_1$ and $W_2$ are independent $L^2(D)$-valued Wiener processes which are detailed in Sec.2.

The problem we are concerned with here is the random dynamics of (1.3) and (1.4). In fact, first (see Sec.3), we give a series of some time uniform *a priori* estimates in different energy spaces which is a key step for the problem we are concerned with. In this part some of technique and Itô formula are used frequently. Being different from the deterministic system we have to further give estimates on the high moment of the solution and maximal estimates on any finite time interval. Here the maximal estimates for stochastic integral (see Lemma 2.2) is an important tool.

Then by a Galerkin method we prove the existence and uniqueness of the solution in $\varepsilon_1$ and $\varepsilon_2$. Usually the Galerkin method for SPDEs leads to martingale solution instead of weak solution. However, in this approach we transform (1.3) and (1.4) to PDEs with random coefficient. Then by the *a priori* estimates obtained in Sec.3, we construct full probability measure set $\tilde{\Omega}$ (see Sec.4), such that for any fixed random parameter $\omega \in \tilde{\Omega}$, the classic Galerkin approximation can be applied to obtain a unique solution for (1.3) and (1.4), which defines a continuous random dynamical system (RDS). A similar method is also applied to the stochastic porous medium equation[12].

Last by the *a priori* estimates in Sec.3, the system is dissipative in space $\varepsilon_2$ in the sense of mean square. Then we can construct a random attractor for the RDS defined in $\varepsilon_2$ endowed with weak topology. Here for the existence of random force we cannot follow the approach of Ref. [8] to obtain a global attractor in $\varepsilon_2$ with usual topology. Although the random attractor is constructed in weak topology, we proved the existence of a stationary measure in $\varepsilon_2$ endowed with usual topology.

The rest of this paper is organized as follows: In Sec.2 we give some functional setting for the problem and some conditions we will need. In Sec.3 a series of *a priori* estimates is given and then the existence of the solution of our stochastic Zakharov equations is given in Sec.4. In Sec.4 we construct a random global attractor in $\varepsilon_2$ endowed with weak topology and proved the existence of a stationary measure in $\varepsilon_2$ endowed with usual topology.

## 2 Preliminary

In this section we give a detailed description of the stochastic Zakharov equations. Let $D = (0,L)$ with $0 < L < \infty$. Consider the following dissipative stochastic Zakharov equations on $D$ :

$$\frac{1}{\lambda^2}\mathrm{d}n_t + [\alpha n_t - \Delta(n + |E|^2)]\mathrm{d}t = f\mathrm{d}t + \mathrm{d}W_1, \tag{2.1}$$

$$\mathrm{i}\mathrm{d}E + [\Delta E - nE + \mathrm{i}\gamma E]\mathrm{d}t = g\mathrm{d}t + \mathrm{d}W_2 \tag{2.2}$$

with Dirichlet boundary conditions

$$n|_{\partial D} = 0, \quad E|_{\partial D} = 0,$$

and initial conditions

$$n(x,0) = n_0(x), \quad n_t(x,0) = n_1(x), \quad E(x,0) = E_0(x).$$

Here $f, g, \alpha, \gamma$ and $\lambda$ are given, with $\alpha > 0$ and $\gamma > 0$. We will work on the usual functional spaces $L^2(D), H^m(D), H_0^1(D)$ and $H^{-1}(D)$. The inner product on $L^2(D)$ will be denoted by

$$\langle u, v\rangle = \mathrm{Re}\int_D u(x)\bar{v}(x)\mathrm{d}x, \quad u, v \in L^2(D),$$

and the norm by $\|\cdot\|_0$. And the norm on $H^m(D)$ is denoted by $\|\cdot\|_m$. Now we define spaces $\varepsilon_i, i = 0,1,2$, as

$$\mathcal{E}_0 = H^{-1}(D) \times L^2(D) \times H_0^1(D),$$
$$\mathcal{E}_1 = L^2(D) \times H_0^1(D) \times (H^2(D) \cap H_0^1(D)),$$
$$\mathcal{E}_2 = H_0^1(D) \times (H^2(D) \cap H_0^1(D)) \times (H^3(D) \cap H_0^1(D)).$$

Endow $\varepsilon_i$ with the usual product norm $\|\cdot\|_{\varepsilon_i}$. Then $\varepsilon_2 \subset \varepsilon_1 \subset \varepsilon_0$ with compact embeddings.

Here we are given a complete probability space $(\Omega, \mathscr{F}, \{\mathscr{F}_t\}_{t\geqslant 0}, \mathbb{P})$. Denote by $\mathbb{E}$ the expectation operator with respect to $\mathbb{P}$. Stochastic terms $W_1(t)$ and $W_2(t)$ are defined on $(\Omega, \mathscr{F}, \{\mathscr{F}_t\}_{t\geqslant 0}, \mathbb{P})$ by

$$W_1(t) = q_1(x)w_1(t), \quad W_2(t) = q_2(x)w_2(t).$$

Here $w_1(t)$ is a standard real-valued Wiener process and $w_2(t)$ is a standard complex-valued Wiener process which is independent of $w_1(t)$ and $q_1(x)$ and $q_2(x)$ are smooth enough.

In our approach we need the following lemmas. Let $(\mathcal{X}, \|\cdot\|_{\mathcal{X}}) \subset (\mathcal{Y}, \|\cdot\|_{\mathcal{Y}}) \subset (\mathcal{Z}, \|\cdot\|_{\mathcal{Z}})$ be three reflective Banach spaces and $\mathcal{X} \subset \mathcal{Y}$ with compact and dense embedding. Define the Banach space

$$\mathcal{G} = \left\{v : v \in L^2(0,T;\mathcal{X}), \frac{\mathrm{d}v}{\mathrm{d}t} \in L^2(0,T;\mathcal{Z})\right\},$$

with norm

$$\|v\|_{\mathcal{G}}^2 = \int_0^T \|v(s)\|_{\mathcal{X}}^2 \mathrm{d}s + \int_0^T \left\|\frac{\mathrm{d}v}{\mathrm{d}s}(s)\right\|_{\mathcal{Z}}^2 \mathrm{d}s, \quad v \in \mathcal{G}.$$

Then we have the following lemma about compactness result[14].

**Lemma 2.1** If $K$ is bounded in $\mathcal{G}$, then $K$ is precompact in $L^2(0,T;\mathcal{Y})$.

We also need some maximal estimates on stochastic integral. Assume $U$ and $H$ are separable Hilbert spaces and $W$ is a $Q$-Wiener process on $U_0$ with $U_0 = Q^{1/2}U$. Let $L_2^0 = L_2(U_0, H)$ be the space of Hilbert-Schmidt operators from $U_0$ to $H$[20]. Then we have the following lemma[20].

**Lemma 2.2** For any $r \geqslant 1$ and any $L_2^0$-valued predictable process $\Phi(t), t \in [0,T]$,

$$\mathbb{E}\int_0^t \Phi(s)\mathrm{d}W(s) = 0$$

and

$$\mathbb{E}\left(\sup_{s\in[0,t]}\left|\int_0^s \Phi(\sigma)\mathrm{d}W(\sigma)\right|^{2r}\right) \leqslant c_r \sup_{s\in[0,t]}\mathbb{E}\left(\left|\int_0^s \Phi(\sigma)\mathrm{d}W(\sigma)\right|^{2r}\right) \leqslant C_r\, \mathbb{E}\left(\int_0^t \|\Phi(s)\|_{L_2^0}^2 \mathrm{d}s\right)^r,$$

with $c_r$ and $C_r$ some positive constants depending on $r$.

## 3 Time Uniform *a Priori* Estimates

In this section we give *a priori* estimates for $(n_t, n, E)$ in different spaces.

### 3.1 *A Priori* Estimates in $\mathcal{E}_0$

In this subsection we assume $(n_1, n_0, E_0) \in \varepsilon_0, f, q_1 \in H^{-1}(D)$, and $g, q_2 \in H_0^1(D)$. Applying the Itô formula to $\| E(t)\|_0^2$, we have

$$\| E(t)\|_0^2 = \| E_0\|_0^2 + 2\int_0^t \langle E_t(s), E(s)\rangle \mathrm{d}s + \| q_2\|_0^2 t$$

$$= \| E_0\|_0^2 - 2\gamma\int_0^t \| E(s)\|_0^2 \mathrm{d}s + 2\int_0^t \langle -\mathrm{i}g, E(s)\rangle \mathrm{d}s + 2\int_0^t \langle E(s), -\mathrm{i}q_2 \mathrm{d}w_2(s)\rangle + \| q_2\|_0^2 t$$

$$\leqslant \| E_0\|_0^2 - \gamma\int_0^t \| E(s)\|_0^2 \mathrm{d}s + \left(\frac{1}{\gamma}\| g\|_0^2 + \| q_2\|_0^2\right) t + 2\int_0^t \langle E(s), -\mathrm{i}q_2 \mathrm{d}w_2(s)\rangle. \tag{3.1}$$

Taking expectation on both sides of the above inequality yields

$$\mathbb{E}\| E(t)\|_0^2 \leqslant \mathbb{E}\| E_0\|_0^2 - \gamma\, \mathbb{E}\int_0^t \| E(s)\|_0^2 \mathrm{d}s + \left(\frac{1}{\gamma}\| g\|_0^2 + \| q_2\|_0^2\right) t.$$

Then by the Gronwall lemma

$$\mathbb{E}\| E(t)\|_0^2 \leqslant \mathrm{e}^{-\gamma t}\, \mathbb{E}\| E_0\|_0^2 + \frac{1}{\gamma^2}\| g\|_0^2 + \frac{1}{\gamma}\| q_2\|_0^2. \tag{3.2}$$

Also by (3.1), we have

$$\sup_{0\leqslant t\leqslant T}\| E(t)\|_0^2 \leqslant \| E_0\|_0^2 + \left(\frac{1}{\gamma}\| g\|_0^2 + \| q_2\|_0^2\right) T + \sup_{0\leqslant t\leqslant T}\left|\int_0^t \langle E(s), -\mathrm{i}q_2 \mathrm{d}w_2(s)\rangle\right|^2 + 1.$$

By Lemma 2.2, for some positive constant $C$,

$$\mathbb{E}\sup_{0\leqslant t\leqslant T}\left|\int_0^t \langle E(s), -\mathrm{i}q_2 \mathrm{d}w_2(s)\rangle\right|^2 \leqslant C\| q_2\|_0^2\, \mathbb{E}\int_0^T \| E(t)\|_0^2 \mathrm{d}t.$$

Then by (3.2) we have for any $T > 0$ a positive constant $C_T > 0$ such that

$$\mathbb{E}\sup_{0\leqslant t\leqslant T}\| E(t)\|_0^2 \leqslant C_T(\mathbb{E}\| E_0\|_0^4 + \| g\|_0^4 + \| q_2\|_0^4 + 1).$$

By the above estimates we further give an estimate of $\| E(t)\|_0^{2p}$ for any $p \geqslant 1$. Applying the Itô formula and Hölder inequality, we have

$$[\| E(t)\|_0^2]^p = \| E_0\|_0^{2p} + p\int_0^t \| E(s)\|_0^{2(p-1)} \frac{\mathrm{d}}{\mathrm{d}t}\| E(s)\|_0^2 \mathrm{d}s + 2p(p-1)\int_0^t \| E(s)\|_0^{2(p-1)}\| q_2\|_0^2 \mathrm{d}s$$

$$\leqslant \| E_0\|_0^{2p} + p\int_0^t \| E(s)\|_0^{2(p-1)}\left(-\gamma\| E(s)\|_0^2 + \frac{1}{\gamma}\| g\|_0^2 + \| q_2\|_0^2\right)\mathrm{d}s +$$

$$2p\int_0^t \| E(s)\|_0^{2(p-1)}\langle E(s), -\mathrm{i}q_2 \mathrm{d}w_2(s)\rangle + 2p(p-1)\int_0^t \| E(s)\|_0^{2(p-1)}\| q_2\|_0^2 \mathrm{d}s$$

$$\leqslant \| E_0\|_0^{2p} - \frac{\gamma p}{2}\int_0^t \| E(s)\|_0^{2p} \mathrm{d}s + 2p\int_0^t \| E(s)\|_0^{2(p-1)}\langle E(s), -\mathrm{i}q_2 dw_2(s)\rangle +$$

$$C(\| g\|_0, \| q_2\|_0^2) t.$$

Taking expectation on both sides of the above inequality, we derive

$$\mathbb{E}\| E(t)\|_0^{2p} \leqslant \mathbb{E}\| E_0\|_0^{2p} - \frac{\gamma p}{2}\, \mathbb{E}\int_0^t \| E(s)\|_0^{2p} \mathrm{d}s + C(\| g\|_0, \| q_2\|_0^2) t.$$

Then the Gronwall lemma yields

$$\mathbb{E}\|E(t)\|_0^{2p} \leqslant e^{-\gamma pt/2}\mathbb{E}\|E_0\|_0^{2p} + C(\|g\|_0, \|q_2\|_0^2). \tag{3.3}$$

Similarly by Lemma 2.2, for some positive constant $C$,

$$\mathbb{E}\sup_{0\leqslant t\leqslant T}\left|\int_0^t p\|E(s)\|_0^{2(p-1)}\langle E(s), -\mathrm{i}q_2\mathrm{d}w_2(s)\rangle\right|^2 \leqslant Cp^2\|q_2\|_0^2\mathbb{E}\int_0^T\|E(s)\|_0^{4p-2}\mathrm{d}s.$$

Then by (3.3) we have for any $T>0$

$$\mathbb{E}\sup_{0\leqslant t\leqslant T}\|E(t)\|_0^{2p} \leqslant C_T(E_0, g, q_2),$$

where $C_T(E_0, g, q_2)$ is a constant depending on $E_0, g, q_2$ and $T$. Then we draw the following lemma.

**Lemma 3.1** Assume $g, E_0, q_2 \in L^2(D)$. Then $E \in L^{2p}(\Omega; L^\infty(0,T;L^2(D))) \cap L^\infty(0,\infty;L^{2p}(\Omega, L^2(D)))$ for any $T>0$ and $p\geqslant 1$.

Let $m = n_t + \varepsilon n$ for some $\varepsilon > 0$. Then

$$\frac{1}{\lambda^2}\mathrm{d}m + \left[\left(\alpha - \frac{\varepsilon}{\lambda^2}\right)m - \varepsilon\left(\alpha - \frac{\varepsilon}{\lambda^2}\right)n - \Delta(n + |E|^2)\right]\mathrm{d}t = f\mathrm{d}t + \mathrm{d}W_1.$$

Applying the Itô formula to $\|m\|_{-1}^2$ we have

$$\begin{aligned}\frac{1}{\lambda^2}\|m(t)\|_{-1}^2 = {} & \frac{1}{\lambda^2}\|m_0\|_{-1}^2 - 2\left(\alpha - \frac{\varepsilon}{\lambda^2}\right)\int_0^t\|m(s)\|_{-1}^2\mathrm{d}s - 2\int_0^t\langle m(s), (n(s) + |E(s)|^2)\rangle\mathrm{d}s + \\ & 2\varepsilon\left(\alpha - \frac{\varepsilon}{\lambda^2}\right)\int_0^t\langle(-\Delta)^{-1}m(s), n(s)\rangle\mathrm{d}s + 2\int_0^t\langle(-\Delta)^{-1}m(s), f\rangle\mathrm{d}s + \\ & 2\int_0^t\langle(-\Delta)^{-1}m(s), q_1\rangle\mathrm{d}w_1(s) + \lambda^2\|q_1\|_{-1}^2 t,\end{aligned}$$

with $m_0(x) = \varepsilon n_0(x) + n_1(x)$. Notice that

$$-2\langle m, n\rangle = -2\langle(n_t + \varepsilon n), n\rangle = -\frac{\mathrm{d}}{\mathrm{d}t}\|n\|_0^2 - 2\varepsilon\|n\|_0^2.$$

Choosing $\varepsilon \in (0, \min\{\alpha\lambda^2/4, \lambda_1/2\alpha\})$, where $\lambda_1$ is the first eigenvalue of the Laplace operator $-\Delta$, results to

$$\left(\alpha - \frac{\varepsilon}{\lambda^2}\right)\|m\|_{-1}^2 - \varepsilon\left(\alpha - \frac{\varepsilon}{\lambda^2}\right)\langle(-\Delta)^{-1}m, n\rangle + \varepsilon\|n\|_0^2 \geqslant \frac{\varepsilon}{2}\left(\frac{1}{\lambda^2}\|m\|_{-1}^2 + \|n\|_0^2\right) + \frac{\alpha}{2}\|m\|_{-1}^2.$$

Then we derive

$$\begin{aligned}\frac{1}{\lambda^2}\|m(t)\|_{-1}^2 + \|n(t)\|_0^2 \leqslant {} & \frac{1}{\lambda^2}\|m_0\|_{-1}^2 + \|n_0\|_0^2 - \varepsilon\int_0^t\left(\frac{1}{\lambda^2}\|m(s)\|_{-1}^2 + \|n(s)\|_0^2\right)\mathrm{d}s \\ & - 2\int_0^t\langle n_t(s) + \varepsilon n(s), |E(s)|^2\rangle\mathrm{d}s + \frac{1}{\alpha}\|f\|_{-1}^2 t \\ & + 2\int_0^t\langle(-\Delta)^{-1}m(s), q_1\rangle\mathrm{d}w_1(s) + \lambda^2\|q_1\|_{-1}^2 t.\end{aligned} \tag{3.4}$$

Note that

$$\langle nE, (E_t + \gamma E)\rangle = \frac{1}{2}\langle n, (E_t\bar{E} + E\bar{E}_t)\rangle + \gamma\langle n, |E|^2\rangle.$$

Then applying the Itô formula to $\|E\|_1^2$ and by using (2.2) we have

$$\|E(t)\|_1^2 = \|E_0\|_1^2 - 2\gamma\int_0^t\|E(s)\|_1^2\mathrm{d}s - \int_0^t\langle n(s), (E_t(s)\bar{E}(s) + E(s)\bar{E}_t(s))\rangle\mathrm{d}s -$$

$$2\gamma\int_0^t\langle n(s),|E(s)|^2\rangle ds-2\int_0^t\langle\nabla E(s),-i\nabla g\rangle ds-2\int_0^t\langle n(s)E(s),-ig\rangle ds-$$

$$2\int_0^t\langle g,-iq_2dw_2(s)\rangle-2\int_0^t\langle(E_t(s)+\gamma E(s)),q_2dw_2(s)\rangle+\|q_2\|_1^2t. \tag{3.5}$$

Notice that the Itô formula yields

$$d\langle n,|E|^2\rangle=\langle dn,|E|^2\rangle+\langle n,d|E|^2\rangle+\mathrm{Tr}(nq_2^2)dt \tag{3.6}$$

which leads us to make the arrangement (3.4) + 2 × (3.5) + (3.6). Setting

$$H_0(t)=\frac{1}{\lambda^2}\|m\|_{-1}^2+\|n\|_0^2+2\|E\|_1^2+2\langle n,|E|^2\rangle$$

and $\beta=\min\{\varepsilon/2,\gamma\}$, then we have

$$H_0(t)-H_0(0)+\beta\int_0^tH_0(s)ds+2\gamma\int_0^t\|E(s)\|_1^2ds+\frac{\varepsilon}{2}\int_0^t\left(\frac{1}{\lambda^2}\|m(s)\|_{-1}^2+\|n(s)\|_0^2\right)ds$$

$$\leqslant\frac{1}{\alpha}\|f\|_{-1}^2t-4\int_0^t\langle\nabla E(s),-i\nabla g\rangle ds-4\int_0^t\langle n(s)E(s),-ig\rangle ds-4\int_0^t\langle g,-iq_2dw_2(s)\rangle$$

$$+2(\beta-\varepsilon-2\gamma)\int_0^t\langle n(s),|E(s)|^2\rangle ds-4\int_0^t\langle(E_t(s)+\gamma E(s)),q_2dw_2(s)\rangle$$

$$+2\int_0^t\langle(-\Delta)^{-1}m(s),q_1\rangle dw_1(s)+(2\|q_2\|_1^2+\lambda^2\|q_1\|_{-1}^2)t+2\int_0^t\mathrm{Tr}(n(s)q_2^2)ds. \tag{3.7}$$

By the Agmon inequality: for $u$ in $H^1$,

$$\|u\|_{L^\infty}\leqslant c\|u\|_1,$$

and by the Gagliardo-Nirenberg inequality: for $u$ in $H^1$,

$$\|u\|_{L^4}\leqslant c\|u\|_0^{3/4}\|u\|_1^{1/4},$$

we estimate the terms in the right-hand side of (3.7) as follows:

$$-4\langle\nabla E(s),-i\nabla g\rangle\leqslant4\|E(s)\|_1\|g\|_1\leqslant\gamma\|E(s)\|_1^2+\frac{4}{\gamma}\|g\|_1^2,$$

$$-4\langle n(s)E(s),-ig\rangle\leqslant4\|g\|_{L^\infty}\|n(s)\|_0\|E(s)\|_0\leqslant4c\|g\|_1\|n(s)\|_0\|E(s)\|_0$$

$$\leqslant\frac{\varepsilon}{8}\|n(s)\|_0^2+\frac{16c^2}{\varepsilon}\|g\|_1^4+\frac{16c^2}{\varepsilon}\|E(s)\|_0^4,$$

$$2(\beta-\varepsilon-2\gamma)\langle n(s),|E(s)|^2\rangle\leqslant2(2\gamma+\varepsilon-\beta)\|n(s)\|_0\|E(s)\|_{L^4}^2$$

$$\leqslant\frac{\varepsilon}{4}\|n(s)\|_0^2+C\|E(s)\|_{L^4}^4$$

$$\leqslant\frac{\varepsilon}{4}\|n(s)\|_0^2+C\|E(s)\|_0^3\|E(s)\|_1$$

$$\leqslant\frac{\varepsilon}{4}\|n(s)\|_0^2+\gamma\|E(s)\|_1^2+C\|E(s)\|_0^6,$$

$$2\mathrm{Tr}(nq_2^2)\leqslant\frac{\varepsilon}{8}\|n\|_0^2+\frac{8}{\varepsilon}\|q_2\|_{L^4}^4\leqslant\frac{\varepsilon}{8}\|n\|_0^2+\frac{8c}{\varepsilon}\|q_2\|_0^3\|q_2\|_1$$

$$\leqslant\frac{\varepsilon}{8}\|n\|_0^2+\frac{4c}{\varepsilon}\|q_2\|_0^6+\frac{4c}{\varepsilon}\|q_2\|_1^2.$$

Consequently, we deduce from all these estimates that

$$H_0(t)-H_0(0)+\beta\int_0^tH_0(s)ds$$

$$\leqslant\frac{1}{\alpha}\|f\|_{-1}^2t+\frac{4}{\gamma}\|g\|_1^2t+\frac{16c^2}{\varepsilon}\|g\|_1^4t+\frac{16c^2}{\varepsilon}\int_0^t\|E(s)\|_0^4\mathrm{d}s+$$

$$C\int_0^t\|E(s)\|_0^6\mathrm{d}s-4\int_0^t\langle g,-\mathrm{i}q_2\mathrm{d}w_2(s)\rangle-4\int_0^t\langle(E_t(s)+\gamma E(s)),q_2\mathrm{d}w_2(s)\rangle+$$

$$2\int_0^t\langle(-\Delta)^{-1}m(s),q_1\rangle\mathrm{d}w_1(s)+\left(2\|q_2\|_1^2+\lambda^2\|q_1\|_{-1}^2+\frac{4c}{\varepsilon}\|q_2\|_0^6+\frac{4c}{\varepsilon}\|q_2\|_1^2\right)t. \tag{3.8}$$

Taking expectation on both sides of the above inequality we derive

$$\mathbb{E}H_0(t)-\mathbb{E}H_0(0)+\beta\,\mathbb{E}\int_0^tH_0(s)\mathrm{d}s\leqslant C\,\mathbb{E}\int_0^t\|E(s)\|_0^6\mathrm{d}s+C(f,g,q_1,q_2)t.$$

By (3.3) and the Gronwall lemma, we have

$$\mathbb{E}H_0(t)\leqslant \mathrm{e}^{-\beta t}\,\mathbb{E}H_0(0)+C(E_0,f,g,q_1,q_2).$$

On the other hand

$$|2\langle n,|E|^2\rangle|\leqslant\frac{1}{2}\|n\|_0^2+\|E\|_1^2+2c^2\|E\|_0^6.$$

Hence

$$H_0(t)\geqslant\frac{1}{\lambda^2}\|m\|_{-1}^2+\frac{1}{2}\|n\|_0^2+\|E\|_1^2-2c^2\|E\|_0^6. \tag{3.9}$$

Therefore

$$\mathbb{E}[1/\lambda^2\|m\|_{-1}^2+1/2\|n\|_0^2+\|E\|_1^2]\leqslant \mathrm{e}^{-\beta t}\,\mathbb{E}H_0(0)+C(E_0,f,g,q_1,q_2). \tag{3.10}$$

Further we give an estimate of $H_0^p(t)$ for any $p\geqslant 1$. First as

$$\begin{aligned}\|nE\|_{-1}^2&=\int_D(-\Delta)^{-1}(nE)n\bar{E}\mathrm{d}x\leqslant\|(-\Delta)^{-1}(nE)\|_{L^\infty}\|n\|_0\|E\|_0\\&\leqslant c\|(-\Delta)^{-1}(nE)\|_1\|n\|_0\|E\|_0\\&\leqslant c\|nE\|_{-1}\|n\|_0\|E\|_0,\end{aligned}$$

by (3.9) we have

$$\|nE\|_{-1}^2\leqslant c^2\|n\|_0^2\|E\|_0^2\leqslant 2c^2(H_0(t)+2c^2\|E\|_0^6)\|E\|_0^2. \tag{3.11}$$

Applying the Itô formula to $H_0^p(t)$, by (2.2) and (3.9), we have

$$\begin{aligned}H_0^p(t)\leqslant\ & H_0^p(0)+p\int_0^tH_0^{p-1}(s)\frac{\mathrm{d}}{\mathrm{d}t}H_0(s)\mathrm{d}s+\\&\frac{1}{2}p(p-1)\int_0^tH_0^{p-2}(s)[16\|E(s)\|_1^2\|q_2\|_1^2+32\|g\|_0^2\|q_2\|_0^2+16\|n(s)E(s)\|_{-1}^2\|q_2\|_1^2\\&+4\|m(s)\|_{-1}^2\|q_1\|_{-1}^2]\mathrm{d}s\\\leqslant\ & H_0^p(0)+p\int_0^tH_0^{p-1}(s)\Big[-\beta H_0(s)+\frac{1}{\alpha}\|f\|_{-1}^2+\frac{4}{\gamma}\|g\|_1^2+\frac{16c^2}{\varepsilon}\|g\|_1^4+\frac{16c^2}{\varepsilon}\|E(s)\|_0^4\\&+C\|E(s)\|_0^6+2\|q_2\|_1^2+\lambda^2\|q_1\|_{-1}^2+\frac{4c}{\varepsilon}\|q_2\|_0^6+\frac{4c}{\varepsilon}\|q_2\|_1^2\Big]\mathrm{d}s+p\int_0^tH_0^{p-1}(s)\cdot\\&[-4\langle g,\mathrm{i}q_2\mathrm{d}w_2(s)\rangle-4\langle(E_t(s)+\gamma E(s)),q_2\mathrm{d}w_2(s)\rangle+2\langle(-\Delta)^{-1}m(s),q_1\mathrm{d}w_1(s)\rangle]\\&+\frac{1}{2}p(p-1)\int_0^tH_0^{p-2}(s)[32\|g\|_0^2\|q_2\|_0^2+16(H_0(s)+2c^2\|E(s)\|_0^6)\|q_2\|_1^2\\&+32c^2(H_0(s)+2c^2\|E(s)\|_0^6)\|E(s)\|_0^2\|q_2\|_1^2+4\lambda^2(H_0(s)+2c^2\|E(s)\|_0^6)\|q_1\|_{-1}^2]\mathrm{d}s.\end{aligned}$$

Taking expectation on both sides of the above inequality, by the Hölder inequality and (3.3), we derive

$$\mathbb{E}H_0^p(t)\leqslant \mathbb{E}H_0^p(0)-\frac{\beta p}{2}\mathbb{E}\int_0^t H_0^p(s)\mathrm{d}s+C(E_0,f,g,q_1,q_2)t.$$

Then the Gronwall lemma yields

$$\mathbb{E}H_0^p(t)\leqslant \mathrm{e}^{-\beta pt/2}\mathbb{E}H_0^p(0)+C(E_0,f,g,q_1,q_2). \tag{3.12}$$

Now we give an estimate of $(n_t,n,E)$ in space $L^2(\Omega,L^\infty(0,T;\mathcal{E}_0))$ for any $T>0$. First by Lemma 2.2, for some positive constant $C$, we have

$$\begin{aligned}-4\mathbb{E}\sup_{0\leqslant t\leqslant T}\int_0^t\langle \mathrm{i}\nabla E,\nabla q_2\mathrm{d}w_2(s)\rangle&\leqslant 2\mathbb{E}\sup_{0\leqslant t\leqslant T}\left|\int_0^t\langle \mathrm{i}\nabla E,\nabla q_2\mathrm{d}w_2(s)\rangle\right|^2+2\\&\leqslant 2C\|q_2\|_1^2\mathbb{E}\int_0^T\|E(t)\|_1^2\mathrm{d}t+2,\end{aligned}$$

$$\begin{aligned}-4\mathbb{E}\sup_{0\leqslant t\leqslant T}\int_0^t\langle \mathrm{i}n(s)E(s),q_2\mathrm{d}w_2(s)\rangle&\leqslant 2\mathbb{E}\sup_{0\leqslant t\leqslant T}\left|\int_0^t\langle \mathrm{i}n(s)E(s),q_2\mathrm{d}w_2(s)\rangle\right|^2+2\\&\leqslant 2C\|q_2\|_1^2\mathbb{E}\int_0^T\|n(E)(t)\|_{-1}^2\mathrm{d}t+2.\end{aligned}$$

In the same way

$$-4\mathbb{E}\sup_{0\leqslant t\leqslant T}\int_0^t\langle -\mathrm{i}g,q_2\mathrm{d}w_2(s)\rangle\leqslant 2CT\|q_2\|_0^2\|g\|_0^2+2,$$

$$\begin{aligned}2\mathbb{E}\sup_{0\leqslant t\leqslant T}\int_0^t\langle(-\Delta)^{-1}m(s),q_1\mathrm{d}w_1(s)\rangle&\leqslant \mathbb{E}\sup_{0\leqslant t\leqslant T}\left|\int_0^t\langle(-\Delta)^{-1}m(s),q_1\mathrm{d}w_1(s)\rangle\right|^2+1\\&\leqslant C\|q_1\|_{-1}^2\mathbb{E}\int_0^T\|m(t)\|_{-1}^2\mathrm{d}t+1.\end{aligned}$$

Then we deduce from all these estimates, Lemma 3.1, and (3.8)~(3.12) that

$$\mathbb{E}\sup_{0\leqslant t\leqslant T}\|(n_t(t),n(t),E(t))\|_{\mathcal{E}_0}^2\leqslant C_T(n_1,n_0,E_0,f,g,q_1,q_2)$$

for positive constant $C_T$ depending on the initial data.

And similarly we can derive the following estimate in space $L^{2p}(\Omega,L^\infty(0,T;\mathcal{E}_0))$:

$$\mathbb{E}\sup_{0\leqslant t\leqslant T}\|(n_t(t),n(t),E(t))\|_{\mathcal{E}_0}^{2p}\leqslant C_T(n_1,n_0,E_0,f,g,q_1,q_2).$$

Then we have the following result.

**Lemma 3.2** Assume $(n_1,n_0,E_0)\in\mathcal{E}_0,f,q_1\in H^{-1}(D)$, and $g,q_2\in H_0^1(D)$. Then for any $T>0$ and $p\geqslant 1,(n_t,n,E)\in L^{2p}(\Omega;L^\infty(0,T;\mathcal{E}_0))\cap L^\infty(0,\infty;L^{2p}(\Omega,\mathcal{E}_0))$.

### 3.2 *A Priori* Estimates in $\mathcal{E}_1$

In this subsection we assume $(n_1,n_0,E_0)\in\mathcal{E}_1,f\in L^2(D),g,q_1\in H_0^1(D)$, and $q_2\in H^2(D)\cap H_0^1(D)$. Applying the Itô formula to $\|E\|_2^2$ and $\|m\|_0^2$, respectively, we have

$$\begin{aligned}\|E(t)\|_2^2=\|E_0\|_2^2-2\gamma\int_0^t\|E(s)\|_2^2\mathrm{d}s+2\int_0^t\langle n(s)E(s),(\Delta E_t(s)+\gamma\Delta E(s))\rangle\mathrm{d}s\\+2\int_0^t\langle g,(\Delta E_t(s)+\gamma\Delta E(s))\rangle\mathrm{d}s+2\int_0^t\langle(\Delta E_t(s)+\gamma\Delta E),q_2\mathrm{d}w_2(s)\rangle+t\|q_2\|_2^2\end{aligned}$$

and

$$\frac{1}{\lambda^2}\| m(t)\|_0^2=\frac{1}{\lambda^2}\| m_0\|_0^2-2\left(\alpha-\frac{\varepsilon}{\lambda^2}\right)\int_0^t\| m(s)\|_0^2\mathrm{d}s+2\varepsilon\left(\alpha-\frac{\varepsilon}{\lambda^2}\right)\int_0^t\langle m(s),n(s)\rangle\mathrm{d}s+$$
$$2\int_0^t\langle\Delta(n(s)+|E(s)|^2)+f,m(s)\rangle\mathrm{d}s+2\int_0^t\langle m(s),q_1\rangle\mathrm{d}w_1(s)+\lambda^2\| q_1\|_0^2 t.$$

Exactly as in Sec. 3.1, choose $\varepsilon\in(0,\min\{\alpha\lambda^2/4,\lambda_1/2\alpha\})$ in order to obtain

$$\frac{1}{\lambda^2}\| m(t)\|_0^2+\| n(t)\|_1^2\leqslant\frac{1}{\lambda^2}\| m_0\|_0^2+\| n_0\|_1^2-\varepsilon\int_0^t\left(\frac{1}{\lambda^2}\| m(s)\|_0^2+\| n(s)\|_1^2\right)\mathrm{d}s+2\int_0^t\langle n_t(s)+$$
$$\varepsilon n(s),\Delta|E(s)|^2\rangle\mathrm{d}s+2\int_0^t\langle m(s),q_1\rangle\mathrm{d}w_1(s)+\frac{1}{\alpha}\| f\|_0^2 t+\lambda^2\| q_1\|_0^2 t.$$

On the other hand, since

$$\Delta|E|^2=2|\nabla E|^2+2\mathrm{Re}(E\Delta\bar{E}),$$

we have

$$-4\frac{\mathrm{d}}{\mathrm{d}t}\langle nE,\Delta E\rangle$$
$$=-2\langle n_t,\Delta|E|^2\rangle+4\langle n_t,|\nabla E|^2\rangle-4\langle nE_t,\Delta E\rangle-4\langle nE,\Delta E_t\rangle+4\mathrm{Tr}(nq_2(-\Delta)q_2).$$

Let us set

$$H_1(t)=\frac{1}{\lambda^2}\| m\|_0^2+\| n\|_1^2+2\| E\|_2^2-4\langle nE,\Delta E\rangle-4\langle g,\Delta E\rangle.$$

Considering $\beta=\min\{\varepsilon/2,\gamma\}$, we deduce that $H_1(t)$ satisfies

$$H_1(t)-H_1(0)+\beta\int_0^t H_1(s)\mathrm{d}s+2\gamma\int_0^t\| E(s)\|_2^2\mathrm{d}s+\frac{\varepsilon}{2}\int_0^t\left(\frac{1}{\lambda^2}\| m(s)\|_0^2+\| n(s)\|_1^2\right)\mathrm{d}s$$
$$\leqslant\frac{1}{\alpha}\| f\|_0^2 t-4\int_0^t\langle n(s)\Delta E(s),E_t(s)\rangle\mathrm{d}s+4\int_0^t\langle m(s),|\nabla E(s)|^2\rangle\mathrm{d}s+4(\gamma+\varepsilon-\beta)\cdot$$
$$\int_0^t\langle n(s)E(s),\Delta E(s)\rangle\mathrm{d}s+4(\gamma-\beta)\int_0^t\langle g,\Delta E(s)\rangle\mathrm{d}s+2\int_0^t\langle m(s),q_1\rangle\mathrm{d}w_1(s)$$
$$+4\int_0^t\langle\Delta E_t(s)+\gamma\Delta E(s),q_2\mathrm{d}w_2(s)\rangle+\lambda^2\| q_1\|_0^2 t+2\| q_2\|_2^2 t+4\int_0^t\mathrm{Tr}(n(s)q_2(-\Delta)q_2)\mathrm{d}s.$$
$$(3.13)$$

From (3.2) we have

$$-4\int_0^t\langle n(s)\Delta E(s),E_t(s)\rangle\mathrm{d}s=4\int_0^t\langle n^2(s)\Delta E(s),-\mathrm{i}E(s)\rangle\mathrm{d}s+4\gamma\int_0^t\langle n(s)\Delta E(s),E(s)\rangle\mathrm{d}s$$
$$+4\int_0^t\langle n(s)\Delta E(s),-\mathrm{i}g\rangle\mathrm{d}s+4\int_0^t\langle n(s)\Delta E(s),-\mathrm{i}q_2\mathrm{d}w_2(s)\rangle.$$

We majorize the right-hand side of (3.13) as follows:

$$4\langle n^2\Delta E,-\mathrm{i}E\rangle\leqslant4\| E\|_{L^\infty}\|\Delta E\|_0\| n\|_{L^4}^2\leqslant4c\| E\|_1\| E\|_2\| n\|_0^{3/2}\| n\|_1^{1/2}$$
$$\leqslant\frac{\gamma}{2}\| E\|_2^2+\frac{\varepsilon}{2}\| n\|_1^2+C(\| E\|_1^8,\| n\|_0^{12}),$$
$$4(2\gamma+\varepsilon-\beta)\langle n\Delta E,E\rangle\leqslant4(2\gamma+\varepsilon-\beta)\| n\|_0\|\Delta E\|_0\| E\|_{L^\infty}$$
$$\leqslant4(2\gamma+\varepsilon-\beta)c\| n\|_0\| E\|_2\| E\|_1$$
$$\leqslant\frac{\gamma}{2}\| E\|_2^2+C(\| n\|_0^4,\| E\|_1^4),$$

$$4\langle n\Delta E,-\mathrm{i}g\rangle\leqslant 4\|g\|_{L^\infty}\|n\|_0\|\Delta E\|_0\leqslant 4c\|g\|_1\|n\|_0\|E\|_2\leqslant\frac{\gamma}{4}\|E\|_2^2+C(\|n\|_0^4,\|g\|_1^4),$$

$$4(\gamma-\beta)\langle g,\Delta E\rangle\leqslant\frac{\gamma}{4}\|E\|_2^2+C(\|g\|_0^2),$$

$$4\langle m,|\nabla E|^2\rangle\leqslant 4\|m\|_0\|\nabla E\|_{L^4}^2\leqslant 4c\|m\|_0\|E\|_1^{3/2}\|E\|_2^{1/2}$$

$$\leqslant\varepsilon/2\lambda^2\|m\|_0^2+\gamma/2\|E\|_2^2+C(\|E\|_1^6),$$

$$4\mathrm{Tr}(nq_2(-\Delta)q_2)\leqslant 4\|q_2\|_{L^\infty}\|n\|_0\|\Delta q_2\|_0\leqslant 4c\|n\|_0\|q_2\|_2^2\leqslant C(\|n\|_0^2,\|q_2\|_2^4).$$

Therefore

$$H_1(t)-H_1(0)+\beta\int_0^t H_1(s)\mathrm{d}s\leqslant 4\int_0^t\langle n(s)\Delta E(s),-\mathrm{i}q_2\mathrm{d}w_2(s)\rangle+2\int_0^t\langle m(s),q_1\rangle\mathrm{d}w_1(s)+$$

$$4\int_0^t\langle\Delta E_t(s)+\gamma\Delta E(s),q_2\mathrm{d}w_2(s)\rangle$$

$$+C(\|E\|_1^8,\|n\|_0^{12},\|f\|_0^2,\|g\|_1^4,\|q_1\|_0^2,\|q_2\|_2^4)t.$$

Taking expectation on both sides of the above inequality and by the Gronwall lemma, we have

$$\mathbb{E}H_1(t)\leqslant \mathrm{e}^{-\beta t}\,\mathbb{E}H_1(0)+C(n_0,n_1,E_0,f,g,q_1,q_2).$$

On the other hand

$$|-4\langle nE,\Delta E\rangle|\leqslant\frac{1}{2}\|E\|_2^2+C(\|E\|_1^4,\|n\|_0^4),$$

$$|-4\langle g,\Delta E\rangle|\leqslant\frac{1}{2}\|E\|_2^2+C(\|g\|_0^2).$$

Then

$$H_1(t)\geqslant\frac{1}{\lambda^2}\|m\|_0^2+\|n\|_1^2+\|E\|_2^2-C(\|E\|_1^4,\|n\|_0^4,\|g\|_0^2). \tag{3.14}$$

Therefore

$$\mathbb{E}\left[\frac{1}{\lambda^2}\|m\|_0^2+\|n\|_1^2+\|E\|_2^2\right]\leqslant \mathrm{e}^{-\beta t}\,\mathbb{E}H_1(0)+C(n_0,n_1,E_0,f,g,q_1,q_2).$$

By the above estimates we further give an estimate of $H_1^p(t)$ for any $p\geqslant 1$. First

$$\|nE\|_0^2=\int_D nEn\bar{E}\mathrm{d}x\leqslant\|nE\|_0\|n\|_0\|E\|_{L^\infty}\leqslant c\|nE\|_0\|n\|_0\|E\|_1.$$

Then

$$\|nE\|_0^2\leqslant c^2\|n\|_0^2\|E\|_1^2\leqslant\frac{c^2}{2}(\|n\|_0^4+\|E\|_1^4).$$

Moreover

$$\|n\Delta E\|_{-1}^2\leqslant c^2\|n\|_0^2\|E\|_2^2.$$

Then applying the Itô formula to $H_1^p(t)$, by (3.2) and (3.14), we have

$$H_1^p(t)\leqslant H_1^p(0)+p\int_0^t H_1^{p-1}(s)\frac{\mathrm{d}}{\mathrm{d}t}H_1(s)\mathrm{d}s+\frac{1}{2}p(p-1)\int_0^t H_1^{p-2}(s)[16\|n(s)\Delta E(s)\|_{-1}^2\|q_2\|_1^2+$$

$$4\|m(s)\|_0^2\|q_1\|_0^2+16\|E(s)\|_2^2\|q_2\|_2^2+16\|n(s)E(s)\|_0^2\|q_2\|_2^2+16\|g\|_1^2\|q_2\|_1^2]\mathrm{d}s$$

$$\leqslant H_1^p(0)+p\int_0^t H_1^{p-1}(s)[-\beta H_1(s)+C(\|E\|_1^8,\|n\|_0^{12},\|f\|_0^2,\|g\|_1^4,\|q_1\|_0^2,\|q_2\|_2^4)]\mathrm{d}s+$$

$$p\int_0^t H_1^{p-1}(s)\cdot[4\langle n(s)\Delta E(s),-\mathrm{i}q_2\mathrm{d}w_2(s)\rangle+4\langle\Delta E_t(s)+\gamma\Delta E(s),q_2\mathrm{d}w_2(s)\rangle+$$

$2\langle m(s), q_1\rangle \mathrm{d}w_1(s)] + \frac{1}{2}p(p-1)\int_0^t H_1^{p-2}(s)[16c^2\|n\|_0^2(H_1(s) +$

$C(\|E\|_1^4, \|n\|_0^4, \|g\|_0^2))\|q_2\|_1^2 + 4\lambda^2(H_1(s) + C(\|E\|_1^4, \|n\|_0^4, \|g\|_0^2))\|q_1\|_0^2 +$

$16(H_1(s) + C(\|E\|_1^4, \|n\|_0^4, \|g\|_0^2))\|q_2\|_2^2 + 8c^2(\|n\|_0^4 + \|E\|_1^4)\|q_2\|_2^2 + 16\|g\|_1^2\|q_2\|_1^2]\mathrm{d}s.$

Taking expectation on both sides of the above inequality, by the Hölder inequality and (3.9) and (3.12), we have

$$\mathbb{E}H_1^p(t) \leqslant \mathbb{E}H_1^p(0) - \frac{\beta p}{2}\mathbb{E}\int_0^t H_1^p(s)\mathrm{d}s + C(n_0, n_1, E_0, f, g, q_1, q_2)t.$$

Using the Gronwall lemma, we derive

$$\mathbb{E}H_1^p(t) \leqslant \mathrm{e}^{-\beta p t/2}\mathbb{E}H_1^p(0) + C(n_0, n_1, E_0, f, g, q_1, q_2).$$

Having the above estimates, by the same approach for the estimates in space $L^{2p}(\Omega, L^\infty(0, T; \mathcal{E}_0))$ in Sec.3.1, we can have for any $p \geqslant 1$ and $T > 0$

$$\mathbb{E}\sup_{0\leqslant t\leqslant T}\|(n_t(t), n(t), E(t))\|_{\mathcal{E}_1}^{2p} \leqslant C_T(n_1, n_0, E_0, f, g, q_1, q_2).$$

Then we draw the following result.

**Lemma 3.3** Assume $(n_1, n_0, E_0) \in \mathcal{E}_1, f \in L^2(D), g, q_1 \in H_0^1(D)$, and $q_2 \in H^2(D) \cap H_0^1(D)$. Then for any $T > 0$ and $p \geqslant 1, (n_t, n, E) \in L^{2p}(\Omega; L^\infty(0, T; \mathcal{E}_1)) \cap L^\infty(0, \infty; L^{2p}(\Omega, \mathcal{E}_1))$.

### 3.3 *A Priori* Estimates in $\mathcal{E}_2$

In this section we always assume $(n_1, n_0, E_0) \in \mathcal{E}_2, f, g, q_1 \in H_0^1(D)$, and $q_2 \in H_0^1(D) \cap H^3(D)$. Applying the Itô formula to $\|E\|_3^2$ and $\|m\|_1^2$, respectively, we have

$$\|E(t)\|_3^2 = \|E_0\|_3^2 - 2\gamma\int_0^t\|E(s)\|_3^2\mathrm{d}s - 2\int_0^t\langle n(s)E(s), \Delta^2E_t(s) + \gamma\Delta^2E(s)\rangle\mathrm{d}s -$$
$$2\int_0^t\langle g, \Delta^2E_t(s) + \gamma\Delta^2E(s)\rangle\mathrm{d}s - 2\int_0^t\langle\Delta^2E_t(s) + \gamma\Delta^2E(s), q_2\mathrm{d}w_2(s)\rangle + \|q_2\|_3^2 t \tag{3.15}$$

and

$$\frac{1}{\lambda^2}\|m(t)\|_1^2 = \frac{1}{\lambda^2}\|m_0\|_1^2 - 2\left(\alpha - \frac{\varepsilon}{\lambda^2}\right)\int_0^t\|m(s)\|_1^2\mathrm{d}s + 2\varepsilon\left(\alpha - \frac{\varepsilon}{\lambda^2}\right)\int_0^t\langle\nabla m(s), \nabla n(s)\rangle\mathrm{d}s -$$
$$2\int_0^t\langle\Delta(n(s) + |E(s)|^2), \Delta m(s)\rangle\mathrm{d}s - 2\int_0^t\langle f, \Delta m(s)\rangle\mathrm{d}s -$$
$$2\int_0^t\langle\Delta m(s), q_1\rangle\mathrm{d}w_1(s) + \lambda^2\|q_1\|_1^2 t.$$

Thanks to the choice of $\varepsilon$ made from the beginning, this reads

$$\frac{1}{\lambda^2}\|m\|_1^2 + \|n\|_2^2 \leqslant \frac{1}{\lambda^2}\|m_0\|_1^2 + \|n_0\|_2^2 - \varepsilon\int_0^t\left(\frac{1}{\lambda^2}\|m(s)\|_1^2 + \|n(s)\|_2^2\right)\mathrm{d}s + \frac{1}{\alpha}\|f\|_1^2 t -$$
$$2\int_0^t\langle\Delta m(s), \Delta|E(s)|^2\rangle\mathrm{d}s - 2\int_0^t\langle\Delta m(s), q_1\rangle\mathrm{d}w_1(s) + \lambda^2\|q_1\|_1^2 t. \tag{3.16}$$

Notice that

$$
\begin{aligned}
&\langle nE, \Delta^2 E_t + \gamma\Delta^2 E\rangle \\
&= \langle \Delta nE, (\Delta E_t + \gamma\Delta E)\rangle + \langle n\Delta E, (\Delta E_t + \gamma\Delta E)\rangle + 2\langle \nabla n\, \nabla E, (\Delta E_t + \gamma\Delta E)\rangle.
\end{aligned}
$$

Now

$$
\begin{aligned}
&\frac{\mathrm{d}}{\mathrm{d}t}\langle \Delta n, \Delta |E|^2\rangle \\
&= \langle \Delta n_t, \Delta |E|^2\rangle + \langle \Delta |E|_t^2, \Delta n\rangle + 2\mathrm{Tr}((-\Delta)n[(-\Delta)^{1/2}q_2]^2) + 2\mathrm{Tr}((-\Delta)nq_2(-\Delta)q_2),
\end{aligned}
$$

where

$$
\frac{1}{2}\langle \Delta |E|_t^2, \Delta n\rangle = \langle E\Delta n, \Delta E_t\rangle + \langle \Delta E\Delta n, E_t\rangle + 2\langle \nabla E_t \Delta n, \nabla E\rangle.
$$

And

$$
\frac{\mathrm{d}}{\mathrm{d}t}\langle n, |\Delta E|^2\rangle = \langle n_t, |\Delta E|^2\rangle + 2\langle n\Delta E, \Delta E_t\rangle + \mathrm{Tr}(n[(-\Delta)q_2]^2).
$$

Moreover

$$
\langle \nabla E_t \Delta n, \nabla E\rangle = -\langle \Delta E_t\, \nabla n, \nabla E\rangle - \langle \nabla E_t\, \nabla n, \Delta E\rangle,
$$

and finally

$$
\begin{aligned}
&\frac{\mathrm{d}}{\mathrm{d}t}\langle \nabla n\, \nabla E, \Delta E\rangle \\
&= \langle \nabla n\, \nabla E, \Delta E_t\rangle + \langle \nabla n\, \nabla E_t, \Delta E\rangle + \langle \nabla n_t\, \nabla E, \Delta E\rangle + \mathrm{Tr}((-\Delta)^{1/2}n(-\Delta)^{1/2}q_2(-\Delta)q_2).
\end{aligned}
$$

Consequently, from all these identities it can be derived that

$$
\begin{aligned}
\langle nE, \Delta^2 E_t + \gamma\Delta^2 E\rangle = {} & \frac{1}{2}\frac{\mathrm{d}}{\mathrm{d}t}\langle \Delta n, \Delta |E|^2\rangle - \frac{1}{2}\langle \Delta n_t, \Delta |E|^2\rangle - \mathrm{Tr}((-\Delta)n[(-\Delta)^{1/2}q_2]^2) - \\
& \mathrm{Tr}((-\Delta)nq_2(-\Delta)q_2) - \langle \Delta E\Delta n, E_t\rangle + 4\frac{\mathrm{d}}{\mathrm{d}t}\langle \nabla n\, \nabla E, \Delta E\rangle - \\
& 4\langle \nabla n\, \nabla E_t, \Delta E\rangle - 4\langle \nabla n_t\, \nabla E, \Delta E\rangle - 4\mathrm{Tr}((-\Delta)^{1/2}n(-\Delta)^{1/2}q_2(-\Delta)q_2) + \\
& 2\langle \nabla E_t\, \nabla n, \Delta E\rangle + \gamma\langle \Delta nE, \Delta E\rangle + \frac{1}{2}\frac{\mathrm{d}}{\mathrm{d}t}\langle n, |\Delta E|^2\rangle - \frac{1}{2}\langle n_t, |\Delta E|^2\rangle - \\
& \frac{1}{2}\mathrm{Tr}(n[(-\Delta)q_2]^2) + \gamma\langle n, |\Delta E|^2\rangle + 2\gamma\langle \nabla n\, \nabla E, \Delta E\rangle.
\end{aligned}
$$

On the other hand, we also notice that

$$
-\langle g, \Delta^2 E_t\rangle = \langle \nabla g, \nabla\Delta E_t\rangle = \frac{\mathrm{d}}{\mathrm{d}t}\langle \nabla g, \nabla\Delta E\rangle.
$$

Then set

$$
\begin{aligned}
H_2(t) = {} & \frac{1}{\lambda^2}\| m\|_1^2 + \| n\|_2^2 + 2\| E\|_3^2 + 2\langle \Delta n, \Delta |E|^2\rangle + 2\langle n, |\Delta E|^2\rangle + \\
& 16\langle \nabla n\, \nabla E, \Delta E\rangle - 4\langle \nabla g, \nabla\Delta E\rangle.
\end{aligned}
$$

Choose $\beta = \min\{\varepsilon/2, \gamma\}$. From (3.15) and (3.16) we deduce that $H_2(t)$ satisfies

$$
\begin{aligned}
& H_2(t) - H_2(0) + \beta\int_0^t H_2(s)\mathrm{d}s + \frac{\varepsilon}{2}\int_0^t \left(\frac{1}{\lambda^2}\| m(s)\|_1^2 + \| n(s)\|_2^2\right)\mathrm{d}s + 2\gamma\int_0^t \| E(s)\|_3^2\mathrm{d}s \\
\leqslant {} & \frac{1}{\alpha}\| f\|_1^2 t + 2(\beta - \varepsilon)\int_0^t \langle \Delta n(s), \Delta |E(s)|^2\rangle\mathrm{d}s + 2(\beta - \varepsilon - 2\gamma)\int_0^t \langle n(s), |\Delta E(s)|^2\rangle\mathrm{d}s +
\end{aligned}
$$

$8(2\beta-2\varepsilon-\gamma)\cdot\int_0^t\langle\nabla n(s)\,\nabla E(s),\Delta E\rangle ds+4(\gamma-\beta)\int_0^t\langle\nabla g,\nabla\Delta E\rangle ds+4\int_0^t\langle\Delta E(s)\cdot\Delta n(s),E_t(s)\rangle ds+8\int_0^t\langle\nabla n(s)\,\nabla E_t(s),\Delta E(s)\rangle ds+16\int_0^t\langle\nabla E(s)\,\nabla m(s),\Delta E(s)\rangle ds+2\int_0^t\langle m(s),|\Delta E(s)|^2\rangle ds-4\gamma\int_0^t\langle\Delta n(s)E(s),\Delta E(s)\rangle ds-4\int_0^t\langle\Delta^2E_t(s)+\gamma\Delta^2E(s),q_2dw_2(s)\rangle-2\int_0^t\langle\Delta m(s),q_1\rangle dw_1(s)+2\|q_2\|_3^2t+2\int_0^t\mathrm{Tr}(n(s)[(-\Delta)q_2]^2)ds+16\int_0^t\mathrm{Tr}((-\Delta)^{1/2}n(s)(-\Delta)^{1/2}q_2(-\Delta)q_2)ds+\lambda^2\|q_1\|_1^2t+4\int_0^t\mathrm{Tr}((-\Delta)n(s)[(-\Delta)^{1/2}q_2]^2)ds+4\int_0^t\mathrm{Tr}((-\Delta)n(s)q_2(-\Delta)q_2)ds.$

It remains to majorize the right-hand side term by term. We have

$$\begin{aligned}2(\beta-\varepsilon)\langle\Delta n,\Delta|E|^2\rangle&\leqslant 4(\varepsilon-\beta)\Big(\|\Delta n\|_0\|\nabla E\|_{L^4}^2+\|\Delta n\|_0\|\Delta E\|_0\|E\|_{L^\infty}\Big)\\&\leqslant 4c(\varepsilon-\beta)(\|n\|_2\|E\|_1^{3/2}\|E\|_2^{1/2}+\|n\|_2\|E\|_2\|E\|_1)\\&\leqslant\frac{\varepsilon}{20}\|n\|_2^2+C(\|E\|_1^6,\|E\|_2^4),\end{aligned}$$

$$\begin{aligned}2(\beta-\varepsilon-2\gamma)\langle n,|\Delta E|^2\rangle&\leqslant 2(2\gamma+\varepsilon-\beta)\|n\|_0\|\Delta E\|_{L^4}^2\\&\leqslant 2c(2\gamma+\varepsilon-\beta)\|n\|_0\|E\|_2^{3/2}\|E\|_3^{1/2}\\&\leqslant\frac{\gamma}{2}\|E\|_3^2+C(\|E\|_2^6,\|n\|_0^2),\end{aligned}$$

$$8(2\beta-2\varepsilon-\gamma)\langle\nabla n\,\nabla E,\Delta E\rangle\leqslant C(\|n\|_1^2,\|E\|_2^4),$$

$$4(\gamma-\beta)\mathrm{Re}\langle\nabla g,\nabla\Delta E\rangle\leqslant\frac{\gamma}{2}\|E\|_3^2+C(\|g\|_1^2).$$

Using (2.2), we have

$$\begin{aligned}4\int_0^t\langle\Delta n(s)\Delta E(s),E_t(s)\rangle ds=&-4\int_0^t\langle\Delta E(s)\Delta n(s),-in(s)E(s)\rangle ds-\\&4\gamma\int_0^t\langle\Delta n(s)\Delta E(s),E(s)\rangle ds-4\int_0^t\langle\Delta n(s)\Delta E(s),-ig\rangle ds-\\&4\int_0^t\langle\Delta n(s)\Delta E(s),-iq_2dw_2(s)\rangle.\end{aligned}$$

For the terms on the right-hand side of the above identity we have

$$\begin{aligned}-4\langle\Delta E\Delta n,-in(s)E\rangle&\leqslant 4\|\Delta E\|_0\|\Delta n\|_0\|n\|_{L^\infty}\|E\|_{L^\infty}\leqslant 4c\|E\|_2\|n\|_2\|n\|_1\|E\|_1\\&\leqslant\frac{\varepsilon}{20}\|n\|_2^2+C(\|E\|_1^4,\|E\|_2^8,\|n\|_1^8),\end{aligned}$$

$$\begin{aligned}-4\gamma\langle\Delta n\Delta E,E\rangle&\leqslant 4\gamma\|n\|_2\|E\|_2\|E\|_{L^\infty}\leqslant 4c\gamma\|n\|_2\|E\|_2\|E\|_1\\&\leqslant\frac{\varepsilon}{20}\|n\|_2^2+C(\|E\|_1^4,\|E\|_2^4),\end{aligned}$$

$$-4\langle\Delta n\Delta E,-ig\rangle\leqslant 4c\|n\|_2\|E\|_2\|g\|_1\leqslant\frac{\varepsilon}{20}\|n\|_2^2+C(\|g\|_1^4,\|E\|_2^4).$$

In the same way we have

$$8\int_0^t \langle \nabla E_t(s)\,\nabla n(s),\Delta E(s)\rangle \mathrm{d}s = -8\int_0^t \langle \nabla\Delta E(s)\,\nabla n(s),-\mathrm{i}\Delta E(s)\rangle \mathrm{d}s + 8\int_0^t \langle (\nabla n(s))^2 E(s),$$

$$-\mathrm{i}\Delta E(s)\rangle \mathrm{d}s + 8\int_0^t \langle n(s)\,\nabla E(s)\,\nabla n(s),-\mathrm{i}\Delta E(s)\rangle \mathrm{d}s -$$

$$8\gamma\int_0^t \langle \nabla E(s),\Delta E(s)\,\nabla n(s)\rangle \mathrm{d}s + 8\int_0^t \langle \nabla g,\Delta E(s)\,\nabla n(s)\rangle \mathrm{d}s +$$

$$8\int_0^t \langle \Delta E(s)\,\nabla n(s),\nabla q_2 \mathrm{d}w_2(s)\rangle.$$

And for the terms on the right-hand side

$$-8\langle \nabla\Delta E(s)\,\nabla n(s),-\mathrm{i}\Delta E(s)\rangle \leqslant 8\|\nabla\Delta E\|_0\|\nabla n\|_{L^4}\|\Delta E\|_{L^4}$$

$$\leqslant 8c\|E\|_3\|n\|_1^{3/4}\|n\|_2^{1/4}\|E\|_2^{3/4}\|E\|_3^{1/4}$$

$$\leqslant \frac{\varepsilon}{20}\|n\|_2^2 + \frac{\gamma}{2}\|E\|_3^2 + C(\|n\|_1^6,\|E\|_2^6),$$

$$8\langle (\nabla n(s))^2 E(s),-\mathrm{i}\Delta E(s)\rangle \leqslant 8\|E\|_{L^\infty}\|\Delta E\|_0\|\nabla n\|_{L^4}^2 \leqslant 8c\|E\|_1\|E\|_2\|n\|_1^{3/2}\|n\|_2^{1/2}$$

$$\leqslant \frac{\varepsilon}{20}\|n\|_2^2 + C(\|E\|_1^4,\|E\|_2^4,\|n\|_1^6),$$

$$8\langle n(s)\,\nabla E(s)\,\nabla n(s),-\mathrm{i}\Delta E(s)\rangle \leqslant 8\|n\|_{L^\infty}\|\nabla E\|_{L^\infty}\|\nabla n\|_0\|\Delta E\|_0 \leqslant C(\|n\|_1^4,\|E\|_2^4),$$

$$-8\gamma\langle \nabla E(s),\Delta E(s)\,\nabla n(s)\rangle \leqslant 8\gamma\|\nabla E\|_{L^\infty}\|\nabla n\|_0\|\Delta E\|_0 \leqslant C(\|n\|_1^2,\|E\|_2^4)$$

and

$$8\langle \nabla g,\Delta E(s)\,\nabla n(s)\rangle \leqslant 8\|\nabla g\|_0\|\nabla n\|_{L^\infty}\|\Delta E\|_0 \leqslant \frac{\varepsilon}{20}\|n\|_2^2 + C(\|g\|_1^4,\|E\|_2^4).$$

Further

$$16\langle \nabla E,\Delta E\,\nabla m\rangle \leqslant 16\|\nabla E\|_{L^\infty}\|\Delta E\|_0\|\nabla m\|_0 \leqslant \frac{\varepsilon}{2\lambda^2}\|m\|_1^2 + C(\|E\|_2^4),$$

$$2\langle m,|\Delta E|^2\rangle \leqslant 2\|m\|_0\|\Delta E\|_{L^4}^2 \leqslant 2c\|m\|_0\|E\|_2^{3/2}\|E\|_3^{1/2} \leqslant \frac{\gamma}{2}\|E\|_3^2 + C(\|m\|_0^2,\|E\|_2^6),$$

$$-4\gamma\langle \Delta nE,\Delta E\rangle \leqslant 4\gamma\|\Delta n\|_0\|E\|_{L^\infty}\|\Delta E\|_0 \leqslant 4c\gamma\|E\|_1\|E\|_2\|n\|_2 \leqslant \frac{\varepsilon}{20}\|n\|_2^2 + C(\|E\|_1^4,\|E\|_2^4),$$

and by the property of trace

$$16\mathrm{Tr}((-\Delta)^{1/2}n(-\Delta)^{1/2}q_2(-\Delta)q_2) \leqslant 16\|n\|_1\|(-\Delta)^{1/2}q_2\|_{L^\infty}\|q_2\|_2$$

$$\leqslant 16c\|n\|_1\|q_2\|_2^2 \leqslant C(\|n\|_1^2,\|q_2\|_2^4),$$

$$2\mathrm{Tr}(n[(-\Delta)q_2]^2) \leqslant 2\|n\|_0\|(-\Delta)q_2\|_{L^4}^2 \leqslant 2c\|n\|_0\|q_2\|_2^{3/2}\|q_2\|_3^{1/2} \leqslant C(\|n\|_0^2,\|q_2\|_2^6,\|q_2\|_3^2),$$

$$4\mathrm{Tr}((-\Delta)n[(-\Delta)^{1/2}q_2]^2) \leqslant 4c\|n\|_2\|q_2\|_1^{3/2}\|q_2\|_2^{1/2} \leqslant \frac{\varepsilon}{20}\|n\|_2^2 + C(\|q_2\|_1^6,\|q_2\|_2^2),$$

$$4\mathrm{Tr}((-\Delta)nq_2(-\Delta)q_2) \leqslant 4c\|n\|_2\|q_2\|_1\|q_2\|_2 \leqslant \frac{\varepsilon}{20}\|n\|_2^2 + C(\|q_2\|_1^4,\|q_2\|_2^4).$$

Then we deduce from all these estimates that

$$H_2(t)-H_2(0)+\beta\int_0^t H_2(s)\mathrm{d}s\leqslant -4\int_0^t\langle\Delta n(s)\Delta E(s),-\mathrm{i}q_2\mathrm{d}w_2(s)\rangle+8\int_0^t\langle\Delta E(s)\nabla n(s),$$
$$-\mathrm{i}\nabla q_2\mathrm{d}w_2(s)\rangle-4\int_0^t\langle\Delta^2 E_t(s)+\gamma\Delta^2 E(s),q_2\mathrm{d}w_2(s)\rangle-$$
$$2\int_0^t\langle\Delta m(s),q_1\rangle\mathrm{d}w_1(s)+$$
$$C(\|E\|_2^8,\|n\|_1^8,\|m\|_0^2,\|f\|_1^2,\|g\|_1^4,\|q_1\|_1^2,\|q_2\|_3^6)t.$$

Taking expectation on both sides of the above inequality and by the Gronwall lemma, we have

$$\mathbb{E}H_2(t)\leqslant \mathrm{e}^{-\beta t}\,\mathbb{E}H_2(0)+C(n_0,n_1,E_0,f,g,q_1,q_2).$$

On the other hand since

$$|2\langle\Delta n,\Delta|E|^2\rangle|\leqslant\frac{1}{2}\|n\|_2^2+C(\|E\|_1^6,\|E\|_2^4),$$
$$|2\langle n,|\Delta E|^2\rangle|\leqslant\frac{1}{2}\|E\|_3^2+C(\|E\|_2^6,\|n\|_0^2),$$
$$|16\langle\nabla n\ \nabla E,\Delta E\rangle|\leqslant C(\|n\|_1^2,\|E\|_2^4),$$
$$|-4\langle\nabla g,\nabla\Delta E\rangle|\leqslant\frac{1}{2}\|E\|_3^2+C(\|g\|_1^2),$$

we have

$$H_2(t)\geqslant\frac{1}{\lambda^2}\|m\|_1^2+\frac{1}{2}\|n\|_2^2+\|E\|_3^2-C(\|E\|_1^6,\|E\|_2^6,\|n\|_0^2,\|n\|_1^2,\|g\|_1^2).$$

Therefore

$$\mathbb{E}\left[\frac{1}{\lambda^2}\|m\|_1^2+\frac{1}{2}\|n\|_2^2+\|E\|_3^2\right]\leqslant \mathrm{e}^{-\beta t}\,\mathbb{E}H_2(0)+C(n_0,n_1,E_0,f,g,q_1,q_2).$$

Also by the same analysis of Secs. 3.1 and 3.2, we have for $p\geqslant 1$ and any $T>0$ there is a positive constant $C_T$ depending on the initial data, such that

$$\mathbb{E}\sup_{0\leqslant t\leqslant T}\|(n_t(t),n(t),E(t))\|_{\mathcal{E}_2}^{2p}\leqslant C_T(n_0,n_1,E_0,f,g,q_1,q_2).$$

Then we can draw the following lemma.

**Lemma 3.4** Assume $(n_1,n_0,E_0)\in\mathcal{E}_2, f,g,q_1\in H_0^1(D)$, and $q_2\in H^3(D)\cap H_0^1(D)$. Then for any $T>0$ and $p\geqslant 1$, $(n_t,n,E)\in L^{2p}(\Omega;L^\infty(0,T;\mathcal{E}_2))\cap L^\infty(0,\infty;L^{2p}(\Omega,\mathcal{E}_2))$.

## 4 Existence and Uniqueness of Solution

By the above *a priori* estimates we prove the existence and uniqueness of solution for Eqs. (2.1) and (2.2) in space $\mathcal{E}_1$ and $\mathcal{E}_2$. First we assume $(n_1,n_0,E_0)\in\mathcal{E}_1, f\in L^2(D), g,q_1\in H_0^1(D)$, and $q_2\in H^2(D)\cap H_0^1(D)$. Let $\{e_i\}_{i=1}^\infty$ be the eigenvectors of Laplace operator $-\Delta$ on $D$ with Dirichlet boundary condition, which is also an orthonormal basis of $L^2(D)$. Let $P^k$ be the projection from $L^2(D)$ onto the space spanned by $\{e_i: i=1,2,\cdots,k\}$. Define the

Galerkin approximation solution $(n_t^k, n^k, E^k)$ solves

$$\frac{1}{\lambda^2}\mathrm{d}n_t^k + [\alpha n_t^k - \Delta(n^k + P^k|E^k|^2)]\mathrm{d}t = f^k\mathrm{d}t + \mathrm{d}W_1^k, \tag{4.1}$$

$$\mathrm{id}E^k + [\Delta E^k - P^k(n^kE^k) + \mathrm{i}\gamma E^k]\mathrm{d}t = g^k\mathrm{d}t + \mathrm{d}W_2^k, \tag{4.2}$$

$$n_t^k(0) = P^kn_1, \quad n^k(0) = P^kn_0, \quad E^k(0) = P^kE_0. \tag{4.3}$$

Here $f^k = P^kf, g^k = P^kg, W_1^k = P^kW_1$ and $W_2^k = P^kW_2$.

We will treat the above equations pathwise by introducing the following random processes which solve

$$\frac{1}{\lambda^2}\mathrm{d}\eta_t + [\alpha\eta_t - \Delta\eta]\mathrm{d}t = \mathrm{d}W_1,$$

$$\mathrm{id}\xi + [\Delta\xi + \mathrm{i}\gamma\xi]\mathrm{d}t = \mathrm{d}W_2$$

with Dirichlet boundary conditions

$$\eta|_{\partial D} = 0, \quad \xi|_{\partial D} = 0,$$

and initial conditions

$$\eta(x,0) = 0, \quad \eta_t(x,0) = 0, \quad \xi(x,0) = 0, \quad x \in D.$$

Following the same approach in Sec.3, we have for any $T > 0$ and almost all $\omega \in \Omega$

$$\eta_t \in C(0,T;H^1(D)), \quad \eta \in C(0,T;H^2(D) \cap H_0^1(D)),$$
$$\xi \in C(0,T;H^3(D) \cap H_0^1(D)) \tag{4.4}$$

and satisfy the following estimates:

$$\mathbb{E}[\|\eta_t(t)\|_1^2 + \|\eta(t)\|_2^2 + \|\xi(t)\|_3^2] \leqslant C \tag{4.5}$$

for some positive constant $C$ independent of $T$. And for any $T > 0$

$$\mathbb{E}\sup_{0\leqslant t\leqslant T}[\|\eta_t(t)\|_1^2 + \|\eta(t)\|_2^2 + \|\xi(t)\|_3^2] \leqslant C_T \tag{4.6}$$

for some positive constant $C_T$. Now let $R^{k,M} = (N_t^{k,M}, N^{k,M}, F^{k,M})$ be the solution of the following equations:

$$\frac{1}{\lambda^2}N_{tt}^{k,M} + \alpha N_t^{k,M} - \Delta N^{k,M} + \chi_M(\|R^{k,M}\|_{\mathcal{E}_1})\Delta P^k(|E^{k,M}|^2) = f^k, \tag{4.7}$$

$$\mathrm{i}F_t^{k,M} + \Delta F^{k,M} - \chi_M(\|R^{k,M}\|_{\mathcal{E}_1})P^k(n^{k,M}E^{k,M}) + \mathrm{i}\gamma F^{k,M} = g^k \tag{4.8}$$

with initial conditions

$$N_t^{k,M}(0) = P^kn_1, N^{k,M}(0) = P^kn_0, F^{k,M}(0) = P^kE_0.$$

Here $n^{k,M} = N^{k,M} + P^k\eta$ and $E^{k,M} = F^{k,M} + P^k\xi$. And $\chi_M \in C_0^\infty(\mathbf{R})$ such that $\chi_M(r) = 1$ for $|r| \leqslant M$ and $\chi_M(r) = 0$ for $|r| \geqslant 2M$.

Notice that (4.7) and (4.8) are random differential equations with Lipschitz nonlinearity in finite dimension. Then for almost all $\omega \in \Omega$, we have a unique solution $(N_t^{k,M}, N^{k,M}, F^{k,M})$ to (4.7) and (4.8). Define stopping time by

$$\tau_M = \inf\{t > 0: \|R^{k,M}\|_{\mathcal{E}_1} \geqslant M\}$$

if the set $\{\|R^{k,M}\|_{\mathcal{E}_1} \geqslant M\}$ is nonempty; otherwise $\tau_M = \infty$. Since $\tau_M$ is increasing in $M$, let $\tau_\infty = \lim_{M\to\infty}\tau_M$ almost surely. And for $t < \tau_M$,

$$(N_t^{k,M}(t), N^{k,M}(t), F^{k,M}(t)) + (P^k\eta_t(t), P^k\eta(t), P^k\xi(t))$$

satisfies (4.1) and (4.2). By the estimates in Sec.3.2 and (4.5) and (4.6) for any $t \geqslant 0$

$$\mathbb{E}\|(N_t^{k,M}(t), N^{k,M}(t), F^{k,M}(t))\|_{\mathcal{E}_1}^2 \leqslant C(n_0, n_1, E_0, f, g, q_1, q_2) \tag{4.9}$$

with $C(n_0, n_1, E_0, f, g, q_1, q_2)$ independent of $T$ and $M$ and for any $T > 0$

$$\mathbb{E} \sup_{0\leqslant t\leqslant T\wedge \tau_M} \|(N_t^{k,M}(t), N^{k,M}(t), F^{k,M}(t))\|_{\mathcal{E}_1}^2 \leqslant C_T(n_0, n_1, E_0, f, g, q_1, q_2), \tag{4.10}$$

with $C_T(n_0, n_1, E_0, f, g, q_1, q_2)$ independent of $M$. On the other hand we have

$$\mathbb{E}\|(N_t^{k,M}(T\wedge\tau_M), N^{k,M}(T\wedge\tau_M), F^{k,M}(T\wedge\tau_M))\|_{\mathcal{E}_1}^2$$

$$\geqslant \mathbb{E}\Big[I(\tau_M \leqslant T)\|(N_t^{k,M}(T\wedge\tau_M), N^{k,M}(T\wedge\tau_M), F^{k,M}(T\wedge\tau_M))\|_{\mathcal{E}_1}^2\Big] \geqslant M^2\,\mathbb{P}(\tau_M \leqslant T).$$

Then by estimate (4.9) we have

$$\mathbb{P}\{\tau_M \leqslant T\} \leqslant \frac{1}{M^2} C(n_0, n_1, E_0, f, g, q_1, q_2)$$

which yields, by the Borel-Cantelli lemma, for any $T > 0$

$$\mathbb{P}\{\tau_\infty > T\} = 1.$$

Then

$$(N_t^k, N^k; F^k) = \lim_{M\to\infty} (N_t^{k,M}(t), N^{k,M}(t), F^{k,M}(t))$$

satisfies the following random differential equations:

$$\frac{1}{\lambda^2} N_{tt}^k + \alpha N_t^k - \Delta(N^k + P^k|E^k|^2) = f^k, \tag{4.11}$$

$$\mathrm{i}F_t^k + \Delta F^k - P^k(n^k E^k) + \mathrm{i}\gamma F^k = g^k, \tag{4.12}$$

with initial conditions

$$N_t^k(0) = P^k n_1, \quad N^k(0) = P^k n_0, \quad F^k(0) = P^k E_0.$$

Then $(N_t^k, N^k, F^k)$ satisfies the estimates (4.9) and (4.10) and for any $t \geqslant 0$

$$(n_t^k(t), n^k(t), E^k(t)) = (N_t^k(t), N^k(t), F^k(t)) + (P^k\eta_t(t), P^k\eta(t), P^k\xi^k(t))$$

is the unique global solution of (4.1) and (4.3).

Now we will consider (4.11) and (4.12) for a fixed $\omega$. First by (4.10) for any $T > 0$

$$P\Big(\bigcap_{L=1}^{\infty}\bigcup_{l=1}^{\infty}\bigcap_{k=l}^{\infty}\Big\{\sup_{0\leqslant t\leqslant T}\|(N_t^k(t), N^k(t), F^k(t))\|_{\mathcal{E}_1}^2 \geqslant L\Big\}\Big) = 0.$$

Let

$$\tilde{\Omega} = \bigcap_{L=1}^{\infty}\bigcup_{l=1}^{\infty}\bigcap_{k=l}^{\infty}\Big\{\sup_{0\leqslant t\leqslant T}\|(N_t^k(t), N^k(t), F^k(t))\|_{\mathcal{E}_1}^2 \geqslant L\Big\}.$$

Then $\mathbb{P}(\Omega\setminus\tilde{\Omega}) = 0$. Now for any fixed $\omega \in \tilde{\Omega}$, there is an $r(\omega)$ with $0 < r(\omega) < \infty$ such that

$$\sup_{0\leqslant t\leqslant T}\|(N_t^k(t), N^k(t), F^k(t))\|_{\mathcal{E}_1}^2 \leqslant r(\omega). \tag{4.13}$$

Then we can extract a subsequence still denoted by $(N_t^k, N^k, F^k)$ such that for any $T > 0$

$N_t^k$ converges to $N_t$ weakly star in $L^\infty(0, T; L^2(D))$,

$N^k$ converges to $N$ weakly star in $L^\infty(0, T; H_0^1(D))$,

$F^k$ converges to $F$ weakly star in $L^\infty(0, T; H^2(D) \cap H_0^1(D))$.

These convergences are sufficient to pass the limit $k\to\infty$ in linear terms, but we need a strong convergence of $F^k$ for nonlinear terms. In fact from (4.12) and estimate (4.13) $F_t^k$ is bounded in $L^\infty(0,T;L^2(D))$. Then by Lemma 2.1 we can further extract a subsequence still denoted by $F^k$ such that

$$F^k \text{ converges to } F \text{ strongly in } L^2(0,T;H_0^1(D)).$$

Then by a standard procedure we can pass the limit $k\to\infty$ to show that $(N_t,N,F)\in L^\infty(0,T;\mathcal{E}_1)$ is a weak solution of

$$\frac{1}{\lambda^2}N_{tt}+\alpha N_t-\Delta(N+|E|^2)=f, \tag{4.14}$$

$$\mathrm{i}F_t+\Delta F-nE+\mathrm{i}\gamma F=g, \tag{4.15}$$

with initial conditions

$$N_t(x,0)=n_1(x),\quad N(x,0)=n_0(x),\quad F(x,0)=E_0(x),\quad x\in D.$$

Then $(n_t,n,F)=(N_t,N,F)+(\eta_t,\eta,\xi)$ is a solution of (2.1) and (2.2) and satisfies the estimates in Sec.3.2.

Now we prove the uniqueness of the solution. In fact for any $\omega\in\widetilde{\Omega}, f\in L^2(D)$, and $\Delta|E|^2\in L^\infty(0,T;L^2(D))$, we have

$$\frac{1}{\lambda^2}N_{tt}-\Delta N\in L^\infty(0,T;L^2(D)).$$

Then by Lemma 4.1, Chapter II of Ref.[25] we have for almost all $\omega\in\Omega$

$$N\in C(0,T;H_0^1(D)),\quad N_t\in C(0,T;L^2(D)).$$

By a similar method, noticing $g\in H_0^1(D)$ and $F_t\in L^\infty(0,T;L^2(D))$ almost surely, we have, by Ref.[14]

$$F\in C(0,T;H^2(D)\cap H_0^1(D)).$$

Then by (4.4) and the definition of $N$ and $F$

$$(n_t,n,E)\in C(0,T;\mathcal{E}_1)$$

almost surely. Now as the noise is additive, we can follow the same approach as in Ref.[5]. The solution $(n_t,n,E)$ is unique in $L^\infty(0,T;\mathcal{E}_1)$ almost surely and is continuous from $[0,T]$ to $\mathcal{E}_1$ almost surely. Then, for $T$ is arbitrary, by the estimates in Sec.3.2, we derive the following result.

**Theorem 4.1** Assume $(n_1,n_0,E_0)\in\mathcal{E}_1, f\in L^2(D), g, q_1\in H_0^1(D)$, and $q_2\in H^2(D)\cap H_0^1(D)$. Then there is a unique solution $(n_t,n,E)\in L^\infty(\mathbf{R}_+;\mathcal{E}_1)$ almost surely for Eqs. (2.1) and (2.2). And $(n_t,n,E)$ is continuous from $\mathbf{R}_+$ to $\mathcal{E}_1$.

And by the same method as above we can derive the following.

**Theorem 4.2** Assume $(n_1,n_0,E_0)\in\mathcal{E}_2, f,g,q_1\in H_0^1(D)$ and $q_2\in H^3(D)\cap H_0^1(D)$. Then there is a unique solution $(n_t,n,E)\in L^\infty(\mathbf{R}_+;\mathcal{E}_2)$ almost surely for Eqs. (2.1) and (2.2). And $(n_t,n,E)$ is continuous from $\mathbf{R}_+$ to $\mathcal{E}_2$.

## 5 Asymptotic Behavior of the Solution

In this section we study the long time behavior of the solution of (2.1) and (2.2). We will prove the existence of a random attractor and a stationary measure.

We construct a random attractor for stochastic Zakharov equations in $\mathcal{E}_2$ equipped with weak topology. First we show that (2.1) and (2.2) defines a RDS.

For this we assume the probability space $(\Omega, \mathscr{F}, \{\mathscr{F}_t\}_{t\geqslant 0}, \mathbb{P})$ is the following canonical probability space. Let $w(t)$ be a standard scalar Wiener process, define

$$\Omega = \{w \in C(\mathbf{R};\mathbf{R}): w(0) = 0\}$$

endowed with compact-open topology, $\mathbb{P}$ is the corresponding Wiener measure, and $\mathscr{F}$ is the $\mathbb{P}$-completeness of Borel $\sigma$-algebra on $\Omega$. Let

$$\theta_t\omega(\cdot) = \omega(\cdot + t) - \omega(t), \quad t \in \mathbf{R}.$$

Then $(\Omega, \mathscr{F}, \mathbb{P}, (\theta_t)_{t\in\mathbf{R}})$ is a metric dynamical system with the filtration

$$\mathscr{F}_t := \vee_{s\leqslant t}\mathscr{F}_s^t, \quad t \in \mathbf{R},$$

where

$$\mathscr{F}_s^t = \sigma\{w(t_2) - w(t_1): s \leqslant t_1 \leqslant t_2 \leqslant t\}$$

is the smallest $\sigma$-algebra generated by random variables $w(t_2) - w(t_1)$ for all $s \leqslant t_1 \leqslant t_2 \leqslant t$. The metric dynamical system $(\Omega, \mathscr{F}, \mathbb{P}, (\theta_t)_{t\in\mathbf{R}})$ defined above is ergodic. Now we introduce the concept of the RDS.

**Definition 5.1** Let $(E, d)$ be a Polish space and $(\Omega, \mathscr{F}, \mathbb{P}, (\theta_t)_{t\in\mathbf{R}})$ be a metric dynamical system. Suppose the mapping

$$\phi: \mathbf{R}^+ \times \Omega \times E \to E$$

is $(\mathscr{B}(\mathbf{R}^+) \times \mathscr{F} \times \mathscr{B}(E), \mathscr{B}(E))$-measurable and satisfies the following properties:

- $\phi(0, \omega)x = x$;
- $\phi(t, \theta_\tau\omega) \circ \phi(\tau, \omega)x = \phi(t+\tau, \omega)x$ for all $t, \tau \in \mathbf{R}^+$, $x \in E$, and $\omega \in \Omega$ (cocycle property).

Then $\phi$ is called a RDS with respect to $(\theta_t)_{t\in\mathbf{R}}$. If $\phi$ is continuous with respect to $x$ for $t \geqslant 0$ and $\omega \in \Omega$, $\phi$ is called a continuous RDS.

For details we refer to Ref. [1]. Now we recall some knowledge of random attractors.

A set-valued map $B: \Omega \to 2^E$ is called a random closed set if $B(\omega)$ is closed and nonempty and $\omega \mapsto d(x, B(\omega))$ is measurable for all $x \in E$. A random set $B$ is said to be tempered if for almost every $\omega \in \Omega$ and all $\gamma > 0$

$$\lim_{t\to\infty} e^{-\gamma t} \operatorname{diam}(B(\theta_{-t}\omega)) = 0,$$

where $\operatorname{diam}(B) = \sup_{x,y\in B} d(x, y)$. Let $\mathscr{D}$ be the collection of all tempered random sets in $E$. And for such a set we write as $\mathscr{D}$-set. Then we give the following definition of the random attractor, see Refs. [2,3,23].

**Definition 5.2** A random set $\mathscr{A}(\omega) \in \mathscr{D}$ is a random (pullback) attractor for RDS $\phi$ if

- $\mathscr{A}(\omega)$ is a random compact set, i.e., $\omega \mapsto \mathrm{d}(x, \mathscr{A}(\omega))$ is measurable for every $x \in E$ and $\mathscr{A}(\omega)$ is compact for almost every $\omega \in \Omega$;
- $\mathscr{A}(\omega)$ is strictly invariant, i.e., $\phi(t, \omega)\mathscr{A}(\omega) = \mathscr{A}(\theta_t\omega)$ $\forall\, t \geqslant 0$ and for almost every $\omega \in \Omega$;

• $\mathscr{A}(\omega)$ attracts all bounded deterministic sets in $E$, i.e., for all bounded set $B\subset E$ and almost every $\omega\in\Omega$ we have

$$\lim_{t\to\infty}d(\phi(t,\theta_{-t}\omega)B,\mathscr{A}(\omega))=0.$$

The following theorem gives a sufficient condition for the existence of a random attractor, see Refs. [2,3,23].

**Theorem 5.3** Suppose $\phi$ is an RDS on a Polish space $(E,d)$ and there exists a random compact set $B(\omega)$ absorbing every bounded deterministic set $D\subset E$. Then, the set

$$\mathscr{A}(\omega)=\bigcap_{\tau\geqslant 0,\,t\geqslant\tau}\overline{\bigcup\phi(t,\theta_{-t}\omega,B(\theta_{-t}\omega))},$$

is a global random attractor for RDS $\phi$.

For our stochastic Zakharov equations we consider $(N_t,N,F)$ which solves (4.14) and (4.15). First by the construction of $(\eta_t,\eta,\xi)$, there is a $\theta_t$ invariant subset $\Omega_0\subset\Omega$ of full probability such that $\{(\eta_t(t,\omega),\eta(t,\omega),\xi(t,\omega))\}_{t\geqslant 0}$ defines a continuous orbit in $\mathcal{E}_2$ for any $\omega\in\Omega_0$. Moreover $\{(\eta_t(t,\omega),\eta(t,\omega),\xi(t,\omega))\}_{t\geqslant 0}$ depends measurably on $\omega$ and satisfies the cocycle property. Then (4.14) and (4.15) is an evolutionary system with nonautonomous coefficient measurably driven by $\theta_t$. And for any $(n_1,n_0,E_0)\in\mathcal{E}_2$ and any $T>0$, by the discussion in Sec.4 there is a unique solution $(N_t,N,F)\in C(0,T;\mathcal{E}_2)$ for all $\omega\in\Omega_0$. Then system (4.14) and (4.15) defines a RDS, denoted by $\Psi(t,\omega)$, on $\mathcal{E}_2$. And $\Psi(t,\omega)$ is weakly continuous on $\mathcal{E}_2$[5] for all $\omega\in\Omega_0$. Then by $(n_t(t,\omega),n(t,\omega),E(t,\omega))=(N_t(t,\omega)+\eta_t(t,\omega),N(t,\omega)+\eta(t,\omega),F(t,\omega)+\xi(t,\omega))$, system (2.1) and (2.2) also defines a weakly continuous RDS on $\mathcal{E}_2$, which we denote by $\Phi(t,\omega)$.

Denote by $B(0,r)$ the ball centers at 0 with radius $r$ in $\mathcal{E}_2$. Then by the estimates in Sec.3.3 and (4.6) there is a random variable $R(\omega)$ such that for any $r>0$, $(n_1,n_0,E_0)\in B(0,r)$, there is a random time $t_r(w)>0$ if $t>t_r(w)$ for almost all $\omega\in\Omega$,

$$\|\Psi(t,\theta_{-t}\omega)(n_1,n_0,E_0)\|_{\mathcal{E}_2}\leqslant R(\omega).$$

Set

$$\mathscr{A}(\omega)=\bigcap_{\tau\geqslant 0,\,t\geqslant\tau}\overline{\bigcup\Psi(t,\theta_{-t}\omega)B(0,R(\omega))}^{\mathcal{E}_2^w}.$$

Here the closure is taken with respect to the weak topology of $\mathcal{E}_2$. Now we show that random attractor $\mathscr{A}(\omega)$ is tempered. By the estimates in Sec.3.3 and (4.6), we have

$$\mathbb{E}\sup_{0\leqslant t\leqslant 1}R^2(\theta_t\omega)<\infty.$$

Then by the Birkhoff ergodic theorem[21],

$$\lim_{s\to\pm\infty}\frac{\sup_{t\in[0,1]}R^2(\theta_{t+s}\omega)}{s}=0$$

on a $\theta$-invariant subset of $\Omega$ with full probability measure. Hence

$$\lim_{t\to\pm\infty}\frac{R^2(\theta_t\omega)}{t}=0$$

on a $\theta$-invariant subset of $\Omega$ with full probability measure, i.e., $R(\omega)$ is tempered. Then we have the following result.

**Theorem 5.4** Assume $f, g, q_1 \in H_0^1(D)$ and $q_2 \in H^3(D) \cap H_0^1(D)$. Stochastic Zakharov equations (2.1) and (2.2) have a global random attractor $\mathscr{A}(\omega)$ which is a tempered random compact set in $\mathcal{E}_2$ endowed with weak topology.

**Remark 5.5** Here we cannot follow the approach of Ref. [8] to prove the existence of a random attractor in $\mathcal{E}_2$ with usual topology for the existence of noise term.

By random dynamical theory, the existence of a random attractor yields the existence of one stationary measure. However, here the random attractor is obtained in weak topology of $\mathcal{E}_2$ which does not yield the existence of a stationary measure in $\mathcal{E}_2$. But in the following we will construct one stationary measure in $\mathcal{E}_2$ by the estimates in Sec. 3.3.

Assume $(n_1, n_0, E_0) \in \mathcal{E}_2$. According to the results in Sec. 4 and estimates in Sec. 3.3, system (2.1) and (2.2) has a unique solution $(n_t, n, E)$ with $(n_t(0), n(0), E(0)) = (n_1, n_0, E_0)$ and for any $t > 0$ satisfies

$$\mathbb{E}[\|n_t(t)\|_1^2 + \|n(t)\|_2^2 + \|E(t)\|_3^2] \leqslant C \tag{5.1}$$

for some positive constant $C > 0$ which is independent of $t > 0$.

Now let $\mu_t$ be the distribution of $(n_t(t), n(t), E(t))$ for $t \geqslant 0$. Following the classical Bogolyubov-Krylov argument[1,21], we define

$$\bar{\mu}_t = \frac{1}{t}\int_0^t \mu_s \mathrm{d}s$$

as

$$\bar{\mu}_t(\Gamma) = \frac{1}{t}\int_0^t \mu_s(\Gamma)\mathrm{d}s$$

for any $\Gamma \in \mathscr{B}(\mathcal{E}_1)$. By (5.1)

$$\int_{\mathcal{E}_1} \|v\|_{\mathcal{E}_2}^2 \bar{\mu}_t(\mathrm{d}v) = \frac{1}{t}\int_0^t E\|n_t(s), n(s), E(s))\|_{\mathcal{E}_2}^2 \mathrm{d}s \leqslant C.$$

By the Chebyshev inequality and because $\mathcal{E}_2$ is a compact embedding into $\mathcal{E}_1$, $\{\bar{\mu}_t\}_{t \geqslant 0}$ is tight in $\mathcal{E}_1$. Then there is a sequence $\{\bar{\mu}_{t_k}\}$ with $t_k \to \infty$ as $k \to \infty$ and a probability measure $\mu$ on $\mathcal{E}_1$ such that

$$\bar{\mu}_{t_k} \to \mu \text{ weakly as } k \to \infty.$$

And $\mu$ is a stationary measure for stochastic Zakharov equations on $\mathcal{E}_1$ by a standard method[21]. Moreover by (5.1) $\mu$ is in fact support on $\mathcal{E}_2$, that is, $\mu$ is a stationary measure for stochastic Zakharov equations on $\mathcal{E}_2$. Then we can draw the following result.

**Theorem 5.6** Assume $f, g, q_1 \in H_0^1(D)$ and $q_2 \in H^3(D) \cap H_0^1(D)$. Stochastic Zakharov equations (2.1) and (2.2) have one stationary measure on $\mathcal{E}_2$.

## References

[1] Arnold L. Random dynamical systems. New York: Springer-Verlag, 1998.

[2] Crauel H, Flandoli F. Attractor for random dynamical systems. Probab Theory Relat Fields, 1994, 100: 365.

[3] Crauel H, Debussche A, Flandoli F. Random attractors. J Dyn Differ Equ, 1997, 9: 307.

[4] E W, Li X, Vanden-Eijnden E. Some Recent Progress in Multiscale Modeling, Multiscale Modelling and Simulation. Lecture Notes in Computer Science Engineering Vol 39. Berlin: Springer, 2004, 3 – 21.

[5] Flahaut I. Attractors for the dissipative Zakharov system. Nonlinear Anal Theory (Methods Appl), 1991, 16: 599.

[6] Guio P, Forme F. Zakharov simulations of Langmuir turbulence: Effects on the ion-acoustic waves in incoherent scattering. Phys Plasmas, 2006, 13: 122902.

[7] Glangetas L, Merle F. Existence of self-similar blow-up solution for Zakharov equation in dimension two. Part I. Commun Math Phys, 1994, 160: 173.

[8] Goubet O, Moise I. Attractors for the dissipative Zakharov system. Nonlinear Anal Theory (Methods Appl) 1998, 31: 823.

[9] Guo B L, Sheng L J. The global existence and uniqueness of classical solutions of periodic initial boundary problems of Zakharov equations. Acta Math Appl Sin, 1982, 5: 310.

[10] Guo B L. On the IBVP for some more extensive Zakharov equations. J Math, 1987, 7: 269(in Chinese).

[11] Imkeller P, Monahan A. In Stochastic Climate Dynamics. Special Issue of Stochastics Dyn, 2002, 2.

[12] Kim J U. On the stochastic porous medium equation. J Differ Equations, 2006, 220: 163.

[13] Li Y S. On the initial boundary value problems for two dimensional systems of Zakharov equations and of complex-Schrödinger-real-Boussinesq equations. J Partial Differ Equ, 1992, 5: 81.

[14] Lions J L, Macenes E. Problémes aux Limites Non Homogénes et Applications. Paris: Dunod, 1968.

[15] Landman M, Papanicolaou G C, Sulem C, et al. Stability of isotropic self-similar dynamics for scalar collapse, Phys Rev A, 1992, 46: 7869.

[16] Makhankov V G. Dynamics of classical solitons (in nonintegrable system). Phys Rep Phys Lett, 1978, 35: 1.

[17] Masselin V. A result on the blow-up rate for the Zakharov system in dimension 3. SIAM J Math Anal, 2001, 33: 440.

[18] Newman D L, Robinson P A, Goldman M V. Field structure of collapsing wave packets in 3D strong Langmuir turbulence. Phys Rev Lett, 1989, 62: 2132.

[19] Papanicolaou G C, Sulem C, Sulem P L, et al. Singular solutions of the Zakharov equations for Langmuir turbulence. Phys Fluids B, 1991, 3: 969.

[20] Da Prato G, Zabczyk J. Stochastic equations in infinite dimensions. Cambridge: Cambridge University Press, 1992.

[21] Da Prato G, Zabczyk J. Ergodicity for infinite dimensional systems. Cambridge: Cambridge University Press, 1996.

[22] Robinson P A, Newman D L, Goldman M V. Three dimensional strong Langmuir turbulence and wave collapse. Phys Rev Lett, 1988, 61: 702.

[23] Schmalfu B. International seminar on applied mathematics-nonlinear dynamics // Reitmann V, Riedrich T, Koksch N. Attractor approximation and global behaviour. TU Dresden, 1992, 185 – 192.

[24] Schochet S H, Weinstein M I. The nonlinear Schrödinger limit of the Zakharov equation governing Langmuir turbulence. Commun Math Phys, 1986, 106: 569.

[25] Temam R. Infinite-dimensional dynamical systems in mechanics and physics. 2nd ed. New York: Springer-Verlag, 1997.

[26] Zakharov V E. Collapse of Langmuir waves. Sov Phys JETP, 1972, 35: 908.

[27] Zakharov V E, Shabat A B. Integrable system ol nonlinear equations in Mathematical Physics. Funct Anal Appl, 1974, 83: 43.

# Attractor and Spatial Chaos for the Brusselator in $\mathbf{R}^N$ *

Guo Boling(郭柏灵) Han Yongqian(韩永前)

**Abstract**

Brusselator is an important dynamical system which can be described by a reaction-diffusion system. In this paper it is proved that this reaction-diffusion system possesses a global attractor $\mathscr{A}$ in the corresponding phase space. The upper and lower bounds of Kolmogorov $\varepsilon$-entropy of $\mathscr{A}$ are obtained.

Moreover we give a more detailed study of spatial chaos of the attractor $\mathscr{A}$ for the Brusselator in $\mathbf{R}^N$. We interpret a group of spatial shifts as a dynamical system which acts on the attractor $\mathscr{A}$. By using the technique of unstable manifolds, it is proved that this dynamical system is chaotic. In order to clarify the nature of this chaos, we construct the Lipschitz-continuous homeomorphic embedding of a typical model dynamical system whose chaotic behavior is evident, into the spatial shifts on the attractor $\mathscr{A}$. This typical dynamical system generalizes the symbolic system. It was first introduced by Zelik.

**Keywords** the brusselator; global attractor; spatial chaos

## 1 Introduction

We consider the following Cauchy problem of a reaction-diffusion system

$$u_t = d\Delta u - (b+1)u + u^2 v + a, \tag{1.1}$$

$$u_t = d\Delta v + bu - u^2 v, \quad x \in \mathbf{R}^N, \quad t > 0. \tag{1.2}$$

$$u(x, t=0) = u_0(x) \geqslant 0, \quad v(x, t=0) = v_0(x) \geqslant 0. \tag{1.3}$$

This system is a model of a certain chemical morphogenetic process due to Turing[23]. It is named Brusselator. Here $d$, $a$ and $b$ are strictly positive constants, $N=1,2,3$.

System (1.1) and (1.2) has been extensively studied[1,3,9,11,13,16,17,19,20,24, etc.], and references therein. In particular, system (1.1) and (1.2) was applied to study oscillatory turing pattern[23], stationary pattern selection and competition[13]. Hollis, Martin and Pierre[11] proved existence of a global bounded solution of (1.1) and (1.2) with initial boundary conditions in the bounded domain of $\mathbf{R}^N$. For fixed $v_0$, the global attractor of $u$ for problem (1.1)~(1.3) was established in [9].

To study the spatial chaos of problem (1.1)~(1.3), the classical Sobolev spaces (such as $H^1(\mathbf{R}^N)$ $(l\geqslant 1)$) do not seem to be adequate because nondecaying initial data and a number of

* Nonlinear Analysis, 2009, 70: 3917-3931.

natural structures from the physical point of view (such as spatially periodic solutions, travelling waves, static states, etc.) are out of consideration. For example, it is obvious that the static state $(u_s, v_s)=(a, b/a)$ is a solution of (1.1) and (1.2), but $u_s, v_s \notin C([0,T];H^1(\mathbf{R}^N))$. Therefore we discuss system (1.1)~(1.3) in some weighted Sobolev spaces such as [7,26,27, etc.].

**Definition 1.1**[27] A function $\phi \in C_{\mathrm{loc}}(\mathbf{R}^N)$ is a weight function with the growth rate $\alpha \geqslant 0$ if the conditions $\phi(x+y) \leqslant C_\phi e^{\alpha|x|}\phi(y)$ and $\phi(x)>0$ are satisfied for every $x, y \in \mathbf{R}^N$.

**Definition 1.2**[27] Let $\phi$ be a weight function with the growth rate $\alpha$. We define the spaces

$$W_\phi^{l,p}(\mathbf{R}^N) = \left\{u \in \mathscr{D}'(\mathbf{R}^N) : \|u\|_{\phi,l,p}^p \equiv \sum_{|\beta| \leqslant l} \int_{\mathbf{R}^N} \phi(x)\,|\partial_x^\beta u(x)|^p \mathrm{d}x < \infty, l \in \mathbf{N}\right\},$$

$$W_{b,\phi}^{l,p}(\mathbf{R}^N) = \left\{u \in \mathscr{D}'(\mathbf{R}^N) : \|u\|_{b,\phi,l,p}^p \equiv \sup_{x_0 \in \mathbf{R}^N} \phi(x_0)\left\|u, B_{x_0}^1\right\|_{l,p}^p < \infty, l \in \mathbf{N}\right\}.$$

We will write $W_b^{l,p}$ instead of $W_{b,1}^{l,p}$.

It is obvious that the static state $(u_s, v_s)=(a, b/a)$ of (1.1) and (1.2) belongs to $W_b^{l,p}(\mathbf{R}^N)$.

The semi-linear system of parabolic equations

$$u_t = a\Delta_x u + \lambda_0 u - f(u, \nabla u) + g(x), \quad x \in \Omega, \tag{1.4}$$

$$u|_{\partial\Omega} = 0, \quad u|_{t=0} = u_0, \tag{1.5}$$

in an unbounded domain $\Omega$ has been studied in [4,7,26,27] and references therein. If the nonlinear function $f(u, \nabla u)$ satisfies the dissipation assumption

$$f(u, \nabla u) \cdot u \geqslant -C, \tag{1.6}$$

the global attractor of (1.4) and (1.5) has been constructed, the upper and lower bounds of Kolmogorov $\varepsilon$-entropy of the infinite-dimensional global attractor for Eqs. (1.4) and (1.5) have been obtained, and the spatial complexity and spatial chaos of the global attractor for Eqs. (1.4) and (1.5) have been studied in [7,26,27] and references therein. Unfortunately, the nonlinearity of Eqs. (1.1) and (1.2) does not satisfy the dissipation assumption (1.6).

The well-posedness of some reaction-diffusion systems in $\mathbf{R}^N$ has been extensively studied by many authors, see [6,10,18, etc.] and references therein.

For system (1.1)~(1.3), the global well-posedness of the solution in $L^\infty([0,\infty); W_b^{2,2}(\mathbf{R}^N))$ has been established in [9]. In this paper we construct the attractor $\mathscr{A}$ of system (1.1)~(1.3) which is bounded in $W_b^{2,2}(\mathbf{R}^N)$ and compact in a local topology of $W_{\mathrm{loc}}^{2,2}(\mathbf{R}^N)$. Although Hausdorff dimension and fractal dimension of this attractor $\mathscr{A}$ may be infinite, the upper and lower bounds of Kolmogorov $\varepsilon$-entropy of the attractor $\mathscr{A}$ are obtained. In order to study the spatial chaos of the attractor $\mathscr{A}$, we interpret a group of spatial shifts as a dynamical system which acts on the attractor $\mathscr{A}$. By using the technique of unstable manifolds, it is proved that this dynamical system is chaotic. In order to clarify the nature of this chaos, we construct the Lipschitz-continuous homeomorphic embedding of a typical model dynamical system whose chaotic behavior is evident, into the spatial shifts on the attractor $\mathscr{A}$. This typical dynamical system generalizes the symbolic system. It was first suggested by Zelik[7,26,27].

Throughout this paper, in order to simplify the exposition, different positive constants might be denoted by the same letter $C$; if necessary, by $C(\cdot,\cdot)$ we denote the constant depending only on the quantities appearing in parentheses.

The plan of this paper is as follows. In Section 2, we derive several *a priori* estimates for the solutions of (1.1)~(1.3), and establish the global well-posedness of the solution again. In Section 3, we construct the global attractor $\mathscr{A}$, and obtain the upper bound of Kolmogorov $\varepsilon$-entropy of attractor $\mathscr{A}$. In Section 4, we study the spatial chaos of attractor $\mathscr{A}$, and obtain the lower bound of Kolmogorov $\varepsilon$-entropy of attractor $\mathscr{A}$.

## 2 *A Priori* Estimates, Existence and Uniqueness of Solution

In this section, our main task is to improve *a priori* estimates in [9]. First we recall the properties of several classes of Sobolev spaces in unbounded domains, which will be used blow. For a detailed study of these spaces[7,26,27].

**Proposition 2.1**[27] Let $\phi$ be a weight function with the growth rate $\alpha$, and $R$ be a positive number. Then the following estimates are valid:

$$C_2\int_{\mathbf{R}^N}\phi(x)\,|u(x)|^p\mathrm{d}x\leqslant\int_{\mathbf{R}^N}\phi(x_0)\int_{B_{x_0}^R}|u(x)|^p\mathrm{d}x\mathrm{d}x_0\leqslant C_1\int_{\mathbf{R}^N}\phi(x)\,|u(x)|^p\mathrm{d}x. \tag{2.1}$$

**Corollary 2.2**[27] The equivalent norm in a weighted Sobolev space $W_\phi^{l,p}(\mathbf{R}^N)$ is given by

$$\|u\|_{\phi,l,p}=\left(\int_{\mathbf{R}^N}\phi(x_0)\,\|u,B_{x_0}^R\|_{l,p}^p\mathrm{d}x_0\right)^{1/p}. \tag{2.2}$$

In particular, norms (2.2) are equivalent for different $R\in\mathbf{R}_+$.

**Corollary 2.3** Let $u\in W_{b,\phi}^{l,p}(\mathbf{R}^n)$ and $\phi$ be a weight function with the growth rate $\alpha<\varepsilon$. Then

$$C_1\|u\|_{\phi,l,p}^p\leqslant\int_{\mathbf{R}^N}\phi(x_0)\int_{\mathbf{R}^N}\mathrm{e}^{-\varepsilon(1+|x-x_0|^2)^{1/2}}\|u,B_x^1\|_{l,p}^p\mathrm{d}x\mathrm{d}x_0\leqslant C_2\|u\|_{\phi,l,p}^p, \tag{2.3}$$

$$C_1\|u\|_{b,\phi,l,p}^p\leqslant\sup_{x_0\in\mathbf{R}^N}\left\{\phi(x_0)\int_{\mathbf{R}^N}\mathrm{e}^{-\varepsilon(1+|x-x_0|^2)^{1/2}}\|u,B_x^1\|_{l,p}^p\mathrm{d}x\right\}\leqslant C_2\|u\|_{b,\phi,l,p}^p. \tag{2.4}$$

**Proof** Estimate (2.3) has been proved in [9] and estimate (2.4) has been proved in [26].

**Lemma 2.4** Let $u_0,v_0\in W_b^{2,2}$. Then for all smooth solutions $(u,v)$ of (1.1)~(1.3), the following estimates are valid:

$$u(x,t)\geqslant\alpha(t)=a(1-\mathrm{e}^{-(b+1)t})/(b+1),\quad\forall x\in\mathbf{R}^N t\geqslant0; \tag{2.5}$$

$$\|v\|_{L^\infty(\mathbf{R}^N\times[0,\infty))}\leqslant C(\|v_0\|_{L^\infty(\mathbf{R}^N)},a,b,d). \tag{2.6}$$

**Proof** See the proof of Lemma 2.4 in [9,14].

**Lemma 2.5** Let $u_0, v_0 \in W_b^{2,2}$. Then for all smooth solutions $(u, v)$ of (1.1)~(1.3), the following estimates are valid:

$$\int_T^{T+1}(\|u(t), B_{x_0}^1\|_{1,2}^2 + \|v(t), B_{x_0}^1\|_{1,2}^2)dt + \|u(T), B_{x_0}^1\|_{0,2}^2 + \|v(T), B_{x_0}^1\|_{0,2}^2$$

$$\leqslant Ce^{-KT}(\|u_0\|_{\phi_\varepsilon^0,0,2}^2 + \|v_0\|_{\phi_\varepsilon^0,0,2}^2) + C, \quad \forall T \geqslant 0, \quad x_0 \in \mathbf{R}^N, \tag{2.7}$$

where positive constants $K = \min\{1, \alpha(1)/2\}$ and $C = C(d, a, b, N)$ are independent of $T$ and $x_0$, $\alpha(1)$ is defined in (2.5), and $\phi_\varepsilon^0(x) = e^{-\varepsilon(1+|x-x_0|^2)^{1/2}}$.

It is new that the constant $C$ of estimate (2.7) is independent of initial data $u_0$ and $v_0$.

**Proof** Summing (1.1) and (1.2), we obtain

$$(u+v)_t - d\Delta(u+v) + u - a = 0. \tag{2.8}$$

Multiplying (2.8) by $\phi_\varepsilon^0(u+v)$ and integrating by parts, we get

$$\frac{1}{2}\frac{d}{dt}(\phi_\varepsilon^0, |u+v|^2) + d(\phi_\varepsilon^0, |\nabla(u+v)|^2) + (\phi_\varepsilon^0, |u+v|^2)$$

$$\leqslant(\phi_\varepsilon^0, |a+v||u+v|) + dN\varepsilon(\phi_\varepsilon^0, |u+v||\nabla(u+v)|), \tag{2.9}$$

where we have used the fact $|\partial_j\phi_\varepsilon^0| \leqslant \varepsilon\phi_\varepsilon^0$. By using the Hölder inequality, we have

$$(\phi_\varepsilon^0, |a+v||u+v|) \leqslant \frac{1}{4}(\phi_\varepsilon^0, |u+v|^2) + (\phi_\varepsilon^0, |a+v|^2) \leqslant 2(\phi_\varepsilon^0, |u+v|^2) + C, \tag{2.10}$$

$$dN\varepsilon(\phi_\varepsilon^0, |u+v||\nabla(u+v)|) \leqslant \frac{d}{2}(\phi_\varepsilon^0, |\nabla(u+v)|^2) + \frac{dN^2\varepsilon^2}{2}(\phi_\varepsilon^0, |u+v|^2). \tag{2.11}$$

Letting $0 < \varepsilon < 1$, putting together (2.9)~(2.11) and applying the Gronwall inequality, we obtain

$$\|u(T) + v(T)\|_{\phi_\varepsilon^0,0,2}^2 + \int_T^{T+1}\|u(t) + v(t)\|_{\phi_\varepsilon^0,1,2}^2 dt +$$

$$\int_0^T e^{(1+dN^2/2)(T-t)}\|u(t) + v(t)\|_{\phi_\varepsilon^0,1,2}^2 dt$$

$$\leqslant Ce^{(1+dN^2/2)T}\|u_0 + v_0\|_{\phi_\varepsilon^0,0,2}^2 + C, \quad \forall T \geqslant 0, \tag{2.12}$$

where $C = C(d, a, N)$.

Thanks to $u, v \geqslant 0$ and the fact that $e^{-\varepsilon(1+|x-x_0|^2)^{1/2}} \geqslant e^{-\varepsilon}$ if $x \in B_{x_0}^1$ uniformly with respect to $x_0 \in \mathbf{R}^N$, (2.12) implies (2.7) for $0 \leqslant T \leqslant 1$. For $T \geqslant 1$, (2.7) can be proved as follows.

Multiplying (1.2) by $\phi_\varepsilon^0 v$ and integrating by parts, we get

$$\frac{1}{2}\frac{d}{dt}(\phi_\varepsilon^0, v^2) + \frac{\alpha^2(1)}{2}(\phi_\varepsilon^0, v^2) + d(\phi_\varepsilon^0, |\nabla v|^2) +$$

$$\frac{1}{2}(\phi_\varepsilon^0, (uv - b)^2) + \frac{1}{2}(\phi_\varepsilon^0, \{u^2 - \alpha^2(1)\}v^2)$$

$$\leqslant \frac{1}{2}(\phi_\varepsilon^0, b^2) + dN\varepsilon(\phi_\varepsilon^0, v|\nabla v|), \quad \forall t \geqslant 1, \tag{2.13}$$

where $\alpha(1)$ is defined in (2.5), and the fact $|\partial_j\phi_\varepsilon^0|\leqslant\varepsilon\phi_\varepsilon^0$ is used. By using the Hölder inequality, we have

$$dN\varepsilon(\phi_\varepsilon^0, v|\nabla v|)\leqslant\frac{d}{2}(\phi_\varepsilon^0,|\nabla v|^2)+\frac{dN^2\varepsilon^2}{2}(\phi_\varepsilon^0, v^2). \tag{2.14}$$

Letting $0<\varepsilon<\min\{\alpha(1)/\sqrt{2dN^2},1\}$, putting together (2.5) and (2.12)~(2.14) and applying the Gronwall inequality, we obtain

$$\begin{aligned}&\|v(T)\|_{\phi_\varepsilon^0,0,2}^2+\int_T^{T+1}\|v(t)\|_{\phi_\varepsilon^0,1,2}^2 dt+\int_1^T e^{\alpha(1)(t-T)/2}\|v(t)\|_{\phi_\varepsilon^0,1,2}^2 dt\\ \leqslant\ & Ce^{-\alpha(1)T/2}(\|u_0\|_{\phi_\varepsilon^0,0,2}^2+\|v_0\|_{\phi_\varepsilon^0,0,2}^2)+C,\quad\forall T\geqslant1,\end{aligned} \tag{2.15}$$

where $C=C(d,a,b,N)$.

Multiplying (2.8) by $\phi_\varepsilon^0(u+v)$, we obtain (2.9) again. By using the Hölder inequality and (2.15), we have

$$\begin{aligned}(\phi_\varepsilon^0,|a+v||u+v|\leqslant&\frac{1}{4}(\phi_\varepsilon^0,|u+v|^2)+2(\phi_\varepsilon^0,a^2+v^2)\\ \leqslant&\frac{1}{4}(\phi_\varepsilon^0,|u+v|^2)+Ce^{-\alpha(1)t/2}(\|u_0\|_{\phi_\varepsilon^0,0,2}^2+\|v_0\|_{\phi_\varepsilon^0,0,2}^2)+C,\quad\forall t\geqslant1.\end{aligned} \tag{2.16}$$

Letting $0<\varepsilon<\min\{1/\sqrt{2dN^2},\alpha(1)/\sqrt{2dN^2},1\}$, putting together (2.9),(2.11),(2.12) and (2.16) and applying the Gronwall inequality, we obtain

$$\begin{aligned}&\|u(T)+v(T)\|_{\phi_\varepsilon^0,0,2}^2+\int_T^{T+1}\|u(t)+v(t)\|_{\phi_\varepsilon^0,1,2}^2 dt+\int_1^T e^{t-T}\|u(t)+v(t)\|_{\phi_\varepsilon^0,1,2}^2 dt\\ \leqslant\ & C(e^{-T}+e^{-\alpha(1)T/2}(\|u_0\|_{\phi_\varepsilon^0,0,2}^2+\|v_0\|_{\phi_\varepsilon^0,0,2}^2)+C,\quad\forall T\geqslant1,\end{aligned} \tag{2.17}$$

where $C=C(d,a,b,N)$.

Thanks to $u,v\geqslant0$ and the fact that $e^{-\varepsilon(1+|x-x_0|^2)^{1/2}}\geqslant e^{-\varepsilon}$ if $x\in B_{x_0}^1$ uniformly with respect to $x_0\in\mathbf{R}^N$, (2.17) implies (2.7) for $T\geqslant1$. This lemma is proved. □

**Lemma 2.6** Let $u_0, v_0\in W_b^{2,2}$. Then the following estimate is valid for all smooth solutions $(u,v)$ of problem (1.1)~(1.3):

$$\begin{aligned}&\|u(T)\|_{W_b^{2,2}}^2+\|v(T)\|_{W_b^{2,2}}^2+\|u_t(T)\|_{L_b^2}^2+\|v_t(T)\|_{L_b^2}^2+\\ &\int_T^{T+1}(\|u(t)\|_{W_b^{3,2}}^2+\|v(t)\|_{W_b^{3,2}}^2+\|u_t(t)\|_{W_b^{1,2}}^2+\|v_t(t)\|_{W_b^{1,2}}^2)dt\\ \leqslant\ & Q(d,a,b,N,\|u_0\|_{W_b^{2,2}}+\|v_0\|_{W_b^{2,2}})e^{-k_0T}+Q_0(d,a,b,n),\quad\forall T\geqslant0\end{aligned} \tag{2.18}$$

where constant $K_0$ is positive, functions $Q$ and $Q_0$ are monotonic.

**Proof** See the proof of Lemma 2.6~2.9 in [9].

In terms of the estimate (2.7), we can improve the estimates of Lemma 2.6~2.9 in [9] such that $Q_0$ of (2.18) is independent of initial data $u_0$ and $v_0$. This improvement is the key to prove the existence of attractor $\mathscr{A}$.

**Theorem 2.7** Let $u_0, v_0\in W_b^{2,2}$. Then there exists a unique solution $(u,v)$ of system

(1.1)~(1.3) such that $u, v \geqslant 0$ and $u, v \in L^{\infty}([0,\infty); W_b^{2,2}) \cap C([0,\infty); W_b^{1,2}) \cap W^{1,\infty}([0,\infty); L_b^2) \cap L_{\mathrm{loc}}^2([0,\infty); W_b^{3,2}) \cap W_{\mathrm{loc}}^{1,2}([0,\infty); W_b^{1,2}) \cap C([0,\infty); W_{\mathrm{loc}}^{2,2}) \cap C^1([0,\infty); L_{\mathrm{loc}}^2)$. Moreover the following estimate holds for every two solutions $(u_1(t), v_1(t))$ and $(u_2(t), v_2(t))$of Eqs. (1.1) and (1.2):

$$\left\| u_1(t) - u_2(t), B_{x_0}^1 \right\|_{0,2} + \left\| v_1(t) - v_2(t), B_{x_0}^1 \right\|_{0,2}$$
$$\leqslant C e^{K_0 t}(\| u_1(0) - u_2(0) \|_{\phi_\varepsilon^0,0,2} + \| v_1(0) - v_2(0) \|_{\phi_\varepsilon^0,0,2}), \quad \forall t \geqslant 0, \tag{2.19}$$

where positive constants $C$ and $K_0$ are independent of $\varepsilon > 0$ and $x_0 \in \mathbf{R}^N$.

**Proof** See the proof of Theorem 1.3 in [9].

**Corollary 2.8** Let $u_0, v_0 \in W_b^{2,2}$ and $(u(t), v(t))$ be the solution of problem (1.1)~(1.3). Then a semigroup $S_t$ can be defined by problem (1.1)~(1.3) in the space $W_b^{2,2} \times W_b^{2,2}$.

$$S_t : W_b^{2,2} \times W_b^{2,2} \rightarrow W_b^{2,2} \times W_b^{2,2}, \quad (u(t), v(t)) = S_t(u_0, v_0). \tag{2.20}$$

Let us formulate some results on the local smoothing property for the solutions of (1.1)~(1.3). These results imply that the absorbing sets of semigroup $S_t$ are compact sets in $W_{\mathrm{loc}}^{2,2}(\mathbf{R}^N)$-topology.

**Lemma 2.9** Let $u_0, v_0 \in W_b^{2,2}$ and $(u(t), v(t))$ be the solution of problem (1.1)~(1.3). Then the following estimate is valid:

$$\| u(T) \|_{W_b^{3,2}}^2 + \| v(T) \|_{W_b^{3,2}}^2 + \| u_t(T) \|_{W_b^{1,2}}^2 + \| v_t(T) \|_{W_b^{1,2}}^2 +$$
$$\int_T^{T+1} (\| u(t) \|_{W_b^{4,2}}^2 + \| v(t) \|_{W_b^{4,2}}^2 + \| u_t(t) \|_{W_b^{2,2}}^2 + \| v_t(t) \|_{W_b^{2,2}}^2) \mathrm{d}t$$
$$\leqslant Q(d, a, b, \| u(0) \|_{W_b^{2,2}}, \| v(0) \|_{W_b^{2,2}}) e^{-K_0 T} + Q_0(d, a, b, N), \quad \forall T \geqslant 1 \tag{2.21}$$

where constant $K_0 > 0$, and functions $Q$ and $Q_o$ are monotonic.

**Proof** See the proof of Lemma 2.10 in [9]. Based on (2.7), the estimate of [9] can be improved such that $Q_0$ is independent of initial data $u_0$ and $v_0$.

We conclude this section by formulating some results on the smoothing property for the difference of the solutions of (1.1)~(1.3), which are of fundamental importance to our study of attractor of semigroup $S_t$.

**Lemma 2.10** Let $(u_1(t), v_1(t))$ and $(u_2(t), v_2(t))$ be two solutions of problem (1.1)~(1.3) with initial data $u_j(0), v_j(0) \in W_b^{2,2}(j = 1, 2)$. Assume that there exists a positive constant $M_0 > 0$ such that $\| u_j(0) \|_{b,2,2} + \| v_j(0) \|_{b,2,2} \leqslant M_0 (j = 1, 2)$. Then the following estimate holds:

$$\left\| u_1(1) - u_2(1), B_{x_0}^1 \right\|_{2,2} + \left\| v_1(1) - v_2(1), B_{x_0}^1 \right\|_{2,2} +$$
$$\left\| \partial_t u_1(1) - \partial_t u_2(1), B_{x_0}^1 \right\|_{0,2} + \left\| \partial_t v_1(1) - \partial_t v_2(1), B_{x_0}^1 \right\|_{0,2}$$
$$\leqslant C(\| u_1(0) - u_2(0) \|_{\phi_\varepsilon^0,0,2} + \| v_1(0) - v_2(0) \|_{\phi_\varepsilon^0,0,2}), \tag{2.22}$$

where the constant $C$ depends only on Eqs. (1.1), (1.2) and $M_0$.

**Proof** Let $U(t)=u_1(t)-u_2(t)$ and $V(t)=v_1(t)-v_2(t)$. Then the functions $U$ and $V$ satisfy the following equations

$$\partial_t U - d\Delta U + (b+1)U - u_1^2 V - (u_1+u_2)v_2 U = 0, \tag{2.23}$$

$$\partial_t V - d\Delta V - bU + u_1^2 V + (u_1+u_2)v_2 U = 0, \tag{2.24}$$

$$U(0) = u_1(0) - u_2(0), \quad V(0) = v_1(0) - v_2(0). \tag{2.25}$$

Multiplying Eqs. (2.23) and (2.24) by $\phi_\varepsilon^0 U(t)$ and $\phi_\varepsilon^0 V(t)$, respectively, integrating over $x\in\mathbf{R}^N$ by parts, and using estimate (2.18), the Hölder inequality and Sobolev embedding theorem $W_b^{2,2}\subset L^\infty$, we have

$$\frac{\mathrm{d}}{\mathrm{d}t}(\phi_\varepsilon^0, |U(t)|^2) + (\phi_\varepsilon^0, |\nabla U(t)|^2) \leqslant C(\phi_\varepsilon^0, |U(t)|^2 + |V(t)|^2),$$

$$\frac{\mathrm{d}}{\mathrm{d}t}(\phi_\varepsilon^0, |V(t)|^2) + (\phi_\varepsilon^0, |\nabla V(t)|^2) \leqslant C(\phi_\varepsilon^0, |U(t)|^2 + |V(t)|^2).$$

Applying the Gronwall inequality, we obtain that

$$\|U(T)\|_{\phi_\varepsilon^0,0,2}^2 + \|V(T)\|_{\phi_\varepsilon^0,0,2}^2 + \int_0^T (\|U(T)\|_{\phi_\varepsilon^0,1,2}^2 + \|V(T)\|_{\phi_\varepsilon^0,1,2}^2)\mathrm{d}t$$
$$\leqslant C\mathrm{e}^{K_0 T}(\|u_1(0)-u_2(0)\|_{\phi_\varepsilon^0,0,2}^2 + \|v_1(0)-v_2(0)\|_{\phi_\varepsilon^0,0,2}^2), \quad \forall T\geqslant 0, \tag{2.26}$$

where positive constants $C$ and $K_0$ are independent of $\varepsilon>0$ and $x_0\in\mathbf{R}^N$.

Multiplying Eqs. (2.23) and (2.24) by $\sum_{j=1}^N \partial_j(\phi_\varepsilon^0 \partial_j U(t))$ and $\sum_{j=1}^N \partial_j(\phi_\varepsilon^0 \partial_j V(t))$, respectively, integrating over $x\in\mathbf{R}^N$ by parts, and using estimate (2.18), the Hölder inequality and Sobolev embedding theorem $W_b^{2,2}\subset L^\infty$, we have

$$\frac{\mathrm{d}}{\mathrm{d}t}(\phi_\varepsilon^0, |\nabla U(t)|^2) + d(\phi_\varepsilon^0, |\Delta U(t)|^2)$$
$$\leqslant C(\phi_\varepsilon^0, |U(t)|^2) + C(\phi_\varepsilon^0, |V(t)|^2) + C(\phi_\varepsilon^0, |\nabla U(t)|^2),$$
$$\frac{\mathrm{d}}{\mathrm{d}t}(\phi_\varepsilon^0, |\nabla V(t)|^2) + d(\phi_\varepsilon^0, |\Delta V(t)|^2)$$
$$\leqslant C(\phi_\varepsilon^0, |U(t)|^2) + C(\phi_\varepsilon^0, |V(t)|^2) + C(\phi_\varepsilon^0, |\nabla V(t)|^2).$$

Using estimate (2.26) and applying the Gronwall inequality, we obtain that

$$\|U(T)\|_{\phi_\varepsilon^0,1,2}^2 + \|V(T)\|_{\phi_\varepsilon^0,1,2}^2 + \int_0^T (\|U(T)\|_{\phi_\varepsilon^0,2,2}^2 + \|V(T)\|_{\phi_\varepsilon^0,2,2}^2)\mathrm{d}t$$
$$\leqslant C\mathrm{e}^{K_1 T}(\|u_1(0)-u_2(0)\|_{\phi_\varepsilon^0,1,2}^2 + \|v_1(0)-v_2(0)\|_{\phi_\varepsilon^0,1,2}^2), \quad \forall T\geqslant 0, \tag{2.27}$$

where positive constants $C$ and $K_1$ depend only on Eqs. (1.1), (1.2) and $M_0$. It follows from estimate (2.26) and Corollary 2.3 that there exists a point $T_0\in[0,1/2]$ such that

$$\|U(T_0)\|_{b,1,2}^2 + \|V(T_0)\|_{b,1,2}^2 \leqslant C(\|u_1(0)-u_2(0)\|_{b,0,2}^2 + \|v_1(0)-v_2(0)\|_{b,0,2}^2).$$

Now applying estimate (2.27) at the initial time moment $t=T_0$ instead of $t=0$, we derive that

$$\|U(T)\|_{\phi_\varepsilon^0,1,2}^2 + \|V(T)\|_{\phi_\varepsilon^0,1,2}^2 + \int_{T_0}^T (\|U(T)\|_{\phi_\varepsilon^0,2,2}^2 + \|V(T)\|_{\phi_\varepsilon^0,2,2}^2)\mathrm{d}t$$
$$\leqslant C(\|U(T_0)\|_{b,1,2}^2 + \|V(T_0)\|_{b,1,2}^2)$$

$$\leqslant C(\| u_1(0) - u_2(0)\|_{b,0,2}^2 + \| v_1(0) - v_2(0)\|_{b,0,2}^2), \quad \forall T \in [T_0, 2]. \quad (2.28)$$

Now let us differentiate Eqs. (2.23) and (2.24) with respect to $t$. Denote $W(t) = \partial_t U(t)$ and $G(t) = \partial_t V(t)$. Then these functions satisfy the equations

$$\partial_t W - d\Delta W + (b+1)W - u_1^2 G - (u_1 + u_2)v_2 W = 2u_1 u_{1t} V + \{(u_1 + u_2)v_2\}_t U, \quad (2.29)$$

$$\partial_t G - d\Delta G - bW + u_1^2 G + (u_1 + u_2)v_2 W = -2u_1 u_{1t} V - \{(u_1 + u_2)v_2\}_t U, \quad (2.30)$$

$$W(0) = u_{1t}(0) - u_{2t}(0), \quad G(0) = v_{1t}(0) - v_{2t}(0). \quad (2.31)$$

Taking the inner products of (2.29) with $\phi_\varepsilon^0 W$ and (2.30) with $\phi_\varepsilon^0 G$, respectively, and using estimates (2.18) and (2.27), the Hölder inequality and Sobolev embedding theorem $W_b^{2,2} \subset L^\infty$, we obtain

$$\begin{aligned}
&\frac{\mathrm{d}}{\mathrm{d}t}(\phi_\varepsilon^0, |W|^2) + 2d(\phi_\varepsilon^0, |\nabla W|^2) \\
\leqslant\; & C(\phi_\varepsilon^0, |W|^2) + C(\phi_\varepsilon^0, |G|^2) + C(\phi_\varepsilon^0, |u_{1t}|^4)^{1/2}(\phi_\varepsilon^0, |U|^4 + |V|^4)^{1/2} + \\
& C(\phi_\varepsilon^0, |u_{2t}|^4 + |v_{2t}|^4)^{1/2}(\phi_\varepsilon^0, |U|^4)^{1/2} \\
\leqslant\; & C(\phi_\varepsilon^0, |W|^2) + C(\phi_\varepsilon^0, |G|^2) + \\
& C(\phi_{\varepsilon/2}^0, |u_{1t}|^2 + |\nabla u_{1t}|^2)(\phi_{\varepsilon/2}^0, |U|^2 + |V|^2 + |\nabla U|^2 + |\nabla V|^2) + \\
& C(\phi_{\varepsilon/2}^0, |u_{2t}|^2 + |v_{2t}|^2 + |\nabla u_{2t}|^2 + |\nabla v_{2t}|^2)(\phi_{\varepsilon/2}^0, |U|^2 + |\nabla U|^2) \\
\leqslant\; & C(\phi_\varepsilon^0, |W|^2) + C(\phi_\varepsilon^0, |G|^2) + C_1(\phi_\varepsilon^0, |\nabla u_{1t}|^2 + |\nabla u_{2t}|^2 + |\nabla v_{2t}|^2) + C_1,
\end{aligned} \quad (2.32)$$

$$\begin{aligned}
&\frac{\mathrm{d}}{\mathrm{d}t}(\phi_\varepsilon^0, |G|^2) + 2d(\phi_\varepsilon^0, |\nabla G|^2) \\
\leqslant\; & C(\phi_\varepsilon^0, |W|^2) + C(\phi_\varepsilon^0, |G|^2) + C_1(\phi_\varepsilon^0, |\nabla u_{1t}|^2 + |\nabla u_{2t}|^2 + |\nabla v_{2t}|^2) + C_1,
\end{aligned} \quad (2.33)$$

$$C_1 = C\mathrm{e}^{K_1 t}(\| u_1(0) - u_2(0)\|_{\phi_{\varepsilon/2}^2,1,2}^2 + \| v_1(0) - v_2(0)\|_{\phi_{\varepsilon/2}^2,1,2}^2), \quad \forall t \geqslant 0,$$

where Sobolev embedding theorem $H^1(\mathbf{R}^N) \subset L^4(\mathbf{R}^N)$ has been used. By Eqs. (2.23) and (2.24) and Sobolev embedding theorem $W_b^{2,2} \subset L^\infty$, we have

$$\begin{aligned}
&\| u_{1t}(0) - u_{2t}(0)\|_{\phi_\varepsilon^0,0,2} + \| v_{1t}(0) - v_{2t}(0)\|_{\phi_\varepsilon^0,0,2} \\
\leqslant\; & C(\| u_1(0) - u_2(0)\|_{\phi_\varepsilon^0,2,2} + \| v_1(0) - v_2(0)\|_{\phi_\varepsilon^0,2,2}).
\end{aligned} \quad (2.34)$$

By (2.18), (2.32)~(2.34), Corollary 2.3 and the Gronwall inequality, we obtain that

$$\begin{aligned}
&\| W(T)\|_{\phi_\varepsilon^0,0,2}^2 + \| G(T)\|_{\phi_\varepsilon^0,0,2}^2 + \int_0^T (\| W(t)\|_{\phi_\varepsilon^0,1,2}^2 + \| G(t)\|_{\phi_\varepsilon^0,1,2}^2)\mathrm{d}t \\
\leqslant\; & C\mathrm{e}^{K_2 T}(\| u_1(0) - u_2(0)\|_{\phi_\varepsilon^0,2,2}^2 + \| v_1(0) - v_2(0)\|_{\phi_\varepsilon^0,2,2}^2), \quad \forall T \geqslant 0, 
\end{aligned} \quad (2.35)$$

where positive constants $C$ and $K_2$ depend only on Eqs. (1.1), (1.2) and $M_0$. By Eqs. (2.23) and (2.24) and estimates (2.18), (2.27) and (2.35), it follows that

$$\|U(T)\|^2_{\phi^0_\varepsilon,2,2}+\|V(T)\|^2_{\phi^0_\varepsilon,2,2}+\int_0^T(\|U(t)\|^2_{\phi^0_\varepsilon,3,2}+\|V(t)\|^2_{\phi^0_\varepsilon,3,2})\mathrm{d}t$$
$$\leqslant Ce^{K_3T}(\|u_1(0)-u_2(0)\|^2_{\phi^0_\varepsilon,2,2}+\|v_1(0)-v_2(0)\|_{\phi^0_\varepsilon,2,2}),\quad \forall T\geqslant 0,\quad (2.36)$$

where positive constants $C$ and $K_3$ depend only on Eqs. (1.1), (1.2) and $M_0$. It follows from estimate (2.28) and Corollary 2.3 that there exists a point $T_1\in[T_0,1)$ such that

$$\|U(T_1)\|^2_{b,2,2}+\|V(T_1)\|^2_{b,2,2}\leqslant C(\|u_1(0)-u_2(0)\|^2_{b,0,2}+\|v_1(0)-v_2(0)\|^2_{b,0,2}).$$

Now applying estimates (2.35) and (2.36) at the initial time moment $t=T_1$ instead of $t=0$, we derive that

$$\|U(T)\|^2_{\phi^0_\varepsilon,2,2}+\|V(T)\|^2_{\phi^0_\varepsilon,2,2}+\|U_t(T)\|^2_{\phi^0_\varepsilon,0,2}+\|V_t(T)\|^2_{\phi^0_\varepsilon,0,2}+$$
$$\int_{T_1}^T(\|U(t)\|^2_{\phi^0_\varepsilon,3,2}+\|V(t)\|^2_{\phi^0_\varepsilon,3,2}+\|U_t(T)\|^2_{\phi^0_\varepsilon,1,2}+\|V_t(T)\|^2_{\phi^0_\varepsilon,1,2})\mathrm{d}t$$
$$\leqslant C(\|U(T_1)\|^2_{b,2,2}+\|V(T_1)\|^2_{b,2,2})$$
$$\leqslant C(\|u_1(0)-u_2(0)\|^2_{b,0,2}+\|v_1(0)-v_2(0)\|^2_{b,0,2}),\quad \forall T\geqslant T_1,\quad (2.37)$$

where the positive constant $C$ depends only on Eqs. (1.1), (1.2) and $M_0$. Thanks to the fact that $e^{-\varepsilon(1+|x-x_0|^2)^{1/2}}\geqslant e^{-\varepsilon}$ if $x\in B^1_{x_0}$ uniformly with respect to $x_0\in\mathbf{R}^N$, estimate (2.37) implies estimate (2.22). This theorem is proved. □

## 3 The Attractor

In this section, we prove the existence of a locally compact attractor $\mathscr{A}$ for the semigroup $S_t$ generated by Eqs. (1.1)~(1.3), and obtain the upper bound for ε-entropy of the attractor.

According to Lemma 2.6, the semigroup $S_t: W_b^{2,2}(\mathbf{R}^N)\times W_b^{2,2}(\mathbf{R}^N)\to W_b^{2,2}(\mathbf{R}^N)\times W_b^{2,2}(\mathbf{R}^N)$ possesses a bounded absorbing set $\mathscr{B}$ in the space $W_b^{2,2}(\mathbf{R}^N)\times W_b^{2,2}(\mathbf{R}^N)$; that is, for every other bounded subset $B\subset W_b^{2,2}(\mathbf{R}^N)\times W_b^{2,2}(\mathbf{R}^N)$ there exists $T=T(B)$ such that $S_tB\subset\mathscr{B}$ if $t\geqslant T$. Nevertheless, in contrast to the case of bounded domains, the compact attractor in $W_b^{2,2}(\mathbf{R}^N)\times W_b^{2,2}(\mathbf{R}^N)$ usually does not exist[26]. Consequently, following[7,26,27], we construct below the attractor $\mathscr{A}$ of semigroup $S_t$. $\mathscr{A}$ attracts bounded subsets of $W_b^{2,2}(\mathbf{R}^N)\times W_b^{2,2}(\mathbf{R}^N)$ only in a local topology of the space $W_{\mathrm{loc}}^{2,2}(\mathbf{R}^N)\times W_{\mathrm{loc}}^{2,2}(\mathbf{R}^N)$; i. e., $\mathscr{A}$ is the $(W_b^{2,2}(\mathbf{R}^N), W_{\mathrm{loc}}^{2,2}(\mathbf{R}^N))$-attractor in the notation of [2,5].

**Definition 3.1**[2,27] A set $\mathscr{A}\subset W_b^{2,2}(\mathbf{R}^N)\times W_b^{2,2}(\mathbf{R}^N)$ is defined as the attractor of the semigroup $S_t$ if the following assumptions hold:

(i) The set $\mathscr{A}$ is compact in $W_{\mathrm{loc}}^{2,2}(\mathbf{R}^N)\times W_{\mathrm{loc}}^{2,2}(\mathbf{R}^N)$ and is bounded in $W_b^{2,2}(\mathbf{R}^n)\times W_b^{2,2}(\mathbf{R}^N)$.

(ii) The set $\mathscr{A}$ is strictly invariant with respect to $S_t$; i. e., $S_t\mathscr{A}=\mathscr{A}$ for $t\geqslant 0$.

(iii) The set $\mathscr{A}$ is an attracting set for $S_t$ in a local topology; i. e., for every neighborhood $\mathscr{O}(\mathscr{A})$ of $\mathscr{A}$ in the topology of the space $W_{\mathrm{loc}}^{2,2}(\mathbf{R}^N)\times W_{\mathrm{loc}}^{2,2}(\mathbf{R}^N)$ and for every bounded subset $B$

in the uniform topology of $W_b^{2,2}(\mathbf{R}^N)\times W_b^{2,2}(\mathbf{R}^N)$, there exists $T=T(\mathcal{O}, B)$ such that $S_tB\subset\mathcal{O}(\mathcal{A})$ if $t\geqslant T$.

**Theorem 3.2** (Global Attractor) The semigroup $S_t$ defined in Corollary 2.8 possesses an attractor $\mathcal{A}$ in the sense of Definition 3.1 that has the structure $\mathcal{A}=\mathcal{K}|_{t=0}$, where we denote by $\mathcal{K}$ the set of all solutions $(u,v)$ of (1.1)~(1.3) that are defined and bounded for all $t\in\mathbf{R}$, $\sup_{t\in\mathbf{R}}\|u(t)\|_{b,2,2}+\|v(t)\|_{b,2,2}<\infty$.

**Proof** According to the existence theorem of attractor for abstract semigroups[2,5,8,21], it is sufficient to verify the following conditions:

(ⅰ) The semigroup $S_t$ possesses a compact absorbing set $K$ in $W_{\rm loc}^{2,2}(\mathbf{R}^N)\times W_{\rm loc}^{2,2}(\mathbf{R}^N)$-topology.

(ⅱ) The operators $S_t$ have closed graphs on $K$ in the $W_{\rm loc}^{2,2}(\mathbf{R}^N)\times W_{\rm loc}^{2,2}(\mathbf{R}^N)$-topology for every fixed $t\geqslant 0$.

Estimates (2.18) and (2.21) imply that the set

$$K=\{w\in W_b^{3,2}(\mathbf{R}^N)\times W_b^{3,2}(\mathbf{R}^N)\mid \|w\|_{W_b^{3,2}\times W_b^{3,2}}\leqslant R\}$$

is an absorbing set of semigroup $S_t$ if $R$ is large enough. Obviously, the absorbing set $K$ is compact in $W_{\rm loc}^{2,2}(\mathbf{R}^N)\times W_{\rm loc}^{2,2}(\mathbf{R}^N)$. Condition (ⅰ) is verified.

Suppose that $w_0^n, w_0\in K$, $w_0=W_{\rm loc}^{2,2}-\lim_{n\to\infty}w_0^n$, and $w=W_{\rm loc}^{2,2}-\lim_{n\to\infty}S_tw_0^n$. It follows from (2.2) that $w_0=L_{\phi_\varepsilon^0}^2-\lim_{n\to\infty}w_0^n$. Using estimate (2.19), we have $S_tw_0=L_{\rm loc}^2-\lim_{n\to\infty}S_tw_0^n$, and $w=S_tw_0$. Thus condition(ⅱ) is verified. Consequently the semigroup $S_t$ possesses an $(W_b^{2,2}, W_{\rm loc}^{2,2})$-attractor $\mathcal{A}=\mathcal{K}|_{t=0}$ by the abstract existence theorem of attractor. This theorem is proved. □

Let us recall the definition of $\varepsilon$-entropy[7,12,22,26,27].

**Definition 3.3** Let $\mathcal{M}$ be a metric space and $\mathcal{K}$ be a precompact subset of it. For a given $\varepsilon>0$, let $N_\varepsilon(\mathcal{K})=N_\varepsilon(\mathcal{K},\mathcal{M})$ be the minimal number of $\varepsilon$-balls in $\mathcal{M}$ that cover the set $\mathcal{K}$. Kolmogorov $\varepsilon$-entropy of $\mathcal{K}$ in $\mathcal{M}$ is the number

$$\mathbb{H}_\varepsilon(\mathcal{K})=\mathbb{H}_\varepsilon(\mathcal{K},\mathcal{M})=\ln N_\varepsilon(\mathcal{K}).$$

**Theorem 3.4** (Upper Bound of $\varepsilon$-Entropy) Let $\mathcal{A}$ be the attractor of semigroup $S_t$ defined in Theorem 3.2, and $\mathrm{vol}_{x_0}(R)=\mathrm{vol}(B_{x_0}^R)$. Then for every $0<\varepsilon\leqslant\varepsilon_0<1$, the following estimates are valid:

$$\mathbb{H}_\varepsilon(\mathcal{A}|_{B_{x_0}^R}, W_b^{2,2}(B_{x_0}^R))\leqslant C\mathrm{vol}_{x_0}\left(R+K\ln\frac{1}{\varepsilon}\right)\ln\frac{1}{\varepsilon},\quad \forall R\in\mathbf{R}_+,\quad \forall x_0\in\mathbf{R}^N, \tag{3.1}$$

where the constants $C$, $K$ and $\varepsilon_0$ are independent of $R$ and $x_0\in\mathbf{R}^N$.

**Proof** The proof of this theorem is based on estimate (2.22) with a special choice of the weight function $\phi$ and is completely analogous to the proof of [26, Theorem 8.1]. For simplicity, the detailed proof is omitted. □

## 4 Unstable Manifolds and Spatial Chaos

Let the group $\{T_h | h\in \mathbf{R}^N\}$ of spatial shifts act on the attractor $\mathscr{A}$ of (1.1)~(1.3),

$$T_h\mathscr{A} = \mathscr{A}, \quad (T_h u)(x) = u(x+h), \quad h\in \mathbf{R}^N, u\in \mathscr{A}.$$

The main task of this section is to study the action of this group on the attractor from the dynamical point of view. The usual way to indicate the chaotic behavior of a dynamical system $T_h: \mathscr{A}\to\mathscr{A}$ is to find a closed invariant subset $M\subset\mathscr{A}$ in the corresponding phase space and construct a homeomorphism $\tau: M\to\mathscr{M}$ such that

$$\tau:(T_h|_M, M)\to(\hat{T}_h,\mathscr{M}), \quad \hat{T}_h = \tau\circ T_h\circ\tau^{-1},$$

where $(\hat{T}_h,\mathscr{M})$ is a model example of a dynamical system whose chaotic behavior is evident. To study the spatial chaos generated by the action of $\{T_h|_h\in\mathbf{R}^N\}$ on the attractor $\mathscr{A}$, we use the technique of unstable manifolds developed in [7,26,27]. Let

$$\mathscr{L} = \begin{pmatrix} d\Delta + b - 1 & a^2 \\ -b & d\Delta - a^2 \end{pmatrix}.$$

Assume that $b>1$ and $a^2<b+1-2\sqrt{b}$, then the spectrum $\sigma(\mathscr{L})$ (in the space $L^2(\mathbf{R}^N)\times L^2(\mathbf{R}^N)$) satisfies the condition $\sigma(\mathscr{L})\cap\{\mathrm{Re}\, z>0\}\neq\emptyset$. Therefore Eqs. (1.1) and (1.2) possess at least one exponentially unstable, spatially homogeneous equilibrium $(u_s, v_s)=(a, b/a)$. As usual, we start by studying the linear nonhomogeneous problem

$$\partial_t w - \mathscr{L}w = h(t), \tag{4.1}$$

which corresponds to the linearization of (1.1) and (1.2) at $(u_s, v_s)=(a, b/a)$. Here $w$ and $h$ are vector-valued functions. For vector-valued functions, to simplify notation, we write below $[E]^2$ instead of $E\times E$ where $E$ can be any functional space. To study unstable manifolds, we need the following functional spaces:

**Definition 4.1**[27] Let $\gamma\in\mathbf{R}$ be arbitrary. Space $\mathbb{L}_\gamma(E)$ is defined by the expression

$$\mathbb{L}_\gamma(E) = \{u\in L^\infty_{\mathrm{loc}}(\mathbf{R}_-, E) \mid |||u|||_{\gamma,E} = \sup_{t\leqslant 0} \mathrm{e}^{-\gamma t}\|u\|_E < \infty\},$$

where $E$ is a certain Banach subspace of distributions $D'(\mathbf{R}^N)$.

**Lemma 4.2** Let $b>1, a^2<b+1-2\sqrt{b}$, and $\gamma>\frac{1}{2}(b-1-a^2+\sqrt{(b-1-a^2)^2-4a^2})>0$. Then for every $h\in\mathbb{L}_\gamma([L^2_b(\mathbf{R}^N)]^2)$, Eq. (4.1) possesses a unique backward solution $w(t)\in\mathbb{L}_\gamma([W_b^{2-\theta,2}(\mathbf{R}^N)]^2)$ $(\forall\theta\in(0,1))$. Thus a linear operator

$$\mathrm{T}_\gamma: \mathbb{L}_\gamma([L^2_b(\mathbf{R}^N)]^2\to\mathbb{L}_\gamma([W_b^{2-\theta,2}(\mathbf{R}^N)]^2), \quad w(t) = (\mathrm{T}_\gamma h)(t), \tag{4.2}$$

is well-defined. Moreover, $\forall\theta\in(0,1)$, the following estimates are valid:

$$|||(\mathrm{T}_\gamma h)|||_{\gamma,[W_b^{2-\theta,2}]^2} \leqslant C\,|||h(s)|||_{\gamma,[L_b^2]^2}, \tag{4.3}$$

$$|||(\mathrm{T}_\gamma h)|||_{\gamma,[W_b^{2+\theta_0-\theta,2}]^2} \leqslant C\,|||h(s)|||_{\gamma,[W_b^{\theta_0,2}]^2}, \quad \forall\theta_0\geqslant 0, \quad h\in\mathbb{L}_\gamma([W_b^{\theta_0,2}]^2), \tag{4.4}$$

$$|||(\mathrm{T}_\gamma h)|||_{\gamma,[W_b^{2+\theta_1-\theta,2}]^2} \leqslant C\,|||h(s)|||^{1-\theta_1}_{\gamma,[L_b^2]^2}\,|||h(s)|||^{\theta_1}_{\gamma,[W_b^{1,2}]^2}, \forall\theta_1\in(0,1), \quad h\in\mathbb{L}_\gamma([W_b^{1,2}]^2). \tag{4.5}$$

**Proof** For every rapidly decreasing function $\psi \in \mathscr{S}(\mathbf{R}^N)$, $\forall \varepsilon > 0$, $\exists r_n > 0$ such that

$$\left|\int_{|x| \geqslant r_n} |\psi(x)|^2 \mathrm{d}x\right| \leqslant \frac{\varepsilon}{2^n}, \quad \forall n = 1, 2, \cdots$$

Then

$$\begin{aligned}\left|\int_{|x| \geqslant r_1} \psi(x) h(x) \mathrm{d}x\right| &\leqslant \sum_{n=1}^{\infty} \left|\int_{r_n \leqslant |x| \leqslant r_{n+1}} \psi(x) h(x) \mathrm{d}x\right| \\ &\leqslant \sum_{n=1}^{\infty} \left\{\int_{r_n \leqslant |x| \leqslant r_{n+1}} |\psi(x)|^2 \mathrm{d}x\right\}^{1/2} \left\{\int_{r_n \leqslant |x| \leqslant r_{n+1}} |h(x)|^2 \mathrm{d}x\right\}^{1/2} \\ &\leqslant \sum_{n=1}^{\infty} \frac{\varepsilon}{2^n} \|h\|_{L_b^2(\mathbf{R}^N)} \leqslant \varepsilon \|h\|_{L_b^2(\mathbf{R}^N)}, \quad \forall h \in L_b^2(\mathbf{R}^N),\end{aligned}$$

and the integration $\int_{\mathbf{R}^N} \psi(x) h(x) \mathrm{d}x \, (\forall \psi \in \mathscr{S}(\mathbf{R}^N), h \in L_b^2(\mathbf{R}^N))$ is well-defined. Therefore, $\forall h \in \mathbb{L}_\gamma([L_b^2(\mathbf{R}^N)]^2)$, Eq. (4.1) possesses a unique backward solution

$$\begin{aligned} w(t) = &\frac{1}{O_+ - O_-} \int_{-\infty}^{t} \mathrm{e}^{\lambda_+(t-s)} \begin{pmatrix} O_+ & -O_+ O_- \\ 1 & -O_- \end{pmatrix} G(t-s) * h(s) \mathrm{d}s + \\ &\frac{1}{O_+ - O_-} \int_{-\infty}^{t} \mathrm{e}^{\lambda_-(t-s)} \begin{pmatrix} -O_- & O_+ O_- \\ -1 & O_+ \end{pmatrix} G(t-s) * h(s) \mathrm{d}s, \quad t \leqslant 0, \end{aligned} \tag{4.6}$$

where $w = \begin{pmatrix} w_1 \\ w_2 \end{pmatrix}$, $h = \begin{pmatrix} h_1 \\ h_2 \end{pmatrix}$, $G * h = \begin{pmatrix} G * h_1 \\ G * h_2 \end{pmatrix}$,

$$\begin{aligned} &G(t) = (4\pi d t)^{-N/2} \exp\left(-\frac{|x|^2}{4dt}\right), \quad O_\pm = -\frac{1}{2b}\left(b - 1 + a^2 \pm \sqrt{(b-1-a^2)^2 - 4a^2}\right), \\ &\lambda_\pm = \frac{1}{2}\left(b - 1 + a^2 \pm \sqrt{(b-1-a^2)^2 - 4a^2}\right). \end{aligned} \tag{4.7}$$

Let us first consider the case $h \in \mathbb{L}_\gamma([L^2(\mathbf{R}^N)]^2)$. Due to the estimate

$$\|(-\Delta)^{(2-\theta)/2} \mathrm{e}^{d\Delta t} g\|_{L^2} \leqslant C t^{-(2-\theta)/2} \|g\|_{L^2}, \quad t > 0, \quad \forall \theta \in (0, 2),$$

we have

$$\mathrm{e}^{-\gamma t} \|w(t)\|_{[W^{2-\theta,2}]^2} \leqslant C \sup_{s \in (-\infty, t]} \mathrm{e}^{-\gamma s} \|h(s)\|_{[L^2]^2}, \quad \forall \theta \in (0, 2). \tag{4.8}$$

Moreover, for any $\theta \in (0, 2)$ we have by the same argument that

$$\mathrm{e}^{-\gamma t} \|w(t)\|_{[W^{2+\theta_0-\theta,2}]^2} \leqslant C \sup_{s \in (-\infty, t]} \mathrm{e}^{-\gamma s} \|h(s)\|_{[W^{\theta_0,2}]^2}, \quad \forall h \in \mathbb{L}_\gamma([W^{\theta_0,2}]^2), \tag{4.9}$$

$$\begin{aligned} \mathrm{e}^{-\gamma t} \|w(t)\|_{[W^{2+\theta_1-\theta,2}]^2} &\leqslant C \sup_{s \in (-\infty, t]} \mathrm{e}^{-\gamma s} \|h(s)\|_{[W^{\theta_1,2}]^2} \\ &\leqslant C \sup_{s \in (-\infty, t]} \mathrm{e}^{-\gamma s} \|h(s)\|_{[L^2]^2}^{1-\theta_1} \|h(s)\|_{[W^{1,2}]^2}^{\theta_1}, \\ &\forall \theta_1 \in (0, 1), \quad h \in \mathbb{L}_\gamma([W^{1,2}]^2). \end{aligned} \tag{4.10}$$

Now we consider the general case $h \in \mathbb{L}_\gamma([L_b^2(\mathbf{R}^N)]^2)$. Let us fix an arbitrary $x_0 \in \mathbf{R}^N$ and introduce a new unknown vector function $w_{x_0}(t) = w(t)\phi_\varepsilon^0$, where $\phi_\varepsilon^0(x) = \mathrm{e}^{-\varepsilon(1+|x-x_0|^2)^{1/2}}$ and $\varepsilon > 0$ is a small parameter that will be specified below. We note that weight functions $\phi_\varepsilon^0$ satisfy the conditions $|\nabla_x \phi_\varepsilon^0| \leqslant C\varepsilon\phi_\varepsilon^0$ and $|\Delta\phi_\varepsilon^0| \leqslant C\varepsilon\phi_\varepsilon^0$. It is easy to verify that the vector functions $w_{x_0}$ satisfy the equations

$$\partial_t w_{x_0} - \mathscr{L} w_{x_0} = \phi_\varepsilon^0 h - 2d(\nabla\phi_\varepsilon^0 \cdot \nabla_x) w - d(\Delta\phi_\varepsilon^0) w = h_{x_0}(t).$$

Estimates (4.8)~(4.10) yield

$$\begin{aligned}
e^{-\gamma t}\|w_{x_0}(t)\|_{[W^{2-\theta,2}]^2} &\leqslant C \sup_{s\in(-\infty,t]} e^{-\gamma s}\|h_{x_0}(s)\|_{[L^2]^2} \\
&\leqslant C_1 \sup_{s\in(-\infty,t]} e^{-\gamma s}\|\phi_\varepsilon^0 h(s)\|_{[L^2]^2} + C_1\varepsilon \sup_{s\in(-\infty,t]} e^{-\gamma s}\|w_{x_0}(s)\|_{[W^{1,2}]^2},
\end{aligned}$$

$$\begin{aligned}
e^{-\gamma t}\|w_{x_0}(t)\|_{[W^{2+\theta_0-\theta,2}]^2} &\leqslant C \sup_{s\in(-\infty,t]} e^{-\gamma s}\|h_{x_0}(s)\|_{[W^{\theta_0,2}]^2} \\
&\leqslant C_1 \sup_{s\in(-\infty,t]} e^{-\gamma s}\|\phi_\varepsilon^0 h(s)\|_{[W^{\theta_0,2}]^2} + C_1\varepsilon \sup_{s\in(-\infty,t]} e^{-\gamma s}\|w_{x_0}(s)\|_{[W^{1+\theta_0,2}]^2},
\end{aligned}$$

$$\begin{aligned}
e^{-\gamma t}\|w_{x_0}(t)\|_{[W^{2+\theta_1-\theta,2}]^2} &\leqslant C \sup_{s\in(-\infty,t]} e^{-\gamma s}\|h_{x_0}(s)\|_{[W^{\theta_1,2}]^2} \\
&\leqslant C_1 \sup_{s\in(-\infty,t]} e^{-\gamma s}\|\phi_\varepsilon^0 h(s)\|_{[L^2]^2}^{1-\theta_1}\|\phi_\varepsilon^0 h(s)\|_{[W^{1,2}]^2}^{\theta_1} + \\
&\quad C_1\varepsilon \sup_{s\in(-\infty,t]} e^{-\gamma s}\|w_{x_0}(s)\|_{[W^{1+\theta_1,2}]^2}.
\end{aligned}$$

Fixing $\varepsilon>0$ small enough such that $C_1\varepsilon\leqslant\frac{1}{2}$, and letting $\theta\in(0,1)$, we derive that

$$\begin{aligned}
\sup_{t\leqslant 0} e^{-\gamma t} \sup_{x_0\in\mathbf{R}^N}\|w(t),B_{x_0}^1\|_{[W^{2-\theta,2}]^2} &\leqslant C \sup_{t\leqslant 0} e^{-\gamma t}\sup_{x_0\in\mathbf{R}^N}\|\phi_\varepsilon^0 w(t)\|_{[W^{2-\theta,2}]^2} \\
&\leqslant C \sup_{t\leqslant 0} e^{-\gamma t}\sup_{x_0\in\mathbf{R}^N}\|w_{x_0}(t)\|_{[W^{2-\theta,2}]^2} \leqslant C ||| h |||_{\gamma,[L_b^2]^2},
\end{aligned} \tag{4.11}$$

$$\sup_{t\leqslant 0} e^{-\gamma t} \sup_{x_0\in\mathbf{R}^N}\|w(t),B_{x_0}^1\|_{[W^{2+\theta_0-\theta,2}]^2} \leqslant C ||| h |||_{\gamma,[W_b^{\theta_0,2}]^2}, \tag{4.12}$$

$$\sup_{t\leqslant 0} e^{-\gamma t} \sup_{x_0\in\mathbf{R}^N}\|w(t),B_{x_0}^1\|_{[W^{2+\theta_1-\theta,2}]^2} \leqslant C ||| h |||_{\gamma,[L_b^2]^2}^{1-\theta_1} ||| h |||_{\gamma,[W_b^{1,2}]^2}^{\theta_1}. \tag{4.13}$$

This lemma is proved. □

**Definition 4.3**[27] Let $\mathbb{B}_{\omega,\xi}=\mathbb{B}_{\omega,\xi}(\mathbf{R}^N,\mathbf{C})(\xi\in\mathbf{R}^N)$ denote the subspace of $L^\infty(\mathbf{R}^N,\mathbf{C})$ that consists of all functions $\psi$ whose Fourier transforms $\hat{\psi}$(in the sense of distributions) satisfy the condition supp $\hat{\psi}\subset\xi+[-\omega,\omega]^N$. We define the space $\mathbb{B}_{\omega,\xi}^{\mathrm{Re}}=\mathbb{B}_{\omega,\xi}^{\mathrm{Re}}(\mathbf{R}^N,\mathbf{R})$ by the expression

$$\mathbb{B}_{\omega,\xi}^{\mathrm{Re}} = \{v\in L^\infty(\mathbf{R}^N): \exists\, u\in\mathbb{B}_{\omega,\xi}(\mathbf{R}^N,\mathbf{C}), v = \mathrm{Re} > u\}.$$

**Proposition 4.4**[27] Let $\sqrt{N}\omega<|\xi|$. Then every entire function $\psi\in\mathbb{B}_{\omega,\xi}$ is uniquely determined by its real part Re $\psi$. Moreover, for every $M>0$, the following estimate is valid:

$$|\psi(y)|\leqslant C_M \sup_{x\in\mathbf{R}^N}(1+|x-y|)^{-M}\|\mathrm{Re}\psi, B_y^1\|_{0,\infty}$$

where the constant $C_M$ is independent of $y\in\mathbf{R}^N$. Consequently, the spaces $\mathbb{B}_{\omega,\xi}^{\mathrm{Re}}$ and $\mathbb{B}_{\omega,\xi}$ are isomorphic. We denote this isomorphism by $\mathscr{R}:\mathbb{B}_{\omega,\xi}\to\mathbb{B}_{\omega,\xi}^{\mathrm{Re}}$.

**Lemma 4.5** Let $b>1, a^2<b+1-2\sqrt{b}, \lambda_\pm$ and $O_\pm$ be defined in (4.7) Then there exist $\gamma>0, \xi_0\in\mathbf{R}^N, \omega>0, e_0=\begin{pmatrix} O_+ \\ 1 \end{pmatrix}$, and the operator $\mathscr{P}_\gamma:\mathbb{B}_{\omega,\xi_0}^{\mathrm{Re}}\to\mathbb{L}_\gamma([W_b^{2,2}]^2)$ such that:

(ⅰ) For every $u_0 \in \mathbb{B}^{\mathrm{Re}}_{\omega,\xi_0}$, the function $w \in \mathbb{L}_\gamma([W_b^{2,2}]^2)$ defined by $w(t) = \mathscr{P}_\gamma(u_0)(t)$ is a solution of (4.1) with $h=0$ and

$$w(0) = \mathscr{P}_\gamma(u_0)(0) = u_0 e_0.$$

(ⅱ) $\gamma > \frac{\lambda_+}{2} > 0$ and $\sqrt{N}\omega < |\xi_0|$.

(ⅲ) For every $M \in \mathbb{R}_+$ and $u_0 \in \mathbb{B}^{\mathrm{Re}}_{\omega,\xi_0}$, the following estimate holds:

$$\left\| \mathscr{P}_\gamma(u_0)(t), B^1_{x_0} \right\|_{[W^{2,2}]^2} \leqslant C_M \mathrm{e}^{\gamma t} \sup_{x \in \mathbf{R}^N} \{ (1 + |x - x_0|)^{-M} \| u_0, B^1_x \|_{L^\infty} \}, \quad \forall t \leqslant 0. \tag{4.14}$$

Moreover the constant $C_M$ is independent of $x_0 \in \mathbf{R}^N$.

**Proof** Applying the $x$-Fourier transform to homogeneous equation (4.1), we have

$$\hat{w}_t = \begin{pmatrix} -d|\xi|^2 - 1 + b & a^2 \\ -b & -d|\xi|^2 - a^2 \end{pmatrix} \hat{w} = \hat{\mathscr{L}} \hat{w}. \tag{4.15}$$

There exists an eigenvalue $\hat{\lambda}(\xi) = \lambda_+ - d|\xi|^2$ of operator $\hat{\mathscr{L}}$ such that $\hat{\mathscr{L}} e_0 = \hat{\lambda}(\xi) e_0$. First we can take $\xi_0 \in \mathbf{R}^N$ such that $\xi_0 \neq 0$ and $\hat{\lambda}(\xi_0) \geqslant \frac{3}{4}\lambda_+$. Then we take $\omega > 0$ such that $\hat{\lambda}(\xi) \geqslant \frac{2}{3}\lambda_+$ for every $\xi \in \{\xi_0 + [-\omega, \omega]^N\}$ and $\sqrt{N}\omega < |\xi_0|$. Now we fixed $\gamma = \frac{2}{3}\lambda_+$. For every function $u_0 \in \mathbb{B}^{\mathrm{Re}}_{\omega,\xi_0}$, there exists a solution $\hat{w}$ of Eq. (4.15) with initial data $\hat{w}(t=0) = \hat{u}_0 e_0$ as follows:

$$\hat{w}(t) = \mathrm{e}^{\lambda_+ t} \mathrm{e}^{-d|\xi|^2 t} \hat{u}_0 e_0. \tag{4.16}$$

Take a cutoff function $\psi \in C_0^\infty(\mathbf{R}^N)$ such that $\psi(\xi) = 1$ if $\xi \in \{\xi_0 + [-\omega, \omega]^N\}$, and $\psi(\xi) = 0$ if $\xi \notin \{\xi_0 + [-2\omega, 2\omega]^N\}$. Then formula (4.16) can be rewritten in the equivalent form

$$\hat{w}(t) = \mathrm{e}^{\lambda_+ t} \mathrm{e}^{-d|\xi|^2 t} \psi(\xi) \hat{u}_0 \mathrm{e}^0. \tag{4.17}$$

Let $K(x) = (F_\xi^{-1}\psi)(x)$ and

$$w(t) = \mathscr{P}_\gamma(u_0)(t) = \mathrm{e}^{\lambda_+ t} K * G(t) * u_0 e_0. \tag{4.18}$$

Here $G(t)$ is defined in (4.7). It is clear that $w(t)$ $(t \leqslant 0)$ is a solution of (4.1) with $h=0$ and $w(0) = \mathscr{P}_\gamma(u_0)(0) = u_0 e_0$. Due to our construction of the cutoff function, it is not difficult to verify that

$$|D_x^L F(x)| \leqslant C_M (1 + |x|)^{-M}, \quad \forall 0 \leqslant L \leqslant 2, \quad M \in \mathbf{R}_+. \tag{4.19}$$

By the representation of $G(t)$ and (4.19), it follows that

$$|D_x^L w(x_0, t)| \leqslant C \mathrm{e}^{\gamma t} \int_{\mathbf{R}^N} (1 + |x_0 - y|)^{-M} |u_0(y)| \mathrm{d}y, \quad \forall 0 \leqslant L \leqslant 2, \quad M \in \mathbf{R}_+. \tag{4.20}$$

By changing the order of integration, we have

$$\begin{aligned} \int_{\mathbf{R}^N} (1 + |x_0 - x|)^{-M} \int_{B_x^1} |u_0(y)| \mathrm{d}y \mathrm{d}x &= \int_{\mathbf{R}^N} |u_0(y)| \int_{B_y^1} (1 + |x_0 - x|)^{-M} \mathrm{d}x \mathrm{d}y \\ &\geqslant C \int_{\mathbf{R}^N} (1 + |x_0 - y|)^{-M} |u_0(y)| \mathrm{d}y, \quad \forall M \in \mathbf{R}_+. \end{aligned} \tag{4.21}$$

Inserting (4.21) into (4.20), we obtain

$$|D_x^L w(x_0,t)| \leqslant C_M e^{\gamma t} \sup_{x\in \mathbf{R}^M} \{(1+|x-x_0|)^{-M} \| u_0, B_x^1 \|_{0,\infty}\}, \quad \forall 0 \leqslant L \leqslant 2, \quad M \in \mathbf{R}_+. \tag{4.22}$$

Estimate (4.14) is an immediate corollary of (4.22). In particular, this estimate implies that the operator $\mathscr{P}_\gamma$ is indeed a bounded operator from $\mathbb{B}^{\mathrm{Re}}_{\omega,\xi_0}$ to $\mathbb{L}_\gamma([W_b^{2,2}]^2)$. This lemma is proved. □

**Theorem 4.6** Let the assumptions of Lemma 4.5 be valid. Then there exist $r>0$ and a $C^1$-map

$$\mathscr{U}_0 : B(0,r,\mathbb{B}_{\omega,\xi_0}) \to \mathscr{A} \tag{4.23}$$

such that the following estimate is valid:

$$\| \mathscr{U}_0(w_0) - \mathscr{R} w_0 e_0 \|_{[W_b^{2,2}]^2} \leqslant C(\| w_0 \|^2_{L_b^\infty} + \| w_0 \|^3_{L_b^\infty}), \quad \forall w_0 \in B(0,r,\mathbb{B}_{w,\xi_0}), \tag{4.24}$$

where $B(0,r,\mathbb{B}_{\omega,\xi_0})$ is $r$-ball in the space $\mathbb{B}_{\omega,\xi_0}$ centered at 0, the constants $\omega$ and $\xi_0$ and vector $e_0$ are defined in Lemma 4.5, and the $\mathscr{R}$ is the isomorphism defined in Proposition 4.4. Moreover this map is a Lipschitz-continuous homeomorphism between $B(0,r,\mathbb{B}_{\omega,\xi_0})$ and its image $\mathscr{U}_0(B(0,r,\mathbb{B}_{\omega,\xi_0}))$ in the sense that

$$C_1 \| w_1 - w_2 \|_{L_b^\infty} \leqslant \| \mathscr{U}_0(w_1) - \mathscr{U}_0(w_2) \|_{[W_b^{2,2}]^2} \leqslant C_2 \| w_1 - w_2 \|_{L_b^\infty}. \tag{4.25}$$

**Proof** We are now going to find the backward solutions of Eqs. (1.1) and (1.2) near the equilibrium $(u_s, v_s) = (a, b/a)$ by using the implicit function theorem. Let us fix $\gamma, \omega$ and $\xi_0$ defined in Lemma 4.5, and consider the equation

$$w + \mathrm{T}_{2\gamma} F(w) = \mathscr{P}_\gamma(\mathscr{R} w_0), \quad w \in \mathbb{L}_\gamma([W_b^{2-\theta,2}]^2), w_0 \in \mathbb{B}_{\omega,\xi_0}. \tag{4.26}$$

Here $w = \begin{pmatrix} w_1 \\ w_2 \end{pmatrix}$, $F(w) = \begin{pmatrix} w_1^2 w_2 + 2a w_1 w_2 + b w_1^2/a \\ -w_1^2 w_2 - 2a w_1 w_2 - b w_1^2/a \end{pmatrix}$ and $0<\theta<\frac{1}{2}$. We note that every solution of (4.26) is simultaneously a solution of Eqs. (1.1) and (1.2). Hence it is sufficient to solve (4.26) in the space $\mathbb{L}_\gamma([W_b^{2-\theta,2}]^2)$. To this end, we introduce the function $\mathscr{F}: \mathbb{L}_\gamma([W_b^{2-\theta,2}]^2) \times \mathbb{B}_{\omega,\xi_0} \to \mathbb{L}_\gamma([W_b^{2-\theta,2}]^2)$ via $\mathscr{F}(w, w_0) = w + \mathrm{T}_{2\gamma} F(w) - \mathscr{P}_\gamma(\mathscr{R} w_0)$. Due to Lemmas 4.2 and 4.5 and the fact that Nemitsky operator $\mathscr{O} w = F(w)$ belongs to the space $C^1(\mathbb{L}_\gamma([W_b^{2-\theta,2}]^2), \mathbb{L}_{2\gamma}([L_b^2]^2))$, which can be directly verified by Sobolev embedding theorems[15], we have that $\mathscr{F}$ belongs to the class $C^1(\mathbb{L}_\gamma([W_b^{2-\theta,2}]^2) \times \mathbb{B}_{\omega,\xi_0}, \mathbb{L}_\gamma([W_b^{2-\theta,2}]^2))$ and $D_u \mathscr{F}(0,0) = Id$. Hence it follows from the implicit function theorem (see [25], for instance) that there exist a neighborhood $B(0,r,\mathbb{B}_{\omega,\xi_0})$ and $C^1$-function $\mathscr{U}: B(0,r,\mathbb{B}_{\omega,\xi_0}) \to \mathbb{L}_\gamma([W_b^{2-\theta,2}]^2)$ such that $\mathscr{F}(\mathscr{U}(w_0), w_0) \equiv 0$ and $|||\, \mathscr{U}(w_0) \,|||_{\gamma,[W_b^{2-\theta,2}]^2} \leqslant 2 \| w_0 \|_{\mathbb{B}_{\omega,\xi_0}}$. Consequently $\mathscr{U}(w_0)(t)$ is a backward solution of problem (1.1) and (1.2). Eq. (4.26) and Lemmas 4.2 and 4.5 imply that

$$|||\, \mathscr{U}(w_0) - \mathscr{P}_\gamma(\mathscr{R} w_0) \,|||_{2\gamma,[W_b^{2-\theta,2}]^2} \leqslant C \,|||\, F(\mathscr{U}(w_0)) \,|||_{2\gamma,[L_b^2]^2}$$

$$\leqslant C(|||\mathscr{U}(w_0)|||^2_{\gamma,[W_b^{2-\theta,2}]^2} + |||\mathscr{U}(w_0)|||^3_{\gamma,[W_b^{2-\theta,2}]^2})$$
$$\leqslant C(\|w_0\|^2_{\mathbb{B}_{\omega,\xi_0}} + \|w_0\|^3_{\mathbb{B}_{\omega,\xi_0}}). \tag{4.27}$$

It follows from estimate (4.5) and Sobolev embedding theorems[15] that

$$|||\mathscr{U}(w_0) - \mathscr{P}_\gamma(\mathscr{R}w_0)|||_{2\gamma,[W_b^{2,2}]^2} \leqslant C\,|||F(\mathscr{U}(w_0))|||^{1-\theta}_{2\gamma,[L_b^2]^2}\,|||F(\mathscr{U}(w_0))|||^{\theta}_{2\gamma,[W_b^{1,2}]^2}$$
$$\leqslant C(|||\mathscr{U}(w_0)|||^2_{\gamma,[W_b^{2-\theta,2}]^2} + |||\mathscr{U}(w_0)|||^3_{\gamma,[W_b^{2-\theta,2}]^2}).$$

Then, by using estimate (4.27), we have

$$|||\mathscr{U}(w_0) - \mathscr{P}_\gamma(\mathscr{R}w_0)|||_{2\gamma,[W_b^{2,2}]^2} \leqslant C(\|w_0\|^2_{\mathbb{B}_{\omega,\xi_0}} + \|w_0\|^3_{\mathbb{B}_{\omega,\xi_0}}). \tag{4.28}$$

We now set $\mathscr{U}_0(u_0) = \mathscr{U}(u_0)|_{t=0}$. Then (4.28) implies estimate (4.24). The assertion $\mathscr{U}_0(B(0,r,\mathbb{B}_{\omega,\xi_0}))\subset\mathscr{A}$ follows immediately from the description of the attractor $\mathscr{A}$ in Theorem 3.2 and the fact that the solution $w(t)=\mathscr{U}(w_0)(t)$ of problem (1.1) and (1.2), which is first defined only for $t\leqslant 0$, can be extended to a complete solution $w(t)$ $(t\in\mathbf{R})$ with $w(0)=\mathscr{U}_0(w_0)$ due to Theorem 2.7.

Thus it remains to verify estimate (4.25), Let functions $w_1$, $w_2\in B(0,r,\mathbb{B}_{\omega,\xi_0})$, $w^j(t)=\mathscr{U}(w_j)(t)(j=1,2)$ be the corresponding backward solutions of (1.1) and (1.2), $v_0=w_1-w_2$ and $V(t)=w^1(t)-w^2(t)$. Then the vector-valued function satisfies

$$V + \mathrm{T}_{2\gamma}(F(w^1) - F(w^2)) - \mathscr{P}_\gamma(\mathscr{R}v_0) = 0. \tag{4.29}$$

It follows from estimate (4.4) and Eq. (4.29) that

$$|||V - \mathscr{P}_\gamma(\mathscr{R}v_0)|||_{\gamma,[W_b^{2,2}]^2}$$
$$\leqslant C\,|||F(w^1) - F(w^2)|||_{2\gamma,[W_b^{1,2}]^2}$$
$$\leqslant C(|||w^1|||_{\gamma,[W_b^{2-\theta,2}]^2} + |||w^2|||_{\gamma,[W_b^{2-\theta,2}]^2} +$$
$$|||w^1|||^2_{\gamma,[W_b^{2-\theta,2}]^2} + |||w^2|||^2_{\gamma,[W_b^{2-\theta,2}]^2})\,|||V|||_{\gamma,[W_b^{2,2}]^2}$$
$$\leqslant C(r+r^2)\,|||V|||_{\gamma,[W_b^{2,2}]^2},\quad \forall w_1,w_2\in B(0,r,\mathbb{B}_{\omega,\xi_0}). \tag{4.30}$$

Decreasing $r$ such that $C(r+r^2)<1$, we derive that

$$(1 - C(r+r^2))\,|||V|||_{\gamma,[W_b^{2,2}]^2} \leqslant |||\mathscr{P}_\gamma(\mathscr{R}v_0)|||_{\gamma,[W_b^{2,2}]^2}$$
$$\leqslant (1 + C(r+r^2))\,|||V|||_{\gamma,[W_b^{2,2}]^2}. \tag{4.31}$$

It follows from (4.31) and (4.14) that

$$|||\mathscr{U}_0(w_1) - \mathscr{U}_0(w_2)|||_{[W_b^{2,2}]^2} \leqslant |||V|||_{\gamma,[W_b^{2,2}]^2}$$
$$\leqslant \frac{|||\mathscr{P}_\gamma(\mathscr{R}v_0)|||_{\gamma,[W_b^{2,2}]^2}}{1 - C(r+r^2)}$$
$$\leqslant \frac{C_2}{1 - C(r+r^2)}\|w_1 - w_2\|_{L_b^\infty}. \tag{4.32}$$

By Sobolev embedding theorem and estimates (4.30)~(4.32), we obtain that

$$\|w_1 - w_2\|_{L_b^\infty} \leqslant C_1 \|\mathcal{U}_0(w_1) - \mathcal{U}_0(w_2)\|_{[W_b^{2,2}]^2} +$$
$$C_1 \|\mathcal{U}_0(w_1) - \mathcal{U}_0(w_2) - (w_1 - w_2)e_0\|_{[W_b^{2,2}]^2}$$
$$\leqslant C_1 \|\mathcal{U}_0(w_1) - \mathcal{U}_0(w_2)\|_{[W_b^{2,2}]^2} + C_1 \left|\!\left|\!\left| V - \mathcal{P}_\gamma(\mathcal{R}v_0) \right|\!\right|\!\right|_{\gamma,[W_b^{2,2}]^2}$$
$$\leqslant C_1 \|\mathcal{U}_0(w_1) - \mathcal{U}_0(w_2)\|_{[W_b^{2,2}]^2} + \frac{C_1 C_2 C(r + r^2)}{1 - C(r + r^2)} \|w_1 - w_2\|_{L_b^\infty}. \quad (4.33)$$

Let us decrease $r$ if necessary such that $\dfrac{C_1 C_2 C(r+r^2)}{1-C(r+r^2)} < \dfrac{1}{2}$. Then estimates (4.32) and (4.33) imply estimate (4.25). This lemma is proved. □

**Corollary 4.7** Let $\{T_h : h \in \mathbf{R}^N\}$ be a group of spatial translations $(T_h u_0)(x) = u_0(x+h)$, $\mathbb{K} = B(0, r, \mathbb{B}_{\omega,\xi_0}(\mathbf{R}^N, \mathbf{C}))$, and constants $r$, $\omega$ and $\xi_0$ be defined in Theorem 4.6. Then $T_h \mathcal{A} = \mathcal{A}$ and $T_h \mathbb{K} = \mathbb{K}$. Moreover the map $\mathcal{U}_0 : \mathbb{K} \to \mathcal{A}$ commutes with this group, $T_h \mathcal{U}_0(u_0) = \mathcal{U}_0(T_h u_0)$ for every $u_0 \in \mathbb{K}$ and $h \in \mathbf{R}^N$.

Assertion $T_h \mathcal{U}_0(u_0) = \mathcal{U}_0(T_h u_0)$ is an immediate corollary of our construction of the map $\mathcal{U}_0$ and the uniqueness part of the implicit function theorem.

It follows from Theorem 4.6 and Corollary 4.7 that the map $\mathcal{U}_0 : \mathbb{K} \to \mathcal{A}$ realizes the homeomorphism $\mathcal{U}_0 : (T_h, \mathbb{K}) \to (T_h, \mathcal{U}_0(\mathbb{K})) \subset (T_h, \mathcal{A})$ for every $h \in \mathbf{R}^N$. The obtained model $(T_h, \mathbb{K})$ is rather complicated by itself. Let us now simplify this model by considering the case of discrete dynamics $(\mathcal{J}_l, l \in \mathbf{Z}^N)$.

**Definition 4.8**[27] Let $\mathbb{D} = \{Z \in \mathbf{C} : |z| \leqslant 1\}$ be a unitary disc on the complex plane and $\mathcal{M} = \mathbb{D}^{\mathbf{Z}^N}$ be the space of all functions $v : \mathbf{Z}^N \to \mathbb{D}$. We endow this space by a Frechet topology generated by the system of seminorms

$$\|v, B_0^R\|_{0,\infty} = \sup_{l \in \mathbf{Z}^N, |l| \leqslant R} |v(l)|, \quad (4.34)$$

and denote the space thus obtained by $\mathcal{M}_{\rm loc}$. We define a group $(\mathcal{J}_l, l \in \mathbf{Z}^N)$ on $\mathcal{M}$ by the standard expression

$$(\mathcal{J}_l v)(m) = v(l+m), \quad l, m \in \mathbf{Z}^N, \quad v \in \mathcal{M}. \quad (4.35)$$

To construct $(\mathcal{J}_l, \mathcal{M})$, we use the Whittaker-Shannon-Kotel′nikov interpolation formula[12,7,26,27] to represent the functions from $\mathbb{B}_{\omega,\xi_0}$.

**Proposition 4.9** Every function $u(x)$ from the class $\mathbb{B}_{\omega',\xi_0}$ can be represented in the form

$$u(x) = \sum_{l \in \mathbf{Z}^N} u(\delta k) g_{\rho,k}(x), \quad \rho > 0, \quad \delta = \frac{\pi}{\omega' + \rho}, \quad (4.36)$$

and

$$g_{\rho,k}(x) = e^{i\xi_0 \cdot (x - \delta k)} \prod_{j=1}^N \frac{\sin\rho(x_j - \delta k_j) \cdot \sin(\omega' + \rho)(x_j - \delta k_j)}{\rho(\omega' + \rho)(x_j - \delta k_j)}. \quad (4.37)$$

Moreover $g_{\rho,k} \in \mathbb{B}_{\omega'+2\rho,\xi_0}$.

We now fix $\omega', \rho > 0$ such that $\omega' + 2\rho < \omega$, and define a map $\kappa : \mathcal{M} \to \mathbb{B}_{\omega,\xi_0}$ via

$$\kappa(v) = \sum_{l \in \mathbf{Z}^N} v(l) g_{\rho,l}(x) \tag{4.38}$$

where the functions $g_{\rho,l}(x)$ are defined in (4.37).

**Lemma 4.10** Let $\omega', \rho > 0$ such that $\omega' + 2\rho < \omega, 0 < \nu < 1$, and the map $\kappa: \mathcal{M} \to \mathbb{B}_{\omega,\xi_0}$ be defined by (4.38). Then the map $\kappa$ is Lipschitz continuous in the metrics of $L^\infty$, and

$$|\kappa(v)(x)| \leqslant C \sup_{l \in \mathbf{Z}^N} |v(l)| \left( \prod_{j=1}^{N} (1 + |x_j - l_j|^2) \right)^{-\nu/2} \tag{4.39}$$

Moreover, for every $R > \sqrt{N}$,

$$\| \kappa(v), B_0^{\delta R} \|_{0,\infty} \geqslant \| v, B_0^R \|_{0,\infty}. \tag{4.40}$$

**Proof** Since $(\omega' + \rho)\delta = \pi$, we have $g_{\rho,k}(\delta l) = \begin{cases} 1 & l = k \\ 0 & l \neq k \end{cases}$, $\kappa(v)(\delta l) = v(l)$ for every $l \in \mathbf{Z}^N$, and $\| v, B_0^R \|_{0,\infty} = \sup_{l \in \mathbf{Z}^N, |l| \leqslant R} |v(l)| = \sup_{l \in \mathbf{Z}^N, |l| \leqslant R} |\kappa(v)(\delta l)| \leqslant \sup_{|x| \leqslant \delta R} |\kappa(v)(x)| = \| \kappa(v), B_0^{\delta R} \|_{0,\infty}$. Estimate (4.40) is proved. The proof of estimate (4.39) is based on the evident estimate $|g_{\rho,l}(x)| \leqslant C \left( \prod_{j=1}^{N} (1 + |x_j - l_j|^2) \right)^{-1}$ for every $l \in \mathbf{Z}^N$ and $x \in \mathbf{R}^N$ which can be verified in a direct way. It follows from estimate (4.39) that the map $\kappa$ is Lipschitz continuous in the metrics of $L^\infty$. □

**Lemma 4.11** Let $\omega', \rho > 0$ such that $\omega' + 2\rho < \omega, \omega, r, \xi_0$ be defined in Theorem 4.6, the map $\kappa: \mathcal{M} \to \mathbb{B}_{\omega,\xi_0}$ be defined by (4.38), $\tilde{\kappa}(v) = \frac{r}{C} \kappa(v)$ for every $v \in \mathcal{M}$, and constant $C$ be defined in (4.39). Then $\tilde{\kappa}(\mathcal{M}) \subset B(0, r, \mathbb{B}_{\omega,\xi_0})$, the set $\tilde{\kappa}(\mathcal{M})$ is invariant with respect to the discrete group $T_{\delta l}$, and this group commutes with the map $\tilde{\kappa}$, i.e. $\tilde{\kappa} \circ \mathcal{J}_l = T_{\delta l} \circ \tilde{\kappa}, l \in \mathbf{Z}^N$, where $\delta > 0$ is the same as in Proposition 4.9.

**Proof** It follows from (4.37)~(4.39) that $\tilde{\kappa}(\mathcal{M}) \subset B(0, r, \mathbb{B}_{\omega,\xi_0})$. For every $v \in \mathcal{M}$, we have

$$\begin{aligned} T_{\delta l} \kappa(v)(x) &= T_{\delta l} \sum_{k \in \mathbf{Z}^N} v(k) g_{\rho,k}(x) = \sum_{k \in \mathbf{Z}^N} v(k) g_{\rho,k}(x + \delta l) = \sum_{k \in \mathbf{Z}^N} v(k) g_{\rho,k-l}(x) \\ &= \sum_{k \in \mathbf{Z}^N} v(k + l) g_{\rho,k}(x) = \sum_{k \in \mathbf{Z}^N} \mathcal{J}_l v(k) g_{\rho,k}(x) = \kappa(\mathcal{J}_l v)(x). \end{aligned}$$

Thus $T_{\delta l} \tilde{\kappa}(\mathcal{M}) = \tilde{\kappa}(\mathcal{M})$ and $\tilde{\kappa} \circ \mathcal{J}_l = T_{\delta l} \circ \tilde{\kappa}$ for every $l \in \mathbf{Z}^N$. □

Combining the embedding constructed in Lemma 4.11 with the embedding constructed in Corollary 4.7, we obtain the following result.

**Theorem 4.12** Let the assumptions of Lemma 4.5 hold. Then the map $\tau = \mathcal{U}_0 \circ \tilde{\kappa}$ is a Lipschitz-continuous (in the metrics of $L^\infty$) isomorphism between $\mathcal{M}$ and $\tau(\mathcal{M}) \subset \mathcal{A}$ that commutes with the spatial shifts,

$$\tau: (\mathcal{J}_l, \mathcal{M}) \to (T_{\delta l}, \tau(\mathcal{M})), \quad \tau \circ \mathcal{J}_l = T_{\delta l} \circ \tau, \quad \forall l \in \mathbf{Z}^N. \tag{4.41}$$

We conclude this paper by formulating some results on the lower bound for the ε-entropy of the attractor $\mathcal{A}$ of Eqs. (1.1) and (1.2).

**Lemma 4.13** Let $u_0^1, u_0^2 \in B(0, \hat{r}, \mathbb{B}_{\omega,\xi_0}^{\mathrm{Re}}), \hat{r} \leqslant r$, and constants $r, \omega$ and $\xi_0$ be defined in Theorem 4.6. Then for every $\rho>0$, we have

$$\|\mathcal{U}_0(u_0^1)-\mathcal{U}_0(u_0^2)\|_{[W_b^{2,2}(B_0^\rho)]^2}\geqslant L\|u_0^1-u_0^2\|_{L^\infty(B_0^\rho)}-C(\hat{r}^2+\hat{r}^3). \tag{4.42}$$

where $L$ is independent of $\rho$.

**Proof** Applying (4.24) we have

$$\|\mathcal{U}_0(u_0^1)-\mathcal{U}_0(u_0^2)\|_{[W_b^{2,2}(B_0^\rho)]^2}$$

$$\geqslant\|u_0^1e_0-u_0^2e_0\|_{[W_b^{2,2}(B_0^\rho)]^2}-(\|\mathcal{U}_0(u_0^1)-u_0^1e_0\|_{[W_b^{2,2}(\mathbf{R}^N)]^2}+\|\mathcal{U}_0(u_0^2)-u_0^2e_0\|_{[W_b^{2,2}(\mathbf{R}^N)]^2})$$

$$\geqslant L\|u_0^1-u_0^2\|_{L^\infty(B_0^\rho)}-C(\hat{r}^2+\hat{r}^3).$$

□

**Theorem 4.14** Let the assumptions of Theorem 4.6 hold. Then for the attractor $\mathscr{A}$ of Eqs. (1.1) and (1.2), we have the following entropy estimate

$$\mathbb{H}_\varepsilon(\mathscr{A},W_b^{2,2}(B_0^\rho))\geqslant C\rho^N\ln\frac{1}{\mathscr{E}}\text{ for }\rho>\rho_0\text{ and }0<\varepsilon\leqslant\varepsilon_0<1. \tag{4.43}$$

Moreover, for every $\delta>0$ there exists $C_\delta>0$ such that

$$\mathbb{H}_{\mathscr{E}}(\mathscr{A},W_b^{2,2}(B_0^1))\geqslant C_\delta\left(\ln\frac{1}{\varepsilon}\right)^{N+1-\delta}. \tag{4.44}$$

**Proof** Let $\varepsilon>0$ be small enough, $\hat{r}=\min\{1,\sqrt{\varepsilon/(6C)}\}$, functions $v_0^1,v_0^2\in B(0,\hat{r},\mathbb{B}_{\omega,\xi_0}^{\mathrm{Re}})$ such that $C(\hat{r}^2+\hat{r}^3)<\frac{\mathscr{E}}{2}$, and

$$\|v_0^1-v_0^2\|_{L^\infty(B_0^\rho)}\geqslant\frac{\mathscr{E}}{L}. \tag{4.45}$$

Then it follows from (4.42) that

$$\|\mathcal{U}_0(v_0^1)-\mathcal{U}_0(v_0^2)\|_{[W_b^{2,2}(B_0^\rho)]^2}\geqslant\frac{\mathscr{E}}{2}. \tag{4.46}$$

Estimates (4.45) and (4.46) and the fact $\mathcal{U}_0(v_0^j)\in\mathscr{A}\ (j=1,2)$ imply that

$$\begin{aligned}\mathbb{H}_{\mathscr{E}}(\mathscr{A},W_b^{2,2}(B_0^\rho))&\geqslant\mathbb{H}_{\mathscr{E}/L}(B(0,\hat{r},\mathbb{B}_{\omega,\xi_0}^{\mathrm{Re}}),C_b(B_0^\rho))\\&=\mathbb{H}_{\mathscr{E}/(\hat{r}L)}(B(0,1,\mathbb{B}_{\omega,\xi_0}^{\mathrm{Re}}),C_b(B_0^\rho)).\end{aligned} \tag{4.47}$$

Estimates (4.43) and (4.44) are immediate results of (4.47) and Propositions 4.10 and 4.11 of [27]. This theorem is proved. □

### References

[1] Auchmuty J G, Nicolis G. Bifurcation analysis of nonlinear reaction-diffusion equations-I. Evolution equations and steady state solutions. Bull Math Biol, 1975, 37: 323 – 365.

[2] Babin A V, Vishik M I. Attractors of evolution equations// Studies in Mathematics and Its Applications. vol. 25. Amsterdam: North-Holland, 1992.

[3] Borckmans P, Wit A De, Dewel G. Competition in ramped turing structures. Physica A, 1992, 188: 137 – 157.

[4] Chepyzhov V V, Vishik M I. Kolmogorov ε-entropy of attractors of reaction-diffusion systems. Mat Sb, 1998, 189 (2): 81 – 110. Translation in Sb Math 1998, 189 (1 – 2): 235 – 263.

[5] Chepyzhov V V, Vishik M I. Attractors for equations of mathematical physics// American Mathematical Society Colloquium Publications, Vol 49. American Mathematical Society, Providence, RI, 2002.

[6] Collet P, Xin J. Global existence and large time asymptotic bounds of $L^\infty$ solutions of thermal diffusive combustion systems on $\mathbf{R}^N$. Ann Sc Norm Super Pisa Cl Sci, 1996, 23 (4): 625-642.

[7] Efendiev M A, Zelik S V. The attractor for a nonlinear reaction-diffusion system in an unbounded domain. Comm Pure Appl Math, 2001, 54 (6): 625-688.

[8] Hale J K. Asymptotic behavior of dissipative systems// Mathematical Surveys and Monographs, Vol 25. Providence, RI: Amer Math Soc, 1998.

[9] Han Yongqian, Guo Boling. Attractors for the brusselator in $\mathbf{R}^N$. J Partial Differential Equations, 2007, 20 (2): 169-182.

[10] Herrero M A, Lacey A A, Velàzquez J L. Global existence for reaction-diffusion systems modelling ignition. Arch Ration Mech Anal, 1998, 142: 219-251.

[11] Hollis S L. Martin R H, Pierre M. Global existence and boundedness in reaction-diffusion systems. SIAM J Math Anal, 1987, 18 (3): 744-761.

[12] Kolmogorov A N, Tihomirov V M. $\varepsilon$-entropy and $\varepsilon$-capacity of sets in functional space. Amer Math Soc Transl, 1961, 17 (2): 277-364.

[13] Lefever R, Herschkowitz-Kaufman M, Turner J W. Dissipative structures in a soluble non-linear reaction-diffusion system. Phys Lett A, 1977, 60 (5): 389-391.

[14] Lightbourne J H, Martin R H. Relatively continuous nonlinear perturbations of analytic semigroups. J Nonlinear Anal-Theory Meth Appl, 1977, 1: 277-292.

[15] Lions J L, Magenes E. Problémes aux limites non homogénes et applications, Vol 1 // Travaux et Recherches Math matiques, Vol 17. Dunod, Paris, 1968.

[16] Maier-Paape S, Lauterbach R. Heteroclinic cycles for reaction diffusion systems by forced symmetry-breaking. Trans Amer Math Soc, 2000, 352 (7): 2937-2991.

[17] Nicolis G, Prigogine Ilya. Self-organization in nonequilibrium systems. New York: John Wiley and Sons Inc, 1977.

[18] Pierre M, Schmitt D. Blowup in reaction-diffusion systems with dissipation of mass. SIAM Rev, 2000, 42 (1): 93-106.

[19] Prigogine E, Nicolis G. Biological order, structure and instabilities. Quart Rev Biophys, 1971, 4: 107-148.

[20] Rothe F. Global solutions of reaction-diffusion systems // Lecture Notes in Math, Vol 1072. Berlin: Springer-Verlag, 1984.

[21] Temam R. Infinite-dimensional dynamical system in mechanics and physics // Applied Mathematical Sciences, Vol 68. New York: Springer, 1988.

[22] Triebel H. Interpolation theory, function spaces, differential operators // North-Holland Mathematical Library, Vol 18. North-Holland, Amsterdam-New York, 1978.

[23] Turing A M. The chemical basis of morphogenesis. Philos Trans R Soc Lond Ser B, 1952, 237: 37-72.

[24] Yang L, Zhabotinsky A M, Epstein I R. Stable squares and other oscillatory turing patterns in a reaction-diffusion model. Phys Rev Lett, 2004, 92 (19).

[25] Zeidler E. Nonlinear functional analysis and its applications. I. Fixed-point Theorems. New York: Springer, 1986.

[26] Zelik S V. The attractor for a nonlinear reaction-diffusion system in the unbounded domain and Kolmogorov's $\varepsilon$-entropy. Math Nachr, 2001, 232: 129-179.

[27] Zelik S V. Attractors of reaction-diffusion systems in unbounded domains and their spatial complexity. Commun Pure Appl Math, 2003, 56: 584-637.

# Global Solutions of a Dynamical Equation in Ferrimagnet *

Guo Boling(郭柏灵) Pu Xueke(蒲学科)

**Abstract**

We consider global solutions of a dynamical equation in ferrimagnet. We show that it admits a global weak solution by using the penalty method. By the energy estimates method we show there exists a unique global smooth solution. Finally we establish the relationship between this equation and wave maps.

**Keywords** dynamical equation in ferrimagnet; penalty method; well-posedness; wave maps

## 1 Introduction

In this paper we consider the following dynamical equation in ferrimagnetic materials of the following form

$$\begin{cases} u_t = u \times \Box u + \lambda u \times (u \times \Box u), & \text{in } \Omega \times \mathbf{R}^+ \\ u(x,0) = u_0, \quad u_t(x,0) = u_1 \end{cases}, \tag{1.1}$$

where $u = (u_1, u_2, u_3) \in \mathbb{S}^2$ (i. e. the unit sphere in $\mathbf{R}^3$) is the unknown vector that maps a regular domain $\Omega \subset \mathbf{R}^d$ to the unit sphere $\mathbb{S}^2 \subset \mathbf{R}^3$, $\times$ denotes the cross product in $\mathbf{R}^3$ and $\lambda > 0$ is a positive damping parameter.

This equation was first derived by Borisov, Kiseliev and Talutz[2] when studying the ferrimagnetic phenomenon in ferrimagnetic materials. Just as the Landau-Lifshitz equation[3-5]

$$u_t = u \times \Delta u - \lambda u \times (u \times \Delta u) \tag{1.2}$$

is important in the study of continuum ferromagnets, this equation plays a key role in the study of ferrimagnetic materials. It aroused the interests of both physicists and mathematicians greatly. Thus we should study it mathematically rigorously. Formally, this equation is very similar to the Landau-Lifshitz equation with the Laplacian operator $\Delta$ replaced by the wave operator $\Box$.

Wave maps are maps from Minkowski space $\mathcal{M}$ into a Riemannian manifold $\mathcal{N} \hookrightarrow \mathbf{R}^k$ that satisfy the wave equation with partial derivatives replaced by covariant derivatives. They are the prototypes of geometric wave equations and are studied by many authors and the interested readers can refer to [8] for more details. It can be also regarded as a wave equation from $\mathcal{M}$ to

* Nonlinear Analysis, 2009, 70: 3894-3901.

$\mathbf{R}^k$, with the range restricted to the manifold $\mathcal{N}$. When the target manifold is the unit sphere $\mathbb{S}^k \hookrightarrow \mathbf{R}^{k+1}$, the wave map reads

$$\Box u = (|\nabla u|^2 - |u_t|^2)u. \tag{WM}$$

We should add here that this is the famous nonlinear $\sigma$-model, and it is important in the study of the dynamics of anti-ferromagnetic materials, see [6,7,9] for more details.

For the Landau-Lifshitz equation (1.2), Alouges and Soyeur[1] established the existence of weak solutions and considered the limit as the damping parameter goes to zero or to infinity. On the other hand, Guo and Hong[4] established the relationship between the Landau-Lifshitz equation and the heat flow of harmonic maps. Then two natural questions arose. One is whether there exist suitable solutions for Eq. (1.1), another one is the relationship between this equation and the wave map (WM) from $\Omega$ to $\mathbb{S}^2$. These questions are our main targets in this paper.

We will consider two important cases. The first one is when the problem is posed periodically. In this case, we denote $M = \Omega = [-L_1, L_1] \times \cdots \times [-L_d, L_d]$, and the problem (1.1) is completed with the periodical condition $u(x-L,t) = u(x+L,t)$, where the vector $L = (L_1, \cdots, L_d)$. We will consider the existence of weak solutions of the periodical problem. For this purpose, we need to make clear what a weak solution is. It is not difficult to verify that when $u$ is regular, by taking cross product with $\lambda u$, Eq. (1.1) is equivalent to

$$\frac{1}{1+\lambda^2}\partial_t u - \frac{\lambda}{1+\lambda^2} u \times \partial_t u - u \times \partial_{tt} u + u \times \Delta u = 0. \tag{1.3}$$

This leads us to introduce the following notion of a weak solution for Eq. (1.1).

**Definition 1.1** Let $u_0 \in H^1(M)$, $|u_0| = 1$ a.e., and $u_1 \in L^2(M)$. We say that $u(x,t)$ is a weak solution of the problem (1.1) provided:

(i) for all $T>0$, $u \in L^2(0,T;H^1(M))$, $u_t \in L^2(0,T;L^2(M))$, with $|u| = 1$ a.e.;

(ii) for all $\Phi \in H^1(M \times [0,T])$, with $\Phi(x,T) = 0$, then there holds:

$$\frac{1}{1+\lambda^2}\int_{M_T} u_t \cdot \Phi \mathrm{d}x\mathrm{d}t + \frac{\lambda}{1+\lambda^2}\int_{M_T}(u \times u_t)\cdot \Phi \mathrm{d}x\mathrm{d}t + \int_M (u \times u_t)\cdot \Phi|_{t=0}\mathrm{d}x +$$
$$\int_{M_T} u \times u_t \cdot \Phi_t \mathrm{d}x\mathrm{d}t - \sum_{i=1}^m \int_{M_T} u \times \frac{\partial u}{\partial x_i}\cdot\frac{\partial \Phi}{\partial x_i} = 0; \tag{1.4}$$

(iii) $u(x,0) = u_0(x)$ in the trace sense.

In this case, we have the following:

**Theorem 1.1** Let $u_0 \in H^1(M, \mathbb{S}^2)$ and $u_1 \in L^2(M, T_u\,\mathbb{S}^2)$. Then there exists a global weak solution of the problem (1.1) with initial data $(u_0, u_1)$.

The second case is when the problem is posed in $\mathbf{R}^d$. For this Cauchy problem, we can only have the local well-posedness result in dimension $d \geqslant 2$, however when we restrict ourselves to dimension $d = 1$, we have the global well-posedness result.

**Theorem 1.2** For any data $(u_0, u_1) \in L^2_{\mathrm{loc}}(\mathbf{R}) \times H^1(\mathbf{R})$, such that $\nabla u_0$ belongs to $H^1(\mathbf{R})$, $|u_0| = 1$ a.e., there exists a unique global smooth solution $u$ of class $H^2$ which preserves the regularity of the initial data.

We remark that the existence of global smooth solutions in the periodical case can be established as the same and we will not handle it explicitly in this paper.

In the final section, we show that when the damping parameter $\lambda\to\infty$, the solution approximates to the wave map that takes values in $\mathbb{S}^2$. Thus we find new interpretations of the wave maps, and this may link the mathematical studies of wave maps to applications in physics.

Throughout this paper, $Du=(u_t, \nabla u)$ always denotes the space-time derivative of $u$, and $T_u\mathbb{S}^2$ denotes the tangent space of $\mathbb{S}^2$ at $u$. We use $\|\cdot\|$ to denote the $L^2$-norm and $\|\cdot\|_X$ to denote the $X$-norm on either $M$ or $\mathbf{R}^d$. $M_T$ always denotes the product $M\times[0,T]$ and $C$ denotes a constant which can vary from line to line.

This paper is organized as follows. In the next section, we prove Theorem 1.1. In Section 3, we prove Theorem 1.2. Finally in Section 4, we show its relationship with wave maps.

## 2 Proof of Theorem 1.1

In this section, we focus ourselves on the existence of global weak solutions of Eq. (1.1). For this purpose, we deduce two equivalent forms of this equation. One is (1.3) and the other one is by taking the cross product of Eq. (1.3) with $u$,

$$\Box u+\frac{\lambda}{1+\lambda^2}\partial_t u+\frac{1}{1+\lambda^2}u\times\partial_t u+(|\partial_t u|^2+|\nabla u|^2)u. \tag{2.1}$$

We then construct the following penalized problem of this equation

$$\partial_{tt}u^k-\Delta u^k+\frac{\lambda}{1+\lambda^2}\partial_t u^k+\frac{1}{1+\lambda^2}u^k\times\partial_t u^k+k(|u^k|^2-1)u^k=0, \tag{2.1$_\mathrm{k}$}$$

completed with initial data $u^k(x,0)=u_0$ and $u_t^k(x,0)=u_1$.

In the following, we use Galerkin's method to construct a weak solution $u^k$ of the penalized problem (2.1$_\mathrm{k}$) and then pass to the limit of $u^k$ to get a global weak solution of Eq. (1.1).

### 2.1 Constructing Weak Solutions $u^k$ of (2.1$_\mathrm{k}$)

Fix an integer $k$. Let $w_i=w_i(x)$ be a smooth orthonormal basis of $L^2(M)$ and an orthogonal basis of $H^1(M)$. Fix now a positive integer $m$. We look for a function $u_m:[0,T]\to H^1(M)$ of the form

$$u_m=\sum_{i=1}^m\varphi_i(t)w_i, \tag{2.2}$$

where $\varphi_i(t)$ are $\mathbf{R}^3$-valued vectors. We hope to select the coefficients $\varphi_i(t)(0\leqslant t\leqslant T, i=1,\cdots,m)$ such that for all $1\leqslant i\leqslant m$,

$$\left\langle\partial_{tt}u_m-\Delta u_m+\frac{\lambda}{1+\lambda^2}\partial_t u_m+\frac{1}{1+\lambda^2}u_m\times\partial_t u_m+k(|u_m|^2-1)u_m, w_i\right\rangle=0, \tag{2.3}$$

where $\langle\cdot,\cdot\rangle$ denotes the inner product in $L^2(M)$ with the initial condition

$$u_m(x,0)=\sum_{i=1}^{m}\varphi_i(0)w_i(x)\to u_0\quad \text{in } H^1(M),$$

$$\frac{\partial u_m}{\partial t}(x,0)=\sum_{i=1}^{m}\varphi_{it}(0)w_i(x)\to u_1\quad \text{in } L^2(M).$$

Inserting the expression (2.2) of $u_m$ into (2.3), we get an ODE system for $\{\varphi_i(t)\}_{i=1,2,\cdots,m}$, and from the standard existence theory of ODEs by Picard's method, we have the local existence of approximate solutions $\{u_m\}$ for (2.3).

Below we establish estimates of $\{u_m\}$ on $[0,T]$ to get a global solution of the penalized problem $(2.1_k)$. Multiplying (2.1) by $\frac{\partial\varphi_i}{\partial t}$ and summing for $1\leqslant i\leqslant m$, we get

$$\frac{1}{2}\frac{\mathrm{d}}{\mathrm{d}t}\int_M|\partial_t u_m|^2+|\nabla u_m|^2\mathrm{d}x+\frac{\lambda}{1+\lambda^2}\int_M|\partial_t u_m|^2\mathrm{d}x+\frac{k}{4}\frac{\mathrm{d}}{\mathrm{d}t}\int_M(|u_m|^2-1)^2=0. \tag{2.4}$$

Integrating over $[0,T]$, we get

$$\int_M|\partial_t u_m|^2+|\nabla u_m|^2\mathrm{d}x+\frac{2\lambda}{1+\lambda^2}\int_{M\times[0,T]}|\partial_t u_m|^2\mathrm{d}x+\frac{k}{2}\int_M(|u_m|^2-1)^2\mathrm{d}x$$
$$\leqslant\int_M|\partial_t u_m(0)|^2+|\nabla u_m(0)|^2\mathrm{d}x+\frac{k}{2}\int_M(|u_m(0)|^2-1)^2. \tag{2.5}$$

Thus for fixed $k$, $\{|u_m|^2-1\}$, $\{|\partial_t u_m|\}$ and $\{|\nabla u_m|\}$ are bounded in $L^\infty(0,T;L^2(M))$. Furthermore, by Young's inequality $\int_M|u_m|^2\leqslant \mathrm{mes}(M)+\frac{1}{2}\int_M(|u_m|^2-1)^2$, we know that $\{u_m\}$ is bounded in $L^\infty(0,T;L^2(M))$. Thus we can select a subsequence of $\{u_m\}$ (still denoted by $\{u_m\}$ for convenience) such that

$$\begin{cases} u_m\to u^k & \text{weakly star in } L^\infty(0,T;H^1(M)) \text{ and weakly in } H^1(M_T)\\ u_m\to u^k & \text{strongly in } L^2(M_T) \text{ and a.e.}\\ |u_m|^2-1\to|u^k|^2-1 & \text{weakly in } L^2(M_T)\end{cases}. \tag{2.6}$$

Taking limit $m\to\infty$ in (2.1), we deduce that for all $\psi$ in span $\{w_i\}_{i=1}^m$ and $\varphi\in C^\infty[0,T]$ with $\varphi(T)=0$, there holds

$$-\int_{M_T}u_t^k\cdot(\varphi'(t)\psi(x))\mathrm{d}x\mathrm{d}t-\int_M\varphi(0)u_1\cdot\psi\mathrm{d}x+\sum_{i=1}^m\int_{M_T}\varphi(t)\frac{\partial u^k}{\partial x_i}\cdot\frac{\partial\psi}{\partial x_i}\mathrm{d}x\mathrm{d}t+$$
$$\frac{\lambda}{1+\lambda^2}\int_{M_T}\partial_t u^k\cdot(\varphi(t)\psi(x))\mathrm{d}x\mathrm{d}t+\frac{1}{1+\lambda^2}\int_M u^k\times u_t^k\cdot(\varphi(t)\psi(x))\mathrm{d}x\mathrm{d}t+$$
$$\int_{M_T}k(|u^k|^2-1)u^k\cdot(\varphi(t)\psi(x))\mathrm{d}x\mathrm{d}t=0.$$

By a density argument, we can show from the above equation that

$$-\int_{M_T}u_t^k\cdot\phi_t\mathrm{d}x\mathrm{d}t-\int_M u_1\cdot\phi(x,0)\mathrm{d}x+\sum_{i=1}^m\int_{M_T}\frac{\partial u^k}{\partial x_i}\cdot\frac{\partial\phi}{\partial x_i}\mathrm{d}x\mathrm{d}t+\frac{\lambda}{1+\lambda^2}\int_{M_T}\partial_t u^k\cdot\phi\mathrm{d}x\mathrm{d}t+$$
$$\frac{1}{1+\lambda^2}\int_M u^k\times u_t^k\cdot\phi\mathrm{d}x\mathrm{d}t+\int_{M_T}k(|u^k|^2-1)u^k\cdot\phi\mathrm{d}x\mathrm{d}t=0 \tag{2.7}$$

for all $\phi(x,t)\in H^1(M_T)$ with $\phi(x,T)=0$.

The above argument shows that the penalized problem has a global weak solution. In what follows we will prove that we can pass to the limit of $u^k$ to get a global weak solution of the problem (1.1).

## 2.2 Passing to the Limit of $u^k$

To pass to the limit of $u^k$, we need to make uniform estimates of $u^k$ in $k$. Passing to the limit $m \to \infty$ in (2.5), we know that

$$\int_M \left|\frac{\partial u^k}{\partial t}\right|^2 + |\nabla u^k|^2 dx + \frac{2\lambda}{1+\lambda^2}\int_{M_T} |\partial_t u^k|^2 dx + \frac{1}{2}\int_M k(|u^k|^2 - 1)^2 dx \leqslant C, \tag{2.8}$$

where $C$ is independent of $k$. Thus

$$\left\|\frac{\partial u^k}{\partial t}\right\|_{L^2(M)}, \quad \|\nabla u^k\|_{L^2(M)} \leqslant C, \tag{2.9}$$

moreover

$$\int_M (|u^k|^2 - 1)^2 dx \leqslant \frac{C}{k}. \tag{2.10}$$

Thus we can extract from $u^k$ a subsequence (still denoted by $\{u^k\}$), such that

$$\begin{cases} u^k \to u & \text{weakly star in } L^\infty(0,T;H^1(M)) \text{ and weakly in } H^1(M_T) \\ |u^k|^2 - 1 \to 0 & \text{strongly in } L^2(M_T) \text{ and a.e.} \end{cases}, \tag{2.11}$$

which also implies that for all $t \in [0,T)$, $|u| = 1$ a.e. in $M$.

Let $\Phi \in C^\infty(\overline{M_T})$ with $\Phi(T) = 0$. Setting $\phi = u^k \times \Phi$, we then have

$$-\int_{M_T} u_t^k \cdot (u^k \times \Phi_t) dx dt - \int_M u_1 \cdot (u_0 \times \Phi(0)) dx + \sum_{i=1}^m \int_{M_T} \frac{\partial u^k}{\partial x_i} \cdot \left(u^k \times \frac{\partial \Phi}{\partial x_i}\right) dx dt +$$
$$\frac{\lambda}{1+\lambda^2}\int_M \partial_t u^k \cdot (u^k \times \Phi) + \int_{M_T} |u^k|^2 \Phi \cdot u_t^k dx dt - \int_{M_T} (u^k \cdot \Phi) u^k \cdot u_t^k dx dt = 0, \tag{2.12}$$

where $\phi \in (H^1(M_T))^3$ with $\phi(x,T) = 0$. On the other hand, by integration by parts we have

$$\begin{aligned}\int_{M_T} (u^k \cdot \Phi) u^k \cdot u_t^k dx dt &= \int_{M_T} (u^k \cdot \Phi) \frac{1}{2} \frac{\partial}{\partial t}(|u^k|^2 - 1) dx dt \\ &= -\frac{1}{2}\int_{M_T} (|u^k|^2 - 1)(u_t^k \Phi + u^k \Phi_t) \to 0,\end{aligned}$$

thanks to (2.11).

Thus passing to the limit in (2.11) ($k \to \infty$), we are led to

$$\int_{M_T} (u \times u_t) \cdot \Phi_t dx + \int_M (u_0 \times u_1) \cdot \Phi(x,0) dx + \frac{1}{1+\lambda^2}\int_{M_T} u_t \cdot \Phi dx dt -$$
$$\sum_{i=1}^m \int_{M_T} \left(u \times \frac{\partial u}{\partial x_i} \cdot \frac{\partial \Phi}{\partial x_i}\right) dx dt + \frac{\lambda}{1+\lambda^2}\int_{M_T} (u \times u_t) \cdot \Phi dx dt. \tag{2.13}$$

By a density argument, the above equality still holds for all $\Phi \in H^1(M_T)$ with $\Phi(x,T) = 0$, from which we finish our proof of Theorem 1.1 that the problem (1.1) has a global weak solution in the distribution sense.

# 3 Proof of Theorem 1.2

In the following, we always denote $Du=(u_t, \nabla u)$ the space-time derivative of $u$.

**Lemma 3.1** Let $u_0 \in L^2_{\text{loc}}$ with $(\nabla u_0, u_1) \in L^2$. Then for any smooth solution $u$ of Eq. (1.1), the following inequality holds for all time

$$\int_{\mathbf{R}^d} |u_t|^2 + |\nabla u|^2 \mathrm{d}x \leqslant \int_{\mathbf{R}^d} |u_1|^2 + |\nabla u_0|^2 \mathrm{d}x, \tag{3.1}$$

which implies that $Du \in L^\infty(0, T; L^2(\mathbf{R}^d))$ for any $T>0$.

**Proof** Taking inner product (2.1) with $u_t$, and integrating over $\mathbf{R}^d$ lead to the result.

□

In what follows, we only consider the case $d=1$ and denote by $\partial$ the first order spatial derivative operator.

**Lemma 3.2** Let $u_0 \in L^2_{\text{loc}}$ with $(\partial u_0, u_1) \in H^1$. Then for any smooth solution $u: \mathbf{R} \to \mathbb{S}^2$ of Eq. (1.1), the following *a priori* estimate holds

$$\sup_{0 \leqslant t \leqslant T} \| D^2 u \|^2_{L^2(\mathbf{R})} \leqslant C, \tag{3.2}$$

for any $T>0$, where $C$ depends on the initial data $u_0$ and $u_1$.

**Proof** We use the equivalent form (2.1) to deduce our result. Applying $\partial$ to Eq. (2.1), we get

$$\Box(\partial u) + \frac{\lambda}{1+\lambda^2}\partial_t \partial u = -\frac{1}{1+\lambda^2}\partial(u \times \partial_t u) - \partial\{(|u_t|^2 - |\nabla u|^2)u\}.$$

Taking inner product with $\partial_t \partial u$, and then integrating over $M$, we get

$$\begin{aligned}
\text{LHS} &= \frac{1}{2}\frac{\mathrm{d}}{\mathrm{d}t}\int |\partial_t \partial u|^2 + |\partial \nabla u|^2 \mathrm{d}x + \frac{\lambda}{1+\lambda^2}\int |\partial_t \partial u|^2 \mathrm{d}x \\
&= -\frac{1}{1+\lambda^2}\int \langle \partial u \times \partial_t u, \partial_t \partial u \rangle \mathrm{d}x - \int \langle (|u_t|^2 - |\nabla u|^2)\partial u, \partial_t \partial u \rangle \mathrm{d}x - \\
&\quad \int \langle \partial(|u_t|^2 - |\nabla u|^2)u, \partial_t \partial u \rangle \mathrm{d}x \\
&=: \mathrm{I} + \mathrm{II} + \mathrm{III},
\end{aligned} \tag{3.3}$$

where $\langle \cdot, \cdot \rangle$ denotes the $\mathbf{R}^3$ inner product. For I and II, we have

$$|\mathrm{I}| \leqslant \frac{1}{1+\lambda^2}\int |Du|^2 |D^2 u| \mathrm{d}x \leqslant \frac{C}{1+\lambda^2}\|D^2 u\|^{\frac{3}{2}} \|Du\|,$$

$$|\mathrm{II}| \leqslant \int |Du|^3 |D^2 u| \mathrm{d}x \leqslant C\|D^2 u\|^2 \|Du\|^2,$$

where we have used the Hölder inequality and the following two interpolation inequalities

$$\|Du\|_{L^4} \leqslant C\|D^2 u\|^{\frac{1}{4}} \|Du\|^{\frac{3}{4}}, \quad \|Du\|_{L^6} \leqslant C\|D^2 u\|^{\frac{1}{3}} \|Du\|^{\frac{2}{3}}.$$

For III, integrating by parts, we have

$$\mathrm{III} = \int \langle \partial(|u_t|^2 - |\nabla u|^2)\partial u, \partial_t u \rangle \mathrm{d}x,$$

thanks to the fact $u \cdot u_t = 0$. Thus,

$$|\mathrm{III}| \leqslant \int |Du|^3 |D^2 u| \mathrm{d}x \leqslant C\|D^2 u\|^2 \|Du\|^2.$$

Adding together the terms I, II and III, we have

$$\text{LHS} \leqslant C\|D^2 u\|^2 \|Du\|^2 + \frac{C}{1+\lambda^2}\|D^2 u\|^{3/2}\|Du\|. \tag{3.4}$$

Applying $\partial_t$ to (2.1), we have

$$\Box(\partial_t u) + \frac{\lambda}{1+\lambda^2}\partial_{tt} u = -\frac{1}{1+\lambda^2}\partial_t(u \times \partial_t u) - \partial_t\{(|u_t|^2 - |\nabla u|^2)u\}.$$

Taking inner product with $\partial_{tt} u$, and then integrating over $\mathbf{R}$, we get

$$\begin{aligned}\text{LHS}_1 &= \frac{1}{2}\frac{\mathrm{d}}{\mathrm{d}t}\int |\partial_{tt} u|^2 + |\partial_t \nabla u|^2 \mathrm{d}x + \frac{\lambda}{1+\lambda^2}\int |\partial_{tt} u|^2 \mathrm{d}x \\ &= -\int \langle(|u_t|^2 - |\nabla u|^2)\partial_t u, \partial_{tt} u\rangle \mathrm{d}x - \int \langle \partial_t(|u_t|^2 - |\nabla u|^2)u, \partial_{tt} u\rangle \mathrm{d}x \\ &=: \mathrm{IV} + \mathrm{V}.\end{aligned}$$

Noting that $u \cdot u_t = 0$ implies $u \cdot \partial_{tt} u = -|\partial_t u|^2$, and using Hölder inequalities and interpolation inequalities we have the same estimates for $\text{LHS}_1$:

$$\text{LHS}_1 \leqslant C\|D^2 u\|^2 \|Du\|^2 + \frac{C}{1+\lambda^2}\|D^2 u\|^{\frac{3}{2}}\|Du\|. \tag{3.5}$$

Adding (3.4) and (3.5) together, we arrive at a Gronwall-type inequality:

$$\frac{\mathrm{d}}{\mathrm{d}t}\|D^2 u\|^2 \leqslant C\|D^2 u\|^2 + \frac{C}{1+\lambda^2}\|D^2 u\|^{\frac{3}{2}}, \tag{3.6}$$

thanks to Lemma 3.1. This inequality implies a global estimate of $\|D^2 u\|_{L^2}^2$, i.e. for any $T>0$, we have that

$$\sup_{0\leqslant t\leqslant T}\|D^2 u\|_{L^2(\mathbf{R})}^2 \leqslant C, \tag{3.7}$$

which completes our proof. □

**Remark** Higher order estimates are standard by using the same procedure as above. In fact, one can show that for any $k\geqslant 2$,

$$\sup_{0\leqslant t\leqslant T}\|D^k u\|_{L^2(\mathbf{R})}^2 \leqslant C, \tag{3.8}$$

where $C$ depends on the $H^k$-norm of the initial data, i.e. the solution can be as smooth as the initial data. On the other hand, the local existence of smooth solutions can be established by the Banach fixed point theorem, of which we omit the details here, and one can refer to Shatah's book[8] for the details. □

When $d=1$, the local theory and our global *a priori* estimates imply that there exists a global smooth solution to Eq. (1.1) as long as the initial data are smooth.

Now $H^2$ uniqueness follows from the $H^2$ *a priori* estimates. Suppose that $u$ and $v$ are both $H^2$-solutions of the equation. Denote $w = u - v$, then

$$\Box w + \frac{\lambda}{1+\lambda^2}w_t + \frac{1}{1+\lambda^2}(u \times w_t + w \times v_t) + [(|u_t|^2 - |\nabla u|^2)u - (|v_t|^2 - |\nabla v|^2)v] = 0. \tag{3.9}$$

Taking inner product with $\partial_t(u-v)$ and integrating over $\boldsymbol{R}$, we get

$$\mathrm{LHS}_w := \frac{1}{2}\frac{\mathrm{d}}{\mathrm{d}t}\|D(u-v)\|^2 + \frac{\lambda}{1+\lambda^2}\|\partial_t(u-v)\|^2$$

$$= \frac{1}{1+\lambda^2}\int\langle w\times\partial_t v, w_t\rangle\mathrm{d}x + \int\langle u_t - v_t, (|u_t|^2 - |\nabla u|^2)u - (|v_t|^2 - |\nabla v|^2)v\rangle\mathrm{d}x$$

$$=: \mathscr{R}_1 + \mathscr{R}_2.$$

For part $\mathscr{R}_1$, we have for $H^2$-solutions

$$\mathscr{R}_1 \leqslant \frac{1}{1+\lambda^2}\|u-v\|_{L^4}\left\|\frac{\partial v}{\partial t}\right\|_{L^4}\|D(u-v)\|_{L^2}$$
$$\leqslant \frac{C}{1+\lambda^2}\|D(u-v)\|_{L^2}\|u-v\|_{H^1(\mathbf{R}^{1+1})}. \tag{3.10}$$

On the other hand, using the fact that $u\cdot u_t=0$ for any $H^2$-solution, we can show that

$$\mathscr{R}_2 = \int\langle u_t, (|v_t|^2 - |\nabla v|^2)u - (|v_t|^2 - |\nabla v|^2)v\rangle\mathrm{d}x -$$
$$\int\langle v_t, (|u_t|^2 - |\nabla u|^2)u - (|u_t|^2 - |\nabla u|^2)v\rangle\mathrm{d}x$$
$$\leqslant C\int|u-v|\,|Du - Dv|(|Du|^2 + |Dv|^2)\mathrm{d}x,$$

which implies

$$\mathscr{R}_2 \leqslant C\|D(u-v)\|_{L^2}\|\,|u-v|(|Du|^2+|Dv|^2)\|_{L^2}.$$

Then by Sobolev's embedding theorem, we have

$$\mathscr{R}_2 \leqslant C\|D(u-v)\|_{L^2}\|u-v\|_{L^6}(\|Du\|_{L^6}^2 + \|Dv\|_{L^6}^2)$$
$$\leqslant C\|D(u-v)\|_{L^2}\|u-v\|_{H^1(\mathbf{R}^{1+1})}. \tag{3.11}$$

Adding together (3.10) and (3.11) we arrive at the following inequality

$$\frac{\mathrm{d}}{\mathrm{d}t}\|D(u-v)\| \leqslant C\|u-v\|_{H^1(\mathbf{R}^{1+1})}. \tag{3.12}$$

On the other hand, we have for $w$ that

$$\frac{1}{2}\frac{\mathrm{d}}{\mathrm{d}t}\|w(t)\|_{L^2}^2 \leqslant \int_{\mathbf{R}^1} w_t w\,\mathrm{d}x \leqslant \|w_t\|_{L^2}\|w\|_{L^2}.$$

therefore,

$$\frac{\mathrm{d}}{\mathrm{d}t}\|(u-v)(t)\|_{L^2} \leqslant 2\|(u-v)_t\|_{L^2} \leqslant 2\|D(u-v)\|_{L^2}. \tag{3.13}$$

Putting together (3.12) and (3.13), we have

$$\frac{\mathrm{d}}{\mathrm{d}t}\|(u-v)(t)\|_{H^1(\mathbf{R}^{1+1})} \leqslant C\|(u-v)(t)\|_{H^1(\mathbf{R}^{1+1})}, \tag{3.14}$$

from which the uniqueness of $H^2$-solutions follows. This shows that in dimension one, there exists a unique global smooth solution to the equation.

## 4 Limiting Process as $\lambda$ Goes to Zero or to Infinity

In this section, we analyze the limiting process when $\lambda$ goes to 0 or to $\infty$. We will confine ourselves to the periodical case, although a similar analysis holds when the problem is posed in $\mathbf{R}^d$. We denote by $u^\lambda$ a weak solution we constructed for (1.1) in Section 2. It satisfies the following properties:

(i) $\nabla u^\lambda \in L^\infty(0,T;L^2(M))$, $\partial_t u^\lambda \in L^\infty(0,T;L^2(M))$, $|u^\lambda|=1$ a.e.;

(ii) for all $\Phi \in H^1(M_T)$, with $\Phi(x,T)=0$, then (1.4) holds for $u^\lambda$;

(iii) $u^\lambda(x,0)=u_0^\lambda(x)$ in the trace sense;

(iv) for all $T>0$, inequality (3.1) holds for $u^\lambda$.

From this we know that $\partial_t u^\lambda \in L^\infty(0,T;L^2(M))$ and $u^\lambda \in L^\infty(0,T;H^1(M))$ are uniformly bounded with respect to $\lambda$, so we can extract a subsequence such that

$$u^\lambda \rightarrow u \quad \text{in } L^2(0,T;H^1(M)) \text{ weakly;}$$
$$\partial_t u^\lambda \rightarrow \partial_t u \quad \text{in } L^2(0,T;L^2(M)) \text{ weakly.}$$

We can now pass to the limit ($\lambda \rightarrow 0$) in each of the terms of the following equation (i.e. the definition of weak solutions of (1.1))

$$\frac{1}{1+\lambda^2}\int_{M_T} u_t^\lambda \cdot \Phi \mathrm{d}x\mathrm{d}t + \frac{\lambda}{1+\lambda^2}\int_{M_T}(u^\lambda \times u_t^\lambda)\cdot \Phi \mathrm{d}x\mathrm{d}t +$$
$$\int_M (u^\lambda \times u_t^\lambda)\cdot \Phi|_{t=0}\mathrm{d}x + \int_{M_T} u^\lambda \times u_t^\lambda \cdot \Phi_t \mathrm{d}x\mathrm{d}t - \sum_{i=1}^m \int_{M_T} u^\lambda \times \frac{\partial u^\lambda}{\partial x_i}\cdot \frac{\partial \Phi}{\partial x_i} = 0 \tag{4.1}$$

to get

$$\int_{M_T} u_t \cdot \Phi \mathrm{d}x\mathrm{d}t + \int_M (u \times u_t)\cdot \Phi|_{t=0}\mathrm{d}x + \int_{M_T} u \times u_t \cdot \Phi_t \mathrm{d}x\mathrm{d}t - \sum_{i=1}^m \int_{M_T} u \times \frac{\partial u}{\partial x_i}\cdot \frac{\partial \Phi}{\partial x_i} = 0 \tag{4.2}$$

for all $\Phi \in H^1(M_T)$ with $\Phi(x,T)=0$. This is just the definition of weak solutions for the ferrimagnetic equation without the $\lambda$-term,

$$u_t = u \times \Box u, \tag{4.3}$$

with the same initial data.

**Proposition 4.1** The weak solution of (1.1) constructed in Section 1, up to a subsequence, weakly converges to a weak solution of (4.3), as $\lambda$ goes to 0.

We turn to the case when $\lambda$ tends to infinity. Again, since $\partial_t u^\lambda \in L^\infty(0,T;L^2(M))$ and $u^\lambda \in L^\infty(0,T;H^1(M))$ are uniformly bounded with respect to $\lambda$, we can extract a subsequence such that

$$u^\lambda \rightarrow u \quad \text{in } L^2(0,T;H^1(M)) \text{ weakly;}$$
$$\partial_t u^\lambda \rightarrow \partial_t u \quad \text{in } L^2(0,T;L^2(M)) \text{ weakly.}$$

Noting that $\frac{\lambda}{1+\lambda^2}\rightarrow 0$ as $\lambda \rightarrow \infty$ we can pass to the limit ($\lambda \rightarrow \infty$) in each term of (4.1) to get

$$\int_M (u \times u_t)\cdot \Phi|_{t=0}\mathrm{d}x + \int_{M_T} u \times u_t \cdot \Phi_t \mathrm{d}x\mathrm{d}t - \sum_{i=1}^m \int_{M_T} u \times \frac{\partial u}{\partial x_i}\cdot \frac{\partial \Phi}{\partial x_i} = 0, \tag{4.4}$$

for all $\Phi \in H^1(M_T)$ with $\Phi(x,T)=0$.

Setting $\Phi = u \times \phi$ in (4.4), for all $\phi \in C^\infty \in (M_T)$ with $\phi(x,T)=0$, one immediately has

$$\int_M u_1 \cdot \phi(x,0) + \int_{M_T} u_t \cdot \phi_t - \int_{M_T} |u_t|^2 u \cdot \phi + \int_{M_T} |\nabla u|^2 u \cdot \phi - \sum_{i=1}^{m} \int_{M_T} \frac{\partial u}{\partial x_i} \cdot \frac{\partial \phi}{\partial x_i}.$$

This shows that the $u$ we obtained as the limit of $u^\lambda$ as $\lambda \to \infty$ is a weak solution of the following wave map

$$u_{tt} - \Delta u = -(|u_t|^2 - |\nabla u|^2)u. \tag{WM}$$

**Proposition 4.2** The weak solution of (1.1) constructed in Section 1, up to a subsequence, weakly converges to a weak solution of (WM), as $\lambda$ goes to infinity.

## References


[1] Alouges F, Soyeur A. On global weak solutions for Landau-Lifshitz equations: Existence and nonuniqueness. Nonlinear Anal TAM, 1992, 18 (11): 1071 - 1084.

[2] Borisov A B, Kiseliev V V, Talutz G G. Solitons in a ferrimagnet. Solid State Commun, 1982, 44 (3): 411 - 412.

[3] Chen Y. The weak solutions to the evolution problems of harmonic maps. Math Z, 1989, 201: 69 - 74.

[4] Guo B, Hong M. The Landau-Lifshitz equation of the ferromagnetic spin chain and harmonic maps. Calc Var, 1993, 1: 311 - 334.

[5] Landau L, Lifshitz E M. On the theory of the dispersion of magnetic permeability in ferromagnetic bodies. Phsy Z Sowj, 1935, 8 (153) (Reproduced in Collected Papers of L. D. Landau, Pergamon Press, New York, 1965: 101 - 114).

[6] Papanicolaou N. Antiferromagnetic domain walls. Phys Rev B, 1995, 51 (21): 15062 - 15073.

[7] Shatah J. Weak solutions and development of singularities in the *SU* (2) $\sigma$-model. Comm Pure Appl Math, 1988, 41: 459 - 469.

[8] Shatah J, Struwe M. Geometric wave equations//Courant Institute of Mathematical Sciences, Vol 2, Amer Math Soc, Providence, RI, 2000.

[9] Shatah J, Zeng C. Schrödinger maps and anti-ferromagnetic chains. Comm Math Phys, 2006, 262: 299 - 315.

# Blowup and Global Existence of the Nonlinear Schrödinger Equations with Multiple Potentials *

Gan Zaihui(甘在会) Guo Boling(郭柏灵) Zhang Jian(张健)

**Abstract**

This paper is concerned with the Cauchy problem for the nonlinear Schrödinger equations with multiple potentials. Under some assumptions on these potentials, we first obtain some sufficient conditions of blowup according to the initial energy and the Cauchy initial data directly. We next establish a sufficient condition of global existence by using potential well method and the relation between the Cauchy data and the ground state. We finally answer the question that how small the Cauchy initial data need to be for global existence by using scaling argument.

**Keywords** Nonlinear Schrödinger equations; multiple potentials; blowup; global existence

## 1 Introduction

We consider the Cauchy problem for the following nonlinear Schrödinger equation with multiple potentials:

$$i\psi_t = -\Delta\psi + |x|^2\psi - K(x)|\psi|^{p-1}\psi - Q(x)|\psi|^{q-1}\psi, \quad t \geqslant 0, \quad x \in \mathbf{R}^N, \tag{1.1}$$

$$\psi(0,x) = \psi_0(x), \quad x \in \mathbf{R}^N, \tag{1.2}$$

where $N \geqslant 2, 1 \leqslant p < q < \frac{N+2}{(N-2)^+}$ ($=\infty$ if $N=2$), $\psi(t,x)$ is a complex-valued function in spacetime $\mathbf{R}\times\mathbf{R}^N$, $K(x)$ and $Q(x)$ are given $C^1$ smooth functions in $\mathbf{R}^N$ and satisfy the following assumptions:

$(H1)$: Both $K(x)$ and $Q(x)$ are bounded in $\mathbf{R}^N$ with $K(x)>0$ and $Q(x)\geqslant 0$.

$(H2)$: $\forall\, x\in\mathbf{R}^N$, $x\,\nabla K(x)\leqslant 0$ and $x\,\nabla Q(x)\leqslant 0$.

When $K(x)=C$ ($C$ is a constant), $Q(x)=0$ and $p=3$, equation (1.1) is the well-known Gross-Pitaevskii equation, which describes the remarkable Bose-Einstein condensate with attractive inter-particle interactions under magnetic trap[1,3,7,12,17]. The harmonic potential $|x|^2$ models a magnetic field whose role is to confine the movement of particles[1,18]. For $K(x)=C$ and $Q(x)=0$, a lot of problems[4,10,13,15,16,20,21] have been studied. Fujiwara[10] studied the smoothness off the time zero of Schrödinger kernel with potentials of quadratic growth, Oh[15] obtained the local well-posedness of nonlinear Schrödinger equations

* Communications on Pure and Applied Analysis 2009, 8 (4):1303－1312.

with subquadratic potentials (including harmonic potential) in corresponding energy space; Zhang[20] and Hon, Zhang and Zhou[13] studied sufficient conditions for global existence of solutions. Moreover, the existence and stability of standing waves with ground state were studied in Zhang[20,21]. A good review can be found in Cazenave[4] as well as Sulem and Sulem[16]. In addition, for equation (1.1) without the term $|x|^2\psi$, when $Q(x)=0$, $V(x)=V(\varepsilon|x|)$ and $p=1+\frac{4}{N}$ with $N\geqslant 2$, Fibich and Wang[9] as well as Liu and Wang[14] studied the stability and instability of the ground-state solitary waves for (1.1), respectively. Moreover, for the nonlinear or quasilinear Schrödinger-type equations, there have been some works[2,5,6,8,11] studying the global solutions. But for general $C^1$ smooth functions $K(x)$ and $Q(x)$, however, little of work about the sufficient conditions of blowup and global existence for solutions to Eq. (1.1) has been carried out.

In the present paper, we first obtain some sufficient conditions for blowup according to the conservation of energy and momentum. Secondly, by defining proper functionals, manifolds and invariant sets, using potential well method and the relation between the initial data and the ground state, we obtain a sufficient conditions for global existence. Finally, we answer the question that how small the initial data need to be for global existence by using scaling argument.

For simplicity, we denote $\int_{\mathbf{R}^N}\cdot\,\mathrm{d}x$ by $\int\cdot\,\mathrm{d}x$ throughout this paper.

## 2 Preliminaries

For (1.1), we define the energy space in the course of nature as

$$\Sigma := H^1(\mathbf{R}^N)\cap\{\varphi:\varphi|x|\in L^2(\mathbf{R}^N)\}. \tag{1.3}$$

By (1.3), $\Sigma$ becomes a Hilbert space, continuously embedded in $H^1(\mathbf{R}^N)$, when endowed with the inner product

$$\langle\varphi,\phi\rangle_\Sigma := \int\nabla\varphi\,\nabla\bar{\phi}\mathrm{d}x+\int\varphi\bar{\phi}\mathrm{d}x+\int|x|^2\varphi\bar{\phi}\mathrm{d}x, \tag{1.4}$$

whose associated norm we denote by $\|\cdot\|_\Sigma$. Moreover, we define the energy functional $E$ on $\Sigma$ by

$$\begin{aligned}E(\varphi) :=& \frac{1}{2}\int|\nabla\varphi|^2\mathrm{d}x+\frac{1}{2}\int|x|^2|\varphi|^2\mathrm{d}x-\\&\frac{1}{p+1}\int K(x)|\varphi|^{p+1}\mathrm{d}x-\frac{1}{q+1}\int Q(x)|\varphi|^{q+1}\mathrm{d}x.\end{aligned} \tag{1.5}$$

From the viewpoint of the Hamiltonian system, $E$ is the generating Hamiltonian of (1.1). Furthermore, for $u\in\Sigma$ and $N\geqslant 2, 1\leqslant p<q<\frac{N+2}{(N-2)^+}(=\infty \text{ if } N=2)$, we define the following functionals:

$$S(u) := \frac{1}{2}\int|\nabla u|^2\mathrm{d}x+\frac{1}{2}\int|x|^2|u|^2\mathrm{d}x+\frac{1}{2}\int|u|^2\mathrm{d}x-$$

$$\frac{1}{p+1}\int K(x)|u|^{p+1}\mathrm{d}x-\frac{1}{q+1}\int Q(x)|u|^{q+1}\mathrm{d}x, \tag{1.6}$$

$$R(u):=\int|\nabla u|^2\mathrm{d}x+\int|x|^2|u|^2\mathrm{d}x+\int|u|^2\mathrm{d}x-\int K(x)|u|^{p+1}\mathrm{d}x-\int Q(x)|u|^{q+1}\mathrm{d}x. \tag{1.7}$$

Since the assumptions $(H1)$, $(H2)$ and the Sobolev embedding theorem, the functionals $S(u)$ and $R(u)$ are well defined.

Now we define a manifold:

$$M:=\{u\in\Sigma\setminus\{0\},R(u)=0\}, \tag{1.8}$$

and a constrained variational problem:

$$d_M:=\inf_{u\in M}S(u). \tag{1.9}$$

By Wang and Zeng[19], we get

**Lemma 2.1** There exists $W(x)\in M$ such that $S(W)=\min_{u\in M}S(u)$.

**Remark 2.1** It is obvious that $R(W)=0$ and

$$\begin{aligned}S(W)&=\frac{p-1}{2(p+1)}\int K(x)|W|^{p+1}\mathrm{d}x+\frac{q-1}{2(q+1)}\int Q(x)|W|^{q+1}\mathrm{d}x\\&<\frac{q-1}{2(q+1)}\left(\int K(x)|W|^{p+1}\mathrm{d}x+\int Q(x)|W|^{q+1}\mathrm{d}x\right)\\&=\frac{q-1}{2(q+1)}\left(\int|\nabla W|^2\mathrm{d}x+\int|x|^2|W|^2\mathrm{d}x+\int|W|^2\mathrm{d}x\right).\end{aligned}$$

By Oh[15] (see also Cazenave[4]), the following local well-posedness for the Cauchy problem (1.1), (1.2) is proposed.

**Proposition 2.1** Let $\psi_0(x)\in\Sigma$. Then there exists a unique solution $\psi(t,x)$ of the Cauchy problem (1.1),(1.2) in $C([0,T),\Sigma)$ for some $T\in(0,\infty]$(maximal existence time) and either $T=\infty$ or else $T<\infty$ and $\lim_{t\to T}\|\psi\|_{\Sigma}=\infty$ Moreover, $\psi$ satisfies the following two conservation laws:

(i) Conservation of momentum

$$\int|\psi|^2\mathrm{d}x=\int|\psi_0|^2\mathrm{d}x,\quad t\in[0,T); \tag{2.1}$$

(ii) Conservation of energy

$$E(\psi)=E(\psi_0),\quad t\in[0,T). \tag{2.2}$$

# 3 Blowing up Solutions

In this section, we will give some sufficient conditions for solutions of (1.1),(1.2) blowing up in finite time according to initial energy.

**Theorem 3.1** Let $N\geqslant2$, $1+\frac{4}{N}\leqslant p<q<\frac{N+2}{(N-2)^+}$ ($=\infty$ if $N=2$), $\psi_0\in\Sigma$ and $\psi\in C([0,T),\Sigma)$ be the solution of the Cauchy problem (1.1),(1.2) on $[0,T)$. If one of the following three conditions holds:

(ⅰ) $E(\psi_0)<0$;

(ⅱ) $E(\psi_0)=0$, $\mathrm{Im}\int x\psi_0\nabla\overline{\psi}_0\mathrm{d}x>0$;

(ⅲ) $E(\psi_0)>0$, and $\mathrm{Im}\int x\psi_0\nabla\overline{\psi}_0\mathrm{d}x\geqslant\left[2E(\psi_0)\int|x|^2|\psi_0|^2\mathrm{d}x\right]^{\frac{1}{2}}$.

Then one has $T<\infty$ and $\lim\limits_{t\to T}\|\psi\|_\Sigma=\infty$.

**Remark 3.1** In fact, in Theorem 3.1, from (1.5) and (2.2), for ∀ function $\varphi_0\in\Sigma\setminus\{0\}$ given, one can always choose a $h$ large enough such that $\psi_0=h\varphi_0$ and $E(\psi_0)<0$. Similarly, there always exists a function $\psi_0\in\Sigma\setminus\{0\}$ such that condition (ⅱ) or (ⅲ) holds.

Now we give a lemma which is key to the proof of Theorem 3.1.

**Lemma 3.2** Suppose that $\psi_0\in\Sigma$. Let $\psi\in C([0,T);\Sigma)$ is a solution of the Cauchy problem (1.1),(1.2) on $[0,T)$. Put

$$J(t)=\int|x|^2|\psi|^2\mathrm{d}x. \tag{3.1}$$

Then one has

$$J''(t)=16E(\psi_0)-16\int|x|^2|\psi|^2\mathrm{d}x-\frac{4(p-1)N-16}{p+1}\int K(x)|\psi|^{p+1}\mathrm{d}x-\frac{4(q-1)N-16}{q+1}\int Q(x)|\psi|^{q+1}\mathrm{d}x+\frac{2}{p+1}\int x\nabla K(x)|\psi|^{p+1}\mathrm{d}x+\frac{2}{q+1}\int x\nabla Q(x)|\psi|^{q+1}\mathrm{d}x.$$

**Proof** From (3.1), one gets

$$\begin{aligned}
J'(t)&=\int|x|^2(\psi_t\overline{\psi}+\psi\overline{\psi}_t)\mathrm{d}x\\
&=2\mathrm{Re}\int|x|^2\overline{\psi}\psi_t\mathrm{d}x=2\mathrm{Im}\int|x|^2\overline{\psi}(\mathrm{i}\psi_t)\mathrm{d}x\\
&=2\mathrm{Im}\int|x|^2\overline{\psi}(-\Delta\psi+|x|^2\psi-K(x)|\psi|^{p-1}\psi-Q(x)|\psi|^{q-1}\psi)\mathrm{d}x\\
&=2\mathrm{Im}\int|x|^2\overline{\psi}(-\Delta\psi)\mathrm{d}x+2\mathrm{Im}\int|x|^4|\psi|^2\mathrm{d}x-\\
&\quad 2\mathrm{Im}\int|x|^2K(x)|\psi|^{p-1}|\psi|^2\mathrm{d}x-2\mathrm{Im}\int|x|^2Q(x)|\psi|^{q-1}|\psi|^2\mathrm{d}x\\
&=2\mathrm{Im}\int|x|^2\overline{\psi}(-\Delta\psi)\mathrm{d}x=-2\mathrm{Im}\int|x|^2\overline{\psi}\Delta\psi\mathrm{d}x\\
&=2\mathrm{Im}\int\nabla(|x|^2\overline{\psi})\nabla\psi\mathrm{d}x=2\mathrm{Im}\int(\nabla|x|^2\overline{\psi}+|x|^2\nabla\overline{\psi})\nabla\psi\mathrm{d}x\\
&=2\mathrm{Im}\int\nabla|x|^2\overline{\psi}\nabla\psi\mathrm{d}x+2\mathrm{Im}\int|x|^2|\nabla\psi|^2\mathrm{d}x\\
&=2\mathrm{Im}\int\nabla|x|^2\overline{\psi}\nabla\psi\mathrm{d}x=2\mathrm{Im}\int2x\overline{\psi}\nabla\psi\mathrm{d}x\\
&=4\mathrm{Im}\int x\overline{\psi}\nabla\psi\mathrm{d}x=-4\mathrm{Im}\int\nabla(x\overline{\psi})\psi\Delta x\\
&=-4\mathrm{Im}\int(\overline{\psi}+x\nabla\overline{\psi})\psi\mathrm{d}x=-4\mathrm{Im}\int|\psi|^2\mathrm{d}x-4\mathrm{Im}\int x\psi\nabla\overline{\psi}\mathrm{d}x
\end{aligned}$$

$$= -4\mathrm{Im}\int x\psi\,\nabla\bar{\psi}\mathrm{d}x, \tag{3.2}$$

where Re$u$ and Im$u$ denote the real part and imaginary part of $u$, respectively.

$$\begin{aligned} J''(t) &= 8\mathrm{Re}\int x\,\nabla\bar{\psi}\mathrm{i}\psi_t\mathrm{d}x + 4\mathrm{Re}\int N\bar{\psi}\mathrm{i}\psi_t\mathrm{d}x \\ &= 8\int|\nabla\psi|^2\mathrm{d}x - 8\int|x|^2|\psi|^2\mathrm{d}x - \frac{4(p-1)}{p+1}N\int K(x)|\psi|^{p+1}\mathrm{d}x - \\ &\quad \frac{4(q-1)}{q+1}N\int Q(x)|\psi|^{q+1}\mathrm{d}x + \\ &\quad \frac{2}{p+1}\int x\,\nabla K(x)|\psi|^{p+1}\mathrm{d}x + \frac{2}{q+1}\int x\,\nabla Q(x)|\psi|^{q+1}\mathrm{d}x \\ &= 16E(\psi_0) - 16\int|x|^2|\psi|^2\mathrm{d}x - \frac{4(p-1)N-16}{p+1}\int K(x)|\psi|^{p+1}\mathrm{d}x - \\ &\quad \frac{4(q-1)N-16}{q+1}\int Q(x)|\psi|^{q+1}\mathrm{d}x + \frac{2}{p+1}\int x\,\nabla K(x)|\psi|^{p+1}\mathrm{d}x + \\ &\quad \frac{2}{q+1}\int x\,\nabla Q(x)|\psi|^{q+1}\mathrm{d}x. \end{aligned} \tag{3.3}$$

□

So far, the proof of Lemma 3.1 is completed. Now we begin to prove Theorem 3.1.

**Proof** We prove Theorem 3.1 by contradiction. Suppose that the maximal existence time $T$ of the solution $\psi$ of the Cauchy problem (1.1), (1.2) is infinity. By Lemma 3.1, $(H1)$, $(H2)$ and $1+\frac{4}{N}\leqslant p<q<\frac{N+2}{(N-2)^+}(=\infty$ if $N=2)$, we have

$$J''(t)\leqslant 16E(\psi_0),\quad 0\leqslant t<\infty. \tag{3.4}$$

By a classical analysis, the following identity is true.

$$J(t) = J(0) + J'(0)t + \int_0^t (t-s)J''(s)\mathrm{d}s,\quad 0\leqslant t<\infty. \tag{3.5}$$

It follows from (3.4) that

$$J(t)\leqslant J(0) + J'(0)t + 8E(\psi_0)t^2,\quad 0\leqslant t<\infty. \tag{3.6}$$

Moreover, $J(t)$ is a nonnegative function,

$$J(0) = \int|x|^2|\psi_0|^2\mathrm{d}x \tag{3.7}$$

and

$$J'(0) = -4\mathrm{Im}\int x\psi_0\,\nabla\bar{\psi}_0\mathrm{d}x. \tag{3.8}$$

In the following, we discuss this theorem through three cases.

(a) If (ⅰ) holds, then from (3.4) we have $J''(t)\leqslant 16E(\psi_0)<0$. Thus $J(t)$ is a concave function of $t$ which implies that there exists $T^*<\infty$ such that $\lim\limits_{t\to T^*}J(t)=0$.

(b) If (ⅱ) holds, then from (3.4) and (3.8) we have $J''(t)\leqslant 16E(\psi_0)=0$ and $J'(0)<0$. Thus there exists $T^*<\infty$ such that $\lim\limits_{t\to T^*}J(t)=0$.

(c) If (ⅲ) holds, then let

$$y(t) = J(0) + J'(0)t + 8E(\psi_0)t^2, \tag{3.9}$$

we have $[J'(0)]^2 - 32J(0)E(\psi_0) \geqslant 0$. Thus there exists at least a $t_1$ such that

$$y(t_1) = J(0) + J'(0)t_1 + 8E(\psi_0)t_1^2 = 0. \tag{3.10}$$

Therefore, from (3.6) and (3.10), there exists $T^* < \infty$ such that $\lim_{t \to T^*} J(t) = 0$.

By (a), (b) and (c), we can get $\lim_{t \to T^*} \int |x|^2 |\psi|^2 dx = 0$, which together with (2.1) leads us to the contradiction. So the maximal existence time $T$ of the solution $\psi$ to the Cauchy problem (1.1), (1.2) is finite. □

This completes the proof of Theorem 3.1.

## 4 Global Existence

In this section, we will discuss the sufficient condition for global existence of the Cauchy problem (1.1), (1.2) by using potential well method and scaling argument.

**Theorem 4.1** Let $N \geqslant 2$ and $1 \leqslant p < q < \frac{N+2}{(N-2)^+}$ ($= \infty$ if $N = 2$). Assume that $\psi_0 \in \Sigma$ and satisfies

$$\int |\nabla \psi_0|^2 dx + \int |x|^2 |\psi_0|^2 dx + \int |\psi_0|^2 dx - \frac{2}{p+1}\int K(x)|\psi_0|^{p+1} dx - \frac{2}{q+1}\int Q(x)|\psi_0|^{q+1} dx < \frac{p-1}{p+1}\int K(x)|W|^{p+1} dx + \frac{q-1}{q+1}\int Q(x)|W|^{q+1} dx, \tag{4.1}$$

and

$$\int |\nabla \psi_0|^2 dx + \int |x|^2 |\psi_0|^2 dx + \int |\psi_0|^2 dx > \int K(x)|\psi_0|^{p+1} dx + \int Q(x)|\psi_0|^{q+1} dx. \tag{4.2}$$

Then the solution $\psi(t,x)$ of the Cauchy problem (1.1), (1.2) globally exists on $t \in [0, \infty)$. Moreover, $\psi(t,x)$ satisfies on $t \in [0, \infty)$,

$$\frac{p-1}{2(p+1)} \|\psi\|_\Sigma^2 \leqslant \frac{q-1}{2(q+1)} \|W\|_\Sigma^2. \tag{4.3}$$

In the following, we answer the question that how small the initial data need to be for global existence of the Cauchy problem (1.1), (1.2).

**Corollary 4.1** For $N \geqslant 2$ and $1 \leqslant p < q < \frac{N+2}{(N-2)^+}$ ($= \infty$ if $N = 2$). If $\psi_0 \in \Sigma$ and satisfies

$$\int |\nabla \psi_0|^2 dx + \int |x|^2 |\psi_0|^2 dx + \int |\psi_0|^2 dx < \frac{p-1}{p+1}\int K(x)|W|^{p+1} dx + \frac{q-1}{q+1}\int Q(x)|W|^{q+1} dx. \tag{4.4}$$

then the solution $\psi(t,x)$ of the Cauchy problem (1.1), (1.2) globally exists. Moreover, on

$t\in[0,\infty)$ one has

$$\frac{p-1}{2(p+1)}\|\psi\|_{\Sigma}^{2}\leqslant\frac{q-1}{2(q+1)}\|W\|_{\Sigma}^{2}. \tag{4.5}$$

**Remark 4.1** For $\forall$ function $\varphi_0\in\Sigma\setminus\{0\}$ given, one can always choose a $\varepsilon$ small enough such that $\psi_0=\varepsilon\psi_0$ and $\psi_0$ satisfies

$$\int|\nabla\psi_0|^2\mathrm{d}x+\int|x|^2|\psi_0|^2\mathrm{d}x+\int|\psi_0|^2\mathrm{d}x$$
$$<\frac{p-1}{p+1}\int K(x)|W|^{p+1}\mathrm{d}x+\frac{q-1}{q+1}\int Q(x)|W|^{q+1}\mathrm{d}x,$$

that is, there always exists a function $\psi_0$ such that (4.4) holds. Moreover, since

$$\frac{p-1}{p+1}\int K(x)|W|^{p+1}\mathrm{d}x+\frac{q-1}{q+1}\int Q(x)|W|^{q+1}\mathrm{d}x>0$$

and

$$\int|\nabla\psi_0|^2\mathrm{d}x+\int|x|^2|\psi_0|^2\mathrm{d}x+\int|\psi_0|^2\mathrm{d}x-\int K(x)|\psi_0|^{p+1}\mathrm{d}x-\int Q(x)|\psi_0|^{q+1}\mathrm{d}x$$
$$\leqslant\int|\nabla\psi_0|^2\mathrm{d}x+\int|x|^2|\psi_0|^2\mathrm{d}x+\int|\psi_0|^2\mathrm{d}x-\frac{2}{p+1}\int K(x)|\psi_0|^{p+1}\mathrm{d}x-\frac{2}{q+1}\int Q(x)|\psi_0|^{q+1}\mathrm{d}x,$$

one can always choose $\psi_0=\varepsilon\psi_0$ such that $\psi_0$ satisfies (4.1) and (4.2).

Before we prove Theorem 4.1 and Corollary 4.1, we first give a lemma.

**Lemma 4.2** Let

$$X:=\{u\in\Sigma,R(u)>0,S(u)<S(W)\} \tag{4.6}$$

Then $X$ is invariant under the flow generated by the Cauchy problem (1.1), (1.2).

**Proof** Let $\psi_0(x)\in X$, and $\psi(t)$ be the solution of the Cauchy problem (1.1),(1.2) with this $\psi_0(x)$. By (1.5), (2.1) and (2.2), one has

$$S(\psi(t))=S(\psi_0),\quad t\in[0,T). \tag{4.7}$$

Thus from $S(\psi_0)<S(W)$, it follows that

$$S(\psi(t))<S(W),\quad t\in[0,T). \tag{4.8}$$

To check $\psi(t)\in X$, we need to prove

$$R(\psi(t))>0,\quad t\in[0,T). \tag{4.9}$$

If (4.9) is not true, by continuity, in views of $R(\psi_0)>0$, there would exist a $t^*>0$ such that

$$R(\psi(t^*))=0. \tag{4.10}$$

It follows that $\psi(t^*)\in M$. Recall (4.8), we also have $S(\psi(t^*))<S(W)$. This is impossible from Lemma 2.1. So $X$ is invariant under the flow generated by the Cauchy problem (1.1), (1.2). □

Now we begin to prove Theorem 4.1 and Corollary 4.1.

**Proof of Theorem 4.1** By (4.1) and (4.2), we have $S(\psi_0)<S(W)$ and $R(\psi_0)>0$. Namely, $\psi_0(x)\in X$. Thus from Lemma 4.1, the solution $\psi(t,x)$ of the Cauchy problem (1.1),(1.2) satisfies

$$\psi(t)\in X,\quad t\in[0,T). \tag{4.11}$$

By (4.11) and Remark 2.1, we get

$$\frac{1}{2}\int|\nabla\psi|^2\mathrm{d}x+\frac{1}{2}\int|x|^2|\psi|^2\mathrm{d}x+\frac{1}{2}\int|\psi|^2\mathrm{d}x-$$
$$\frac{1}{p+1}\int K(x)|\psi|^{p+1}\mathrm{d}x-\frac{1}{q+1}\int Q(x)|\psi|^{q+1}\mathrm{d}x$$
$$<\frac{p-1}{2(p+1)}\int K(x)|W|^{p+1}\mathrm{d}x+\frac{q-1}{2(q+1)}\int Q(x)|W|^{q+1}\mathrm{d}x.$$
$$<\frac{q-1}{2(q+1)}\left(\int K(x)|W|^{p+1}\mathrm{d}x+\int Q(x)|W|^{q+1}\mathrm{d}x\right),\tag{4.12}$$

and

$$\int|\nabla\psi|^2\mathrm{d}x+\int|x|^2|\psi|^2\mathrm{d}x+\int|\psi|^2\mathrm{d}x>\int K(x)|\psi|^{p+1}\mathrm{d}x+\int Q(x)|\psi|^{q+1}\mathrm{d}x.\tag{4.13}$$

For $N\geqslant 2$ and $1<p<q<\frac{N+2}{(N-2)^+}$ ($=\infty$ if $N=2$), from (4.13) we get

$$\begin{aligned}&\frac{1}{p+1}\int K(x)|\psi|^{p+1}\mathrm{d}x+\frac{1}{q+1}\int Q(x)|\psi|^{q+1}\mathrm{d}x\\&<\frac{1}{p+1}\left(\int|\nabla\psi|^2\mathrm{d}x+\int|x|^2|\psi|^2\mathrm{d}x+\int|\psi|^2\mathrm{d}x\right).\end{aligned}\tag{4.14}$$

Thus it follows from (4.12) and (4.14) that

$$\begin{aligned}&\frac{p-1}{2(p+1)}\left(\int|\nabla\psi|^2\mathrm{d}x+\int|x|^2|\psi|^2\mathrm{d}x+\int|\psi|^2\mathrm{d}x\right)\\&<\frac{q-1}{2(q+1)}\left(\int K(x)|W|^{p+1}\mathrm{d}x+\int Q(x)|W|^{q+1}\mathrm{d}x\right)\\&=\frac{q-1}{2(q+1)}\left(\int|\nabla W|^2\mathrm{d}x+\int|x|^2|W|^2\mathrm{d}x+\int|W|^2\mathrm{d}x\right).\end{aligned}\tag{4.15}$$

Thus we established the boundedness of $\psi(t,x)$ in $\Sigma$ for $t\in[0,T)$. So it must have $T=\infty$. That is the solution $\psi(t,x)$ of the Cauchy problem (1.1), (1.2) exists globally on $t\in[0,\infty)$. Moreover, from (4.15), $\psi(t,x)$ satisfies (4.3).

**Proof of Corollary 4.1** By (4.4), it is clear that $\psi_0(x)$ satisfies (4.1). Now we show $\psi_0(x)$ also satisfies (4.2). We proceed as follows. If $\psi_0(x)$ doesn't satisfy (4.2), then one would have

$$\begin{aligned}&\int K(x)|\psi_0|^{p+1}\mathrm{d}x+\int Q(x)|\psi_0|^{q+1}\mathrm{d}x\\&\geqslant\int|\nabla\psi_0|^2\mathrm{d}x+\int|x|^2|\psi_0|^2\mathrm{d}x+\int|\psi_0|^2\mathrm{d}x.\end{aligned}\tag{4.16}$$

Thus there would exist a $0<\mu\leqslant 1$ such that

$$\begin{aligned}&\int K(x)|\mu\psi_0|^{p+1}\mathrm{d}x+\int Q(x)|\mu\psi_0|^{q+1}\mathrm{d}x\\&=\int|\nabla(\mu\psi_0)|^2\mathrm{d}x+\int|x|^2|\mu\psi_0|^2\mathrm{d}x+\int|\mu\psi_0|^2\mathrm{d}x.\end{aligned}\tag{4.17}$$

Recall (1.7), it follows that

$$R(\mu\psi_0)=0. \tag{4.18}$$

On the other hand, by (1.6) one has

$$S(\mu\psi_0)\leqslant\frac{1}{2}\int|\nabla(\mu\psi_0)|^2\mathrm{d}x+\frac{1}{2}\int|x|^2|\mu\psi_0|^2\mathrm{d}x+\frac{1}{2}\int|\mu\psi_0|^2\mathrm{d}x. \tag{4.19}$$

From (4.4), it follows that

$$S(\mu\psi_0)<\frac{1}{2}\mu^2\left[\frac{p-1}{p+1}\int K(x)|W|^{p+1}\mathrm{d}x+\frac{q-1}{q+1}\int Q(x)|W|^{q+1}\mathrm{d}x\right]$$
$$=\mu^2 S(W).$$

Noting that $0<\mu\leqslant 1$ and (4.18) we get $\mu\psi_0\in M$ and $S(\mu\psi_0)<S(W)$. This is impossible by Lemma 2.1. Hence we see that $\psi_0(x)$ satisfies (4.2). From Theorem 4.1, the solution $\psi(t,x)$ of the Cauchy problem (1.1), (1.2) globally exists and on $t\in[0,\infty)$, one has (4.5).

## References


[1] Bradley C C, Sackett C A, Hulet R G. Bose-Einstein condensation of Lithium: observation of limited condensate number. Phys Rev Lett, 1997, 78: 985 – 989.

[2] Blue P, Colliander J. Global well-posedness in Sobolev space implies global existence for weighted $L^2$ initial data for $L^2$-critical NLS. Commun Pure Applied Anal, 2006, 5: 691 – 708.

[3] Caradoc-Davies B M, Ballagh R J, Burnett K. Coherent dynamics of vortex formation in trapped Bose-Einstein condensates. Phys Rev Lett, 1999, 83: 895 – 898.

[4] Cazenave T. An Introduction to Nonlinear Schrödinger Equations. Textos de Métodos Matemáticos 22. Instituto de Mathemática-UFRJ, Rio de Janeiro, 1989.

[5] Chen J. A variational argument to finding global solutions of a quasilinear Schrödinger equation. Commun Pure Applied Anal, 2008, 7: 83 – 88.

[6] Chen J, Guo B. Sharp global existence and blowing up results for inhomogeneous Schrödinger equations. Discrete and Continuous Dynamical Systems Series B, 2007, 8: 357 – 367.

[7] Dalfovo F, Giorgini S, Pitaevskii L P, et al. Theory of Bose-Einstein condensation in trapped gases. Reviews of Modern Physics, 1999, 71: 463 – 512.

[8] Silva D De, Pavlovic N, Staffilani G, et al. Global well-posedness for the $L^2$ critical nonlinear Schrödinger equation in higher dimensions. Commun Pure Applied Anal, 2007, 6: 1023 – 1041.

[9] Fibich G, Wang X P. Stability of solitary waves for nonlinear Schrödinger equations with inhomogeneous nonlinearities. Phys D, 2003, 175: 96 – 108.

[10] Fujiwara D. Remarks on convergence of the Feynman path integrals. Duke Math J, 1980, 47: 559 – 600.

[11] Gan Z H, Guo B L. Sharp threshold of global existence for the generalized Davey-Stewartson System in $R^2$. Commun Pure Applied Anal, 2009, 8(3): 913 – 922.

[12] García-Ripoll J J, Péerez-García V M. Extended parametric resonances in nonlinear Schrödinger systems. Phys Rev Lett, 1999, 83: 1715 – 1718.

[13] Hon Y C, Zhang J, Zhou X. Threshold of global existence for 3-D attractive nonlinear Schrödinger equations under confined potentials. Comput math Appl, 2002, 44: 317 – 322.

[14] Liu Y, Wang X P. Instability of standing waves of the Schrödinger equation with inhomogeneous

nonlinearity. Trans Amer Math Soc, 2006, 358: 2105 - 2122.

[15] Oh Y G. Cauchy problem and Ehrenfest's law of nonlinear Schrödinger equations with potentials. J Diff Eqs, 1989, 81: 255 - 274.

[16] Sulem C, Sulem P L. The nonlinear Schrödinger equation, self-focusing and wave collapse. Springer, 1999.

[17] Tsurumi T, Wadati M. Collapses of wavefunctions in multi-dimensional coupled nonlinear Schrödinger equations under harmonic potential. J Phys Soc Jpn, 1997, 66: 3031 - 3034.

[18] Wadati M, Tsurumi T. Critical number of atoms for the magnetically trapped Bose-Einstein condensate with negative s-wave scattering length. Phys Lett A, 1998, 247: 287 - 293.

[19] Wang X F, Zeng B. On concentration of positive bound states of nonlinear Schrödinger equations with competing potential functions. SIAM J Math Anal, 1997, 28: 633 - 655.

[20] Zhang J. Stability of attractive Bose-Einstein condensates. J Stat Phys, 2000, 101: 731 - 746.

[21] Zhang J. Stability of standing waves for nonlinear Schrödinger equations with unbounded potentials. Z Angew Math Phys, 2000, 51: 498 - 503.

# Instability of Standing Wave, Global Existence and Blowup for the Klein-Gordon-Zakharov System with Different-degree Nonlinearities *

Gan Zaihui(甘在会) Guo Boling(郭柏灵) Zhang Jian(张 健)

**Abstract**

This paper discusses the Klein-Gordon-Zakharov system with different-degree nonlinearities in two and three space dimensions. Firstly, we prove the existence of standing wave with ground state by applying an intricate variational argument. Next, by introducing an auxiliary functional and an equivalent minimization problem, we obtain two invariant manifolds under the solution flow generated by the Cauchy problem to the aforementioned Klein-Gordon-Zakharov system. Furthermore, by constructing a type of constrained variational problem, utilizing the above two invariant manifolds as well as applying potential well argument and concavity method, we derive a sharp threshold for global existence and blowup. Then, combining the above results, we obtain two conclusions of how small the initial data are for the solution to exist globally by using dilation transformation. Finally, we prove a modified instability of standing wave to the system under study.

**Keywords** Klein-Gordon-Zakharov system; standing wave; global existence; blowup; instability

## 1 Introduction

In the present paper, we consider the rescaled Klein-Gordon-Zakharov system in two and three space dimensions:

$$c^{-2}u_{tt} - \Delta u + c^2 u = -nu - |u|^2 u, \quad t > 0, \quad x \in \mathbf{R}^N, \tag{1.1}$$

$$n_t + \nabla \cdot V = 0, \quad t > 0, \quad x \in \mathbf{R}^N, \tag{1.2}$$

$$\alpha^{-2} V_t + \nabla n + \nabla |u|^2 = 0, \quad t > 0, \quad x \in \mathbf{R}^N, \tag{1.3}$$

$$\begin{cases} u(0,x) = u_0(x), & u_t(0,x) = u_1(x), \quad x \in \mathbf{R}^N, \\ n(0,x) = n_0(x), & V(0,x) = v_0(x), \quad x \in \mathbf{R}^N, \end{cases} \tag{1.4}$$

where $N = 2, 3$, $(c, \alpha)$ are two parameters (see [10] for the rescaling with physical constants), $u = u(t,x)$ is an unknown complex vector-valued function, $V = V(t,x)$ is an unknown real vector-valued function and $n = n(t,x)$ is an unknown real scalar-valued function. When $N = 3$ and with the absence of $-|u|^2 u$, the system (1.1)~(1.3) describes the interaction between Langmuir waves and ion sound waves in a plasma (see Bergé, Bidégaray and Colin[2], Dendy[3], Masmoudi and Nakanishi[10], Thornhill and ter Haar[17] as

* This work is supported by National Natural Science Foundation of PR China (10801102, 10771151).
J Differential Equations, 2009, 246: 4097-4128.

well as Zakharov[20]), $u:\mathbf{R}_+\times\mathbf{R}^3\to\mathbf{R}^3$ is the electric field, $n:\mathbf{R}_+\times\mathbf{R}^3\to\mathbf{R}$ is the density fluctuation of ions, $c^2$ is the plasma frequency and $\alpha$ the ion sound speed. In this paper, we assume $0<\alpha<c$ which is physically nature since $c^2/\alpha^2$ is the same order as the mass ratio of the ions and the electrons (see Bellan[1], Masmoudi and Nakanishi[11]), and take function $u$ as complex scalar-valued, because it does not matter what kind of value of the function $u$ takes in our argument[13].

In the theoretical investigations of the dynamics of strong Langmuir turbulence in plasma physics, various types of Zakharov equations play a very important role (see Guo[6], Ozawa, Tsutaya and Tsutsumi[13,14], Thornhill and ter Haar[17] as well as Tsutaya[18]). When $N=3$, there have been some works for the Klein-Gordon-Zakharov system (1.1)~(1.3) with the absence of the term $-|u|^2u$ (see [4,5,11,13,18,21]):

$$c^{-2}u_{tt}-\Delta u+c^2u=-nu,\quad t>0,\quad x\in\mathbf{R}^3,\tag{1.5}$$

$$\alpha^{-2}n_{tt}-\Delta n=\Delta|u|^2,\quad t>0,\quad x\in\mathbf{R}^3,\tag{1.6}$$

$$\begin{cases}u(0,x)=u_0(x),\quad u_t(0,x)=u_1(x),\quad x\in\mathbf{R}^3,\\ n(0,x)=n_0(x),\quad n_t(0,x)=n_1(x),\quad x\in\mathbf{R}^3,\end{cases}\tag{1.7}$$

In the case of $c=1$ and $\alpha=1$, Ozawa, Tsutaya and Tsutsumi[14] studied the Cauchy problem (1.5)~(1.7) by using the method of normal forms introduced by Shatah[16]. They proved that for small initial data, there exists the unique global solution to (1.5)~(1.6) and showed that the solution approaches asymptotically the free solution as $t\to\infty$. Tsutaya[18] proved that there exist global solutions to (1.5)~(1.7) for small data using the invariant Sobolev space but without applying the null condition technique, and thus improved the regularity requirements on the initial data. On the other hand, when $c=1$ and $0<\alpha<1$, Ozawa, Tsutaya and Tsutsumi[13] showed that the discrepancy between the propagation speeds in the system (1.5) and (1.6) yields the time-local well-posedness in $H^1(\mathbf{R}^3)\oplus L^2(\mathbf{R}^3)\oplus L^2(\mathbf{R}^3)\oplus\dot{H}^{-1}(\mathbf{R}^3)$, and by combining this result and the energy conservation law, they also obtained the unique global solution to (1.5)~(1.7) in the energy space for small initial data. Gan and Zhang[4,5] proved the existence of standing wave with ground state of (1.5)~(1.6) by applying an intricate variational argument and obtained the instability of the standing wave by applying Payne and Sattinger's potential well argument (see Payne and Sattinger[15]) and Levine's concavity method (see Levine[8]). Zhang and Gan[21] derived a sharp condition of global existence for the Cauchy problem (1.5)~(1.7). Recently, Li[9] obtained exact explicit travelling wave solution of the system (1.5)~(1.6) for $c=1$ and $\alpha>0$ with arbitrary space dimension. Ohta and Todorova[12] proved the strong instability of standing wave solutions to the system (1.5)~(1.6) when $c=1$, $\alpha>0$ and $N=2,3$. When $0<\alpha<c$, Masmoudi and Nakanishi[11] proved existence and uniqueness of solutions to (1.5)~(1.6) in the energy space $H^1(\mathbf{R}^3)\times L^2(\mathbf{R}^3)$ on some time interval which is uniform with respect to two large parameters $c$ and $\alpha$.

When $N=2$ and $c=\alpha=1$, Guo and Yuan[7] proved without assuming that the Cauchy data are small, the existence and uniqueness of the global smooth solution for the Cauchy problem of the Klein-Gordon-Zakharov system (1.1)~(1.4) via the so-called continuous method and delicate *a priori* estimates and studied the asymptotic behavior of the solutions to the Klein-Gordon-Zakharov system (1.1)~(1.3) with a small parameter approaching zero.

To our knowledge, for the Klein-Gordon-Zakharov system (1.1)~(1.3) (system with two different-degree nonlinearities, two parameters ($c$ , $\alpha$) and $N=2,3$), no results are known so far on the existence and instability of standing wave as well as sharp threshold of global existence and blowup. In the present paper, we are interested in the sharp threshold of global existence, existence and instability of standing wave for the system (1.1)~(1.3) with $\alpha<c$. More specifically, by applying and generalizing the methods studying nonlinear wave equations (see for example [8,15,16]) and the Klein-Gordon-Zakharov system with same-degree nonlinearity (see for example [4,13,14,18,21]), we shall derive a sharp threshold of global existence for the Cauchy problem (1.1)~(1.4). On the other hand, using the sharp threshold, we answer the question of how small the Cauchy data ($u_0$, $u_1$, $n_0$, $v_0$) are for the global solutions of the Cauchy problem (1.1)~(1.4) to exist in the light of the relation between $\int_{\mathbf{R}^N} n_0|u_0|^2\mathrm{d}x$ and 0. Finally, combining the above results, we shall obtain the modified instability of standing wave for the Klein-Gordon-Zakharov system (1.1)~(1.3).

Now we recall that the local well-posedness for the Klein-Gordon-Zakharov system (1.5)~(1.6) with initial data (1.7) was performed by Ohta and Todorova[12], Ozawa, Tsutaya and Tsutsumi[13] when $\alpha\neq c, N=2,3$ and Masmoudi and Nakanishi[11] when $\alpha<c, N=3$. Using the idea of the papers of Ozawa, Tsutaya and Tsutsumi[13], Guo and Yuan[7] as well as Masmoudi and Nakanishi[11], we can prove the local well-posedness of the Klein-Gordon-Zakharov system (1.1)~(1.3) in the energy space $H^1(\mathbf{R}^N)\times L^2(\mathbf{R}^N)\times L^2(\mathbf{R}^N)\times L^2(\mathbf{R}^N)$ with $\alpha<c$, that is, for any $(u_0, u_1, n_0, v_0)\in H^1(\mathbf{R}^N)\times L^2(\mathbf{R}^N)\times L^2(\mathbf{R}^N)\times L^2(\mathbf{R}^N)$ there exists a unique solution

$$(u, u_t, n, V)\in C([0,T_{\max});H^1(\mathbf{R}^N)\times L^2(\mathbf{R}^N)\times L^2(\mathbf{R}^N)\times L^2(\mathbf{R}^N))$$

of (1.1)~(1.3) with the following conserved energy

$$\varepsilon(u, u_t, n, V)=\varepsilon(u_0, u_1, n_0, v_0) \tag{1.8}$$

for all $t\in[0,T_{\max})$, where $\varepsilon(u,u_t,n,V)$ is defined by

$$\varepsilon(u,u_t,n,V)=c^{-2}\int_{\mathbf{R}^N}|u_t|^2\mathrm{d}x+\int_{\mathbf{R}^N}|\nabla u|^2\mathrm{d}x+c^2\int_{\mathbf{R}^N}|u|^2\mathrm{d}x+\frac{1}{2}\int_{\mathbf{R}^N}|u|^4\mathrm{d}x+$$
$$\frac{1}{2}\int_{\mathbf{R}^N}|n|^2\mathrm{d}x+\frac{1}{2}\alpha^{-2}\int_{\mathbf{R}^N}|V|^2\mathrm{d}x+\int_{\mathbf{R}^N}n|u|^2\mathrm{d}x. \tag{1.9}$$

Since the system (1.1)~(1.3) includes different-degree nonlinearities and derivative nonlinearity, in order to get the results in this paper aforementioned, we must improve some techniques different from those proposed in the papers studying the Klein-Gordon-Zakharov system with quadratic nonlinearity and the nonlinear coupled system of the Klein-Gordon and wave equations without derivative and different-degree nonlinearities[4,5,12,21,22]. On the one hand, for the derivative nonlinearity, we must introduce a homogeneous Sobolev space $\dot{H}^{-1}(\mathbf{R}^N)$ which is defined by

$$\dot{H}^{-1}(\mathbf{R}^N) = \{ n \mid \exists\, v:\mathbf{R}^N \to \mathbf{R}^N \text{ such that } n = -\nabla \cdot v, \\ v \in L^2(\mathbf{R}^N) \text{ and } \| n \|_{\dot{H}^{-1}(\mathbf{R}^N)} = \| v \|_{L^2(\mathbf{R}^N)} \}. \tag{1.10}$$

Unlike the nonlinear Klein-Gordon system without derivative nonlinearity, we cannot obtain the blowup in finite time in $L^2(\mathbf{R}^N) \times H^1(\mathbf{R}^N)$. Instead by defining

$$F_{(c,\alpha)}(t) := 2c^{-2}\int_{\mathbf{R}^N} |u|^2 \mathrm{d}x + \alpha^{-2}\int_{\mathbf{R}^N} |g|^2 \mathrm{d}x, \tag{1.11}$$

where $g$ will be defined by assumption (A1) at the end of this section, we can only obtain that the solution of the Cauchy problem (1.1)~(1.4) blows up in finite time in $H^1(\mathbf{R}^N) \times \dot{H}^{-1}(\mathbf{R}^N)$. On the other hand, the different-degree nonlinearities in the system (1.1)~(1.3) yield the following:

In Lemma 3.2, the zero point $\lambda$ of $K(\phi_\lambda, \psi_\lambda) = 0$ is not unique, which brings about much more complexities to show the existence and instability of the standing wave for the system (1.1)~(1.3).

For simplicity, throughout this paper, we denote various positive constants by $C$, $\int_{\mathbf{R}^N} \cdot\, \mathrm{d}x$ by $\int \cdot\, \mathrm{d}x$ and make the following assumption:

(A1) There exists a real vector-valued function $g(t,x) \in L^2(\mathbf{R}^N)$ such that

$$g_t(t,x) = V(t,x).$$

In addition, for simplicity, we introduce a modified $H^1(\mathbf{R}^N)$-norm as

$$\| u \|^2_{H^1_c(\mathbf{R}^N)} = c^2\int |u|^2 \mathrm{d}x + \int |\nabla u|^2 \mathrm{d}x. \tag{1.12}$$

**Remark 1.1** In fact, from the mathematical point of view, when $\alpha \geqslant c > 0$, the results in the present paper are still true. But it has been pointed out in Masmoudi and Nakanishi[11] that the system (1.5)~(1.6) does not have the null form structure as in the sixteenth reference of [11] and this suggests that when $\alpha = c$ the system (1.5)~(1.6) may be locally ill-posed in $H^1(\mathbf{R}^N) \times L^2(\mathbf{R}^N)$. Thus, as the system (1.5)~(1.6), in view of physical relativity and the local well-posedness, we only consider the case $0 < \alpha < c$ for the system (1.1)~(1.3) in the present paper.

## 2 Preliminaries

In this section, we first define several functionals and manifolds, then consider two constrained minimization problems and finally give some elementary results.

For $(\phi, \psi) \in H^1(\mathbf{R}^N) \times L^2(\mathbf{R}^N)$, we define two functionals and a manifold as

$$J(\phi,\psi):=\int|\nabla\phi|^2\mathrm{d}x+c^2\int|\phi|^2\mathrm{d}x+\frac{1}{2}\int|\phi|^4\mathrm{d}x+\frac{1}{2}\int|\psi|^2\mathrm{d}x+\int\psi|\phi|^2\mathrm{d}x, \tag{2.1}$$

$$K(\phi,\psi):=2\int|\nabla\phi|^2\mathrm{d}x+2c^2\int|\phi|^2\mathrm{d}x+2\int|\phi|^4\mathrm{d}x+\int|\psi|^2\mathrm{d}x+3\int\psi|\phi|^2\mathrm{d}x, \tag{2.2}$$

$$\mathscr{M}:=\{(\phi,\psi)\in H^1(\mathbf{R}^N)\times L^2(\mathbf{R}^N)\mid K(\phi,\psi)=0,(\phi,\psi)\neq(0,0)\}. \tag{2.3}$$

In addition, we define a constrained variational problem

$$d_{\mathscr{M}}:=\inf_{(\phi,\psi)\in\mathscr{M}}J(\phi,\psi). \tag{2.4}$$

**Remark 2.1** Since functionals $J(u)$ and $K(u)$ include two different high-degree nonlinearities (higher than two-degree), in order to prove $d_{\mathscr{M}}>0$, it is more difficult than functionals $J(u)$ and $K(u)$ without different-degree nonlinearities. We will prove $d_{\mathscr{M}}>0$ in Section 3.

Besides functionals $J(\varphi,\psi)$ and $K(\phi,\psi)$ defined by (2.1) and (2.2), we also introduce another functional

$$\begin{aligned}I(\phi,\psi):&=J(\phi,\psi)-\frac{1}{v+1}K(\phi,\psi)\\&=\frac{v-1}{v+1}\int|\nabla\phi|^2\mathrm{d}x+\frac{v-1}{v+1}c^2\int|\phi|^2\mathrm{d}x+\frac{v-3}{2(v+1)}\int|\phi|^4\mathrm{d}x+\\&\quad\frac{v-1}{2(v+1)}\int|\psi|^2\mathrm{d}x+\frac{v-2}{v+1}\int\psi|\phi|^2\mathrm{d}x\end{aligned} \tag{2.5}$$

and set

$$\mathscr{M}^-:=\{(\phi,\psi)\in H^1(\mathbf{R}^N)\times L^2(\mathbf{R}^N)\mid K(\phi,\psi)\leqslant 0,(\phi,\psi)\neq(0,0)\}, \tag{2.6}$$

where $v>1$ is a constant. In addition, we define a constrained variational problem

$$d_{\mathscr{M}^-}:=\inf_{(\phi,\psi)\in\mathscr{M}^-}I(\phi,\psi). \tag{2.7}$$

In the following, we give some elementary results according to the above definitions.

**Lemma 2.1** If $K(\phi,\psi)\leqslant 0$ and $(\phi,\psi)\neq(0,0)$, then $I(s^{\frac{1}{2}}\phi,s\psi)$ is an increasing function of $s\in(0,\infty)$, where functionals $K(\phi,\psi)$ and $I(\phi,\psi)$ are defined by (2.2) and (2.5).

**Proof** From (2.5) it follows that

$$\begin{aligned}I(s^{\frac{1}{2}}\phi,s\psi)=&\frac{v-1}{v+1}s\int|\nabla\phi|^2\mathrm{d}x+\frac{v-1}{v+1}sc^2\int|\phi|^2\mathrm{d}x+\frac{v-3}{2(v+1)}s^2\int|\phi|^4\mathrm{d}x+\\&\frac{v-1}{2(v+1)}s^2\int|\psi|^2\mathrm{d}x+\frac{v-2}{v+1}s^2\int\psi|\phi|^2\mathrm{d}x.\end{aligned} \tag{2.8}$$

By $K(\phi,\psi)\leqslant 0$ and (2.2), we have

$$K(\phi,\psi)=2\int|\nabla\phi|^2\mathrm{d}x+2c^2\int|\phi|^2\mathrm{d}x+2\int|\phi|^4\mathrm{d}x+\int|\psi|^2\mathrm{d}x+3\int\psi|\phi|^2\mathrm{d}x\leqslant 0$$

which yields that

$$2\int|\nabla\phi|^2\mathrm{d}x+2c^2\int|\phi|^2\mathrm{d}x\leqslant-2\int|\phi|^4\mathrm{d}x-\int|\psi|^2\mathrm{d}x-3\int\psi|\phi|^2\mathrm{d}x. \tag{2.9}$$

By the inequality

$$\left|\int \psi |\phi|^2 \mathrm{d}x\right| \leqslant \frac{1}{2}\int |\psi|^2 \mathrm{d}x + \frac{1}{2}\int |\phi|^4 \mathrm{d}x, \tag{2.10}$$

$(\phi,\psi)\neq(0,0)$ and (2.9) imply that

$$0 < 2\int |\nabla\phi|^2 \mathrm{d}x + 2c^2\int |\phi|^2 \mathrm{d}x \leqslant \frac{1}{2}\left(\int |\psi|^2 \mathrm{d}x - \int |\phi|^4 \mathrm{d}x\right). \tag{2.11}$$

From $K(\phi,\psi)\leqslant 0, v>1, (\phi,\psi)\neq(0,0)$, (2.10) and (2.11), for $s\in(0,\infty)$ we get

$$\begin{aligned}\frac{\mathrm{d}I(s^{\frac{1}{2}}\phi, s\psi)}{\mathrm{d}s} &= \frac{v-1}{v+1}\int |\nabla\phi|^2 \mathrm{d}x + \frac{v-1}{v+1}c^2\int |\phi|^2 \mathrm{d}x + \frac{v-3}{v+1}s\int |\phi|^4 \mathrm{d}x + \\ &\quad \frac{v-1}{v+1}s\int |\psi|^2 \mathrm{d}x + \frac{2(v-2)}{v+1}s\int \psi|\phi|^2 \mathrm{d}x \\ &\geqslant \frac{v-1}{v+1}\int |\nabla\phi|^2 \mathrm{d}x + \frac{v-1}{v+1}c^2\int |\phi|^2 \mathrm{d}x + \frac{1}{v+1}s\left(\int |\psi|^2 \mathrm{d}x - \int |\phi|^4 \mathrm{d}x\right) \\ &> 0,\end{aligned}$$

which concludes our proof. □

Using Lemma 2.1, we can get the following conclusion.

**Proposition 2.1** $\mathcal{M}$ and $\mathcal{M}^-$ are non-empty, and

$$d_{\mathcal{M}} = \inf_{(\phi,\psi)\in\mathcal{M}} J(\phi,\psi) = d_{\mathcal{M}^-} = \inf_{(\phi,\psi)\in\mathcal{M}^-} I(\phi,\psi). \tag{2.12}$$

Furthermore, $I(\varphi,\psi)>d_{\mathcal{M}}$ if $K(\phi,\psi)<0$.

**Proof** **Step 1** First we prove that $\mathcal{M}^-$ is non-empty. Choose any $(\phi,\psi)\in H^1(\mathbf{R}^N)\times L^2(\mathbf{R}^N)$ with $(\phi,\psi)\neq(0,0)$ and consider

$$\begin{aligned}P(s) &= K(s^{\frac{1}{2}}\phi, s\psi) \\ &= 2s\left(\int |\nabla\phi|^2 \mathrm{d}x + c^2\int |\phi|^2 \mathrm{d}x\right) + 2s^2\int |\phi|^4 \mathrm{d}x + s^2\int |\psi|^2 \mathrm{d}x + 3s^2\int \psi|\phi|^2 \mathrm{d}x,\end{aligned}$$

thus if we take $2\int |\phi|^4 \mathrm{d}x + \int |\psi|^2 \mathrm{d}x + 3\int \psi|\phi|^2 \mathrm{d}x < 0$, then we see that $P(s)<0$ for sufficiently large $s>1$ and hence $(s^{\frac{1}{2}}\phi, s\psi)\in\mathcal{M}^-$.

**Step 2** Next, to prove that $\mathcal{M}$ is non-empty, we choose $(\phi,\psi)\in H^1(\mathbf{R}^N)\times L^2(\mathbf{R}^N)$ such that $K(\phi,\psi)<0$ and consider $K(s^{\frac{1}{2}}\phi, s\psi)$. Now for $s=1$, $K(\phi,\psi)<0$; for $s>0$ close to zero, $K(s^{\frac{1}{2}}\phi, s\psi)>0$ from the expression for $K(s^{\frac{1}{2}}\phi, s\psi)$. Therefore by continuity, there exists an $s_0\in(0,1)$ such that $K(s_0^{\frac{1}{2}}\phi, s_0\psi)=0$, i.e., $(s_0^{\frac{1}{2}}\phi, s_0\psi)\in\mathcal{M}$.

**Step 3** Furthermore, we show that (2.12) holds. By Lemma 2.1, $I(s^{\frac{1}{2}}\phi, s\psi)$ is an increasing function of $s\in(0,\infty)$ provided that $K(\phi,\psi)\leqslant 0$ and $(\phi,\psi)\neq(0,0)$. Since $K(s_0^{\frac{1}{2}}\phi, s_0\psi)=0$, by (2.5) we have

$$d_{\mathcal{M}} \leqslant J(s_0^{\frac{1}{2}}\phi, s_0\psi) = I(s_0^{\frac{1}{2}}\phi, s_0\psi) + \frac{1}{v+1}K(s_0^{\frac{1}{2}}\phi, s_0\psi) = I(s_0^{\frac{1}{2}}\phi, s_0\psi),$$

which together with Lemma 2.1 implies that for $s_0\in(0,1)$,

$$d_{\mathcal{M}} \leqslant \inf_{(s_0^{\frac{1}{2}}\phi, s_0\psi)\in\mathcal{M}^-} J(s_0^{\frac{1}{2}}\phi, s_0\psi) = \inf_{(s_0^{\frac{1}{2}}\phi, s_0\psi)\in\mathcal{M}^-} I(s_0^{\frac{1}{2}}\phi, s_0\psi) \leqslant \inf_{(\phi,\psi)\in\mathcal{M}^-} I(\phi,\psi) = d_{\mathcal{M}^-}. \tag{2.13}$$

But by definition (2.4)

$$d_{\mathscr{M}} = \inf_{(\phi,\psi)\in\mathscr{M}} J(\phi,\psi) = \inf_{(\phi,\psi)\in\mathscr{M}} I(\phi,\psi) \geqslant \inf_{(\phi,\psi)\in\overline{\mathscr{M}}} I(\phi,\psi) = d_{\overline{\mathscr{M}}}. \tag{2.14}$$

Thus (2.13) and (2.14) imply (2.12).

Finally, we show: If $K(\phi,\psi)<0$, then $I(\phi,\psi)>d_{\mathscr{M}}$. Since $K(\phi,\psi)<0$, from Step 2 it follows that there exists an $s\in(0,1)$ such that $K(s^{\frac{1}{2}}\phi,s\psi)=0$ and $(\phi,\psi)\neq(0,0)$. From (2.5) we have

$$J(s^{\frac{1}{2}}\phi,s\psi) = I(s^{\frac{1}{2}}\phi,s\psi) \geqslant d_{\mathscr{M}}$$

which yields from Lemma 2.1 that

$$d_{\mathscr{M}} \leqslant I(s^{\frac{1}{2}}\phi,s\psi) < I(\phi,\psi).$$

Therefore, if $K(\phi,\psi)<0$, then $I(\phi,\psi)>d_{\mathscr{M}}$.

So far, the proof of Proposition 2.1 is completed. □

## 3 Existence of Standing Wave

In this section, we discuss the existence of standing wave for the Klein-Gordon-Zakharov system (1.1)~(1.3) and prove $d_{\mathscr{M}}>0$ which was defined in Section 2. First we give the definitions of standing wave and ground state.

**Definition 3.1** (Standing wave) If a pair of real functions $(\phi,\psi)=(\phi(x),\psi(x))$, $x\in\mathbf{R}^N$, verifies the system

$$\begin{cases} -\Delta\phi + c^2\phi = -\phi\psi - |\phi|^2\phi, & x\in\mathbf{R}^N \\ \psi + |\phi|^2 = 0, & x\in\mathbf{R}^N \end{cases}, \tag{E}$$

and

$$(\phi,\psi)\in H^1(\mathbf{R}^N)\times L^2(\mathbf{R}^N)\setminus\{(0,0)\},$$

then

$$u(t,x) = \phi(x), \quad n(t,x) = \psi(x), \quad t\geqslant 0, \quad x\in\mathbf{R}^N,$$

satisfy (1.1)~(1.3), and are called standing wave solution (or standing wave) of the system (1.1)~(1.3).

**Definition 3.2** (Ground state) A solution $(\phi,\psi)$ of (E) is termed as a ground state (solution) if it has some minimal action $J(\phi,\psi)$ (defined by (2.1)) among all positive solutions of (E).

**Remark 3.1** Since the system (1.1)~(1.3) is equivalent to

$$\begin{cases} c^{-2}u_{tt} - \Delta u + c^2 u = -nu - |u|^2 u \\ \alpha^{-2}n_{tt} - \Delta n = \Delta |u|^2 \end{cases}, \tag{KGZ}$$

Definitions 3.1 and 3.2 are consistent with the related definitions for single nonlinear Klein-Gordon equation. Moreover, from the physical viewpoint, an important role is played by the ground state solution of (E).

In the following, we give two lemmas.

**Lemma 3.1** $J(\phi,\psi)$ is bounded below on $\mathscr{M}$ and $J(\phi,\psi)>0$ for all $(\phi,\psi)\in\mathscr{M}$.

**Proof** From (2.1)~(2.4), we get on $\mathscr{M}$

$$J(\phi,\psi)=\frac{1}{3}\int|\nabla\phi|^2\mathrm{d}x+\frac{1}{3}c^2\int|\phi|^2\mathrm{d}x+\frac{1}{6}\int|\psi|^2\mathrm{d}x-\frac{1}{6}\int|\phi|^4\mathrm{d}x \tag{3.1}$$

and

$$\int\psi|\phi|^2\mathrm{d}x<0. \tag{3.2}$$

By the inequality

$$\left|\int\psi|\phi|^2\mathrm{d}x\right|\leqslant\frac{1}{2}\int|\psi|^2\mathrm{d}x+\frac{1}{2}\int|\phi|^4\mathrm{d}x, \tag{3.3}$$

$K(\phi,\psi)=0$ and $(\phi,\psi)\neq(0,0)$, it follows that

$$\begin{aligned}2\int|\nabla\phi|^2\mathrm{d}x+2c^2\int|\phi|^2\mathrm{d}x+2\int|\phi|^4\mathrm{d}x+\int|\psi|^2\mathrm{d}x&=-3\int\psi|\phi|^2\mathrm{d}x\\&\leqslant\frac{3}{2}\int|\psi|^2\mathrm{d}x+\frac{3}{2}\int|\phi|^4\mathrm{d}x\end{aligned}$$

which implies

$$0<2\int|\nabla\phi|^2\mathrm{d}x+2\int|\phi|^2\mathrm{d}x\leqslant\frac{1}{2}\int|\psi|^2\mathrm{d}x-\frac{1}{2}\int|\phi|^4\mathrm{d}x, \tag{3.4}$$

that is,

$$\int|\psi|^2\mathrm{d}x>\int|\phi|^4\mathrm{d}x. \tag{3.5}$$

Hence if $K(\phi,\psi)=0$ and $(\phi,\psi)\neq(0,0)$, then (3.5) is always true. By (3.1) and (3.5) we get $J(\phi,\psi)>0$ for all $(\phi,\psi)\in\mathscr{M}$. □

**Lemma 3.2** For $(\phi,\psi)\in H^1(\mathbf{R}^N)\times L^2(\mathbf{R}^N)$, $(\phi,\psi)\neq(0,0)$ and $\lambda>0$, let $\phi_\lambda(x)=\lambda\phi(x)$, $\psi_\lambda(x)=\lambda\psi(x)$. Thus one of the following two conclusions holds:

(Ⅰ) If $K(\phi_\lambda,\psi_\lambda)$ has two different positive zero points $\lambda_1$ and $\lambda_2$ (depending on $(\phi,\psi)$) and $\lambda_2>\lambda_1$, then $K(\phi_{\lambda_1},\psi_{\lambda_1})=0$, $K(\phi_{\lambda_2},\psi_{\lambda_2})=0$ and one gets the following eight possibilities.

(Ⅰ-1) $K(\phi_\lambda,\psi_\lambda)>0$ for $\lambda\in(0,\lambda_1)$,
$K(\phi_\lambda,\psi_\lambda)<0$ for $\lambda\in(\lambda_1,\lambda_2)$,
$K(\phi_\lambda,\psi_\lambda)>0$ for $\lambda\in(\lambda_2,\infty)$,
$\forall\lambda\in(0,\lambda_2]$, $J(\phi_{\lambda_1},\psi_{\lambda_1})\geqslant J(\phi_\lambda,\psi_\lambda)$,
$\forall\lambda\in[\lambda_2,\infty)$, $J(\phi_{\lambda_2},\psi_{\lambda_2})\leqslant J(\phi_\lambda,\psi_\lambda)$.

(Ⅰ-2) $K(\phi_\lambda,\psi_\lambda)>0$ for $\lambda\in(0,\lambda_1)\cup(\lambda_1,\lambda_2)$,
$K(\phi_\lambda,\psi_\lambda)>0$ for $\lambda\in(\lambda_2,\infty)$,
$\forall\lambda\in(0,\lambda_2]$, $J(\phi_{\lambda_2},\psi_{\lambda_2})\geqslant J(\phi_\lambda,\psi_\lambda)$,
$\forall\lambda\in(\lambda_2,\infty)$, $J(\phi_\lambda,\psi_\lambda)\geqslant J(\phi_{\lambda_2},\psi_{\lambda_2})$.

(Ⅰ-3) $K(\phi_\lambda,\psi_\lambda)>0$ for $\lambda\in(0,\lambda_1)\cup(\lambda_1,\lambda_2)$,

$$K(\phi_\lambda,\psi_\lambda)<0 \quad \text{for } \lambda\in(\lambda_2,\infty),$$
$$\forall\lambda\in(0,\infty),\quad J(\phi_{\lambda_2},\psi_{\lambda_2})\geqslant J(\phi_\lambda,\psi_\lambda).$$

(Ⅰ-4) $K(\phi_\lambda,\psi_\lambda)>0$ for $\lambda\in(0,\lambda_1)$,
$K(\phi_\lambda,\psi_\lambda)<0$ for $\lambda\in(\lambda_1,\lambda_2)\cup(\lambda_2,\infty)$,
$\forall\lambda\in(0,\infty)$, $J(\phi_{\lambda_1},\psi_{\lambda_1})\geqslant J(\phi_\lambda,\psi_\lambda)$.

(Ⅰ-5) $K(\phi_\lambda,\psi_\lambda)<0$ for $\lambda\in(0,\lambda_1)$,
$K(\phi_\lambda,\psi_\lambda)>0$ for $\lambda\in(\lambda_1,\lambda_2)\cup(\lambda_2,\infty)$,
$\forall\lambda\in(0,\infty)$, $J(\phi_{\lambda_1},\psi_{\lambda_1})\leqslant J(\phi_\lambda,\psi_\lambda)$.

(Ⅰ-6) $K(\phi_\lambda,\psi_\lambda)<0$ for $\lambda\in(0,\lambda_1)$,
$K(\phi_\lambda,\psi_\lambda)>0$ for $\lambda\in(\lambda_1,\lambda_2)$,
$K(\phi_\lambda,\psi_\lambda)<0$ for $\lambda\in(\lambda_2,\infty)$,
$\forall\lambda\in(0,\lambda_2]$, $J(\phi_{\lambda_1},\psi_{\lambda_1})\leqslant J(\phi_\lambda,\psi_\lambda)$,
$\forall\lambda\in[\lambda_2,\infty)$, $J(\phi_{\lambda_2},\psi_{\lambda_2})\geqslant J(\phi_\lambda,\psi_\lambda)$.

(Ⅰ-7) $K(\phi_\lambda,\psi_\lambda)<0$ for $\lambda\in(0,\lambda_1)\cup(\lambda_1,\lambda_2)$,
$K(\phi_\lambda,\psi_\lambda)>0$ for $\lambda\in(\lambda_2,\infty)$,
$\forall\lambda\in(0,\infty)$, $J(\phi_{\lambda_2},\psi_{\lambda_2})\leqslant J(\phi_\lambda,\psi_\lambda)$.

(Ⅰ-8) $K(\phi_\lambda,\psi_\lambda)<0$ for $\lambda\in(0,\lambda_1)\cap(\lambda_1,\lambda_2)\cup(\lambda_2,\infty)$,
$\forall\lambda\in(0,\lambda_1)$, $J(\phi_{\lambda_1},\psi_{\lambda_1})\leqslant J(\phi_\lambda,\psi_\lambda)$,
$\forall\lambda\in[\lambda_1,\infty)$, $J(\phi_{\lambda_1},\psi_{\lambda_1})\geqslant J(\phi_\lambda,\psi_\lambda)$.

(Ⅱ) If $K(\phi_\lambda,\psi_\lambda)$ has two same positive zero points $\lambda^*$, then one gets $K(\phi_{\lambda^*},\psi_{\lambda^*})=0$ and the following four possibilities hold.

(Ⅱ-1) $K(\phi_\lambda,\psi_\lambda)<0$ for $\lambda\in(0,\lambda^*)$,
$K(\phi_\lambda,\psi_\lambda)>0$ for $\lambda\in(\lambda^*,\infty)$,
$\forall\lambda\in(0,\infty)$, $J(\phi_{\lambda^*},\psi_{\lambda^*})\leqslant J(\phi_\lambda,\psi_\lambda)$.

(Ⅱ-2) $K(\phi_\lambda,\psi_\lambda)<0$ for $\lambda\in(0,\lambda^*)$,
$K(\phi_\lambda,\psi_\lambda)<0$ for $\lambda\in(\lambda^*,\infty)$,
$\forall\lambda\in(0,\lambda^*)$, $J(\phi_{\lambda^*},\psi_{\lambda^*})\leqslant J(\phi_\lambda,\psi_\lambda)$,
$\forall\lambda\in(\lambda^*,\infty)$, $J(\phi_{\lambda^*},\psi_{\lambda^*})\geqslant J(\phi_\lambda,\psi_\lambda)$.

(Ⅱ-3) $K(\phi_\lambda,\psi_\lambda)>0$ for $\lambda\in(0,\lambda^*)$,
$K(\phi_\lambda,\psi_\lambda)<0$ for $\lambda\in(\lambda^*,\infty)$,
$\forall\lambda\in(0,\infty)$, $J(\phi_{\lambda^*},\psi_{\lambda^*})\geqslant J(\phi_\lambda,\psi_\lambda)$.

(Ⅱ-4) $K(\phi_\lambda,\psi_\lambda)>0$ for $\lambda\in(0,\lambda^*)$,

$$K(\phi_\lambda,\psi_\lambda)>0 \quad \text{for } \lambda\in(\lambda^*,\infty),$$
$$\forall\lambda\in(0,\lambda^*),\quad J(\phi_{\lambda^*},\psi_{\lambda^*})\geqslant J(\phi_\lambda,\psi_\lambda),$$
$$\forall\lambda\in(\lambda^*,\infty),\quad J(\phi_{\lambda^*},\psi_{\lambda^*})\leqslant J(\phi_\lambda,\psi_\lambda).$$

**Proof** By (2.1) and (2.2), we have

$$J(\phi_\lambda,\psi_\lambda)=\lambda^2\left(\int|\nabla\phi|^2dx+c^2\int|\phi|^2dx\right)+\frac{1}{2}\lambda^4\int|\phi|^4dx+\frac{1}{2}\lambda^2\int|\psi|^2dx+\lambda^3\int\psi|\phi|^2dx, \tag{3.6}$$

$$K(\phi_\lambda,\psi_\lambda)=2\lambda^2\left(\int|\nabla\phi|^2dx+c^2\int|\phi|^2dx\right)+2\lambda^4\int|\phi|^4dx+\lambda^2\int|\psi|^2dx+3\lambda^3\int\psi|\phi|^2dx. \tag{3.7}$$

From the definition of $\mathcal{M}$($\mathcal{M}$ is not an empty set which has been shown in Proposition 2.1), there must exist a $\gamma>0$ such that $K(\phi_\gamma,\psi_\gamma)=0$ and $(\phi_\gamma,\psi_\gamma)\neq(0,0)$. Thus by (3.7), one of the following two cases is true.

In the following, we first prove (Ⅰ).

**Proof of** (Ⅰ) If $K(\phi_\lambda,\psi_\lambda)$ has two positive zero points $\lambda_1$ and $\lambda_2$ with $\lambda_2>\lambda_1$, then from (3.7) we obtain $K(\phi_{\lambda_1},\psi_{\lambda_1})=0$, $K(\phi_{\lambda_2},\psi_{\lambda_2})=0$ and $(\phi_{\lambda_1},\psi_{\lambda_1})\neq(0,0)$, $(\phi_{\lambda_2},\psi_{\lambda_2})\neq(0,0)$. First of all, we show (Ⅰ-1) holds. By the expression (3.7), it follows that the possibility holds: $K(\phi_\lambda,\psi_\lambda)>0$ for $\lambda\in(0,\lambda_1)$, $K(\phi_\lambda,\psi_\lambda)<0$ for $\lambda\in(\lambda_1,\lambda_2)$ and $K(\phi_\lambda,\psi_\lambda)>0$ for $\lambda\in(\lambda_2,\infty)$. Since

$$\frac{d}{d\lambda}J(\phi_\lambda,\psi_\lambda)-2\lambda\left(\int|\nabla\phi|^2dx+c^2\int|\phi|^2dx+\frac{1}{2}\int|\psi|^2dx\right)+2\lambda^3\int|\phi|^4dx+3\lambda^2\int\psi|\phi|^2dx$$
$$=\lambda^{-1}K(\phi_\lambda,\psi_\lambda), \tag{3.8}$$

noting that $K(\phi_{\lambda_1},\psi_{\lambda_1})=0$ and $K(\phi_{\lambda_2},\psi_{\lambda_2})=0$, it follows that for $\forall\lambda\in(0,\lambda_2]$, $J(\phi_{\lambda_1},\psi_{\lambda_1})\geqslant J(\phi_\lambda,\psi_\lambda)$ and for $\forall\lambda\in[\lambda_2,\infty)$, $J(\phi_{\lambda_2},\psi_{\lambda_2})\leqslant J(\phi_\lambda,\psi_\lambda)$.

Similarly, we can prove possibilities (Ⅰ-2)~(Ⅰ-8). □

Next we prove (Ⅱ).

**Proof of** (Ⅱ) If $K(\varphi_\lambda,\psi_\lambda)$ has two same positive zero points $\lambda^*$, then from (3.7) one gets $K(\phi_{\lambda^*},\psi_{\lambda^*})=0$ and $(\phi_{\lambda^*},\psi_{\lambda^*})\neq(0,0)$. We first show (Ⅱ-1) holds. Applying (3.7), it follows that the possibility holds: $K(\phi_\lambda,\psi_\lambda)<0$ for $\lambda\in(0,\lambda^*)$, $K(\phi_\lambda,\psi_\lambda)>0$ for $\lambda\in(\lambda^*,\infty)$. Thus from (3.8), we get for $\forall\lambda\in(0,\infty)$, $J(\phi_{\lambda^*},\psi_{\lambda^*})\leqslant J(\phi_\lambda,\psi_\lambda)$.

Similarly, we can show possibilities (Ⅱ-2)~(Ⅱ-4). □

So far, the proof of Lemma 3.2 is completed. □

In the following, we give the existence of standing waves for the system (1.1)~(1.3) whose proof strongly depends upon Lemma 3.2.

**Remark 3.2** In Lemma 3.2 and its proof, with the absence of $|\phi|^4$ in $K(\phi,\psi)$, $K(\phi_\lambda,\psi_\lambda)=0$ exists only one standing point $\lambda^*$ which depends only on $(\phi,\psi)$. In this case, action $J(\phi_\lambda,\psi_\lambda)$ can be determined uniquely by the standing point $\lambda^*$. However, with the presence of $|\phi|^4$ in $K(\phi,\psi)$, owing to the interaction of multiple nonlinearities in $K(\phi,\psi)$, $K(\phi_\lambda,\psi_\lambda)=0$ exist two different positive standing points $\lambda_1$, $\lambda_2$ or two same positive standing points $\lambda_1=\lambda_2=\lambda^*$ which depend only on $(\phi,\psi)$. In this case, action $J(\phi_\lambda,\psi_\lambda)$ cannot be determined uniquely by the standing points, but varies from the value of $K(\phi_\lambda,\psi_\lambda)$ on the intervals $(0,\lambda_1)$, $(\lambda_1,\lambda_2)$ and $(\lambda_2,\infty)$ or $(0,\lambda^*)$ and $(\lambda^*,\infty)$. Therefore, from the physical viewpoint, the above mathematical discussions can be viewed that: the more the standing points, the more complex the action becomes. In other words, the form of the functional $K(\phi,\psi)$ decides directly the characteristic of action $J(\phi,\psi)$.

**Theorem 3.1** There exists $(P,Q)\in\mathscr{M}$ such that

$$J(P,Q)=\inf_{(\phi,\psi)\in\mathscr{M}} J(\phi,\psi)=d_{\mathscr{M}}.$$

**Proof** Considering the constrained variational problem (2.4) and applying Lemma 3.2 we may let $\{((\phi_n)_\xi,(\psi_n)_\xi):n\in\mathbf{N}\}\subset\mathscr{M}$ be a minimizing sequence for (2.4), then one gets

$$\begin{cases}K((\phi_n)_\xi,(\psi_n)_\xi)=0\\ J((\phi_n)_\xi,(\psi_n)_\xi)\to\inf\limits_{(\phi,\psi)\in\mathscr{M}}J(\phi,\psi) \quad \text{as } n\to\infty\end{cases},\tag{3.9}$$

where

$$\begin{cases}\xi=\lambda_1 \text{ or } \xi\in[\lambda_2,\infty), & \text{if (I-1) in Lemma 3.2 holds}\\ \xi=\lambda_2 \text{ or } \xi\in[\lambda_2,\infty), & \text{if (I-2) in Lemma 3.2 holds}\\ \xi=\lambda_2, & \text{if (I-3) in Lemma 3.2 holds}\\ \xi=\lambda_1, & \text{if (I-4) in Lemma 3.2 holds}\\ \xi\in(0,\infty), & \text{if (I-5) in Lemma 3.2 holds}\\ \xi=\lambda_2 \text{ or } \xi\in(0,\lambda_2], & \text{if (I-6) in Lemma 3.2 holds}\\ \xi\in(0,\infty), & \text{if (I-7) in Lemma 3.2 holds}\\ \xi=\lambda_1 \text{ or } \xi\in(0,\lambda_1], & \text{if (I-8) in Lemma 3.2 holds}\\ \xi\in(0,\infty), & \text{if (II-1) in Lemma 3.2 holds}\\ \xi=\lambda^* \text{ or } \xi\in(0,\lambda^*), & \text{if (II-2) in Lemma 3.2 holds}\\ \xi=\lambda^*, & \text{if (II-3) in Lemma 3.2 holds}\\ \xi=\lambda^* \text{ or } \xi\in[\lambda^*,\infty), & \text{if (II-4) in Lemma 3.2 holds}\end{cases}$$

Let $\phi^*$ and $\psi^*$ denote the Schwarz spherical rearrangements of functions $\phi$ and $\psi$, respectively. We recall that $\phi^*$ and $\psi^*$ are spherically symmetric, non-increasing (with respect to $|x|$) functions, and the symmetrization has the following properties:

$$\int|\nabla\phi^*|^2\mathrm{d}x\leqslant\int|\nabla\phi|^2\mathrm{d}x,\quad \int|\nabla\psi^*|^2\mathrm{d}x\leqslant\int|\nabla\psi|^2\mathrm{d}x,\tag{3.10}$$

$$\int|\phi^*|^\sigma\mathrm{d}x=\int|\phi|^\sigma\mathrm{d}x,\quad \int|\psi^*|^\sigma\mathrm{d}x=\int|\psi|^\sigma\mathrm{d}x \quad \text{for } \sigma>1.\tag{3.11}$$

Furthermore, it is evident that

$$(\phi_\lambda)^*=(\phi^*)_\lambda,\quad (\psi_\lambda)^*=(\psi^*)_\lambda,\tag{3.12}$$

where as in Lemma 3.2, $\phi_\lambda(x)=\lambda\phi(x)$, $\psi_\lambda(x)=\lambda\psi(x)$.

Now we consider the minimizing sequence $\{((\phi_n)_\xi, (\psi_n)_\xi): n \in \mathbf{N}\}$ and choose suitable $((\phi_n)_\xi)^*$, $((\psi_n)_\xi)^*$ such that

$$K((((\phi_n)_\xi)^*)_{\eta_n}, \quad (((\psi_n)_\xi)^*)_{\eta_n}) = 0, \tag{3.13}$$

where

$$\begin{cases} \eta_n \in (0,\lambda_2] \text{ or } \eta_n = \lambda_2, & \text{if (I-1) in Lemma 3.2 holds} \\ \eta_n \in (0,\lambda_2] \text{ or } \eta_n = \lambda_2, & \text{if (I-2) in Lemma 3.2 holds} \\ \eta_n \in (0,\infty), & \text{if (I-3) in Lemma 3.2 holds} \\ \eta_n \in (0,\infty), & \text{if (I-4) in Lemma 3.2 holds} \\ \eta_n = \lambda_1, & \text{if (I-5) in Lemma 3.2 holds} \\ \eta_n \in (\lambda_2,\infty) \text{ or } \eta_n = \lambda_1, & \text{if (I-6) in Lemma 3.2 holds} \\ \eta_n = \lambda_2, & \text{if (I-7) in Lemma 3.2 holds} \\ \eta_n \in [\lambda_1,\infty) \text{ or } \eta_n = \lambda_1, & \text{if (I-8) in Lemma 3.2 holds} \\ \eta_n = \lambda^*, & \text{if (II-1) in Lemma 3.2 holds} \\ \eta_n \in [\lambda^*,\infty) \text{ or } \eta_n = \lambda^*, & \text{if (II-2) in Lemma 3.2 holds} \\ \eta_n \in (0,\infty), & \text{if (II-3) in Lemma 3.2 holds} \\ \eta_n \in (0,\lambda^*] \text{ or } \eta_n = \lambda^*, & \text{if (II-4) in Lemma 3.2 holds} \end{cases}. \tag{3.14}$$

Let

$$P_n = (((\phi_n)_\xi)^*)_{\eta_n}, \quad Q_n = (((\psi_n)_\xi)^*)_{\eta_n}. \tag{3.15}$$

From (3.12), we also have

$$P_n = (((\phi_n)_\xi)_{\eta_n})^*, \quad Q_n = (((\psi_n)_\xi)_{\eta_n})^*$$

and therefore by (3.1), (3.10) and (3.11), we have from $K((\phi_n)_\xi,(\psi_n)_\xi)=0$ that

$$J(P_n, Q_n) \leqslant J[((\phi_n)_\xi)_{\eta_n}, ((\psi_n)_\xi)_{\eta_n}] \leqslant J((\phi_n)_\xi, (\psi_n)_\xi). \tag{3.16}$$

The right-hand side inequality in (3.16) is a consequence of Lemma 3.2. Thus

$$\{(P_n, Q_n): n \in \mathbf{N}\} \subset \mathscr{M}$$

and by (3.16) we have

$$J(P_n, Q_n) \leqslant J((\phi_n)_\xi, (\psi_n)_\xi).$$

Therefore $\{(P_n, Q_n): n \in \mathbf{N}\}$ is also a minimizing sequence for (2.4).

From (3.1), (3.9) and Sobolev's embedding theorem, one knows that $\| P_n \|_{H^1(\mathbf{R}^N)}$ and $\| Q_n \|_{L^2(\mathbf{R}^N)}$ are both bounded for all $n \in \mathbf{N}$. Then there exists a subsequence

$$\{(P_n)_k : k \in \mathbf{N}\} \subset \{P_n : n \in \mathbf{N}\}$$

such that

$$(P_n)_k \rightharpoonup P_\infty \quad \text{weakly in } H^1(\mathbf{R}^N) \text{ as } n \to \infty.$$

Now for

$$\{(Q_n)_k : k \in \mathbf{N}\} \subset \{Q_n : n \in \mathbf{N}\}$$

there also exists a subsequence

$$\{((Q_n)_k)_m : m \in \mathbf{N}\} \subset \{(Q_n)_k : k \in \mathbf{N}\}$$

such that

$$((Q_n)_k)_m \rightharpoonup Q_\infty \quad \text{weakly in } L^2(\mathbf{R}^N) \text{ as } n \to \infty. \tag{3.17}$$

It is of course that

$$((P_n)_k)_m \rightharpoonup P_\infty \quad \text{weakly in } H^1(\mathbf{R}^N) \text{ as } n \to \infty. \tag{3.18}$$

Thus we extract a subsequence $\{(((P_n)_k)_m, ((Q_n)_k)_m) : m \in \mathbf{N}\}$ from $\{(P_n, Q_n) : n \in \mathbf{N}\}$ such that (3.17) and (3.18) hold. For simplicity, we still denote $\{(((P_n)_k)_m, ((Q_n)_k)_m) : m \in \mathbf{N}\}$ by $\{(P_n, Q_n) : n \in \mathbf{N}\}$.

Note the compactness lemma in [19]: For $2 < \delta < \frac{N+2}{N-2}$ (when $N=2, \frac{N+2}{N-2} = \infty$), the embedding $H^1_{\text{radial}}(\mathbf{R}^N) \to L^\delta(\mathbf{R}^N)$ is compact, where

$$H^1_{\text{radial}}(\mathbf{R}^N) = \{f(x) \in H^1(\mathbf{R}^N), f(x) = f(|x|) \text{ is a function of } |x| \text{ alone}\}.$$

Thus from (3.18), one gets

$$P_n \to P_\infty \quad \text{strongly in } L^4(\mathbf{R}^N) \text{ as } n \to \infty. \tag{3.19}$$

Now we assert that $(P_\infty, Q_\infty) \neq (0,0)$, which will be proved by contradiction. Otherwise, if $(P_\infty, Q_\infty) \equiv (0,0)$, then from (3.17) and (3.19) one obtains

$$\begin{aligned} &P_n \to 0 \quad \text{strongly in } L^4(\mathbf{R}^N) \text{ as } n \to \infty, \\ &Q_n \rightharpoonup 0 \quad \text{weakly in } L^2(\mathbf{R}^N) \text{ as } n \to \infty. \end{aligned} \tag{3.20}$$

Moreover, from Hölder's inequality

$$\left| \int Q_n |P_n|^2 \mathrm{d}x \right| \leqslant \left( \int |Q_n|^2 \mathrm{d}x \right)^{\frac{1}{2}} \left( \int |P_n|^4 \mathrm{d}x \right)^{\frac{1}{2}}$$

and (3.20), we get

$$\int Q_n |P_n|^2 \mathrm{d}x \to 0 \quad \text{as } n \to \infty. \tag{3.21}$$

Since $(P_n, Q_n) \in \mathscr{M}$, $K(P_n, Q_n) = 0$ and (3.21) imply that

$$2\int |\nabla P_n|^2 \mathrm{d}x + 2c^2 \int |P_n|^2 \mathrm{d}x + 2\int |P_n|^4 \mathrm{d}x + \int |Q_n|^2 \mathrm{d}x \to 0 \quad \text{as } n \to \infty$$

which together with (3.20) yields that

$$2\int |\nabla P_n|^2 \mathrm{d}x + 2c^2 \int |P_n|^2 \mathrm{d}x + \int |Q_n|^2 \mathrm{d}x \to 0 \quad \text{as } n \to \infty. \tag{3.22}$$

On the other hand, from $(P_n, Q_n) \in \mathscr{M}$ it follows that

$$2\int |\nabla P_n|^2 \mathrm{d}x + 2c^2 \int |P_n|^2 \mathrm{d}x + \int |Q_n|^2 \mathrm{d}x = -3\int Q_n |P_n|^2 \mathrm{d}x. \tag{3.23}$$

Noting Young inequality

$$\left| \int Q_n |P_n|^2 \mathrm{d}x \right| \leqslant \frac{1}{3} \int |Q_n|^2 \mathrm{d}x + \frac{3}{4} \int |P_n|^4 \mathrm{d}x$$

and (3.23), we get

$$2\int|\nabla P_n|^2\mathrm{d}x+2c^2\int|P_n|^2\mathrm{d}x+\int|Q_n|^2\mathrm{d}x\leqslant\int|Q_n|^2\mathrm{d}x+\frac{9}{4}\int|P_n|^4\mathrm{d}x,$$

namely,

$$2\int|\nabla P_n|^2\mathrm{d}x+2c^2\int|P_n|^2\mathrm{d}x\leqslant\frac{9}{4}\int|P_n|^4\mathrm{d}x,$$

which together with Sobolev's embedding theorem implies that

$$\begin{aligned}2\int|\nabla P_n|^2\mathrm{d}x+2c^2\int|P_n|^2\mathrm{d}x&\leqslant\frac{9}{4}\int|P_n|^4\mathrm{d}x\\&\leqslant\frac{9}{4}C\Big(\int|\nabla P_n|^2\mathrm{d}x+c^2\int|P_n|^2\mathrm{d}x\Big)^2\\&\leqslant C\Big(2\int|\nabla P_n|^2\mathrm{d}x+2c^2\int|P_n|^2\mathrm{d}x\Big)^2.\end{aligned}\tag{3.24}$$

From (3.24), we have

$$2\int|\nabla P_n|^2\mathrm{d}x+2c^2\int|P_n|^2\mathrm{d}x\geqslant C.$$

Thus we get

$$2\int|\nabla P_n|^2\mathrm{d}x+2c^2\int|P_n|^2\mathrm{d}x+\int|Q_n|^2\mathrm{d}x\geqslant C,$$

which is contradictory with (3.22). So $(P_\infty,Q_\infty)\neq(0,0)$.

Now we take

$$P=\overline{P}_\xi=((P_\infty)_\xi)_\eta,\quad Q=\overline{Q}_\xi=((Q_\infty)_\xi)_\eta\tag{3.25}$$

with

$$\begin{cases}\xi=\lambda_1\text{ or }\xi\in[\lambda_2,\infty), & \text{if (I-1) in Lemma 3.2 holds}\\ \xi=\lambda_2\text{ or }\xi\in(\lambda_2,\infty), & \text{if (I-2) in Lemma 3.2 holds}\\ \xi=\lambda_2, & \text{if (I-3) in Lemma 3.2 holds}\\ \xi=\lambda_1, & \text{if (I-4) in Lemma 3.2 holds}\\ \xi\in(0,\infty), & \text{if (I-5) in Lemma 3.2 holds}\\ \xi=\lambda_2\text{ or }\xi\in(0,\lambda_2], & \text{if (I-6) in Lemma 3.2 holds}\\ \xi\in(0,\infty), & \text{if (I-7) in Lemma 3.2 holds}\\ \xi=\lambda_1\text{ or }\xi\in(0,\lambda_1], & \text{if (I-8) in Lemma 3.2 holds}\\ \xi\in(0,\infty), & \text{if (II-1) in Lemma 3.2 holds}\\ \xi=\lambda^*\text{ or }\xi\in(0,\lambda^*), & \text{if (II-2) in Lemma 3.2 holds}\\ \xi=\lambda^*, & \text{if (II-3) in Lemma 3.2 holds}\\ \xi=\lambda^*\text{ or }\xi\in[\lambda^*,\infty), & \text{if (II-4) in Lemma 3.2 holds}\end{cases},$$

such that

$$K(\overline{P}_\xi,\overline{Q}_\xi)=K[((P_\infty)_\xi)_\eta,((Q_\infty)_\xi)_\eta]=0,\tag{3.26}$$

where

$$\begin{cases} \eta \in (0,\lambda_2] \text{ or } \eta = \lambda_2, & \text{if ( I -1) in Lemma 3.2 holds} \\ \eta \in (0,\lambda_2] \text{ or } \eta = \lambda_2, & \text{if ( I -2) in Lemma 3.2 holds} \\ \eta \in (0,\infty), & \text{if ( I -3) in Lemma 3.2 holds} \\ \eta \in (0,\infty), & \text{if ( I -4) in Lemma 3.2 holds} \\ \eta = \lambda_1, & \text{if ( I -5) in Lemma 3.2 holds} \\ \eta \in (\lambda_2,\infty) \text{ or } \eta = \lambda_1, & \text{if ( I -6) in Lemma 3.2 holds} \\ \eta = \lambda_2, & \text{if ( I -7) in Lemma 3.2 holds} \\ \eta \in [\lambda_1,\infty) \text{ or } \eta = \lambda_1, & \text{if ( I -8) in Lemma 3.2 holds} \\ \eta = \lambda^*, & \text{if ( II -1) in Lemma 3.2 holds} \\ \eta \in [\lambda^*,\infty) \text{ or } \eta = \lambda^*, & \text{if ( II -2) in Lemma 3.2 holds} \\ \eta \in (0,\infty), & \text{if ( II -3) in Lemma 3.2 holds} \\ \eta \in (0,\lambda^*] \text{ or } \eta = \lambda^*, & \text{if ( II -4) in Lemma 3.2 holds} \end{cases}.$$

By (3.17), (3.18) and (3.19), thus one gets

$$\begin{cases} ((P_n)_\xi)_\eta \to \overline{P}_\xi & \text{strongly in } L^4(\mathbf{R}^N) \text{ as } n \to \infty, \\ ((P_n)_\xi)_\eta \to \overline{P}_\xi & \text{weakly in } H^1(\mathbf{R}^N) \text{ as } n \to \infty, \\ ((Q_n)_\xi)_\eta \to \overline{Q}_\xi & \text{weakly in } L^2(\mathbf{R}^N) \text{ as } n \to \infty. \end{cases} \tag{3.27}$$

Since $K((P_n)_\xi, (Q_n)_\xi)=0$, Lemma 3.2 shows that

$$J[((P_n)_\xi)_\eta,((Q_n)_\xi)_\eta] \leqslant J[(P_n)_\xi,(Q_n)_\xi]. \tag{3.28}$$

Hence, using (3.26) and (3.27) one has

$$J(\overline{P}_\xi,\overline{Q}_\xi) \leqslant \varliminf_{n\to\infty} J[((P_n)_\xi)_\eta,((Q_n)_\xi)_\eta] \leqslant \varliminf_{n\to\infty} J[(P_n)_\xi,(Q_n)_\xi] = \inf_{(\phi,\psi)\in\mathscr{M}} J(\phi,\psi). \tag{3.29}$$

As $(\overline{P}_\xi,\overline{Q}_\xi)\neq(0,0)$ and $K(\overline{P}_\xi,\overline{Q}_\xi)=0$, one has $(\overline{P}_\xi,\overline{Q}_\xi)\in\mathscr{M}$. Therefore, from (3.25) and (3.29), $(P,Q)=(\overline{P}_\xi,\overline{Q}_\xi)$ solves the minimization problem

$$J(P,Q) = \min_{(\phi,\psi)\in\mathscr{M}} J(\phi,\psi). \qquad \square \tag{3.30}$$

From Lemma 3.1 and Theorem 3.1, we get an important conclusion—Proposition 3.1 which is key throughout this paper.

**Proposition 3.1** $d_{\mathscr{M}}>0$.

At the end of this section, we get a conclusion on the existence of a standing wave for the Klein-Gordon-Zakharov system (1.1)~(1.3), which will be helpful to prove the modified instability of the standing wave for the system (1.1)~(1.3) in Section 5.

**Theorem 3.2** Assume $(P,Q)$ is a solution of the problem (3.30). Then $Q=-|P|^2$ and $(P,Q)$ is a ground state solution of the following system:

$$\begin{cases} -\Delta P + c^2 P = -PQ - |P|^2 P, & x \in \mathbf{R}^N \\ Q + |P|^2 = 0, & x \in \mathbf{R}^N \end{cases}. \tag{3.31}$$

Moreover, $(u(t,x),n(t,x))=(P(x),Q(x))$ is a standing wave solution for the system (1.1)~(1.3).

**Proof** Since $(P,Q)$ is a solution of the problem (3.30), there exists a Lagrange multiplier $\Lambda$ such that

$$\delta_P[J + \Lambda K] = 0, \quad \delta_Q[J + \Lambda K] = 0, \tag{3.32}$$

where $\delta_\phi G$ denotes the variation of $G(\phi, \psi)$ about $\phi$. By the formula

$$\delta_\phi G(\phi, \psi) = \frac{\partial}{\partial \eta} G(\phi + \eta\delta\phi, \psi)|_\eta = 0,$$

we get

$$\delta_\phi[J + \Lambda K] = (2 + 4\Lambda)\int(-\Delta\phi \cdot \delta\phi + c^2\phi\delta\phi)\mathrm{d}x + (2 + 8\Lambda)\int(|\phi|^2\phi\delta\phi)\mathrm{d}x + (2 + 6\Lambda)\int\phi\psi\delta\phi\mathrm{d}x, \tag{3.33}$$

$$\delta_\psi[J + \Lambda K] = (1 + 2\Lambda)\int\psi\delta\psi\mathrm{d}x + (1 + 3\Lambda)\int|\phi|^2\delta\psi\mathrm{d}x \tag{3.34}$$

where $\delta\phi$ denotes the variation of $\phi$. By (3.32) one has

$$(2 + 4\Lambda)\int(|\nabla P|^2 + c^2|P|^2)\mathrm{d}x + (2 + 8\Lambda)\int|P|^4\mathrm{d}x + (2 + 6\Lambda)\int|P|^2Q\mathrm{d}x = 0, \tag{3.35}$$

$$(1 + 2\Lambda)\int|Q|^2\mathrm{d}x + (1 + 3\Lambda)\int|P|^2Q\mathrm{d}x = 0, \tag{3.36}$$

From $K(P,Q)=0$ it follows that

$$2\int|\nabla P|^2\mathrm{d}x + 2c^2\int|P|^2\mathrm{d}x + 2\int|P|^4\mathrm{d}x + \int|Q|^2\mathrm{d}x + 3\int Q|P|^2\mathrm{d}x = 0. \tag{3.37}$$

By (3.35), (3.36) and (3.37), one gets $\Lambda=0$.

Thus (3.35) and (3.36) yield that

$$2\int(\nabla P\,\nabla\bar{P})\mathrm{d}x + 2c^2\int P\bar{P}\mathrm{d}x + 2\int|P|^2P\bar{P}\mathrm{d}x + 2\int P\bar{P}Q\mathrm{d}x = 0, \tag{3.38}$$

$$\int Q^2\mathrm{d}x + \int|P|^2Q\mathrm{d}x = 0. \tag{3.39}$$

Thus (3.38) and (3.39) conclude that

$$\begin{cases} -\Delta P + c^2P = -PQ - |P|^2P, & x \in \mathbf{R}^N \\ Q + |P|^2 = 0, & x \in \mathbf{R}^N, \end{cases}$$

that is,

$$\begin{cases} -\Delta P + c^2P = 0, & x \in \mathbf{R}^N \\ Q = -|P|^2, & x \in \mathbf{R}^N. \end{cases}$$

Putting $(u(t,x), n(t,x)) = (P(x), Q(x))$ into the system (1.1)~(1.3), we can obtain easily that $(P, Q)$ is a standing wave solution for the system (1.1)~(1.3).

This completes the proof of Theorem 3.2. □

## 4 Blowup and Global Existence

In this section, we discuss global existence and blowup for the solution to the Cauchy problem (1.1)~(1.4). Firstly, we give a sharp threshold of global existence and blowup for the solution $(u(t,x),n(t,x),V(t,x))$ to the Cauchy problem (1.1)~(1.4) in terms of the relationship between the initial energy $\varepsilon(u_0,u_1,n_0,v_0)$ and $d_{\mathcal{M}}$, namely

**Theorem 4.1** Let $(u_0,u_1,n_0,v_0)\in H^1(\mathbf{R}^N)\times L^2(\mathbf{R}^N)\times L^2(\mathbf{R}^N)\times L^2(\mathbf{R}^N)$ and satisfy

$$\varepsilon(u_0,u_1,n_0,v_0)<d_{\mathcal{M}}. \tag{4.1}$$

Then:

(i) If

$$K(u_0,n_0)<0, \tag{4.2}$$

then the solution $(u(t,x),n(t,x),V(t,x))$ of the Cauchy problem (1.1)~(1.4) blows up infinite time, that is, there exists $T>0$ such that

$$\lim_{t\to T}(\|u\|^2_{L^2(\mathbf{R}^N)}+\|n\|^2_{\dot{H}^{-1}(\mathbf{R}^N)})=\infty.$$

(ii) If

$$K(u_0,n_0)>0, \tag{4.3}$$

then the solution $(u(t,x),n(t,x),V(t,x))$ of the Cauchy problem (1.1)~(1.4) exists globally on $t\in[0,\infty)$. Moreover, $(u(t,x),n(t,x),V(t,x))$ satisfies

$$c^{-2}\|u_t\|^2_{L^2(\mathbf{R}^N)}+\frac{1}{2}\alpha^{-2}\|V\|^2_{L^2(\mathbf{R}^N)}+\|u\|^2_{H^1_c(\mathbf{R}^N)}+\frac{1}{2}\|n\|^2_{L^2(\mathbf{R}^N)}<d_{\mathcal{M}}, \tag{4.4}$$

or

$$c^{-2}\|u_t\|^2_{L^2(\mathbf{R}^N)}+\frac{1}{2}\alpha^{-2}\|V\|^2_{L^2(\mathbf{R}^N)}+\frac{4}{3}\|u\|^2_{H^1_c(\mathbf{R}^N)}+\frac{1}{6}\|n\|^2_{L^2(\mathbf{R}^N)}<2d_{\mathcal{M}}+\frac{1}{6}c^*d^2_{\mathcal{M}}. \tag{4.5}$$

Here and hereafter, for simplicity, we denote $\|u\|^p_{L^p(\mathbf{R}^N)}=\int|u|^p\,\mathrm{d}x$ and $c^*$ is the positive constant which satisfies the Sobolev's inequality

$$\|u\|^4_{L^4(\mathbf{R}^N)}\leqslant C\|u\|^4_{H^1_c(\mathbf{R}^N)}.$$

**Remark 4.1** Unlike the Klein-Gotdon-Zakhatov system (1.5)~(1.6), for the Klein-Gordon-Zakharov system (1.1)~(1.3), we know that the initial energy $\varepsilon(u_0,u_1,n_0,v_0)$ is always nonnegative from the inequality

$$\left|\int n_0|u_0|^2\mathrm{d}x\right|\leqslant\frac{1}{2}\int|n_0|^2\mathrm{d}x+\frac{1}{2}\int_{\mathbf{R}^2}|u_0|^4\mathrm{d}x.$$

But $d_{\mathcal{M}}>0$ (Proposition 3.1) can also guarantee the set

$$\{(u_0,u_1,n_0,v_0)\in H^1(\mathbf{R}^N)\times L^2(\mathbf{R}^N)\times L^2(\mathbf{R}^N)\times L^2(\mathbf{R}^N):\varepsilon(u_0,u_1,n_0,v_0)<d_{\mathcal{M}}\}$$

is not empty.

From Theorem 4.1, we can obtain two corollaries based on the relationship between $\int n_0|u_0|^2\mathrm{d}x$ and 0.

When $\int n_0|u_0|^2\mathrm{d}x<0$, we get

**Corollary 4.1** (Small data criterion I) If $(u_0, u_1, n_0, v_0) \in H^1(\mathbf{R}^N) \times L^2(\mathbf{R}^N) \times L^2(\mathbf{R}^N) \times L^2(\mathbf{R}^N)$ and satisfies

$$\| u_0 \|^2_{H^1_c(\mathbf{R}^N)} + \frac{1}{2} c^* \| u_0 \|^4_{H^1_c(\mathbf{R}^N)} + \frac{1}{2} \| n_0 \|^2_{L^2(\mathbf{R}^N)} + \frac{1}{2} \alpha^{-2} \| v_0 \|^2_{L^2(\mathbf{R}^N)} + c^{-2} \| u_1 \|^2_{L^2(\mathbf{R}^N)} < d_{\mathcal{M}}, \tag{4.6}$$

and

$$\int n_0 | u_0 |^2 \mathrm{d}x < 0, \tag{4.7}$$

then the solution $(u(t,x), n(t,x), V(t,x))$ of the Cauchy problem (1.1)~(1.4) exists globally. Here, $c^*$ is the same as that appeared in Theorem 4.1.

When $\int n_0 | u_0 |^2 \mathrm{d}x > 0$, we get

**Corollary 4.2** (Small data criterion II) If $(u_0, u_1, n_0, v_0) \in H^1(\mathbf{R}^N) \times L^2(\mathbf{R}^N) \times L^2(\mathbf{R}^N) \times L^2(\mathbf{R}^N)$ and satisfies

$$\| u_0 \|^2_{H^1_c(\mathbf{R}^N)} + c^* \| u_0 \|^4_{H^1_c(\mathbf{R}^N)} + \| n_0 \|^2_{L^2(\mathbf{R}^N)} + \frac{1}{2} \alpha^{-2} \| v_0 \|^2_{L^2(\mathbf{R}^N)} + c^{-2} \| u_1 \|^2_{L^2(\mathbf{R}^N)} < d_{\mathcal{M}}, \tag{4.8}$$

and

$$\int n_0 | u_0 |^2 \mathrm{d}x > 0, \tag{4.9}$$

then the solution $(u(t,x), n(t,x), V(t,x))$ of the Cauchy problem (1.1)~(1.4) exists globally. Here, $c^*$ is the same as that appeared in Theorem 4.1.

**Remark 4.2** In fact, on the one hand, Corollaries 4.1 and 4.2 give two sufficient conditions of global existence for the solution to the Cauchy problem (1.1)~(1.4). On the other hand, the two corollaries answer the question: how small are the initial data $(u_0, u_1, n_0, v_0)$, the solution to the Cauchy problem (1.1)~(1.4) exists globally?

**Remark 4.3** For the Klein-Gordon-Zakharov system (1.1)~(1.3) with the absence of the high-degree nonlinearity $-|u|^2 u$, one can obtain the same result as Theorem 4.1 and using the result, one can also get that when the initial energy $\varepsilon(u_0, u_1, n_0, v_0) < 0$, the solution of the Cauchy problem (1.1)~(1.4) blows up in finite time. The similar result cannot be obtained in the present paper on account of the presence of the high-degree nonlinearity $-|u|^2 u$ such that the initial energy $\varepsilon(u_0, u_1, n_0, v_0)$ is always nonnegative from Remark 4.1. But from Corollaries 4.1 and 4.2, we get two results on the global existence with small data of the Cauchy problem for the Klein-Gordon-Zakharov system (1.1)~(1.3). This kind of results here are more meticulous than those appeared in the Klein-Gordon-Zakharov system without the high-degree nonlinearity $-|u|^2 u$.

Before we prove Theorem 4.1, Corollaries 4.1 and 4.2, we first give some preliminaries.

Rewrite the energy functional $\varepsilon$ as

$$\varepsilon(\phi_1, \phi_2, \psi_1, \psi_2) = c^{-2} \int | \phi_2 |^2 \mathrm{d}x + \frac{1}{2} \alpha^{-2} \int | \psi_2 |^2 \mathrm{d}x + J(\phi_1, \psi_1). \tag{4.10}$$

Now we define a set $S$ as

$$S := \{(\phi_1,\phi_2,\psi_1,\psi_2) \in H^1(\mathbf{R}^N) \times L^2(\mathbf{R}^N) \times L^2(\mathbf{R}^N) \times L^2(\mathbf{R}^N) : \varepsilon(\phi_1,\phi_2,\psi_1,\psi_2) < d_{\mathcal{M}}\}.$$

Furthermore, we introduce two invariant sets for the Cauchy problem (1.1)~(1.4) as

$$S_1 := \{(\phi_1,\phi_2,\psi_1,\psi_2) \in S \mid K(\phi_1,\psi_1) > 0\} \cup \{(0,\phi_2,0,\psi_2) \in S\},$$

$$S_2 := \{(\phi_1,\phi_2,\psi_1,\psi_2) \in S \mid K(\phi_1,\psi 1) < 0\}.$$

The fundamental property of the two sets $S_1$ and $S_2$ is the following.

**Proposition 4.1** $S_1$ and $S_2$ are invariant regions under the solution flow generated by the Cauchy problem (1.1)~(1.4).

**Proof** First we show $S_1$ is an invariant region. Let $(u_0, u_1, n_0, v_0) \in S_1$ and assume that there exists a $t^*$ such that $(u(t^*), u_t(t^*), n(t^*), V(t^*)) \notin S_1$. Then $(u(t^*), n(t^*)) \neq (0,0)$ and $K(u(t^*), n(t^*)) \leqslant 0$, i.e., $(u(t^*), n(t^*)) \in \mathcal{M}^-$. Let

$$s = \inf\{0 \leqslant t \leqslant t^* \mid (u(t), u_t(t), n(t), V(t)) \notin S_1\}, \tag{4.11}$$

then $K(u(t), n(t)) \geqslant 0$ for $0 \leqslant t < s$. Let $\{s_k\}$ be the minimizing sequence for problem (4.11), then $(u(s_k), n(s_k)) \in \mathcal{M}^-$ and by weak lower semi-continuity of $K(u(\cdot), n(\cdot))$, we have

$$K(u(s), n(s)) \leqslant \liminf_{k \to \infty} K(u(s_k), n(s_k)) \leqslant 0,$$
$$(u(s), n(s)) \neq (0,0). \tag{4.12}$$

On the other hand, from (1.8), (1.9) and (4.10) it follows that

$$\begin{aligned} I(u(s), n(s)) &= \liminf_{t \to s^-} I(u(t), n(t)) \\ &\leqslant \liminf_{t \to s^-} \left( I(u(t), n(t)) + \frac{1}{v+1} K(u(t), n(t)) \right) \\ &= \liminf_{t \to s^-} J(u(t), n(t)) \\ &\leqslant \liminf_{t \to s^-} \varepsilon(u(t), u_t(t), n(t), V(t)) \\ &< d_{\mathcal{M}}, \end{aligned}$$

which contradicts (4.12) from Proposition 2.1 and (2.7). Thus $S_1$ is invariant.

Next we show $S_2$ is also an invariant region.

Let $(u_0, u_1, n_0, v_0) \in S_2$ and assume that there exists a $\tau$ such that $(u(\tau), u_t(\tau), n(\tau), V(\tau)) \notin S_2$, from (1.8), (1.9), (2.5) and (4.10) it follows that $I(u(\tau), n(\tau)) \leqslant d_{\mathcal{M}}$. Let

$$s = \inf\{0 \leqslant t \leqslant \tau \mid (u(t), u_t(t), n(t), V(t)) \notin S_2\}. \tag{4.13}$$

then $I(u(s), n(s)) \leqslant d_{\mathcal{M}}$ and $I(u(t), n(t)) > d_{\mathcal{M}}$ for au $0 < t < s$ from Proposition 2.1.

On the other hand, from (2.5) we have

$$\begin{aligned} K(u(s), n(s)) &= \liminf_{t \to s^-} (v+1)(J(u(t), n(t)) - I(u(t), n(t))) \\ &\leqslant \liminf_{t \to s^-} (v+1)(\varepsilon(u(t), u_t(t), n(t), V(t)) - d_{\mathcal{M}}) \\ &\leqslant (v+1)(\varepsilon(u_0, u_1, n_0, v_0) - d_{\mathcal{M}}) \\ &< 0, \end{aligned}$$

which contradicts $I(u(s),n(s))\leqslant d_{\mathcal{M}}$ from Proposition 2.1 (if $K(\phi,\psi)<0$, then $I(\phi,\psi)>d_{\mathcal{M}}$). Therefore $S_2$ is also invariant. □

Now we begin to prove Theorem 4.1, Corollaries 4.1 and 4.2 by utilizing Payne and Sattinger's potential well argument[15] and Levine's concavity method[8] as well as introducing suitable dilation transformation.

First we prove Theorem 4.1.

**Proof of Theorem 4.1** From (1.9), (2.1) and (4.1), it follows that

$$J(u_0,n_0)\leqslant \varepsilon(u_0,u_1,n_0,v_0) < d_{\mathcal{M}}. \tag{4.14}$$

We first prove (ⅰ) of Theorem 4.1.

Ⅰ. (4.2), (4.14) and Proposition 4.1 imply that $(u(t,x),u_t(t,x),n(t,x),V(t,x))\in S_2$. Thus from (4.10) we get

$$K(u(t,x),n(t,x)) < 0 \quad \text{for } t\in[0,T) \tag{4.15}$$

and

$$J(u(t,x),n(t,x)) < d_{\mathcal{M}}. \tag{4.16}$$

Since $(u(t,x),n(t,x),V(t,x))$ is a solution of the Cauchy problem (1.1)~(1.4) on $[0,T)$, by assumption (A1), we put

$$F_{(c,\alpha)}(t) = 2c^{-2}\int |u|^2\mathrm{d}x + \alpha^{-2}\int |g|^2\mathrm{d}x. \tag{4.17}$$

Thus we get

$$F'_{(c,\alpha)}(t) = \int[2c^{-2}(u_tu^* + uu_t^*) + 2\alpha^{-2}gg_t]\mathrm{d}x, \tag{4.18}$$

where $u^*$ is the complex conjugate of $u$. By computing $F''_{(c,\alpha)}(t)$ we obtain

$$\begin{aligned}
F''_{(c,\alpha)}(t) &= 4c^{-2}\int |u_t|^2\mathrm{d}x + 4c^{-2}\mathrm{Re}\int u_{tt}u^*\mathrm{d}x + 2\alpha^{-2}\int |g_t|^2\mathrm{d}x + 2\alpha^{-2}\int gg_{tt}\mathrm{d}x \\
&= 4c^{-2}\int |u_t|^2\mathrm{d}x + 2\alpha^{-2}\int |V|^2\mathrm{d}x - 4c^2\int |u|^2\mathrm{d}x - 4\int |\nabla u|^2\mathrm{d}x - \\
&\quad 4\int |u|^4\mathrm{d}x - 4\int n|u|^2\mathrm{d}x + 2\int g(-\nabla n - \nabla|u|^2)\mathrm{d}x \\
&= 4c^{-2}\int |u_t|^2\mathrm{d}x + 2\alpha^{-2}\int |V|^2\mathrm{d}x - 4c^2\int |u|^2\mathrm{d}x - 4\int |\nabla u|^2\mathrm{d}x - \\
&\quad 4\int |u|^4\mathrm{d}x - 4\int n|u|^2\mathrm{d}x + 2\int \nabla\cdot g(n + |u|^2)\mathrm{d}x \\
&= 4c^{-2}\int |u_t|^2\mathrm{d}x + 2\alpha^{-2}\int |V|^2\mathrm{d}x - 4c^2\int |u|^2\mathrm{d}x - 4\int |\nabla u|^2\mathrm{d}x - \\
&\quad 4\int |u|^4\mathrm{d}x - 4\int n|u|^2\mathrm{d}x - 2\int |n|^2\mathrm{d}x - 2\int n|u|^2\mathrm{d}x \\
&= 4c^{-2}\int |u_t|^2\mathrm{d}x + 2\alpha^{-2}\int |V|^2\mathrm{d}x - 4c^2\int |u|^2\mathrm{d}x - 4\int |\nabla u|^2\mathrm{d}x - \\
&\quad 4\int |u|^4\mathrm{d}x - 2\int |n|^2\mathrm{d}x - 6\int n|u|^2\mathrm{d}x \\
&= 2\left(2c^{-2}\int |u_t|^2\mathrm{d}x + \alpha^{-2}\int |V|^2\mathrm{d}x\right) - 2K(u,n).
\end{aligned} \tag{4.19}$$

In view of (4.15), one has

$$F''_{(c,\alpha)}(t) > 0 \quad \text{for } t \in [0, T). \tag{4.20}$$

By (1.8)~(1.10), (4.17) and (4.19), one gets

$$F''_{(c,\alpha)}(t) = 5\left(2c^{-2}\int |u_t|^2 dx + \alpha^{-2}\int |V|^2 dx\right) + 2c^2\int |u|^2 dx + 2\int |\nabla u|^2 dx + \int |n|^2 dx - \int |u|^4 dx - 6\varepsilon(u_0, u_1, n_0, v_0). \tag{4.21}$$

Because $K(u, n) < 0$, we have

$$2c^2\int |u|^2 dx + 2\int |\nabla u|^2 dx + 2\int |u|^4 dx + \int |n|^2 dx < -3\int n|u|^2 dx \leqslant \frac{3}{2}\int |u|^4 dx + \frac{3}{2}\int |n|^2 dx,$$

which implies that

$$0 < 2c^2\int |u|^2 dx + 2\int |\nabla u|^2 dx < \frac{1}{2}\int |n|^2 dx - \frac{1}{2}\int |u|^4 dx. \tag{4.22}$$

Since $F_{(c,\alpha)}(t)$ is a convex function of $t$ by (4.20), it follows that if there exists a time $t_1$ such that $F'_{(c,\alpha)}(t) > 0$, then $F_{(c,\alpha)}(t)$ is increasing for all $t > t_1$ (within the interval of existence). Now we give a remark before we continue to prove (ⅰ) of Theorem 4.1.

**Remark 4.4** If $F_{(c,\alpha)}(t)$ is increasing for all $t > t_1$, then we can conclude that $\int |u|^2 dx$ is increasing for all $t > t_1$. In fact, let

$$G_c(t) = c^{-2}\int |u|^2 dx.$$

We get

$$G'_c(t) = c^{-2}\int (u_t u^* + u u_t^*) dx,$$

$$\begin{aligned}
G''_c(t) &= 2c^{-2}\int |u_t|^2 dx - 2c^2\int |u|^2 dx - 2\int |\nabla u|^2 dx - 2\int |u|^4 dx - 2\int n|u|^2 dx \\
&= 2c^{-2}\int |u_t|^2 dx - K(u,n) + \int n|u|^2 dx + \int |n|^2 dx \\
&= 2c^{-2}\int |u_t|^2 dx - K(u,n) - \left(-\int n|u|^2 dx\right) + \int |n|^2 dx \\
&\geqslant 2c^{-2}\int |u_t|^2 dx - K(u,n) + \int |n|^2 dx - \left(\frac{1}{2}\int |n|^2 dx + \frac{1}{2}|u|^4 dx\right) \\
&= 2c^{-2}\int |u_t|^2 dx - K(u,n) + \frac{1}{2}\int |n|^2 dx - \frac{1}{2}\int |u|^4 dx.
\end{aligned}$$

By (4.22) and $K(u, n) < 0$, the above inequality implies $G''_c(t) > 0$. So $G'_c(t)$ is strictly increasing for all $t > 0$. Thus if we choose $(u_0, u_1)$ properly such that

$$G'_c(0) = c^{-2}\int (u_1 u_0^* + u_0 u_1^*) dx \geqslant 0, \tag{R1}$$

then for all $t > 0$, $G'_c(t) > 0$. Therefore, $G_c(t) = \int |u|^2 dx$ is strictly increasing for all $t > 0$. Without loss of generality, for simplicity, we omit the condition (R1) in the present paper and assume that if $F_{(c,\alpha)}(t)$ is increasing for all $t>t_1$, then $\int |u|^2 dx$ is increasing for all $t>t_1$.

Now we continue to our proof. From Remark 4.4 and (4.22), we get the quantity

$$2c^2\int |u|^2 dx + 2\int |\nabla u|^2 dx + \int |n|^2 dx - \int |u|^4 dx - 6\varepsilon(u_0, u_1, n_0, v_0)$$

will eventually become positive and will remain positive thereafter. Thus form (4.21) and (4.22), we would have

$$F''_{(c,\alpha)}(t) \geqslant 5\left(2c^{-2}\int |u_t|^2 dx + \alpha^{-2}\int |V|^2 dx\right). \tag{4.23}$$

In view of (4.17), (4.18) and (4.23), using Hölder's inequality, we get

$$\begin{aligned} F_{(c,\alpha)}(t)F''_{(c,\alpha)}(t) &\geqslant 5\left(2c^{-2}\int |u|^2 dx + \alpha^{-2}\int |g|^2 dx\right)\left(2c^{-2}\int |u_t|^2 dx + \alpha^{-2}\int |V|^2 dx\right) \\ &= 5\Big(4c^{-4}\int |u^*|^2 dx\int |u_t|^2 dx + 2c^{-2}\alpha^{-2}\int |u|^2 dx\int |V|^2 dx + \\ &\quad 2c^{-2}\alpha^{-2}\int |u_t|^2 dx\int |g|^2 dx + \alpha^{-4}\int |g|^2 dx\int |V|^2 dx\Big) \\ &\geqslant 5\left(4\left(c^{-2}\int |u^* u_t| dx\right)^2 + 2c^{-2}\alpha^{-2}\left(\int |uV| dx\right)^2\right) + \\ &\quad 5\left(2c^{-2}\alpha^{-2}\left(\int |u_t g| dx\right)^2 + \left(\alpha^{-2}\int |gV| dx\right)^2\right) \\ &\geqslant \frac{5}{4}(F'_{(c,\alpha)}(t))^2. \end{aligned} \tag{4.24}$$

Since

$$\left(F^{-\frac{1}{4}}_{(c,\alpha)}(t)\right)'' = -\frac{1}{4}F^{-\frac{9}{4}}_{(c,\alpha)}(t)\left(F_{(c,\alpha)}(t)F''_{(c,\alpha)}(t) - \frac{5}{4}(F'_{(c,\alpha)}(t))^2\right), \tag{4.25}$$

from (4.24) we get

$$\left(F^{-\frac{1}{4}}_{(c,\alpha)}(t)\right)'' \leqslant 0. \tag{4.26}$$

Therefore $F^{-\frac{1}{4}}_{(c,\alpha)}(t)$ is concave for sufficiently large $t$ and there exists a finite time $T^*$ such that

$$\lim_{t\to T^*} F^{-\frac{1}{4}}_{(c,\alpha)}(t) = 0.$$

In other words,

$$\lim_{t\to T^*} F_{(c,\alpha)}(t) = \infty.$$

Thus one has $T<\infty$ and

$$\lim_{t\to T}\left(2c^{-2}\int |u|^2 dx + \alpha^{-2}\int |g|^2 dx\right) = \infty.$$

From assumption (A1), (1.1)~(1.3) and (1.10), we get

$$\lim_{t\to T}\left(2c^{-2}\int |u|^2 dx + \int \left| |\alpha\nabla|^{-1} n\right|^2 dx\right) = \infty.$$

The proof of (i) of Theorem 4.1 will be completed once we have shown that $F'_{(c,\alpha)}(t)>0$ for some $t$. We prove it by contradiction. Suppose

$$F'_{(c,\alpha)}(t) \leqslant 0 \quad \text{for all } t. \tag{4.27}$$

Since $F_{(c,\alpha)}(t)$ is convex from (4.20), $F_{(c,\alpha)}(t)$ must tend to a finite, nonnegative limit $L$ as $t \to \infty$. By Proposition 4.1, we assert $L > 0$. Therefore one has, as $t \to \infty$, $F_{(c,\alpha)} \to L > 0$, $F'_{(c,\alpha)}(t) \to 0$ and $F''_{(c,\alpha)}(t) \to 0$. Thus (4.19) and $K(u,n) < 0$ yield

$$\lim_{t\to\infty} 2\left(2c^{-2}\int |u_t|^2 \mathrm{d}x + \alpha^{-2}\int |V|^2 \mathrm{d}x\right) = 0 \tag{4.28}$$

and

$$\lim_{t\to\infty} K(u,n) = 0. \tag{4.29}$$

Now for any fixed $t > 0$, because of $K(u,n) < 0$, there exists a $0 < \lambda < 1$ such that $K(\lambda^{\frac{1}{2}}u, \lambda n) = 0$ and $(\lambda^{\frac{1}{2}}u, \lambda n) \neq (0,0)$. Therefore by (2.2) we obtain

$$-\frac{3}{2}\lambda^2\int n|u|^2\mathrm{d}x = \lambda\left(c^2\int |u|^2\mathrm{d}x + \int |\nabla u|^2\mathrm{d}x\right) + \lambda^2\int |u|^4\mathrm{d}x + \frac{1}{2}\lambda^2\int |n|^2\mathrm{d}x \tag{4.30}$$

and

$$\int n|u|^2\mathrm{d}x < 0. \tag{4.31}$$

Thus by (2.4), we get

$$J(\lambda^{\frac{1}{2}}u, \lambda n) \geqslant d_{\mathcal{M}}. \tag{4.32}$$

From (4.30) and (4.31) it follows that

$$\begin{aligned} J(\lambda^{\frac{1}{2}}u, \lambda n) &= \lambda\left(c^2\int |u|^2\mathrm{d}x + \int |\nabla u|^2\mathrm{d}x\right) + \frac{1}{2}\lambda^2\int |u|^4\mathrm{d}x + \frac{1}{2}\lambda^2\int |n|^2\mathrm{d}x + \lambda^2\int n|u|^2\mathrm{d}x \\ &= -\frac{3}{2}\lambda^2\int n|u|^2\mathrm{d}x - \frac{1}{2}\lambda^2\int |u|^4\mathrm{d}x + \lambda^2\int n|u|^2\mathrm{d}x \\ &= -\frac{1}{2}\lambda^2\int n|u|^2\mathrm{d}x - \frac{1}{2}\lambda^2\int |u|^4\mathrm{d}x. \end{aligned} \tag{4.33}$$

So

$$\begin{aligned} &J(u,n) - J(\lambda^{\frac{1}{2}}u, \lambda n) \\ &= c^2\int |u|^2\mathrm{d}x + \int |\nabla u|^2\mathrm{d}x + \frac{1}{2}\int |u|^4\mathrm{d}x + \frac{1}{2}\int |n|^2\mathrm{d}x + \int n|u|^2\mathrm{d}x + \\ &\quad \frac{1}{2}\lambda^2\int n|u|^2\mathrm{d}x + \frac{1}{2}\lambda^2\int |u|^4\mathrm{d}x \\ &= c^2\int |u|^2\mathrm{d}x + \int |\nabla u|^2\mathrm{d}x + \int |u|^4\mathrm{d}x + \frac{1}{2}\int |n|^2\mathrm{d}x + \int n|u|^2\mathrm{d}x - \frac{1}{2}\int |u|^4\mathrm{d}x + \\ &\quad \frac{1}{2}\lambda^2\int n|u|^2\mathrm{d}x + \frac{1}{2}\lambda^2\int |u|^4\mathrm{d}x. \end{aligned} \tag{4.34}$$

On the other hand, from (4.29) and (3.5) we get as $t \to \infty$,

$$\frac{1}{2}\int |u|^4\mathrm{d}x < \frac{1}{2}\int |n|^2\mathrm{d}x \tag{4.35}$$

which together with (2.2) and (4.29) yields

$$\int |u|^4\mathrm{d}x \leqslant -\int n|u|^2\mathrm{d}x \leqslant \frac{1}{2}\int |u|^4\mathrm{d}x + \frac{1}{2}\int |n|^2\mathrm{d}x. \tag{4.36}$$

By (4.35) and (4.36) we get for $0<\lambda<1$,

$$\frac{1}{2}(\lambda^2-1)\int|u|^4\mathrm{d}x\geqslant\frac{1}{2}(\lambda^2-1)\left(-\int n|u|^2\mathrm{d}x\right).$$

So

$$\frac{1}{2}\lambda^2\int|u|^4\mathrm{d}x-\frac{1}{2}\int|u|^4\mathrm{d}x+\frac{1}{2}\lambda^2\int n|u|^2\mathrm{d}x\geqslant\frac{1}{2}\int n|u|^2\mathrm{d}x. \tag{4.37}$$

Thus from (4.29), (4.34) and (4.37) we obtain as $t\to\infty$,

$$\begin{aligned}&J(u,n)-J(\lambda^{\frac{1}{2}}u,\lambda n)\\&\geqslant c^2\int|u|^2\mathrm{d}x+\int|\nabla u|^2\mathrm{d}x+\int|u|^4\mathrm{d}x+\frac{1}{2}\int|n|^2\mathrm{d}x+\int n|u|^2\mathrm{d}x+\frac{1}{2}\int n|u|^2\mathrm{d}x\\&=c^2\int|u|^2\mathrm{d}x+\int|\nabla u|^2\mathrm{d}x+\int|u|^4\mathrm{d}x+\frac{1}{2}\int|n|^2\mathrm{d}x+\frac{3}{2}\int n|u|^2\mathrm{d}x\\&=\frac{1}{2}K(u,n).\end{aligned} \tag{4.38}$$

All in all, by (4.29), (4.32) and (4.38), we may conclude that as $t\to\infty$,

$$J(u,n)\geqslant J(\lambda^{\frac{1}{2}}u,\lambda n)\leqslant d_{\mathscr{M}}, \tag{4.39}$$

which contradicts $J(u,n)<d_{\mathscr{M}}$ from (4.16). So the supposition (4.27) is not true. Thus $F'_{(c,\alpha)}(t)>0$ for some $t$.

So far, we complete the proof of (ⅰ) of Theorem 4.1. Next we prove (ⅱ).

Ⅱ. (4.3), (4.14) and Proposition 4.1 imply that $(u,u_t,n,V)\in S_1$, namely,

(1) $(u,n)=(0,0),\varepsilon(0,u_t,0,V)<d_{\mathscr{M}}$, or

(2) $K(u,n)>0,\varepsilon(u,u_t,n,V)<d_{\mathscr{M}}$ for $t\in[0,T)$.

If (1) holds, the result is evidently true.

If (2) holds, then the following two cases hold:

(a) $\int n|u|^2\mathrm{d}x\geqslant 0$, or

(b) $\int n|u|^2\mathrm{d}x<0$, and

$$\int n|u|^2\mathrm{d}x>-\frac{2}{3}c^2\int|u|^2\mathrm{d}x-\frac{2}{3}\int|\nabla u|^2\mathrm{d}x-\frac{1}{3}\int|n|^2\mathrm{d}x-\frac{2}{3}\int|u|^4\mathrm{d}x.$$

For the case (a) $\int n|u|^2\mathrm{d}x\geqslant 0$, from (1.8), (1.9) and (4.14) we obtain

$$\begin{aligned}&c^{-2}\int|u_t|^2\mathrm{d}x+\frac{1}{2}\alpha^{-2}\int|V|^2\mathrm{d}x+c^2\int|u|^2\mathrm{d}x+\int|\nabla u|^2\mathrm{d}x+\frac{1}{2}\int|u|^4\mathrm{d}x+\frac{1}{2}\int|n|^2\mathrm{d}x\\&\leqslant\varepsilon(u_0,u_1,n_0,v_0)<d_{\mathscr{M}}.\end{aligned} \tag{4.40}$$

Thus we established the boundedness of $u_t$ in $L^2(\mathbf{R}^N)$, $V$ in $L^2(\mathbf{R}^N)$, $u$ in $H_c^1(\mathbf{R}^N)$ and $n$ in $L^2(\mathbf{R}^N)$ for $t\in[0,T)$. Hence it must be $T=\infty$. Then the solution $(u,n,V)$ of the Cauchy problem (1.1) ~ (1.4) exists globally on $t\in[0,\infty)$. Furthermore, (4.40) implies the estimate (4.4).

For the case (b) $\int n|u|^2\mathrm{d}x<0$ and

$$\int n|u|^2\mathrm{d}x > -\frac{2}{3}c^2\int |u|^2\mathrm{d}x - \frac{2}{3}\int |\nabla u|^2\mathrm{d}x - \frac{1}{3}\int |n|^2\mathrm{d}x - \frac{2}{3}\int |u|^4\mathrm{d}x,$$

from (1.9), (1.10) and the inequality

$$-\int n|u|^2\mathrm{d}x \leqslant \frac{1}{2}\int |n|^2\mathrm{d}x + \frac{1}{2}\int |u|^4\mathrm{d}x,$$

we get

$$\begin{aligned}
&c^{-2}\int |u_t|^2\mathrm{d}x + c^2\int |u|^2\mathrm{d}x + \int |\nabla u|^2\mathrm{d}x + \frac{1}{2}\int |u|^4\mathrm{d}x + \\
&\frac{1}{2}\int |n|^2\mathrm{d}x + \frac{1}{2}\alpha^{-2}\int |V|^2\mathrm{d}x - \left(-\int n|u|^2\mathrm{d}x\right) \\
\geqslant\ & c^{-2}\int |u_t|^2\mathrm{d}x + c^2\int |u|^2\mathrm{d}x + \int |\nabla u|^2\mathrm{d}x + \frac{1}{2}\alpha^{-2}\int |V|^2\mathrm{d}x.
\end{aligned}$$

Thus by (1.9) and (4.14), we obtain

$$\begin{aligned}
&c^{-2}\int |u_t|^2\mathrm{d}x + c^2\int |u|^2\mathrm{d}x + \int |\nabla u|^2\mathrm{d}x + \frac{1}{2}\alpha^{-2}\int |V|^2\mathrm{d}x \\
\leqslant\ & \varepsilon(u, u_t, n, V) \\
=\ & \varepsilon(u_0, u_1, n_0, v_0) < d_{\mathscr{M}}.
\end{aligned}\tag{4.41}$$

On the other hand, from $K(u,n)>0$ and $J(u,n)<d_{\mathscr{M}}$, we get

$$\int n|u|^2\mathrm{d}x > -\frac{2}{3}c^2\int |u|^2\mathrm{d}x - \frac{2}{3}\int |\nabla u|^2\mathrm{d}x - \frac{1}{3}\int |n|^2\mathrm{d}x - \frac{2}{3}\int |u|^4\mathrm{d}x, \tag{4.42}$$

and

$$c^2\int |u|^2\mathrm{d}x + \int |\nabla u|^2\mathrm{d}x + \frac{1}{2}\int |u|^4\mathrm{d}x + \frac{1}{2}\int |n|^2\mathrm{d}x + \int n|u|^2\mathrm{d}x < d_{\mathscr{M}}. \tag{4.43}$$

Thus from (4.42) and (4.43) it follows that

$$\frac{1}{3}c^2\int |u|^2\mathrm{d}x + \frac{1}{3}\int |\nabla u|^2\mathrm{d}x + \frac{1}{6}\int |n|^2\mathrm{d}x - \frac{1}{6}\int |u|^4\mathrm{d}x < d_{\mathscr{M}},$$

namely,

$$\frac{1}{3}c^2\int |u|^2\mathrm{d}x + \frac{1}{3}\int |\nabla u|^2\mathrm{d}x + \frac{1}{6}\int |n|^2\mathrm{d}x < d_{\mathscr{M}} + \frac{1}{6}\int |u|^4\mathrm{d}x$$

which together with Sobolev's inequality

$$\int |u|^4\mathrm{d}x \leqslant c^*\left(c^2\int |u|^2\mathrm{d}x + \int |\nabla u|^2\mathrm{d}x\right)^2$$

yields that

$$\frac{1}{3}c^2\int |u|^2\mathrm{d}x + \frac{1}{3}\int |\nabla u|^2\mathrm{d}x + \frac{1}{6}\int |n|^2\mathrm{d}x < d_{\mathscr{M}} + \frac{1}{6}c^*\left(c^2\int |u|^2\mathrm{d}x + \int |\nabla u|^2\mathrm{d}x\right)^2. \tag{4.44}$$

By (4.41), we obtain

$$c^2\int |u|^2\mathrm{d}x + \int |\nabla u|^2\mathrm{d}x < d_{\mathscr{M}}. \tag{4.45}$$

Thus it follows from (4.44) and (4.45) that

$$\frac{1}{3}c^2\int |u|^2\mathrm{d}x + \frac{1}{3}\int |\nabla u|^2\mathrm{d}x + \frac{1}{6}\int |n|^2\mathrm{d}x < d_{\mathscr{M}} + \frac{1}{6}c^*d_M^2, \tag{4.46}$$

where in (4.44) and (4.46), $c^*$ is the same as that appeared in Theorem 4.1.

Therefore by (4.41) and (4.46), we get

$$c^{-2}\int |u_t|^2 dx + \frac{4}{3}c^2\int |u|^2 dx + \frac{4}{3}\int |\nabla u|^2 dx + \frac{1}{2}\alpha^{-2}\int |V|^2 dx + \frac{1}{6}\int |n|^2 dx$$

$$< 2d_{\mathcal{M}} + \frac{1}{6}c^* d_M^2. \tag{4.47}$$

Thus we established the boundedness of $u_t$ in $L^2(\mathbf{R}^N)$, $V$ in $L^2(\mathbf{R}^N)$, $u$ in $H_c^1(\mathbf{R}^N)$ and $n$ in $L^2(\mathbf{R}^N)$ for $t\in[0,T)$. Hence it must be $T=\infty$. Then the solution $(u, n, V)$ of the Cauchy problem (1.1) ~ (1.4) exists globally on $t\in[0,\infty)$. Furthermore, (4.37) implies the estimate (4.5). From the discussions of the case (a) and the case (b), we complete the proof of (ii) of Theorem 4.1.

From the above arguments, the proof of Theorem 4.1 is completed. □

Next, we show Corollaries 4.1 and 4.2 by introducing proper dilation transformation and applying the result in Theorem 4.1. First we show Corollary 4.1.

**Proof of Corollary 4.1** By (1.9), (4.6), (4.7) and Sobolev's inequality

$$\int |u|^4 dx \leqslant c^*\left(c^2\int |u|^2 dx + \int |\nabla u|^2 dx\right)^2,$$

we get $(u_0, n_0)\neq(0,0)$ and

$$\begin{aligned}\varepsilon(u_0,u_1,n_0,v_0) &= c^{-2}\int |u_1|^2 dx + \frac{1}{2}\alpha^{-2}\int |v_0|^2 dx + c^2\int |u_0|^2 dx + \int |\nabla u_0|^2 dx + \\ &\quad \frac{1}{2}\int |u_0|^4 dx + \frac{1}{2}\int |n_0|^2 dx + \int n_0 |u_0|^2 dx \\ &\leqslant c^{-2}\int |u_1|^2 dx + \frac{1}{2}\alpha^{-2}\int |v_0|^2 dx + c^2\int |u_0|^2 dx + \\ &\quad \int |\nabla u_0|^2 dx + \frac{1}{2}c^*\left(c^2\int |u_0|^2 dx + \int |\nabla u_0|^2 dx\right)^2 + \\ &\quad \frac{1}{2}\int |n_0|^2 dx + \int n_0 |u_0|^2 dx \\ &< d_{\mathcal{M}}.\end{aligned} \tag{4.48}$$

Now we show that

$$K(u_0, n_0) > 0. \tag{4.49}$$

In the following, we prove (4.49) by contradiction. If (4.49) is not true, one would have

$$K(u_0, n_0) \leqslant 0. \tag{4.50}$$

Thus there exists $0<\lambda\leqslant 1$ such that

$$K(\lambda^{\frac{1}{2}}u_0, \lambda n_0) = 0 \quad \text{and} \quad (\lambda^{\frac{1}{2}}u_0, \lambda n_0) \neq (0,0)$$

since $(u_0, n_0)\neq(0,0)$. So $(\lambda^{\frac{1}{2}}u_0, \lambda n_0)\in\mathcal{M}$. By (2.3) and (2.4) we get

$$J(\lambda^{\frac{1}{2}}u_0, \lambda n_0) \geqslant d_{\mathcal{M}}. \tag{4.51}$$

On the other hand, for $0<\lambda\leqslant 1$, $(\lambda^{\frac{1}{2}}u_0, \lambda n_0)$ and $(u_1, v_0)$ still satisfy (4.6) and (4.7). It follows from (4.6), (4.7), Sobolev's inequality and $0<\lambda\leqslant 1$ that

$$J(\lambda^{\frac{1}{2}}u_0,\lambda n_0)$$
$$=\lambda\left(c^2\int|u_0|^2dx+\int|\nabla u_0|^2dx\right)+\frac{1}{2}\lambda^2\int|n_0|^2dx+\frac{1}{2}\lambda^2\int|u_0|^4dx+\lambda^2\int n_0|u_0|^2dx$$
$$\leqslant c^2\int|u_0|^2dx+\int|\nabla u_0|^2dx+\frac{1}{2}\int|n_0|^2dx+\frac{1}{2}c^*\left(c^2\int|u_0|^2dx+\int|\nabla u_0|^2dx\right)^2$$
$$<d_{\mathscr{M}}, \tag{4.52}$$

which is contradictory to (4.51). Thus (4.49) is true. Therefore by (ⅱ) of Theorem 4.1, we get the conclusion of Corollary 4.1. □

Now we show Corollary 4.2.

**Proof of Corollary 4.2** From (1.9), (4.8) and (4.9), using Sobolev's inequality

$$\int|u|^4dx\leqslant c^*\left(c^2\int|u|^2dx+\int|\nabla u|^2dx\right)^2,$$

where $c^*$ is the same as that appeared in Theorem 4.1 and the inequality

$$\left|\int n|u|^2dx\right|\leqslant\frac{1}{2}\int|n|^2dx+\frac{1}{2}\int|u|^4dx,$$

we get

$$\begin{aligned}\varepsilon(u_0,u_1,n_0,v_0)=&\ c^{-2}\int|u_1|^2dx+c^2\int|u_0|^2dx+\int|\nabla u_0|^2dx+\frac{1}{2}\int|u_0|^4dx+\\&\frac{1}{2}\int|n_0|^2dx+\frac{1}{2}\alpha^{-2}\int|v_0|^2dx+\int n_0|u_0|^2dx\\\leqslant&\ c^{-2}\int|u_1|^2dx+c^2\int|u_0|^2dx+\int|\nabla u_0|^2dx+\frac{1}{2}\int|n_0|^2dx+\\&\frac{1}{2}\alpha^{-2}\int|v_0|^2dx+\frac{1}{2}c^*\left(c^2\int|u_0|^2dx+\int|\nabla u_0|^2dx\right)^2+\\&\frac{1}{2}\int|n_0|^2dx+\frac{1}{2}\int|u_0|^4dx\\\leqslant&\ c^{-2}\int|u_1|^2dx+c^2\int|u_0|^2dx+\int|\nabla u_0|^2dx+\int|n_0|^2dx+\\&c^*\left(c^2\int|u_0|^2dx+\int|\nabla u_0|^2dx\right)^2+\frac{1}{2}\alpha^{-2}\int|v_0|^2dx\\<&\ d_{\mathscr{M}}.\end{aligned} \tag{4.53}$$

Moreover, it follows from (4.9) that

$$K(u_0,n_0)=2c^2\int|u_0|^2dx+2\int|\nabla u_0|^2dx+2\int|u_0|^4dx+\int|n_0|^2dx+3\int n_0|u_0|^2dx$$
$$>0. \tag{4.54}$$

From (4.53) and (4.54), using (ⅱ) of Theorem 4.1, we get the conclusion of Corollary 4.2. □

## 5 Modified Instability of Standing Wave

In this section, from Lemma 3.2 and Theorem 3.2, on the characterization of standing wave of the Klein-Gordon-Zakharov system (1.1)~(1.3) with minimal action, we can get a result on the modified instability of standing wave for the system (1.1)~(1.3).

**Theorem 5.1** (Modified instability of standing wave) Let $Q = -|P|^2$ and $(P, Q)$ be a ground state solution of (3.31). Put

$$P = \zeta \overline{P}, \quad Q = \zeta \overline{Q}, \tag{5.1}$$

where $\zeta$ is taken by

$$\begin{cases} \zeta = \lambda_1 \text{ or } \zeta \in [\lambda_2, \infty), & \text{if (I-1) in Lemma 3.2 holds} \\ \zeta = \lambda_2, & \text{if (I-3) in Lemma 3.2 holds} \\ \zeta = \lambda_1, & \text{if (I-4) in Lemma 3.2 holds} \\ \zeta \in (0, \lambda_1), & \text{if (I-5) in Lemma 3.2 holds} \\ \zeta = \lambda_2 \text{ or } \zeta \in (0, \lambda_2], & \text{if (I-6) in Lemma 3.2 holds} \\ \zeta \in (\lambda_2, \infty), & \text{if (I-7) in Lemma 3.2 holds} \\ \zeta = \lambda_1 \text{ or } \zeta \in (0, \lambda_1], & \text{if (I-8) in Lemma 3.2 holds} \\ \zeta \in (\lambda^*, \infty), & \text{if (II-1) in Lemma 3.2 holds} \\ \zeta = \lambda^* \text{ or } \zeta \in (0, \lambda^*), & \text{if (II-2) in Lemma 3.2 holds} \\ \zeta = \lambda^*, & \text{if (II-3) in Lemma 3.2 holds} \end{cases}$$

Then for any $\varepsilon > 0$, there exists $(u_0, u_1, n_0, v_0) \in H^1(\mathbf{R}^N) \times L^2(\mathbf{R}^N) \times L^2(\mathbf{R}^N) \times L^2(\mathbf{R}^N)$ such that

$$\| u_0 - \overline{P} \|_{H^1(\mathbf{R}^N)} < \varepsilon, \quad \| n_0 - \overline{Q} \|_{L^2(\mathbf{R}^N)} < \varepsilon \tag{5.2}$$

and the following property holds:

The solution $(u, n, V)$ of the Cauchy problem for the Klein-Gordon-Zakharov system (1.1)~(1.3) corresponding to the initial data

$$\begin{cases} u(0,x) = u_0(x), & u_t(0,x) = 0, \\ n(0,x) = n_0(x), & V(0,x) = 0, \end{cases} \tag{5.3}$$

is defined on $T \in (0, \infty)$ such that

$$(u, n, u_t, V) \in C([0,T); H^1(\mathbf{R}^N) \times L^2(\mathbf{R}^N) \times L^2(\mathbf{R}^N) \times L^2(\mathbf{R}^N))$$

and

$$\lim_{t \to T} (c^{-2} \| u \|^2_{L^2(\mathbf{R}^N)} + \alpha^{-2} \| n \|^2_{\dot{H}^{-1}(\mathbf{R}^N)}) = \infty. \tag{5.4}$$

**Remark 5.1** For the Klein-Gordon-Zakharov system with the absence of the high-degree nonlinearity $-|u|^2 u$, instability of standing wave for it has been studied by Gan and Zhang[4] as well as Ohta and Todorova[12]. But for the Klein-Gordon-Zakharov system with the presence of the high-degree nonlinearity $-|u|^2 u$ such as the system (1.1)~(1.3), to our knowledge, no results are known on this problem. In the present paper, we get a modified instability result of standing wave for the system (1.1)~(1.3), which depends strongly on Lemma 3.2. In other words, to some extent, the form of the functional $K(\phi, \psi)$ decides directly the characteristic of standing wave for the nonlinear wave equation or system.

**Proof of Theorem 5.1** From the initial data (5.3), (1.9), (1.10) and (2.1), it follows that

$$\varepsilon(u_0, u_1, n_0, v_0) = J(u_0, n_0). \tag{5.5}$$

Now we take

$$u_0(x) = \mu\bar{P}(x), \quad n_0(x) = \mu\bar{Q}(x), \tag{5.6}$$

with $\mu>0$ and

$$\begin{cases} \mu \in (\lambda_1, \lambda_2), & \text{if (I-1) in Lemma 3.2 holds} \\ \mu \in (\lambda_2, \infty), & \text{if (I-3) in Lemma 3.2 holds} \\ \mu \in (\lambda_1, \infty), & \text{if (I-4) in Lemma 3.2 holds} \\ \mu = \lambda_1, & \text{if (I-5) in Lemma 3.2 holds} \\ \mu \in (\lambda_2, \infty), & \text{if (I-6) in Lemma 3.2 holds} \\ \mu = \lambda_2, & \text{if (I-7) in Lemma 3.2 holds} \\ \mu \in (\lambda_1, \infty), & \text{if (I-8) in Lemma 3.2 holds} \\ \mu = \lambda^*, & \text{if (II-1) in Lemma 3.2 holds} \\ \mu \in (\lambda^*, \infty) \text{ or } \mu = \lambda^*, & \text{if (II-2) in Lemma 3.2 holds} \\ \mu \in (\lambda^*, \infty), & \text{if (II-3) in Lemma 3.2 holds} \end{cases}$$

For any $\varepsilon>0$, one can always choose suitable $\zeta$ and $\mu$ such that

$$\begin{cases} \| u_0 - \bar{P} \|_{H^1(\mathbf{R}^N)} = |\mu - 1| \, \|\bar{P}\|_{H^1(\mathbf{R}^N)} < \varepsilon, \\ \| n_0 - \bar{Q} \|_{L^2(\mathbf{R}^N)} = |\mu - 1| \, \|\bar{Q}\|_{L^2(\mathbf{R}^N)} < \varepsilon. \end{cases} \tag{5.7}$$

From Lemma 3.2 and (5.6), it follows that

$$\begin{cases} K(u_0, n_0) < K(\zeta\bar{P}, \zeta\bar{Q}) = 0, \\ J(u_0, n_0) < J(\zeta\bar{P}, \zeta\bar{Q}) = d_{\mathcal{M}}. \end{cases} \tag{5.8}$$

Again, we note (5.5) and obtain

$$\varepsilon(u_0, u_1, n_0, v_0) = J(u_0, n_0) < d_{\mathcal{M}}. \tag{5.9}$$

In view of 1 of Theorem 4.1, we complete the proof of Theorem 5.1. □

**References**

[1] Bellan P M. Fundamentals of Plasmas Physics. Cambridge: Cambridge Univ. Press, 2006.

[2] Bergé L, Bidégaray B, Colin T. A perturbative analysis of the time-envelope approximation in strong Langmuir turbulence. Phys D, 1996, 95: 351－379.

[3] Dendy R O. Plasma Dynamics. Oxford: Oxford Univ Press, 1990.

[4] Gan Z, Zhang J. Instability of standing waves for Klein-Gordon-Zakharov equations with different propagation speeds in three space dimensions. J Math Anal Appl, 2005, 307: 219－231.

[5] Gan Z. Orbital instability of standing waves for the Klein-Gordon-Zakharov system. Adv Nonlinear Stud, 2008, 8: 413－428.

[6] Guo B. Global smooth solution for the system of Zakharov equations in nonhomogeneous medium. Northeast Math J, 1990, 6: 379－390.

[7] Guo B, Yuan G. Global smooth solution for the Klein-Gordon-Zakharov equations. J Math Phys, 1995, 36

(8): 4119 - 4124.

[8] Levine H A. Instability and non-existence of global solutions to nonlinear wave equations of the form $Pu_{tt} = -Au + F(u)$. Trans Amer Math Soc 1974, 192: 1 - 21.

[9] Li J. Exact explicit travelling wave solutions for $(n+1)$-dimensional Klein-Gordon-Zakharov equations. Chaos Solitons Fractals, 2007, 34: 867 - 871.

[10] Masmoudi N, Nakanishi K. From the Klein-Gordon-Zakharov system to the nonlinear Schrödinger equation. J Hyperbolic Differ Equ, 2005, 2 (4): 975 - 1008.

[11] Masmoudi N, Nakanishi K. Energy convergence for singular limits of Zakharov type system. Invent Math, 2008, 172: 535 - 583.

[12] Ohta M, Todorova G. Strong instability of standing waves for the nonlinear Klein-Gordon equation and the Klein-Gordon-Zakharov system. SIAM J Math Anal, 2007, 38 (6): 1912 - 1931.

[13] Ozawa T, Tsutaya K, Tsutsumi Y. Well-posedness in energy space for the Cauchy problem of the Klein-Gordon-Zakharov equations with different propagation speeds in three space dimensions. Math Ann, 1999, 313: 127 - 140.

[14] Ozawa T, Tsutaya K, Tsutsumi Y. Normal form and global solutions for the Klein-Gordon-Zakharov equations. Ann Inst H Poincaré Anal Non Linéaire, 1995, 12 (4): 459 - 503.

[15] Payne L E. Sattinger D H. Saddle points and instability of nonlinear hyperbolic equations. Israel J Math, 1975, 22 (3 - 4): 273 - 303.

[16] Shatah J. Normal forms and quadratic nonlinear Klein-Gordon equations. Comm Pure Appl Math, 1985, 38: 685 - 696.

[17] Thornhill S G, ter Haar D. Langmuir turbulence and modulational instability. Phys Rep 1987, 43: 43 - 99.

[18] Tsutaya K. Global existence of small amplitude solutions for the Klein-Gordon-Zakharov equations. Nonlinear Anal, 1996, 27 (12): 1373 - 1380.

[19] Strauss W. Existence of solitary waves in high-dimensions. Comm Math Phys, 1977, 55: 149 - 164.

[20] Zakharov V E. Collapse of Langmuir waves. Sov Phys JETP, 1972, 35: 908 - 914.

[21] Zhang J, Gan Z. Sharp conditions of global existence for Klein-Gordon-Zakharov equations in three space dimensions. Adv Math, 2005, 34 (2): 241 - 244.

[22] Zhang J. On the standing wave in coupled nonlinear Klein-Gordon equations. Math Methods Appl Sci, 2003, 26: 11 - 25.

# Well Posedness for the Stochastic Cahn-Hilliard Equation Driven by Lévy Space-time White Noise *

Guo Boling(郭柏灵) Wang Guolian(王国联) Wang Shu(王 术)

**Abstract**

In this paper we study a stochastic Cahn-Hilliard equation driven by Lévy space-time white noise with Neumann boundary conditions. We establish the global existence and uniqueness of a mild solution under some regularity and boundedness on the coefficients.

## 1 Introduction

We consider the following stochastic version of the Cahn-Hilliard partial differential equation with initial and homogeneous Neumann boundary conditions:

$$(P)\begin{cases}\dfrac{\partial u}{\partial t}+\Delta^2 u=\Delta f(u)+\sigma(u)\dot{F}, & \text{in } [0,T]\times D\\ u(0,\cdot)=u_0 & \\ \dfrac{\partial u}{\partial n}=\dfrac{\partial \Delta u}{\partial n}=0 & \text{on } [0,T]\times\partial D\end{cases},$$

where $\Delta$ denotes the Laplace operator, $D:=[0,\pi]^d$ ($d\leqslant 3$ is the space dimension), $u_0$ is the initial function, $\sigma(\cdot):\mathbf{R}\to\mathbf{R}$ is a continuous function satisfying some assumptions given below, and $\dot{F}$ is a $d+1$-parameter Lévy space-time white noise which actually consists of a Brownian sheet, denoted by $\dot{W}$, on $D\times[0,\infty)$ and a Poisson space-time white noise on $D\times[0,\infty)$ (the detail will be given below in this section).

The deterministic Cahn-Hilliard equation (i.e. $\sigma=0$) was proposed in [3,4] to describe the evolution of the phase separation of a binary alloy when temperature has been quenched from a value $T_0$, above the critical temperature $T_c$, to a temperature less than $T_c$. The function $u$ represents a scaled concentration, and $-\Delta u+f(u)$ represents the chemical potential. The function $f$ is the derivative of the homogeneous free energy $E$. In its original form, $E$ contains a logarithmic term, which makes the study of this equation delicate. In some circumstances, $E$ can be approximated by a polynomial of even degree with a strictly positive dominant coefficient. The Neumann boundary conditions reflect the conservation of mass and insulation from outside. This case has been extensively studied (see for instance [8, 9, 6] and

* Differential and Integral Equations, 2009, 22,(5-6):543-560.

references therein).

The stochastic Cahn-Hilliard equation, such as in the case $\dot{F} = \dot{W}$ (Gaussian space-time white noise), has been studied by several authors (see e.g. Da Prato and Debussche[13] and Cardon-Weber[5]). In the case when $\dot{F}$ is a Lévy space-time white noise, the local well posedness of the solution to problem $(P)$ has already been proved by Bo and Wang[2] under some mild assumptions on the coefficients, where they failed to solve the global solution and left it as a challenging problem.

Enlightened by the approach used by Cardon-Weber[5], Gyöngy[10] and Da Prato and Gatarek[14], and noticing the Burkhölder-Davis-Gundy inequality which was proved in Knoche[11], we prove the global well posedness of the solution to the problem $(P)$ in a weak sense, and our local well posedness is better than Bo and Wang's[2].

For this aim, we first introduce Lévy space-time white noise and stochastic integration[2, 16, 1].

Let $(\Omega, \mathscr{F}, \{\mathscr{F}_t\}_{t\geqslant 0}, P)$ be a filtered probability space with filtration $\{\mathscr{F}_t\}_{t\geqslant 0}$ satisfying the usual conditions and $(E_i, \mathcal{E}_i, \mu_i)$ $(i = 1, 2)$ be two $\sigma$-finite measurable spaces. We call $N:(E_1, \mathcal{E}_1, \mu_1)\times(E_2, \mathcal{E}_2, \mu_2)\times(\Omega, \mathscr{F}, P)\to\mathbf{N}\cup\{0\}\cup\{\infty\}$ a Poisson noise on $(E_1, \mathcal{E}_1, \mu_1)$ if, for any $\mathbf{A}\in\mathcal{E}_1, B\in\mathcal{E}_2$ and $n\in\mathbf{N}\cup\{0\}\cup\{\infty\}$,

$$P(\omega\in\Omega: N(\mathbf{A}, B, \omega) = n) = \frac{\mathrm{e}^{-\mu_1(\mathbf{A})\mu_2(B)}(\mu_1(\mathbf{A})\mu_2(B))^n}{n!}.$$

In particular, when $(E_1, \mathcal{E}_1, \mu_1) = ([0,\infty)\times D, \mathscr{B}([0,\infty)\times D), \mathrm{d}t\times\mathrm{d}x)$, we define the compensated random martingale measure

$$M(B, A, t) := N([0, t)\times A, B) - t\mu_1(A)\mu_2(B),$$

by assuming that $\mu_1([0, t]\times A)\mu_2(B) < \infty$ for all $(t, A, B)\in[0,\infty)\times\mathscr{B}(D)\times\mathcal{E}_2$. Furthermore, let $p:[0,\infty)\times D\times E_2\times\Omega\to\mathbf{R}$ be an $\{\mathscr{F}_t\}_{t\geqslant 0}$-predictable function satisfying

$$\mathbb{E}\left(\int_0^t\int_A\int_B |p(s, x, y)|^2\mu_2(\mathrm{d}y)\mathrm{d}x\mathrm{d}s\right) < \infty, \tag{1.1}$$

for all $t>0$ and $(A, B)\in\mathscr{B}(D)\times\mathcal{E}_2$. We define a stochastic integral process

$$\left\{R_t := \int_0^{t+}\int_A\int_B p(s, x, y)M(\mathrm{d}y, \mathrm{d}x, \mathrm{d}s) : t\geqslant 0\right\},$$

which is a square integrable $\{\mathscr{F}_t\}_{t\geqslant 0}$ martingale. Let $\langle\cdot,\cdot\rangle$ and $[\cdot,\cdot]$ denote respectively the quadratic variation and the conditional quadratic variation of a process (refer to page 97 of Protter[15] for definitions). Then

$$\langle R, R\rangle_t = \int_0^t\int_A\int_B |p(s, x, y)|^2\mu_2(\mathrm{d}y)\mathrm{d}x\mathrm{d}s,$$

$$[R, R]_t = \int_0^{t+}\int_A\int_B |p(s, x, y)|^2 N(\mathrm{d}s\times\mathrm{d}x, \mathrm{d}y).$$

It is well known that a Lévy space-time white noise has the following structure:

$$\dot{F}(x, t) = \dot{W}(x, t) + \int_{U_0} h_1(t, x, y)\dot{M}(\mathrm{d}y, x, t) + \int_{E_2/U_0} h_2(t, x, y)\dot{N}(\mathrm{d}y, x, t),$$

for some $U_0 \in \varepsilon_2$ such that $\mu_2(E_2/U_0) < \infty$, where $h_1, h_2: [0, \infty) \times D \times E_2 \to \mathbf{R}$ are measurable functions, and $\dot{W}(t,x)$ is a Gaussian space-time white noise on $D \times [0, \infty)$; i.e., $\mathbb{E}[\dot{W}(x,t)\dot{W}(x',t')] = \delta(t-t')\delta(x-x')$, $\dot{M}$ and $\dot{N}$ are the Radon-Nikodym derivatives defined by

$$\dot{M}(\mathrm{d}y, x, t) = \frac{M(\mathrm{d}y, \mathrm{d}x, \mathrm{d}t)}{\mathrm{d}t \times \mathrm{d}x}, \quad \dot{N}(\mathrm{d}y, x, t) = \frac{N(\mathrm{d}y, \mathrm{d}x, \mathrm{d}t)}{\mathrm{d}t \times \mathrm{d}x},$$

for $(t, x, y) \in [0, \infty) \times D \times E_2$.

Now we give a definition of the solution in a weak sense (as in Walsh[17]). We say that $u$ is a weak solution of problem $(P)$ if, for all $\phi \in C^4(D)$ with

$$\frac{\partial \phi}{\partial n} = \frac{\partial \Delta \phi}{\partial n} = 0 \quad \text{on } \partial D,$$

$u$ satisfies

$$\begin{aligned}\int_D u(t,x)\phi(x)\mathrm{d}x = &\int_D u_0(x)\phi(x)\mathrm{d}x - \int_0^t\int_D \Delta^2\phi(x)u(s,x)\mathrm{d}x\mathrm{d}s + \\ &\int_0^t\int_D \Delta\phi(x)f(u(s,x))\mathrm{d}x\mathrm{d}s + \int_0^t\int_D \phi(x)\sigma(u(s,x))W(\mathrm{d}x,\mathrm{d}s) + \\ &\int_0^{t+}\int_D\int_{U_0} \phi(x)\sigma(u(s-,x))h_1(s,x,z)M(\mathrm{d}z,\mathrm{d}x,\mathrm{d}s) + \\ &\int_0^{t+}\int_D\int_{E_2/U_0} \phi(x)\sigma(u(s-,x))h_2(s,x,z)N(\mathrm{d}s \times \mathrm{d}x,\mathrm{d}z).\end{aligned} \tag{1.2}$$

The paper is organized as follows. Section 2 contains some preliminaries on some useful estimates. In Section 3, we give the local well posedness of the solution to the problem $(P)$ under some mild assumptions. In the last section, the main results and their proofs are presented.

## 2 Preliminaries

In this section, we first give some estimates about the Green's function $G$ generated by the operator $\frac{\partial}{\partial t} + \Delta^2$ on the domain $[0, \infty) \times D$. The estimates are from Eidelman and Ivasisen[7] on Green's functions on smooth domains and were extended to the domain $D$ by Cardon-Weber[5].

Let $A$ denote the operator $-\Delta$ on the domain $\mathscr{D}(A) = \{u \in H^2(D): \frac{\partial u}{\partial n} = 0 \text{ on } \partial D\}$. The following family $(\varepsilon_k)_{k \in \mathbf{N}^d}$ is a basis of eigenfunctions of $A$ in $L^2(D)$. If $d = 1$,

$$\varepsilon_k(x) = \cos(kx)\sqrt{\frac{2}{\pi}} \text{ if } k \neq 0, \text{ and } \varepsilon_0 = \frac{1}{\sqrt{\pi}};$$

and for $d \in \{1,2,3\}$,

$$\varepsilon_k(x) = \prod_{i=1}^{d} \varepsilon_{k_i}(x_i),$$

associated with the eigenvalues $\lambda_k = \sum_{i=1}^{d} k_i^2 = |k|^2$. The semigroup $S(t)$ generated by $-A^2$ is denoted by $S(t)=\mathrm{e}^{-tA^2}$; that is, for $z \in L^2(D)$,

$$S(t)z = \sum_{k \in \mathbf{N}^d} \mathrm{e}^{-\lambda_k^2 t} \langle z, \varepsilon_k \rangle \varepsilon_k,$$

where $\langle \cdot, \cdot \rangle$ denotes the usual scalar product in $L^2(D)$; this is a convolution semigroup with the Green's function $G$ defined by

$$G(t,x,y) = \sum_{\mathbf{N}^d} \mathrm{e}^{-\lambda_k^2 t} \varepsilon_k(x) \varepsilon_k(y).$$

We lists some useful properties about the Green's function (see Cardon-Weber[5] for details) to the following work.

**Lemma 2.1** There exist $C>0$ and $c>0$ such that, for $t \in (0, T]$, $x, y \in D$, $\alpha$ a dimensional exponent satisfying $|\alpha| \leqslant 3$,

$$|G(t,x,y)| \leqslant \frac{C}{t^{d/4}} \exp\left(-c \frac{|x-y|^{4/3}}{|t|^{1/3}}\right); \tag{2.1}$$

$$|\partial_x^\alpha G(t,x,y)| \leqslant \frac{C}{t^{(d+|\alpha|)/4}} \exp\left(-c \frac{|x-y|^{4/3}}{|t|^{1/3}}\right); \tag{2.2}$$

$$|\partial_t G(t,x,y)| \leqslant \frac{C}{t^{(d+4)/4}} \exp\left(-c \frac{|x-y|^{4/3}}{|t|^{1/3}}\right); \tag{2.3}$$

The following result (Gyöngy[10]) gives precise estimates of the regularizing effect of convolution with $G, \Delta G$ and $G^2$.

**Lemma 2.2** Let $J$ be defined, for all $v \in L^1([0,T], L^\rho(D))$, $0 \leqslant t_0 \leqslant t \leqslant T$, and $x \in D$, by

$$J(v)(t_0,t,x) = \int_{t_0}^{t} \int_D H(t-s,x,y) v(s,y) \mathrm{d}y \mathrm{d}s.$$

Then, for any $\rho \in [1,\infty]$, $q \in [\rho,\infty]$ and $\frac{1}{r} = \frac{1}{q} - \frac{1}{\rho} + 1 \in [0,1]$, $J$ is a bounded operator from $L^1([0,T], L^\rho(D))$ to $L^\infty([0,T], L^q(D))$ such that:

(i) If $H(t,x,y) = G(s,x,y)$, there exists a constant $C>0$ such that

$$\|J(v)(t_0,t,\cdot)\|_q \leqslant C \int_{t_0}^{t} (t-s)^{\frac{d}{4r}-\frac{d}{4}} \|v(s,\cdot)\|_\rho \mathrm{d}s. \tag{2.4}$$

(ii) If $H(t,x,y) = \Delta G(s,x,y)$ (if $d=3$, we also need $r<3$; and if $d=2, r \neq \infty$), there exists a constant $C>0$ such that

$$\|J(v)(t_0,t,\cdot)\|_q \leqslant C \int_{t_0}^{t} (t-s)^{\frac{d}{4r}-\frac{d+2}{4}} \|v(s,\cdot)\|_\rho \mathrm{d}s. \tag{2.5}$$

(iii) If $H(t,x,y) = G^2(s,x,y)$ (if $d=3$, we also need $r<\frac{3}{2}$; and if $d=2, r \neq \infty$), there exists a constant $C>0$ such that

$$\|J(v)(t_0,t,\cdot)\|_q \leqslant C \int_{t_0}^{t} (t-s)^{\frac{d}{4r}-\frac{d}{2}} \|v(s,\cdot)\|_\rho \mathrm{d}s. \tag{2.6}$$

By the way, we give the Burkhölder-Davis-Gundy inequality from Knoche[11], which is very useful like the Burkhölder inequality.

**Lemma 2.3** Let $p:[0,\infty)\times D\times E_2\times\Omega\rightarrow\mathbf{R}$ be $\{\mathscr{F}_t\}_{t\geq 0}$-predictable and satisfy (1.1). We denote by $X$ the integral process

$$\left\{X_t:=\int_0^{t+}\int_D\int_{E_2}p(s,y,z)M(dz,dy,ds):t\geqslant 0\right\}.$$

Then, for any $T>0$ and $q\geqslant 2$, there exists a constant $C(q,T)>0$ such that

$$\sup_{t\in[0,T]}(\mathbb{E}|X_t|^q\leqslant C(q,T)\left(\int_0^T\int_D\int_{E_2}(\mathbb{E}[|p(s,y,z)|^q])^{\frac{2}{q}}\mu_2(dz)dyds\right)^{\frac{q}{2}}.\quad(2.7)$$

Now we give an equivalent form of equation (1.2) represented by the Green function's $G$ as follows, which will be used to prove the existence and uniqueness of the solution to the problem $(P)$,

$$\begin{aligned}u(t,x)=&\int_D G(t,x,y)u_0(y)dy+\int_0^t\int_D\Delta G(t-s,x,y)f(u(s,y))dyds+\\&\int_0^t\int_D\int_{E_2/U_0}G(t-s,x,y)\sigma(u(s,y))h_2(s,y,z)\mu_2(dz)dyds+\\&\int_0^t\int_D G(t-s,x,y)\sigma(u,(s,y))W(dy,ds)+\\&\int_0^{t+}\int_D\int_{E_2}G(t-s,x,y)\sigma(u(s-,y))(h_1(s,y,z)I_{U_0}(z)+\\&h_2(s,y,z)I_{E_2/U_0}(z))\mu_2(dz)dyds.\end{aligned}\quad(2.8)$$

# 3 Local Solution

In this section, we apply the estimates of the Green's function to the existence and uniqueness of solutions for a similar equation with truncated coefficients. This yields the existence and uniqueness of the solution on the time interval $[0,\tau)$, where $\tau$ is a stopping time.

Let $n>0$ and denote by $K_n$, defined on $[0,\infty)$, a $C^1$ function such that $|K_n'|\leqslant 2$ and

$$K_n(x)=\begin{cases}1, & \text{if } x<n\\0, & \text{if } x\geqslant n+1.\end{cases}$$

Let

$$g(t,y)=\int_{E_2/U_0}h_2(t,y,z)\mu_2(dz)$$

and

$$h(t,y,z)=h_1(s,y,z)I_{U_0}(z)+h_2(s,y,z)I_{E_2/U_0}(z).$$

Then we prove the local well posedness for the equation obtained from the problem $(P)$ by truncating the operators determined by $f$ and $\sigma$,

$$
\begin{aligned}
u_n(t,x) = & \int_D G(t,x,y)u_0(y)\mathrm{d}y + \\
& \int_0^t\int_D \Delta G(t-s,x,y)K_n(\|u_n(s,\cdot)\|_q)f(u_n(s,y))\mathrm{d}y\mathrm{d}s + \\
& \int_0^t\int_D G(t-s,x,y)K_n(\|u_n(s,\cdot)\|_q)\sigma(u_n(s,y))g(s,y)\mathrm{d}y\mathrm{d}s + \\
& \int_0^t\int_D G(t-s,x,y)K_n(\|u_n(s,\cdot)\|_q)\sigma(u_n(s,y))W(\mathrm{d}y,\mathrm{d}s) + \\
& \int_0^{t+}\int_D\int_{E_2} G(t-s,x,y)K_n(\|u_n(s,\cdot)\|_q\sigma(u_n(s-,y))h(s,y,z)M(\mathrm{d}z,\mathrm{d}y,\mathrm{d}s).
\end{aligned}
\tag{3.1}
$$

We quote the definition of the local solution to problem $(P)$ coming from [2]. Define

$$\tau_n = \inf\{t \geqslant 0: \|u_n(t,\cdot)\|_q \geqslant n\}.$$

Let $\tau = \lim_{n\to\infty}\tau_n$, and define $u(t,\cdot) := u_n(t,\cdot)$ on $\{t<\tau_n<\tau\}$. Then $u(t,\cdot)$ is a solution to (3.1) on $\{t<\tau\}$. This solution is called a local solution.

Before giving the local solution, we give some reasonable assumptions about the problem $(P)$:

(a) $f$ is a polynomial of degree 3 with positive dominant coefficients;

(b) $\sigma$ is Lipschitzian and has linear growth; i.e., there exists a constant $C>0$ such that

$$|\sigma(x)| \leqslant C(1+|x|), \quad \text{for all } x \in \mathbf{R}.$$

For $g, h$ and $\mu_2$,

(c) $\sup_{t\in[0,T]}\|g(t,\cdot)\|_q < \infty$;

(d) $G(t-s,x,y)h(s,y,z)$ is $L^q([0,t]\times D^2\times E_2)$, $\mathrm{d}s\times\mathrm{d}x\times\mathrm{d}y\times\mu_2(E_2)$ integrable, for $0\leqslant t\leqslant T$;

(e) $\mu_2(E_2)<\infty$.

**Theorem 3.1** Suppose the assumptions (a)~(e) are satisfied. Let $d\in\{1,2,3\}$ and $q\in(4,\infty)$. Then, for every $\mathscr{F}_0$-measurable $u_0: D\times\Omega\to\mathbf{R}$ with $\mathbb{E}(\|u_0(\cdot)\|_q^q)<\infty$, there exists a unique local solution $\{u(t,x):(t,x)\in[0,T]\times D\}$ to the problem $(P)$ and there exists a stopping time $\tau$ such that $\mathbb{E}(\|u(t\wedge\tau,\cdot)\|_q^q)<\infty$, for any $t\in[0,T]$.

We want to use a fixed-point argument to prove Theorem 3.1. So we first introduce our working space.

Let $\mathscr{H}$ be all $L^q(D)$-valued $\mathscr{F}_t$ adapted r.c.l.l. processes $u(t,x)$ such that the norm

$$\|u\|_{\mathscr{H}} = (\sup_{t\in[0,T]}(\mathbb{E}(\|u(t,\cdot)\|_q^q))^{1/q} \tag{3.2}$$

is finite. Then, under the norm (3.2), $\mathscr{H}$ is a Banach space.

For convenience, let us define operators $A_i (i=0,1,2,3,4)$ such that

$$A_0(u_0)(t,x) = \int_D G(t,x,y)u_0(y)\mathrm{d}y,$$

$$A_1(u)(t,x) = \int_0^t\int_D \Delta G(t-s,x,y)K_n(\|u(s,\cdot)\|_q)f(u(s,y))\mathrm{d}y\mathrm{d}s,$$

$$A_2(u)(t,x)=\int_0^t\int_D G(t-s,x,y)K_n(\|u(s,\cdot)\|_q)\sigma(u(s,y))W(\mathrm{d}y\mathrm{d}s),$$

$$A_3(u)(t,x)=\int_0^{t+}\int_D\int_{E_2} G(t-s,x,y)K_n(\|u(s,\cdot)\|_q)\times$$
$$\sigma(u(s^-,y))h(s,y,z)M(\mathrm{d}z,\mathrm{d}y,\mathrm{d}s),$$

$$A_4(u)(t,x)=\int_0^t\int_D G(t-s,x,y)K_n(\|u(s,\cdot)\|_q)\sigma(u(s,y))g(s,y)\mathrm{d}y\mathrm{d}s,$$

$$Y(u)(t,x):=A_0(u_0)(t,x)+\sum_{i=1}^{4}A_i(u)(t,x).$$

In order to prove the local well posedness of the solution to the problem $(P)$, i.e., Theorem 3.1, we first give the following propositions:

**Proposition 3.1** Under the conditions of Theorem 3.1, for any $u\in\mathscr{H}, A_i(u)\in\mathscr{H}$.

**Proof** We always assume that $u\in\mathscr{H}$ in the following proof. Using (2.1) and Young's inequality for $1/q=1+1/q-1$, we have

$$\begin{aligned}\|A_0(u_0)\|_q&=\left\|\int_D G(t,\cdot,y)u_0(y)\mathrm{d}y\right\|_q\\&\leqslant Kt^{-\frac{d}{4}}\left\|\int_D \exp\left(-C\frac{|\cdot-y|^{\frac{4}{3}}}{t^{\frac{1}{3}}}\right)u_0(y)\mathrm{d}y\right\|_q\\&\leqslant Kt^{-\frac{d}{4}}\left\|\exp\left(-C\frac{|\cdot-y|^{\frac{4}{3}}}{t^{\frac{1}{3}}}\right)\right\|_1\|u_0(\cdot)\mathrm{d}y\|_q\leqslant K\|u_0(\cdot)\mathrm{d}y\|_q.\end{aligned}\tag{3.3}$$

Thus, $A_0(u_0)\in\mathscr{H}$, under the condition $\mathbb{E}(\|u_0\|_q^q)<\infty$.

As for $A_1(u)$, we apply (2.5) with $\rho=q/3$; we have $1/r=1-2/q$, which satisfies $r_1\neq\infty$ if $d=3$ and $r_1<3$ if $d=3$. Then

$$\begin{aligned}\|A_1(u)\|_q&\leqslant\int_0^t|t-s|^{\frac{d}{4r}-\frac{d+2}{4}}\|f(u(s,\cdot))K_n(\|u(s,\cdot)\|_q)\|_\rho\mathrm{d}s\\&\leqslant C(n+1)^3T^{1+d/(4r_1)-(d+2)/4},\end{aligned}$$

so that $A_1(u)\in\mathscr{H}$.

As for $A_2(u)$, applying Burkhölder's inequality, assumption (b), and (2.6) with $1/r=1/(q/2)-1/\rho+1$, we have

$$\begin{aligned}\|A_2(u)\|_{\mathscr{H}}^q&=\sup_{t\in[0,T]}\mathbb{E}\left(\left\|\int_0^t\int_D G(t-s,x,y)\sigma(u(s,y))K_n(\|u(s,\cdot)\|_q)W(\mathrm{d}y,\mathrm{d}s)\right\|_q^q\right)\\&\leqslant C\sup_{t\in[0,T]}\mathbb{E}\left\|\int_0^t\int_D G^2(t-s,x,y)\sigma^2(u(s,y))K_n^2(\|u(s,\cdot)\|_q)\mathrm{d}y\mathrm{d}s\right\|_{\frac{q}{2}}^{\frac{q}{2}}\\&\leqslant C\sup_{t\in[0,T]}\left(\int_0^t(t-s)^{\frac{d}{4r}-\frac{d}{2}}\|\sigma^2(u(s,y))K_n^2(\|u(s,\cdot)\|_q)\|_\rho\mathrm{d}s\right)^{\frac{q}{2}}\\&\leqslant C\sup_{t\in[0,T]}\left(\int_0^t(t-s)^{\frac{d}{4r}-\frac{d}{2}}\|(1+u^2(s,\cdot))K_n^2(\|u(s,\cdot)\|_q)\|_\rho\mathrm{d}s\right)^{\frac{q}{2}}\end{aligned}$$

$$\leqslant C\sup_{t\in[0,T]}\left(\int_0^t (t-s)^{\frac{d}{4r}-\frac{d}{2}}(1+(n+1)^2)\mathrm{d}s\right)^{\frac{q}{2}}\leqslant C(T)(n+1)^q. \tag{3.4}$$

In the last inequality, one can set $\rho=q/2$, and $r=1$. Thus we have $A_2(u)\in\mathscr{H}$.

As for $A_3(u)$, let $q>4$. By assumptions (b), (d) and Lemma 2.3, using the Hölder inequality, we have

$$\|A_3(u)\|_{\mathscr{H}}^q=\sup_{t\in[0,T]}\mathbb{E}\left(\left\|\int_0^{t+}\int_D\int_{E_2}G(t-s,x,y)\sigma(u(s^-,y))\times\right.\right.$$

$$\left.\left.K_n(\|u(s,\cdot)\|_q)h(s,y,z)M(\mathrm{d}z,\mathrm{d}y,\mathrm{d}s)\right\|_q^q\right)$$

$$\leqslant C\sup_{t\in[0,T]}\int_D\mathbb{E}\left(\left|\int_0^{t+}\int_D\int_{E_2}G(t-s,x,y)\sigma(u(s,y))K_n(\|u(s,\cdot)\|_q)h(s,y,z)\times\right.\right.$$

$$\left.\left.M(\mathrm{d}z,\mathrm{d}y,\mathrm{d}s)\right|^q\right)\mathrm{d}x$$

$$\leqslant C\sup_{t\in[0,T]}\int_D\left(\left|\int_0^t\int_D\int_{E_2}(\mathbb{E}|G(t-s,x,y)\sigma(u(s,y)K_n(\|u(s,\cdot)\|_q)\times\right.\right.$$

$$\left.\left.h(s,y,z)|^q)^{\frac{2}{q}}\right)\mu_2(\mathrm{d}z)\mathrm{d}y\mathrm{d}s\right|^{\frac{q}{2}}\mathrm{d}x$$

$$\leqslant C\sup_{t\in[0,T]}\int_D\left|\int_0^t\int_D\int_{E_2}|G(t-s,x,y)h(s,y,z)|^2\times\right.$$

$$\left.K_n(\|u(s,\cdot)\|_q)(\mathbb{E}(1+|u(s,y)|)^q)^{2/q}\mu_2(\mathrm{d}z)\mathrm{d}y\mathrm{d}s\right|^{\frac{q}{2}}\mathrm{d}x$$

$$\leqslant C(T)\|n+1\|^q\,\mathbb{C}\sup_{t\in[0,T]}\int_D\left|\int_0^t\int_D\int_{E_2}|G(t-s,x,y))h(s,y,z)|^q)\times\right.$$

$$\left.\mu_2(\mathrm{d}z)\mathrm{d}y\mathrm{d}s\right|\mathrm{d}x. \tag{3.5}$$

Thus, we have $A_3(u)\in\mathscr{H}$.

Applying (2.4) with $\rho=q/2$ and $\frac{1}{r}=\frac{1}{q}-\frac{1}{\rho}+1$, we have

$$\|A_4(u)\|_q\leqslant C\int_0^t(t-s)^{\frac{d}{4r}-\frac{d}{4}}\|g(s,\cdot)\sigma(u(s,\cdot))K_n(\|u(s,\cdot)\|_q)\|_\rho\mathrm{d}s$$

$$\leqslant C\int_0^t(t-s)^{\frac{d}{4r}-\frac{d}{4}}(\|g(s,\cdot)\|_{q/2}+\|g(s,\cdot)u(s,\cdot)\|_{q/2})K_n(\|u(s,\cdot)\|_q)\mathrm{d}s$$

$$\leqslant C\int_0^t(t-s)^{\frac{d}{4r}-\frac{d}{4}}(C(q)\|g(s,\cdot)\|_q+(n+1)\|g(s,\cdot)\|_q)\mathrm{d}s$$

$$\leqslant C\sup_{t\in[0,T]}\|g(t,\cdot)\|_q\int_0^t(t-s)^{\frac{d}{4r}-\frac{d}{4}}\mathrm{d}s. \tag{3.6}$$

Since $q>4$ and $\frac{d}{4r}-\frac{d}{4}>-1$, it follows that $\|A_4(u)\|_{\mathscr{H}}<\infty$.

This completes the proof of this proposition.

**Proposition 3.2** Under the conditions of Theorem 3.1, the operator $Y$ is a contraction on $\mathscr{H}$; i.e., there exists a constant $\kappa\in(0,1)$ such that

$$\|Y(u)-Y(v)\|_{\mathscr{H}}\leqslant\kappa\|u-v\|_{\mathscr{H}}$$

for any $u,v\in\mathscr{H}$.

**Proof** Let $u,v\in\mathscr{H}$; we have the following estimates. One can see [2, 5] that

$$\|K_n(\|u(s,\cdot)\|_q)f(u(s,\cdot))-K_n(\|v(s,\cdot)\|)f(v(s,\cdot))\|_\rho\leqslant C_n\|u(s,\cdot)-v(s,\cdot)\|_q,$$

$$\|K_n(\|u(s,\cdot)\|_q)\sigma(u(s,\cdot))-K_n(\|v(s,\cdot)\|)\sigma(v(s,\cdot))\|_\rho\leqslant C_n\|u(s,\cdot)-v(s,\cdot)\|_q.$$

Let $u\in\mathscr{H}$. If $\rho\in[1,q]$ and $r\neq\infty$ if $d=3$ and $r<3$ if $d=3$ such that $1/q=1/r+1/\rho-1$, then, for $A_1$, we have

$$\begin{aligned}\|A_1(u)-A_1(v)\|_{\mathscr{H}}^q&\leqslant C_n\sup_{t\in[0,T]}\mathbb{E}\int_0^t|t-s|^{\frac{d}{4r}-\frac{d+2}{4}}\|u(s,\cdot)-v(s,\cdot)\|_q\mathrm{d}s\\&\leqslant C_n\sup_{t\in[0,T]}\left[\left(\int_0^t|t-s|^{\frac{d}{4r}-\frac{d+2}{4}}\mathrm{d}s\right)^{q-1}\times\right.\\&\left.\int_0^t|t-s|^{\frac{d}{4r}-\frac{d+2}{4}}\mathbb{E}\|u(s,\cdot)-v(s,\cdot)\|_q^q\mathrm{d}s\right]\leqslant C_nT^{q\left(\frac{2-d}{4}+\frac{d}{4r}\right)}\|u-v\|_{\mathscr{H}}^q.\end{aligned}$$

Choose $T_1$ such that $C_nT^{q\left(\frac{2-d}{4}+\frac{d}{4r}\right)}<1$. Then the map $A_1$ is a contraction on $\mathscr{H}$.

Let $u\in\mathscr{H}$; Burkhölder's inequality, (2.1), (2.6) and assumption (b) imply,

$$\begin{aligned}&\|A_2(u)-A_2(v)\|_{\mathscr{H}}^q\\&\leqslant C(q)\sup_{t\in[0,T]}\int_D\mathbb{E}\left(\left|\int_0^t\int_D G^2(t-s,x,y)[K_n(\|u(s,\cdot)\|_q)\sigma(u(s,\cdot))-\right.\right.\\&\left.\left.K_n(\|v(s,\cdot)\|)\sigma(v(s,\cdot))]^2\mathrm{d}y\mathrm{d}s\right|^{q/2}\right)\mathrm{d}x\\&\leqslant C(n,q)\sup_{t\in[0,T]}\mathbb{E}\left(\left|\int_0^t(t-s)^{\frac{d}{4r}-\frac{d}{2}}\|u(s,\cdot)-v(s,\cdot)\|_q^2\mathrm{d}s\right|^{q/2}\right)\\&\leqslant C(n,q)T^{q(-d/2+d/(4r))}\|u-v\|_{\mathscr{H}}^q.\end{aligned}$$

Choose $T_2$ such that $C(n,q)T^{q(-d/2+d/(4r))}<1$. Then the map $A_2$ is a contraction on $\mathscr{H}$.

Let $q>4$. By assumptions (b), (d) and Lemma 2.3, using the Hölder inequality, we have

$$\begin{aligned}&\|A_3(u)-A_3(v)\|_{\mathscr{H}}^q=\sup_{t\in[0,T]}\mathbb{E}(\|A_3(u(t,\cdot))-A_3(v(t,\cdot))\|_q^q)\\&\leqslant C\sup_{t\in[0,T]}\int_D\mathbb{E}\left(\left|\int_0^{t+}\int_D\int_{E_2}G(t-s,x,y)h(s,y,z)(K_n(\|u(s,\cdot)\|_q)\sigma(u(s^-,y))-\right.\right.\\&\left.\left.K_n(\|v(s,\cdot)\|_q)\sigma(v(s^-,y)))M(\mathrm{d}z,\mathrm{d}y,\mathrm{d}s)\right|^q\right)\mathrm{d}x\\&\leqslant C(q)\sup_{t\in[0,T]}\int_D\left(\left|\int_0^t\int_D\int_{E_2}(\mathbb{E}|G(t-s,x,y)h(s,y,z)(K_n(\|u(s,\cdot)\|_q)\sigma(u(s^-,y))-\right.\right.\\&\left.\left.K_n(\|v(s,\cdot)\|_q)\sigma(v(s^-,y)))|^q)^{\frac{2}{q}}\right)\mu_2(\mathrm{d}z)\mathrm{d}y\mathrm{d}s\right|^{\frac{q}{2}}\mathrm{d}x\end{aligned}$$

$$\leqslant C(q) \sup_{t\in[0,T]}\int_D \left|\int_0^t\int_D\int_{E_2} |G(t-s,x,y)h(s,y,z)|^2 \times (\mathbb{E}(|(K_n(\|u(s,\cdot)\|_q)\sigma(u(s^-,y)) - K_n(\|v(s,\cdot)\|_q)\sigma(v(s^-,y)))|)^q)^{2/q}\mu_2(\mathrm{d}z)\mathrm{d}y\mathrm{d}x\right|^{\frac{q}{2}}\mathrm{d}x$$

$$\leqslant C(q) \sup_{t\in[0,T]}\int_D\left(\int_0^t\int_D\int_{E_2} \mathbb{E}(|K_n(\|u(s,\cdot)\|_q)u(s,y)) - K_n(\|v(s,\cdot)\|_q)v(s,y)|)^q \times \mu_2(\mathrm{d}z)\mathrm{d}y\mathrm{d}s\left(\int_0^t\left|\int_D\int_{E_2}|G(t-s,x,y)h(s,y,z)|^{\frac{2q}{q-2}}\mu_2(\mathrm{d}z)\mathrm{d}y\mathrm{d}s\right|^{\frac{q}{2}}\right)^{\frac{q-2}{2}}\right)\mathrm{d}x$$

$$\leqslant C(n,q)T\mu_2(E_2)\|u-v\|_{\mathscr{H}}^q\int_0^t\int_D\int_{E_2}|G(t-s,x,y)h(s,y,z)|^q\mu_2(\mathrm{d}z)\mathrm{d}y\mathrm{d}s.$$

Choose $T_3$ such that

$$C(n,q)T\mu_2(E_2)\int_0^t\int_D\int_{E_2}|G(t-s,x,y)h(s,y,z)|^q\mu_2(\mathrm{d}z)\mathrm{d}y\mathrm{d}s<1.$$

Then the map $A_3$ is a contraction on $\mathscr{H}$.

Applying (2.4), with $\rho=q/2$ and $\frac{1}{r}=\frac{1}{q}-\frac{1}{\rho}+1$, we have

$$\|A_4(u)-A_4(v)\|_{\mathscr{H}}^q\leqslant C\sup_{t\in[0,T]}\int_0^t(t-s)^{\frac{d}{4r}-\frac{d}{4}}\|g(s,\cdot)(\sigma(u(s,\cdot))K_n(\|u(s,\cdot)- \sigma(v(s,\cdot))K_n(\|v(s,\cdot)\|_q))\|_\rho\mathrm{d}s$$

$$\leqslant C\int_0^t(t-s)^{\frac{d}{4r}-\frac{d}{4}}C(n,q)\|g(s,\cdot)\|_q\|u(s,y)-v(s,y))\|_q^q\mathrm{d}s$$

$$\leqslant C\sup_{t\in[0,T]}\|g(t,\cdot)\|_qT^{\frac{d}{4r}-\frac{d}{4}+1}\|u-v\|_{\mathscr{H}}^q.$$

Since $q>4$ and $\frac{d}{4r}-\frac{d}{4}>-1$, choose $T$ such that

$$C\sup_{t\in[0,T]}\|g(t,\cdot)\|_qT^{\frac{d}{4r}-\frac{d}{4}+1}<1.$$

Then the map $A_4$ is a contraction on $\mathscr{H}$.

Finally, let $T=\frac{1}{4}\min\{T_1,T_2,T_3,T_4\}<1$. The map $Y$ is a contraction on $\mathscr{H}$, thus we complete the proof of this proposition. □

Using Proposition 3.1, Proposition 3.2, and the fixed-point argument on the space $\mathscr{H}$ endowed with the norm 3.2, one can get Theorem 3.1.

**Remark 3.1** Bo and Wang[2] have proved that Theorem 3.1 holds for $q\in(8,\infty)$ under the same assumptions in Theorem 3.1. We can improve the result if we use the norm (3.2), but Bo and Wang define a norm $\|\cdot\|_{\mathscr{H}}$ (depending on $(\lambda,q)$) on $\mathscr{H}$ by $\|u\|_{\mathscr{H}}=(\sup_{t\in[0,T]}e^{-\lambda t}(\mathbb{E}(\|u(t,\cdot)\|_q^q))^{1/q}$.

## 4 Global Solution

By uniqueness of the solution to (3.1), the local property of stochastic integrals yields, for

$m>n$, $u_m(t,\cdot)=u_n(t,\cdot)$ if $t<\tau_n$, so that we can define a process $u$ by $u(t,\cdot)=u_n(t,\cdot)$ on $t\leqslant\tau_n$. Set $\tau_\infty=\lim_{n\to\infty}\tau_n$; clearly $u$ is a solution to (3.1) on the interval $[0,\tau_\infty)$, and is unique. We just need to prove that $\tau_\infty=+\infty$ almost surely, and use an argument similar to that of reference [13,5].

In order to obtain the global solution, we need stronger conditions than the assumptions given in Section 3:

(b)′ $\sigma:\mathbf{R}\to\mathbf{R}$ is a bounded and Lipschitzian function;

(d)′ $G(t-s,x,y)h(s,y,z)$ is $L^{2q}([0,t]\times D^2\times E_2)$, $ds\times dx\times dy\times\mu_2(E_2)$ integrable, for $0\leqslant t\leqslant T$.

Now we can give our main result, the global well posedness of the problem ($P$), in the present paper.

**Theorem 4.1** Suppose the assumptions (a), (b)′, (c), (d)′ and (e) are satisfied. Let $d\in\{1,2,3\}$ and $q\in(4,\infty)$. Then for every $\mathscr{F}_0$-measurable $u_0$: $D\times\Omega\to\mathbf{R}$ with $\mathbb{E}(\|u_0(\cdot)\|_q^q)<\infty$, for any $T>0$, there exists a unique solution $\{u(t,x):(t,x)\in[0,T]\times D\}$ to the problem ($P$) such that $\mathbb{E}(\|u(t,\cdot)\|_q^q)<\infty$, for any $t\in[0,T]$.

Before proving this theorem, we first prove two *a priori* estimates in the following two propositions.

Let $H$, $L$ be defined respectively by

$$H(u)(t,x)=A_1(u)(t,x)$$

$$L(u)(t,x)=A_2(u)(t,x)+A_3(u)(t,x)+A_4(u)(t,x).$$

It is obvious that, for all $L^q(D)$-valued $\mathscr{F}_t$ adapted r.c.l.l. processes $u(t,x)$, the following norm

$$\|u\|_{\mathscr{H}_1}=(\sup_{t\in[0,T]}(\mathbb{E}(\|u(t,\cdot)\|_q^\beta))^{1/\beta} \tag{4.1}$$

is finite. Thus $\mathscr{H}_1$ is a Banach space.

**Proposition 4.1** Let the assumptions (a), (b)′, (c), (d)′ and (e) hold; then we have

$$\sup_n\sup_{t\in[0,T]}\sup_{x\in D}\mathbb{E}(|L(u_n)(t,x)|^\delta)<+\infty,$$

for $\delta\in[1,2q]$.

**Proof** Assume the assumptions (a), (b)′, (c), (d)′ and (e) hold. We have the following facts, for $u\in\mathscr{H}_1$.

As for $A_2(u)(t,x)$, similar to (3.4), we have

$$\begin{aligned}
&\|A_2(u)\|_{\mathscr{H}_1}^\beta\\
&\leqslant C\sup_{t\in[0,T]}\int_D\mathbb{E}\left(\left|\int_0^t\int_D G(t-s,x,y)\sigma(u(s,y))K_n(\|u(s,\cdot)\|_q)W(dy,ds)\right|^\beta\right)dx\\
&\leqslant C\sup_{t\in[0,T]}\int_D\mathbb{E}\left(\left|\int_0^t\int_D G^2(t-s,x,y)\sigma^2(u(s,y))K_n^2(\|u(s,\cdot)\|_q)dyds\right|\right)^{\frac{\beta}{2}}dx\\
&\leqslant C(T)\|\sigma\|_\infty^\beta\leqslant\infty.
\end{aligned}$$

As for $A_3(u)(t,x)$, similar to (3.5), for $\beta\leqslant 2q$, we have

$$\begin{aligned}
\|A_3(u)\|_{\mathscr{H}_1}^{\beta} &= \sup_{t\in[0,T]} \mathbb{E}(\|A_3(u(t,\cdot))\|_q^{\beta}) \\
&\leqslant C \sup_{t\in[0,T]}\int_D \mathbb{E}\left(\left|\int_0^{t+}\int_D\int_{E_2} G(t-s,x,y)\sigma(u(s,y))h(s,y,z)M(\mathrm{d}z,\mathrm{d}y,\mathrm{d}s)\right|^{\beta}\right)\mathrm{d}x \\
&\leqslant C \sup_{t\in[0,T]}\int_D \left|\int_0^{t}\int_D\int_{E_2} (\mathbb{E}|G(t-s,x,y)\sigma(u(s,y))h(s,y,z)|^{\beta})^{\frac{2}{\beta}}\mu_2(\mathrm{d}z)\mathrm{d}y\mathrm{d}s\right|^{\frac{\beta}{2}}\mathrm{d}x \\
&\leqslant C(T)\|\sigma\|_{\infty}^{\beta} \sup_{t\in[0,T]}\int_D \left|\int_0^{t}\int_D\int_{E_2} |G(t-s,x,y)h(s,y,z)|^2\mu_2(\mathrm{d}z)\mathrm{d}y\mathrm{d}s\right|^{\frac{\beta}{2}}\mathrm{d}x \\
&\leqslant C(T)\|\sigma\|_{\infty}^{\beta} \sup_{t\in[0,T]}\int_D\int_0^{t}\int_D\int_{E_2} |G(t-s,x,y)h(s,y,z)|^{2q}\mu_2(\mathrm{d}z)\mathrm{d}y\mathrm{d}s\mathrm{d}x.
\end{aligned}$$

For $A_4(u)(t,x)$, similar to (3.6), we have

$$\|A_4(u)\|_{\mathscr{H}_1}^{\beta} \leqslant C \sup_{t\in[0,T]}\|g(s,\cdot)\|_q^{\beta}\|\sigma\|_{\infty}^{\beta}\left(\int_0^t (t-s)^{\frac{d}{4r}-\frac{d}{4}}\mathrm{d}s\right)^{\beta}.$$

Since $q>4$ and $\frac{d}{4r}-\frac{d}{4}>-1$, it follows that $\|A_4(u)\|_{\mathscr{H}_1}^{\beta}<\infty$. The above estimates tell us that

$$\sup_n \sup_{t\in[0,T]} \sup_{x\in D} \mathbb{E}(|L(u_n)(t,x)|^{\delta}) < \infty,$$

where $\delta\in[1,2q]$. Thus we complete the proof of this proposition. □

We just need to prove a uniform upper estimate for $H(u_n)$. To this end we note first that $v_n = u_n - L(u_n)$ is a solution of the following system:

$$\begin{aligned}
&\frac{\partial v_n}{\partial t}(t,x) + \Delta^2 v_n(t,x) - \\
&\qquad \Delta[K_n(\|v_n(t,\cdot)+L(u_n)(t,\cdot)\|_q)f(v_n(t,x)+L(u_n)(t,x))] = 0,
\end{aligned} \tag{4.2}$$

with the initial boundary conditions

$$v_n(0,\cdot) = u_0(\cdot), \quad \frac{\partial v_n}{\partial n} = \frac{\partial \delta v_n}{\partial n} = 0.$$

Again the above problem can be made rigorous as problem $(P)$. For any $\phi\in C^4(D)$ such that $\phi$ satisfies $\frac{\partial\phi}{\partial n}=\frac{\partial\Delta\phi}{\partial n}=0$ on $\partial D$, $u$ satisfies

$$\begin{aligned}
\int_D[v_n(t,x)-u_0(x)]\phi(x)\mathrm{d}x &= -\int_0^t\int_D v_n(s,x)\Delta^2\phi(x)\mathrm{d}x\mathrm{d}s + \\
&\int_0^t\int_D K_n(\|v_n(s,\cdot)+L(u_n)(s,\cdot)\|_q)f(v_n(s,x)+L(u_n)(s,x))\Delta\phi(x)\mathrm{d}x\mathrm{d}s.
\end{aligned}$$

Using the Galerkin method, we can prove that, for every $T>0$, $v_n$ is the weak solution on $[0,T]$ to the above system, and $v_n\in L^2([0,T],W^{2,2}(D))$, where $W^{2,2}(D)$ denotes the Sobolev space.

**Proposition 4.2** Let the assumptions (a), (b)′, (c), (d)′ and (e) hold; then we have, for $k\in[1,q/3]$,

$$\sup_n \mathbb{E}(\sup_{t\in[0,T]}\|H(u_n)(t,\cdot)\|_q^k) < \infty.$$

**Proof** Recall that, for $A = -\Delta$, $\alpha \neq 0$, and $u \in \mathrm{Dom}(A^{\alpha})$, by convention, denote by $\mathbf{N}^{d,*}$ the set $\mathbf{N}^d/\{0\}$, then

$$A^{\alpha} = \sum_{k \in \mathbf{N}^{d,*}} \lambda_k^{\alpha} \langle \varepsilon_k, u \rangle \varepsilon_k.$$

$A^{\alpha}u$ exists for every $u$ such that $\sum_{k \in \mathbf{N}^{d,*}} \lambda_k^{2\alpha} \langle \varepsilon_k, u \rangle^2 < \infty$. In what follows, for a function $u$: $[0, T] \times D \to \mathbf{R}$, we set

$$m(u)(t) = \langle \varepsilon_0, u(t, \cdot) \rangle = \pi^{-\frac{d}{2}} \int_D u(t, x) \mathrm{d}x \quad \text{and} \quad \tilde{u}(t, y) = u(t, y) - m(u)(t).$$

Notice that $A^{\alpha}\tilde{u}(t, y) = A^{\alpha}u(t, y)$ for $\alpha \neq 0$. Apply $A^{-1}$ to equation (4.2) and take its scalar product in $L^2(D)$ with $\tilde{v}_n(t, \cdot)$; this leads to

$$\|A^{-\frac{1}{2}} \tilde{v}_n(t, \cdot)\|_2^2 - \|A^{-\frac{1}{2}} \tilde{v}_n(0, \cdot)\|_2^2 + \int_0^t \|A^{\frac{1}{2}} \tilde{v}_n(s, \cdot)\|_2^2 \mathrm{d}s +$$

$$\int_0^t K_n(\|v_n(s, \cdot) + L(u_n)(s, \cdot)\|_q) \int_D f(v_n(s, x) + L(u_n)(s, x)) \tilde{v}_n(t, x) \mathrm{d}x \mathrm{d}s = 0.$$

Let us deal with the last term in the above inequality.

$$\int_D f(v_n(s, x) + L(u_n)(s, x)) \tilde{v}_n(t, x) \mathrm{d}x$$

$$= \int_D f(v_n(s, x) + L(u_n)(s, x))(v_n(s, x) + L(u_n)(s, x)) \mathrm{d}x -$$

$$\int_D f(v_n(s, x) + L(u_n)(s, x))(L(u_n)(s, x) + m(v_n)(s)) \mathrm{d}x = \text{I} + \text{II}.$$

The polynomial $f$ is of degree 3 with positive dominant coefficient, hence, $\lim_{|x| \to +\infty} f(x) = +\infty$, and there exists $a$, $C$ and $c > 0$ such that

$$\text{I} \geqslant \frac{7}{8} a \|v_n(s, \cdot) + L(u_n)(s, \cdot)\|_4^4 - c.$$

In addition,

$$\text{II} \leqslant \frac{5}{4} \int_D |v_n(s, x) + L(u_n)(s, x)|^3 (|L(u_n)(s, x)| + |m(v_n)(s)|) \mathrm{d}x +$$

$$C \int_D (|L(u_n)(s, x)| + |m(v_n)(s)|) \mathrm{d}x$$

$$\leqslant \frac{5}{8} a \|v_n(s, \cdot) + L(u_n)(s, \cdot)\|_4^4 + C(|m(v_n)(s)|^4 + \|L(u_n)(s, \cdot)\|_4^4).$$

Since $K_n$ is a positive bounded function,

$$\|A^{-\frac{1}{2}} \tilde{v}_n(t, \cdot)\|_2^2 - \|A^{-\frac{1}{2}} \tilde{v}_n(0, \cdot)\|_2^2 + \int_0^t \|A^{\frac{1}{2}} \tilde{v}_n(s, \cdot)\|_2^2 \mathrm{d}s +$$

$$\frac{a}{4} \int_0^t K_n(\|v_n(s, \cdot) + L(u_n)(s, \cdot)\|_q) \|v_n(s, \cdot) + L(u_n)(s, \cdot)\|_4^4 \mathrm{d}s$$

$$\leqslant \int_0^t C(1 + |m(v_n)(s)|^4 + \|L(u_n)(s, \cdot)\|_4^4) \mathrm{d}s.$$

Taking the scalar product of the solution to the equation with the function $\varepsilon_0$, we obtain

$$\frac{\partial}{\partial t} \langle v_n(t, \cdot), \varepsilon_0 \rangle = 0, \langle v_n(0, \cdot), \varepsilon_0 \rangle = \langle u_0(\cdot), \varepsilon_0 \rangle,$$

hence, $m(u_0)=\langle \varepsilon, u_0\rangle$. Since $A^{-\frac{1}{2}}\tilde{u}=A^{-\frac{1}{2}}u_0$, it follows that, for every $T>0$, we have

$$\|A^{-\frac{1}{2}}\tilde{v}_n(T,\cdot)\|_2^2-\|A^{-\frac{1}{2}}u_0(\cdot)\|_2^2+\int_0^T\|A^{\frac{1}{2}}v_n(s,\cdot)\|_2^2\mathrm{d}s+$$

$$\frac{a}{4}\int_0^T K_n(\|v_n(s,\cdot)+L(u_n)(s,\cdot)\|_q)\|v_n(s,\cdot)+L(u_n)(s,\cdot)\|_4^4\mathrm{d}s$$

$$\leqslant\int_0^T C(1+|m(u_0)(s)|^4+\|L(u_n)(s,\cdot)\|_4^4)\mathrm{d}s.$$

This yields

$$\int_0^t K_n(\|v_n(s,\cdot)+L(u_n)(s,\cdot)\|_q)\|v_n(s,\cdot)+L(u_n)(s,\cdot)\|_4^4\mathrm{d}s$$
$$\leqslant C_T(1+|m(u_0)|^4+\|L(u_n)(s,\cdot)\|_\infty^4)\mathrm{d}s+\|A^{-\frac{1}{2}}\tilde{v}_n(0,\cdot)\|_2^2. \tag{4.3}$$

We now take the scalar product in $L^2(D)$ of equation (4.2) with $v_n$; using Green's formula, we obtain

$$\frac{1}{2}\frac{\partial}{\partial t}\|v_n\|_2^2+\|\Delta v_n(t,\cdot)\|_2^2$$
$$=K_n(\|v_n(t,\cdot)+L(u_n(t,\cdot))\|_q)\int_D f(v_n(t,x)+L(u_n)(t,x))\Delta v_n(t,x)\mathrm{d}x.$$

Noticing (4.3), we have

$$\|v_n\|_2^2+\int_0^t[\|\Delta[v_n(t,\cdot)]\|_2^2+m(v_n(s,\cdot))^2]\mathrm{d}s$$

$$\leqslant\|u_0\|_2^2+C_T(1+\|L(u_n)\|_\infty^6)+C_T(1+\|L(u_n)\|_\infty^6)[\|A^{-\frac{1}{2}}u_0\|_2^2+m(u_0)^4].$$

Proposition 4.1 tells us that, for $k\in[1,q/3]$,

$$\sup_n\mathbb{E}(\sup_{t\in[0,T]}\|v_n(t,\cdot)\|_2^{2k})<\infty,$$
$$\sup_n\mathbb{E}\Big(\int_0^T[\|\Delta[v_n(t,\cdot)\|_2^2+m(v_n(s,\cdot))^2]\mathrm{d}s]^k\Big)<\infty.$$

Furthermore, by Sobolev's embedding theorem there exists $C>0$ such that, for $r\geqslant2, d<4$, if $u\in W^{2,2}(D)$,

$$\|u\|_{L^r(D)}\leqslant C\|u\|_{W^{2,2}(D)}.$$

Thus, for $2\leqslant r<\infty$,

$$\sup_n\mathbb{E}\Big(\Big[\int_0^T\|v_n(t,\cdot)\|_r^2\mathrm{d}t\Big]^k\Big)<\infty.$$

For $2\leqslant r<\infty$,

$$\sup_n\mathbb{E}(\sup_{t\in[0,T]}\|u_n(t,\cdot)\|_2^{2k})<\infty,\quad\sup_n\mathbb{E}\Big(\Big[\int_0^T\|\Delta u_n(t,\cdot)\|_r^2\mathrm{d}s\Big]^k\Big)<\infty.$$

Let us use the interpolation method to prove that $u_n$ belongs almost surely to $L^a([0,T],L^q(D))$, with $2\leqslant q\leqslant r, a\geqslant1\vee\frac{2q}{r}$. The Hölder inequality implies that, if $q=2(1-\lambda)+r\lambda$, for $\lambda\in[0,1]$, that

$$\int_0^T\|u_n(t,\cdot)\|_q^a\mathrm{d}t\leqslant\int_0^T\|u_n(t,\cdot)\|_2^{2a(1-\lambda)/q}\|u_n(t,\cdot)\|_r^{ar\lambda/q}\mathrm{d}t.$$

Taking $\lambda = 2q/(ar)$, we obtain

$$\int_0^T \| u_n(t,\cdot)\|_q^a \mathrm{d}t \leqslant \sup_{t\in[0,T]} \| u_n(t,\cdot)\|_2^{2a(1-\lambda)/q} \int_0^T \| u_n(t,\cdot)\|_r^2 \mathrm{d}t$$

for $q\in[2,\infty)$ and $a\in[q,\infty)$ and

$$\sup_n \mathbb{E}\left(\left[\int_0^T \| u_n(t,\cdot)\|_q^a \mathrm{d}t\right]^k\right) < \infty,$$

$$\| H(u_n)(t,\cdot)\|_q \leqslant C\int_0^t (t-s)^{\frac{d}{4r}-\frac{d+2}{4}} (\| u_n(s,\cdot)\|_q^3 + 1)\mathrm{d}s.$$

Let $\gamma$, $\gamma' \in (1,\infty)$ be conjugate exponents, with $\gamma$ close enough to unity to ensure $\left(\frac{d}{4r} - \frac{d+2}{4}\right)\gamma > -1$; the Hölder inequality implies that

$$\| H(u_n)(t,\cdot)\|_q \leqslant C\left[\int_0^t (t-s)^{\left(\frac{d}{4r}-\frac{d+2}{4}\right)\gamma} \mathrm{d}s\right]^{\frac{1}{\gamma}} \left[\int_0^t (\| u_n(s,\cdot)\|_q^3 + 1)^{\gamma'} \mathrm{d}s\right]^{\frac{1}{\gamma'}}.$$

We obtain

$$\sup_n \mathbb{E}(\sup_{t\in[0,T]} \| H(u_n)(t,\cdot)\|_q^k) < \infty.$$

Thus we end the proof of this proposition. □

**The proof of Theorem (4.1)** Noticing Proposition 4.1, Proposition 4.2, and the estimates (3.3), we have

$$\sup_n \mathbb{E}(\sup_{t\in[0,T]} \| u_n(t,\cdot)\|_q^k) < \infty.$$

We can now conclude that $\tau_\infty = +\infty$ almost surely; indeed, for every $T>0$,

$$P(\tau_n \leqslant T) = P(\sup_{t\leqslant T} \| u_n(t,\cdot)\|_q \geqslant n) \leqslant \mathbb{E}(\sup_{t\leqslant T} \| u_n(t,\cdot)\|_q^{2k}) n^{-2k},$$

so that $\lim_{n\to\infty} P(\tau_n \leqslant T) = 0$, Therefore, we can construct the solution to the problem $(P)$ on every interval $[0,T]$. Then we obtain the main theorem in this paper. □

## References


[1] Albeverio J, Wu J, Zhang T. Parabolic SPDEs driven by Poisson white noise. Stoch Proc Appl, 1998, 74: 21-36.

[2] Bo L, Wang Y. Stochastic Cahn-Hilliard partial differential equations with Lévy spacetime white noise. Stochastics and Dynamics, 2006, 6: 229-244.

[3] Cahn J W. On spinodal decomposition. Acta Metallurgica, 1961, 9: 795-801.

[4] Cahn J W, Hilliard J E. Free energy for a nonuniform system. I. Interfacial free energy. J Chem Phys, 1958, 2: 258-267.

[5] Cardon-Weber C. Cahn-Hilliard stochastic equation: existence of the solution and of its density. Bernoulli, 2001, 7: 777-816.

[6] Debussche A, Dettori L. On the Cahn-Hilliard equation with a Logarithmic free energy. Nonlinear Anal, 1995, 24: 1497-1514.

[7] Eidelman S D, Ivasisen N V. Investigation of the Green matrix for a homogeneous parabolic boundary value problem. Trans Moscow Math Soc, 1970, 23: 179-242.

[8] Elliott C M, Zhang S. On the Cahn-Hilliard equation. Arch Rat Mech Anal, 1986, 96: 339-357.

[9] Grant C P. Spinodal decomposition for the Cahn-Hilliard equation. Comm Partial Differential Equations, 1993, 18: 453 - 490.

[10] Gyöngy J. Existence and uniqueness results for semilinear stochastic partial differential equations. Stochastic Process Appl, 1998, 73: 271 - 299.

[11] Knoche C. Mild solutions of SPDEs driven by Poisson noise in infinite dimensions and their dependence on initial conditions. PhD thesis, Universität Bielefeld, 2005.

[12] Novick A, Segel L A. Nonlinear aspects of the Cahn-Hilliard equation. Physica D, 1984, 10: 277 - 298.

[13] Da Prato G, Debussche A. Stochastic Cahn-Hilliard equation. Nonlinear Anal, 1996, 26: 241 - 263.

[14] Da Prato G, Gatarek D. Stochastic Burgers equation with correlated noise. Stochastics Stochastics, 1992, 52: 29 - 41.

[15] Protter P. Stochastic integration and differential equations. Springer, 1990.

[16] Truman A, Wu J. Stochastic Burgers equation with Lévy space-time white noise // Probabilistic Method in Fluids (World Scientific, 2003), 298 - 323.

[17] Walsh J. An introduction to stochastic partial differential equations. Ecole d'Eté de Probabilités de Saint-Flour XIV // Lecture Notes in Math, Vol 1180, Springer, 1986.

# Existence and Stability of Steady Waves for the Hasegawa-Mima Equation *

Guo Boling(郭柏灵) Huang Daiwen(黄代文)

**Abstract**

By introducing a compactness lemma and considering a constrained variational problem, we obtain a set $G_{\mathbf{R}^2}$ of steady waves for Hasegawa-Mima equation, which describes the motion of drift waves in plasma. Moreover, we prove that $G_{\mathbf{R}^2}$ is a stable set for the initial value problem of the equation, in the sense that a solution $\psi(t)$ which starts near $G_{\mathbf{R}^2}$ will remain near it for all time.

## 1 Introduction

It is commonly believed that drift waves and drift-wave turbulence play a major role in the understanding of anomalous transport at the plasma edge of a tokamak fusion reactor. One-field equation describing the electrostatic potential fluctuations in this regime is Hasegawa-Mima equation:

$$\frac{\partial(\Delta\psi-\psi)}{\partial t}+\frac{\partial\psi}{\partial x}\frac{\partial(\Delta\psi-\psi)}{\partial y}-\frac{\partial\psi}{\partial y}\frac{\partial(\Delta\psi-\psi)}{\partial x}=0, \tag{1.1}$$

where $(x,y)\in\mathbf{R}^2$, $\psi$ describes the electrostatic fluctuation and $\psi(x,y)\to 0$ as $|x|+|y|\to+\infty$. The derivation of (1.1) can be found in [1].

There are many works about analytical mathematical study for (1.1); see, for example, [2, 3] and references therein. In [2], Grauer proved that the energy for a perturbed Hasegawa-Mima equation saturates at a finite level, which was observed by numerical simulations. Guo and Han in [3] studied the global well-posedness of Cauchy problem for (1.1). One of their results is that the solution $\psi(t)$ of (1.1) with $\psi(0)\in W^{2,2}(\mathbf{R}^2)\cap W^{2,\infty}(\mathbf{R}^2)$ exists globally and is unique. However, the global well-posedness of (1.1) with $\psi(0)\in W^{2,q}(\mathbf{R}^2)$ is still not attacked, where $W^{2,q}(\mathbf{R}^2)$ is the usual Sobolev space with norm $\|\cdot\|_{2,q}$ and $1<q<2$. A natural problem is whether the solution of (1.1) with the initial data $\psi(0)$ is close to a steady wave $\psi_0$ for all time or not, if $\psi(0)$ is sufficiently close to $\psi_0$ in $W^{2,q}(\mathbf{R}^2)$. The problem is concerned with the existence and stability of steady waves of (1.1).

Here, we are interested in studying the above problem. $\psi_0$ is a steady wave for (1.1) if and only if there exists some function $f$ such that $\psi_0=f(\Delta\psi_0-\psi_0)$. In order to prove the

* Boundary Value Problems, 2009, 509801.

existence and nonlinear stability of steady waves for (1.1), we consider the existence and property of critical points of the so-called energy-Casimir functional:

$$I_{\mathbf{R}^2}(\psi) = \Phi_{\mathbf{R}^2}(\psi) + \Psi_{\mathbf{R}^2}(\psi), \tag{1.2}$$

where

$$\left.\begin{aligned} \Phi_{\mathbf{R}^2}(\psi) &= \frac{1}{2}\int_{\mathbf{R}^2}(|\nabla\psi|^2 + |\psi|^2), \\ \Psi_{\mathbf{R}^2}(\psi) &= \int_{\mathbf{R}^2} F(\Delta\psi - \psi), \end{aligned}\right\} \tag{1.3}$$

are two conserved quantities of (1.1), called the total energy and the generalized enstrophy respectively. Here $\int_{\mathbf{R}^2}\cdot\, dx dy$ is denoted by $\int_{\mathbf{R}^2}\cdot$, and $F$ is an arbitrary $C^1$ function. The critical points $\psi_0$ of $I_{\mathbf{R}^2}$ are steady waves of (1.1), given by $\psi_0 = f(\Delta\psi_0 - \psi_0)$, where $f = F'$.

A usual approach to prove the existence of stable critical points of $I_{\mathbf{R}^2}$ is to find extremum points of it, which is the well-known Liapunov method. If $\psi_0$ is a global or local extremum point of $I_{\mathbf{R}^2}$ in an appropriate defined function space $X$, then it follows that $\psi_0$ is a steady nonlinearly stable solutions of (1.1); see, for example, [4]. There are two examples for $F$ such that $I_{\mathbf{R}^2}$ have a global extremum. One is $F$ satisfying that $F''(x) \geqslant 0$ for any $x \in \boldsymbol{R}$ and there exists $\psi_0$ such that $\psi_0 = F'(\Delta\psi_0 - \psi_0)$. In this case,

$$I_{\mathbf{R}^2}(\psi) - I_{\mathbf{R}^2}(\psi_0) \geqslant \frac{1}{2}\int_{\mathbf{R}^2}(|\nabla(\psi - \psi_0)|^2 + |\psi - \psi_0|^2), \tag{1.4}$$

which implies that $\psi_0$ is a global minimizer of $I_{\mathbf{R}^2}$. Therefore, the steady wave $\psi_0$ is nonlinearly stable in the following sense: for any $\varepsilon > 0$, there exists $\delta > 0$ such that if $\|\psi(0) - \psi_0\|_X < \delta$ and $\psi(t) \in C([0, T), X)$ is a solution of (1.1) with initial data $\psi(0)$, then for any $t \in [0, T)$, $\|\psi(t) - \psi_0\|_1 < \varepsilon$, where $\|\cdot\|_1$ is the norm of $H^1(\mathbf{R}^2)$ and $T > 0$. The other example is $F$, which has the properties: $F''(x) \leqslant -c$ for large enough $c > 0$ and there exists $\psi_0$ such that $F'(\Delta\psi_0 - \psi_0) = \psi_0$. In this example,

$$\begin{aligned} I_{\mathbf{R}^2}(\psi) - I_{\mathbf{R}^2}(\psi_0) &\leqslant \frac{1}{2}\int_{\mathbf{R}^2}(|\nabla(\psi - \psi_0)|^2 + |\psi - \psi_0|^2) - c\int_{\mathbf{R}^2}|\Delta(\psi - \psi_0) - (\psi - \psi_0)|^2 \\ &\leqslant -c'\int_{\mathbf{R}^2}|\Delta(\psi - \psi_0) - (\psi - \psi_0)|^2, \end{aligned} \tag{1.5}$$

where $c' > 0$. So $\psi_0$ is a global maximizer of the functional $I_{\mathbf{R}^2}$ and nonlinearly stable in the above sense.

However, for some $F$, all critical points of $I_{\mathbf{R}^2}$ are neither global nor local extremum points of $I_{\mathbf{R}^2}$. Among them, some critical points are saddle points regarded as an unstable equilibria or transient excited state of (1.1). In the present paper, we consider the existence and stability defined later of saddle points of $I_{\mathbf{R}^2}$ for $F(x) = -(1/q)|x|^q$, where $q = p/(p-1)$, $2 < p < \infty$, and $x \in \mathbf{R}$. Since $F''(x) < 0$ and there does not exist a positive constant $c > 0$ such that $F''(x) < -c$ for any $x \neq 0$, $F$ is not within the range of the above two examples. As shown in Proposition A.1 (Proposition A.1, Definition A.2, Proposition A.3, and Remark A.4 are given in the appendix), the functional (1.2) is neither bounded from above nor from below in $W^{2,q}(\mathbf{R}^2)$, that is, it is impossible to prove the existence of critical points by finding global extremum points of $I_{\mathbf{R}^2}$. However, through studying the constrained variational problem

$$M_{\mathbf{R}^2}=\inf_{\tilde{\psi}\in W^{2,q}(\mathbf{R}^2),\ \|\tilde{\psi}\|_1=1}\int_{\mathbf{R}^2}|\Delta\tilde{\psi}-\tilde{\psi}|^q, \qquad (M_{\mathbf{R}^2})$$

we obtain the existence of critical points of (1.2). In fact, if $\tilde{\psi}_0$ is a minimizer for $(M_{\mathbf{R}^2})$, then, according to Lagrange Multiplier Principle, with the transform $\psi_0=M_{\mathbf{R}^2}^{1/(2-q)}\tilde{\psi}_0$, $\psi_0$ is a steady wave of (1.1) in $\mathbf{R}^2$. With Definition A.2 and Proposition A.3, $\psi_0$ is a ground state and saddle point of $I_{\mathbf{R}^2}$. Let $Z_{\mathbf{R}^2}$ be the set of all minimizers for $(M_{\mathbf{R}^2})$, that is,

$$Z_{\mathbf{R}^2}=\left\{\tilde{\psi}_0;\int_{\mathbf{R}^2}(|\nabla\tilde{\psi}_0|^2+|\tilde{\psi}_0|^2)=1,\int_{\mathbf{R}^2}(|\Delta\tilde{\psi}_0-\tilde{\psi}_0|^q)=M_{\mathbf{R}^2}\right\}, \tag{1.6}$$

and let $G_{\mathbf{R}^2}$ be the set of steady waves of (1.1) corresponding to minimizers of $(M_{\mathbf{R}^2})$, that is,

$$G_{\mathbf{R}^2}=\{\psi_0;\psi_0=M_{\mathbf{R}^2}^{1/(2-q)}\tilde{\psi}_0,\tilde{\psi}_0\in Z_{\mathbf{R}^2}\}. \tag{1.7}$$

As is presented in Remark A.4, $G_{\mathbf{R}^2}$ is the set of all ground states of the functional $I_{\mathbf{R}^2}$. Although the elements of $G_{\mathbf{R}^2}$ are saddle points of $I_{\mathbf{R}^2}$ regarded as an unstable state of (1.1), we prove that $G_{\mathbf{R}^2}$ is a stable set in the sense of Definition 1.1, that is, a solution $\psi(t)$ of (1.1) which starts near $G_{\mathbf{R}^2}$ will remain near it for all time.

**Definition 1.1** Let $E$ be a function space with norm $\|\cdot\|_E$, and $T\in(0,\infty]$. A set $G\subset E$ is called $E$-stable with respect to (1.1) in $\mathbf{R}^2$, if for any $\varepsilon>0$, there exists a $\delta>0$ such that if $\psi\in C([0,T),E)$ is a solution to (1.1) with initial data $\psi(0)$ satisfying $\inf_{\psi_0\in G}\|\psi(0)-\psi_0\|_E\leqslant\delta$, then for any $t\in[0,T)$, $\inf_{\psi_0\in G}\|\psi(t,\cdot)-\psi_0\|_E\leqslant\varepsilon$.

One gives some explanations for the above definition as follows. If $G$ has only one element $\psi_0$, then the steady wave $\psi_0$ is nonlinearly stable in the usual sense. But, in general, the elements of $G$ might not be unique. For example, as shown in Theorem 1.3, $G_{\mathbf{R}^2}=\{M_{\mathbf{R}^2}^{1/(2-q)}\tilde{\psi}_0(\cdot+y),-M_{\mathbf{R}^2}^{1/(2-q)}\tilde{\psi}_0(\cdot+y);y\in\mathbf{R}^2\}$. In this case, if $\psi(0)$ is sufficiently close to $G_{\mathbf{R}^2}$, then, for any $t>0$, the form of $\psi(t)$ is almost similar to that of $M_{\mathbf{R}^2}^{1/(2-q)}\tilde{\psi}_0$ or $-M_{\mathbf{R}^2}^{1/(2-q)}\tilde{\psi}_0$.

Now one turns to describe his two main results.

**Theorem 1.2** If $\{\tilde{\psi}_m\}$ is a minimizing sequence for $(M_{\mathbf{R}^2})$, then there exist $\{y_{m_k}\}\subset\mathbf{R}^2$ and a subsequence $\{\tilde{\psi}_{m_k}\}$ such that $\{\tilde{\psi}_{m_k}(\cdot+y_{m_k})\}$ is a convergent sequence in $W^{2/q}(\mathbf{R}^2)$. In particular, the minimization problem $(M_{\mathbf{R}^2})$ has a minimizer $\tilde{\psi}_0$.

**Theorem 1.3** According to Definition 1.1, the ground state set $G_{\mathbf{R}^2}$ is $W^{2,q}(\mathbf{R}^2)$-stable with respect to (1.1) in $\mathbf{R}^2$. Moreover, there is a unique, up to translation, positive radially symmetric $C^2$ minimizer $\tilde{\psi}_0$ of $(M_{\mathbf{R}^2})$, and

$$G_{\mathbf{R}^2}=\{M_{\mathbf{R}^2}^{1/(2-q)}\tilde{\psi}_0(\cdot+y),-M_{\mathbf{R}^2}^{1/(2-q)}\tilde{\psi}_0(\cdot+y);y\in\mathbf{R}^2\}. \tag{1.8}$$

The important step to obtain Theorems 1.1 and 1.2 is to prove that the infimum is achieved. If $\{\tilde{\psi}_m\}$ is a minimizing sequence of $(M_{\mathbf{R}^2})$, then

$$\| \tilde{\psi}_m \|_1 = 1, \quad \int_{\mathbf{R}^2} | \Delta \tilde{\psi}_m - \tilde{\psi}_m |^q \to M_{\mathbf{R}^2}, \quad m \to +\infty. \tag{1.9}$$

Going if necessary to a subsequence, we may assume $\tilde{\psi}_m \rightharpoonup \tilde{\psi}_0$ weakly in $W^{2,q}(\mathbf{R}^2)$, so that

$$\int_{\mathbf{R}^2} | \Delta \tilde{\psi}_0 - \tilde{\psi}_0 |^q \leqslant \liminf \int_{\mathbf{R}^2} | \Delta \tilde{\psi}_m - \tilde{\psi}_m |^q = M_{\mathbf{R}^2}. \tag{1.10}$$

Thus $\tilde{\psi}_0$ is a minimizer of $(M_{\mathbf{R}^2})$ provided $\| \tilde{\psi}_0 \|_1 = 1$. Since $W^{2,q}(\mathbf{R}^2) \hookrightarrow H^1(\mathbf{R}^2)$ is not compact, we cannot derive $\| \tilde{\psi}_0 \|_1 = 1$ from $\| \tilde{\psi}_m \|_1 = 1$ and $\tilde{\psi}_m \rightharpoonup \tilde{\psi}_0$ weakly in $W^{2,q}(\mathbf{R}^2)$. Therefore, we cannot directly derive the existence of minimizer from any minimizing sequence. However, if we obtain the result that for any minimizing sequence $\{\tilde{\psi}_m\}$ there exist $\{y_{m_k}\} \subset \mathbf{R}^2$ and a subsequence $\{\tilde{\psi}_{m_k}\}$ such that $\{\tilde{\psi}_{m_k}(\cdot + y_{m_k})\}$ is a convergent sequence in $W^{2,q}(\mathbf{R}^2)$, which is the first part of Theorem 1.2, then we prove that the infimum is achieved.

In order to prove Theorem 1.2, we construct Lemma 2.1, which is used to study the behavior at infinity of the minimizing sequence $\{\tilde{\psi}_m\}$ and to overcome the loss of compactness of $(M_{\mathbf{R}^2})$. Theorem 1.2 is proved by two steps. Firstly, using Lemma 2.4, we prove that for any minimizing sequence $\{\tilde{\psi}_m\}$, there exist a subsequence $\{\tilde{\psi}_{m_k}\}$ and $\{y_{m_k}\} \subset \mathbf{R}^2$ such that $\tilde{\psi}_{m_k}(\cdot + y_{m_k}) \rightharpoonup \tilde{\psi}_0 \neq 0$ weakly in $W^{2,q}(\mathbf{R}^2)$, which denotes $0 \leqslant \alpha_\infty < 1$. Here $\alpha_\infty$ is a quantity related to $\{\tilde{\psi}_{m_k}(\cdot + y_{m_k})\}$ defined in Lemma 2.1. Secondly, according to Lemmas 2.1 and 2.3 based on Ekeland Principle, we know that if $\alpha_\infty > 0$, then $\alpha_\infty \geqslant 1$. Therefore, putting together the results of the above steps, we obtain $\alpha_\infty = 0$, which implies that there exists a sequence $y_{m_k} \subset \mathbf{R}^2$ such that the sequence $\{\tilde{\psi}_{m_k}(\cdot + y_{m_k})\}$ is convergent in $W^{2,q}(\mathbf{R}^2)$. Applying Theorem 1.2, we prove that $G_{\mathbf{R}^2}$ is a stable set with respect to (1.1), which is the first part of Theorem 1.3. The second part about the structure of $G_{\mathbf{R}^2}$ is obtained by studying the properties of the elliptic equation satisfied by $M_{\mathbf{R}^2}^{1/(2-q)} \tilde{\psi}_0$. Our method in proving the existence and stability of steady waves for (1.1) is different from that in [5]. In [5], Albert considered constrained variational problems with concentration-compactness Lemma introduced by Lions[6, 7] and proved the existence and stability of solitary waves to Kdv equation and some nonlocal equations.

The paper is organized as follows. In Section 2, we establish three lemmas for proving Theorem 1.2. In Section 3, we give the proofs of Theorems 1.2 and 1.3. In Section 4, we consider the existence and stability of steady waves for Hasegawa-Mima equation in general periodic domains and give the application of Lemmas 2.1 and 2.3 to study the existence and stability of steady waves for two-dimensional incompressible fluid in an infinite strip channel. Two propositions about the properties of the functional $I_{\mathbf{R}^2}$ for $F(x) = -(1/q)|x|^q$ and the definition of the ground state of $I_{\mathbf{R}^2}$ are presented in the appendix.

## 2 Three Lemmas

At first, we give some notations used later. Let $L^q(\Omega)$ be the usual Lebesgue space with norm $|\cdot|_q$, where $\Omega$ is an unbounded domain in $\mathbf{R}^2$ and $q=p/(p-1), 2<p<\infty$. The space $W_0^{2,q}(\Omega)$ is the completion of $D(\Omega)$ with respect to $\|\cdot\|_{2,q}$, where $D(\Omega)$ is the set of all $C^\infty$-functions with compact support in $\Omega$. The space $H_0^1(\Omega)$ is the completion of $D(\Omega)$ with respect to $\|\cdot\|_1$.

Now we give Lemma 2.1, which is used to study the behavior at infinity of minimizing sequence $\{\tilde{\psi}_m\}$ for $(M_{\mathbf{R}^2})$.

**Lemma 2.1** (Compactness Lemma) Let $\{\tilde{\psi}_m\}\subset W_0^{2,q}(\Omega)$ be a sequence such that $\tilde{\psi}_m\to\tilde{\psi}$ weakly in $W_0^{2,q}(\Omega)$ and define

$$\left.\begin{aligned}\alpha_\infty &:= \lim_{R\to\infty}\limsup_{m\to\infty}\int_{\Omega\cap\{|x|>R\}}(|\nabla\tilde{\psi}_m|^2+|\tilde{\psi}_m|^2),\\ \beta_\infty &:= \lim_{R\to\infty}\limsup_{m\to\infty}\int_{\Omega\cap\{|x|>R\}}|\Delta\tilde{\psi}_m-\tilde{\psi}_m|^q.\end{aligned}\right\}\tag{2.1}$$

Then one has

(1) $\displaystyle\limsup_{m\to\infty}\int_\Omega(|\nabla\tilde{\psi}_m|^2+|\tilde{\psi}_m|^2)=\int_\Omega(|\nabla\tilde{\psi}|^2+|\tilde{\psi}|^2)+\alpha_\infty$,

(2) $\displaystyle\limsup_{m\to\infty}\int_\Omega|\Delta\tilde{\psi}_m-\tilde{\psi}_m|^q\geqslant\int_\Omega|\Delta\tilde{\psi}-\tilde{\psi}|^q+\beta_\infty$,

(3) $(\alpha_\infty)^{q/2}M_\Omega\leqslant\beta_\infty$, where

$$M_\Omega=\inf_{\tilde{\psi}\in W_0^{2,q}(\Omega),\|\tilde{\psi}\|_1=1}\int_\Omega|\Delta\tilde{\psi}-\tilde{\psi}|^q.\tag{$M_\Omega$}$$

**Proof** (i) For any $R>0$, let $B_R=\{x;|x|<R\}, B_R^c=\{x;|x|\geqslant R\}$:

$$\begin{aligned}&\limsup_{m\to\infty}\int_\Omega(|\nabla\tilde{\psi}_m|^2+|\tilde{\psi}_m|^2)\\ &=\limsup_{m\to\infty}\left[\int_{\Omega\cap B_R}(|\nabla\tilde{\psi}_m|^2+|\tilde{\psi}_m|^2)+\int_{\Omega\cap B_R^c}(|\nabla\tilde{\psi}_m|^2+|\tilde{\psi}_m|^2)\right].\end{aligned}\tag{2.2}$$

Since the imbedding $W^{2,q}(\Omega\cap B_R)\hookrightarrow H^1(\Omega\cap B_R)$ is compact and $\tilde{\psi}_m\to\tilde{\psi}$ weakly in $W_0^{2,q}(\Omega)$,

$$\lim_{m\to\infty}\int_{\Omega\cap B_R}(|\nabla\tilde{\psi}_m|^2+|\tilde{\psi}_m|^2)=\int_{\Omega\cap B_R}(|\nabla\tilde{\psi}|^2+|\tilde{\psi}|^2).\tag{2.3}$$

Combining (2.2) with (2.3), we get

$$\begin{aligned}&\limsup_{m\to\infty}\int_\Omega(|\nabla\tilde{\psi}_m|^2+|\tilde{\psi}_m|^2)\\ &=\int_{\Omega\cap B_R}(|\nabla\tilde{\psi}|^2+|\tilde{\psi}|^2)+\limsup_{m\to\infty}\int_{\Omega\cap B_R^c}(|\nabla\tilde{\psi}_m|^2+|\tilde{\psi}_m|^2).\end{aligned}\tag{2.4}$$

Letting $R\to\infty$ in the above formula, we obtain (1).

(ii) Using the weakly lower semicontinuity of a norm, we have

$$\liminf_{m\to\infty}\int_{\Omega\cap B_R}|\Delta\tilde{\psi}_m-\tilde{\psi}_m|^q\geqslant\int_{\Omega\cap B_R}|\Delta\tilde{\psi}-\tilde{\psi}|^q,\quad\text{for any } R>0.\tag{2.5}$$

Applying (2.5), we have

$$\limsup_{m\to\infty}\int_\Omega|\Delta\tilde{\psi}_m-\tilde{\psi}_m|^q$$

$$
\begin{aligned}
&\geqslant \liminf_{m\to\infty}\int_{\Omega\cap B_R} |\Delta\tilde{\psi}_m - \tilde{\psi}_m|^q + \limsup_{m\to\infty}\int_{\Omega\cap B_R^c} |\Delta\tilde{\psi}_m - \tilde{\psi}_m|^q \\
&\geqslant \int_{\Omega\cap B_R} |\Delta\tilde{\psi} - \tilde{\psi}|^q + \limsup_{m\to\infty}\int_{\Omega\cap B_R^c} |\Delta\tilde{\psi}_m - \tilde{\psi}_m|^q.
\end{aligned} \tag{2.6}
$$

Letting $R\to\infty$ in (2.6), we deduce (2).

(iii) Applying elementary inequalities, we can prove that the norm $|(\Delta-1)\cdot|_q$ is equivalent to $\|\cdot\|_{2,q}$ in $W_0^{2,q}(\Omega)$. Let $\varphi_R \in C^\infty(\mathbf{R}^2)$ such that

$$
\varphi_R(x) = \begin{cases} 0, & |x| < R \\ 1, & |x| > R+1 \end{cases} \tag{2.7}
$$

and $0 < \varphi_R(x) \leqslant 1$ on $\mathbf{R}^2$. It follows from the definition of $M_\Omega$ and the convexity of the function $g(x) = |x|^q (x\in\mathbf{R})$ that

$$
\begin{aligned}
&M_\Omega\Big[\int_\Omega (|\nabla\tilde{\psi}_m\varphi_R|^2 + |\tilde{\psi}_m\varphi_R|^2)\Big]^{q/2} \\
&\leqslant \int_\Omega |\Delta(\tilde{\psi}_m\varphi_R) - \tilde{\psi}_m\varphi_R|^q \\
&= \int_\Omega |\varphi_R(\Delta\tilde{\psi}_m - \tilde{\psi}_m) + \tilde{\psi}_m\Delta\varphi_R + 2\nabla\tilde{\psi}_m\cdot\nabla\varphi_R|^q \\
&= \int_\Omega \left|(1-\varepsilon)\frac{\varphi_R}{1-\varepsilon}(\Delta\tilde{\psi}_m - \tilde{\psi}_m) + \varepsilon\left(\tilde{\psi}_m\Delta\frac{\varphi_R}{\varepsilon} + 2\nabla\tilde{\psi}_m\cdot\nabla\frac{\varphi_R}{\varepsilon}\right)\right|^q \\
&\leqslant (1-\varepsilon)\int_\Omega\left(\frac{\varphi_R}{1-\varepsilon}\right)^q |\Delta\tilde{\psi}_m - \tilde{\psi}_m|^q + \varepsilon\int_\Omega\left|\tilde{\psi}_m\Delta\frac{\varphi_R}{\varepsilon} + 2\nabla\tilde{\psi}_m\cdot\nabla\frac{\varphi_R}{\varepsilon}\right|^q,
\end{aligned} \tag{2.8}
$$

where $0<\varepsilon<1$. Since $\tilde{\psi}_m \rightharpoonup \tilde{\psi}$ weakly in $W_0^{2,q}(\Omega)$ and the embedding $W^{2,q}(\Omega\cap\{x: R<|x|<R+1\}) \hookrightarrow H^1(\Omega\cap\{x: R<|x|<R+1\})$ is compact,

$$
\tilde{\psi}_m\Delta\frac{\varphi_R}{\varepsilon} \to \tilde{\psi}\Delta\frac{\varphi_R}{\varepsilon}, \nabla\tilde{\psi}_m\nabla\frac{\varphi_R}{\varepsilon} \to \nabla\tilde{\psi}\nabla\frac{\varphi_R}{\varepsilon} \text{ in } L^q(\Omega\cap\{x: R<|x|<R+1\}), \tag{2.9}
$$

which implies

$$
\lim_{R\to\infty}\limsup_{m\to\infty}\int_\Omega\left|\tilde{\psi}_m\Delta\frac{\varphi_R}{\varepsilon} + 2\nabla\tilde{\psi}_m\cdot\nabla\frac{\varphi_R}{\varepsilon}\right|^q = 0. \tag{2.10}
$$

With the definitions of $\alpha_\infty, \beta_\infty$,

$$
\left.\begin{aligned}
&\lim_{R\to\infty}\limsup_{m\to\infty}\Big[\int_\Omega (|\nabla\tilde{\psi}_m\varphi_R|^2 + |\tilde{\psi}_m\varphi_R|^2)\Big]^{q/2} = (\alpha_\infty)^{q/2}, \\
&\lim_{R\to\infty}\limsup_{m\to\infty}\int_\Omega \varphi_R^q |\Delta\tilde{\psi}_m - \tilde{\psi}_m|^q = \beta_\infty.
\end{aligned}\right\} \tag{2.11}
$$

We derive from (2.8)~(2.11)

$$
M_\Omega(\alpha_\infty)^{q/2} \leqslant (1-\varepsilon)^{1-q}\beta_\infty. \tag{2.12}
$$

Since $\varepsilon$ is arbitrary, we have

$$
M_\Omega(\alpha_\infty)^{q/2} \leqslant \beta_\infty. \tag{2.13}
$$

□

**Remark 2.1** In [8], Huang and Li have used a concentration-compactness principle at infinity, similar to Lemma 2.1, to study the existence of positive solutions for some quasilinear equations on unbounded domains in $\mathbf{R}^N$.

In the following, we give Lemma 2.3, which is used to find a Palais-Smale sequence $\{\tilde{\psi}_m^1\}$ of $I_{\mathbf{R}^2}$ through the minimizing sequence $\{\tilde{\psi}_m\}$ for $(M_{\mathbf{R}^2})$. Firstly, we give some notations. Let

$$\left.\begin{aligned} I_\Omega(\psi) &= \frac{1}{2}\int_\Omega(|\nabla\psi|^2+|\psi|^2)-\frac{1}{q}\int_\Omega|\Delta\psi-\psi|^q,\\ J_\Omega(\tilde{\psi}) &= \frac{\int_\Omega|\Delta\tilde{\psi}-\tilde{\psi}|^q}{\left[\int_\Omega(|\nabla\tilde{\psi}|^2+|\tilde{\psi}|^2)\right]^{q/2}}. \end{aligned}\right\} \tag{2.14}$$

By the definition of Fréchet derivative, it is easy to verify that $I_\Omega\in C^1(W_0^{2,q}(\Omega),\mathbf{R})$, $J_\Omega\in C^1(W_0^{2,q}(\Omega)\setminus\{0\},\mathbf{R})$ and

$$\left.\begin{aligned} \langle I_\Omega'(\psi),h\rangle &= \int_\Omega(\nabla\psi\nabla h+\psi h)-\int_\Omega|\Delta\psi-\psi|^{q-2}(\Delta\psi-\psi)(\Delta h-h),\\ \langle J_\Omega'(\tilde{\psi}),h\rangle &= \frac{q\|\tilde{\psi}\|_1^{q-2}\left[\|\tilde{\psi}\|_1^2\int_\Omega|A\tilde{\psi}|^{q-2}A\tilde{\psi}Ah-\|\tilde{\psi}\|_{2,q}^q\int_\Omega(\nabla\tilde{\psi}\nabla h+\tilde{\psi}h)\right]}{\left[\int_\Omega(|\nabla\tilde{\psi}|^2+|\tilde{\psi}|^2)\right]^q}. \end{aligned}\right\} \tag{2.15}$$

where $\psi,\tilde{\psi},h\in W_0^{2,q}(\Omega)$ and $A\tilde{\psi}=\Delta\tilde{\psi}-\tilde{\psi}$.

**Lemma 2.2** If $\{\tilde{\psi}_m\}$ is a minimizing sequence of $(M_\Omega)$, then there is a minimizing sequence $\{\tilde{\psi}_m^1\}$ such that $\|\tilde{\psi}_m^1-\tilde{\psi}_m\|_{2,q}<1/m$, $J_\Omega(\tilde{\psi}_m^1)\to M_\Omega$, $J_\Omega'(\tilde{\psi}_m^1)\to 0$ in $W^{-2,q'}(\Omega)$ as $m\to\infty$, and

$$I_\Omega(\psi_m)\to\left(\frac{1}{2}-\frac{1}{q}\right)M_\Omega^{2/(2-q)},\quad I_\Omega'(\psi_m)\to 0\quad\text{in } W^{-2,q'}(\Omega),\ \text{as } m\to\infty, \tag{2.16}$$

where $\psi_m=M_\Omega^{1/(2-q)}\tilde{\psi}_m^1$, and $W^{-2,q'}(\Omega)$ is the dual space of $W_0^{2,q}(\Omega)$, $1/q+1/q'=1$. Moreover, if $\alpha_\infty,\beta_\infty$ are quantities related to $\tilde{\psi}_m^1$ in Lemma 2.1, then

$$M_\Omega\cdot\alpha_\infty=\beta_\infty. \tag{2.17}$$

**Proof** Using the definitions of $M_\Omega$ and $J_\Omega$, we have

$$M_\Omega=\inf_{0\neq\tilde{\psi}\in W_0^{2,q}(\Omega)}J_\Omega(\tilde{\psi})=\lim_{m\to\infty}J_\Omega(\tilde{\psi}_m). \tag{2.18}$$

Applying the Ekeland Variational Principle (cf. [9, page 51]) to (2.18), we get a Palais-Smale sequence $\{\tilde{\psi}_m^1\}$, which satisfies

$$\left\{\begin{aligned} &\|\tilde{\psi}_m^1-\tilde{\psi}_m\|_{2,q}<\frac{1}{m},\quad J_\Omega(\tilde{\psi}_m^1)\to M_\Omega,\\ &J_\Omega'(\tilde{\psi}_m^1)\to 0\quad\text{in } W^{-2,q'}(\Omega),\ \text{as } m\to\infty. \end{aligned}\right\} \tag{2.19}$$

Then, according to the definitions of $I_\Omega$ and $\psi_m$, for any $h\in W_0^{2,q}(\Omega)$, we get

$$\begin{aligned} I_\Omega(\psi_m) &= \frac{1}{2}M_\Omega^{2/(2-q)}\int_\Omega(|\nabla\tilde{\psi}_m^1|^2+|\tilde{\psi}_m^1|^2)-\frac{1}{q}M_\Omega^{q/(2-q)}\int_\Omega|\Delta\tilde{\psi}_m^1-\tilde{\psi}_m^1|^q\\ &\to\frac{1}{2}M_\Omega^{2/(2-q)}-\frac{1}{q}M_\Omega^{q/(2-q)+1}\\ &=\left(\frac{1}{2}-\frac{1}{q}\right)M_\Omega^{2/(2-q)}, \end{aligned} \tag{2.20}$$

$$\langle I'_\Omega(\psi_m), h\rangle = M_\Omega^{1/(2-q)}\int_\Omega (\nabla\tilde{\psi}_m^1\,\nabla h + \tilde{\psi}_m^1 h) - M_\Omega^{(q-1)/(2-q)}\int_\Omega |A\tilde{\psi}_m^1|^{q-2}A\tilde{\psi}_m^1 Ah$$
$$= M_\Omega^{(q-1)/(2-q)}\left[M_\Omega\int_\Omega (\nabla\tilde{\psi}_m^1\,\nabla h + \tilde{\psi}_m^1 h) - \int_\Omega |A\tilde{\psi}_m^1|^{q-2}A\tilde{\psi}_m^1 Ah\right]$$
$$\to 0(\text{because of } \langle J'_\Omega(\tilde{\psi}_m^1), h\rangle \to 0), \quad \text{as } m\to\infty.$$

Since (2.20) implies

$$I'_\Omega(M_\Omega^{1/(2-q)}\tilde{\psi}_m^1) \to 0 \quad \text{in } W^{-2,q'}(\Omega), \text{ as } m\to\infty, \tag{2.21}$$

it follows that

$$\langle I'_\Omega(M_\Omega^{1/(2-q)}\tilde{\psi}_m^1), M_\Omega^{1/(2-q)}\tilde{\psi}_m^1\varphi_R\rangle \to 0, \quad \text{as } m\to\infty \text{ uniformly for } R\geqslant 1, \tag{2.22}$$

where $\varphi_R$ is the function defined in the proof of Lemma 2.1. With the definition of $I_\Omega$, we have

$$\langle I'_\Omega(M_\Omega^{1/(2-q)}\tilde{\psi}_m^1), M_\Omega^{1/(2-q)}\tilde{\psi}_m^1\varphi_R\rangle$$
$$= M_\Omega^{2/(2-q)}\int_\Omega (\nabla\tilde{\psi}_m^1\,\nabla(\tilde{\psi}_m^1\varphi_R) + \tilde{\psi}_m^1\tilde{\psi}_m^1\varphi_R) -$$
$$M_\Omega^{q/(2-q)}\int_\Omega |\Delta\tilde{\psi}_m^1 - \tilde{\psi}_m^1|^{q-2}(\Delta\tilde{\psi}_m^1 - \tilde{\psi}_m^1)[\Delta(\tilde{\psi}_m^1\varphi_R) - \tilde{\psi}_m^1\varphi_R]. \tag{2.23}$$

Using (2.22), and letting $R\to\infty$ after $m\to\infty$ in (2.23), we obtain

$$M_\Omega\cdot\alpha_\infty = \beta_\infty. \tag{2.24}$$

□

At last, we give another lemma for proving Theorem 1.2.

**Lemma 2.4** If $\{\tilde{\psi}_m\}$ is bounded in $W^{2,q}(\mathbf{R}^2)$, and

$$\sup_{y\in\mathbf{R}^2}\int_{B(y,R)} |\tilde{\psi}_m|^q \to 0, \quad \text{for some } R>0, \tag{2.25}$$

then $\tilde{\psi}_m\to 0$ in $H^1(\mathbf{R}^2)$.

**Proof** Applying interpolation inequalities, for $\tilde{\psi}\in W^{2,q}(\mathbf{R}^2)$, we have

$$\|\nabla\tilde{\psi}\|^2_{L^2(B(y,R))} \leqslant c\|\tilde{\psi}\|^{2(1-\gamma_1)}_{L^q(B(y,R))}\|\tilde{\psi}\|^{2\gamma_1}_{W^{2,q}(B(y,R))}, \tag{2.26}$$

$$\|\tilde{\psi}\|^2_{L^2(B(y,R))} \leqslant c\|\tilde{\psi}\|^{2(1-\gamma_2)}_{L^q(B(y,R))}\|\tilde{\psi}\|^{2\gamma_2}_{W^{2,q}(B(y,R))}, \tag{2.27}$$

where $\gamma_1 = 1/q$, $\gamma_2 = 1/q - 1/2$, and $c$ is a positive constant. Letting $B_1 = B(0,R)$, $B_2 = B(y_2,R)$ where $y_2\in\partial B(0,R)$, $B_3 = B(y_3,R)$, $B_4 = B(y_4,R)$ where $\{y_3,y_4\} = \partial B_1\cap\partial B_2$, $\cdots$, we cover $\mathbf{R}^2$ by the above balls of radius $R$ such that each point of $\mathbf{R}^2$ is contained in at most 3 balls. Therefore, combining (2.26) with (2.27), we obtain

$$\left.\begin{aligned}\|\nabla\tilde{\psi}_m\|^2_{L^2(\mathbf{R}^2)} &\leqslant 3c\left(\sup_{y\in\mathbf{R}^2}\int_{B(y,R)}|\tilde{\psi}_m|^q\right)^{2(1-\gamma_1)}\|\tilde{\psi}_m\|^{2\gamma_1}_{W^{2,q}(\mathbf{R}^2)},\\ \|\tilde{\psi}_m\|^2_{L^2(\mathbf{R}^2)} &\leqslant 3c\left(\sup_{y\in\mathbf{R}^2}\int_{B(y,R)}|\tilde{\psi}_m|^q\right)^{2(1-\gamma_2)}\|\tilde{\psi}_m\|^{2\gamma_2}_{W^{2,q}(\mathbf{R}^2)}.\end{aligned}\right\} \tag{2.28}$$

According to the assumptions of Lemma 2.4 and the above two inequalities, $\tilde{\psi}_m\to 0$ in $H^1(\mathbf{R}^2)$.

□

# 3 Proof of Theorems 1.2 and 1.3

Now we turn to prove our main results.

**Proof of Theorem 1.2** Using Lemma 2.3 and $I_{\mathbf{R}^2}\in C^1(W^{2,q}(\mathbf{R}^2),\mathbf{R})$, for any minimizing sequence $\{\tilde{\psi}_m\}$ of $(M_{\mathbf{R}^2})$, we have

$$\left.\begin{aligned} &I_{\mathbf{R}^2}(M_{\mathbf{R}^2}^{1/(2-q)}\tilde{\psi}_m)\to\left(\frac{1}{2}-\frac{1}{q}\right)M_{\mathbf{R}^2}^{2/(2-q)},\\ &I'_{\mathbf{R}^2}(M_{\mathbf{R}^2}^{1/(2-q)}\tilde{\psi}_m)\to 0 \quad \text{in } W^{-2,q'}(\mathbf{R}^2) \text{ as } m\to\infty.\end{aligned}\right\} \tag{3.1}$$

Moreover, $\{\tilde{\psi}_m\}$ is bounded in $W^{2,q}(\mathbf{R}^2)$.

Lemma 2.4 implies that there exists $\{y_{m_k}\}\subset\mathbf{R}^2$ such that

$$\tilde{\psi}^1_{m_k}:=\tilde{\psi}_{m_k}(\cdot+y_{m_k})\to\tilde{\psi}_0\neq 0 \quad \text{weakly in } W^{2,q}(\mathbf{R}^2). \tag{3.2}$$

In fact, since $\|\tilde{\psi}_m\|_1=1$, by Lemma 2.4, there exists $\nu>0$ such that

$$\liminf_{m\to\infty}\sup_{y\in\mathbf{R}^2}\int_{B(y,R)}|\tilde{\psi}_m|^q>\nu. \tag{3.3}$$

Then, the above inequality and the boundedness of $\{\tilde{\psi}_m\}$ in $W^{2,q}(\mathbf{R}^2)$ imply that there exist a subsequence $\{\tilde{\psi}_{m_k}\}$ and $y_{m_k}\in\mathbf{R}^2$ such that

$$\int_{B(y_{m_k},R)}|\tilde{\psi}_{m_k}|^q>\frac{\nu}{2} \tag{3.4}$$

and $\tilde{\psi}_{m_k}(\cdot+y_{m_k})\to\tilde{\psi}_0$ weakly in $W^{2,q}(\mathbf{R}^2)$. Letting $\tilde{\psi}^1_{m_k}:=\tilde{\psi}_{m_k}(\cdot+y_{m_k})$, we know that $\{\tilde{\psi}^1_{m_k}\}$ is a minimizing sequence of $(M_{\mathbf{R}^2})$ and $\int_{B(0,R)}|\tilde{\psi}^1_{m_k}|^q>\nu/2$. Since the embedding $W^{2,q}(B(0,R))\hookrightarrow L^q(B(0,R))$ is compact, $\tilde{\psi}^1_{m_k}\to\tilde{\psi}_0$ in $L^q(B(0,R))$, which implies $\|\tilde{\psi}_0\|^q_{L^q(B(0,R))}\geqslant\nu/2$. So $\tilde{\psi}^1_{m_k}\to\tilde{\psi}_0\neq 0$ weakly in $W^{2,q}(\mathbf{R}^2)$.

Let $\Omega=\mathbf{R}^2$ and $\alpha_\infty,\beta_\infty$ be quantities related to $\tilde{\psi}^1_{m_k}$ in Lemma 2.1. With Lemma 2.1, we have

$$\int_{\mathbf{R}^2}(|\nabla\tilde{\psi}_0|^2+|\tilde{\psi}_0|^2)+\alpha_\infty=1. \tag{3.5}$$

In order to prove Theorem 1.2, we have to show $\alpha_\infty=0$. Since $\tilde{\psi}_0\neq 0$ in $W^{2,q}(\mathbf{R}^2)$ implies $0\leqslant\alpha_\infty<1$, arguing by contradiction, we assume $0<\alpha_\infty<1$. Using Lemmas 2.1 and 2.3, we have

$$M_{\mathbf{R}^2}(\alpha_\infty)^{q/2}\leqslant\beta_\infty=M_{\mathbf{R}^2}\alpha_\infty. \tag{3.6}$$

Since Sobolev Theorem implies $M_{\mathbf{R}^2}>0$, we derive from (3.6) $\alpha_\infty\geqslant 1$, which contradicts the assumption that $0<\alpha_\infty<1$. Thus $\tilde{\psi}_0$ is a minimizer for the minimization problem $(M_{\mathbf{R}^2})$. Since $W^{2,q}(\mathbf{R}^2)$ is a uniformly convex space, $\|\tilde{\psi}^1_{m_k}\|_{2,q}\to\|\tilde{\psi}_0\|_{2,q}$ and $\tilde{\psi}^1_{m_k}\to\tilde{\psi}_0$ weakly in $W^{2,q}(\mathbf{R}^2)$, we know that $\tilde{\psi}^1_{m_k}\to\tilde{\psi}_0$ in $W^{2,q}(\mathbf{R}^2)$. □

**Proof of Theorem 1.3** We divide the proof into two steps.

**Step 1** We prove that $G_{\mathbf{R}^2}$ is a stable set. Assume that the ground state set $G_{\mathbf{R}^2}$ is not $W^{2,q}(\mathbf{R}^2)$-stable. Then there exist $\varepsilon_0>0$, $\psi_m^0 \subset W^{2,q}(\mathbf{R}^2)$ and $t_m\in[0,T)$ such that

$$\inf_{\psi_0\in G_{\mathbf{R}^2}} \|\psi_m^0-\psi_0\|_E \leqslant \frac{1}{m}, \tag{3.7}$$

$$\inf_{\psi_0\in G_{\mathbf{R}^2}} \|\psi_m(t_m)-\psi_0\|_E \geqslant \varepsilon_0, \tag{3.8}$$

where $\psi_m\in C([0,T),E)$ is a solution to (1.1) with $\psi_m(0)=\psi_m^0$. Let $\tilde{\psi}_m = M_{\mathbf{R}^2}^{-1/(2-q)}\psi_m^0$. Equation (3.7) implies that

$$\int_{\mathbf{R}^2}(|\nabla\tilde{\psi}_m|^2+|\tilde{\psi}_m|^2)\to 1,\quad \int_{\mathbf{R}^2}|\Delta\tilde{\psi}_m-\tilde{\psi}_m|^q\to M_{\mathbf{R}^2}. \tag{3.9}$$

Then there exists $\{r_m\}\subset\mathbf{R}$, $r_m\to 1$ as $m\to+\infty$ such that

$$\{r_m\tilde{\psi}_m\} \text{ is a minimizing sequence for } (M_{\mathbf{R}^2}). \tag{3.10}$$

Using (1.3), we have

$$\|\psi_m^0\|_1 = \|\psi_m(t_m)\|_1,\quad \|\psi_m^0\|_{2,q} = \|\psi_m(t_m)\|_{2,q}. \tag{3.11}$$

With (3.10) and (3.11), we know that $\{r_m M_{\mathbf{R}^2}^{-1/(2-q)}\psi_m(t_m)\}$ is a minimizing sequence for $(M_{\mathbf{R}^2})$. By Theorem 1.2, there exist $y_{m_k}\in\mathbf{R}^2$ and $\tilde{\psi}_{m_k}^1\in Z_{\mathbf{R}^2}$ such that

$$\begin{aligned}
&\left\|r_{m_k}M_{\mathbf{R}^2}^{-1/(2-q)}\psi_{m_k}(t_{m_k})(\cdot+y_{m_k})-\tilde{\psi}_{m_k}^1\right\|_{2,q}\\
=&\left\|r_{m_k}M_{\mathbf{R}^2}^{-1/(2-q)}\psi_{m_k}(t_{m_k})-\tilde{\psi}_{m_k}^1(\cdot-y_{m_k})\right\|_{2,q}\\
\leqslant&\frac{\varepsilon_0}{2M_{\mathbf{R}^2}^{1/(2-q)}},
\end{aligned}\tag{3.12}$$

for sufficiently large $m_k$. Since $r_m\to 1$ and $\|\psi_k(t_m)\|_{2,q}$ is bounded, we derive from (3.8) and (3.12)

$$\begin{aligned}
\varepsilon_0&\leqslant\left\|\psi_{m_k}(t_{m_k})-M_{\mathbf{R}^2}^{1/(2-q)}\tilde{\psi}_{m_k}^1(\cdot-y_{m_k})\right\|_{2,q}\\
&\leqslant\left\|\psi_{m_k}(t_{m_k})-r_{m_k}\psi_{m_k}(t_{m_k})\right\|_{2,q}+\left\|r_{m_k}\psi_{m_k}(t_{m_k})-M_{\mathbf{R}^2}^{1/(2-q)}\tilde{\psi}_{m_k}^1(\cdot-y_{m_k})\right\|_{2,q}\\
&\leqslant\left\|\psi_{m_k}(t_{m_k})-r_{m_k}\psi_{m_k}(t_{m_k})\right\|_{2,q}+M_{\mathbf{R}^2}^{1/(2-q)}\left\|r_{m_k}M_{\mathbf{R}^2}^{-1/(2-q)}\psi_{m_k}(t_{m_k})-\tilde{\psi}_{m_k}^1(\cdot-y_{m_k})\right\|_{2,q}\\
&\leqslant\frac{3}{4}\varepsilon_0,
\end{aligned}\tag{3.13}$$

for sufficiently large $m_k$. Equation (3.13) is a contradiction. Therefore, the ground state set $G_{\mathbf{R}^2}$ is $W^{2,q}(\mathbf{R}^2)$-stable with respect to (1.1).

**Step 2** Show the structure of $G_{\mathbf{R}^2}$. If $\tilde{\psi}_0$ is a minimizer for $(M_{\mathbf{R}^2})$, then $|\tilde{\psi}_0|$ is also a minimizer for $(M_{\mathbf{R}^2})$. We can assume that $\psi_0$ is a nonnegative minimizer for $(M_{\mathbf{R}^2})$. Let $\psi_0 = M_{\mathbf{R}^2}^{1/(2-q)}\tilde{\psi}_0$, then $I'_{\mathbf{R}^2}(\psi_0)=0$ in $W^{-2,q'}(\mathbf{R}^2)$, that is,

$$-\Delta\psi_0+\psi_0=\psi_0^{p-1},\quad \psi_0(x)\to 0\quad\text{as } |x|\to+\infty, \tag{3.14}$$

where $p>2$ is given in Section 2. In fact, using the implicit function theorem, we know that $M=\{\tilde{\psi}:\tilde{\psi}\in W^{2,q}(\mathbf{R}^2),\int_{\mathbf{R}^2}|\nabla\tilde{\psi}|^2+|\tilde{\psi}|^2=1\}$ is a $C^1$-submanifold of $W^{2,q}(\mathbf{R}^2)$. Then, according to the Lagrange Multiplier Principle, there exists $\lambda \in \mathbf{R}$ such that

$$-q\Psi'_{\mathbf{R}^2}(\tilde{\psi}_0) = 2\lambda\Phi'_{\mathbf{R}^2}(\tilde{\psi}_0) \quad \text{in } W^{-2,q'}(\mathbf{R}^2), \tag{3.15}$$

where $\Psi_{\mathbf{R}^2}(\tilde{\psi}_0) = -(1/q)\int_{\mathbf{R}^2}|\Delta\tilde{\psi}_0 - \tilde{\psi}_0|^q$, $\Phi_{\mathbf{R}^2}(\tilde{\psi}_0) = (1/2)\int_{\mathbf{R}^2}(|\nabla\tilde{\psi}_0|^2 + |\tilde{\psi}_0|^2)$. Since $\tilde{\psi}_0$ is a minimizer for $(M_{\mathbf{R}^2})$, with the definition of $M_{\mathbf{R}^2}$, we obtain $\lambda = (q/2)M_{\mathbf{R}^2}$. So, we conclude $I'_{\mathbf{R}^2}(\psi_0) = 0$.

Applying elliptic regularity theory ([10], Lemma 1.30), we prove that $\psi_0 \in C^2(\mathbf{R}^2)$. According to the strong maximum principle, $\psi_0$ is positive. Using the moving plane method[11], we show that $\psi_0$ is radially symmetric. Moreover, by the uniqueness result in [12], $\psi_0$ is unique up to translations. Therefore, $\tilde{\psi}_0$ is a unique, up to translation, positive radially symmetric $C^2$ minimizer of $(M_{\mathbf{R}^2})$, and

$$G_{\mathbf{R}^2} = \{M_{\mathbf{R}^2}^{1/(2-q)}\tilde{\psi}_0(\cdot + y), -M_{\mathbf{R}^2}^{1/(2-q)}\tilde{\psi}_0(\cdot + y); y \in \mathbf{R}^2\}. \tag{3.16}$$

□

**Remark 3.1** The existence of nontrivial critical points of (1.2) for $F(x) = -(1/q)|x|^q$ can also be proved by the following method. Since $W_r^{2,q}(\mathbf{R}^2) \hookrightarrow H_r^1(\mathbf{R}^2)$ is compact, where $W_r^{2,q}(\mathbf{R}^2)$ and $H_r^1(\mathbf{R}^2)$ are the set of the radially symmetric functions of $W^{2,q}(\mathbf{R}^2)$ and $H^1(\mathbf{R}^2)$, respectively, with Principle of Symmetric Criticality proposed by Palais[13] and Fountain Theorem ([10], Chapter 3), we can prove that there are infinitely many solutions to (1.2). However, we do not know how to consider the stability of these solutions. So, we cannot study the existence of nontrivial critical points of (1.2) with this method. Through considering the constrained variational problem $(M_{\mathbf{R}^2})$, we obtain the existence of a set $G_{\mathbf{R}^2}$ of steady waves for (1.1) and know that $G_{\mathbf{R}^2}$ is $W^{2,q}(\mathbf{R}^2)$-stable with respect to (1.1).

## 4 Hasegawa-Mima Equation in Periodic Domains

In this section, we consider the existence and stability of steady waves for Hasegawa-Mima equation in periodic domains:

$$\frac{\partial(\Delta\psi - \psi)}{\partial t} + \frac{\partial\psi}{\partial x}\frac{\partial(\Delta\psi - \psi)}{\partial y} - \frac{\partial\psi}{\partial y}\frac{\partial(\Delta\psi - \psi)}{\partial x} = 0, \tag{4.1}$$

where $(x, y) \in \Omega$, $\Omega$ is a periodic domain defined later, $\psi(x, y) \to 0$ as $|x| + |y| \to +\infty$, and $\psi|_{(x,y)\in\partial\Omega} = 0$. Moreover, as a byproduct, we prove the existence and stability of steady two-dimensional incompressible flows in infinite strip channel.

At first, we give the definition and two examples of periodic domain.

**Definition 4.1** (periodic domain) If $\Omega$ is a domain in $\mathbf{R}^2$, and there are a partition $\{\Omega_n\}$ of $\Omega$ and points $\{y_n\}$ in $\mathbf{R}^2$ satisfying the following conditions: (1) $\{y_n\}$ forms a subgroup of $\mathbf{R}^2$, (2) $\Omega_0$ is a bounded domain in $\mathbf{R}^2$, (3) $\Omega_n = y_n + \Omega_0$, then $\Omega$ is called a periodic domain.

It is clear that $\mathbf{R}^2$ is a periodic domain. The other example of periodic domain is

$$Q = \{(x, y) \in \mathbf{R} \times \mathbf{R} : |y| < 1\}.$$

Similarly to the method used in studying (1.1), in order to prove the existence and stability of steady waves for (4.1), we consider the following minimization problem:

$$M_\Omega = \inf_{\tilde{\psi}\in W_0^{2,q}(\Omega),\ \|\tilde{\psi}\|_1=1} \int_\Omega |\Delta\tilde{\psi} - \tilde{\psi}|^q. \qquad (M_\Omega)$$

Let

$$\left.\begin{aligned} Z_\Omega &= \left\{\tilde{\psi}_0;\int_\Omega(|\nabla\tilde{\psi}_0|^2 + |\tilde{\psi}_0|^2) = 1,\ \int_\Omega(|\Delta\tilde{\psi}_0 - \tilde{\psi}_0|^q) = M_\Omega\right\},\\ G_\Omega &= \{\psi_0;\psi_0 = M_\Omega^{1/(2-q)}\tilde{\psi}_0, \tilde{\psi}_0 \in Z_\Omega\}. \end{aligned}\right\} \qquad (4.2)$$

If $\tilde{\psi}_0 \in Z_\Omega$, then, according to Lagrange Multiplier Principle, $\psi_0 = M_\Omega^{1/(2-q)}\tilde{\psi}_0$ is a steady wave for (4.1). Moreover, as presented in Proposition A.3, $\psi_0$ is a ground state and saddle point of $I_\Omega$.

Through considering the above minimization problem ($M_\Omega$), we obtain the following result similar to Theorems 1.2 and 1.3.

**Theorem 4.2** If $\{\tilde{\psi}_m\}$ is a minimizing sequence for ($M_\Omega$), then there exist a subsequence $\{\tilde{\psi}_{m_k}\}$ and $\{y_{n_{m_k}}\}$ such that $\tilde{\psi}_{m_k}(\cdot + y_{n_{m_k}}) \to \tilde{\psi} \neq 0$ weakly in $W_0^{2,q}(\Omega)$. Moreover, the ground state set $G_\Omega$ is $W_0^{2,q}(\Omega)$-stable with respect to (4.1).

**Proof** Without loss of generality, we assume that $\Omega = \bigcup_{n\in\mathbf{Z}}\Omega_n$, where $\Omega_i$, and $\Omega_j$ are disjoint if $i \neq j$, and there exists $\Omega_0 \subset \mathbf{R}^2$ such that $\Omega_n = y_n + \Omega_0$. Similar to Lemma 2.3, using $I_\Omega \in C^1(W_0^{2,q}(\Omega),\mathbf{R})$, we know that the minimizing sequence $\{\tilde{\psi}_m\}$ satisfies

$$\left.\begin{aligned} I_\Omega(M_\Omega^{1/(2-q)}\tilde{\psi}_m) &\to \left(\frac{1}{2} - \frac{1}{q}\right)M_\Omega^{2/(2-q)},\\ I'_\Omega(M_\Omega^{1/(2-q)}\tilde{\psi}_m) &\to 0 \quad \text{in } W^{-2,q'}(\Omega), \text{ as } m \to \infty. \end{aligned}\right\} \qquad (4.3)$$

If we let $\psi_m = M_\Omega^{1/(2-q)}\tilde{\psi}_m$, then (4.3) implies

$$\int_\Omega(|\nabla\psi_m|^2 + |\psi_m|^2) = \int_\Omega|\Delta\psi_m - \psi_m|^q + o(1). \qquad (4.4)$$

Letting $d_m = \max_{n\in\mathbf{Z}}\left[\int_{\Omega_n}(|\nabla\psi_m|^2 + |\psi_m|^2)\right]^{1/2}$, applying Sobolev inequality, (4.3) and (4.4), we have

$$\begin{aligned} o(1) + I_\Omega(\psi_m) &= \left(\frac{1}{2} - \frac{1}{q}\right)\sum_{n\in\mathbf{Z}}\int_{\Omega_n}(|\nabla\psi_m|^2 + |\psi_m|^2)\\ &\geqslant \left(\frac{1}{2} - \frac{1}{q}\right)\max_{n\in\mathbf{Z}}\left[\int_{\Omega_n}(|\nabla\psi_m|^2 + |\psi_n|^2)\right]^{(2-q)/2}\sum_{n\in\mathbf{Z}}\left[\int_{\Omega_n}(|\nabla\psi_m|^2 + |\psi_m|^2)\right]^{q/2}\\ &\geqslant \left(\frac{1}{2} - \frac{1}{q}\right)d_m^{2-q}C\sum_{n\in\mathbf{Z}}\int_{\Omega_n}|\Delta\psi_m - \psi_m|^q\\ &\geqslant \left(\frac{1}{2} - \frac{1}{q}\right)d_m^{2-q}C\sum_{n\in\mathbf{Z}}\int_{\Omega_n}(|\Delta\psi_m|^2 + \|\psi_m\|^2) + o(1), \end{aligned} \qquad (4.5)$$

for some constant $C>0$ independent of $m$. Since $\int_\Omega(|\nabla\tilde{\psi}_m|^2 + |\tilde{\psi}_m|^2) = 1$ for all $m$, (4.5) implies that there exists $\mu>0$ such that $d_m \geqslant \mu$ for $m = 1,2,\cdots$ Therefore, for each $m$, there exists $\Omega_{n_m}$ such that

$$\left[\int_{\Omega_{n_m}}(|\nabla\psi_m|^2+|\psi_m|^2)\right]^{1/2}\geqslant\frac{\mu}{2}. \tag{4.6}$$

Equation (4.6) means that there exists a point $y_{n_m}$ such that

$$\left[\int_{\Omega_{n_m}}(|\nabla\psi_m|^2+|\psi_m|^2)\right]^{1/2}=\left[\int_{\Omega_0}(|\nabla\psi_m(x+y_{n_m})|^2+|\psi_m(x+y_{n_m})|^2)\right]^{1/2}\geqslant\frac{\mu}{2}. \tag{4.7}$$

Since $\{\psi_m(x+y_{n_m})\}$ is bounded in $W_0^{2,q}(\Omega)$, there exists a subsequence $\{\psi_{m_k}(x+y_{n_{m_k}})\}$ such that

$$\psi_{m_k}(x+y_{n_{m_k}})\rightarrow\psi\neq0\quad\text{weakly in } W_0^{2,q}(\Omega). \tag{4.8}$$

So $\tilde{\psi}_{m_k}(x+y_{n_{m_k}})=M_\Omega^{-1/(2-q)}\psi_{m_k}(x+y_{n_{m_k}})\rightarrow M_\Omega^{-1/(2-q)}\psi=\tilde{\psi}\neq0$ weakly in $W_0^{2,q}(\Omega)$.

Let $\alpha_\infty,\beta_\infty$ be quantities related to $\{\tilde{\psi}_{m_k}(x+y_{n_{m_k}})\}$ in Lemma 2.1 Similarly to the proof of the first part of Theorem 1.3, we can prove the second part of Theorem 4.2. □

In the following, we give an application of Lemmas 2.1 and 2.3 to study the existence and stability of steady two-dimensional incompressible waves. The well-known vorticity equation governing the two-dimensional incompressible flow in an infinite strip channel is

$$\frac{\partial\Delta\psi}{\partial t}+\frac{\partial\psi}{\partial x}\frac{\partial\Delta\psi}{\partial y}-\frac{\partial\psi}{\partial y}\frac{\partial\Delta\psi}{\partial x}=0,\quad(x,y)\in Q=\mathbf{R}\times(0,1), \tag{4.9}$$

where $\psi$ is a stream function.

If the stream function $\psi$ satisfies

$$\psi(x,0)=\psi(x,1)=0\quad\text{for any } x\in\mathbf{R},\quad\psi(x,y)\rightarrow0\quad\text{as } |x|\rightarrow+\infty, \tag{4.10}$$

then there are two conserved quantities of (4.9) and (4.10): the total energy $\Phi_Q(\psi)=(1/2)\int_Q|\nabla\psi|^2$, the generalized enstrophy $\Psi_Q(\psi)=\int_Q F(\Delta\psi)$, where $F$ is an arbitrary $C^1$ function. Letting $F(x)=-(1/q)|x|^q$ for $x\in\mathbf{R}$, where $q=p/(p-1)$ and $2<p<+\infty$, we consider the minimization problem:

$$M_Q:=\inf\left\{\int_Q(|\Delta\tilde{\psi}|^q);\psi\in W_0^{2,q}(Q),\int_Q|\nabla\tilde{\psi}|^2=1\right\}. \tag{$M_Q$}$$

Let $Z_Q=\left\{\tilde{\psi};\int_Q|\Delta\tilde{\psi}|^q=M_Q,\text{ and}\int_Q|\nabla\tilde{\psi}|^2=1\right\}$, $G_Q=\{\psi;\psi=M_Q^{1/(2-q)}\tilde{\psi},\tilde{\psi}\in Z_Q\}$. Then we have the following result, which is a corollary of Theorem 4.2.

**Corollary 4.3** The ground state set $G_Q$ is not empty. Moreover, it is $W_0^{2,q}(Q)$-stable with respect to (4.9) and (4.10).

# Appendix

**Proposition A.1** The functional

$$I_\Omega(\psi)=\frac{1}{2}\int_\Omega(|\nabla\psi|^2+|\psi|^2)-\frac{1}{q}\int_\Omega|\Delta\psi-\psi|^q \tag{A.1}$$

is neither bounded from above nor from below in $W_0^{2,q}(\Omega)$.

**Proof** For given $\psi \neq 0$ in $W_0^{2,q}(\Omega)$, from $I_\Omega(t\psi) = (t^2/2)\int_\Omega (|\nabla\psi|^2 + |\psi|^2) - (|t|^q/q)\int_\Omega |\Delta\psi - \psi|^q$ and $1<q<2$, we have

$$I_\Omega(t\psi) \to +\infty, \quad \text{as } |t| \to +\infty. \tag{A.2}$$

In the following, we prove that $I_\Omega$ is not bounded from below in $W_0^{2,q}(\Omega)$.

Let $E_r = \{\psi; \psi \in W_0^{2,q}(\Omega)$, and $\psi$ is radially symmetric$\}$, $\{e_i\}_{i=1}^{+\infty}$ be a Schauder basis for $E_r$ (cf. [14]), $X_i = \text{span}\{e_i\}$, and $\gamma_k = \overline{\bigoplus_{i=k}^{+\infty} X_i}$. Firstly, we claim

$$\alpha_k = \sup_{u\in\gamma_k,\ \|\psi\|_{2,q}=1} \|\psi\|_1 \to 0, \quad \text{as } k \to +\infty. \tag{A.3}$$

It is clear that $\alpha_k$ is a nonnegative decreasing sequence in $\mathbf{R}$. Let $\alpha_k \to \alpha \geqslant 0$, as $k \to +\infty$. For given $k$, there exists $\psi_k$ such that $\|\psi_k\|_{2,q} = 1$ and $\|\psi_k\|_1 \geqslant \alpha_k/2$. The fact that $\|\psi_k\|_{2,q} = 1$ and $\psi_k \in \gamma_k$ implies that $\psi_k \rightharpoonup \psi = 0$ in $E_r$ as $k \to +\infty$. Since $E_r \hookrightarrow H_{0,r}^1(\Omega)$ is compact, $\psi_k \to 0$ in $H_{0,r}^1(\Omega)$ as $k \to +\infty$, where $H_{0,r}^1(\Omega) = \{\psi : \psi \in H_0^1(\Omega)$, and $u$ is radially symmetric$\}$. So $\alpha_k \to \alpha = 0$. With the definition of $\alpha_k$,

$$I_\Omega(\psi) \leqslant \frac{1}{2}\alpha_k^2 \|\psi\|_{2,q}^2 - \frac{1}{q}\|\psi\|_{2,q}^q \quad \text{on the subspace } \gamma_k. \tag{A.4}$$

Let $\psi \in \gamma_k$, $\|\psi\|_{2,q} = \alpha_k^{-2/(2-q)}$,

$$I_\Omega(\psi) \leqslant \frac{1}{2}\alpha_k^2 \alpha_k^{-4/(2-q)} - \frac{1}{q}\alpha_k^{-2q/(2-q)} = \left(\frac{1}{2} - \frac{1}{q}\right)\alpha_k^{-2q/(2-q)} \to -\infty, \quad \text{as } k \to +\infty. \tag{A.5}$$

□

Let $N = \{\psi; \psi \in W_0^{2,q}(\Omega), \langle I'_\Omega(\psi), \psi\rangle = 0\}$, which is usually called Nehari manifold, $N_1 = \{\psi; \psi \in W_0^{2,q}(\Omega), \langle I'_\Omega(\psi), h\rangle = 0$ for any $h \in W_0^{2,q}(\Omega)\}$, which is the set of all critical points for $I_\Omega$. It is clear that $N \supset N_1$.

**Definition A.2** A function $\psi_0$ is called a ground state for $I_\Omega$, if $\psi_0$ is a critical point of $I_\Omega$ and $I_\Omega(\psi_0) \geqslant I_\Omega(\psi)$, that is, $\|\psi_0\|_1^2 \leqslant \|\psi\|_1^2$, for any $\psi \in N_1$.

**Proposition A.3** If $\tilde{\psi}_0$ is a minimizer for $(M_\Omega)$, that is, $\tilde{\psi} \in Z_\Omega$, then $\psi_0 = M_\Omega^{1/(2-q)}\tilde{\psi}_0$ is a ground state solution of $I_\Omega$. Moreover, $\psi_0$ is a "Mountain Pass type" critical point (saddle point) of $I_\Omega$.

**Proof** As is presented in the proof of Theorem 1.3, $\psi_0$ is a critical point of $I_\Omega$ and

$$I_\Omega(\psi_0) = \frac{1}{2}\int_\Omega (|\nabla\psi_0|^2 + |\psi_0|^2) - \frac{1}{q}\int_\Omega |\Delta\psi_0 - \psi_0|^q = \left(\frac{1}{2} - \frac{1}{q}\right)M_\Omega^{2/(2-q)}. \tag{A.6}$$

The definition of $M_\Omega$ implies that $\|h\|_{2,q}^q \geqslant M_\Omega \|h\|_1^q$ for all $h \in W_0^{2,q}(\Omega)$. If $\psi \in N$, which implies that $\|\psi\|_1^2 = \|\psi\|_{2,q}^q$, then $\|\psi\|_1^2 \geqslant M_\Omega^{2/(2-q)}$. Therefore, for any $\psi \in N$,

$$I_\Omega(\psi) = \left(\frac{1}{2} - \frac{1}{q}\right)\|\psi\|_1^2 \leqslant \left(\frac{1}{2} - \frac{1}{q}\right)M_\Omega^{2/(2-q)} = I_\Omega(\psi_0). \tag{A.7}$$

Since $N_1 \subset N$, by Definition A.2, $\psi_0$ is a ground state solution of $I_\Omega$.

For $0\neq\psi\in E$ and $t\geqslant 0$, let

$$g(t)=I_\Omega(t\psi)=\frac{t^2}{2}\int_\Omega(|\nabla\psi|^2+|\psi|^2)-\frac{t^q}{q}\int_\Omega|\Delta\psi-\psi|^q. \tag{A.8}$$

Then there exists a unique $t(\psi)>0$ such that $I_\Omega(t(\psi)\psi)=\inf_{t\geqslant 0}I_\Omega(t\psi)$ and $(dg(t)/dt)|_{t=t(\psi)}=0$, that is, $t(\psi)\int_\Omega(|\nabla\psi|^2+|\psi|^2)-t(\psi)^{q-1}\int_\Omega|\Delta\psi-\psi|^q=0$, which implies $t(\psi)\psi\in N$. Let

$$c_1=\sup_{\psi\in N}I_\Omega(\psi),\quad c_2=\sup_{0\neq\psi\in E}\inf_{t\geqslant 0}I_\Omega(t\psi). \tag{A.9}$$

Therefore, $I_\Omega(\psi_0)=c_1=c_2$. It follows from the definition of $c_2$ that $\psi_0$ is a "Mountain Pass type" critical point (saddle point) of $I_\Omega$. □

**Remark A.4** From the proof of Proposition A.3, we obtain

$$G_\Omega=\{\psi_0;\psi_0=M_\Omega^{1/(2-q)}\tilde{\psi}_0,\tilde{\psi}_0\in Z_\Omega\}=\{\psi_0;\psi_0\in N_1,\|\psi_0\|_1^2\leqslant\|\psi\|_1^2\text{ for any }\psi\in N_1\}. \tag{A.10}$$

In fact, if $\psi_0=M_\Omega^{1/(2-q)}\tilde{\psi}_0$, where $\tilde{\psi}_0\in Z_\Omega$, then $\psi_0\in N_1$ and $\|\psi_0\|_1^2\leqslant\|\psi\|_1^2$ for any $\psi\in N_1$. On the other hand, if $\psi_0\in N_1$ and $\|\psi_0\|_1^2\leqslant\|\psi\|_1^2$ for any $\psi\in N_1$, then $\|\psi_0\|_1^2=\|\psi_0\|_{2,q}^q=M_\Omega^{2/(2-q)}$. Letting $\tilde{\psi}_0=M_\Omega^{-1/(2-q)}\psi_0$, we know $\tilde{\psi}_0\in Z_\Omega$.

## References


[1] Hasegawa A, Mima K. Pseudo-three-dimensional turbulence in magnetized nonuniform plasma. The Physics of Fluids, 1978, 21 (1): 87-92.

[2] Grauer R. An energy estimate for a perturbed Hasegawa-Mima equation. Nonlinearity, 1998, 11 (3): 659-666.

[3] Guo B. Han Y. Existence and uniqueness of global solution of the Hasegawa-Mima equation. Journal of Mathematical Physics, 2004, 45 (4): 1639-1647.

[4] Holm D D, Marsden J E, Ratiu T, et al. Nonlinear stability of fluid and plasma equilibria. Physics Reports, 1985, 123 (1-2): 1-116.

[5] Albert J P. Concentration compactness and the stability of solitary-wave solutions to nonlocal equations// Applied Analysis, Vol 221 of Contemporary Mathematics, American Mathematical Society, Providence, RI, USA, 1999, 1-29.

[6] Lions P L. The concentration-compactness principle in the calculus of variations. The locally compact case. Ⅰ. Annales de l'Institut Henri Poincaré. Analyse Non Linéaire, 1984, 1 (2): 109-145.

[7] Lions P L. The concentration-compactness principle in the calculus of variations. The locally compact case. Ⅱ. Annales de l'Institut Henri Poincaré. Analyse Non Linéaire, 1984, 1 (4): 223-283.

[8] Huang D, Li Y. A concentration-compactness principle at infinity and positive solutions of some quasilinear elliptic equations in unbounded domains. Journal of Mathematical Analysis and Applications, 2005, 304 (1): 58-73.

[9] Struwe M. Variational methods: Applications to Nonlinear Partial Differential Equations and Hamiltonian System// Vol 34 of Results in Mathematics and Related Areas (3). 2nd edition. Berlin: Springer, 1996.

[10] Willem M. Minimax theorems // Vol 24 of progress in nonlinear differential equations and their applications, Birkhauser, Boston, Mass, USA, 1996.

[11] Gidas B, Ni W M, Nirenberg L. Symmetry of positive solutions of nonlinear elliptic equations in $\mathbf{R}^n$ // Mathematical Analysis and Applications, Part A, Vol 7 of Advances in Mathematics: Supplementary Studies. New York: Academic Press, 1981, 369-402.

[12] Kwong M K. Uniqueness of positive solutions of $\Delta u - u + u^p = 0$ in $\mathbf{R}^n$. Archive for Rational Mechanics and Analysis, 1989, 105 (3): 243-266.

[13] Palais R S. The principle of symmetric criticality. Communications in Mathematical Physics, 1979, 69 (1): 19-30.

[14] Triebel H. Interpolation theory, function spaces, differential operators // Vol 18 of North-Holland Mathematical Library. Amsterdam: North-Holland, 1978.

# Blowup Profile to the Solutions of Two-coupled Schrödinger Equations *

Chen Jianqing(陈建清) Guo Boling(郭柏灵)

**Abstract**

The model of the following two-coupled Schrödinger equations, $\mathrm{i}u_t+\frac{1}{2}\Delta u=(g_{11}|u|^{2p}+g|u|^{p-1}|v|^{p+1})u$, $(t,x)\in\mathbf{R}_+\times\mathbf{R}^N$, and $\mathrm{i}v_t+\frac{1}{2}\Delta v=(g|u|^{p+1}|v|^{p-1}+g_{22}|v|^{2p})v$, $(t,x)\in\mathbf{R}_+\times\mathbf{R}^N$, is proposed in the study of the Bose-Einstein condensates [Mitchell, et al. Self-traping of partially spatially incoherent light. Phys Rev Lett 1996, 77: 490]. We prove that for suitable initial data and $p$ the solution blows up exactly like $\delta$ function. As a by-product, we prove that similar phenomenon occurs for the critical two-coupled Schrödinger equations with harmonic potential [Perez-Garcia V M, Beitia T B, Sybiotic solitons in heteronuclear multicomponent Bose-Einstein condensates. Phys Rev A, 2005, 72, 033620.], $\mathrm{i}u_t+\frac{1}{2}\Delta u=\frac{\Omega}{2}|x|^2u+(g_{11}|u|^{2p}+g|u|^{p-1}|v|^{p+1})u$, $x\in\mathbf{R}^N$, and $\mathrm{i}v_t+\frac{1}{2}\Delta v=\frac{\Omega}{2}|x|^2v+(g|u|^{p+1}|v|^{p-1}+g_{22}|v|^{2p})v$, $x\in\mathbf{R}^N$.

## 1 Introduction and Main Results

This paper contains two parts. In the first part, we study the following model

$$\mathrm{i}u_t+\frac{1}{2}\Delta u=(g_{11}|u|^{2p}+g|u|^{p-1}|v|^{p+1})u,\quad x\in\mathbf{R}^N$$

$$\mathrm{i}v_t+\frac{1}{2}\Delta v=(g|u|^{p+1}|v|^{p-1}+g_{22}|v|^{2p})v,\quad x\in\mathbf{R}^N,\tag{SE}$$

with initial data

$$u(0,x)=\phi(x),\quad v(0,x)=\psi(x).\tag{IV}$$

In here $\mathrm{i}=\sqrt{-1}$, $g_{11}>0$, $g_{22}>0$, and $g<0$, $0<p<2/(N-2)^+$, where $2/(N-2)^+=+\infty$ when $N=1,2$ and $(N-2)^+=N-2$ when $N\geqslant 3$.

System (SE) appears in several branches of physics, such as in the study of interactions of waves with different polarizations[3] or in the description of nonlinear modulations of two monochromatic waves[22]. We refer to Refs. [1,6,7,13,18, and 19] for a comprehensive list of references. Physically, the solution $(u,v)$ denotes the beam in Kerr-like photorefractive media. The positive constants $g_{11}$ and $g_{22}$ are called defocusing case. The coupling constant

* Journal of Mathematical Physics, 2009, 50, 023505.

$g<0$ means that the interaction is attractive. The model (SE) also appears in two component Bose-Einstein condensates in the limit of strong transverse confinement[11] with the scattering lengths $a_{ij}(i,j=1,2)$ and $a_{11},a_{22}>0$ and $a_{12},a_{21}<0$, and $g_{ij}=2a_{ij}\alpha^{i+j-2}/a_0$ and $g_{12}=g_{21}=g$.

For the system (SE), it is natural to study the solitary wave solutions. To obtain solitary wave solutions of (SE), one may set $u(x)=e^{i\lambda t}u_1(x)$ and $v(x)=e^{i\mu t}v_1(x)$ and transforms (SE) into a coupled elliptic system[9]. Results in this aspect have been studied extensively during the past several years. We refer to Refs. [9, 10, and 28] for various existence and qualitative properties of solitary wave solutions. The main purpose of the present paper is to study the existence of blow-up solutions of Cauchy problem (SE) and the blow-up profile to the blow-up solutions. From the physical point of view, the important case of (SE) is $p=1$ and $N=2$ or $N=3$. But we would rather to study (SE) in a more general case. Before stating the main results, we list some notations used in this paper. By $H^1$ we denote the usual Sobolev space $H^1(\mathbf{R}^N)$ with usual norm $\|\cdot\|_{H^1}^2=\int_{\mathbf{R}^N}(|\cdot|^2+|\nabla\cdot|^2)$. $\|\cdot\|_{L^q}$ denotes the norm in $L^q(\mathbf{R}^N)(1\leqslant q\leqslant+\infty)$ space. $\Sigma=\{u\in H^1;\ |x|u\in L^2(\mathbf{R}^N)\}$ with norm $\|u\|_{\Sigma}^2=\int_{\mathbf{R}^N}(|\nabla u|^2+|u|^2)\mathrm{d}x$ and $\Sigma\times\Sigma$ is simply denoted by $\Sigma^2$. $S=S(\mathbf{R}^N)$ is the Schwartz space of rapidly decreasing $C^\infty$-functions and $S'$ is the dual of $S$.

In Sec. 2, we give some conditions on the existence of blow-up solutions of (SE) and (IV). Our idea is taken from Glassey[12]. More precisely, for a local solution $(u(t),v(t))$ of (SE) and (IV), we define $h(t)=\int|x|^2(|u(t)|^2+|v(t)|^2)$. Then

$$h''(t)=8E(u,v)+\frac{2Np-4}{p+1}\int(g_{11}|u|^{2p+2}+2g|u|^{p+1}|v|^{p+1}+g_{22}|v|^{2p+2}),$$

where

$$E(u,v)=\int\Big[\frac{1}{4}(|\nabla u|^2+|\nabla v|^2)+\frac{1}{2p+2}(g_{11}|u|^{2p+2}+2g|u|^{p+1}|v|^{p+1}+g_{22}|v|^{2p+2})\Big]$$

is the energy of the solution $(u(t),v(t))$. Combining this with the classical inequality $\|f\|_{L^2}^2\leqslant\|xf\|_{L^2}\|\nabla f\|_{L^2}$, we obtain the following theorem.

**Theorem 1.1** Let $\frac{2}{N}\leqslant p<2/(N-2)^+$ and $(\phi,\psi)\in\Sigma^2$. Suppose that $\begin{pmatrix} g_{11}, g \\ g, g_{22}\end{pmatrix}$ is negative definite and one of the following conditions holds.

(i) $E_0=E(\phi,\psi)<0$.

(ii) $E_0=0$ and $\operatorname{Im}\int(\bar{\phi}x\nabla\phi+\bar{\psi}x\nabla\psi)<0$.

(iii) $E_0>0$ and $\operatorname{Im}\int(\bar{\phi}x\nabla\phi+\bar{\psi}x\nabla\psi)<-2\sqrt{h(0)E_0}$.

Then the solution $(u(t),v(t))$ of (SE) and (IV) blows up in a finite time, i.e., there is $T_*>0$ such that

$$\lim_{t\to T_*^-}\int(|\nabla u|^2+|\nabla v|^2)=+\infty.$$

**Remark 1.1** From the conserved energy identity (2.2) and Gagliardo-Nirenberg inequality, we know that if $0<p<2/N$, then for any initial data $(\phi,\psi)\in\Sigma^2$ the solutions of (SE) and (IV) exist globally in time. However, when $p\geqslant 2/N$, there exist blow-up solutions of (SE) and (IV) as stated in Theorem 1.1. Therefore $p=2/N$ is the critical exponent in the sense of the existence and nonexistence of blow-up solutions.

Theorem 1.1 gives a sufficient condition to the existence of blow-up solutions of (SE) and (IV). A natural but interesting question is the following.

How does the blow-up solution behave near the blow-up time?

Recently, many mathematicians and physicists have studied the behavior of the blow-up solutions near blow-up time for the single Schrödinger equation[17],

$$i\varphi_t+\Delta\varphi+|\varphi|^{4/N}\varphi=0,\quad(t,x)\in\mathbf{R}_+\times\mathbf{R}^N,\tag{1.1}$$

with initial data $\varphi(0,x)=\varphi_0$, see Refs. [2,26,27,15,16, and 21] and the references therein. It is proven that the existence of blow-up solution of (1.1) and their asymptotical behavior near blow-up time have close relationship with the ground state solution of the following elliptic equation:

$$-\Delta R+R=|R|^{4/N}R,\quad R\in H^1,\quad\text{and}\quad R(x)>0.\tag{1.2}$$

For example, Weinstein[26,27] proved that if $\|\varphi_0\|_{L^2}=\|R\|_{L^2}$ [where $R(x)$ is the ground state solution of (1.2)] and the solution $\varphi(t,x)$ of (1.1) blows up at $T_0$, then

$$\lim_{t\to T_0}\lambda(t)^{N/2}u(t,\lambda(t)(x+y(t)))=R(x)\quad\text{in}\quad H^1,$$

where $y(t)\in\mathbf{R}^N$ and $\lambda(t)\in\mathbf{R}$ are some suitable functions. Many refined blow-up behavior of the blow-up solution has been proven by Merle[15,16]. During this procedure, the uniqueness of the ground state solution of (1.2) played an essential role.

However, for the problem (SE) we consider here, there are some results[9,10,28] on the ground state solution of the elliptic system,

$$-\lambda u+\frac{1}{2}\Delta u=(g_{11}|u|^{2p}+g|u|^{p-1}|v|^{p+1})u,\quad x\in\mathbf{R}^N$$

$$-\mu v+\frac{1}{2}\Delta v=(g|u|^{p+1}|v|^{p-1}+g_{22}|v|^{2p})v,\quad x\in\mathbf{R}^N$$

$$u(x)>0,\quad v(x)>0,\quad u(x),\quad v(x)\to 0\quad(|x|\to\infty).\tag{1.3}$$

But by Ref. [20] there is no uniqueness of the ground state solution of (1.3). So the methods of Weinstein[27] and Merle[15,16] cannot be used to study the asymptotical behavior of the blow-up solutions of (SE). However, borrowing some idea from Nawa[21], we can obtain some partial asymptotical results on the blow-up solutions of (SE) and (IV) in the case of $p=2/N$. For $a\in\mathbf{R}^N$ and $T>0$, we define

$$G(t;a,T)=e^{-i|x-a|^2/2(T-t)},\quad G_0(a,T)=G(0;a,T)=e^{-i|x-a|^2/2T}.$$

Denote

$$\mathcal{H}_0 = \{(\phi,\psi) \in \Sigma \times \Sigma; E(\phi,\psi) = 0\},$$

$$\mathcal{H}_1 = \{(\phi,\psi) \in \Sigma \times \Sigma; E(\phi,\psi) = 0 \text{ and (SE) with initial data } (u(0),v(0)) = (\phi,\varphi) \text{ has a global solution}\},$$

$$\mathcal{B}_0 = \{(\Phi,\Psi) \in \Sigma^2 \setminus (0,0); \exists a \in \mathbf{R}^N, T > 0, \text{ and } (\phi,\psi) \in \mathcal{H}_0 \text{ such that } (\Phi,\Psi) = (G_0(a,T)\phi, G_0(a,T)\psi) = (\mathrm{e}^{-\mathrm{i}|x-a|^2/2T}\phi(x), \mathrm{e}^{-\mathrm{i}|x-a|^2/2T}\psi(x))\},$$

$$\mathcal{B}_1 = \{(\Phi,\Psi) \in \Sigma^2 \setminus (0,0); \exists a \in \mathbf{R}^N, T > 0, \text{ and } (\phi,\psi) \in \mathcal{H}_1 \text{ such that } (\Phi,\Psi) = (G_0(a,T)\phi, G_0(a,T)\psi) = (\mathrm{e}^{-\mathrm{i}|x-a|^2/2T}\phi(x), \mathrm{e}^{-\mathrm{i}|x-a|^2/2T}\psi(x))\}.$$

Our next result is as follows.

**Theorem 1.2** Let $p = 2/N$ and $\begin{pmatrix} g_{11}, g \\ g, g_{22} \end{pmatrix}$ be negative definite. The solution $(U(t), V(t))$ of (SE) blows up at $t = T_m$ like the $\delta$ function, i.e.,

(1) $\lim_{t \to T_m} \int |x-a|^2(|U(t)|^2 + |V(t)|^2) = 0$;

(2) $\lim_{t \to T_m} (|U(t,\cdot)|^2 + |V(t,\cdot)|^2) = \delta(\cdot - a)\int(|\Phi|^2 + |\Psi|^2)$ in $S'$;

if and only if the initial data $(U(0), V(0)) = (\Phi, \Psi)$ are an element of $\mathcal{B}_1$ such that

$$(\Phi,\Psi) = (G_0(a,T_m)\phi, G_0(a,T_m)\psi), \quad (\phi,\psi) \in \mathcal{H}_1$$

and $\delta(\cdot - a)$ is the Dirac measure at $x = a$.

The proof of Theorem 1.2 is contained in Sec.4 after some preliminaries presented in Sec.3. The final section is the second part of this paper. We use the results obtained in Sec.4 to study the blow-up profile for the blow-up solutions of the following two-coupled Schrödinger equations with harmonic potential:

$$\mathrm{i}u_t + \frac{1}{2}\Delta u = \frac{\Omega}{2}|x|^2 u + (g_{11}|u|^{2p} + g|u|^{p-1}|v|^{p+1})u, \quad x \in \mathbf{R}^N$$

$$\mathrm{i}v_t + \frac{1}{2}\Delta v = \frac{\Omega}{2}|x|^2 v + (g|u|^{p+1}|v|^{p-1} + g_{22}|v|^{2p})v, \quad x \in \mathbf{R}^N$$

$$u(0,x) = \Phi(x), \quad v(0,x) = \Psi(x). \quad \text{(SEH)}$$

Recalling that for the single Schrödinger equation with harmonic potential,

$$\mathrm{i}u_t + \frac{1}{2}\Delta u = \frac{\Omega}{2}|x|^2 u - \lambda |u|^{4/N}u, \tag{1.4}$$

the blow-up solutions of the Cauchy problem (1.4) with suitable initial data have been studied in Refs. [4 and 29]. We point out that Carles[4] studied (1.4) by establishing a relationship between the solutions of (1.4) and the following equation:

$$\mathrm{i}u_t + \frac{1}{2}\Delta u = -\lambda |u|^{1+4/N}u. \tag{1.5}$$

Using the ideas of Carles[4,24], we are able to establish some blow-up properties to the blow-up solutions of (SEH). The main theorem is the following.

**Theorem 1.3** If the initial data $(\Phi, \Psi)$ are element of $\mathcal{B}_1$ such that

$$(\Phi,\Psi) = (G_0(a,T_m)\phi, G_0(a,T_m)\psi), \quad (\phi,\psi) \in \mathcal{H}_1$$

and $\delta(\cdot - a)$ is the Dirac measure at $x = a$, then the solution $(U_1(t), V_1(t))$ of (SEH) blows up at $t = T_m$. Moreover, the following properties hold.

(i-1)

$$\lim_{t \to \frac{\arctan(\Omega T_m)}{\Omega}} \int |x - a\cos(\Omega t)|^2(|U_1(t)|^2 + |V_1(t)|^2) = 0.$$

(i-2)

$$\lim_{t\to T_m} \frac{1}{(1+\Omega^2t^2)^2}\left[\left|U_1\left(\frac{\arctan(\Omega t)}{\Omega}, \frac{x}{(1+\Omega^2t^2)^{\frac{1}{2}}}\right)\right|^2 + \left|V_1\left(\frac{\arctan(\Omega t)}{\Omega}, \frac{1}{(1+\Omega^2t^2)^{1/2}}\right)\right|^2\right]$$
$$= \delta(\cdot - a)\int(|\Phi|^2 + |\Psi|^2) \text{ in } S'.$$

Notations: As above and henceforth, $\int_{\mathbf{R}^N} \cdot \, dx$ is simply denoted by $\int \cdot$.

## 2 Existence of Blow-up Solutions

In this section, we are interested in under what conditions, the solutions of (SE) and (IV) blow up in a finite time. First, from the results of Cazenave[5], we have immediately the following local existence result.

**Lemma 2.1** Let $0 < p < 2/(N-2)^+$ and $(\phi, \psi) \in H^1 \times H^1$. Then there is a $T_0 = T(\|\phi\|, \|\psi\|)$ and a unique solution $(u(t), v(t)) \in C([0, T_0], H^1 \times H^1)$ satisfying the conserved particle number,

$$\int(|u(t)|^2 + |v(t)|^2) = \int(|\phi|^2 + |\psi|^2), \tag{2.1}$$

and the conserved energy,

$$E(u,v) = \int\left[\frac{1}{4}(|\nabla u|^2 + |\nabla v|^2) + \frac{1}{2p+2}(g_{11}|u|^{2p+2} + 2g|u|^{p+1}|v|^{p+1} + g_{22}|v|^{2p+2})\right]$$
$$\equiv E(\phi, \psi). \tag{2.2}$$

Next using the results of Cazenave[5] we have that if $(\phi, \psi) \in \Sigma^2$, then the solution $(u(t), u(t)) \in C([0, T_0), \Sigma^2)$. Moreover, we have the following theorem.

**Theorem 2.2** Let $0 < p < 2/(N-2)^+$, $(\phi, \psi) \in \Sigma^2$, and $(u(t), v(t)) \in C([0, T_0), \Sigma^2)$ be the solution of (SE) and (IV). Define $h(t) = \int |x|^2(|u(t)|^2 + |v(t)|^2)$. Then $h(t)$ is well defined for $t \in [0, T_0)$. Moreover,

$$h'(t) = 2\mathrm{Im}\int(\bar{u}x\,\nabla u + \bar{v}x\,\nabla v), \tag{2.3}$$

$$h''(t) = 2\int\left(|\nabla u|^2 + |\nabla v|^2 + \frac{Np}{p+1}(g_{11}|u|^{2p+2} + 2g|u|^{p+1}|v|^{p+1} + g_{22}|v|^{2p+2})\right). \tag{2.4}$$

**Proof** We only prove (2.3) and (2.4) formally. Since $(u(t), v(t)) \in \Sigma^2$, we have that

$$h'(t) = 2\mathrm{Re}\int |x|^2(\bar{u}u_t + \bar{v}v_t) = -\mathrm{Im}\int |x|^2(\bar{u}\Delta u + \bar{v}\Delta v) = 2\mathrm{Im}\int (\bar{u}x\nabla u + \bar{v}x\nabla v),$$

$$h''(t) = -2\mathrm{Im}\int (u_t(N\bar{u} + 2x\nabla\bar{u}) + v_t(N\bar{v} + 2x\nabla\bar{v})) = -2\mathrm{Re}\int\Big[\Big(\frac{1}{2}\Delta u - u(g_{11}|u|^{2p} + g|u|^{p-1}|v|^{p+1})\Big)(N\bar{u} + 2x\nabla\bar{u}) + \Big(\frac{1}{2}\Delta v - v(g|u|^{p+1}|v|^{p-1} + g_{22}|v|^{2p})\Big)(N\bar{u} + 2x\nabla\bar{u})\Big].$$

Note that

$$\mathrm{Re}\int \frac{N}{2}\bar{u}\Delta u = -\frac{N}{2}\int |\nabla u|^2, \quad \mathrm{Re}\int \Delta u x\nabla\bar{u} = \frac{N-2}{2}\int |\nabla u|^2.$$

Then

$$\mathrm{Re}\int \frac{1}{2}\Delta u(N\bar{u} + 2x\nabla\bar{u}) = -\int |\nabla u|^2.$$

Similarly,

$$\mathrm{Re}\int \frac{1}{2}\Delta v(N\bar{v} + 2x\nabla\bar{v}) = -\int |\nabla v|^2,$$

$$\int |u|^{2p}u(N\bar{u} + 2x\nabla\bar{u}) = \int\Big(N|u|^{2p+2} - \frac{N}{p+1}|u|^{2p+2}\Big),$$

$$\int |u|^{p-1}|v|^{p+1}u(N\bar{u} + 2x\nabla\bar{u}) = \int\Big(N|u|^{p+1}|v|^{p+1} + \frac{2}{p+1}|v|^{p+1}x\nabla|u|^{p+1}\Big),$$

$$\int |u|^{p+1}|v|^{p-1}v(N\bar{v} + 2x\nabla\bar{v}) = \int\Big(N|u|^{p+1}|v|^{p+1} + \frac{2}{p+1}|u|^{p+1}x\nabla|v|^{p+1}\Big),$$

and

$$\int |v|^{2p}v(N\bar{v} + 2x\nabla\bar{v}) = \int\Big(N|v|^{2p+2} - \frac{N}{p+1}|v|^{2p+2}\Big).$$

Therefore

$$h''(t) = 2\int\Big(|\nabla u|^2 + |\nabla v|^2 + \frac{Np}{p+1}(g_{11}|u|^{2p+2} + 2g|u|^{p+1}|v|^{p+1} + g_{22}|v|^{2p+2})\Big).$$

The proof is complete. □

**Proof of Theorem 1.1** Assuming that the solutions exist globally in time, we obtain from Lemma 2.1 and Theorem 2.2 that

$$h''(t) = 8E(u,v) + \frac{2Np-4}{p+1}\int (g_{11}|u|^{2p+2} + 2g|u|^{p+1}|v|^{p+1} + g_{22}|v|^{2p+2}). \tag{2.5}$$

Since $p \geqslant 2/N$ and $\begin{pmatrix} g_{11}, g \\ g, g_{22} \end{pmatrix}$ is negative definite, we obtain that

$$h''(t) \leqslant 8E(u,v) = 8E(\phi,\psi) = 8E_0.$$

By a classical analysis, we have that

$$h(t) = h(0) + h'(0)t + \int_0^t (t-\tau)h''(\tau)\mathrm{d}\tau, \quad 0 \leqslant t < +\infty.$$

It follows that

$$h(t) \leqslant h(0) + h'(0)t + 4E_0 t^2, \quad 0 \leqslant t < +\infty. \tag{2.6}$$

Therefore one of the (ⅰ)~(ⅲ) of Theorem 1.1 implies that for some $T_0 > 0$,

$$\lim_{t \to T_0^-} h(t) = 0,$$

which together with (2.1) lead to a contradiction. □

**Remark 2.1** When $p = 1$ and $N = 2$, Theorem 1.1 has been proven in Refs. [25 and 23]. For general $N$ and $0 < p < 2/(N-2)^+$, Ma and Zhao[14] and Fanelli and Montefusco[8] gave the existence of the blow-up solutions for different combinations of the coefficients of $g$, $g_{11}$, and $g_{22}$. Theorem 1.1 seems to give a slight general condition on the existence of blow-up solutions of (SE) and (IV).

## 3 Some Preliminary Properties

In this section, we study some properties on the solutions of (SE) and (IV) in the case of $p = 2/N$. From now on we assume $p = 2/N$ and write the model as

$$\begin{aligned} & \mathrm{i}u_t + \frac{1}{2}\Delta u = (g_{11}|u|^{4/N} + g|u|^{2/N-1}|v|^{2/N+1})u, \quad x \in \mathbf{R}^N \\ & \mathrm{i}v_t + \frac{1}{2}\Delta v = (g|u|^{2/N+1}|v|^{2/N-1} + g_{22}|v|^{4/N})v, \quad x \in \mathbf{R}^N, \end{aligned} \tag{SE}$$

and

$$u(0,x) = \phi(x), \quad v(0,x) = \psi(x) \tag{IV}$$

Note from (2.5) that for $p = 2/N$, $h''(t) = 8E(u,v)$. Theorem 2.2 implies that one of the conditions (ⅰ)~(ⅲ) can produce the blow-up solutions of (SE) and (IV). For $\omega > 0$, we define a new map

$$J(\omega):(u,v) \in \Sigma^2 \to J(\omega)(u,v) \in \Sigma^2,$$

such that

$$\begin{aligned} [J(\omega)(u,v)](t,x) &:= ([J(\omega)u](t,x), [J(\omega)v](t,x)) \\ &= \left((1-\omega t)^{-N/2} G\left(t;0,\frac{1}{\omega}\right) u\left(\frac{t}{1-\omega t}, \frac{x}{1-\omega t}\right),\right. \\ &\quad \left.(1-\omega t)^{-N/2} G\left(t;0,\frac{1}{\omega}\right) v\left(\frac{t}{1-\omega t}, \frac{x}{1-\omega t}\right)\right). \end{aligned}$$

Then we have the following proposition.

**Proposition 3.1** Let $p = 2/N$. If $(u,v) \in C([0,T],\Sigma^2)$ is a solution of (SE) with $(u(0),v(0)) = (\phi,\psi)$, then $J(\omega)(u,v)$ is also a solution of (SE). Moreover,

$$J(\omega)(u,v) \in C\left(\left[0,\frac{T}{1+\omega T}\right],\Sigma^2\right).$$

**Proof** The proofs follow from direct computations, which are the same as in Ref. [26]. We omit the details here. □

Next, we define another transform. For any $a \in \mathbf{R}^N$ and $T > 0$, using the notation

$$G(t;a,T) = \mathrm{e}^{-\mathrm{i}|x-a|^2/2(T-t)},$$

we define

$$J(a,T):(u,v)\in\Sigma^2\to J(a,T)(u,v)\in\Sigma^2,$$

such that

$$\begin{aligned}[J(a,T)(u,v)](t,x)&:=([J(a,T)u](t,x),[J(a,T)v](t,x))\\&=\left((T-t)^{-N/2}G(t;a,T)u\left(\frac{t}{T(T-t)},\frac{x-a}{T-t}\right),\right.\\&\quad\left.(T-t)^{-N/2}G(t;a,T)v\left(\frac{t}{T(T-t)},\frac{x-a}{T-t}\right)\right).\end{aligned}$$

Define another map

$$j(a,T):(\phi,\psi)\in\Sigma\to j(a,T)(\phi,\psi),$$

such that

$$j(a,T)(\phi,\psi)=\left(T^{-N/2}G(0;a,T)\phi\left(\frac{x-a}{T}\right),T^{-N/2}G(0;a,T)\psi\left(\frac{x-a}{T}\right)\right).$$

Then $J(a,T)$ and $j(a,T)$ have inverse map. Indeed denote $w(t,x)=[J(a,T)u](t,x)$, then we have from direct computations that

$$u(s,y)=[J^{-1}(a,T)w](s,y)=\left(\frac{T}{1+Ts}\right)^{N/2}e^{iT|y|^2/2(1+Ts)}w\left(\frac{T^2s}{1+Ts},a+\frac{Ty}{1+Ts}\right),\tag{3.1}$$

and $u(0,y)=(T)^{N/2}e^{iT|y|^2/2}w(0,a+Ty)$, where we have denoted $s=t/T(T-t)$ and $y=(x-a)/(T-t)$.

**Proposition 3.2** Let $T_0$ be any positive constant. Let $(u(t),v(t))$ be a solution of (SE) with initial condition $(u(0),v(0))=(\phi(x),\psi(x))$ such that $(u(t),v(t))\in C([0,T_0],\Sigma^2)$. Then $J(a,T)^{-1}$ transforms $(u(t),v(t))$ into a solution $(U(t),V(t))\in C([0,T_0/T(T-T_0)];\Sigma^2)$ with initial data

$$(U(0),V(0))=j^{-1}(a,T)(u(0),v(0))=(j^{-1}(a,T)u(0),j^{-1}(a,T)v(0)).\tag{3.2}$$

For $J(a,T)$, we have a reverse result.

**Proof** Keep the expression of $J(a,T)^{-1}$[see (3.1)] in mind. We have from the choice of $s$ and $y$ that if $(u(t),v(t))\in C([0,T_0],\Sigma^2)$ is a solution of (SE), then clearly $s\in[0,T_0/T(T-T_0)]$. Hence

$$(U(t),V(t))\in C\left(\left[0,\frac{T_0}{T(T-T_0)}\right],\Sigma^2\right),$$

where

$$U(t)=\left(\frac{T}{1+Tt}\right)^{N/2}e^{iT|x|^2/2(1+Tt)}u\left(\frac{T^2t}{1+Tt},a+\frac{Tx}{1+Tt}\right)$$

and

$$V(t)=\left(\frac{T}{1+Tt}\right)^{N/2}e^{iT|x|^2/2(1+Tt)}v\left(\frac{T^2t}{1+Tt},a+\frac{Tx}{1+Tt}\right).$$

Now we verify that $(U(t),V(t))$ is a solution of (SE). By direct calculation, we have that

$$U_t = \left(\frac{T}{1+Tt}\right)^{N/2} e^{iT|x|^2/2(1+Tt)}\left(-\frac{NTu}{2(1+Tt)} - \frac{iT^2|x|^2u}{2(1+Tt)^2} + u_t\frac{T^2}{(1+Tt)^2} - \frac{T^2x\nabla u}{(1+Tt)^2}\right),$$

$$\Delta U = \left(\frac{T}{1+Tt}\right)^{N/2} e^{iT|x|^2/2(1+Tt)}\left(-\frac{T^2|x|^2u}{(1+Tt)^2} - \frac{2iT^2x\nabla u}{(1+Tt)^2} + \frac{iTNu}{1+Tt} - \frac{T^2\Delta u}{(1+Tt)^2}\right).$$

Similarly,

$$V_t = \left(\frac{T}{1+Tt}\right)^{N/2} e^{iT|x|^2/2(1+Tt)}\left(-\frac{NTv}{2(1+Tt)} - \frac{iT^2|x|^2v}{2(1+Tt)^2} + v_t\frac{T^2}{(1+Tt)^2} - \frac{T^2x\nabla v}{(1+Tt)^2}\right),$$

$$\Delta V = \left(\frac{T}{1+Tt}\right)^{N/2} e^{iT|x|^2/2(1+Tt)}\left(-\frac{T^2|x|^2v}{(1+Tt)^2} - \frac{2iT^2x\nabla v}{(1+Tt)^2} + \frac{iTNv}{1+Tt} - \frac{T^2\Delta v}{(1+Tt)^2}\right).$$

Thus

$$\begin{aligned} iU_t + \frac{1}{2}\Delta U &= \left(\frac{T}{1+Tt}\right)^{N/2+2} e^{iT|x|^2/2(1+Tt)}\left(iu_t + \frac{1}{2}\Delta u\right) \\ &= (g_{11}|U|^{4/N} + g|U|^{2/N-1}|V|^{1+2/N})U \end{aligned}$$

and

$$\begin{aligned} iV_t + \frac{1}{2}\Delta V &= \left(\frac{T}{1+Tt}\right)^{N/2+2} e^{iT|x|^2/2(1+Tt)}\left(iv_t + \frac{1}{2}\Delta v\right) \\ &= (g|U|^{1+2/N}|V|^{2/N-1} + g_{22}|V|^{4/N})V. \end{aligned}$$

Therefore $(U(t), V(t))$ is a solution of (SE) with $(U(0), V(0))$ satisfying (3.2).

The proofs on the transform of $J(a, T)$ are similar. We omit the details. □

## 4 Blow-up Profile of the Blow-up Solutions

After developing some properties of the solutions of (SE) in the case of $p = 2/N$, attentions are now turned to the blow-up profile of the blow-up solutions in this case. The main purpose of this section is to prove Theorem 1.2. We need some further lemmas.

**Lemma 4.1** Let $p = 2/N$ and $(u(t), v(t)) \in C([0, T), \Sigma^2)$ be a solution of (SE) with $(u(0), v(0)) = (\phi, \psi)$. Put

$$(U(t), V(t)) = ([J^{-1}(a,T)u](t), [J^{-1}(a,T)v](t)).$$

Then $(U(t), V(t)) \in C([0, +\infty), \Sigma^2)$ is also a solution of (SE) with

$$(U(0), V(0)) = (j^{-1}(a,T)\phi, j^{-1}(a,T)\psi).$$

Moreover, we have the following two identities:

$$\begin{aligned} &E(U(0), V(0)) \\ &= T^2E(\phi,\psi) + \frac{T}{2}\mathrm{Im}\int(\bar{\phi}(x)(x-a)\nabla\phi + \bar{\psi}(x)(x-a)\nabla\psi) + \\ &\frac{1}{4}\int|x-a|^2(|\phi(x)|^2 + |\psi(x)|^2), \end{aligned} \tag{4.1}$$

$$\begin{aligned} &\int|x-a|^2(|u(t,x)|^2 + |v(t,x)|^2) \\ &= \frac{T^2}{(1+Ts)^2}\Big(\int|y|^2(|U(0)|^2 + |V(0)|^2) + \\ &2s\mathrm{Im}\int[\bar{U}(0)y\nabla U(0) + \bar{V}(0)y\nabla V(0)] + 4E(U(0), V(0))s^2\Big). \end{aligned} \tag{4.2}$$

**Proof** According to Proposition 3.2, we only need to prove (4.1) and (4.2). Recalling that

$$U(s,y)=\left(\frac{T}{1+Ts}\right)^{N/2}e^{iT|y|^2/2(1+Ts)}u\left(\frac{T^2s}{1+Ts},a+\frac{Ty}{1+Ts}\right)$$

and

$$U(0,y)=T^{N/2}e^{iT|y|^2/2}\phi(a+Ty),$$

we have from direct calculations that

$$\int|\nabla U(0,y)|^2dy=\int|x-a|^2|\phi(x)|^2+T^2\int|\nabla\phi(x)|^2+2T\mathrm{Im}\int\bar{\phi}(x)(x-a)\nabla\phi(x).$$

Similarly,

$$\int|\nabla V(0,y)|^2dy=\int|x-a|^2|\psi(x)|^2+T^2\int|\nabla\psi(x)|^2+2T\mathrm{Im}\int\bar{\psi}(x)(x-a)\nabla\psi(x).$$

Thus

$$\begin{aligned}&E(U(0,y),V(0,y))\\&=\frac{T}{2}\mathrm{Im}\int(\bar{\phi}(x)(x-a)\nabla\phi(x)+\bar{\psi}(x)(x-a)\nabla\psi(x))+T^2E(\phi,\psi)+\\&\frac{1}{4}\int|x-a|^2(|\phi(x)|^2+|\psi(x)|^2).\end{aligned}$$

Therefore (4.1) holds.

To see (4.2), note again that

$$U(s,y)=\left(\frac{T}{1+Ts}\right)^{N/2}e^{iT|y|^2/2(1+Ts)}u\left(\frac{T^2s}{1+Ts},a+\frac{Ty}{1+Ts}\right).$$

It is deduced from direct calculations that

$$\int|y|^2|U(s,y)|^2dy=\left(\frac{T}{1+Ts}\right)^{-2}\int|x-a|^2|u(t,x)|^2dx,$$

which means that

$$\int|x-a|^2|u(t,x)|^2dx=\left(\frac{T}{1+Ts}\right)^{2}\int|y|^2|U(s,y)|^2dy.$$

Similarly,

$$\int|x-a|^2|v(t,x)|^2dx=\left(\frac{T}{1+Ts}\right)^{2}\int|y|^2|V(s,y)|^2dy.$$

Therefore

$$\int|x-a|^2(|u(t,x)|^2+|v(t,x)|^2)dx=\left(\frac{T}{1+Ts}\right)^{2}\int|y|^2(|U(s,y)|^2+|V(s,y)|^2)dy. \tag{4.3}$$

Notice that $(U(s,y),V(s,y))$ is a solution of (SE) with initial data $(U(0,y),V(0,y))$. Denote

$$H(s)=\int|y|^2(|U(s,y)|^2+|V(s,y)|^2).$$

Then similar to the proof of Theorem 1.1, we have that

$$H(s)=H(0)+H'(0)s+4E(U(0),V(0))s^2,$$

which is to say

$$\int |y|^2(|U(s,y)|^2 + |V(s,y)|^2)\mathrm{d}y$$

$$= \int |y|^2(|U(0)|^2 + |V(0)|^2) + 2s\,\mathrm{Im}\int [\bar{U}(0)y\nabla U(0) + \bar{V}(0)y\nabla V(0)] + 4E(U(0),V(0))s^2. \tag{4.4}$$

This proves (4.2). □

**Lemma 4.2** The condition $(\Phi,\Psi)\in\mathscr{B}_0$ is equivalent to one of the following three conditions.

(ⅰ)′ $E(\Phi,\Psi)<0$.

(ⅱ)″ $E(\Phi,\Psi)=0$ and $\mathrm{Im}\int(x-a)(\bar{\Phi}\nabla\phi+\bar{\Psi}\nabla\psi)<0$.

(ⅲ)′ $E(\Phi,\Psi)>0$ and

$$\mathrm{Im}\int(x-a)(\bar{\Phi}\nabla\phi+\bar{\Psi}\nabla\psi) < -\sqrt{E(\Phi,\Psi)\int|x-a|^2(|\Phi(x)|^2+|\Psi(x)|^2)}.$$

**Proof** We only sketch the argument. Note that for $(\Phi,\Psi)\in\mathscr{B}_0$, there is $a\in\mathbf{R}^N$, $T>0$, and $(\phi,\psi)\in\mathscr{H}_0$[remember $E(\phi,\psi)=0$] such that

$$(\Phi,\Psi) = (\mathrm{e}^{-\mathrm{i}|x-a|^2/2T}\phi(x), \mathrm{e}^{-\mathrm{i}|x-a|^2/2T}\psi(x)).$$

Hence we deduce from direct calculations that

$$\int|\nabla\phi(x)|^2 = \frac{1}{T^2}\int|x-a|^2|\Phi(x)|^2 + \int|\nabla\Phi(x)|^2 + \frac{2}{T}\mathrm{Im}\int\bar{\Phi}(x)(x-a)\nabla\Phi(x).$$

Similarly,

$$\int|\nabla\psi(x)|^2 = \frac{1}{T^2}\int|x-a|^2|\Psi(x)|^2 + \int|\nabla\Psi(x)|^2 + \frac{2}{T}\mathrm{Im}\int\bar{\Psi}(x)(x-a)\nabla\Psi(x).$$

Thus

$$0 = E(\phi,\psi)$$

$$= E(\Phi,\Psi) + \frac{1}{4}\left(\frac{1}{T^2}\int|x-a|^2(|\Phi(x)|^2+|\Psi(x)|^2) + \frac{2}{T}\mathrm{Im}\int(x-a)(\bar{\Phi}(x)\nabla\Phi(x) + \bar{\Psi}(x)\nabla\Psi(x))\right).$$

This proves the lemma. □

**Lemma 4.3** Let $(\phi,\psi)\in\mathscr{H}_1$ and $T>0$ be any fixed positive number. Put

$$(\Phi,\Psi) = j(a,T)(\phi,\psi) = \left(T^{-N/2}\mathrm{e}^{-\mathrm{i}|x-a|^2/2T}\phi\left(\frac{x-a}{T}\right), T^{-N/2}\mathrm{e}^{-\mathrm{i}|x-a|^2/2T}\psi\left(\frac{x-a}{T}\right)\right).$$

Then the solution $(U(t),V(t))$ of (SE) with initial data $(U(0),V(0))=(\Phi,\Psi)$ blows up at $t=T$ and satisfies the following.

(1) $\lim\limits_{t\to T}\int|x-a|^2(|U(t)|^2+|V(t)|^2)=0$.

(2) $\lim\limits_{t\to T}(|U(t,\cdot)|^2+|V(t,\cdot)|^2) = \delta(\cdot-a)\int(|\Phi|^2+|\Psi|^2)$ in $S'$.

**Proof** We first solve the equation (SE) with initial data $(\phi,\psi)\in\mathscr{H}_1$. Then according to the definition of $\mathscr{H}_1$, there is a unique solution $(u(t),v(t))$ of (SE) such that $(u(t),v(t))$

$\in C([0,+\infty),\Sigma^2)$. Now Put

$$(\widetilde{U}(t),\widetilde{V}(t)) = ([J(a,T)u](t),[J(a,T)v](t)).$$

By Proposition 3.2 $(\widetilde{U}(t),\widetilde{V}(t))\in C([0,T];\Sigma^2)$ is a solution of (SE) with

$$(\widetilde{U}(0),\widetilde{V}(0)) = ([j(a,T)u](0),[j(a,T)v](0)).$$

It follows from the uniqueness of the solutions of (SE) that

$$(U(t),V(t)) = (\widetilde{U}(t),\widetilde{V}(t)) = ([J(a,T)u](t),[J(a,T)v](t)).$$

Recalling the notations of $s$ and $y$, we obtain from (3.1) that

$$u(s,y) = \left(\frac{T}{1+Ts}\right)^{N/2} e^{iT|y|^2/2(1+Ts)} U\left(\frac{T^2 s}{1+Ts}, a+\frac{Ty}{1+Ts}\right)$$

and

$$u(0,y) = T^{N/2} e^{iT|y|^2/2} U(0,a+Ty).$$

Similar to those calculations used in the proof of Lemma 4.1, we obtain from $E(u(0),v(0)) = 0$ that

$$\begin{aligned}&\int |x-a|^2(|u(t,x)|^2+|v(t,x)|^2)\mathrm{d}x\\ &=\left(\frac{T}{1+Ts}\right)^2\int |y|^2(|U(s,y)|^2+|V(s,y)|^2)\mathrm{d}y\\ &=\left(\frac{T}{1+Ts}\right)^2\Big[\int |y|^2(|U(0)|^2+|V(0)|^2)+2s\,\mathrm{Im}\int(\overline{U}(0)y\,\nabla U(0)+\overline{V}(0)y\,\nabla V(0))+\\ &\quad 4E(U(0),V(0))s^2\Big].\end{aligned}$$

It is now deduced from

$$\begin{aligned}0=E(\phi,\psi) &= E(\Phi,\Psi)+\frac{1}{4}\left(\frac{1}{T^2}\int(|x-a|^2|\phi(x)|^2+|x-a|^2|\psi(x)|^2)+\right.\\ &\left.\frac{2}{T}\mathrm{Im}\int(x-a)(\overline{\Phi}(x)\,\nabla\Phi(x)+\overline{\Psi}(x)\,\nabla\Psi(x))\right).\end{aligned}$$

that (1) holds.

Next to see that (2) holds, it suffices to prove that for any $\rho\in S$,

$$\lim_{t\to T}\int\rho(x)(|U(t,x)|^2+|V(t,x)|^2) = \rho(a)\int(|U(0)|^2+|V(0)|^2).$$

Note that

$$\begin{aligned}&\left|\int(\rho(x)-\rho(a))(|U(t,x)|^2+|V(t,x)|^2)\right|\\ &\leqslant\int_{|x-a|>\varepsilon}|\rho(x)-\rho(a)|(|U(t,x)|^2+|V(t,x)|^2)+\\ &\quad\int_{|x-a|\leqslant\varepsilon}|\rho(x)-\rho(a)|(|U(t,x)|^2+|V(t,x)|^2)\\ &\leqslant M\int_{|x-a|>\varepsilon}(|U(t,x)|^2+|V(t,x)|^2)+\\ &\quad\int_{|x-a|\leqslant\varepsilon}(\rho(x)-\rho(a))(|U(t,x)|^2+|V(t,x)|^2)\end{aligned}$$

$$\leqslant \frac{M}{\varepsilon^2}\int |x-a|^2(|U(t,x)|^2+|V(t,x)|^2)+$$

$$\sup_{|x-a|\leqslant\varepsilon}|\rho(x)-\rho(a)|\int(|U(t,x)|^2+|V(t,x)|^2).$$

First letting $t\to T$, we get that

$$(*)\leqslant \sup_{|x-a|\leqslant\varepsilon}|\rho(x)-\rho(a)|\int(|U(0,x)|^2+|V(0,x)|^2).$$

Second letting $\varepsilon\to 0$, we get that $(*)\to 0$. This proves (2). □

Now we are in a position to prove Theorem 1.2.

**Proof of Theorem 1.2** Let $(\Phi,\Psi)$ be an element of $\mathscr{B}_1$ such that

$$(\Phi,\Psi)=(G_0(a,T_m)\phi,G_0(a,T_m)\psi)=(\mathrm{e}^{-\mathrm{i}|x-a|^2/2T_m}\phi(x),\mathrm{e}^{-\mathrm{i}|x-a|^2/2T_m}\psi(x)),$$

where $(\phi,\psi)\in\mathscr{H}_1$. Define $(\phi_0,\psi_0)\in\Sigma^2$ such that

$$\phi(x)=T_m^{-N/2}\phi_0\left(\frac{x-a}{T_m}\right),\quad \psi(x)=T_m^{-N/2}\psi_0\left(\frac{x-a}{T_m}\right).$$

Thus

$$\phi_0(y)=T_m^{N/2}\phi(a+T_my),\quad \psi_0(y)=T_m^{N/2}\psi(a+T_my).$$

From $E(\phi,\psi)=0$, we have that

$$E(\phi_0,\psi_0)=T_m^2E(\phi,\psi)=0.$$

By the uniqueness of the solutions of the Cauchy problem of (SE), we have that $(\phi_0,\psi_0)\in\mathscr{H}_1$. Hence according to Lemma 4.3 and

$$(\Phi,\Psi)=\left(T_m^{-N/2}\mathrm{e}^{-\mathrm{i}|x-a|^2/2T_m}\phi_0\left(\frac{x-a}{T_m}\right),T_m^{-N/2}\mathrm{e}^{-\mathrm{i}|x-a|^2/2T_m}\psi_0\left(\frac{x-a}{T_m}\right)\right).$$

We get that the solution of (SE) with initial data $(\Phi,\Psi)$ satisfies (1) and (2).

Conversely, suppose that the solution $(U(t),V(t))$ blows up at $t=T_m$ and satisfies (1) and (2). Put

$$(u(t),v(t))=(J^{-1}(a,T_m)U(t),J^{-1}(a,T_m)V(t)).$$

Then $(u(t),v(t))\in C([0,+\infty),\Sigma^2)$ and

$$\begin{aligned}E(u(t),v(t))&=E(u(0),v(0))=E(j^{-1}(a,T_m)U(t),j^{-1}(a,T_m)V(t))\\&=T_m^2E(U(0),V(0))+\frac{1}{4}\int|x-a|^2(|U(0)|^2+|V(0)|^2)+\\&\quad\frac{T_m}{2}\mathrm{Im}\int(x-a)(\overline{U}(0)\nabla U(0)+\overline{V}(0)\nabla V(0))\\&=\lim_{t\to T_m}\int|x-a|^2(|U(t,x)|^2+|V(t,x)|^2)=0\end{aligned}$$

by (2) and (1) of Lemma 4.3. From $(U(0),V(0))=(j(a,T_m)u(0),j(a,T_m)v(0))$ and the expression of $j(a,T_m)$, we know that $(U(0),V(0))\in\mathscr{B}_1$. The proof of Theorem 1.2 is complete. □

## 5 System with Harmonic Potential

This section is the second part of this paper. We use the results obtained in the first part to study the blow-up properties of the blow-up solutions of (SEH). Our ideas are due to Carles[4,24]. More precisely, Consider the Cauchy problem

$$iu_t + \frac{1}{2}\Delta u = \frac{\Omega}{2}|x|^2 u - \lambda |u|^{4/N}u, \quad t \geqslant 0, \quad x \in \mathbf{R}^N$$
$$u(0,x) = \varphi_0(x), \quad \omega > 0, \quad \lambda > 0, \tag{5.1}$$

and

$$iu_t + \frac{1}{2}\Delta u = -\lambda |u|^{4/N}u, \quad t \geqslant 0, \quad x \in \mathbf{R}^N$$
$$u(0,x) = \varphi_0(x), \quad \lambda > 0. \tag{5.2}$$

The following results have been proven by Carles[4].

(Ⅰ) Assume that $u(t) \in C([0,t_0'),\Sigma)(t_0' \in (0,\pi/2\Omega])$ is a solution of (5.1). Let

$$v(t,x) = \frac{1}{(1+(\Omega t)^2)^{N/4}} e^{i|x|^2\Omega^2 t/2(1+\Omega^2 t^2)} u\left(\frac{\arctan(\Omega t)}{\Omega}, \frac{x}{(1+\Omega^2 t^2)^{1/2}}\right).$$

Then $v(t) \in C([0,\tan(\Omega t_0')/\Omega),\Sigma)$ is a solution of (5.2). In particular, if $u(t) \in C([0,\pi/2\Omega),\Sigma)$, then $v(t) \in C([0,+\infty),\Sigma)$ is a global solution of (5.2).

(Ⅱ) Assume that $v(t) \in C([0,t_0),\Sigma)$ $(t_0 > 0)$ is a solution of (5.2). Let

$$u(t,x) = \frac{1}{(\cos(\Omega t))^{N/2}} e^{-i\Omega |x|^2 \tan(\Omega t)/2} v\left(\frac{\tan(\Omega t)}{\Omega}, \frac{x}{\cos(\Omega t)}\right).$$

Then $u(t) \in C([0,\arctan(\Omega t_0)/\Omega),\Sigma)$ is a solution of Cauchy problem (5.1). In particular, if $v(t) \in C([0,+\infty),\Sigma)$, then $u(t) \in C([0,\pi/2\Omega),\Sigma)$ is a local solution of (5.1).

From these results, we have immediately the following relations between solutions of (SE) with initial data $(\Phi, \Psi)$ and (SEH).

**Lemma 5.1**

(Ⅰ1) Assume that $(U(t),V(t)) \in C([0,t_0),\Sigma^2)$ $(t_0 > 0)$ is a solution of (SE) with $(U(0),V(0)) = (\Phi,\Psi)$. Let

$$U_1(t,x) = \frac{1}{(\cos(\Omega t))^{N/2}} e^{-i\Omega |x|^2 \tan(\Omega t)/2} U\left(\frac{\tan(\Omega t)}{\Omega}, \frac{x}{\cos(\Omega t)}\right)$$

and

$$V_1(t,x) = \frac{1}{(\cos(\Omega t))^{N/2}} e^{-i\Omega |x|^2 \tan(\Omega t)/2} V\left(\frac{\tan(\Omega t)}{\Omega}, \frac{x}{\cos(\Omega t)}\right).$$

Then

$$(U_1(t), V_1(t)) \in C\left(\left[0, \frac{\arctan(\Omega t_0)}{\Omega}\right), \Sigma^2\right)$$

is a solution of (SEH). In particular, if $(U(t),V(t)) \in C([0,+\infty),\Sigma)$, then

$$(U_1(t), V_1(t)) \in C\left(\left[0, \frac{\pi}{2\Omega}\right), \Sigma\right)$$

is a local solution of (SEH).

(Ⅰ2) Assume that $(U_1(t),V_1(t)) \in C([0,t_0'),\Sigma)$ $(t_0' \in (0,\pi/2\Omega])$ is a solution of (SEH). Let

$$U(t,x) = \frac{1}{(1+(\Omega t)^2)^{N/4}} e^{i|x|^2\Omega^2 t/2(1+\Omega^2 t^2)} U_1\left(\frac{\arctan(\Omega t)}{\Omega}, \frac{x}{(1+\Omega^2 t^2)^{1/2}}\right)$$

and

$$V(t,x)=\frac{1}{(1+(\Omega t)^2)^{N/4}}e^{i|x|^2\Omega^2 t/2(1+\Omega^2 t^2)}V_1\left(\frac{\arctan(\Omega t)}{\Omega},\frac{x}{(1+\Omega^2 t^2)^{1/2}}\right).$$

Then

$$(U(t),V(t))\in C\left(\left[0,\frac{\tan(\Omega t_0')}{\Omega}\right),\Sigma^2\right)$$

is the solution of (SE) with initial data $(\Phi,\Psi)$. In particular, if $(U_1(t),V_1(t))\in C([0,\pi/2\Omega),\Sigma^2)$, then

$$(U(t),V(t))\in C([0,+\infty),\Sigma^2)$$

is a global solution of (SE) with initial data $(\Phi,\Psi)$.

The following proposition is a direct consequence of Lemma 5.1.

**Proposition 5.2** Assume that $(U(t),V(t))\in C([0,T),\Sigma^2)$ is a blow-up solution of (SE) with initial data $(\Phi,\Psi)$ ($T$ is the maximal existence time). Then

$$(U_1(t),V_1(t))\in C\left(\left[0,\frac{\arctan(\Omega T)}{\Omega}\right),\Sigma^2\right)$$

is a blow-up solution of (SEH), where $[0,\arctan(\Omega T_0)/\Omega)$ is the maximal interval of existence time.

**Proof of Theorem 1.3** Assume that in (SEH) the initial data $(\Phi,\Psi)\in\mathscr{B}_1$. According to Theorem 1.2, we know that (SE) with initial data $(\Phi,\Psi)$ has a solution $(U(t),V(t))$ in $\Sigma^2$ which blows up at $t=T_m$ and satisfies (1) and (2) of Theorem 1.2. Let

$$U_1(t,x)=\frac{1}{(\cos(\Omega t))^{N/2}}e^{-i\Omega|x|^2\tan(\Omega t)/2}U\left(\frac{\tan(\Omega t)}{\Omega},\frac{x}{\cos(\Omega t)}\right)\tag{5.3}$$

and

$$V_1(t,x)=\frac{1}{(\cos(\Omega t))^{N/2}}e^{-i\Omega|x|^2\tan(\Omega t)/2}V\left(\frac{\tan(\Omega t)}{\Omega},\frac{x}{\cos(\Omega t)}\right).\tag{5.4}$$

Then by Proposition 5.2, we have that

$$(U_1(t),V_1(t))\in C\left(\left[0,\frac{\arctan(\Omega T_m)}{\Omega}\right),\Sigma^2\right)$$

is a blow-up solution of (SEH). Substitute the expression of $(U_1(t),V_1(t))$ into (1) of Theorem 1.2. We get from direct calculations that (i−1) of Theorem 1.3 holds. Since

$$U(t,x)=\frac{1}{(1+(\Omega t)^2)^{N/4}}e^{i|x|^2\Omega^2 t/2(1+\Omega^2 t^2)}U_1\left(\frac{\arctan(\Omega t)}{\Omega},\frac{x}{(1+\Omega^2 t^2)^{1/2}}\right)$$

and

$$V(t,x)=\frac{1}{(1+(\Omega t)^2)^{N/4}}e^{i|x|^2\Omega^2 t/2(1+\Omega^2 t^2)}V_1\left(\frac{\arctan(\Omega t)}{\Omega},\frac{x}{(1+\Omega^2 t^2)^{1/2}}\right),$$

we get that (i−2) of Theorem 1.3 holds. The proof is complete. □

## References

[1] Akhmediev N, Ankiewicz A. Partially coherent solitons on a finite background. Phys Rev Lett, 1999, 82, 2661.

[2] Berestycki H, Cazenave T. Instabilite des etats stationnaires dans les equations de Schrödinger et de Klein-

Gordon non linearires. C R Acad Sci Paris, Seire I, 1981, 293, 489.
[3] Berkhoer A L, Zakharov V E. Self exitation of waves with different polarizations in nonlinear media. Sov Phys JETP, 1970, 31, 486.
[4] Carles R. Critical nonlinear Schrödinger equations with and without harmonic potential. Math Models Meth Appl Sci, 2002, 12, 1513.
[5] Cazenave T. Semilinear Schrödinger Equations. Courant Institute of Mathematical Sciences. AMS, Providence, RI, 2005, Vol. 10.
[6] Cipolatti R, Zumpichiatti W. Orbital stable standing waves for a system of coupled nonlinear Schrödinger equations. Nonlinear Anal Theory, Methods Appl, 2000, 42, 445.
[7] Chow K W. Periodic solutions for a system of four coupled nonlinear Schrödinger equations. Phys Lett A, 2001, 285, 319.
[8] Fanelli L, Montefusco E. On the blow-up threshold for weakly coupled nonlinear Schrödinger equations. J Phys A: Math Theor, 2007, 40, 14139.
[9] Lin T C, Wei J. Spikes in two-component systems of nonlinear Schrödinger equations with trapping potential. J Differ Equations, 2006, 229, 538.
[10] Maia L A, Montefusco E, Pellacci B. Positive solutions for a weakly coupled Schrödinger system. J Differ Equations, 2006, 229, 743.
[11] Perez-Garcia V M, Beitia T B. Sybiotic solitons in heteronuclear multicomponent Bose-Einstein condensates. Phys Rev A, 2005, 72, 033620.
[12] Glassey R T. On the blowing-up of solutions to the Cauchy problem for the nonlinear Schrödinger equation. J Math Phys 1977, 18, 1794.
[13] Hioe F T. Solitary waves for $N$ coupled Schrödinger equations. Phys Rev Lett, 1999, 82, 1152.
[14] Ma L, Zhao L. Sharp thresholds of blow-up and global existence for the coupled nonlinear Schrödinger equations. e-print arXiv:07043921.
[15] Merle F. Determination of blow-up solutions with minimal mass or nonlinear Schrödinger equations with critical power. Duke Math J, 1993, 69, 427.
[16] Merle F. Asymptotics for $L^2$ minimal blow-up solutions of critical nonlinear Schrödinger equation. Ann Inst Henri Poincare, Anal Non Linéaire 1996, 13, 553.
[17] Merle F, Raphael P. Profiles and quantization of the blow-up mass for critical nonlinear Schrödinger equation. Commun Math Phys, 2005, 253, 675.
[18] Mitchell M, Chen Z, Shih M, et al. Self-traping of partially spatially incoherent light. Phys Rev Lett, 1996, 77, 490.
[19] Mitchell M, Segev M. Self-traping of incoherent white light. Nature (London), 1997, 387, 858.
[20] Montefusco E, Pellaccci B, Squassina M. Semiclassical states for weakly coupled nonlinear Schrödinger system. J Eur Math Soc, 2008, 10, 47.
[21] Nawa H, Tsutsumi M. On blow-up for the pseudo-comformally in variant nonlinear Schrödinger equation. Funkc Ekvac, 1989, 32, 417.
[22] Newboult G K, Parker D F, Faulkner T R. Coupled nonlinear Schrödinger equations arising in the study of monomode step-index optical fibers. J Math Phys, 1989, 30, 930.
[23] Roberts D C, Newell A C. Finite-time collapse of $N$ classical fields described by coupled nonlinear Schrödinger equations. Phys Rev E, 2006, 74, 047602.
[24] Rybin A V, Varzugin G G, Lindberg M, et al. Similarity solutions and collapse in the attractive Gross-Pitaevskii equation. Phys Rev E, 2000, 62, 6224.

[25] Tsutsumi Y, Zhang J. Instability of optical solitons for two wave interaction model in cubic nonlinear media. Adv Math Sci Appl, 1998, 8, 691.

[26] Weinstein M I. Nonlinear Schrödinger equations and sharp interpolation estimates. Commun Math Phys, 1983, 87, 567.

[27] Weinstein M I. On the structure and formation singularities in solutions to nonlinear dispersive evolution equations. Commun Partial Differ Equ, 1986, 11, 545.

[28] Wei J, Weth T. Nonradial bound states for a system of coupled Schrödinger equations. Atti Accad Naz Lincei Cl Sci Fis, Mat Nat, Rend Lincei, Mat Appl, 2007, 18, 279.

[29] Zhang J. Stability of attractive Bose-Einstein condensates. J Stat Phys, 2000, 102, 731.

# Stochastic Landau-Lifshitz Equation *

Guo Boling(郭柏灵) Pu Xueke(蒲学科)

**Abstract**

In this paper, we establish the existence of global weak solutions to the stochastic Landau-Lifshitz equation in $\mathbf{R}^d$ for any $d>0$. When restricted to the real line $\mathbf{R}^1$, we show that the Cauchy problem of this equation admits a unique global smooth solution with the difference method.

## 1 Introduction

As is well known, the Landau-Lifshitz-Gilbert equation[6,11] plays an important role in understanding magnetic phenomena emerging from magnetic materials. Many mathematicians have contributed their efforts to the study of this equation and there has been a broad study from both the mathematical and physical point of view regarding existence and uniqueness of strong or weak solutions and regularities as well as its link with harmonic maps, see [1,2,7,12,15,18] to list only a few; the list can be expanded as one wishes.

On the other hand, when thermal activation is introduced in the Landau-Lifshitz-Gilbert equation by a stochastic thermal field $H_{\text{th}}$, which is added to the effective field $H_{\text{eff}}$, we are led to the following stochastic Landau-Lifshitz-Gilbert equation[4]:

$$\frac{\mathrm{d}M}{\mathrm{d}t}=-\gamma M\wedge(H_{\text{eff}}+H_{\text{th}})-\alpha\gamma M\wedge[M\wedge(H_{\text{eff}}+H_{\text{th}})]. \tag{1.1}$$

When $\alpha=0$, we call this equation the stochastic Landau-Lifshitz equation. Physical evidence shows that the stochastic integral should be understood in the sense of Stratonovich instead of Itô[13] to yield correct thermal-equilibrium properties, which we will illustrate mathematically rigorously under the assumption that the magnetization field $M$ is of unit length. For weak damping $\alpha\ll 1$ we can drop the fluctuating field from the relaxation term of the above equation to arrive at

$$\frac{\mathrm{d}M}{\mathrm{d}t}=-\gamma M\wedge(H_{\text{eff}}+H_{\text{th}})-\alpha\gamma M\wedge(M\wedge H_{\text{eff}}), \tag{1.2}$$

which will be called the stochastic Landau-Lifshitz-Gilbert equation in the following. In fact, the two models are completely equivalent in the average sense when the condition of

* Differential and Integral Equations, 2009, 22 (3-4): 251-274.

thermodynamic consistency is applied[4], thus in the following we consider the formally simpler one (1.2). One more thing that should be illustrated here is that since a large number of microscopic degrees of freedom contribute to this mechanism, the thermal field is assumed to be a Gaussian random process with zero mean value. This observation shows that the average effects of the thermal field taken over different realizations vanish. In this paper, we restrict ourselves to the case when the effective field $H_{\text{eff}}=\Delta M$.

Let us put forward the stochastic Landau-Lifshitz equation mathematically. Let $(\Omega,\mathscr{F},(\mathscr{F}_t)_t,\mathbb{P})$ be a complete filtration probability space, and $W_t=(W_t^{(1)},W_t^{(2)},W_t^{(3)})$ be a three dimensional $\mathscr{F}_t$-adapted Brownian motion on $(\Omega,\mathscr{F},\mathbb{P})$ with the following statistical properties:

$$\mathbb{E}W_t^{(i)}=0,\quad \mathbb{E}[W_t^{(i)}W_s^{(j)}]=\delta_{ij}t\wedge s,\quad \text{for } i,j=1,2,3. \tag{1.3}$$

For fundamentals of stochastic differential equations, we refer the reader to [3, 9, 14, 13] for more details.

When $\alpha=0$, setting all the coefficients to be one, we can rewrite the above stochastic Landau-Lifshitz equation (1.1) as

$$\mathrm{d}u=u\wedge(\Delta u\mathrm{d}t+\circ\,\mathrm{d}M_t), \tag{1.4}$$

where $\circ d$ means that the stochastic integral is understood in the sense of Stratonovich, and $u=(u_1,u_2,u_3)$ is the magnetization vector from $\mathbf{R}^d\to\mathbb{S}^2$. This is a system of Langevin equations with multiplicative noise, because the multiplicative factor $u\wedge\circ$ is a function of the unknown $u$[8]. The reason why we use Stratonovich integral is illustrated in Section 2. However, it is convenient for us to use the equivalent Itô equation. By the relationship between Itô integral and Stratonovich integral[5]

$$X\circ\mathrm{d}Y=X\mathrm{d}Y+\frac{1}{2}\mathrm{d}X\mathrm{d}Y, \tag{1.5}$$

for two Itô processes $X$, $Y$, the equivalent Itô equation may be written as

$$\mathrm{d}u=u\wedge\Delta u\mathrm{d}t-u\mathrm{d}t+u\wedge\mathrm{d}W. \tag{SLL}$$

We call this equation (SLL) the stochastic Landau-Lifshitz equation, which can be completed with a given initial data $u(t=0,x)=u_0$.

In what follows this paper is concerned with the mathematical study of the equation (SLL). We will focus on two aspects of its solutions, namely, weak solution and smooth solution. For this purpose, we define the weak solution as follows.

**Definition 1.1** (Weak Solution) We call an $\mathscr{F}_t$-adapted stochastic process $u(t)$ with $|u(t)|=1$ for almost every $x\in\mathbf{R}^d$ and $\mathbb{E}\|\nabla u(t)\|_{L^2}^2<\infty$ for all $t\in[0,T]$ a weak solution of the stochastic Landau-Lifshitz equation (SLL), provided that

$$\langle u(t),v(t)\rangle=\langle u_0,v(0)\rangle-\int_0^t\langle u\wedge\nabla u,\nabla v\rangle\mathrm{d}s-\int_0^t\langle u(s),v(s)\rangle\mathrm{d}s+\int_0^t\langle v(s),u(s)\wedge\mathrm{d}W(s)\rangle\quad \text{a.s.}, \tag{1.6}$$

for any $v(s)\in L^\infty(0,T;H^1(\mathbf{R}^d))$.

This definition requires only that $\nabla u(t)\in L^2$, and $|u(t)|=1$ for almost every $x\in\mathbf{R}^d$. In this case, we seek a solution of this equation in the Sobolev space $\dot{H}^1(\mathbf{R}^d)$, and we can regard the solution $(u(t))_{t\geqslant 0}$ as a stochastic process in infinite-dimensional space with continuous parameters $t\geqslant 0$.

By smooth solutions[16] we mean that the process $(u(t))_{t\geqslant 0}$ is in any Sobolev space $H^k$ for $k=1,2,\cdots$. For the existence part, the method is mainly based on the vanishing viscosity method and the finite difference method. We consider the viscous equation

$$du=-\varepsilon u\wedge u\wedge\Delta u dt+u\wedge\Delta u dt-u dt+u\wedge dW \qquad (\mathrm{SLL}_\varepsilon\text{-}1)$$

with damping parameter $\varepsilon>0$. Equivalently, we rewrite this equation as

$$du=\varepsilon\Delta u dt+\varepsilon|\nabla u|^2 u dt+u\wedge\Delta u dt-u dt+u\wedge dW, \qquad (\mathrm{SLL}_\varepsilon\text{-}2)$$

whose equivalence when $u(t)$ is regular is stated in the Appendix. One can see immediately that these two equations are two variants of the stochastic Landau-Lifshitz-Gilbert equation (1.2).

Our main results are the existence of global weak solutions on $\mathbf{R}^d$ and the existence of a unique global smooth solution on the real line $\mathbf{R}^1$ for the stochastic Landau-Lifshitz equation for any initial data (without a smallness restriction), see Theorem 3.1 and Theorem 5.1 for a clearer statement. As a corollary of Section 4 and Section 5, we get the existence of global in time smooth solutions for the Cauchy problem of the stochastic Landau-Lifshitz-Gilbert equation (1.2) on $\mathbf{R}^1$ for $\alpha>0$ fixed. See the corollary for details.

This paper is organized as follows: in Section 2, we show mathematically why the Itô integral is not appropriate for our model and thus we should choose the Stratonovich integral to keep our model physically meaningful and to yield correct thermal equilibrium properties. In Section 3, we study the existence of weak solutions to our equation (SLL) by difference method in $\mathbf{R}^d$ for any dimension $d=1,2,\cdots$. In Section 4, we study the local in time existence of smooth solutions for the approximating equation ($\mathrm{SLL}_\varepsilon$-2) by a finite difference method and finally, in Section 5, based on the local theory we get the existence of smooth solutions of the Cauchy problem of the equation (SLL) on $\mathbf{R}^1$. As a corollary stated at the end of this paper, we also get the existence of global smooth solutions of the Cauchy problem of the stochastic Landau-Lifshitz-Gilbert equation on $\mathbf{R}^1$.

**Convention** the letters $C$, $c$ will denote some positive constants, which are unimportant and may change from line to line.

**Notation** we use $\|\cdot\|_m$ and $\langle u,v\rangle_m=\langle(-\Delta)^{m/2}u,(-\Delta)^{m/2}v\rangle_0$ respectively to denote the norm and the inner product of the homogeneous Hilbert space $\dot{H}^m$. When $m=0$ we use $\|\cdot\|$ and $\langle\cdot,\cdot\rangle$ for simplicity.

## 2 Why the Itô Integral Is not Appropriate

In this section, we show why the Itô integral is not appropriate for our model. By physical intuition, we require that the magnetization field $u(t)$ lies on the unit sphere $\mathbb{S}^2$ for all time when it initially lies on the sphere. Starting from this observation, we convince the reader that the Itô integral is not appropriate for our model and we should choose the Stratonovich integral for the study of the model.

Assume that the Itô integral is appropriate for our model. We can rewrite the equation as

$$du = u \wedge \Delta u dt + B(u)dW,$$

where $B(u)$ is antisymmetric:

$$B(u) = \begin{pmatrix} 0 & -u_3 & u_2 \\ u_3 & 0 & -u_1 \\ -u_2 & u_1 & 0 \end{pmatrix}.$$

By the Itô formula, we have

$$|u|^2(t) = |u_0|^2 + 2\int_0^t \langle u, u \wedge \Delta u\rangle ds + 2\int_0^t \langle u, u \wedge dW\rangle + \int_0^t \mathrm{Tr}(BB^*)ds = \int_0^t \mathrm{Tr}(BB^*)ds,$$

where $|u|^2 = |u_1|^2 + |u_2|^2 + |u_3|^2$ denotes the vector norm in $\mathbf{R}^3$. One can compute directly that $\mathrm{Tr}(B(u)B^*(u)) = 2|u|^2$. Inserting this into the Itô formula, we have

$$|u|^2(t) = |u_0|^2 + 2\int_0^t |u|^2 ds,$$

which implies that $|u|^2(t) = |u_0|^2 e^{2t} = e^{2t}$. This shows that, in the Itô sense, the solution is not on the sphere $\mathbb{S}^2$ although the magnetization field initially lies on the unit sphere.

On the contrary, when the integral is understood in the Stratonovich sense, correspondingly for (SLL), one can show by the same procedure as above that the solution $u(t)$ lies on the unit sphere $\mathbb{S}^2$ when the magnetization field initially lies on the unit sphere. The above analysis shows from the mathematical point of view that the Stratonovich integral is appropriate for the study of the stochastic Landau-Lifshitz equation.

## 3 Global Weak Solution for (SLL)

In this section, we establish the existence of global weak solutions to the stochastic Landau-Lifshitz equation (SLL). More precisely, we have the following theorem.

**Theorem 3.1** Let $u_0$ be such that $u_0 \in \mathbb{S}^2$ for almost every $x \in \mathbf{R}^d$, $u_0 \in \dot{H}^1(\mathbf{R}^d)$ and is measurable with respect to $\mathscr{F}$. There exists a global weak solution $u(t)$ of the stochastic Landau-Lifshitz equation, which is an adapted process satisfying the stochastic Landau-Lifshitz equation in the weak sense in Definition 1.1. The solution remains on the unit sphere $\mathbb{S}^2$ as time evolves, and keeps $\mathbb{E}\|\nabla u\|^2$ bounded for all time.

We utilize the difference method to derive the existence of weak solutions of this equation. Let $u^h(x_i)$ denote the magnetization vector located at the grid point $x_i = \lambda_i h$ in $\mathbf{R}^d$, for $\lambda_i = (\lambda_i^{(1)}, \cdots, \lambda_i^{(d)}) \in \mathbf{Z}^d$. Denoting $h_j = (0, \cdots, 0, \underbrace{h}_{j^{\text{th}}}, 0, \cdots, 0)$, we use the forward difference quotient

$$D_j^+ u^h(x_i) = \frac{u^h(t, x_i + h_j) - u^h(t, x_i)}{h} \tag{3.1}$$

to approximate the space derivatives $\partial_{x_j} u$. Similarly one can define the backward difference quotient $D_j^- u^h(x_i)$. The Laplacian operator $\Delta$ is approximated by

$$\tilde{\Delta} u^h(x_i) = \sum_{j=1}^d D_j^+ D_j^- u^h(x_i) = \sum_{j=1}^d D_j^- D_j^+ u^h(x_i).$$

Denote

$$\| u^h \|_{L_h^p}^p = \sum_{i \in \mathbf{Z}^d} | u^h(x_i) |^p h^d,$$

when $p = \infty$, $\| u^h \|_{L_h^\infty} = \sup_{i \in \mathbf{Z}^d} | u^h(x_i) |$. Higher-order Sobolev norms can be defined by higher-order difference quotients similarly. Rewriting the equation (SLL) in the discrete form for the unknowns $\{ u^h(t, x_i) \}_i$, we obtain

$$du^h(t, x_i) = u^h(t, x_i) \wedge \tilde{\Delta} u^h(t, x_i) dt - u^h(t, x_i) dt + B(u^h(t, x_i)) dW_t. \tag{3.2}$$

This is a standard stochastic differential system (with respect to $t$) of the following form

$$dX_t = \sigma(t, X_t) dt + b(t, X_t) dW_t, \tag{3.3}$$

whose coefficients

$$u^h \mapsto u^h \wedge \tilde{\Delta} u^h - u^h(t, x_i), \quad u^h \mapsto B(u^h(t, x_i))$$

are locally Lipschitz. Then by standard SDE theory, we know that there exists a local solution for $\{ u^h(t, x_i) \}_{i \in \mathbf{Z}^d}$.

- **Zeroth estimates** Using the Itô formula, we deduce that

$$\begin{aligned} d| u^h(x_j) |^2 = & 2( u^h(t, x_i), u^h(t, x_i) \wedge \tilde{\Delta} u^h(t, x_i)) dt - \\ & 2( u^h(t, x_i), u^h(t, x_i)) dt + 2( u^h(t, x_i), u^h(t, x_i) \wedge dW) + \\ & \mathrm{Tr}[ B(u^h(t, x_i)) B^*(u^h(t, x_i))] dt = 0, \end{aligned}$$

where $B(u)$ is defined in Section 2 and

$$\mathrm{Tr}[ B(u) B^*(u)] = 2| u |^2. \tag{3.4}$$

This shows that $u^h(t, x_i)$ remains of unit length for almost surely $\omega \in \Omega$.

- **First estimates** Applying $D_j^+$ to the equation (3.2), we have

$$dD_j^+ u^h(t, x_i) = D_j^+[ u^h(t, x_i) \wedge \tilde{\Delta} u^h(t, x_i)] dt - D_j^+ u^h(t, x_i) dt + D_j^+ u^h(t, x_i) \wedge dW(t).$$

Using the Itô formula and summation we have

$$\begin{aligned} \sum_{j, x_i} d| D_j^+ u^h(t, x_i) |^2 = & 2 \sum_{j, x_i} \langle D_j^+ u^h(t, x_i), D_j^+[ u^h(t, x_i) \wedge \tilde{\Delta} u^h(t, x_i)] \rangle dt - \\ & 2 \sum_{j, x_i} \langle D_j^+ u^h(t, x_i), D_j^+ u^h(t, x_i) \rangle dt + \\ & 2 \sum_{j, x_i} \langle D_j^+ u^h(t, x_i), D_j^+ u^h(t, x_i) \wedge dW \rangle + \\ & \sum_{j, x_i} \mathrm{Tr}[ B(D_j^+ u^h(t, x_i)) B^*(D_j^+ u^h(t, x_i))] dt = 0, \end{aligned}$$

thanks to the well-known integration by parts formula in the discrete form

$$\sum_{x_i} v^h(x_i) \cdot D_j^+ u^h(x_i) = -\sum_{x_i} D_j^- v^h(x_i) \cdot u^h(x_i). \tag{3.5}$$

This implies that

$$\| u^h \|^2_{\dot{H}^1_h} = \| u_0^h \|^2_{\dot{H}^1_h}, \tag{3.6}$$

for almost surely $\omega \in \Omega$.

In the above, we constructed a solution $u^h$ defined on the mesh $\mathbf{Z}_h^d := \{x_i : x_i = \lambda_i h, \lambda_i \in \mathbf{Z}^d\}$ in $\mathbf{R}^d$ which satisfies the stochastic equation (3.2). To extend the solution to the whole space, we use an interpolation process introduced in [10]. Denote by $C_i$ the cell

$$C_i = \{x_{i;1}, x_{i;1} + h_1\} \times \cdots \times \{x_{i;d}, x_{i;d} + h_d\} \tag{3.7}$$

of the $i^{\text{th}}$ lattice $\{x_i\}$. For any $x \in C_i$, let $q_h$, $p_h$, $r_h^{(m)}$ be the interpolation operators introduced below.

- $q_h u^h(x) = u^h(x_i)$.
- $$p_h u^h(x) = u^h(x_i) + \sum_{j=1}^{d} D_j^+ u^h(x_i) \frac{h_j}{h} \cdot (x - x_i) + \cdots + \sum_{j=1}^{d} D_1^+ \cdots D_{j-1}^+ D_{j+1}^+ \cdots D_d^+ u^h(x_i) \prod_{l=1, l\neq j}^{d} \frac{h_l}{h} \cdot (x - x_i) + D_1^+ \cdots D_d^+ u^h(x_i) \prod_{l=1}^{d} \frac{h_l}{h} \cdot (x - x_i).$$
- $$r_h^{(m)} u^h(x) - u^h(x_i) + \sum_{j=1, j\neq m}^{d} D_j^+ u^h(x_i) \frac{h_j}{h} \cdot (x - x_i) + D_1^+ \cdots D_{m-1}^+ D_{m+1}^+ \cdots D_d^+ u^h(x_i) \prod_{l=1, l\neq m}^{d} \frac{h_l}{h} \cdot (x - x_i).$$

For these operators, we have $\frac{\partial}{\partial x_m} p_h u^h = r_h^{(m)}(D_m^+ u^h)$. More importantly, if one of the interpolates $p_h u^h$, $q_h u^h$ or $r_h^{(m)}$ converges strongly (respectively weakly) in $L^2$ as $h \to 0$, then the other two also converge to the same limit in $L^2$ strongly (respectively weakly)[10].

We can see from the above estimates that

$$p_h u^h(x) \to u(x) \quad \text{in} \quad L^2(\Omega; L^\infty(0, T; \dot{H}^1)) \quad \text{weak } *$$
$$p_h u^h(x) \to u(x) \quad \text{in} \quad L^2(\Omega; L^2(0, T; L^2_{\text{loc}})) \quad \text{strongly},$$

where the second one is due to the Sobolev compactness embedding theorem.

To show that the sequence $\{u^h(t, x_i)\}_h$ converges to a weak solution $u(t)$ of the stochastic equation, we consider the interpolation sequence $\{p_h u^h\}_h$. Indeed, for this sequence,

$$\langle p_h u^h(x), p_h v^h(x) \rangle$$
$$= \langle p_h u_0^h(x), p_h v_0^h(x) \rangle + \int_0^t \langle p_h D_i^+(u^h \wedge D_i^- u^h), p_h v^h(x) \rangle \mathrm{d}s -$$

$$\int_0^t \langle p_h u^h(x), p_h v^h(x)\rangle \mathrm{d}s + \int_0^t \langle p_h v^h(x), p_h u^h(x)\mathrm{d}W\rangle \tag{3.8}$$

holds for almost surely $\omega \in \Omega$, and for any sequence $v^h$ such that $p_h v^h$ converges to $v$ in $L^\infty(0,T;H^1)$, thanks to (3.2) and the definition of the interpolation operators.

We now consider the convergence of the second term and the fourth term on the right; the other terms are similar.

**Convergence of the second term** We need only to show that $p_h D_i^+(u^h \wedge D_i^- u^h)$ converges to $\partial_{x_i}(u \wedge \partial_{x_i})$ in $L^\infty(0,T;H_{\mathrm{loc}}^{-1})$ weak*. Since $u^h \wedge D_i^- u^h$ is in a bounded subset of $L^\infty(0,T;L_h^2)$, $p_h(u^h \wedge D_i^- u^h)$, $q_h(u^h \wedge D_i^- u^h)$ and $r_h^{(m)}(u^h \wedge D_i^- u^h)$ converge to the same limit in $L^\infty(0,T;L^2)$. Rewrite $q_h(u^h \wedge D_i^- u^h) = q_h u^h \wedge q_h D_i^- u^h$. By the previous strong convergence result, $q_h u^h \to u$ for almost every $x \in \mathbf{R}^d$ up to a subsequence by the Riesz theorem and $q_h D_i^- u^h \to \partial_{x_i} u$ weak*, one then has

$$p_h(u^h \wedge D_i^+ u^h) \to u \wedge \frac{\partial u}{\partial x_i} \text{ in } L^\infty(0,T;L^2) \text{ weak } *;$$

$$\frac{\partial}{\partial x_i} p_h(u^h \wedge D_i^+ u^h) \to \frac{\partial}{\partial x_i}\left(u \wedge \frac{\partial u}{\partial x_i}\right) \text{ in } L^\infty(0,T;L^2) \text{ weak } *.$$

This implies that $r_h^{(i)} D_i^+(u^h \wedge D_i^- u^h)$ and thus $p_h D_i^+(u^h \wedge D_i^- u^h)$ converges to the same limit $\frac{\partial}{\partial x_i}\left(u \wedge \frac{\partial u}{\partial x_i}\right)$ in $L^\infty(0,T;H_{\mathrm{loc}}^{-1})$ weak *.

**Convergence of the fourth term** Denote

$$\begin{aligned}\mathscr{Q}_h - \mathscr{Q} &:= \int_0^t \langle p_h v^h, p_h u^h \wedge \mathrm{d}W_s\rangle - \int_0^t \langle v, u \wedge \mathrm{d}W_s\rangle \\ &= \int_0^t \langle v, (p_h u^h - u) \wedge \mathrm{d}W_s\rangle + \int_0^t \langle p_h v^h - v, p_h u^h \mathrm{d}W_s\rangle =: \mathscr{M}_h + \mathscr{N}_h.\end{aligned}$$

We use the BDG inequality to deduce that

$$\begin{aligned}\mathbb{E}\sup_{0\leqslant s\leqslant t}|\mathscr{Q}_h - \mathscr{Q}|^2 &\leqslant \mathbb{E}\sup_{0\leqslant s\leqslant t}|\mathscr{M}_h|^2 + \mathbb{E}\sup_{0\leqslant s\leqslant t}|\mathscr{N}_h|^2 \\ &\leqslant \sup_{0\leqslant s\leqslant t}\mathbb{E}\int_0^t \left|\int_{\mathbf{R}^d} v \wedge (p_h u^h - u)\mathrm{d}x\right|^2 \mathrm{d}s + \\ &\quad \sup_{0\leqslant s\leqslant t}\mathbb{E}\int_0^t \left|\int_{\mathbf{R}^d}(p_h v^h - v) \wedge p_h u^h \mathrm{d}x\right|^2 \mathrm{d}s \to 0,\end{aligned}$$

where the first one is due to the weak convergence of $p_h u^h - u$ to 0 in $L^2(0,t;\dot{H}^1)$ for almost surely $\omega \in \Omega$, and $v$ is bounded in $L^2(0,t;L^2)$ and the second one is due to the fact that $p_h u^h$ is bounded in $L^2(0,t;\dot{H}^1)$ for almost surely $\omega \in \Omega$ and $p_h v_h - v$ is convergent strongly in $L^2(0,t;H^1)$. This shows that $\mathbb{E}\sup_{0\leqslant s\leqslant t}|\mathscr{Q}_h - \mathscr{Q}|^2 \to 0$, which implies that there exists a subsequence of $\{p_h u^h(x)\}_h$ such that

$$\int_0^t \langle p_h v^h, p_h u^h \wedge \mathrm{d}W_s\rangle \to \int_0^t \langle v, u \wedge \mathrm{d}W_s\rangle, \tag{3.9}$$

up to a subsequence for almost surely $\omega \in \Omega$ by the Risez theorem; i.e., every sequence $a_n \to a$

in norm has a subsequence $a_{n'}$ such that $a_{n'}\to a$ almost everywhere.

Then passing to the limit of (3.8) as $h\to 0$, we have that

$$\langle u,v\rangle=\langle u_0,v_0\rangle+\int_0^t\langle u\wedge\Delta u,v\rangle ds-\int_0^t\langle u,v\rangle ds+\int_0^t\langle v,u\wedge dW\rangle \quad (3.10)$$

for any $v\in L^\infty(0,T;H^1)$. This shows that $u$ constructed as the limit of the approximating sequence $u^h$ is a weak solution of the stochastic Landau-Lifshitz equation and concludes the proof of the theorem.

## 4 Local Theory for $\varepsilon>0$

In this section, we establish the local in time existence of solutions to the periodic approximating stochastic equation on the finite interval $\Lambda=(-D,D)$

$$du=\varepsilon\Delta u dt+\varepsilon|\nabla u|^2 u dt+u\wedge\Delta u dt-u dt+u\wedge dW, \qquad (\mathrm{SLL}_\varepsilon\text{-}2)$$

for fixed $\varepsilon>0$ with initial data $u_0$. To this end, we consider the following difference equation:

$$du^h(t,x_i)=\varepsilon D^+D^-u^h(t,x_i)dt+\varepsilon|D^+u^h(t,x_i)|^2u^h(t,x_i)dt+ \\ u^h(t,x_i)\wedge D^+D^-u^h(t,x_i)dt-u^h(t,x_i)dt+u^h(t,x_i)\wedge dW, \quad (4.1)$$

where $i\in\mathscr{J}:=\{0,\pm1,\cdots,\pm J\}$, $h=D/J$ is the step size and $D^+,D^-$ are one-dimensional difference quotient operators defined in (3.1). The periodic condition implies that $u^h(t,x_{-J})=u^h(t,x_J)$ for any $t\geqslant 0$. This is a system of stochastic ordinary differential equations with respect to $t$, and the unknowns are $\{u^h(t,x_i)\}_{i\in\mathscr{J}}$. The coefficients of this system are locally Lipschitz thus, by the standard SODE theory, we know that there exists a local solution $\{u^h(t,x_i)\}_{i\in\mathscr{J}}$ on $[0,T_h]$ of the stochastic system (4.1). In the following, we will give some *a priori* estimates independent of the step size $h>0$, then letting $h\to 0$, one can get a local in time solution of the approximating stochastic Landau-Lifshitz equation ($\mathrm{SLL}_\varepsilon$-2) on some public interval $[0,T_0]$.

Before starting our exposition, we introduce some notation. Defining the first-order difference quotient as in (3.1), one can then define the discrete function $\delta u^h$ as

$$\delta u^h:=\{D^+u^h(x_i):i\in\mathscr{J}\},$$

the $L_h^p$-norm of which can be defined by

$$\|\delta u^h\|_{L_h^p}^p=\sum_{j\in\mathscr{J}}|D^+u^h(x_i)|^p h.$$

Applying the difference operator $D^+$ to the first-order difference quotient, one can define the second-order difference quotient of the discrete function $\{u^h(x_i):i\in\mathscr{J}\}$ as

$$D^+D^-u^h(x_i)=\frac{u^h(x_{i+1})-2u^h(x_i)+u^h(x_{i-1})}{h^2}, \text{for all } -J+1\leqslant i\leqslant J-1.$$

Since we are given the periodic boundary condition, the definition can be easily extended to all $i\in\mathscr{J}$. One needs only to use $u^h(x_{-J+1})$ to replace $u^h(x_{J+1})$ and $u^h(x_{J-1})$ to replace $u^h(x_{-J-1})$ in the above formula, so that the second-order difference quotient is well defined for all $i\in\mathscr{J}$. By induction, one can define the $k^{\text{th}}$-order difference quotient for all $i\in\mathscr{J}$ and all

$k\in\mathbf{N}$ and similarly the discrete function $\delta^k u^h$ and their $L_h^p$-norms. For the interpolation inequality in the discrete form, we have the following:

**Lemma 4.1** (Theorem 3, [17]). For any discrete function $u^h=\{u_j\,|\,j=0,\pm1,\cdots,\pm J\}$ on the finite interval $\Lambda$, there holds

$$\|\delta^k u^h\|_{L_h^p}\leqslant c\|u^h\|_{L_h^2}^{1-\frac{k+\frac{1}{2}-\frac{1}{p}}{n}}\left(\|\delta^n u^h\|_{L_h^2}+\frac{\|u^h\|_{L_h^2}}{D^n}\right)^{\frac{k+\frac{1}{2}-\frac{1}{p}}{n}},\tag{4.2}$$

where $2\leqslant p\leqslant\infty$, $0\leqslant k\leqslant n$ for any $n\in\mathbf{N}$ and $c$ is a constant independent of $u^h$.

**Lemma 4.2** Let $u^h(t,x_i)$ be a local solution of the equation (4.1) with $\mathscr{F}_0$ measurable initial data $u_0\in\mathbb{S}^2$ and $\nabla u_0(x)\in L^2$. Then there exist a $T_0>0$ and a constant $C>0$ such that

$$\sup_{0\leqslant t\leqslant T_0}\mathbb{E}\|u^h(t)\|_{L_h^2}^2+\sup_{0\leqslant t\leqslant T_0}\mathbb{E}\|\delta u^h(t)\|_{L_h^2}^2\leqslant C,\tag{4.3}$$

where $T_0$ and $C$ are independent of $h$.

**Proof** Applying the Itô formula for $F(x)=|x|^2$, we have

$$\begin{aligned}d|u^h(t,x_i)|^2=&2\varepsilon u^h(t,x_i)\cdot D^+D^-u^h(t,x_i)dt+2\varepsilon u^h(t,x_i)\cdot|D^+u^h(t,x_i)|^2u^h(t,x_i)dt+\\&2u^h(t,x_i)\cdot u^h(t,x_i)\wedge D^+D^-u^h(t,x_i)dt-2u^h(t,x_i)\cdot u^h(t,x_i)dt+\\&2u^h(t,x_i)\cdot u^h(t,x_i)\wedge dW+\mathrm{Tr}(B(u^h(t,x_i))B^*(u^h(t,x_i)))dt.\end{aligned}$$

One can easily observe that the last four terms vanish. Multiplying the step size $h$, summing over all indices $i$ and integrating over $[0,\ t]$, we obtain

$$\begin{aligned}\sum_i|u^h(t,x_i)|^2h=&\sum_i|u_0^h(t,x_i)|^2h+2\varepsilon\int_0^t\sum_i u^h(t,x_i)\cdot D^+D^-u^h(t,x_i)d\tau+\\&2\varepsilon\int_0^t\sum_i u^h(t,x_i)\cdot|D^+u^h(t,x_i)|^2u^h(t,x_i)h\,d\tau.\end{aligned}$$

We thus have for almost surely $\omega\in\Omega$

$$\|u^h\|_{L_h^2}^2+2\varepsilon\int_0^t\|\delta u^h\|_{L_h^2}^2d\tau\leqslant\|u_0^h\|_{L_h^2}^2+2C\varepsilon\int_0^t\|u^h\|_{L_h^2}^4+\|\delta u^h\|_{L_h^2}^4d\tau,\tag{4.4}$$

thanks to the interpolation inequality in Lemma 4.1 for $p=\infty$.

On the other hand, applying the difference operator $D^+$ to the equation (4.1), and then applying the same procedure as above, one has

$$\begin{aligned}&\sum_i|D^+D^-u^h(t,x_i)|^2h\\&\leqslant\sum_i|D^+D^-u_0^h(t,x_i)|^2h-2\varepsilon\int_0^t\sum_i|D^+D^-u^h(t,x_i)|^2h\,d\tau+\\&\frac{2\varepsilon}{4}\int_0^t\sum_i|D^+D^-u^h(t,x_i)|^2h\,d\tau+2\varepsilon\int_0^t\max_i|u^h(t,x_i)|^2\sum_i|D^+u^h|^4h\,d\tau.\end{aligned}$$

Applying the interpolation inequality for $p=\infty$ and $p=4$, we have

$$\begin{aligned}\|\delta u^h\|_{L_h^2}^2+2\varepsilon\int_0^t\|\delta^2u^h\|_{L_h^2}^2d\tau\leqslant&\|\delta u_0^h\|_{L_h^2}^2+\frac{2\varepsilon}{4}\int_0^t\|\delta^2u^h\|_{L_h^2}^2d\tau+\\&2\varepsilon\int_0^t\|u^h\|_{L_h^2}^{\frac{3}{2}}(\|\delta^2u^h\|_{L_h^2}+\|u^h\|_{L_h^2})^{\frac{1}{2}}\cdot\|\delta u^h\|_{L_h^2}^3(\|\delta^2u^h\|_{L_h^2}+\|\delta u^h\|_{L_h^2})d\tau.\end{aligned}$$

Then, applying the Hölder inequality, one finally obtains for almost surely$\omega\in\Omega$

$$\|\delta u^h\|_{L_h^2}^2+\varepsilon\int_0^t\|\delta^2 u^h\|_{L_h^2}^2\mathrm{d}\tau$$

$$\leqslant\|\delta u_0^h\|_{L_h^2}^2+C\varepsilon\int_0^t(1+\|u^h\|_{L_h^2}^{18}+\|\delta u^h\|_{L_h^2}^{18})\mathrm{d}\tau. \tag{4.5}$$

This together with (4.4) implies that

$$\|u^h\|_{L_h^2}^2+\|\delta u^h\|_{L_h^2}^2+\varepsilon\int_0^t\|\delta^2 u^h\|_{L_h^2}^2\mathrm{d}\tau$$

$$\leqslant\|\delta u_0^h\|_{L_h^2}^2+C\varepsilon\int_0^t(1+\|u^h\|_{L_h^2}^{18}+\|\delta u^h\|_{L_h^2}^{18})\mathrm{d}\tau. \tag{4.6}$$

which shows that there exist constants $C$ and $T_0$ such that for almost surely $\omega\in\Omega$

$$\|u^h\|_{L_h^2}^2+\|\delta u^h\|_{L_h^2}^2\leqslant C,\quad \text{for any } t\in[0,T_0],$$

$$\int_0^{T_0}\|\delta u^h\|_{L_h^2}^2\mathrm{d}\tau\leqslant C.$$

Then taking expectations of the above inequality finishes the proof of the lemma. □

**Lemma 4.3** Let $u^h(t,x_i)$ be a solution of the equation (4.1) with $\mathscr{F}_0$ measurable initial data $u_0\in\mathbb{S}$ and $\nabla u_0(x)\in H^1$. Then there exist a $T_0>0$ and a constant $C>0$ such that

$$\sup_{0\leqslant t\leqslant T_0}\mathbb{E}\|\delta^2 u^h(t)\|_{L_h^2}^2+\int_0^{T_0}\mathbb{E}\|\delta^3 u^h(t)\|_{L_h^2}^2\mathrm{d}t\leqslant C,$$

where $T_0$ and $C$ are independent of $h$.

**Proof** The proof is similar to the one above. Applying the difference operator $D^+D^-$ to the equation (4.1), and then applying the Itô formula to the function $F(x)=|x|^2$, one obtains

$$\begin{aligned}\mathrm{d}|D^+D^-u^h(t,x_i)|^2=\;&2\varepsilon D^+D^-u^h(t,x_i)\cdot D^+D^-D^+D^-u^h(t,x_i)\mathrm{d}t+\\&2\varepsilon D^+D^-u^h(t,x_i)\cdot D^+D^-(|D^+u^h(t,x_i)|^2u^h(t,x_i))\mathrm{d}t+\\&2D^+D^-u^h(t,x_i)\cdot D^+D^-(u^h(t,x_i)\wedge D^+D^-u^h(t,x_i))\mathrm{d}t-\\&2D^+D^-u^h(t,x_i)\cdot D^+D^-u^h(t,x_i)\mathrm{d}t+\\&2D^+D^-u^h(t,x_i)\cdot D^+D^-u^h(t,x_i)\wedge\mathrm{d}W+\\&\mathrm{Tr}(B(D^+D^-u^h(t,x_i))B^*(D^+D^-u^h(t,x_i)))\mathrm{d}t.\end{aligned}$$

Also, the sum of the last three terms vanishes. Summation over all indices $i$ gives that

$$\begin{aligned}&\sum_i|D^+D^-u^h(t,x_i)|^2h\\&=\sum_i|D^+D^-u_0^h(t,x_i)|^2h-2\varepsilon\int_0^t\sum_i|D^+D^-D^+u^h(t,x_i)|^2h\mathrm{d}\tau-\\&2\varepsilon\int_0^t\sum_i D^-D^+D^-u^h(t,x_i)\cdot D^-(|D^+u^h(t,x_i)|^2u^h(t,x_i))h\mathrm{d}\tau+\\&2\int_0^t\sum_i D^-D^+D^-u^h(t,x_i)\cdot D^-(u^h(t,x_i)\wedge D^+D^-u^h(t,x_i))h\mathrm{d}\tau.\end{aligned}$$

Observe that

$$\int_0^t \sum_i D^- D^+ D^- u^h(t,x_i) \cdot D^-(|D^+ u^h(t,x_i)|^2 u^h(t,x_i)) h\,d\tau$$

$$\leqslant \frac{1}{4}\int_0^t \sum_i |D^- D^+ D^- u^h(t,x_i)|^2 h\,d\tau + c\int_0^t \max_i |D^+ u^h(t,x_i)|^2 \times$$

$$\sum_i |D^+ D^- u^h(t,x_i)|^2 h + \sum_i |D^+ u^h(t,x_i)|^6 h\,d\tau$$

and

$$2\int_0^t \sum_i D^- D^+ D^- u^h(t,x_i) \cdot D^-(u^h(t,x_i) \wedge D^+ D^- u^h(t,x_i)) h\,d\tau$$

$$\leqslant \frac{2\varepsilon}{4}\int_0^t \sum_i |D^- D^+ D^- u^h(t,x_i)|^2 h\,d\tau +$$

$$\frac{2}{\varepsilon}\int_0^t |D^- u^h(t,x_i) \wedge D^+ D^- u^h(t,x_i)|^2 h\,d\tau$$

$$\leqslant \frac{2\varepsilon}{4}\int_0^t \sum_i |D^- D^+ D^- u^h(t,x_i)|^2 h\,d\tau +$$

$$\frac{2}{\varepsilon}\int_0^t \max_i |D^- u^h(t,x_i)|^2 \sum_i |D^+ D^- u^h(t,x_i)|^2 h\,d\tau.$$

By the previous Lemma 4.2 and the interpolation inequality in Lemma 4.1, we have

$$\sup_{i,0\leqslant t\leqslant T_0} |u^h(t,x_i)| \leqslant C. \tag{4.7}$$

Combining all the above estimates, one finally obtains

$$\|\delta^2 u^h\|_{L_h^2}^2 + \varepsilon\int_0^t \|\delta^3 u^h\|_{L_h^2}^2 \leqslant \|\delta^2 u_0^h\|_{L_h^2}^2 + c\int_0^t (1 + \|\delta^2 u^h\|_{L_h^2}^2) d\tau.$$

Taking expectations, and using the Gronwall inequality, we know that there exist some constants $C$ and $T_0$ independent of the step size $h$ such that

$$\sup_{0\leqslant t\leqslant T_0} \mathbb{E}\|\delta^2 u^h\|_{L_h^2}^2 + \int_0^{T_0} \mathbb{E}\|\delta^3 u^h\|_{L_h^2}^2 d\tau \leqslant C. \tag{4.8}$$

One finishes the proof of the lemma. □

Similarly, one can establish the higher-order estimates of the solutions and we are led to the following.

**Lemma 4.4** Let $u^h(t,x_i)$ be a solution of the equation (4.1) with the $\mathscr{F}_0$-measurable initial data $u_0^h(x_i) \in \mathbb{S}^2$ and $\nabla u_0(x) \in H^k$ for any $k \geqslant 2$. There exist a $T_0 > 0$ and a constant $C > 0$ such that

$$\sup_{0\leqslant t\leqslant T_0} \mathbb{E}\|\delta^{k+1} u^h(t)\|_{L_h^2}^2 \leqslant C,$$

where $T_0$ and $C$ are independent of $h$.

Gathering together the above lemmas, we see that the *a priori* estimates are independent of the step size $h > 0$, thus one can let $h \to 0$ to get a local in time solution $u(t,x)$ of the periodic approximating stochastic Landau-Lifshitz equation ($\mathrm{SLL}_\varepsilon$-2) on the finite interval $\Lambda$ on some public interval $[0, T_0]$ for $\varepsilon > 0$ fixed. We summarize as follows.

**Proposition 4.1** Let $\varepsilon > 0$ be fixed. The periodic approximating stochastic Landau-

Lifshitz equation ($SLL_\varepsilon$-2) on the finite interval $\Lambda$ admits a local in time solution $u(t,x)$. This solution is constructed as the limit of solutions of the difference schemes (4.1) of the approximating stochastic system ($SLL_\varepsilon$-2) as $h\to 0$.

# 5 Global Solution for (SLL)

In this section, we derive some global *a priori* estimates independent of the viscous parameter $\varepsilon>0$ of the solutions constructed in Section 4 in dimension $d=1$. Then passing to the limit $\varepsilon\to 0$, we get a global smooth solution of the equation (SLL). We summarize our result in the following theorem.

**Theorem 5.1** Let $\nabla u_0\in H^m$ for $m\geqslant 1$ with $u_0\in\mathbb{S}^2$ and $\mathscr{F}_0$ measurable. There exists a unique continuous $\mathscr{F}_t$-adapted solution $(u(t))_{t\geqslant 0}$ such that $u(t)\in\mathbb{S}^2$ for almost every $x\in\mathbf{R}^1$ and $\nabla u(t)\in L^\infty(0,T;H^m)$ for almost surely $\omega\in\Omega$ for the Cauchy problem of the stochastic Landau-Lifshitz equation (SLL) on the real line $\mathbf{R}^1$; furthermore the solution is global in time.

**Lemma 5.1** Let $u(t)$ be a smooth solution of the approximating stochastic Landau-Lifshitz equation ($SLL_\varepsilon$-2) with $\mathscr{F}_0$-measurable smooth initial data $u_0$; then the following estimates hold:

$$|u(t,x)|=1 \quad \text{for any} \quad (t,x)\in[0,T]\times\Lambda \tag{5.1}$$

$$\sup_{0\leqslant t\leqslant T}\mathbb{E}\|\nabla u\|^2_{L^2(\Lambda)}\leqslant C, \tag{5.2}$$

where $C$ may depend on the initial data, but is independent of $\varepsilon, D$.

**Proof** By the Itô formula, we have that

$$\|\nabla u(t)\|^2=\|\nabla u_0\|^2+\mathscr{L}_{11}+\mathscr{L}_{12}+\mathscr{L}_{13}, \tag{5.3}$$

where

$$\mathscr{L}_{11}=2\int_0^t\langle\nabla u,\nabla(-\varepsilon u\wedge u\wedge\Delta u+u\wedge\Delta u-u)\rangle ds,$$

$$\mathscr{L}_{12}=2\int_0^t\langle\nabla u,\nabla u\wedge dW\rangle,\quad \mathscr{L}_{13}=\int_0^t\|\mathrm{Tr}(BB^*)\|_1^2 ds.$$

Let us compute one by one.

$$\begin{aligned}\mathscr{L}_{11}&=2\int_0^t\langle\Delta u,\varepsilon u\wedge u\wedge\Delta u\rangle ds-2\int_0^t\|\nabla u\|^2 ds\\&=-2\varepsilon\int_0^t\|u\wedge\Delta u\|^2 ds-2\int_0^t\|\nabla u\|^2 ds;\end{aligned}$$

$$\mathscr{L}_{12}=0;\quad \mathscr{L}_{13}=2\int_0^t\|\nabla u\|^2 ds.$$

Then (5.3) becomes

$$\|\nabla u(t)\|^2=\|\nabla u_0\|^2-2\varepsilon\int_0^t\|u\wedge\Delta u\|^2 ds. \tag{5.4}$$

This leads to

$$\|\nabla u(t)\|^2\leqslant\|\nabla u_0\|^2,\quad \text{a.s. } \omega\in\Omega, \tag{5.5}$$

and thus

$$\mathbb{E}\|\nabla u(t)\|^2 \leqslant \mathbb{E}\|\nabla u_0\|^2, \tag{5.6}$$

which finishes the proof of the lemma. □

**Remark** Indeed we can expect more about the estimates. By (5.5), one has $\sup_{0\leqslant t\leqslant T}\|\nabla u(t)\|^2 \leqslant \|\nabla u_0\|^2$, then taking expectations implies that

$$\mathbb{E}\sup_{0\leqslant t\leqslant T}\|\nabla u(t)\|^2 \leqslant \|\nabla u_0\|^2. \tag{5.7}$$

Similar remarks hold in the following lemmas.

**Lemma 5.2** Let $u(t)$ be a smooth solution of the approximating stochastic Landau-Lifshitz equation ($\mathrm{SLL}_\varepsilon$-2) with $\mathscr{F}_0$-measurable initial data $u_0$; then for any $T>0$ the following estimates hold:

$$\sup_{0\leqslant t\leqslant T}\mathbb{E}\|u_{xx}\|^2_{L^2(\Lambda)} \leqslant C, \tag{5.8}$$

where $C$ may depend on the initial data, but is independent of $\varepsilon, D$.

**Proof** Since $|u(t,x)|=1$ for any $(t,x)\in[0,T]\times\Lambda$, $u$, $u_x$, $u\wedge u_x$ form an orthogonal basis in $\mathbf{R}^3$. Write

$$u_{xx} = \alpha u + \beta u_x + \gamma u\wedge u_x. \tag{5.9}$$

By simple computation, one has

$$\alpha = -|u_x|^2,\quad \beta = \frac{u_x\cdot u_{xx}}{|u_x|^2},\quad \gamma = \frac{(u\wedge u_x)\cdot u_{xx}}{|u_x|^2}.$$

By the Itô formula, one has for ($\mathrm{SLL}_\varepsilon$-2) that

$$\|\partial_x^2 u(t)\|^2 = \|\partial_x^2 u_0\|^2 + \mathscr{L}_{21} + \mathscr{L}_{22} + \mathscr{L}_{23}, \tag{5.10}$$

where

$$\mathscr{L}_{21} = 2\int_0^t\langle\partial_x^2 u, \partial_x^2(\varepsilon\Delta u + \varepsilon|\nabla u|^2 u + u\wedge\Delta u - u)\rangle\mathrm{d}s,$$

$$\mathscr{L}_{22} = 2\int_0^t\langle\partial_x^2 u, B(\partial_x^2 u)\rangle\mathrm{d}W,\quad \mathscr{L}_{23} = \int_0^t\|\mathrm{Tr}(BB^*)\|_2^2\mathrm{d}s.$$

For the term $\mathscr{L}_{23}$, we have

$$\mathscr{L}_{23} = 2\int_0^t\|u_{xx}\|^2\mathrm{d}s.$$

For the term $\mathscr{L}_{21}$, we have

$$\begin{aligned}\mathscr{L}_{21} = &-2\varepsilon\int_0^t\|u_{xxx}\|^2\mathrm{d}s - 2\int_0^t\langle u_{xxx}, u_x\wedge u_{xx}\rangle\mathrm{d}s - \\ &2\varepsilon\int_0^t\langle u_{xxx}, |u_x|^2u_x + 2(u_x\cdot u_{xx})u\rangle\mathrm{d}s - 2\int_0^t\|u_{xx}\|^2\mathrm{d}s \\ =:&\mathscr{R}_1 + \mathscr{R}_2 + \mathscr{R}_3 + \mathscr{R}_4.\end{aligned}$$

For these terms, we have

$$\mathscr{R}_2 = -2\int_0^t\int_{-D}^D\left[u_x\wedge\left(-|u_x|^2u + \frac{(u\wedge u_x)\cdot u_{xx}}{|u_x|^2}u\wedge u_x\right)\right]\cdot u_{xxx}\mathrm{d}x\mathrm{d}s$$

$$= -2\int_0^t\int_{-D}^D |u_x|^2(u\wedge u_x)\cdot u_{xxx}\mathrm{d}x + \int_{-D}^D \frac{(u\wedge u_x)\cdot u_{xx}}{|u_x|^2}|u_x|^2 u\cdot u_{xxx}\mathrm{d}x\mathrm{d}s$$

$$= -5\int_0^t\int_{-D}^D |u_x|^2(u\wedge u_x)\cdot u_{xxx}\mathrm{d}x\mathrm{d}s,$$

where, in the last step, we have used the relationship

$$u\cdot u_{xxx} = -\frac{3}{2}(|u_x|^2)_x;$$

$$\mathscr{R}_3 = 2\varepsilon\int_0^t\int_{-D}^D |u_x|^2|u_{xx}|^2\mathrm{d}x\mathrm{d}s + 16\varepsilon\int_0^t\int_{-D}^D |u_x\cdot u_{xx}|^2\mathrm{d}x\mathrm{d}s.$$

On the other hand, applying the operator $\partial_x$ to the stochastic Landau-Lifshitz equation (SLL$_\varepsilon$-2), we have

$$\mathrm{d}u_x = \varepsilon\partial_x u_{xx}\mathrm{d}t + \varepsilon\partial_x(|u_x|^2 u)\mathrm{d}t + \partial_x(u\wedge u_{xx})\mathrm{d}t - \partial_x u\mathrm{d}t + \partial_x u\wedge \mathrm{d}W. \tag{5.11}$$

Let $F(p) = |p|^4 = (|p_1|^2 + |p_2|^2 + |p_3|^2)^2$ for a vector $p\in\mathbf{R}^3$. By the Itô formula we have

$$\|\partial_x u\|_{L^4}^4 = \int_{-D}^D F(\partial_x u(t))\mathrm{d}x = 4\int_0^t\langle|\partial_x u|^2\partial_x u, \mathrm{d}u_x\rangle\mathrm{d}s + 4\int_0^t\int_{-D}^D|\partial_x u|^4\mathrm{d}x\mathrm{d}s$$

$$= 4\int_0^t\int_{-D}^D(-\varepsilon|\partial_x u|^2|\partial_{xx}u|^2\mathrm{d}x - 2\varepsilon|\partial_x u\cdot\partial_{xx}u|^2 + \varepsilon|u_x|^6 -$$

$$|u_x|^2(u\wedge u_x)\cdot u_{xxx})\mathrm{d}x\mathrm{d}s + 4\int_0^t\int_{-D}^D|u_x|^2 u_x\cdot u_x\wedge\mathrm{d}x\mathrm{d}W,$$

which can be rewritten clearly as

$$\|\partial_x u\|_{L^4}^4 = 4\int_0^t\int_{-D}^D(-\varepsilon|\partial_x u|^2|\partial_{xx}u|^2 - 2\varepsilon|\partial_x u\cdot\partial_{xx}u|^2 + \varepsilon|u_x|^6 - |u_x|^2(u\wedge u_x)\cdot u_{xxx})\mathrm{d}x\mathrm{d}s. \tag{5.12}$$

Putting together (5.10)~(5.12), we have

$$4\|u_{xx}(t)\|^2 + 8\varepsilon\int_0^t\|u_{xxx}(s)\|^2\mathrm{d}s + 20\varepsilon\int_0^t\int_{-D}^D|u_x(s)|^6\mathrm{d}x\mathrm{d}s$$

$$= 5\|u_x(t)\|_{L^4}^4 - 5\|u_{0x}\|_{L^4}^4 + 4\|u_{0xx}\| + 28\varepsilon\int_0^t\int_{-D}^D|u_x|^2|u_{xx}|^2\mathrm{d}x\mathrm{d}s + 104\varepsilon\int_0^t\int_{-D}^D|u_x\cdot u_{xx}|^2\mathrm{d}x\mathrm{d}s. \tag{5.13}$$

Denote $\mathscr{A} = 28\int_{-D}^D|u_x|^2|u_{xx}|^2\mathrm{d}x + 104\int_{-D}^D|u_x\cdot u_{xx}|^2\mathrm{d}x$. We can estimate this term by

$$\mathscr{A} \leqslant 132\int_{-D}^D|u_x|^2|u_{xx}|^2\mathrm{d}x \leqslant \delta\int|u_x|^6 + c(\delta)\int|u_{xx}|^3\mathrm{d}x$$

$$\leqslant \delta\int_{-D}^D|u_x|^6\mathrm{d}x + c(\delta)\|u_x\|^{\frac{5}{4}}\|u_{xxx}\|^{\frac{7}{4}}$$

$$\leqslant \delta\int_{-D}^D|u_x|^6\mathrm{d}x + \delta\|u_{xxx}\|^2 + c(\delta)\|u_x\|^{10}.$$

Let $\delta = 4$. From (5.13), we have

$$4\| u_{xx}(t)\|^2 + 4\varepsilon\int_0^t \| u_{xxx}(s)\|^2 ds + 16\varepsilon\int_0^t\int_{-D}^D | u_x(s)|^6 dx ds$$
$$= 5\| u_x(t)\|_{L^4}^4 - 5\| u_{0x}\|_{L^4}^4 + 4\| u_{0xx}\|^2 + c(\delta)\int_0^t \| u_x\|^{10} ds$$
$$\leqslant 2\| u_{xx}(t)\|^2 + c\| u_x(t)\|^6 + 4\| u_{0xx}\|^2 + c(\delta)\int_0^t \| u_x\|^{10} ds, \tag{5.14}$$

where in the last step we have used the interpolation inequality in dimension one

$$\| u_x\|_{L^4}^4 \leqslant c\| u_x\|^3 \| u_{xx}\|.$$

Thanks to Lemma 5.1, one then derives from (5.14) that for almost surely $\omega\in\Omega$,

$$\| u_{xx}\|_{L^2(\Lambda)}^2 \leqslant C(T, \| u_0\|_{H^2}), \tag{5.15}$$

which follows

$$\sup_{0\leqslant t\leqslant T} \mathbb{E}\| u_{xx}\|_{L^2(\Lambda)}^2 \leqslant C, \tag{5.16}$$

where the constant $C$ may depend on $\| u_0\|_{H^2}$ but is independent of $\varepsilon$ and $D$. □

**Lemma 5.3** Let $u(t)$ be a smooth solution of the approximating stochastic Landau-Lifshitz equation ($\mathrm{SLL}_\varepsilon$-2); then the following estimates hold

$$\sup_{0\leqslant t\leqslant T} \mathbb{E}\| \partial_x^3 u\|_{L^2(\Lambda)}^2 \leqslant C, \tag{5.17}$$

where $C$ depends on the initial data, but independent of $\varepsilon$, $D$.

**Proof** The proof is similar to that in Lemma 5.2, thus we only sketch the proof. Applying the operator $\mathscr{L}=\partial_x^3$ to the equation ($\mathrm{SLL}_\varepsilon$-2), we have

$$d(\partial_x^3 u) = \varepsilon\partial_x^5 u(t)dt + \varepsilon\partial_x^3(| u_x|^2 u)dt + \partial_x^3(u \wedge \Delta u)dt - \partial_x^3 u dt + \partial_x^3 u \wedge dW.$$

Then, using the Itô formula for $F(u) = | u|^2$, we can derive that

$$\| \partial_x^3 u\|^2 = \| \partial_x^3 u_0\|^2 + \mathscr{L}_{31} + \mathscr{L}_{32} + \mathscr{L}_{33}, \tag{5.18}$$

where

$$\mathscr{L}_{31} = 2\varepsilon\int_0^t \langle\partial_x^3 u(s), \partial_x^5 u(s)\rangle ds + 2\varepsilon\int_0^t \langle\partial_x^3 u, \partial_x^3(| u_x|^2 u)\rangle ds +$$
$$2\int_0^t \langle\partial_x^3 u, \partial_x^3(u \wedge u_{xx})\rangle ds - 2\int_0^t \langle\partial_x^3 u, \partial_x^3 u\rangle ds$$
$$=: \mathscr{R}_1 + \mathscr{R}_2 + \mathscr{R}_3 + \mathscr{R}_4,$$
$$\mathscr{L}_{32} = 2\int_0^t \langle\partial_x^3 u, \partial_x^3 u \wedge dW\rangle = 0,$$
$$\mathscr{L}_{33} = \int_0^t \| \mathrm{Tr}(BB^*)\|_{\dot{H}^3} ds = 2\int_0^t \| \partial_x^3 u\|^2 ds.$$

For the terms on the right of $\mathscr{L}_{31}$, we have

$$\mathscr{R}_2 = -2\varepsilon\int_0^t\int (| u_x|^2 u)_{xx} \cdot u_{xxxx} dx ds$$
$$\leqslant C\varepsilon\int_0^t (| u_{xx}| + | u_{xx}|^2 + | u_{xxx}|)| u_{xxxx}| dx ds$$
$$\leqslant C\varepsilon t + \frac{\varepsilon}{2}\int_0^t\int | u_{xxxx}|^2 dx ds + C\varepsilon\int_0^t\int | u_{xxx}|^2 dx ds + C\varepsilon\int_0^t\int | u_{xx}|^4 dx ds$$

$$\leqslant \varepsilon\int_0^t\int |u_{xxxx}|^2 dx ds + C\varepsilon\int_0^t\int |u_{xxx}|^2 dx ds + \int_0^t \|u_{xx}\|^{\frac{14}{3}} ds,$$

thanks to the interpolation inequality

$$\|u_{xx}\|_{L^4} \leqslant c\|u_{xxxx}\|^{\frac{1}{8}}\|u_{xx}\|^{\frac{7}{8}},$$

and Lemma 5.2.

On the other hand, we can express $u_{xxx}$ in terms of the orthogonal basis $\{u, u_x, u\wedge u_x\}$ in $\mathbf{R}^3$,

$$u_{xxx} = \alpha u + \beta u_x + \gamma u \wedge u_x,$$

with

$$\alpha = -3u_x \cdot u_{xx}, \quad \beta = \frac{u_x \cdot u_{xxx}}{|u_x|^2}, \quad \gamma = \frac{(u\wedge u_x)\cdot u_{xxx}}{|u_x|^2}.$$

Then $\mathcal{R}_3$ can be estimated as

$$\begin{aligned}
\mathcal{R}_3 &= -2\int_0^t\int (u_x \wedge u_{xxx})\cdot u_{xxxx} dx ds \\
&= -2\int_0^t\int \left[-3(u_x\cdot u_{xx})(u_x\wedge u) + \frac{(u\wedge u_x)\cdot u_{xxx}}{|u_x|^2}|u_x|^2 u\right]\cdot u_{xxxx} dx ds \\
&= 6\int_0^t\int [(u_x\cdot u_{xx})(u_x\wedge u)]_x \cdot u_{xxx} dx ds + \\
&\quad 8\int_0^t\int [(u\wedge u_x)\cdot u_{xxx}]u_x\cdot u_{xxx} dx ds + \\
&\quad 6\int_0^t\int [(u\cdot u_x)\cdot u_{xxx}]|u_{xx}|^2 dx ds \\
&\leqslant c\int_0^t\int |u_{xxx}|^2 dx ds + c\int_0^t\int |u_{xx}|^4 dx ds \\
&\leqslant c\int_0^t\int |u_{xxx}|^2 dx ds + c\int_0^t \|u_{xx}\|^6 ds,
\end{aligned}$$

where in the third equality we have used the fact that

$$u\cdot u_{xxxx} = -3|u_{xx}|^2 - 4u_x\cdot u_{xxx},$$

and in the last step we have used the interpolation inequality

$$\|u_{xx}\|_{L^4} \leqslant C\|u_{xx}\|^{\frac{3}{4}}\|u_{xxx}\|^{\frac{1}{4}}.$$

Putting together (5.5) and the estimates for $\mathcal{R}_2, \mathcal{R}_3$ we are led to

$$\|u_{xxx}\|^2 + \varepsilon\int_0^t \|u_{xxxx}\|^2 \leqslant \|u_{0xxx}\|^2 + c\int_0^t \|u_{xxx}\|^2 ds + C, \tag{5.19}$$

which shows that

$$\|u_{xxx}\|^2 \leqslant C(T, \|u_{0xxx}\|^2), \quad \text{a.s. } \omega\in\Omega. \tag{5.20}$$

Taking expectations then finishes the proof of the lemma. □

**Lemma 5.4** Let $u(t)$ be a smooth solution of the approximating stochastic Landau-Lifshitz equation (SLL$_\varepsilon$-2); then the following estimates hold for any $k\geqslant 2$:

$$\sup_{0\leqslant t\leqslant T}\mathbb{E}\|\partial_x^k u\|_{L^2(\Lambda)}^2\leqslant C, \tag{5.21}$$

where $C$ may depend on the initial data, but is independent of $\varepsilon, D$.

**Proof** Applying the operator $\mathscr{L}=(-\Delta)^{k/2}$ to the equation, and then using the Itô formula as above, one can get the estimates. □

These estimates and the local (in time) theory obtained in the previous Section 4 then finish the proof of Theorem 5.1: the existence of global (in time) smooth solutions of the Cauchy problem of the stochastic Landau-Lifshitz equation (SLL) on $\mathbf{R}^1$. Indeed since the above estimates are independent of $\varepsilon>0$ and $D>0$, one can let $\varepsilon\to 0$ and $D\to\infty$ to get a local in time smooth solution of the Cauchy problem (SLL) on the real line $\mathbf{R}^1$, thanks to the local in time existence result of the approximating system established in Proposition 4.1. The globality in time of the solution is a standard application of the continuity method and the global estimates established in the above lemmas. This finishes the proof of Theorem 5.1 once we show its uniqueness. However, uniqueness is implied by the regularity we have proved in the above lemmas. We sketch it in the following.

Let $u_1$ and $u_2$ be two solutions with the same initial data. Setting $w=u-v$, then one has

$$dw=(u\wedge\Delta w+w\wedge\Delta v)dt-wdt+w\wedge dW. \tag{5.22}$$

By the Itô formula, and integration in the $x$-variable, we have

$$\|w(t)\|^2=\int_0^t\langle w(\tau),u(\tau)\wedge\Delta w(\tau)\rangle d\tau\leqslant C\int_0^t\|w(\tau)\|^2+\|\nabla w(\tau)\|^2 d\tau,$$

as well as

$$\begin{aligned}\|\nabla w(t)\|_{L^2}^2&=-\langle w(t),\Delta w(t)\rangle=-2\int_0^t\langle w(\tau)\wedge\Delta v(\tau),\Delta w(\tau)\rangle d\tau\\&\leqslant C\int_0^t\|w(\tau)\|^2+\|\nabla w(\tau)\|^2 d\tau.\end{aligned}$$

Putting together the two inequalities and using the Gronwall inequality the uniqueness of the solution of the (SLL) equation follows.

As a byproduct, we have the existence of smooth solutions for the Cauchy problem of the stochastic Landau-Lifshitz-Gilbert equation (1.2) on $\mathbf{R}^1$ with damping parameter $\alpha>0$ fixed. Noting that $\alpha$ plays the same role as the viscosity parameter $\varepsilon$ plays in ($SLL_\varepsilon$-2), we regard them as the same.

**Corollary 5.1** Let the Gilbert damping parameter $\alpha>0$ be fixed. For any $\mathscr{F}_0$ measurable smooth initial data $u_0$, the stochastic Landau-Lifshitz-Gilbert equation (1.2) admits a global in time smooth solution, which is a continuous $\mathscr{F}_t$-adapted process in the Sobolev space $H^k$ for any $k\in\mathbf{N}$.

**Proof** Indeed, for fixed $\varepsilon=\alpha>0$, Proposition 4.1 gives the existence of local smooth solutions to the periodic problem on the domain $\Lambda$. Since the estimates established in this section are independent of $D$ and global, one can let $D\to\infty$ to get a global in time smooth solution to the Cauchy problem of the stochastic Landau-Lifshitz-Gilbert equation (1.2) on the real line $\mathbf{R}^1$. □

One can also let $D>0$ be fixed, and let $\varepsilon \to 0$ to get a global in time smooth solution for both the stochastic Landau-Lifshitz equation and the stochastic Landau-Lifshitz-Gilbert equation in $\Lambda$ under periodic boundary conditions. We omit the statement of the details here.

## Appendix: The Equivalence Between the Two Approximating Equations ($SLL_\varepsilon$-1) and ($SLL_\varepsilon$-2)

To be complete, we show that the following two equations are equivalent with the same initial data $u_0 \in \mathbb{S}^2$:

$$du = u \wedge u \wedge \Delta u dt + u \wedge \Delta u dt - u dt + u \wedge dW, \tag{A.1}$$

$$du = \Delta u dt + |\nabla u|^2 u dt + u \wedge \Delta u dt - u dt + u \wedge dW. \tag{A.2}$$

The equivalence for a general viscous parameter $\varepsilon>0$ can be shown similarly.

- (A.1)⇒(A.2)

As in Section 2, we can show that, for any $(t,x)$, $|u(t,x)|=1$ holds, which implies that $u\cdot\Delta u = -|\nabla u|^2$. Then by the formula $a\wedge(b\wedge c)=(a\cdot c)b-(a\cdot b)c$, we have

$$u \wedge (u \wedge \Delta u) = -|\nabla u|^2 u - \Delta u,$$

which shows that (A.1) implies (A.2) for regular enough solutions.

- (A.2)⇒(A.1)

We only need to show that the solution of (A.2) remains on the sphere $\mathbb{S}^2$ as time evolves. In fact, by the Itô formula,

$$\begin{aligned} d|u|^2 &= 2u\cdot\Delta u dt + 2u\cdot(|\nabla u|^2 u)dt + 2u\cdot(u\wedge\Delta u) - \\ &\quad 2u\cdot u dt + 2u\cdot(u\wedge dW) + \mathrm{Tr}(BB^*)dt \\ &= 2u\cdot\Delta u dt + 2|\nabla u|^2|u|^2, \end{aligned}$$

which can be rewritten as

$$d|u|^2 = \Delta|u|^2 dt - 2|\nabla u|^2 dt + 2|\nabla u|^2|u|^2 dt,$$

by the fact $\Delta|u|^2 = 2|\nabla u|^2 + 2u\cdot\Delta u.$

Denote $U(t)=|u(t)|^2-1$. One has

$$dU(t) = \Delta U(t)dt + 2|\nabla u(t)|^2 U(t)dt$$

with initial data $U(0)=0$. Applying the Itô formula to this equation, one gets

$$|U(t)|^2 = |U(0)|^2 + 2\int_0^T U(t)\cdot\Delta U(t)dt + 4\int_0^T |\nabla u|^2|U(t)|^2 dt.$$

This implies that

$$|U(t)|^2 + 2\int_0^T |\nabla U(t)|^2 dt = 4\sup_{t,x}|\nabla u|^2\int_0^T |U(t)|^2 dt,$$

which implies $U(t)\equiv 0$. This shows that $|u(t)|\equiv 1$ for all $T>0$ and for almost every $\omega\in\Omega$, thus the equivalence of the two approximating stochastic equations follows.

### References

[1] Carbou G, Fabrie P. Regular Solutions for Landau-Lifschitz equations in a bounded domain. Differential

Integral Equations, 2001, 14: 213-229.

[2] Ding S, Guo B. Hausdorff measure of the singular set of Landau-Lifshitz Equations with a nonlocal term. Commun Math Phys, 2004, 250: 95-117.

[3] Friedman A. Stochastic differential equations and applications, Vol 1. New York: Academic Press, 1975.

[4] Garcia-Palacios L, Lazaro F J. Langevin-dynamics study of the dynamical properties of small magnetic particles. Phys Rev B, 1998, 58: 14937-14958.

[5] Gong G. Introduction to stochastic differential equations (In Chinese). Science Press Publication, 1987.

[6] Guo B, Ding S. Landau-Lifshitz equations// Frontiers of Research with the Chinese Academy of Sciences. Vol 1. Singapore: World Scientific, 2008.

[7] Guo B, Hong M. The Landau-Lifshitz equation of the ferromagnetic spin chain and harmonic maps. Calc Var, 1993, 1: 311-334.

[8] van Kampen N G. Stochastic process in physics and chemistry. Amsterdam: North-Holland, 1981.

[9] Krylov N V, Rozovskii B L. Stochastic differential equations. Journal of Soviet Mathematics, 1983, 14: 1233-1277. (Translated from the Russian edition published in 1979.).

[10] Ladyzenskaya. The boundary value problems of mathematical physics// Applied Mathematical Science, Vol 49. Berlin: Springer-Verlag, 1985.

[11] Landau L, Lifshitz E M. On the theory of the dispersion of magnetic permeability in ferromagnetic bodies. Phsy Z Sowj, 1935, 8: 153 (Reproduced in Collected Papers of Landau L D, Pergamon Press, New York, 1965: 101-114).

[12] Moser R. Partial regularity for the Landau-Lifshitz equation in small dimensions. MPI Preprint 26, 2002.

[13] Øksendal B. Stochastic differential equations: an introduction with applications. Sixth edition. Springer-Verlag, 2003.

[14] Da Prato G, Zabczyk J. Stochastic equations in infinite dimensions// Encyclopedia of Mathematics and its applications, Vol 44. Cambridge: Cambridge University Press, 1992.

[15] Sulem P L, Sulem C, Bardos C. On the continous limit for a system of classical spins. Comm Math Phys, 1986, 107: 431-454.

[16] Zhang X C. Smooth solutions of non-linear stochastic partial differential equations. arXiv: 0801.3883v1 (2008).

[17] Zhou Y L. Applications of discrete functional analysis to the finite difference method. International Academic Publishers, 1990.

[18] Zhou Y L, Guo B L, Tan S B. Existence and uniqueness of smooth solution for system of ferromagnetic chain. Scientia Sinica, Ser A, 1991, 34: 257-266.

# Sharp Threshold of Global Existence for the Generalized Davey-Stewartson System in $\mathbf{R}^2$ *

Gan Zaihui(甘在会) Guo Boling(郭柏灵) Zhang Jian(张 健)

**Abstract**

This paper is concerned with the generalized Davey-Stewartson system in $\mathbf{R}^2$ which appears as mathematical models for the evolution of shallow-water waves having one predominant direction of travel. We obtain a sharp threshold of blowing up and global existence to the Cauchy problem of the system by constructing a type of cross-constrained variational problem and establishing so-called cross-invariant manifolds of the evolution flow. Furthermore, we answer the question: How small are the initial data, the global solutions to the Cauchy problem of the system exist.

**Keywords** Davey-Stewartson system; cross-invariant manifold; sharp threshold; global existence; blowup

## 1 Introduction

In this paper, we consider the Cauchy problem for the generalized Davey-Stewartson system in $\mathbf{R}^2$ as follows:

$$i\phi_t + \Delta\phi + |\phi|^2\phi + E_1(|\phi|^2)\phi = 0, \quad t \geqslant 0, \quad x \in \mathbf{R}^2, \tag{1.1}$$

$$\phi(0,x) = \phi_0(x), \quad x \in \mathbf{R}^2. \tag{1.2}$$

Here $\phi = \phi(t,x)$ is a complex-valued function of $(t,x) \in \mathbf{R}_+ \times \mathbf{R}^2$, $E_1$ is the singular integral operator with symbol $\sigma_1(\xi) = \xi_1^2/|\xi|^2$, $\xi \in \mathbf{R}^2$, $E_1(\psi) = F^{-1}(\xi_1^2/|\xi|^2)F\psi$, $F^{-1}$ and $F$ denote the Fourier inverse transform and Fourier transform on $\mathbf{R}^2$, respectively[6,7,13,14]. In fluid mechanics, system (1.1) describes the evolution of shallow-water waves having one predominant direction of travel[5,6,7]. In the earlier work[13,14,16,17], people pay more attention to the Cauchy problem of the following generalized Davey-Stewartson system:

$$\begin{aligned} &i\phi_t + \Delta\phi + |\phi|^{p-1}\phi + E_1(|\phi|^2)\phi = 0, \quad t \geqslant 0, \quad x \in \mathbf{R}^N \\ &\phi(0,x) = \phi_0(x), \quad x \in \mathbf{R}^N. \end{aligned} \tag{1.3}$$

Ghidaglia and Saut[13] as well as Guo and Wang[14] establish the local existence of the solutions for (1.3) in the energy space $H^1(\mathbf{R}^N)$. Ohta[16] proved that if $p > 3$, then there exist blow-up solutions of (1.3) arbitrarily close to the standing wave in the case of $N = 2$. Moreover, Ohta[14] also proved that if $p \geqslant 1 + \frac{4}{N}$, then the ground state $\phi_\omega$ is unstable for any $\omega \in (0,\infty)$

* Communications on Pure and Applied Analysis, 2009, 8 (3): 913-922.

in the case of $N=2$ or $N=3$.

From the view-point of mathematics, the following problems are very important. Under what conditions, will the water waves become unstable to collapse (blowup)? And under what conditions, will the water waves be stable for all time (global existence)? Especially the sharp thresholds for global existence and collapse are pursued strongly. For the study of global solutions for the nonlinear or quasilinear Schrödinger equations, there have been some works[2,3,4,8,9]. But to our knowledge, few results are known on the study of the global solutions to the generalized Davey-Stewartson system, especially on the sharp threshold of global existence for it except when $N>2$ and $p=3$, Shu and Zhang[19] obtained a sharp criterion for blowup and global existence to the solution of (1.3). In the present paper, we consider the Cauchy problem (1.1), (1.2) and search the sharp threshold for blowup and global existence of its solutions. We first construct three proper functionals, then set a cross-constrained variational problem. By establishing the invariant sets under the flow generated by the Cauchy problem (1.1), (1.2), we obtain a new blowup theorem and establish the sharp threshold for global existence and blowup to the solutions of (1.1), (1.2) by applying the potential well argument (Pagne and Sattinger[18]), the concavity method (Levine[15]) and the scaling argument. These arguments originate from Beresticki and Cazenave[1], Weinstein[20] and Zhang[21,22,23]. In addition, we answer the question: How small are the initial data, the global solutions to the Cauchy problem (1.1), (1.2) exist?

For simplicity, throughout this paper we denote $\int_{\mathbf{R}^2}\cdot\,\mathrm{d}x$ by $\int\cdot\,\mathrm{d}x$.

This paper is organized as follows: In the second section, we give some preliminaries. In the third section, we discuss the cross-constrained variational problem and invariant manifolds. In the last section, we give a new blow-up theorem and a sharp threshold of global existence.

## 2 Preliminaries

In this section, we will recall some known facts and give some elementary results which will be used and play an important role later.

From Ghidaglia and Saut[13] as well as Guo and Wang[14], we can get the local well-posedness of the Cauchy problem (1.1), (1.2) in energy class $H^1(\mathbf{R}^2)$.

**Proposition 2.1** Let $\phi_0\in H^1(\mathbf{R}^2)$. Then there exists a unique solution $\phi$ of the Cauchy problem (1.1), (1.2) in $C([0,T);H^1(\mathbf{R}^2))$ for some $T\in[0,\infty)$ (maximal existence time), and $\phi(t,\cdot)$ satisfies the following two conservation laws of mass and energy for all $t\in[0,T)$:

$$\int|\phi|^2\mathrm{d}x=\int|\phi_0|^2\mathrm{d}x, \tag{2.1}$$

$$E(\phi)=\frac{1}{2}\int|\nabla\phi|^2\mathrm{d}x-\frac{1}{4}\int|\phi|^4\mathrm{d}x-\frac{1}{4}\int|\phi|^2E_1(|\phi|^2)\mathrm{d}x=E(\phi_0). \tag{2.2}$$

In addition, we have the following alternatives: $T=\infty$ (global existence) or else $T<\infty$ and

$$\lim_{t\to T}\|\phi\|_{H^1(\mathbf{R}^2)}=\infty \text{ (blowup)}.$$

**Proposition 2.2** (Ohta[16,17]) Let $\phi_0\in H^1(\mathbf{R}^2)$, $|x|\phi_0\in L^2(\mathbf{R}^2)$ and $\phi(t,x)$ be a solution of the Cauchy problem (1.1),(1.2) in $C([0,T);H^1(\mathbf{R}^2))$. Put

$$J(t):=\int|x|^2|\phi|^2\mathrm{d}x, \tag{2.3}$$

then one has

$$J''(t)=8\int|\nabla\phi|^2\mathrm{d}x-4\int|\phi|^4\mathrm{d}x-4\int|\phi|^2E_1(|\phi|^2)\mathrm{d}x. \tag{2.4}$$

**Proposition 2.3** (Cipolatti[5,6], Ohta[16,17]) For all $\phi\in S(\mathbf{R}^2,\mathbf{R})$ (the Schwartz space of rapidly decreasing functions), the following results hold:

$$\int|\phi|^2E_1(|\phi|^2)\mathrm{d}x\leqslant\int|\phi|^4\mathrm{d}x, \tag{2.5}$$

$$\int\phi x\cdot\nabla\phi\mathrm{d}x=-\int|\phi|^2\mathrm{d}x,\quad \int|\phi|^2\phi x\cdot\nabla\phi\mathrm{d}x=-\frac{1}{2}\int|\phi|^4\mathrm{d}x, \tag{2.6}$$

$$\int E_1(|\phi|^2)\phi x\cdot\nabla\phi\mathrm{d}x=-\frac{1}{2}\int|\phi|^2E_1(|\phi|^2)\mathrm{d}x. \tag{2.7}$$

## 3 The Cross-constrained Variational Problem and the Cross-invariant Manifolds

For $u\in H^1(\mathbf{R}^2)$ and $\omega>0$, we define the following functionals:

$$I(u):=\frac{1}{2}\int|\nabla u|^2\mathrm{d}x+\frac{\omega}{2}\int|u|^2\mathrm{d}x-\frac{1}{4}\int|u|^4\mathrm{d}x-\frac{1}{4}\int|u|^2E_1(|u|^2)\mathrm{d}x, \tag{3.1}$$

$$S(u):=\int|\nabla u|^2\mathrm{d}x+\omega\int|u|^2\mathrm{d}x-\int|u|^4\mathrm{d}x-\int|u|^2E_1(|u|^2)\mathrm{d}x, \tag{3.2}$$

$$Q(u):=\int|\nabla u|^2\mathrm{d}x-\frac{1}{2}\int|u|^4\mathrm{d}x-\frac{1}{2}\int|u|^2E_1(|u|^2)\mathrm{d}x. \tag{3.3}$$

By (2.5) and Sobolev's embedding theorem, the above functionals are well-defined. In addition, we define two manifolds as follows:

$$\Sigma_1:=\{u\in H^1(\mathbf{R}^2)\setminus\{0\},S(u)=0\}, \tag{3.4}$$

$$\Sigma_2:=\{u\in H^1(\mathbf{R}^2),Q(u)=0,S(u)<0\}. \tag{3.5}$$

Furthermore, we define two constrained variational problems:

$$d_{\Sigma_1}:=\inf_{u\in\Sigma_1}I(u), \tag{3.6}$$

$$d_{\Sigma_2}:=\inf_{u\in\Sigma_2}I(u). \tag{3.7}$$

Thus the following results hold.

**Lemma 3.1** $d_{\Sigma_1}>0$.

**Proof** Let $u\in\Sigma_1$. From $S(u)=0$, it follows that

$$\int|u|^4\mathrm{d}x+\int|u|^2E_1(|u|^2)\mathrm{d}x=\int|\nabla u|^2\mathrm{d}x+\omega\int|u|^2\mathrm{d}x. \tag{3.8}$$

By Sobolev's embedding inequality and (2.5), we have

$$\int|u|^4\mathrm{d}x+\int|u|^2E_1(|u|^2)\mathrm{d}x\leqslant2\int|u|^4\mathrm{d}x\leqslant C\left(\int|\nabla u|^2\mathrm{d}x+\omega\int|u|^2\mathrm{d}x\right)^2. \tag{3.9}$$

Here and later on, $C$ denotes various positive constants. From $u\neq 0$, (3.8) and (3.9) yield that

$$\int |u|^4 dx + \int |u|^2 E_1(|u|^2) dx \geqslant c > 0. \tag{3.10}$$

By (3.1), (3.2), (3.4) and (3.6), on $\Sigma_1$ we have

$$I(u) = \frac{1}{4}\left(\int |u|^4 dx + \int |u|^2 E_1(|u|^2) dx\right). \tag{3.11}$$

Thus (3.10) and (3.11) imply that $d_{\Sigma_1} \geqslant c > 0$. □

**Lemma 3.2** There exists $u \in H^1(\mathbf{R}^2) \setminus \{0\}$ such that $S(u)=0$ and $Q(u)=0$.

**Proof** From Ohta[16,17], it follows that there exists $u \in H^1(\mathbf{R}^2) \setminus \{0\}$ such that $u$ is a solution of the following Euler-Lagrangian equation

$$-\Delta u + \omega u - |u|^2 u - E_1(|u|^2) u = 0. \tag{3.12}$$

Thus $S(u)=0$, which is obtained by multiplying (3.12) by $u$ and then integrating with respect to $x$ over $\mathbf{R}^2$. Moreover, from (3.12) and Proposition 2.3, we have the Pohozaev identity

$$\omega \int |u|^2 dx - \frac{1}{2}\int |u|^4 dx - \frac{1}{2}\int |u|^2 E_1(|u|^2) dx = 0, \tag{3.13}$$

which is obtained by multiplying (3.12) by $x \cdot \nabla u$ and then integrating with respect to $x$ over $\mathbf{R}^2$. Noting that $S(u)=0$, thus we get $Q(u)=0$. □

**Lemma 3.3** $\Sigma_2$ is not empty.

**Proof** From Lemma 3.2, it follows that there exists $u \in H^1(\mathbf{R}^2) \setminus \{0\}$ such that $S(u)=0$ and $Q(u)=0$. Let $u_\lambda = \lambda u(\lambda x)$ for $\lambda > 0$, then $u_\lambda \in H^1(\mathbf{R}^2) \setminus \{0\}$. By (3.2) and (3.3), we get

$$S(u_\lambda) = \lambda^2\left(\int |\nabla u|^2 dx - \int |u|^4 dx - \int |u|^2 E_1(|u|^2) dx\right) + \omega \int |u|^2 dx, \tag{3.14}$$

$$Q(u_\lambda) = \lambda^2\left(\int |\nabla u|^2 dx - \frac{1}{2}\int |u|^4 dx - \frac{1}{2}\int |u|^2 E_1(|u|^2) dx\right). \tag{3.15}$$

Thus $S(u)=0$ implies that there exists a $\lambda^* > 1$ such that $S(u_{\lambda^*}) < 0$.

On the other hand, from $\lambda^* > 1$ and $Q(u)=0$, we still have $Q(u_{\lambda^*})=0$. So $u_{\lambda^*} \in \Sigma_2$. This proves that $\Sigma_2$ is not empty. □

**Lemma 3.4** $d_{\Sigma_2} > 0$.

**Proof** Let $u \in \Sigma_2$. It follows from $S(u) < 0$ that $u \neq 0$. By $Q(u)=0$ we obtain

$$I(u) = \frac{\omega}{2}\int |u|^2 dx. \tag{3.16}$$

Thus (3.16) and $u \neq 0$ yield that $d_{\Sigma_2} \geqslant 0$. In the following, we prove $d_{\Sigma_2} \neq 0$ by contradiction.

If $d_{\Sigma_2} = 0$, then from (3.7) there would exist a sequence $\{u_n, n \in \mathbf{Z}^+\} \subset \Sigma_2$ such that $Q(u_n)=0, S(u_n)<0$ and $I(u_n) \to 0$ as $n \to \infty$. Thus (3.16) implies that

$$\int |u_n|^2 dx \to 0 \quad \text{as} \quad n \to \infty. \tag{3.17}$$

According to $S(u_n)<0$, by Gagliardo-Nirenberg inequality and (2.5) we obtain

$$\int|\nabla u_n|^2\mathrm{d}x+\omega\int|u_n|^2\mathrm{d}x<\int|u_n|^4\mathrm{d}x+\int|u_n|^2E_1(|u_n|^2)\mathrm{d}x \\ \leqslant C\left(\int|\nabla u_n|^2\mathrm{d}x\right)\left(\int|u_n|^2\mathrm{d}x\right). \tag{3.18}$$

For $C$ in (3.18), by (3.17) we have that when $n$ sufficiently large,

$$\int|\nabla u_n|^2\mathrm{d}x+\omega\int|u_n|^2\mathrm{d}x\geqslant C\left(\int|\nabla u_n|^2\mathrm{d}x\right)\left(\int|u_n|^2\mathrm{d}x\right) \tag{3.19}$$

It is obvious that (3.18) and (3.19) are contradictory. Since we have showed $d_{\Sigma_2}\geqslant 0$, thus we get $d_{\Sigma_2}>0$. □

Now we define

$$d:=\min\{d_{\Sigma_1},d_{\Sigma_2}\}. \tag{3.20}$$

From Lemma 3.1 and Lemma 3.4, we get $d>0$. Furthermore, we can get a more meticulous result.

**Lemma 3.5** $d_{\Sigma_2}\geqslant d_{\Sigma_1}$.

**Proof** Let $u\in\Sigma_2$, then $S(u)<0$ and $Q(u)=0$. Put

$$u_\lambda=\lambda u(\lambda x). \tag{3.21}$$

By (3.2) and (3.3) we get

$$S(u_\lambda)=\lambda^2\left(\int|\nabla u|^2\mathrm{d}x-\int|u|^4\mathrm{d}x-\int|u|^2E_1(|u|^2)\mathrm{d}x\right)+\omega\int|u|^2\mathrm{d}x, \tag{3.22}$$

$$Q(u_\lambda)=\lambda^2\left(\int|\nabla u|^2\mathrm{d}x-\frac{1}{2}\int|u|^4\mathrm{d}x-\frac{1}{2}\int|u|^2E_1(|u|^2)\mathrm{d}x\right). \tag{3.23}$$

Thus $S(u)<0$ implies that there exists a unique $0<\lambda^*<1$ such that $S(u_{\lambda^*})=0$. It is clear that $u\neq 0$ and $u_{\lambda^*}\neq 0$. By (3.6) it follows that

$$I(u_{\lambda^*})\geqslant d_{\Sigma_1}. \tag{3.24}$$

On the other hand, $Q(u)=0$ and (3.23) imply that for any $\lambda>0, Q(u_\lambda)=0$. Thus it follows from $Q(u)=0$ and $Q(u_\lambda)=0$ that for any $\lambda>0$,

$$I(u_\lambda)=\frac{\omega}{2}\int|u_\lambda|^2\mathrm{d}x=\frac{\omega}{2}\int|u|^2\mathrm{d}x=I(u), \tag{3.25}$$

especially for $\lambda=\lambda^*$, it is also true for $I(u_{\lambda^*})=I(u)$, which together with (3.24) and (3.25) yields

$$I(u)=I(u_\lambda)=I(u_{\lambda^*})\geqslant d_{\Sigma_1}.$$

Therefore, by (3.7) we obtain

$$\inf_{u\in\Sigma_2}I(u)=d_{\Sigma_2}\geqslant d_{\Sigma_1}.$$

So far, the proof of Lemma 3.5 is completed. □

Now we define

$$K:=\{u\in H^1(\mathbf{R}^2),\quad I(u)<d_{\Sigma_1},\quad S(u)<0,\quad Q(u)<0\}$$

Thus we have

**Proposition 3.6** $K$ is not empty.

**Proof** From Lemma 3.2, there exists $u\in H^1(\mathbf{R}^2)\setminus\{0\}$ such that $S(u)=0$ and $Q(u)=0$. Put $u_\beta(x)=\beta u(x)$, then by (3.1), (3.2) and (3.3), one has

$$S(u_\beta) = \beta^2\Big(\int|\nabla u|^2\mathrm{d}x + \omega\int|u|^2\mathrm{d}x\Big)-\beta^4\Big(\int|u|^4\mathrm{d}x+\int|u|^2E_1(|u|^2)\mathrm{d}x\Big),$$

$$Q(u_\beta) = \beta^2\int|\nabla u|^2\mathrm{d}x - \frac{1}{2}\beta^4\Big(\int|u|^4\mathrm{d}x+\int|u|^2E_1(|u|^2)\mathrm{d}x\Big),$$

$$I(u_\beta) = \frac{1}{2}\beta^2\Big(\int|\nabla u|^2\mathrm{d}x + \omega\int|u|^2\mathrm{d}x\Big)-\frac{1}{4}\beta^4\Big(\int|u|^4\mathrm{d}x+\int|u|^2E_1(|u|^2)\mathrm{d}x\Big).$$

Since $d_{\Sigma_1}>0$, for $\beta>1$ large enough, in view of the above identities as well as $S(u)=0$ and $Q(u)=0$, one has that $S(u_\beta)<0, Q(u_\beta)<0$ and $I(u_\beta)<d_{\Sigma_1}$. Thus $u_\beta\in K$.

This completes the proof of Proposition 3.1. □

Furthermore, we have

**Proposition 3.7** $K$ is an invariant manifold of the Cauchy problem (1.1),(1.2).

**Proof** Let $\phi_0\in K$. By Proposition 2.1, there exists a unique

$$\phi(t,x)\in C([0,T);H^1(\mathbf{R}^2))$$

with $T\leqslant\infty$ such that $\phi(t,x)$ is a solution of the Cauchy problem (1.1),(1.2). From (1.1) and (1.2), it follows that

$$I(\phi(t,\cdot)) = I(\phi_0),\quad t\in[0,T). \tag{3.26}$$

Thus $I(\phi_0)<d_{\Sigma_1}$ implies that $I(\phi(t,\cdot))<d_{\Sigma_1}$ for any $t\in[0,T)$.

In the following, we first show $S(\phi(t,\cdot))<0$ for $t\in[0,T)$ by contradiction. Otherwise, from continuity, there were a $t^*\in(0,T)$ such that $S(\phi(t^*,\cdot))=0$. By (3.26), it has $\phi(t^*,\cdot)\neq0$. Again from (3.4) and (3.6), it follows that $I(\phi(t^*,\cdot))\geqslant d_{\Sigma_1}$. This is contradictory to $I(\phi(t^*,\cdot))<d_{\Sigma_1}$ for $t\in[0,T)$. Therefore $S(\phi(t,\cdot))<0$ for all $t\in[0,T)$.

Next, we prove $Q(\phi(t,\cdot))<0$ for $t\in[0,T)$ by contradiction. Otherwise, from continuity, there were a $\bar{t}\in(0,T)$ such that $Q(\phi(\bar{t},\cdot))=0$. Since we have showed $S(\phi(\bar{t},\cdot))<0$, it follows that $\phi(\bar{t},\cdot)\in\Sigma_2$. Thus (3.7), (3.20) and Lemma 3.5 imply that $I(\phi(\bar{t},\cdot))\geqslant d_{\Sigma_2}\geqslant d_{\Sigma_1}$. This contradicts $I(\phi(\bar{t},\cdot))<d_{\Sigma_1}$ for $t\in[0,T)$. Therefore $Q(\phi(t,\cdot))<0$ for all $t\in[0,T)$.

From the above arguments, we have proved $\phi(t,\cdot)\in K$ for any $t\in[0,T)$. □

By the same arguments as Proposition 3.2, the following result holds.

Define

$$K_+ := \{u\in H^1(\mathbf{R}^2),\quad I(u)<d_{\Sigma_1},\quad S(u)<0,\quad Q(u)>0\},$$

$$\mathbf{R}_- := \{u\in H^1(\mathbf{R}^2),\quad I(u)<d_{\Sigma_1},\quad S(u)<0\},$$

$$\mathbf{R}_+ := \{u\in H^1(\mathbf{R}^2),\quad I(u)<d_{\Sigma_1},\quad S(u)>0\}.$$

Then $K_+$, $\mathbf{R}_-$ and $\mathbf{R}_+$ are all invariant manifolds of the Cauchy problem (1.1),(1.2).

**Remark 3.8** Generally, we call the variational problem (3.7) cross-constrained variational problem since there are two constrained conditions in (3.7). Similarly, we call $\mathbf{R}_-$ and $\mathbf{R}_+$ are invariant manifolds of the Cauchy problem (1.1), (1.2). In the course of nature, we call $K$ and $K_+$ cross-invariant manifolds of the Cauchy problem (1.1), (1.2).

By the definitions of $K$, $K_+$ and $\mathbf{R}_+$, (3.6) and Lemma 3.5, the following result is true.

**Proposition 3.9**

$$\{\phi \in H^1(\mathbf{R}^2)\setminus\{0\},\quad I(\phi) < d_{\Sigma_1}\} = \mathbf{R}_+ \cup K_+ \cup K.$$

## 4 Sharp Threshold for Global Existence and Blowup

In this section, we discuss the sharp sufficient condition for blowup and global existence. Firstly, we give a sufficient condition for global existence of the solutions to the Cauchy problem (1.1), (1.2).

**Theorem 4.1** If $\phi_0 \in K_+ \cup \mathbf{R}_+$, then the solution $\phi(t,x)$ of the Cauchy problem (1.1), (1.2) exists globally on $t\in[0,\infty)$.

**Proof** We prove this theorem by two steps.

**Step 1** we prove the case $\phi_0 \in K_+$.

**Step 2** we prove the case $\phi_0 \in \mathbf{R}_+$.

Firstly, we consider Step 1. $\phi_0 \in K_+$.

From $\phi_0 \in K_+$, Proposition 3.3 implies that the solution $\phi(t,x)$ of the Cauchy problem (1.1), (1.2) satisfies that $\phi(t,\cdot)\in K_+$ for $t\in[0,T)$. For fixed $t\in[0,T)$, we denote $\phi(t,\cdot)=\phi$. Thus we have $I(\phi)<d_{\Sigma_1}$, $Q(\phi)>0$ and $S(\phi)<0$. So $\phi\neq 0$. According to (3.1) and (3.3), $I(\phi)<d_{\Sigma_1}$ and $Q(\phi)>0$ yield that

$$\frac{\omega}{2}\int|\phi|^2\mathrm{d}x < d_{\Sigma_1}. \tag{4.1}$$

In the following, we prove that $\int|\nabla\phi|^2\mathrm{d}x$ is bounded. Putting $\phi_\lambda = \lambda^{\frac{1}{2}}\phi(\lambda x)$, we have

$$Q(\phi_\lambda) = \lambda\int|\nabla\phi|^2\mathrm{d}x - \frac{1}{2}\int|\phi|^4\mathrm{d}x - \frac{1}{2}\int|\phi|^2E_1(|\phi|^2)\mathrm{d}x. \tag{4.2}$$

By $Q(\phi)>0$, it follows that when $\lambda\to 0$, $Q(\phi_\lambda)<0$ and when $\lambda\to 1$, $Q(\phi_\lambda)>0$. Thus by continuity, there exists a $0<\lambda^*<1$ such that $Q(\phi_{\lambda^*})=0$. Noting (3.1) and (3.3), we have

$$I(\phi_{\lambda^*}) = \frac{\omega}{2}\int|\phi_{\lambda^*}|^2\mathrm{d}x = \frac{\omega}{2}\lambda^{*-1}\int|\phi|^2\mathrm{d}x. \tag{4.3}$$

From (4.1), it follows that

$$I(\phi_{\lambda^*}) < \lambda^{*-1}d_{\Sigma_1}. \tag{4.4}$$

Also, by (3.2) and (3.3), it follows from $Q(\phi_{\lambda^*})=0$ and $S(\phi)<0$ that

$$S(\phi_{\lambda^*}) = \omega\int|\phi_{\lambda^*}|^2\mathrm{d}x - \int|\nabla\phi_{\lambda^*}|^2\mathrm{d}x = \omega\lambda^{*-1}\int|\phi|^2\mathrm{d}x - \lambda^*\int|\nabla\phi|^2\mathrm{d}x. \tag{4.5}$$

which has two possibilities:

(i) $S(\phi_{\lambda^*})<0$;

(ii) $S(\phi_{\lambda^*})\geqslant 0$.

We first consider (i) $S(\phi_{\lambda^*})<0$. In this case, noting that $Q(\phi_{\lambda^*})=0$, we get $\phi_{\lambda^*}\in\Sigma_2$. Thus by (3.7) and Lemma 3.5, one has

$$I(\phi_{\lambda^*}) \geqslant d_{\Sigma_2} \geqslant d_{\Sigma_1} > I(\phi), \tag{4.6}$$

which implies that

$$I(\phi) - I(\phi_{\lambda^*}) < 0. \tag{4.7}$$

That is,

$$\left(\frac{1}{2}-\frac{1}{2}\lambda^*\right)\int|\nabla\phi|^2\mathrm{d}x + \left(\frac{1}{2}-\frac{1}{2}\lambda^{*-1}\right)\omega\int|\phi|^2\mathrm{d}x < 0. \tag{4.8}$$

Since $0<\lambda^*<1$, one has $\frac{\lambda^{*-1}-1}{1-\lambda^*}>0, \lambda^{*-1}-1>0$ and $1-\lambda^*>0$. By (4.1), we get

$$\int|\nabla\phi|^2\mathrm{d}x < \frac{\lambda^{*-1}-1}{1-\lambda^*}\omega\int|\phi|^2\mathrm{d}x < 2\frac{\lambda^{*-1}-1}{1-\lambda^*}d_{\Sigma_1}. \tag{4.9}$$

Now we consider (ii) $S(\phi_{\lambda^*})\geqslant 0$. In this case, from (4.4) it follows that

$$I(\phi_{\lambda^*}) - \frac{1}{4}S(\phi_{\lambda^*}) < \lambda^{*-1}d_{\Sigma_1}. \tag{4.10}$$

Thus by (4.3) and (4.5), one gets

$$\frac{1}{4}\int|\nabla\phi_{\lambda^*}|^2\mathrm{d}x + \frac{1}{4}\omega\int|\phi_{\lambda^*}|^2\mathrm{d}x < \lambda^{*-1}d_{\Sigma_1}. \tag{4.11}$$

Namely,

$$\lambda^*\int|\nabla\phi|^2\mathrm{d}x + \lambda^{*-1}\omega\int|\phi|^2\mathrm{d}x < 4\lambda^{*-1}d_{\Sigma_1},$$

which together with $0<\lambda^*<1$ implies that

$$\int|\nabla\phi|^2\mathrm{d}x < 4\lambda^{*-2}d_{\Sigma_1}. \tag{4.12}$$

By (4.9) and (4.12), $0<\lambda^*<1$ implies that $\int|\nabla\phi|^2\mathrm{d}x$ is bounded for any $t\in[0,T)$. Thus by (4.1), we have $\phi$ is bounded in $H^1(\mathbf{R}^2)$. So Proposition 2.1 implies that the solution $\phi$ of the Cauchy problem (1.1),(1.2) globally exists on $t\in[0,\infty)$ for $\phi_0\in K_+$.

In the following, we deal with Step 2. $\phi_0\in\mathbf{R}_+$.

Let $\phi_0\in\mathbf{R}_+$. Proposition 3.3 implies that the solution $\phi(t,x)$ of the Cauchy problem (1.1), (1.2) satisfies that $\phi(t,\cdot)\in\mathbf{R}_+$ for $t\in[0,T)$. So we have $I(\phi)<d_{\Sigma_1}$ and $S(\phi)>0$. By $S(\phi)>0$ and (3.2), we get

$$-\int|\phi|^4\mathrm{d}x - \int|\phi|^2E_1(|\phi|^2)\mathrm{d}x > -\left(\int|\nabla\phi|^2\mathrm{d}x + \omega\int|\phi|^2\mathrm{d}x\right). \tag{4.13}$$

From (3.1) and $I(\phi)<d_{\Sigma_1}$, one has

$$\frac{1}{2}\int|\nabla\phi|^2\mathrm{d}x + \frac{\omega}{2}\int|\phi|^2\mathrm{d}x - \frac{1}{4}\int|\phi|^4\mathrm{d}x - \frac{1}{4}\int|\phi|^2E_1(|\phi|^2)\mathrm{d}x < d_{\Sigma_1}. \tag{4.14}$$

Therefore (4.13) and (4.14) conclude that

$$\frac{1}{4}\int|\nabla\phi|^2\mathrm{d}x + \frac{\omega}{4}\int|\phi|^2\mathrm{d}x < d_{\Sigma_1}, \tag{4.15}$$

which implies that $\phi$ is bounded in $H^1(\mathbf{R}^2)$ for any $t\in[0,T)$. Thus by Proposition 2.1, we obtain that the solution $\phi$ of the Cauchy problem (1.1), (1.2) exists globally on $t\in[0,\infty)$ for $\phi_0\in\mathbf{R}_+$.

By the arguments of Step 1 and Step 2, we have completed the proof of Theorem 4.1. □

**Theorem 4.2** If $\phi_0(x)\in K$ and $|x|\phi_0\in L^2(\mathbf{R}^2)$, then the solution $\phi(t,x)$ of the Cauchy problem (1.1), (1.2) blows up in finite time.

**Proof** According to Ginibre and Velo[10,11], from $|x|\phi_0(x)\in L^2(\mathbf{R}^2)$, one has $|x|\phi\in L^2(\mathbf{R}^2)$. By $\phi_0\in K$, Proposition 3.2 implies that $\phi(t,\cdot)\in K$ for $t\in[0,T)$. Thus we have $I(\phi)<d_{\Sigma_1}$, $S(\phi)<0$ and $Q(\phi)<0$. Now we put

$$J(t)=\int|x|^2|\phi|^2\mathrm{d}x. \tag{4.16}$$

Proposition 2.2 and (3.3) imply that

$$J''(t)=8Q(\phi(t,\cdot)),\quad t\in[0,T). \tag{4.17}$$

Fixed $t\in[0,T)$ and denote $\phi(t,\cdot)=\phi$. Thus $\phi$ satisfies that $I(\phi)<d_{\Sigma_1}$, $Q(\phi)<0$ and $S(\phi)<0$. By (2.2) (energy conservation) and (3.3), we have

$$E(\phi)=E(\phi_0)=\frac{1}{2}Q(\phi)=\frac{1}{2}Q(\phi_0)<0$$

since $Q(\phi)<0$. Thus (4.17) concludes

$$J''(t)=16E(\phi_0)<0,$$

which yields the result of Theorem 4.2. □

Using Theorem 4.1, 4.2 and Proposition 3.4, we can get

**Theorem 4.3** If $\phi_0$ satisfies $I(\phi_0)<d_{\Sigma_1}$ and $|x|\phi_0\in L^2(\mathbf{R}^2)$, then the solution $\phi(t,x)$ of the Cauchy problem (1.1), (1.2) blows up in finite time if and only if $\phi_0\in K$.

**Remark 4.4** From Proposition 3.4, Theorem 4.3 shows that Theorem 4.1 is sharp for global existence of the solution to the Cauchy problem (1.1), (1.2) provided that $|x|\phi_0\in L^2(\mathbf{R}^2)$.

In addition, using Theorem 4.1 and scaling argument, we can get another sufficient condition for the global existence of the solution to the Cauchy problem (1.1), (1.2).

**Corollary 1** (Small Data Criterion) If $\phi_0\in H^1(\mathbf{R}^2)$ and satisfies

$$\int|\nabla\phi_0|^2\mathrm{d}x+\omega\int|\phi_0|^2\mathrm{d}x<2d_{\Sigma_1}, \tag{4.18}$$

then the solution $\phi(t,x)$ of the Cauchy problem (1.1), (1.2) exists globally on $t\in[0,\infty)$.

**Proof** According to (3.1) and (4.18), we get $I(\phi_0)<d_{\Sigma_1}$. In addition, we assert that $S(\phi_0)>0$. Otherwise, from (3.2) there were a $0<\lambda\leqslant1$ such that $S(\lambda\phi_0)=0$. Thus by (3.6), one has

$$I(\lambda\phi_0)\geqslant d_{\Sigma_1}. \tag{4.19}$$

On the other hand,

$$\int|\nabla(\lambda\phi_0)|^2\mathrm{d}x+\omega\int|\lambda\phi_0|^2\mathrm{d}x=\lambda^2\left(\int|\nabla\phi_0|^2\mathrm{d}x+\omega\int|\phi_0|^2\mathrm{d}x\right)\leqslant2d_{\Sigma_1}, \tag{4.20}$$

which implies that

$$I(\lambda\phi_0) < d_{\Sigma_1}. \tag{4.21}$$

It is obvious that (4.19) and (4.21) are contradictory. So $S(\phi_0)>0$ holds. From Proposition 3.3, $I(\phi_0)<d_{\Sigma_1}$ and $S(\phi_0)>0$ yield that $\phi_0\in\mathbf{R}_+$, thus Theorem 4.1 implies the result of Corollary 1. □

**Remark 4.5** In fact, Corollary 1 gives an answer to the question: How small are the initial data, the solution of the Cauchy problem (1.1), (1.2) exists globally?

**References**


[1] Berestycki H, Cazenave T. Instabilité des états stationnaires dans les équations de Schrödinger et de Klein-Gordon non linéairees. C R Acad Sci Paris, 1981, 293: 489-492.

[2] Blue P, Colliander J. Global well-posedness in Sobolev space implies global existence for weighted $L^2$ initial data for $L^2$-critical NLS. Commun Pure Applied Anal, 2006, 5: 691-708.

[3] Chen J. A variational argument to finding global solutions of a quasilinear Schrödinger equation. Commun Pure Applied Anal, 2008, 7: 83-88.

[4] Chen J, Guo B. Sharp global existence and blowing up results for inhomogeneous Schrödinger equations. Discrete and Continuous Dynamical Systems Series B, 2007, 8: 357-367.

[5] Cipolatti R. On the existence of standing waves for a Davey-Stewartson system. Comm P D E, 1992, 17: 967-988.

[6] Cipolatti R. On the instability of ground states for a Davey-Stewartson system. Ann Inst Henri Poincaré, Phys Théor, 1993, 58: 85-104.

[7] Davey A, Stewartson K. On three-dimensional packets of surface waves. Proc R Soc London A, 1974, 338: 101-110.

[8] De Silva D, Pavlovič N, Staffilani G, et al. Global well-posedness for the $L^2$ critical nonlinear Schrödinger equation in higher dimensions. Commun Pure Applied Anal, 2007, 6: 1023-1041.

[9] Gan Z H, Guo B L, Zhang J. Blowup and Global Existence of the Nonlinear Schrödinger Equations with Multiple Potentials. Commun Pure Applied Anal, 2009, 8: 1303-1312.

[10] Ginibre J, Velo G. On a class of nonlinear Schrödinger equations. J Funct Anal, 1979, 32: 1-71.

[11] Ginibre J, Velo G. The global Cauchy problem for the nonlinear Schrödinger equation. Ann Inst H Poincaré, Anal Non Linéaire, 1985,2: 309-327.

[12] Glassey R T. On the blowing up of solutions to the Cauchy problem for nonlinear Schrödinger equations. J Math Phys, 1977, 18: 1794-1797.

[13] Ghidaglia J M, Saut J C. On the initial value problem for the Davey-Stewartson systems. Nonlinearity, 1990, 3: 475-506.

[14] Guo B L, Wang B X. The Cauchy problem for Davey-Stewartson systems. Commun Pure Appl Math, 1999, 52: 1477-1490.

[15] Levine H A. Instability and non-existence of global solutions to nonlinear wave equations of the form $Pu_{tt} = -Au + F(u)$. Trans Amer Math Soc, 1974, 192: 1-21.

[16] Ohta M. Blow-up solutions and strong instability of standing waves for the generalized Davey-Stewartson system in $\mathbf{R}^2$. Ann Inst Henri Poincaré, 1995, 63: 111-117.

[17] Ohta M. Instability of standing waves for the generalized Davey-Stewartson system. Ann Inst Henri

Poincaré, 1995, 62: 69 - 80.

[18] Payne L E, Sattinger D H. Saddle points and instability of nonlinear hyperbolic equations. Israel J Math, 1975, 22: 273 - 303.

[19] Shu J, Zhang J. Sharp conditons of global existence for the generalized Davey-Stewartson system. IMA Journal of Applied Mathematics, 2007, 72: 36 - 42.

[20] Weinstein M I. Nonlinear Schrödinger equations and sharp interpolations estimates. Commun Math Phys, 1983, 87: 567 - 576.

[21] Zhang J. Sharp conditions of global existence for nonlinear Schrödinger and Klein-Gordon equations. Nonlinear Anal, 2002, 48: 191 - 207.

[22] Zhang J. Sharp threshold for blowup and global existence in nonlinear Schrödinger equations under a harmonic potential. Commun in PDE, 2005, 30: 1429 - 1443.

[23] Zhang J. Cross-constrained variational problem and nonlinear Schrödinger equation // Foundations of Computational Mathematics. Hong Kong: World Sci Publ, 2002: 457 - 469.

# 3D Stochastic Primitive Equations of the Large-scale Ocean: Global Well-posedness and Attractors *

Guo Boling(郭柏灵) Huang Daiwen(黄代文)

**Abstract**

In this paper, we consider the global well-posedness and long-time dynamics for the three-dimensional viscous primitive equations describing the large-scale oceanic motion under a random forcing, which is an additive white in time noise. We firstly prove the existence and uniqueness of global strong solutions to the initial boundary value problem for the stochastic primitive equations. Subsequently, by studying the asymptotic behavior of strong solutions, we obtain the existence of random attractors for the corresponding random dynamical system.

## 1 Introduction

In order to understand the mechanism of long-term weather prediction and climate changes, one can study the equations governing the motions of the ocean and atmosphere. Under the Boussinesq and hydrostatic approximations, the large-scale motion of the ocean can be well modeled by 3D viscous primitive equations. After a series of pioneering articles by J. L. Lions, R. Temam and S. Wang[24-27], there are many works about the global well-posedness and long-time dynamics of the deterministic primitive oceanic or atmospheric equations[13,43] and references therein. Moreover, some other models, such as the 2D and 3D quasi-geostrophic models, have been the subject of analytical mathematical study[3,5,10,11,14,18,28,29] and references therein.

In the past two decades, there are several research works about the well-posedness of the deterministic 3D primitive equations of the large-scale ocean. In [25], Lions, Temam and Wang obtained the existence of global weak solutions for the primitive equations. In [22], Guillén-Gonzélez, et al. obtained the global existence of strong solutions to the primitive equations with small initial data. Moreover, they proved the local existence of strong solutions to the equation. Hu et al. studied the local existence of strong solutions to the primitive equations under the small depth hypothesis in [23]. In [13], Cao and Titi developed a beautiful approach to proving that the $L^6$-norm of the fluctuation $\tilde{v}$ of horizontal velocity is uniformly in $t$ bounded, and obtained the global well-posedness for the 3D viscous primitive equations.

---

* Commun Math Phys, 2009, 286: 697-723.

Despite the great success of the above developments, a comprehensive understanding of the long-time dynamics of atmospheric and oceanic motions seems to be still incomplete. For example, the non-trivial effects of randomness are not included among all the former literature for the 3D primitive equations. Atmospheric or oceanic motions exhibit fluctuations over a broad range of spatial and temporal scales ranging from centimeters to thousands of kilometers and from seconds to decades and beyond. Such fluctuations can be caused by internal instability processes, as well as by external forcing. For the atmosphere, the main external forcing coming from the solar radiation is stochastic. The external forcing of the ocean comes mainly from the atmosphere. As usual, the atmospheric forcing field must be regarded as random[20,32,33,35,38].

Therefore, in the study of the primitive equations of the large-scale ocean or atmosphere, taking the stochastic external factors into account is reasonable and necessary. There are many works about mathematical study of some stochastic climate models[15-17,30-32]. [17] is one of the first works on a 3D stochastic quasi-geostrophic flow model. However, to our knowledge, no one has considered the long-time dynamics for the 3D stochastic primitive equations of the large-scale ocean or atmosphere. The understanding of the long-time behavior of dynamical systems is one of the most important problems of modern mathematical physics. One way to solve the problem is to study their attractors[4,9,40].

In the present paper, we are interested in considering the existence of random attractors of the 3D stochastic primitive equations (2.1)~(2.4), which, when $\Psi(t,x,y,z)\equiv 0$ in (2.1), is the 3D deterministic primitive equations considered in [13]. In order to do that, we should study the global well-posedness and long-time behavior of global strong solutions to the initial boundary value problem of the 3D stochastic primitive equations, which is denoted by IBVP and will be given in Sec. 2. Our results are as follows.

**Theorem 1.1** [Global well-posedness of IBVP] If $Q\in H^1(\Omega)$ and $U_{t_0}=(v_{t_0}, T_{t_0})\in V$, then, for any given $\mathcal{T}>t_0$, there exists a unique strong solution $U$ of the system (2.9)~(2.15) on the interval $[t_0,\mathcal{T}]$; moreover, the strong solution $U$ is dependent continuously on the initial data. The definitions of the space $V$ and the strong solution to the system (2.9)~(2.15), and some assumptions on the stochastic forcing term are given in Sec. 3.

**Theorem 1.2** [Existence of the random pull-back attractor for the random dynamical system (2.9)~(2.14)] The system (2.9)~(2.14) possesses a unique random pull-back attractor $\mathscr{A}(\omega)$ that captures all the trajectories started at time $-\infty$ and evolved, under the action of the shift $\vartheta_t$, from $-\infty$ to the present time $t=0$. The attractor $\mathscr{A}(\omega)$ has the following properties:

(i) (weak compact) $\mathscr{A}(\omega)$ is bounded and weakly closed in $V$;

(ii) (invariant) for any $t\geqslant 0$, $\psi(t,\omega)\mathscr{A}(\omega)=\mathscr{A}(\vartheta_t\omega)$;

(iii) (attracting) for any deterministic bounded set $B$ in $V$, the sets $\psi(t,\vartheta_{-t}\omega)B$ converge to $\mathscr{A}(\vartheta_t\omega)$ with respect to $V$-weak topology as $t\to+\infty$, i.e.,

$$\lim_{t\to+\infty} d_V^w(\psi(t,\vartheta_{-t}\omega)B,\mathscr{A}(\omega))=0,\ P-\text{a.s.},$$

where the definitions of the random attractor, $\vartheta_t, t\in\mathbf{R}$ and $\psi(t,\omega), t\geqslant 0$ will be given in Subsec. 6.1, and the distance $d_V^w$ is induced by the $V$-weak topology.

There are two main difficulties in considering the long-time dynamics of the 3D stochastic primitive equations.

First, since the random forcing term $\Psi(t,x,y,z)$ is so irregular in time that it is still unknown how to handle $\int_{t_0}^{\tau}\Psi(s,x,y,z)v(s,x,y,z)\mathrm{d}s$, it is impossible to obtain the necessary energy inequalities with usual energy methods. So, in order to prove the local well-posedness of IBVP and make *a priori* estimates of local strong solutions, we should introduce an auxiliary Ornstein-Uhlenbeck process $Z$ which is a stationary ergodic solution to a stochastic Stokes equation. Our main idea used in proving the global well-posedness of IBVP is inspired by [13,22]. The proof of local well-posedness of IBVP is similar to that of the deterministic primitive equations. However, after introducing $Z$, *a priori* estimates of local strong solutions to the stochastic primitive equations are more complicated than the deterministic ones. In order to prove the global well-posedness, we should obtain energy inequalities for the difference $w=v-Z$ between the solution $v$ of IBVP and $Z$. In the course of making *a priori* estimates of $w$, it is most important to find a properly regular condition of $Z$ for the proof of the global well-posedness of IBVP. We need $Z\in L^{\infty}(\mathbf{R};(H^{2+2\gamma}(\Omega))^2)$ for some $\gamma>0$. We give some examples of $\Psi$ in Subsec. 3.2 which enable $Z$ to satisfy the condition.

Second, the random dynamical system, corresponding to the boundary value problem of the 3D stochastic primitive equations, is non-autonomous. We should use a pull-back procedure[6,7] and construct a random attractor which attracts trajectories started at time $-\infty$ and evolved, under the action of the shift $\vartheta_t$, from $-\infty$ to the present time $t=0$. By studying the long-time behavior of strong solutions, we prove the existence of a random compact absorbing set, and obtain a random pullback attractor $\mathscr{A}(\omega)$. In order to study the long-time behavior of strong solutions to IBVP, we must make two key estimates. Firstly, we must make estimates about the $L^3$-norm of the fluctuation $\tilde{w}$ of $w$ before we study the long-time behavior of strong solutions, without which we can only prove the global well-posedness of IBVP. Secondly, we ought to make estimates about the $L^4$-norm of $\tilde{w}$. If we only made estimates about the $L^6$-norm of $\tilde{w}$, we could not study long-time behavior of strong solutions to IBVP.

The paper is organized as follows. In Sec. 2, the 3D stochastic primitive equations are introduced. Our working spaces and a new formulation of the initial boundary value problem for the stochastic primitive equations with an additive noise are given in Sec. 3. We obtain the existence and make *a priori* estimates of local strong solutions to IBVP in Sec. 4. We prove the main results of our paper in Sec. 5,6 .

## 2 The 3D Stochastic Primitive Equations

The 3D viscous primitive equations of the large-scale ocean under a stochastic forcing, in a

Cartesian coordinate system, are written as

$$\frac{\partial v}{\partial t}+(v\cdot\nabla)v+\theta\frac{\partial v}{\partial z}+fk\times v+\nabla p-\frac{1}{Re_1}\Delta v-\frac{1}{Re_2}\frac{\partial^2 v}{\partial z^2}=\Psi(t,x,y,z),\quad(2.1)$$

$$\frac{\partial p}{\partial z}+T=0,\quad(2.2)$$

$$\nabla\cdot v+\frac{\partial\theta}{\partial z}=0,\quad(2.3)$$

$$\frac{\partial T}{\partial t}+v\cdot\nabla T+\theta\frac{\partial T}{\partial z}-\frac{1}{Rt_1}\Delta T-\frac{1}{Rt_2}\frac{\partial^2 T}{\partial z^2}=Q,\quad(2.4)$$

where the unknown functions are $v,\theta,p,T$, $v=(v^{(1)},v^{(2)})$ horizontal velocity, $\theta$ vertical velocity, $p$ pressure, $T$ temperature, $f=f_0(\beta+y)$ Coriolis parameter, $k$ vertical unit vector, $Re_1$, $Re_2$ Reynolds numbers, $Rt_1$, $Rt_2$ the horizontal and vertical heat diffusivity respectively, $Q(x,y,z)$ a given heat source, $\nabla=(\partial_x,\partial_y)$, $\Delta=\partial_x^2+\partial_y^2$. Here, we take $\Psi(t,x,y,z)$ as a stochastic forcing field given in Subsec. 3.2.

The space domain for (2.1)~(2.4) is

$$\Omega=\{(x,y,z):(x,y)\in M \text{ and } z\in(-g(x,y),0)\},$$

where $M$ is a smooth bounded domain in $\mathbf{R}^2$ and $g$ is a sufficiently smooth function. Here we assume $g=1$, that is, $\Omega=M\times(-1,0)$. The boundary value conditions for (2.1)~(2.4) are given by

$$\frac{\partial v}{\partial z}=0,\quad\theta=0,\quad\frac{\partial T}{\partial z}=-\alpha_u T\quad\text{on } M\times\{0\}=\Gamma_u,\quad(2.5)$$

$$\frac{\partial v}{\partial z}=0,\quad\theta=0,\quad\frac{\partial T}{\partial z}=0\quad\text{on } M\times\{-1\}=\Gamma_b,\quad(2.6)$$

$$v\cdot\boldsymbol{n}=0,\quad\frac{\partial v}{\partial\boldsymbol{n}}\times\boldsymbol{n}=0,\quad\frac{\partial T}{\partial\boldsymbol{n}}=0\quad\text{on }\partial M\times[-1,0]=\Gamma_l,\quad(2.7)$$

where $\alpha_u$ is a positive constant and $\boldsymbol{n}$ is the norm vector to $\Gamma_l$.

**Remark 2.1** For more details about the derivation of the 3D deterministic primitive equations, we refer the reader to [25,34] and references therein. Similar to [13], the salinity diffusion equation is omitted here. However, our results are valid when the salinity is taken into account and the boundary value conditions $\left.\frac{\partial v}{\partial z}\right|_{\Gamma_u}=0$, $\left.\frac{\partial T}{\partial z}\right|_{\Gamma_u}=-\alpha_u T$ are replaced by $\left.\frac{\partial v}{\partial z}\right|_{\Gamma_u}=\tau$, $\left.\frac{\partial T}{\partial z}\right|_{\Gamma_u}=-\alpha_u(T-T^*)$ for smooth enough $\tau$, $T^*$ satisfying the compatible boundary conditions: $\tau\cdot\boldsymbol{n}|_{\partial M}=0$, $\frac{\partial\tau}{\partial\boldsymbol{n}}\times\boldsymbol{n}|_{\partial M}=0$, $\frac{\partial T^*}{\partial\boldsymbol{n}}|_{\partial M}=0$.

In this article, we assume that the constants $Re_1$, $Re_2$, $Rt_1$, $Rt_2$ are all equal to 1. Without this assumption, we can also obtain the results of this paper. Integrating (2.3) from $-1$ to $z$ and using (2.5),(2,6) we have

$$\theta(t,x,y,z)=\Phi(v)(t,x,y,z)=-\int_{-1}^{z}\nabla\cdot v(t,x,y,z')\mathrm{d}z',\quad(2.8)$$

moreover, $\int_{-1}^{0}\nabla\cdot v\,\mathrm{d}z=0$. Supposing that $p_b$ is a certain unknown function at $\Gamma_b$, and integrating (2.2) from $-1$ to $z$, we rewrite (2.1)~(2.4) as

$$\frac{\partial v}{\partial t} + (v \cdot \nabla)v + \Phi(v)\frac{\partial v}{\partial z} + fk \times v + \nabla p_b - \int_{-1}^{z} \nabla T dz' - \Delta v - \frac{\partial^2 v}{\partial z^2} = \psi, \quad (2.9)$$

$$\frac{\partial T}{\partial t} + (v \cdot \nabla)T + \Phi(v)\frac{\partial T}{\partial z} - \Delta T - \frac{\partial^2 T}{\partial z^2} = Q, \quad (2.10)$$

$$\int_{-1}^{0} \nabla \cdot v dz = 0. \quad (2.11)$$

The boundary value conditions for (2.9)~(2.11) are given by

$$\frac{\partial v}{\partial z} = 0, \quad \frac{\partial T}{\partial z} = -\alpha_u T \quad \text{on } \Gamma_u, \quad (2.12)$$

$$\frac{\partial v}{\partial z} = 0, \quad \frac{\partial T}{\partial z} = 0 \quad \text{on } \Gamma_b, \quad (2.13)$$

$$v \cdot \boldsymbol{n} = 0, \quad \frac{\partial v}{\partial \boldsymbol{n}} \times \boldsymbol{n} = 0, \quad \frac{\partial T}{\partial \boldsymbol{n}} = 0 \quad \text{on } \Gamma_l, \quad (2.14)$$

and the initial value conditions are

$$U|_{t=t_0} = (v|_{t=t_0}, T|_{t=t_0}) = U_{t_0} = (v_{t_0}, T_{t_0}). \quad (2.15)$$

We call (2.9) ~ (2.15) the initial boundary value problem of the 3D viscous stochastic primitive equations, which is denoted by IBVP.

# 3 New Formulation of IBVP

## 3.1 Somc Function Spaces

$L^p(\Omega)$ is the usual Lebesgue space with the norm $|\cdot|_p, 1 \leqslant p \leqslant \infty$. $H^m(\Omega)$ is the usual Sobolev space ($m$ is a positive integer) with the norm

$$\|h\|_m = \left[\int_\Omega \Big(\sum_{1\leqslant k\leqslant m}\ \sum_{i_j=1,2,3;j=1,\cdots,k} |\nabla_{i_1}\cdots\nabla_{i_k} h|^2 + |h|^2\Big)\right]^{\frac{1}{2}}, \text{where } \nabla_1 = \frac{\partial}{\partial x}, \nabla_2 = \frac{\partial}{\partial y}$$

and $\nabla_3 = \frac{\partial}{\partial z}$. $\int_\Omega \cdot \, d\Omega$ and $\int_M \cdot \, dM$ are denoted by $\int_\Omega \cdot$ and $\int_M \cdot$ respectively.

We define our working spaces for IBVP. Let

$$\mathscr{V}_1 := \left\{ v \in (C^\infty(\Omega))^2; \frac{\partial v}{\partial z}\Big|_{\Gamma_u,\Gamma_b} = 0, v \cdot \boldsymbol{n}\Big|_{\Gamma_l} = 0, \frac{\partial v}{\partial \boldsymbol{n}} \times \boldsymbol{n}\Big|_{\Gamma_l} = 0, \int_{-1}^{0} \nabla \cdot v dz = 0 \right\},$$

$$\mathscr{V}_2 := \left\{ T \in C^\infty(\Omega); \frac{\partial T}{\partial z}\Big|_{\Gamma_u} = -\alpha_u T, \frac{\partial T}{\partial z}\Big|_{\Gamma_b} = 0, \frac{\partial T}{\partial \boldsymbol{n}}\Big|_{\Gamma_l} = 0 \right\},$$

$V_1$ = the closure of $\mathscr{V}_1$ with respect to the norm $\|\cdot\|_1$,

$V_2$ = the closure of $\mathscr{V}_2$ with respect to the norm $\|\cdot\|_1$,

$H_1$ = the closure of $\mathscr{V}_1$ with respect to the norm $\|\cdot\|_2$,

$$V = V_1 \times V_2, \quad H = H_1 \times L^2(\Omega).$$

The inner products and norms on $V$, $H$ are given by

$$(U, U_1)_V = (v, v_1)_{V_1} + (T, T_1)_{V_2},$$

$$(U,U_1)=(v^{(1)},(v_1)^{(1)})+(v^{(2)},(v_1)^{(2)})+(T,T_1),$$

$$\|U\|=(U,U)_V^{\frac{1}{2}}=(v,v)_{V_1}^{\frac{1}{2}}=(T,T)_{V_2}^{\frac{1}{2}}=\|v\|+\|T\|,\quad |U|_2=(U,U)^{\frac{1}{2}},$$

where $U=(v,T),U_1=(v_1,T_1)\in V$, and $(\cdot,\cdot)$ denotes the inner product in $L^2(\Omega)$.

## 3.2 The White Noise and Ornstein Uhlenbeck Process

At first, we define the functionals $a:V\times V\to\mathbf{R}$, $a_1:V_1\times V_1\to\mathbf{R}$, $a_2:V_2\times V_2\to\mathbf{R}$, and their corresponding linear operators $A:V\to V'$, $A_1:V_1\to V'_1$, $A_2:V_2\to V'_2$ by $a(U,U_1)=(AU,U_1)=a_1(v,v_1)+a_2(T,T_1)$, where

$$a_1(v,v_1)=(A_1v,v_1)=\int_\Omega\left(\nabla v\cdot\nabla v_1+\frac{\partial v}{\partial z}\cdot\frac{\partial v_1}{\partial z}\right),$$

$$a_2(T,T_1)=(A_2T,T_1)=\int_\Omega\left(\nabla T\cdot\nabla T_1+\frac{\partial T}{\partial z}\frac{\partial T_1}{\partial z}\right)+\alpha_u\int_{\Gamma_u}TT_1.$$

**Lemma 3.1** (1) $a$ is coercive and continuous, and $A:V\to V'$ is isomorphism. Moreover,

$$\left.\begin{aligned}&a(U,U_1)\leqslant c\|v\|\|v_1\|+c\|T\|\|T_1\|\leqslant c\|U\|\|U_1\|,\\&a(U,U)\geqslant c\|v\|^2+c\|T\|^2\geqslant c\|U\|^2.\end{aligned}\right\}\tag{3.1}$$

In this article, $c$ will denote positive constants and can be determined in concrete conditions.

(2) The isomorphism $A:V\to V'$ can be extended to a self-adjoint unbounded linear operator on $H$ with a compact inverse $A^{-1}:H\to H$ and with the domain of definition of the operator $D(A)=V\cap[(H^2(\Omega))^2\times H^2(\Omega)]$.

**Proof of Lemma 3.1** By $\|v\|^2_{L^2(M)}\leqslant C_M\|\nabla v\|^2_{L^2(M)}$ (cf. [21,p.55]), we prove (3.1). Since the operator $A$ is similar to the usual positive symmetric Laplacian operator on $H_0^1$, the other part of Lemma 3.1 can be proven in the usual argument. Therefore we omit the details of the proof. For more details, the reader can refer to [25, Lemma 2.4]. □

We denote by $0<\lambda_1<\lambda_2\leqslant\cdots$ the eigenvalues of $A_1$ and by $e_1, e_2,\cdots$ a corresponding complete orthonormal system of eigenvectors. We remark that $\|v\|^2\geqslant\lambda_1|v|_2^2$ for any $v\in V_1$. We denote by $e^{t(-A_1)}, t\geqslant 0$, the semigroup on $H_1$ generated by $-A_1$.

In the present paper, we assume that the stochastic forcing $\Psi(t,x,y,z)$ is an additive white in time noise with the form

$$\Psi(t,x,y,z)=G\frac{\partial W}{\partial t},\tag{3.2}$$

where the derivative is in the Itô integral sense, the random process $W$ is a two-sided in time cylindrical Wiener process in $H_1$ with the form $W(t)=\sum_{i=1}^{+\infty}\omega_i(t,\omega)e_i$, and $G$ is a Hilbert-Schmidt operator from $H_1$ to $H^{1+2\gamma_0}(\Omega)\times H^{1+2\gamma_0}(\Omega)$ for some $\gamma_0>0$, i.e. $\sum_{i=1}^{+\infty}\|Ge_i\|^2_{1+2\gamma_0}<+\infty$. Here, $\omega_1,\omega_2,\cdots$ is a sequence of independent standard one-dimensional Brownian motions on a complete probability space $(\Omega,\mathscr{F},P)$ with expectation denoted by $E$, and $H^{1+2\gamma_0}(\Omega)$

is the usual non-integer order Sobolev space.

Now, we define an auxiliary Ornstein-Uhlenbeck process. For any $\alpha \geqslant 0$, let

$$Z_\alpha(t) = W_{A_1}^\alpha(t) = \int_{-\infty}^t e^{(t-s)(-A_1-\alpha)} G \mathrm{d}W(s) \tag{3.3}$$

be the solution of the stochastic Stokes equation

$$\mathrm{d}Z = (-A_1 Z - \alpha Z)\mathrm{d}t + G\mathrm{d}W(t)$$

with $Z(0) = \int_{-\infty}^0 e^{s(A_1+\alpha)} G\mathrm{d}W(s)$. In the following, denote $Z_\alpha(t)$ *by* $Z(t)$ for simplicity. The damping term $\alpha Z$ is not necessary to prove the global well-posedness of IBVP, but that term turns out to be very useful to study the asymptotic behavior of strong solutions to IBVP. So we introduce $\alpha Z$ from the beginning.

**Lemma 3.2**[36,37] If $Z(t)$ is the solution for the above equation, then the process $Z(t)$ is a stationary ergodic solution with continuous trajectories, taking values in $D(A_1^{1+\gamma})$, for any $\gamma < \gamma_0$.

**Remark 3.3** An example for $\Psi$ is $\Psi = \frac{\partial W}{\partial t}$, where $W$ is a two-sided in time finite-dimensional Brownian motion with the form

$$W = \sum_{i=1}^m \delta_i \omega_i(t,\omega) e_i.$$

In the above formula, $\omega_1, \cdots, \omega_m$ are independent standard one-dimensional Brownian motions on a complete probability space $(\Omega, \mathscr{F}, P)$ (with expectation denoted by $E$), and $\delta_i$ are real coefficients. In this case, $Z$ is a stationary ergodic solution with continuous trajectories which take values in $D(A_1^k)$ for any $k \in \mathbf{N}$.

**Remark 3.4** Another example for $\Psi$ is $\Psi = \frac{\partial W}{\partial t}$, where $W$ is a two-sided in time infinite dimensional Brownian motion with the form

$$W(t) = \sum_{i=1}^{+\infty} \mu_i \omega_i(t,\omega) e_i.$$

Here $\omega_1, \omega_2, \cdots$ is a sequence of independent standard one-dimensional Brownian motions on a complete probability space $(\Omega, \mathscr{F}, P)$ (with expectation denoted by $E$) and the coefficients $\mu_i$ satisfy $\sum_{i=1}^{+\infty} \lambda_i^{1+2\gamma_0} \mu_i^2 < +\infty$, for some $\gamma_0 > 0$. In this case,

$$E|A_1^{1+\gamma} Z(t)|_2^2 = E\sum_{i=1}^{+\infty} \left| \int_{-\infty}^t \lambda_i^{1+\gamma} e^{(t-s)(-\lambda_i-\alpha)} \mathrm{d}\omega_i \right|^2$$
$$= \sum_{i=1}^{+\infty} \left| \int_{-\infty}^t \lambda_i^{2+2\gamma} e^{2(t-s)(-\lambda_i-\alpha)} \mu_i^2 \mathrm{d}s \right|^2 = \sum_{i=1}^{+\infty} \frac{\lambda_i^{2+2\gamma}\mu_i^2}{2(\lambda_i+\alpha)} < +\infty,$$

for any $\gamma < \gamma_0$.

**Remark 3.5** If $\Psi(t,x,y,z)$ is independent of the variable $z$, then the process $Z$ is also independent of $z$. In order to prove the global well-posedness of IBVP, we only need lower regular $G$ such that $Z \in L^\infty(\mathbf{R}; H^1(M)) \cap L^2(\mathbf{R}; H^2(M))$. But we need $Z \in L^\infty(\mathbf{R}; H^2(M))$ for proving the existence of random attractors for the 3D stochastic primitive equations.

### 3.3 New Formulation of the System (2.9)~(2.15)

**Definition 3.6** For any $\mathcal{T}>t_0$, a stochastic process $U(t,\omega)=(v,T)$ is called a strong solution (weak solution) to (2.9)~(2.15) in $[t_0,\mathcal{T}]$, if, for $P$-a.e. $\omega\in\Omega$, $U$ satisfies

$$\int_\Omega v(t)\cdot\varphi_1-\int_{t_0}^{\mathcal{T}}\int_\Omega\left\{[(v\cdot\nabla)\varphi_1+\Phi(v)\partial_z\varphi_1]\cdot v-\left[(fk\times v)\cdot\varphi_1+\left(\int_{-1}^{z}T\mathrm{d}z'\right)\nabla\cdot\varphi_1\right]\right\}+$$

$$\int_{t_0}^{\mathcal{T}}\int_\Omega v\cdot A_1\varphi_1=\int_\Omega v_{t_0}\cdot\varphi_1+\int_\Omega[GW(t,\omega)-GW(t_0,\omega)]\cdot\varphi_1,$$

$$\int_\Omega T(t)\varphi_2-\int_{t_0}^{\mathcal{T}}\int_\Omega\{[(v\cdot\nabla)\varphi_2+\Phi(v)\partial_z\varphi_2]T-TA_2\varphi_2\}=\int_\Omega T_{t_0}\varphi_2+\int_{t_0}^{\mathcal{T}}\int_\Omega Q\varphi_2,$$

for all $t\in[t_0,\mathcal{T}]$ and $\varphi=(\varphi_1,\varphi_2)\in D(A_1)\times D(A_2)$, moreover $U\in L^\infty(t_0,\mathcal{T};V)\cap L^2(t_0,\mathcal{T};(H^2(\Omega))^3)$ ($U\in L^\infty(t_0,\mathcal{T};H)\cap L^2(t_0,\mathcal{T};V)$) and $U$ is progressively measurable in these topologies.

Assume that $Y$ is the solution of the initial boundary value problem

$$\begin{cases}\dfrac{\partial Y}{\partial t}+\nabla p_{b_1}-\Delta Y-\dfrac{\partial^2 Y}{\partial z^2}=0,\\ \left.\dfrac{\partial Y}{\partial z}\right|_{\Gamma_u,\Gamma_b}=0,\quad \left.Y\cdot\boldsymbol{n}\right|_{\Gamma_l}=0,\quad \left.\dfrac{\partial Y}{\partial \boldsymbol{n}}\times\boldsymbol{n}\right|_{\Gamma_l}=0,\\ \displaystyle\int_{-1}^{0}\nabla\cdot Y\mathrm{d}z=0,\\ Y(t_0,\omega)=v_{t_0}-Z_{t_0}.\end{cases}$$

If $v_{t_0}\in V_1$, then, for any $\mathcal{T}>t_0$ and $P$-a.e. $\omega\in\Omega$,

$$Y\in L^\infty(t_0,\mathcal{T};V_1)\cap L^2(t_0,\mathcal{T};(H^2(\Omega))^2),\tag{3.4}$$

see, e.g., [22].

Let $u(t)=v(t)-Z(t)-Y$. A stochastic process $U(t,\omega)=(v,T)$ is a strong solution to (2.9)~(2.15) on $[t_0,\mathcal{T}]$, if and only if $(u,T)$ is a strong solution to the following problem on $[t_0,\mathcal{T}]$:

$$\frac{\partial u}{\partial t}+[(u+Z+Y)\cdot\nabla](u+Z+Y)+\Phi(u+Z+Y)\frac{\partial(u+Z+Y)}{\partial z}+$$

$$fk\times(u+Z+Y)-\alpha Z+\nabla p_{b_2}-\int_{-1}^{z}\nabla T\mathrm{d}z'-\Delta u-\frac{\partial^2 u}{\partial z^2}=0,\tag{3.5}$$

$$\frac{\partial T}{\partial t}+[(u+Z+Y)\cdot\nabla]T+\Phi(u+Z+Y)\frac{\partial T}{\partial z}-\Delta T-\frac{\partial^2 T}{\partial z^2}=Q,\tag{3.6}$$

$$\int_{-1}^{0}\nabla\cdot u\mathrm{d}z=0,\tag{3.7}$$

$(u,T)$ satisfies the boundary value conditions (2.12)~(2.14), (3.8)

$$(u|_{t=t_0},T|_{t=t_0})=(0,T_{t_0}).\tag{3.9}$$

**Definition 3.7** If $Z$, $Y$ are defined above, $T_{t_0}\in V_2$, and let $\mathcal{T}$ be a fixed positive time. For $P$-a.e. $\omega\in\Omega$, $(u,T)$ is called a strong solution of the system (3.5)~(3.9) on the time interval $[t_0, \mathcal{T}]$ if it satisfies (3.5)~(3.7) in the weak sense such that

$$u \in L^\infty(t_0,\mathcal{T};V_1) \cap L^2(t_0,\mathcal{T};(H^2(\Omega))^2),$$
$$T \in L^\infty(t_0,\mathcal{T};V_2) \cap L^2(0,\mathcal{T};H^2(\Omega)),$$
$$\frac{\partial u}{\partial t} \in L^1(0,\mathcal{T};(L^2(\Omega))^2), \quad \frac{\partial T}{\partial t} \in L^1(0,\mathcal{T};L^2(\Omega)).$$

**Remark 3.8** (ⅰ) In order to consider the local well-posedness for IBVP, we need to study the initial boundary value problem (3.5)~(3.9). However, only by study of global well-posedness and long-time dynamics of the system (4.18)~(4.22) given in Subsec. 4.2, instead of (3.5)~(3.9), we can obtain the global well-posedness and long-time behavior of strong solutions to IBVP.

(ⅱ) For almost all given paths of the process $Z(t)$, we can study Eqs. (3.5)~(3.7) and (4.18)~(4.20) as deterministic evolution equations.

# 4 Existence and *a priori* Estimates of Local Strong Solutions to IBVP

## 4.1 Local Existence of Strong Solutions

By Remark 3.8, in order to prove the existence of local strong solutions of IBVP, we only need to prove that for (3.5)~(3.9). For convenience, we recall some interpolation inequalities used later[1,21].

(ⅰ) For $h_1 \in H^1(M)$,

$$\|h_1\|_{L^4} \leqslant c\|h_1\|_{L^2}^{\frac{1}{2}}\|h_1\|_{H^1}^{\frac{1}{2}}, \tag{4.1}$$

$$\|h_1\|_{L^5} \leqslant c\|h_1\|_{L^3}^{\frac{3}{5}}\|h_1\|_{H^1}^{\frac{2}{5}}, \tag{4.2}$$

$$\|h_1\|_{L^6} \leqslant c\|h_1\|_{L^4}^{\frac{2}{3}}\|h_1\|_{H^1}^{\frac{1}{3}}. \tag{4.3}$$

(ⅱ)

$$|h_2|_4 \leqslant c|h_2|_2^{\frac{1}{4}}\|h_2\|_1^{\frac{3}{4}}, \quad \text{for } h_2 \in H^1(\Omega). \tag{4.4}$$

**Proposition 4.1**[Existence of local strong solutions to (3.5)~(3.9)] If $Q \in H^1(\Omega)$ and $T_{t_0} \in V_2$, then, for $P$-a.e. $\omega \in \Omega$, there exists $\mathcal{T}^* > t_0$ such that $(u, T)$ is a strong solution of the system (3.5)~(3.9) on the interval $[t_0, \mathcal{T}^*)$.

**Lemma 4.2** If $v_1 \in V_1$, $v_2 \in V_1$ or $V_2$, $v_3 \in L^2(\Omega) \times L^2(\Omega)$ or $L^2(\Omega)$, then

(ⅰ)

$$\left|\int_\Omega v_3 \cdot (v_1 \cdot \nabla) v_2\right| \leqslant c(|v_1|_4^2 + |v_1|_4^8)|\nabla v_2|_2^2 + \varepsilon\left[|v_3|_2^2 + \int_\Omega (|\nabla v_{2z}|^2 + |\Delta v_2|^2)\right],$$

(ⅱ) $$\left|\int_\Omega \Phi(v_1) v_{2z} \cdot v_3\right| \leqslant c|\nabla v_1|_2^{\frac{1}{2}}(|\nabla v_1|_2^2 + |\Delta v_1|_2^2)^{\frac{1}{4}}|v_{2z}|_2^{\frac{1}{2}}|\nabla v_{2z}|_2^{\frac{1}{2}}|v_3|_2.$$

In this article, $\varepsilon$ is a small enough positive constant.

**Proof of Lemma 4.2** By the Hölder inequality, (4.4) and the Young inequality,

$$\int_\Omega |v_1||\nabla v_2||v_3| \leqslant c|v_1|_4^2\left(\int_\Omega |\nabla v_2|^4\right)^{\frac{1}{2}} + \varepsilon|v_3|_2^2$$

$$\leqslant c(|v_1|_4^2 + |v_1|_4^8)|\nabla v_2|_2^2 + \varepsilon\left[|v_3|_2^2 + \int_\Omega (|\nabla v_{2z}|^2 + |\Delta v_2|^2)\right].$$

By the Hölder inequality, the Minkowski inequality and (4.1), we obtain

$$\int_M \left(\int_{-1}^0 |\nabla v_1| dz \int_{-1}^0 |v_{2z}||v_3| dz\right)$$

$$\leqslant \int_M \left[\left(\int_{-1}^0 |\nabla v_1|^2 dz\right)^{\frac{1}{2}}\left(\int_{-1}^0 |v_{2z}|^2 dz\right)^{\frac{1}{2}}\left(\int_{-1}^0 |v_3|^2 dz\right)^{\frac{1}{2}}\right]$$

$$\leqslant c\left[\int_{-1}^0 (|\nabla v_1|_{L^2(M)}|\nabla v_1|_{H^1(M)})\int_{-1}^0 (|v_{2z}|_{L^2(M)}|v_{2z}|_{H^1(M)})\right]^{\frac{1}{2}}|v_3|_2. \qquad \square$$

**Proof of Proposition 4.1** In the sequel, $\omega \in \Omega$ is fixed. We shall prove Proposition 4.1 by the Faedo-Galerkin method. Since the procedure is similar to the proof of the existence of local strong solutions to the 3D Navier-Stokes system in [41], we only give *a priori* estimates of approximate solutions. Let $(u_m, T_m)$ be an approximate solution for the problem (3.5)~(3.9), where $(u_m, T_m) = \sum_{i=1}^m \alpha_{i,m}(t)\phi_i(x)$ and $\{\phi_m\}$ is an completely orthonormal basis of $V$. Let $e = Z + Y$, then $(u_m, T_m)$ satisfies

$$\int_\Omega h_m \cdot \frac{\partial u_m}{\partial t} + \int_\Omega h_m \cdot \left\{[(u_m + e)\cdot\nabla](u_m + e) + \Phi(u_m + e)\frac{\partial(u_m + e)}{\partial z}\right\} - \int_\Omega h_m \cdot \alpha Z +$$

$$\int_\Omega h_m \cdot [fk \times (u_m + e)] - \int_\Omega h_m \cdot \int_{-1}^z \nabla T_m dz' + \int_\Omega h_m \cdot A_1 u_m = 0, \tag{4.5}$$

$$\int_\Omega q_m \frac{\partial T_m}{\partial t} + \int_\Omega q_m \left\{[(u_m + e)\cdot\nabla]T_m + \Phi(u_m + e)\frac{\partial T_m}{\partial z}\right\} + \int_\Omega q_m A_2 T_m = \int_\Omega q_m Q, \tag{4.6}$$

$$u_m(t_0) = 0, \quad T_m(t_0) = T_{0m} \rightarrow T_{t_0} \quad \text{in } V_2, \tag{4.7}$$

where $h_m \in V_{1m}, q_m \in V_{2m}$, and the $m$-dimensional spaces $V_{1m} \times V_{2m} = \mathrm{span}\{\phi_1, \cdots, \phi_m\}$.

$L^2$ estimates about $T_m, u_m$. Letting $q_m = T_m$ in (4.6), by integration by parts and Lemma 3.1, we get

$$\frac{d|T_m|_2^2}{dt} + c\|T_m\|^2 \leqslant c|Q|_2^2, \tag{4.8}$$

which implies $T_m$ is uniformly in $m$ bounded in $L^\infty(t_0, \mathcal{T}; L^2(\Omega)) \cap L^2(t_0, \mathcal{T}; V_2)$ for any $\mathcal{T} > t_0$.

Similarly to Lemma 4.2, by (4.4) and (4.1),

$$-\int_\Omega u_m \cdot [(e\cdot\nabla)u_m + (u_m\cdot\nabla)e + (e\cdot\nabla)e] \leqslant \varepsilon\|u_m\|^2 + c(|e|_4^8 + \|e\|^4)|u_m|_2^2 + c\|e\|^2, \tag{4.9}$$

$$-\int_\Omega u_m \cdot \left[\Phi(u_m)\frac{\partial e}{\partial z} + \Phi(e)\frac{\partial u_m}{\partial z} + \Phi(e)\frac{\partial e}{\partial z}\right] \leqslant \varepsilon\|u_m\|^2 + c\|e\|^2\|e\|_2^2|u_m|_2^2 + c\|e\|^2. \tag{4.10}$$

Letting $h_m = u_m$ in (4.5), and noticing that

$$\int_\Omega \left[ (u_m \cdot \nabla) u_m + \Phi(u_m) \frac{\partial u_m}{\partial z} \right] \cdot u_m = 0,$$

by (4.9), (4.10), Lemma 3.1, the Hölder inequality and the Young inequality, we have

$$\frac{d |u_m|_2^2}{dt} + c\|u_m\|^2 \leqslant c(|e|_4^8 + \|e\|^4 + \|e\|^2\|e\|_2^2)|u_m|_2^2 + c(\|e\|^2 + |T_m|_2^2 + |Z|_2^2). \tag{4.11}$$

By Lemma 3.2, (3.4) and (4.8), we conclude that $u_m$ is uniformly in $m$ bounded in $L^\infty(t_0, \mathcal{T}; H_1) \cap L^2(t_0, \mathcal{T}; V_1)$ for any $\mathcal{T} > t_0$.

$H^1$ estimates about $u_m, T_m$. Similarly to Lemma 4.2, by (4.1), (4.4) and the Young inequality,

$$\left| \int_\Omega A_1 u_m \cdot [(u_m + e)\cdot \nabla](u_m + e) \right|$$

$$\leqslant c(|u_m|_2^2\|u_m\|^6 + |e|_2^2\|e\|^6 + \|e\|^{\frac{1}{2}}\|e\|_2^{\frac{3}{2}})\|u_m\|^2 + c|e|_2^{\frac{1}{2}}\|e\|^2\|e\|_2^{\frac{3}{2}} + \varepsilon |A_1 u_m|_2^2, \tag{4.12}$$

$$\left| \int_\Omega A_1 u_m \cdot \Phi(u_m + e) \frac{\partial(u_m + e)}{\partial z} \right|$$

$$\leqslant c\|e\|^2\|e\|_2^2 + c\|e\|^2\|e\|_2^2\|u_m\|^2 + c|A_1 u_m|_2^2\|u_m\| + \varepsilon |A_1 u_m|_2^2. \tag{4.13}$$

Letting $h_m = A_1 u_m$ in (4.5), by the Hölder inequality, the Young inequality, (4.12) and (4.13), we obtain

$$\frac{d(A_1 u_m, u_m)}{dt} + \frac{1}{2}|A_1 u_m|_2^2 \leqslant c|A_1 u_m|_2^2\|u_m\| + c\|e\|^2\|e\|_2^2 + c\|T_m\|^2 + c(|e|_2^2 + |Z|_2^2) +$$

$$c(|u_m|_2^2\|u_m\|^6 + \|e\|^2\|e\|_2^2 + |e|_2^2\|e\|^6 + \|e\|^{\frac{1}{2}}\|e\|_2^{\frac{3}{2}})\|u_m\|^2. \tag{4.14}$$

Similarly to (4.12) and (4.13),

$$\left| \int_\Omega A_2 T_m [(u_m + e)\cdot \nabla] T_m \right| \leqslant c(|u_m|_2^2\|u_m\|^6 + |e|_2^2\|e\|^6)\|T_m\|^2 + \varepsilon |A_2 T_m|_2^2, \tag{4.15}$$

$$\left| \int_\Omega A_2 T_m \Phi(u_m + e) \frac{\partial T_m}{\partial z} \right| \leqslant c(\|e\|^2\|e\|_2^2 + \|u_m\|^2\|u_m\|_2^2)\|T_m\|^2 + \varepsilon |A_2 T_m|_2^2. \tag{4.16}$$

Letting $q_m = A_2 T_m$ in (4.6), by the Hölder inequality, the Young inequality, (4.15) and (4.16), we have

$$\frac{d\|T_m\|^2}{dt} + \frac{1}{2}|A_2 T_m|_2^2$$

$$\leqslant c(|u_m|_2^2\|u_m\|^6 + |e|_2^2\|e\|^6)\|T_m\|^2 + c(\|e\|^2\|e\|_2^2 + \|u_m\|^2\|u_m\|_2^2)\|T_m\|^2 + |Q|_2^2. \tag{4.17}$$

Since $u_m(t_0) = 0$ and (4.14), by Lemma 3.1, there exists $\mathcal{T}_1 > t_0$ independent of $m$ such that $\| u_m(t) \|$ is small enough, for all $t \in [t_0, \mathcal{T}_1]$. From (4.14) and (4.17), using Lemma 3.1 and Lemma 3.2, by a standard argument used in proving the local well-posedness for the 3D incompressible Navier-Stokes equations, we can prove Proposition 4.1. □

### 4.2 *A Priori* Estimates of Local Strong Solutions to (2.9)~(2.15)

In this subsection, $\omega \in \Omega$ is fixed. By Proposition 4.1, we have proven the local existence of strong solutions to (2.9)~(2.15). Assume $(v, T)$ is a strong solution of the system (2.9)~(2.15) on the interval $[t_0, \mathcal{T}^*)$. In order to obtain the global existence of strong solutions to the system (2.9)~(2.15), we should make *a priori* estimates of local strong solutions $(v, T)$, i.e., we shall prove that if $\mathcal{T}_* < +\infty$, then $\lim\limits_{t \to \mathcal{T}_*^-} \sup(\| v \| + \| T \|) < +\infty$. In order to do that, we consider properties of the barotropic flow $\overline{w}$ and baroclinic flow $\widetilde{w}$ defined in the following, where $w = v - Z$ and $(w, T)$ satisfies

$$\frac{\partial w}{\partial t} + [(w + Z) \cdot \nabla](w + Z) + \Phi(w + Z) \frac{\partial (w + Z)}{\partial z} + fk \times (w + Z) - \alpha Z +$$

$$\nabla p_{b_3} - \int_{-1}^{z} \nabla T \mathrm{d}z' - \Delta w - \frac{\partial^2 w}{\partial z^2} = 0, \tag{4.18}$$

$$\frac{\partial T}{\partial t} + [(w + Z) \cdot \nabla] T + \Phi(w + Z) \frac{\partial T}{\partial z} - \Delta T - \frac{\partial^2 T}{\partial z^2} = Q, \tag{4.19}$$

$$\int_{-1}^{0} \nabla \cdot w \mathrm{d}z = 0, \tag{4.20}$$

$(w, T)$ satisfies the boundary value conditions (2.12)~(2.14), (4.21)

$$(w|_{t=t_0}, T|_{t=t_0}) = (v_{t_0} - Z_{t_0}, T_{t_0}). \tag{4.22}$$

By the definition of $(w, T)$, we can define strong solutions to (4.18)~(4.22) similar to that in Definition 3.7. So, $(w, T)$ is a strong solution of the system (4.18)~(4.22) on the interval $[t_0, \mathcal{T}^*)$.

We define the baroclinic flow $\widetilde{w}$ and find the equations satisfied by $\widetilde{w}$ and $\overline{w}$ as that in [13]. For any $w \in V_1$, denote the fluctuation of the horizontal velocity by $\widetilde{w} = w - \overline{w}$, where $\overline{w} = \int_{-1}^{0} w \mathrm{d}z$. We notice that

$$\overline{\widetilde{w}} = \int_{-1}^{0} \widetilde{w} \mathrm{d}z = 0, \quad \nabla \cdot \overline{w} = 0. \tag{4.23}$$

By integration by parts and (4.23),

$$\int_{-1}^{0} \Phi(w) \frac{\partial w}{\partial z} \mathrm{d}z = \int_{-1}^{0} w \nabla \cdot w \mathrm{d}z = \int_{-1}^{0} \widetilde{w} \nabla \cdot \widetilde{w} \mathrm{d}z, \tag{4.24}$$

$$\int_{-1}^{0} (w \cdot \nabla) w \mathrm{d}z = \int_{-1}^{0} (\widetilde{w} \cdot \nabla) \widetilde{w} \mathrm{d}z + (\overline{w} \cdot \nabla) \overline{w}. \tag{4.25}$$

By integrating the momentum equation (4.18) with respect to $z$ from $-1$ to 0 and using (4.21), (4.24) and (4.25), we get

$$\frac{\partial \overline{w}}{\partial t} - \Delta \overline{w} + \nabla p_{b_3} + \overline{(\widetilde{w} + \widetilde{Z}) \nabla \cdot (\widetilde{w} + \widetilde{Z}) + [(\widetilde{w} + \widetilde{Z}) \cdot \nabla](\widetilde{w} + \widetilde{Z})} +$$

$$[(\overline{w}+\overline{Z})\cdot\nabla](\overline{w}+\overline{Z})+fk\times(\overline{w}+\overline{Z})-\alpha\overline{Z}-\int_{-1}^{0}\int_{-1}^{z}\nabla T\mathrm{d}z'\mathrm{d}z=0\quad\text{in}\quad M,\tag{4.26}$$

$$\nabla\cdot\overline{w}=0\quad\text{in}\quad M,\tag{4.27}$$

$$\overline{w}\cdot\boldsymbol{n}=0,\quad\frac{\partial\overline{w}}{\partial\boldsymbol{n}}\times\boldsymbol{n}=0\quad\text{on}\quad\partial M.\tag{4.28}$$

Subtracting (4.26) from ( 4.18), we know that $\widetilde{w}$ satisfies

$$\frac{\partial\widetilde{w}}{\partial t}-\Delta\widetilde{w}-\frac{\partial^2\widetilde{w}}{\partial z^2}+[(\widetilde{w}+\widetilde{Z})\cdot\nabla](\widetilde{w}+\widetilde{Z})+\Phi(\widetilde{w}+\widetilde{Z})\frac{\partial(\widetilde{w}+\widetilde{Z})}{\partial z}+$$
$$\overline{[(\widetilde{w}+\widetilde{Z})\cdot\nabla](\overline{w}+\overline{Z})+[(\overline{w}+\overline{Z})\cdot\nabla](\widetilde{w}+\widetilde{Z})}+fk\times(\widetilde{w}+\widetilde{Z})-$$
$$\overline{(\widetilde{w}+\widetilde{Z})\nabla\cdot(\widetilde{w}+\widetilde{Z})+[(\widetilde{w}+\widetilde{Z})\cdot\nabla](\widetilde{w}+\widetilde{Z})}-\alpha\widetilde{Z}-\int_{-1}^{z}\nabla T\mathrm{d}z'+$$
$$\int_{-1}^{0}\int_{-1}^{z}\nabla T\mathrm{d}z'\mathrm{d}z=0\quad\text{in}\quad\Omega,\tag{4.29}$$

$$\frac{\partial\widetilde{w}}{\partial z}=0\quad\text{on}\quad\Gamma_u,\quad\frac{\partial\widetilde{w}}{\partial z}=0\quad\text{on}\quad\Gamma_b,\quad\widetilde{w}\cdot\boldsymbol{n}=0,\quad\frac{\partial\widetilde{w}}{\partial\boldsymbol{n}}\times\boldsymbol{n}=0\quad\text{on}\quad\Gamma_l.\tag{4.30}$$

$L^2$ estimates about $T,w$. Taking the inner product of Eq. (4.19) with $T$ in $L^2(\Omega)$, by integration by parts, $T(t,x,y,z)=-\int_z^0\frac{\partial T}{\partial z'}\mathrm{d}z'+T|_{z=0}$, using the Hölder inequality and the Cauchy-Schwarz inequality, we obtain

$$\frac{\mathrm{d}|T|_2^2}{\mathrm{d}t}+\int_\Omega|\nabla T|^2+\int_\Omega\left|\frac{\partial T}{\partial z}\right|^2+\alpha_u|T|_{z=0}|_2^2\leqslant c|Q|_2^2.\tag{4.31}$$

By (4.31) and the Gronwall inequality,

$$|T(t)|_2^2\leqslant \mathrm{e}^{-ct}\left|T_{t_0}\right|_2^2+c|Q|_2^2,\tag{4.32}$$

where $t\geqslant t_0$. From (4.31) and (4.32), for $\mathcal{T}>t_0$ given, there exists a positive constant $C_1(\mathcal{T},\|U_0\|,\|Q\|_1)$ such that

$$\int_{t_0}^{\mathcal{T}}\left[\int_\Omega\left(|\nabla T|^2+\left|\frac{\partial T}{\partial z}\right|^2+|T|^2\right)+|T|_{z=0}|_2^2\right]+|T(t)|_2^2\leqslant C_1,\tag{4.33}$$

where $t\in[t_0,\mathcal{T})$and $\int_{t_0}^{\mathcal{T}}\cdot\,\mathrm{d}s$ is denoted by$\int_{t_0}^{\mathcal{T}}$. Throughout this paper, $C_i(\cdot,\cdot)$ denote the constants dependent only on the quantities appearing in parentheses, $i\in\mathbf{N}$.

Choosing $w$ as a test function in (4.18), similarly to (4.11), we obtain

$$\frac{\mathrm{d}|w|_2^2}{\mathrm{d}t}+\int_\Omega|\nabla w|^2+\int_\Omega\left|\frac{\partial w}{\partial z}\right|^2$$
$$\leqslant c(|Z|_4^8+\|Z\|^4+\|Z\|^2\|Z\|_2^2)|w|_2^2+c(\|Z\|^2+|T|_2^2+|Z|_2^2).\tag{4.34}$$

By $\lambda_1|w|_2^2\leqslant\|w\|^2$ and the Gronwall inequality, for $t\geqslant t_0$, we derive from (4.34)

$$|w(t)|_2^2\leqslant \mathrm{e}^{\int_{t_0}^t[-\lambda_1+c(|Z|_4^8+\|Z\|^4+\|Z\|^2\|Z\|_2^2)]\mathrm{d}\tau}|w(t_0)|_2^2+$$
$$c\int_{t_0}^t\mathrm{e}^{\int_\sigma^t[-\lambda_1+c(|Z|_4^8+\|Z\|^4+\|Z\|^2\|Z\|_2^2)]\mathrm{d}\tau}(\|Z\|^2+|T|_2^2+|Z|_2^2)\mathrm{d}\sigma.\tag{4.35}$$

By Lemma 3.2 and (4.35), for $\mathcal{T}>t_0$ given, there exists a positive constant $C_2(\mathcal{T},\|U_0\|,\|Q\|_1,\|Z_{t_0}\|_2)$ such that

$$\int_{t_0}^{\mathcal{T}}\int_\Omega\left(|\nabla w|^2+\left|\frac{\partial w}{\partial z}\right|^2\right)+|w(t)|_2^2\leqslant C_2,\quad \text{for any } t\in[t_0,\mathcal{T}). \tag{4.36}$$

By the Minkowski inequality and the Hölder inequality, we derive from (4.36),

$$\int_{t_0}^{\mathcal{T}}\int_M(|\nabla\overline{w}|^2+|\overline{w}|^2)+\|\overline{w}(t)\|_{L^2}^2\leqslant C_2,\quad \forall\, t\in[t_0,\mathcal{T}). \tag{4.37}$$

$L^4$ estimates about $T$. We take the inner product of Eq. (4.19) with $|T|^2T$ in $L^2(\Omega)$ and obtain

$$\begin{aligned}&\frac{1}{4}\frac{\mathrm{d}|T|_4^4}{\mathrm{d}t}+3\int_\Omega|\nabla T|^2|T|^2+3\int_\Omega\left|\frac{\partial T}{\partial z}\right|^2|T|^2+\alpha_u\int_M|T|_{z=0}|^4\\&=\int_\Omega Q|T|^2T-\int_\Omega\left\{[(w+Z)\cdot\nabla]T+\Phi(w+Z)\frac{\partial T}{\partial z}\right\}|T|^2T.\end{aligned} \tag{4.38}$$

Since $T^4(t,x,y,z)=-\int_z^0\frac{\partial T^4}{\partial z'}\mathrm{d}z'+T^4|_{z=0}$, by the Hölder inequality and the Cauchy-Schwarz inequality, we get

$$|T|_4^4\leqslant c\left(\int_\Omega|T|^2\left|\frac{\partial T}{\partial z}\right|^2\right)+\frac{1}{2}\int_\Omega T^4+|T|_{z=0}|_4^4. \tag{4.39}$$

By integration by parts, Hölder inequality and Young inequality, we derive from (4.38) and (4.39),

$$\frac{\mathrm{d}|T|_4^4}{\mathrm{d}t}+3\int_\Omega|\nabla T|^2|T|^2+3\int_\Omega\left|\frac{\partial T}{\partial z}\right|^2|T|^2+\alpha_u\int_M|T|_{z=0}|^4\leqslant c|Q|_4^4. \tag{4.40}$$

By the Gronwall inequality, (4.39) and (4.40), we obtain

$$|T(t)|_4^4\leqslant \mathrm{e}^{-ct}\left|T_{t_0}\right|_4^4+c|Q|_4^4\leqslant C_3, \tag{4.41}$$

where $t\geqslant t_0$, $C_3$ is a positive constant.

**Remark 4.3** In the following, the result of $L^3$ estimates about $\widetilde{w}$ will be used in studying the long-time behavior of strong solutions to the stochastic primitive equations. Since $L^3$ estimates about $\widetilde{w}$ are similar to $L^4$ estimates about $\widetilde{w}$, we make $L^3$ estimates about $\widetilde{w}$ here.

$L^3$ estimates about $\widetilde{w}$. By integration by parts,

$$-\int_\Omega\left[(\widetilde{w}\cdot\nabla)\widetilde{w}-\left(\int_{-1}^z\nabla\cdot\widetilde{w}\mathrm{d}z'\right)\frac{\partial\widetilde{w}}{\partial z}\right]\cdot|\widetilde{w}|\widetilde{w}=0,$$

$$-\int_\Omega\{[(\overline{w}+\overline{Z})\cdot\nabla]\widetilde{w}\}\cdot|\widetilde{w}|\widetilde{w}=\frac{1}{3}\int_\Omega|\widetilde{w}|^3\nabla\cdot(\overline{w}+\overline{Z})=0,$$

$$\begin{aligned}&-\int_\Omega\{[(\widetilde{w}+\widetilde{Z})\cdot\nabla](\overline{w}+\overline{Z})\}\cdot|\widetilde{w}|\widetilde{w}\\&=\int_\Omega(\overline{w}+\overline{Z})\cdot[(\widetilde{w}+\widetilde{Z})\cdot\nabla]|\widetilde{w}|\widetilde{w}+\int_\Omega|\widetilde{w}|\widetilde{w}\cdot(\overline{w}+\overline{Z})\nabla\cdot(\widetilde{w}+\widetilde{Z}),\end{aligned}$$

$$\int_\Omega\overline{(\widetilde{w}+\widetilde{Z})\nabla\cdot(\widetilde{w}+\widetilde{Z})+[(\widetilde{w}+\widetilde{Z})\cdot\nabla](\widetilde{w}+\widetilde{Z})}\cdot|\widetilde{w}|\widetilde{w}$$

$$= \int_\Omega \int_{-1}^{0} (\widetilde{w} + \widetilde{Z})^{(1)} (\widetilde{w} + \widetilde{Z}) \mathrm{d}z \cdot \partial_x (|\widetilde{w}|\widetilde{w}) + \int_\Omega \int_{-1}^{0} (\widetilde{w} + \widetilde{Z})^{(2)} (\widetilde{w} + \widetilde{Z}) \mathrm{d}z \cdot \partial_y (|\widetilde{w}|\widetilde{w}), \tag{4.42}$$

where $\widetilde{w} + \widetilde{Z} = ((\widetilde{w} + \widetilde{Z})^{(1)}, (\widetilde{w} + \widetilde{Z})^{(2)})$. Taking the inner product of Eq. (4.29) with $|\widetilde{w}|\widetilde{w}$ in $L^2(\Omega) \times L^2(\Omega)$, by (4.42), we get

$$\begin{aligned}
&\frac{1}{3}\frac{\mathrm{d}|\widetilde{w}|_3^3}{\mathrm{d}t} + \int_\Omega \left(|\nabla \widetilde{w}|^2 |\widetilde{w}| + \frac{4}{9}|\nabla |\widetilde{w}|^{\frac{3}{2}}|^2\right) + \int_\Omega \left(|\partial_z \widetilde{w}|^2 |\widetilde{w}| + \frac{4}{9}|\partial_z |\widetilde{w}|^{\frac{3}{2}}|^2\right) \\
&= \int_\Omega (\overline{w} + \overline{Z}) \cdot [(\widetilde{w} + \widetilde{Z}) \cdot \nabla] |\widetilde{w}|\widetilde{w} + \int_\Omega |\widetilde{w}|\widetilde{w} \cdot (\overline{w} + \overline{Z}) \nabla \cdot (\widetilde{w} + \widetilde{Z}) - \\
&\int_\Omega [(\widetilde{Z} \cdot \nabla)\widetilde{w} + (\widetilde{w} \cdot \nabla)\widetilde{Z} + (\widetilde{Z} \cdot \nabla)\widetilde{Z}] \cdot |\widetilde{w}|\widetilde{w} - \\
&\int_\Omega \left[\Phi(\widetilde{Z}) \frac{\partial \widetilde{w}}{\partial z} + \Phi(\widetilde{w}) \frac{\partial \widetilde{Z}}{\partial z} + \Phi(\widetilde{Z}) \frac{\partial \widetilde{Z}}{\partial z}\right] \cdot |\widetilde{w}|\widetilde{w} - \\
&\int_\Omega \{[(\overline{w} + \overline{Z}) \cdot \nabla]\widetilde{Z}\} \cdot |\widetilde{w}|\widetilde{w} + \int_\Omega \int_{-1}^{0} (\widetilde{w} + \widetilde{Z})^{(1)} (\widetilde{w} + \widetilde{Z}) \mathrm{d}z \cdot \partial_x (|\widetilde{w}|\widetilde{w}) + \\
&\int_\Omega \int_{-1}^{0} (\widetilde{w} + \widetilde{Z})^{(2)} (\widetilde{w} + \widetilde{Z}) \mathrm{d}z \cdot \partial_y (|\widetilde{w}|\widetilde{w}) - \int_\Omega (fk \times \widetilde{Z}) \cdot |\widetilde{w}|\widetilde{w} - \\
&\int_\Omega \left(\int_{-1}^{z} T \mathrm{d}z' - \int_{-1}^{0}\int_{-1}^{z} T \mathrm{d}z' \mathrm{d}z\right) \nabla \cdot |\widetilde{w}|\widetilde{w} + \int_\Omega \alpha \widetilde{Z} \cdot |\widetilde{w}|\widetilde{w}.
\end{aligned} \tag{4.43}$$

By Hölder inequality, we derive from (4.43),

$$\begin{aligned}
&\frac{1}{3}\frac{\mathrm{d}|\widetilde{w}|_3^3}{\mathrm{d}t} + \int_\Omega \left(|\nabla \widetilde{w}|^2 |\widetilde{w}| + \frac{4}{9}|\nabla |\widetilde{w}|^{\frac{3}{2}}|^2\right) + \int_\Omega \left(|\partial_z \widetilde{w}|^2 |\widetilde{w}| + \frac{4}{9}|\partial_z |\widetilde{w}|^{\frac{3}{2}}|^2\right) \\
&\leqslant c(\|\overline{w}\|_{L^4} + \|\overline{Z}\|_{L^4}) \left(\int_\Omega |\nabla \widetilde{w}|^2 |\widetilde{w}|\right)^{\frac{1}{2}} \left\{ \left[\int_M \left(\int_{-1}^{0} |\widetilde{w}|^3 \mathrm{d}z\right)^2\right]^{\frac{1}{4}} + \right. \\
&\left. \left[\int_M \left(\int_{-1}^{0} |\widetilde{w}|^2 \mathrm{d}z\right)^2\right]^{\frac{1}{8}} \left[\int_M \left(\int_{-1}^{0} |\widetilde{Z}|^4 \mathrm{d}z\right)^2\right]^{\frac{1}{8}} \right\} + \\
&c\left(\int_\Omega |\nabla \widetilde{w}|^2 |\widetilde{w}|\right)^{\frac{1}{2}} |\widetilde{w}|_2^{\frac{1}{2}} \left\{ \left[\int_M \left(\int_{-1}^{0} |\widetilde{Z}|^2 \mathrm{d}z\right)^4\right]^{\frac{1}{4}} + \|\overline{|T|}\|_{L^4} \right\} + \\
&c\left(\int_\Omega |\nabla \widetilde{w}|^2 |\widetilde{w}|\right)^{\frac{1}{2}} \left[\int_M \left(\int_{-1}^{0} |\widetilde{w}|^2 \mathrm{d}z\right)^{\frac{5}{2}}\right]^{\frac{1}{2}} + I_1 + I_2 + I_3.
\end{aligned} \tag{4.44}$$

$$I_1 = \int_\Omega \left\{ -[(\overline{w} + \overline{Z}) \cdot \nabla]\widetilde{Z} + (\overline{w} + \overline{Z}) \nabla \cdot \widetilde{Z} - [(\widetilde{w} + \widetilde{Z}) \cdot \nabla]\widetilde{Z} - \Phi(\widetilde{Z}) \frac{\partial \widetilde{Z}}{\partial z} \right\} \cdot |\widetilde{w}|\widetilde{w},$$

$$I_2 = -\int_\Omega \left[(\widetilde{Z} \cdot \nabla)\widetilde{w} + \Phi(\widetilde{Z}) \frac{\partial \widetilde{w}}{\partial z} + \Phi(\widetilde{w}) \frac{\partial \widetilde{Z}}{\partial z}\right] \cdot |\widetilde{w}|\widetilde{w},$$

$$I_3 = -\int_\Omega (fk \times \widetilde{Z} - \alpha \widetilde{Z}) \cdot |\widetilde{w}|\widetilde{w}.$$

By Minkowski inequality, (4.1) and Hölder inequality,

$$\left[\int_M \left(\int_{-1}^{0} |\widetilde{w}|^3 \mathrm{d}z\right)^2\right]^{\frac{1}{2}} \leqslant c|\widetilde{w}|_3^{\frac{3}{2}} \left[\int_{-1}^{0} \left(\|\nabla |\widetilde{w}|^{\frac{3}{2}}\|_{L^2}^2 + \||\widetilde{w}|^{\frac{3}{2}}\|_{L^2}^2\right) \mathrm{d}z\right]^{\frac{1}{2}}. \tag{4.45}$$

By Minkowski inequality, Hölder inequality and (4.2),

$$\int_M\left(\int_{-1}^0|\widetilde{w}|^2\mathrm{d}z\right)^{\frac{5}{2}}\leqslant\left(\int_{-1}^0\|\widetilde{w}\|_{L^3}^{\frac{6}{5}}\|\widetilde{w}\|_{H^1}^{\frac{4}{5}}\mathrm{d}z\right)^{\frac{5}{2}}\leqslant c\|\widetilde{w}\|^2|\widetilde{w}|_3^3. \qquad (4.46)$$

By Hölder inequality, Minkowski inequality, (4.1), $|\widetilde{w}|_4\leqslant|\widetilde{w}|_3^{\frac{1}{2}}\|\widetilde{w}\|^{\frac{1}{2}}$, $|\widetilde{w}|_6^6=\int_\Omega|\widetilde{w}|^{\frac{3}{2}\cdot 4}\leqslant|\widetilde{w}|_3^{\frac{3}{2}}\||\widetilde{w}|^{\frac{3}{2}}\|^3$ and Young inequality, we have

$$I_1\leqslant c(\|\overline{w}\|_{L^4}+\|\overline{Z}\|_{L^4})\left[\int_{-1}^0\left(\int_M|\nabla\widetilde{Z}|^4\right)^{\frac{1}{2}}\mathrm{d}z\right]^{\frac{1}{2}}|\widetilde{w}|_4^2+c|\nabla\widetilde{Z}|_2|\widetilde{w}|_6^3+$$

$$c|\widetilde{Z}|_6|\nabla\widetilde{Z}|_2|\widetilde{w}|_6^2+c\left[\int_{-1}^0\left(\int_M|\nabla\widetilde{Z}|^4\right)^{\frac{1}{2}}\mathrm{d}z\right]^{\frac{1}{2}}\left[\int_{-1}^0\left(\int_M\left|\frac{\partial\widetilde{Z}}{\partial z}\right|^4\right)^{\frac{1}{2}}\mathrm{d}z\right]^{\frac{1}{2}}|\widetilde{w}|_4^2$$

$$\leqslant\varepsilon\int_\Omega(|\nabla|\widetilde{w}|^{\frac{3}{2}}|^2+|\partial_z|\widetilde{w}|^{\frac{3}{2}}|^2)+c(1+|w|_2^{\frac{3}{2}}\|w\|^{\frac{3}{2}}\|Z\|^{\frac{3}{2}}\|Z\|_2^{\frac{3}{2}}+$$

$$|Z|_2^{\frac{3}{2}}\|Z\|^3\|Z\|_2^{\frac{3}{2}}+\|Z\|^4+|Z|_6^6+\|w\|^2)|\widetilde{w}|_3^3+c\|w\|^{\frac{3}{2}}+$$

$$c\|Z\|^3+c\|Z\|^2\|Z\|_2^2+c\|w\|^2. \qquad (4.47)$$

By integration by parts, the Hölder inequality and (4.45),

$$I_2\leqslant\int_\Omega\left[\left(\int_{-1}^z|\widetilde{w}|\right)\left|\frac{\partial\widetilde{Z}}{\partial z}\right||\nabla\widetilde{w}||\widetilde{w}|+|\nabla\widetilde{Z}||\widetilde{w}|^3+\left(\int_{-1}^z|\widetilde{w}|\right)|\nabla\widetilde{Z}|\left|\frac{\partial\widetilde{w}}{\partial z}\right||\widetilde{w}|\right]+$$

$$\int_\Omega\left[|\widetilde{Z}||\nabla\widetilde{w}||\widetilde{w}|^2+\left(\int_{-1}^z|\nabla\widetilde{Z}|\right)\left|\frac{\partial\widetilde{w}}{\partial z}\right||\widetilde{w}|^2\right]$$

$$\leqslant\varepsilon\int_\Omega(|\nabla\widetilde{w}|^2|\widetilde{w}|+|\partial_z\widetilde{w}|^2|\widetilde{w}|+|\nabla|\widetilde{w}|^{\frac{3}{2}}|^2+|\partial_z|\widetilde{w}|^{\frac{3}{2}}|^2)+$$

$$(1+|Z|_4^8+\|Z\|^4+\|Z\|\|Z\|_2^3+\|Z\|^2\|Z\|_2^2)|\widetilde{w}|_3^3. \qquad (4.48)$$

Since $I_3\leqslant c|Z|_3^3+c|\widetilde{w}|_3^3$, by (4.1) and the Young inequality, we derive from (4.44)~(4.48),

$$\frac{\mathrm{d}|\widetilde{w}|_3^3}{\mathrm{d}t}+\int_\Omega\left(|\nabla\widetilde{w}|^2|\widetilde{w}|+\frac{4}{9}|\nabla|\widetilde{w}|^{\frac{3}{2}}|^2\right)+\int_\Omega\left(|\partial_z\widetilde{w}|^2|\widetilde{w}|+\frac{4}{9}|\partial_z|\widetilde{w}|^{\frac{3}{2}}|^2\right)$$

$$\leqslant c(1+\|\overline{w}\|_{L^2}^2\|\overline{w}\|_{H^1}^2+\|\widetilde{w}\|^2+|w|_2^{\frac{3}{2}}\|w\|^{\frac{3}{2}}\|Z\|^{\frac{3}{2}}\|Z\|_2^{\frac{3}{2}}+|Z|_2^{\frac{3}{2}}\|Z\|^3\|Z\|_2^{\frac{3}{2}}+$$

$$\|Z\|^2+|Z|_4^8+\|Z\|^4+\|Z\|\|Z\|_2^3+\|Z\|^2\|Z\|_2^2+|Z|_6^6)|\widetilde{w}|_3^3+c\|w\|^{\frac{3}{2}}+$$

$$c|w|_2^2\|w\|^2+c\|w\|^2+c\|Z\|^3+c\|Z\|^8+c|T|_4^4+c\|Z\|^2\|Z\|_2^2. \qquad (4.49)$$

By the Gronwall inequality, Lemma 3.2, (4.36), (4.37) and (4.41), for $\mathcal{T}>t_0$ given, there exists a positive constant $C_4(\mathcal{T},\|U_0\|,\|Q\|_1,\|Z_{t_0}\|_2)$ such that

$$|\widetilde{w}(t)|_3^3\leqslant C_4,\quad\text{for any } t\in[t_0,\tau). \qquad (4.50)$$

**Remark 4.4** In estimating $\int_\Omega\Phi(\widetilde{w})\frac{\partial\widetilde{Z}}{\partial z}\cdot|\widetilde{w}|\widetilde{w}$ in $I_2$, we can not diminish the exponent 3 in the coefficient $\|Z\|\|Z\|_2^3$ of $|\widetilde{w}|_3^3$. We guess that it is impossible to prove the global well-posedness of IBVP when $Z\in L^\infty(\mathbf{R};V_1)\cap L^2(\mathbf{R};(H^2(\Omega))^2)$.

$L^4$ estimates about $\widetilde{w}$. Taking the inner product of Eq. (4.29) with $|\widetilde{w}|^2\widetilde{w}$ in $L^2(\Omega)\times L^2(\Omega)$, similarly to (4.44), by integration by parts and the Hölder inequality, we obtain

$$\frac{1}{4}\frac{d|\widetilde{w}|_4^4}{dt}+\int_\Omega\left(|\nabla\widetilde{w}|^2|\widetilde{w}|^2+\frac{1}{2}|\nabla|\widetilde{w}|^2|^2\right)+\int_\Omega\left(|\partial_z\widetilde{w}|^2|\widetilde{w}|^2+\frac{1}{2}|\partial_z|\widetilde{w}|^2|^2\right)$$

$$\leqslant c(\|\overline{w}\|_{L^4}+\|\overline{Z}\|_{L^4})\left(\int_\Omega|\nabla\widetilde{w}|^2|\widetilde{w}|^2\right)^{\frac{1}{2}}\left\{\left[\int_M\left(\int_{-1}^0|\widetilde{w}|^4dz\right)^2\right]^{\frac{1}{4}}+\right.$$

$$\left.\left[\int_M\left(\int_{-1}^0|\widetilde{w}|^4dz\right)^2\right]^{\frac{1}{8}}\left[\int_M\left(\int_{-1}^0|\widetilde{Z}|^4dz\right)^2\right]^{\frac{1}{8}}\right\}+$$

$$c\left(\int_\Omega|\nabla\widetilde{w}|^2|\widetilde{w}|^2\right)^{\frac{1}{2}}\left[\int_M\left(\int_{-1}^0|\widetilde{w}|^2dz\right)^2\right]^{\frac{1}{4}}\left\{\left[\int_M\left(\int_{-1}^0|\widetilde{Z}|^2dz\right)^4\right]^{\frac{1}{4}}+c\|\overline{|T|}\|_{L^4}\right\}+$$

$$c\left(\int_\Omega|\nabla\widetilde{w}|^2|\widetilde{w}|^2\right)^{\frac{1}{2}}\left[\int_M\left(\int_{-1}^0|\widetilde{w}|^2dz\right)^3\right]^{\frac{1}{2}}+J_1+J_2+J_3, \tag{4.51}$$

$$J_1=-\int_\Omega\left\{[(\overline{w}+\overline{Z})\cdot\nabla]\widetilde{Z}-(\overline{w}+\overline{Z})\nabla\cdot\widetilde{Z}+[(\widetilde{w}+\widetilde{Z})\cdot\nabla]\widetilde{Z}+\Phi(\widetilde{Z})\frac{\partial\widetilde{Z}}{\partial z}\right\}\cdot|\widetilde{w}|^2\widetilde{w},$$

$$J_2=-\int_\Omega\left[(\widetilde{Z}\cdot\nabla)\widetilde{w}+\Phi(\widetilde{Z})\frac{\partial\widetilde{w}}{\partial z}+\Phi(\widetilde{w})\frac{\partial\widetilde{Z}}{\partial z}\right]\cdot|\widetilde{w}|^2\widetilde{w},$$

$$J_3=-\int_\Omega(fk\times\widetilde{Z}-\alpha\widetilde{Z})\cdot|\widetilde{w}|^2\widetilde{w}.$$

By the Minkowski inequality, (4.1) and the Hölder inequality,

$$\left(\int_M\left(\int_{-1}^0|\widetilde{w}|^4dz\right)^2\right]^{\frac{1}{2}}\leqslant c|\widetilde{w}|_4^2\left[\int_{-1}^0(\|\nabla|\widetilde{w}|^2\|_{L^2(M)}^2+\||\widetilde{w}|^2\|_{L^2(M)}^2)dz\right]^{\frac{1}{2}}. \tag{4.52}$$

By the Minkowski inequality, (4.3) and the Hölder inequality,

$$\int_M\left(\int_{-1}^0|\widetilde{w}|^2dz\right)^3\leqslant\left[\int_{-1}^0\left(\int_M|\widetilde{w}|^6\right)^{\frac{1}{3}}dz\right]^3\leqslant c\|\widetilde{w}\|^2|\widetilde{w}|_4^4. \tag{4.53}$$

By the Hölder inequality, the Minkowski inequality, (4.1), (4.4), $|\widetilde{w}|_6^6\leqslant|\widetilde{w}|_4^3\||\widetilde{w}|^2\|^{\frac{3}{2}}$, $|\widetilde{w}|_8^8=|\widetilde{w}|_4^2\||\widetilde{w}|^2\|^3$ and the Young inequality, we get

$$J_1\leqslant\varepsilon\int_\Omega(|\nabla|\widetilde{w}|^2|^2+|\partial_z|\widetilde{w}|^2|^2)+c|w|_2^2\|w\|^2+c|Z|_2^2\|Z\|^2+\|Z\|^{\frac{8}{5}}\|Z\|_2^{\frac{8}{5}}+$$

$$c|Z|_4^4+c(1+\|Z\|^{\frac{4}{3}}\|Z\|_2^{\frac{4}{3}}+\|Z\|^4+\|Z\|^{\frac{2}{3}}\|Z\|_2^{\frac{2}{3}}+\|Z\|^{\frac{8}{5}}\|Z\|_2^{\frac{8}{5}})|\widetilde{w}|_4^4. \tag{4.54}$$

Similarly to (4.48), by (4.52),

$$J_2\leqslant\varepsilon\int_\Omega(|\nabla\widetilde{w}|^2|\widetilde{w}|^2+|\partial_z\widetilde{w}|^2|\widetilde{w}|^2+|\nabla|\widetilde{w}|^2|^2+|\partial_z|\widetilde{w}|^2|^2)+$$

$$(1+|Z|_4^8+\|Z\|^4+\|Z\|\|Z\|_2^3+\|Z\|^2\|Z\|_2^2)|\widetilde{w}|_4^4. \tag{4.55}$$

By (4.1), the Hölder inequality, the Minkowski inequality and the Young inequality, we derive from (4.51)~(4.55)

$$\frac{d|\widetilde{w}|_4^4}{dt}+\int_\Omega\left(|\nabla\widetilde{w}|^2|\widetilde{w}|^2+\frac{1}{2}|\nabla|\widetilde{w}|^2|^2\right)+\int_\Omega\left(|\partial_z\widetilde{w}|^2|\widetilde{w}|^2+\frac{1}{2}|\partial_z|\widetilde{w}|^2|^2\right)$$

$$\leqslant c(1+|w|_2^2\|w\|^2+|Z|_2^2\|Z\|^2+|Z|_8^8+\|w\|^2+\|Z\|^4+\|Z\|^{\frac{4}{3}}\|Z\|_2^{\frac{4}{3}}+$$

$$\|Z\|^{\frac{2}{3}}\|Z\|_2^{\frac{2}{2}}+\|Z\|^{\frac{8}{5}}\|Z\|_2^{\frac{8}{5}}+\|Z\|\|Z\|_2^{\frac{3}{2}}+\|Z\|^2\|Z\|_2^{\frac{2}{2}})|\tilde{w}|_4^4+$$

$$c|w|_2^2\|w\|^2+c|Z|_2^2\|Z\|^2+c|Z|_4^4+c|T|_4^4+\|Z\|^{\frac{8}{5}}\|Z\|_2^{\frac{8}{5}}. \qquad (4.56)$$

By the Gronwall inequality, Lemma 3.2, (4.36), (4.37) and (4.41), for $\mathcal{T}>t_0$ given, there exists a positive constant $C_5(\mathcal{T},\|U_0\|,\|Q\|_1,\|Z_{t_0}\|_2)$ such that

$$\int_{t_0}^{\mathcal{T}}\int_\Omega\left[\left(|\nabla\tilde{w}|^2|\tilde{w}|^2+\frac{1}{2}|\nabla|\tilde{w}|^2|^2\right)+\left(|\partial_z\tilde{w}|^2|\tilde{w}|^2+\frac{1}{2}|\partial_z|\tilde{w}|^2|^2\right)\right]+|\tilde{w}(t)|_4^4\leqslant C_5, \qquad (4.57)$$

where $t_0\leqslant t<\mathcal{T}$.

$L^2$ estimates about $\nabla\overline{w}$. By Lemma 4.2,

$$\left|\int_M[(\overline{w}+\overline{Z})\cdot\nabla](\overline{w}+\overline{Z})\cdot\Delta\overline{w}\right|\leqslant(\|\overline{w}\|_{L^2}\|\overline{w}\|_{H^1}+\|\overline{Z}\|_{L^4}^2)\|\nabla\overline{Z}\|_{L^2}\|\nabla\overline{Z}\|_{H^1}+$$

$$c(\|\overline{w}\|_{H^1}^2+\|\overline{w}\|_{L^2}^2\|\overline{w}\|_{H^1}^2+\|\overline{Z}\|_{H^1}^2+\|\overline{Z}\|_{H^1}^4)\|\nabla\overline{w}\|_{L^2}^2+\varepsilon\|\Delta\overline{w}\|_{L^2}^2. \qquad (4.58)$$

By the Hölder inequality and the Minkowski inequality,

$$\left|\int_M\int_{-1}^0[\tilde{w}\nabla\cdot\tilde{w}+(\tilde{w}\cdot\nabla)\tilde{w}]\mathrm{d}z\cdot\Delta\overline{w}\right|\leqslant c\int_\Omega|\nabla\tilde{w}|^2|\tilde{w}|^2+\varepsilon\|\Delta\overline{w}\|_{L^2}^2. \qquad (4.59)$$

By the Hölder inequality, $|Z|_\infty\leqslant c\|Z\|_2$, the Minkowski inequality and (4.1),

$$\left|\int_M\int_{-1}^0[\tilde{Z}\nabla\cdot\tilde{w}+(\tilde{Z}\cdot\nabla)\tilde{w}+\tilde{w}\nabla\cdot\tilde{Z}+(\tilde{w}\cdot\nabla)\tilde{Z}+\tilde{Z}\nabla\cdot\tilde{Z}+(\tilde{Z}\cdot\nabla)\tilde{Z}]\mathrm{d}z\cdot\Delta\overline{w}\right|$$

$$\leqslant c\|Z\|_2^2\|w\|^2+\|Z\|\|Z\|_2|w|_2\|w\|+|Z|_2\|Z\|^2\|Z\|_2+\varepsilon\|\Delta\overline{w}\|_{L^2}^2. \qquad (4.60)$$

By integration by parts, (4.27) and (4.28), we have (for more detail[13])

$$\int_M\nabla p_{b_3}\cdot\Delta\overline{w}=0,\quad -\int_M\int_{-1}^0\int_{-1}^z\nabla T\mathrm{d}z'\mathrm{d}z\cdot\Delta\overline{w}=0. \qquad (4.61)$$

Taking the inner product of Eq. (4.26) with $-\Delta\overline{w}$ in $L^2(M)\times L^2(M)$, by $(fk\times\overline{w})\cdot\Delta\overline{w}=0$, the Hölder inequality, the Young inequality, (4.58)~(4.61), and choosing $\varepsilon$ small enough, we obtain

$$\frac{\mathrm{d}\|\nabla\overline{w}\|_{L^2}^2}{\mathrm{d}t}+\|\Delta\overline{w}\|_{L^2}^2$$

$$\leqslant c(\|\overline{w}\|_{H^1}^2+\|\overline{w}\|_{L^2}^2\|\overline{w}\|_{H^1}^2+\|\overline{Z}\|_{H^1}^2+\|\overline{Z}\|_{H^1}^4)\|\nabla\overline{w}\|_{L^2}^2+c\int_\Omega|\nabla\tilde{w}|^2|\tilde{w}|^2+$$

$$c\|Z\|_2^2\|w\|^2+\|Z\|\|Z\|_2|w|_2\|w\|+|Z|_2\|Z\|^2\|Z\|_2^2+c|Z|_2^2. \qquad (4.62)$$

By the Gronwall inequality, Lemma 3.2, (4.36), (4.37) and (4.57), for $\mathcal{T}\geqslant t_0$ given, there exists a positive constant $C_6(\mathcal{T},\|U_0\|,\|Q\|_1,\|Z_{t_0}\|_2)$ such that

$$\|\nabla\overline{w}(t)\|_{L^2}^2\leqslant C_6,\quad \text{for any } t\in[t_0,\mathcal{T}). \qquad (4.63)$$

$L^2$ estimates about $\partial_z w$. Taking the derivative, with respect to $z$, of Eq. (4.18), we get

$$\frac{\partial w_z}{\partial t}-\Delta w_z-\frac{\partial^2 w_z}{\partial z^2}+[(w+Z)\cdot\nabla](w_z+Z_z)+\Phi(w+Z)\frac{\partial(w_z+Z_z)}{\partial z}+$$

$$[(w_z + Z_z) \cdot \nabla](w + Z) - (w_z + Z_z)\nabla \cdot (w + Z) + fk \times (w_z + Z_z) - \nabla T - \alpha Z_z = 0. \tag{4.64}$$

By integrations by parts, the Hölder inequality, the Minkowski inequality, the Sobolev incquality and (4.4),

$$-\int_\Omega \left\{[(w + Z) \cdot \nabla]Z_z + \Phi(w + Z)\frac{\partial Z_z}{\partial z}\right\} \cdot w_z$$

$$\leqslant |w + Z|_4 |\nabla Z_z|_2 |w_z|_4 + \int_\Omega \left|\int_{-1}^z (w + Z)\mathrm{d}z'\right| (|Z_{zz}||\nabla w_z| + |\nabla Z_z||w_{zz}|)$$

$$\leqslant \varepsilon \|w_z\|^2 + c\|Z\|_2^2 + c|w + Z|_4^8 |w_z|_2^2 + c\|w + Z\|^2 \|Z\|_{2+2\gamma}^2, \tag{4.65}$$

where $0 < \gamma < \gamma_0$, $\gamma_0$ is given in the definition of $\Psi$. By integration by parts, the Hölder inequality, (4.4) and the Young inequality,

$$-\int_\Omega \{[(w_z + Z_z) \cdot \nabla](w + Z) - (w_z + Z_z)\nabla \cdot (w + Z)\} \cdot w_z$$

$$\leqslant \varepsilon \|w_z\|^2 + c\|Z\|_2^2 + c|w + Z|_4^8 |Z_z|_2^2 + c|w + Z|_4^8 |w_z|_2^2. \tag{4.66}$$

Taking the inner product of Eq. (4.64) with $w_z$ in $L^2(\Omega) \times L^2(\Omega)$, by integration by parts, the Hölder inequality, the Poincaré inequality, (4.65), (4.66), and choosing $\varepsilon$ small enough, we obtain

$$\frac{\mathrm{d}|w_z|_2^2}{\mathrm{d}t} + \int_\Omega |\nabla w_z|^2 + \int_\Omega \left|\frac{\partial w_z}{\partial z}\right|^2$$

$$\leqslant c(|\overline{w}|_{H^1}^8 + |\widetilde{w}|_4^8 + |Z|_4^8)(|w_z|_2^2 + |Z_z|_2^2) +$$

$$c\|Z\|_2^2 + c\|w + Z\|^2 \|Z\|_{2+2\gamma}^2 + c|T|_2^2. \tag{4.67}$$

By the Gronwall inequality, Lemma 3.2, (4.36), (4.37), (4.57) and (4.63), for $\mathcal{T} > t_0$ given, there exists a positive constant $C_7(\mathcal{T}, \|U_0\|, \|Q\|_1, \|Z_{t_0}\|_2)$ such that

$$\int_{t_0}^{\mathcal{T}} \|w_z\|^2 + |w_z(t)|_2^2 \leqslant C_7, \quad \text{for any } t \in [t_0, \mathcal{T}). \tag{4.68}$$

$L^2$ estimates about $\nabla w$. Similarly to Lemma 4.2, by (4.4), we have

$$\left|\int_\Omega \{[(w + Z) \cdot \nabla](w + Z)\} \cdot \Delta w\right|$$

$$\leqslant \varepsilon(|\Delta w|_2^2 + |\nabla w_z|_2^2) + c(|w|_4^8 + |Z|_4^8)(|\nabla w|_2^2 + |\nabla Z|_2^2) + c\|Z\|_2^2, \tag{4.69}$$

$$\left|\int_\Omega [\Phi(w + Z)(w_z + Z_z)] \cdot \Delta w\right|$$

$$\leqslant \varepsilon|\Delta w|_2^2 + c\|Z\|_2^2 + c(|w_z|_2^2 + |\nabla w_z|_2^2 + |w_z|_2^2 |\nabla w_z|_2^2)(|\nabla w|_2^2 + |\nabla Z|_2^2) +$$

$$c(|Z_z|_2^2 + |\nabla Z_z|_2^2 + |Z_z|_2^2 |\nabla Z_z|_2^2)(|\nabla w|_2^2 + |\nabla Z|_2^2). \tag{4.70}$$

Taking the inner product of Eq. (4.18) with $-\Delta w$ in $L^2(\Omega) \times L^2(\Omega)$, by the Hölder inequality, $(fk \times w) \cdot \Delta w = 0$, $\int_\Omega \nabla p_{b_3} \cdot \Delta w = 0$, (4.69), (4.70), and choosing $\varepsilon$ small enough, we get

$$\frac{\mathrm{d}|\nabla w|_2^2}{\mathrm{d}t} + \int_\Omega |\Delta w|^2 + \int_\Omega |\nabla w_z|^2$$

$$\leqslant c(|w|_4^8+|Z|_4^8+|w_z|_2^2+|\nabla w_z|_2^2+|w_z|_2^2|\nabla w_z|_2^2+|Z_z|_2^2+|\nabla Z_z|_2^2+ |Z_z|_2^2|\nabla Z_z|_2^2)|\nabla w|_2^2+c|\nabla T|_2^2+c(|w|_4^8+|Z|_4^8+|w_z|_2^2+|\nabla w_z|_2^2+ |w_z|_2^2|\nabla w_z|_2^2+|Z_z|_2^2+|\nabla Z_z|_2^2+|Z_z|_2^2|\nabla Z_z|_2^2)|\nabla Z|_2^2+c\|Z\|_2^2. \qquad (4.71)$$

By the Gronwall inequality, Lemma 3.2, (4.33), (4.57), (4.63) and (4.68), for $\mathcal{T}>t_0$ given, there exists a positive constant $C_8(\mathcal{T},\|U_0\|,\|Q\|_1,\|Z_{t_0}\|_2)$ such that

$$\int_{t_0}^{\mathcal{T}}|\Delta w|_2^2+|\nabla w(t)|_2^2\leqslant C_8,\quad \text{for any } t\in[t_0,\mathcal{T}). \qquad (4.72)$$

$L^2$ estimates about $\nabla T, T_z$. Similarly to Lemma 4.2, by (4.4),

$$\left|\int_\Omega[(w+Z)\cdot\nabla]T(\Delta T+T_{zz})\right| \leqslant c(|w|_4^8+|Z|_4^8)|\nabla T|_2^2+\varepsilon(|\Delta T|_2^2+|\nabla T_z|_2^2+|T_{zz}|_2^2), \qquad (4.73)$$

$$\left|\int_\Omega\Phi(w+Z)T_z(\Delta T+T_{zz})\right| \leqslant\varepsilon(|\Delta T|_2^2+|T_{zz}|_2^2+|\nabla T_z|_2^2)+ c[(|\nabla w|_2^4+|\nabla w|_2^2|\Delta w|_2^2)+c(|\nabla Z|_2^4+|\nabla Z|_2^2|\Delta Z|_2^2)]|T_z|_2^2. \qquad (4.74)$$

Taking the inner product of (4.19) with $-(\Delta T+T_{zz})$ in $L^2(\Omega)$, by the Hölder inequality, the Young inequality, (4.73), (4.74), and choosing $\varepsilon$ small enough, we reach

$$\frac{\mathrm{d}(|\nabla T|_2^2+|T_z|_2^2+\alpha_u|T|_{z=0}|_2^2)}{\mathrm{d}t}+|\Delta T|_2^2+|T_{zz}|_2^2+(|\nabla T_z|_2^2+\alpha_u|\nabla T|_{z=0}|_2^2) \leqslant c(|w|_4^8+|Z|_4^8)|\nabla T|_2^2+c[(|\nabla w|_2^4+|\nabla w|_2^2|\Delta w|_2^2)+ c(|\nabla Z|_2^4+|\nabla Z|_2^2|\Delta Z|_2^2)]|T_z|_2^2+c|Q|_2^2. \qquad (4.75)$$

By the Gronwall inequality, Lemma 3.2, (4.33), (4.57), (4.63), (4.68) and (4.72), for $\mathcal{T}>t_0$ given, there exists a positive constant $C_9(\mathcal{T},\|U_0\|,\|Q\|_1,\|Z_{t_0}\|_2)$ such that

$$|\nabla T(t)|_2^2+|T_z(t)|_2^2\leqslant C_9,\ \text{for any } t\in[t_0,\mathcal{T}). \qquad (4.76)$$

# 5 The Global Well-Posedness of IBVP

**Proof of Theorem 1.1** **Step 1** The existence of global strong solutions. By Proposition 4.1, we can use the method of contradiction to prove Theorem 1.1. Indeed, let $(u,T)$ be a strong solution to the system (3.5)~(3.9) on the maximal interval $[0,\mathcal{T}_*)$, i.e., $(w,T)$ be a strong solution to the system (4.18)~(4.22) on the maximal interval $[0,\mathcal{T}_*)$. If $\mathcal{T}_*<+\infty$, then $\lim\limits_{t\to\mathcal{T}_*^-}\sup(\|w\|+\|T\|)=+\infty$, which is impossible from (4.33), (4.36), (4.68), (4.72) and (4.76).

**Step 2** The uniqueness of global strong solutions. Let $(w_1,T_1)$ and $(w_2,T_2)$ be two strong solutions of (4.18)~(4.22) on the time interval $[t_0,\mathcal{T}]$ with $p'_{b_3}, p''_{b_3}$ and initial data $((w_{t_0})_1,(T_{t_0})_1),((w_{t_0})_2,(T_{t_0})_2)$ respectively.

Define $w = w_1 - w_2$, $T = T_1 - T_2$, $p_{b_3} = p'_{b_3} - p''_{b_3}$. Then $w, T, p_{b_3}$ satisfy

$$\frac{\partial w}{\partial t} - \Delta w - \frac{\partial^2 w}{\partial z^2} + [(w_1 + Z)\cdot\nabla]w + (w\cdot\nabla)(w_2 + Z) + \Phi(w_1 + Z)\frac{\partial w}{\partial z} + \Phi(w)\frac{\partial(w_2 + Z)}{\partial z} + fk\times w + \nabla p_{b_3} - \int_{-1}^{z}\nabla T\mathrm{d}z' = 0, \tag{5.1}$$

$$\frac{\partial T}{\partial t} - \Delta T - \frac{\partial^2 T}{\partial z^2} + [(w_1 + Z)\cdot\nabla]T + (w\cdot\nabla)T_2 + \Phi(w_1 + Z)\frac{\partial T}{\partial z} + \Phi(w)\frac{\partial T_2}{\partial z} = 0, \tag{5.2}$$

$$\int_{-1}^{0}\nabla\cdot w\mathrm{d}z = 0, \tag{5.3}$$

$$w|_{t=t_0} = (w_{t_0})_1 - (w_{t_0})_2,\quad T|_{t=t_0} = (T_{t_0})_1 - (T_{t_0})_2, \tag{5.4}$$

$(w, T)$ satisfies the boundary value conditions $(2.12)\sim(2.14)$. (5.5)

We take the inner product of (5.1) with $w$ in $L^2(\Omega)\times L^2(\Omega)$ and obtain

$$\frac{1}{2}\frac{\mathrm{d}|w|_2^2}{\mathrm{d}t} + \int_\Omega|\nabla w|^2 + \int_\Omega|w_z|^2$$
$$= -\int_\Omega\left\{[(w_1 + Z)\cdot\nabla]w + \Phi(w_1 + Z)\frac{\partial w}{\partial z}\right\}\cdot w - \int_\Omega(fk\times w + \nabla p_{b_3})\cdot w - \int_\Omega\left[(w\cdot\nabla)(w_2 + Z) + \Phi(w)\frac{\partial(w_2 + Z)}{\partial z}\right]\cdot w + \int_\Omega\left(\int_{-1}^{z}\nabla T\mathrm{d}z'\right)\cdot w. \tag{5.6}$$

By integration by parts and Lemma 4.2,

$$\left|\int_\Omega[(w\cdot\nabla)(w_2 + Z)]\cdot w\right| \leqslant \varepsilon\int_\Omega(|\nabla w|^2 + |w_z|^2) + c(|Z|_4^8 + |w_2|_4^8)|w|_2^2,$$

$$\left|\int_\Omega\Phi(w)\frac{\partial(w_2 + Z)}{\partial z}\cdot w\right|$$
$$\leqslant\varepsilon\int_\Omega|\nabla w|^2 + c(|w_{2z}|_2^2|\nabla w_{2z}|_2^2 + |w_{2z}|_2^4 + |Z_z|_2^2|\nabla Z_z|_2^2 + |Z_z|_2^4)|w|_2^2. \tag{5.7}$$

By integration by parts, the Hölder inequality and the Young inequality, we derive from (5.6) and (5.7),

$$\frac{1}{2}\frac{\mathrm{d}|w|_2^2}{\mathrm{d}t} + \int_\Omega|\nabla w|^2 + \int_\Omega|w_z|^2 \leqslant 2\varepsilon\int_\Omega(|\nabla w|^2 + |w_z|^2) + \varepsilon|\nabla T|_2^2 + c(1 + |Z|_4^8 + |w_2|_4^8 + |w_{2z}|_2^2|\nabla w_{2z}|_2^2 + |w_{2z}|_2^4 + |Z_z|_2^2|\nabla Z_z|_2^2 + |Z_z|_2^4)|w|_2^2. \tag{5.8}$$

Similarly to (5.8), we get

$$\frac{1}{2}\frac{\mathrm{d}|T|_2^2}{\mathrm{d}t} + \int_\Omega|\nabla T|^2 + \int_\Omega|T_z|^2 + \alpha_u|T|_{z=0}|_2^2 \leqslant \varepsilon\int_\Omega(|\nabla w|^2 + |w_z|^2) + \varepsilon\int_\Omega(|\nabla T|^2 + |T_z|^2) + c(|T_2|_4^8 + |T_{2z}|_2^2|\nabla T_{2z}|_2^2 + |T_{2z}|_2^4)(|w|_2^2 + |T|_2^2). \tag{5.9}$$

From (5.8) and (5.9), and choosing $\varepsilon$ small enough, we obtain

$$\frac{d(|w|_2^2+|T|_2^2)}{dt}+\int_\Omega(|\nabla w|^2+|w_z|^2)+\int_\Omega(|\nabla T|^2+|T_z|^2)+\alpha_u|T|_{z=0}|_2^2$$
$$\leqslant c(1+|T_2|_4^8+|Z|_4^8+|w_2|_4^8+|w_{2z}|_2^2|\nabla v_{2z}|_2^2+|w_{2z}|_2^4+|Z_z|_2^2|\nabla Z_z|_2^2+$$
$$|Z_z|_2^4)|w|_2^2+c[|T_2|_4^8+(|T_{2z}|_2^2+1)|\nabla T_{2z}|_2^2+|T_{2z}|_2^4+|T_{2z}|_2^2]|T|_2^2. \quad (5.10)$$

By the Gronwall inequality, the result of Step 1 and (5.10), we prove the uniqueness. □

# 6 The Existence of Random Attractors

## 6.1 Preliminaries for Random Attractors

We recall some definitions and results from the theory of random dynamical systems[2,6,7,39].

Let $(X,d)$ be a complete separable metric space (a Polish space), $(\Omega,\mathscr{F},P)$ a complete probability space, $\{\vartheta_t:\Omega\to\Omega,t\in\mathbf{R}\}$ a family of measure preserving transformation such that $\vartheta_0=\mathrm{id}_\Omega$ and $\vartheta_{t+s}=\vartheta_t\circ\vartheta_s$ for all $t,s\in\mathbf{R}$. $\{\vartheta_t\}$ is called a metric dynamical system on $\Omega$, which represents the noise driving a random dynamical system. Here we assume $\vartheta_t$ is ergodic under $P$.

**Definition 6.1** [Random dynamical system] A measurable map $\psi:\mathbf{R}^+\times\Omega\times X\to X(t,\omega,U)\mapsto\psi(t,\omega)U$ is called a random dynamical system if $\psi$ satisfies the cocycle property: $\psi(0,\omega)=\mathrm{id}_X,\psi(t+s,\omega)=\psi(t,\vartheta_s\omega)\psi(s,\omega)$ for all $t,s\in\mathbf{R}^+$ and $P$-a.s. $\omega\in\Omega$. If $\psi(t,\omega):X\to X$ is continuous, then $\psi$ is called a continuous random dynamical system.

Random dynamical systems with continuous time are generated by infinite dimensional stochastic evolution equations under an additive noise with a unique global solution, as well as by differential equations with random coefficients or stochastic differential equations.

**Definition 6.2** [Random compact set] Let $K:\Omega\to 2^X$, $2^X$ be the set of all subsets of $X$. $K$ is called a random compact set if $K(\omega)$ is compact $P$-a.s. and the map $\omega\to d(U,K(\omega))$ is measurable for any $U\in X$, where $d(U,K(\omega))=\inf_{U_1\in K(\omega)}d(U,U_1)$.

**Definition 6.3** Let $A(\omega)$, $B(\omega)$ be two random sets.

(i) $A(\omega)$ attracts $B(\omega)$ if $\lim_{t\to+\infty}d(\psi(t,\vartheta_{-t}\omega)B(\vartheta_{-t}\omega),A(\omega))=0$, $P$-a.s.

(ii) $A(\omega)$ absorbs $B(\omega)$ if there exists $t_B(\omega)$ such that for all $t\geqslant t_B(\omega)$,
$$\psi(t,\vartheta_{-t}\omega)B(\vartheta_{-t}\omega)\subset A(\omega),P\text{-a.s.}$$

**Definition 6.4** [Random attractor] A random set $\mathscr{A}(\omega)$ is said to be a random attractor for the random dynamical system $\psi$ if $P$-a.s.

(i) $\mathscr{A}(\omega)$ is a random compact set.

(ii) $\mathscr{A}(\omega)$ is invariant, that is, $\psi(t,\omega)\mathscr{A}(\omega)=\mathscr{A}(\vartheta_t\omega)$, for $\forall\, t\geqslant 0$.

(iii) $\mathscr{A}(\omega)$ attracts all deterministic bound sets $B\subset X$, i.e.
$$\lim_{t\to+\infty}d(\psi(t,\vartheta_{-t}\omega)B,A(\omega))=0,\quad P\text{-a.s.}$$

**Remark 6.5** $\psi(t,\vartheta_{-t}\omega)U$ can be interpreted as the position at $t=0$ of the trajectory which was $U$ at time $-t$, that is, when $t$ is moving, the trajectory $\psi(t,\vartheta_{-t}\omega)U$ is always at the position at $t=0$. Therefore, the random attractor is also called the random pull-back attractor.

**Theorem 6.6**[6,7] If there exists a random compact set $K(\omega)$ absorbing every bounded non-random set $B\subset X$, the continuous random dynamical system $\psi$ possesses a random pull-back attractor $\mathscr{A}(\omega)$, where

$$\mathscr{A}(\omega)=\overline{\bigcup_{B\subset X}\Lambda_B(\omega)},\quad \Lambda_B(\omega)=\bigcap_{s\geqslant 0}\overline{\bigcup_{t\geqslant s}\psi(t,\vartheta_{-t}\omega)B}.$$

Moreover, $\mathscr{A}(\omega)\subset K(\omega)$, $\mathscr{A}(\omega)$ is unique.

## 6.2 Proof of Theorem 1.2

Now we construct a random dynamical system modeling the boundary value problem of 3D stochastic primitive equations (2.9)~(2.14). Let

$$\Omega=\{\omega:\omega\in C(\mathbf{R},l^2),\omega(0)=0\},$$

$\mathscr{F}$ the Borel sigma-algebra induced by the compact open-topology of $\Omega$, $P$ a Wiener measure on $(\Omega,\mathscr{F})$. Write

$$(\omega_1(t,\omega),\cdots,\omega_k(t,\omega),\cdots)=\omega(t).$$

Define

$$\vartheta_t\omega(s)=\omega(t+s)-\omega(t). \tag{6.1}$$

Then $\vartheta_t$ satisfies $\vartheta_{t+s}=\vartheta_t\circ\vartheta_s$ for all $t,s\in\mathbf{R}$, and $\vartheta_t$ is ergodic under $P$. By Theorem 1.1, let

$$U(t,\omega)=S(t,s;\omega)U_s,\quad (w(t,\omega),T(t,\omega))=\phi(t,s;\omega)(w_s,T_s),$$

where $U(t,\omega)=(v(t,\omega),T(t,\omega))$ is a strong solution to (2.9)~(2.15) on $[s,t]$ with the initial data $U(s)=U_s$ and $(w,T)$ is a strong solution to (4.18)~(4.22) on $[s,t]$ with the initial data $(w_s,T_s)$. Then, for $s\leqslant r\leqslant t$, we have $S(t,s;\omega)=S(t,r;\omega)S(r,s;\omega)$. Due to (6.1), for any $s,t\in\mathbf{R}^+$, $U_0\in V$, we have $P$-a.s.,

$$S(t+s,0;\omega)U_0=S(t,0;\vartheta_s\omega)S(s,0;\omega)U_0.$$

Define $\psi:\mathbf{R}^+\times\Omega\times V\to V$, $\psi(t,\omega)U_0=S(t,0;\omega)U_0$. By Theorem 1.1, the following Proposition 6.8 will prove that $\psi$ is a continuous random dynamical system on $V$ with weak topology over $(\Omega,\mathscr{F},P,\{\vartheta_t\}_{t\in\mathbf{R}})$ and models the random dynamical system generated by (2.9)~(2.14).

In the course of proving Theorem 1.2 by Theorem 6.6, the key step is the proof of the existence of a bounded absorbing set $K(\omega)$, which is compact in $V$ with weak topology.

**Proposition 6.7** [Existence of bounded absorbing sets for the random dynamical system (2.9)~(2.14)] If $Q\in H^1(\Omega)$ and $B_\rho=\{U;\|U\|\leqslant\rho,U\in V\}$, then there exist $r_0(\omega,\|Q\|_1)$ and $t(\omega,\rho)\leqslant-1$ such that for any $t_0\leqslant t(\omega,\rho)$, $U_{t_0}\in B_\rho$,

$$\|S(0,t_0;\omega)U_{t_0}\|\leqslant r_0(\omega).$$

By Remark 6.5, for any bounded set $B\subset V$, there exists $-t_0(B)>0$ big enough such that

$$\psi(-s,\vartheta_s\omega)B=S(0,s;\omega)B\subset B_{r_0(\omega)},\quad\text{for any } s\leqslant t_0.$$

**Proof of Proposition 6.7** From the ergodicity of the process $Z$ with values in $D(A_1^{1+\gamma})$, we have[37]

$$\lim_{s\to-\infty}\frac{1}{-s}\int_s^0(|Z(\tau)|_4^8+\|Z(\tau)\|^4+\|Z(\tau)\|^2\|Z(\tau)\|_2^2)\mathrm{d}\tau$$
$$=E(|Z(0)|_4^8+\|Z(0)\|^4+\|Z(0)\|^2\|Z(0)\|_2^2).$$

By (3.3), it is known that $E(|Z(0)|_4^8+\|Z(0)\|^4+\|Z(0)\|^2\|Z(0)\|_2^2)\to 0$ as $\alpha\to+\infty$. Therefore, by choosing large enough $\alpha$, we have

$$\lim_{s\to-\infty}\frac{1}{-s}\int_s^0[-\lambda_1+c(|Z(\tau)|_4^8+\|Z(t)\|^4+\|Z(\tau)\|^2\|Z(\tau)\|_2^2)]\mathrm{d}\tau\leqslant-\frac{\lambda_1}{2}.$$

This implies the existence of $s_0(\omega)$ such that $s<s_0(\omega)$,

$$\int_s^0[-\lambda_1+c(|Z(\tau)|_4^8+\|Z(\tau)\|^4+\|Z(\tau)\|^2\|Z(\tau)\|_2^2)]\mathrm{d}\tau\leqslant-\frac{\lambda_1}{4}(-s).\quad(6.2)$$

By the assumption (3.2), (4.33) and a similar argument as in [19], $\|Z(\tau)\|^2+|T(\tau)|_2^2+|Z(\tau)|_2^2$ has at most polynomial growth as $\tau\to-\infty$. Equations (4.35) and (6.2) imply the existence of $t_0(\rho,\omega)$ and a.s. the finite random variable $R_0(\omega)$ such that a.s.

$$|w(t)|_2^2\leqslant R_0(\omega),\quad \text{for any } s\leqslant t\leqslant 0,\quad(6.3)$$

where $s\leqslant t_0$, $(w,T)$ is the strong solution of (4.18)~(4.22) with initial data $(w_s, T_s)=(v_s-Z_s, T_s)$ and $(v_s, T_s)\in B_\rho$. Integrating (4.34) from $t$ to $t+1$ for $s\leqslant t\leqslant-1$, by (6.3), we know there exists a.s. a finite random variable $R_1(\omega)$ such that

$$c_2\int_t^{t+1}\left[\int_\Omega\left(|\nabla w|^2+\left|\frac{\partial w}{\partial z}\right|^2+|w|^2\right)\right]\leqslant R_1(\omega).\quad(6.4)$$

By the uniform Gronwall Lemma (cf. [40, p. 91]), $|\widetilde{w}|_3^3\leqslant|\widetilde{w}|_2^{\frac{3}{2}}\|\widetilde{w}\|_2^{\frac{3}{2}}$, (4.41), (6.3), (6.4) and Lemma 3.2, we derive from (4.49)

$$|\widetilde{w}(t+1)|_3^3\leqslant R_2(\omega),\quad(6.5)$$

where $R_2(\omega)$ is an a.s. finite random variable and $t\in[s,-1]$.

By the uniform Gronwall Lemma, (4.41), (6.3)~(6.5), Lemma 3.2 and $|\widetilde{w}|_4^4\leqslant|\widetilde{w}|_3^2\|\widetilde{w}\|^2$, from (4.56), we get

$$|\widetilde{w}(t+2)|_4^4\leqslant R_3(\omega),\quad(6.6)$$

where $R_3(\omega)$ is an a.s. finite random variable and $t\in[s,-2]$. For $t\in[s,-2]$, from (4.56) and (6.6),

$$\int_{t+2}^{t+3}\left[\int_\Omega\left(|\nabla\widetilde{w}|^2|\widetilde{w}|^2+\frac{1}{2}|\nabla|\widetilde{w}|^2|^2\right)+\int_\Omega\left(|\partial_z\widetilde{w}|^2|\widetilde{w}|^2+\frac{1}{2}|\partial_z|\widetilde{w}|^2|^2\right)\right]$$
$$\leqslant R_3(\omega)^2+R_3(\omega)=R_4(\omega).\quad(6.7)$$

From (4.62), by the uniform Gronwall Lemma, (6.3), (6.4) and (6.7),

$$\|\nabla\overline{w}(t+3)\|_{L^2}^2\leqslant R_5(\omega),\quad(6.8)$$

where $R_5(\omega)$ is an a.s. finite random variable and $t\in[s,-3]$.

Similarly, there exist a.s. finite random variables $R_6(\omega)$, $R_7(\omega)$, $R_8(\omega)$ such that

$$|w_z(t+4)|_2^2\leqslant R_6(\omega),\quad\text{for any } t\in[s,-4],\quad(6.9)$$

$$|\nabla w(t+5)|_2^2 \leqslant R_7(\omega), \quad \text{for any } t\in[s,-5], \tag{6.10}$$

$$|\nabla T(t+6)|_2^2 + |T_z(t+6)|_2^2 \leqslant R_8(\omega), \quad \text{for any } t\in[s,-6], \tag{6.11}$$

From (4.32), (6.3), (6.9) ~ (6.11), we know that then there exist $r_0(\omega)$ and $t(\omega,\rho)\leqslant -1$ such that for any $t_0\leqslant t(\omega,\rho)$, $U_{t_0}\in B_\rho$,

$$\|S(0,t_0;\omega)U_{t_0}\|\leqslant r_0(\omega).$$ □

In order to prove Theorem 1.2, we need the following property about the family of mapping $\{S(t,s;\omega)\}_{t\geqslant s}$.

**Proposition 6.8** For any $t\geqslant s$, the mapping $S(t,s;\omega)$ is weakly continuous from $V$ to $V$.

**Proof of Proposition 6.8** Let $\{U_n\}$ be a sequence in $V$ such that $U_n\to U$ weakly in $V$. Then $\{U_n\}$ is bounded in $V$. By the *a priori* estimates in Sec. 4 and the proof of Proposition 6.7, we know that, for any $t\geqslant s$, $\{S(t,s;\omega)U_n\}$ is bounded in $V$. So we extract a subsequence $\{S(t,s;\omega)U_{n_k}\}$ such that $S(t,s;\omega)U_{n_k}\to\mathscr{U}$ weakly in $V$. Since the embedding $V\hookrightarrow L^2(\Omega)\times L^2(\Omega)\times L^2(\Omega)$ is compact, $U_{n_k}\to U$ strongly in $L^2(\Omega)\times L^2(\Omega)\times L^2(\Omega)$. By (5.10), we obtain that $S(t,s;\omega)U_{n_k}\to S(t,s;\omega)U$ strongly in $L^2(\Omega)\times L^2(\Omega)\times L^2(\Omega)$. Then $\mathscr{U}=S(t,s;\omega)U$. Therefore, the sequence $\{S(t,s;\omega)U_n\}$ has a subsequence $\{S(t,s;\omega)U_{n_k}\}$ such that $S(t,s;\omega)U_{n_k}\to S(t,s;\omega)U$ weakly in $V$. □

**Proof of Theorem 1.2** Applying Theorem 6.6, by Proposition 6.7 and Proposition 6.8, we prove Theorem 1.2. □

### References


[1] Adams R A. Sobolev Space. New York: Academic Press, 1975.

[2] Arnold L. Random Dynamical System // Springer Monograghs in Mathematics. Berlin: Springer-Verlag, 1998.

[3] Bourgeois A J, Beale J T. Validity of the quasigeostrophic model for large-scale flow in the atmosphere and ocean. SIAM J Math Anal, 1994, 25: 1023 – 1068.

[4] Babin A V, Vishik M I. Attractors of Evolution Equations (in Russian). Moscow: Nauka, 1989 (English translation. Amsterdam: North Holland, 1992).

[5] Cordoba D. Nonexistence of simple hyperbolic blow-up for the quasi-geostrophic equation. Ann of Math, 1998, 148: 1135 – 1152.

[6] Crauel H, Debussche A, Flandoli F. Random attractors. J Dyn Diff Eq, 1997,29 (2): 307 – 341.

[7] Crauel H, Flandoli F. Attractors of random dynamics systems. Prob Th Rel Fields, 1994, 100: 365 – 393.

[8] Charney J G, Fjortaft R, Von Neumann J. Numerical integration of the barotropic vorticity equation. Tellus, 1950, 2: 237 – 254.

[9] Constantin P, Foias C, Temam R. Attractors representing turbolent flows. Memoirs of AMS, 1985, 53 (314).

[10] Constantin P, Majda A, Tabak E. Formation of strong fronts in the 2-D quasigeostrophic thermal active scalar. Nonlinearity, 1994, 7: 1495 – 1533.

[11] Constantin P, Majda A, Tabak E. Singular front formation in a model for quasigeostrophic flow. Phys Fluids, 1994, 6: 9－11.

[12] Charney J G, Philips N A. Numerical integration of the quasi-geostrophic equations for barotropic simple baroclinic flows. J Meteor, 1953, 10: 71－99.

[13] Cao C, Titi E S. Global well-posedness of the three-dimensional viscous primitive equations of large-scale ocean and atmosphere dynamics. Ann of Math, 2007, 166: 245－267.

[14] Constantin P, Wu J. Behavior of solutions of 2D quasi-geostrophic equations. SIAM J Math Anal, 1999, 30: 937－948.

[15] Duan J, Gao H, Schmalfuss B. Stochastic dynamics of a coupled atmosphere-ocean model. Stoch and Dynam, 2002,2(3): 357－380.

[16] Duan J, Kloeden P E, Schmalfuss B. Exponential stability of the quasi-geostrophic equation under random perturbations. Prog in Probability, 2001, 49: 241－256.

[17] Duan J, Schmalfuss B. The 3D quasi-geostrophic fluid dynamics under random forcing on boundary. Comm in Math Sci, 2003, 1: 133－151.

[18] Embid P F, Majda A. Averaging over fast gravity waves for geophysical flows with arbitrary potential vorticity. Comm in PDE, 1996, 21: 619－658.

[19] Flandoli F. Dissipativity and invariant measures for stochastic Navier-Stokes equations. Nonliear Diff Eq Appl, 1994, 1: 403－423.

[20] Frankignoul C, Hasselmann K. Stochastic climate models, Part II: Application to sea-surface temperature anomalies and thermocline variability. Tellus, 1977, 29: 289－305.

[21] Galdi G P. An Introduction to the Mathematical Theory of the Navier-Stokes Equations, Vol I. Berlin-Heidelberg-New York: Springer-Verlag, 1994.

[22] Guillén-González F, Masmoudi N. Rodríguez-Bellido M A. Anisotropic estimates and strong solutions for the primitive equations. Diff Int Equ, 2001, 14: 1381－1408.

[23] Hu C, Temam R, Ziane M. The primitive equations of the large scale ocean under the small depth hypothesis. Disc and Cont Dyn Sys, 2003, 9(1): 97－131.

[24] Lions J L, Temam R, Wang S. New formulations of the primitive equations of atmosphere and applications. Nonlinearity, 1992, 5: 237－288.

[25] Lions J L, Temam R, Wang S. On the equations of the large scale ocean. Nonlinearity, 1992, 5: 1007－1053.

[26] Lions J L, Temam R, Wang S. Models of the coupled atmosphere and ocean(CAO I). Computational Mechanics Advance, 1993, 1: 1－54.

[27] Lions J L, Temam R, Wang S. Mathematical theory for the coupled atmosphere-ocean models (CAO III). J Math Pures Appl, 1995, 74: 105－163.

[28] Kiselev A, Nazarov F, Volberg A. Global well-posedness for the critical 2D dissipative quasi-geostrophic equation. Invent Math, 2007, 167: 445－453.

[29] Majda A. Introduction to PDEs and Waves for the Atmosphere and Ocean. Courant Lecture Notes in Mathematics, 2003, 9.

[30] Majda A, Eijnden E V. A mathematical framework for stochastic climate models. Comm Pure Appl Math, 2001, 54: 891－974.

[31] Majda A, Wang X. The emergence of large-scale coherent structure under small-scale random bombardments. Comm Pure Appl Math, 2001, 59: 467－500.

[32] Müller P. Stochastic forcing of quasi-geostrophic eddies//Adler R J, müller P, Rozovskii B. Stochastic

Modelling in Physical Oceanography. Basel: Birkhäuser, 1996.
[33] Mikolajewicz U, Maier-Reimer E. Internal secular variability in an OGCM. Climate Dyn, 1990, 4: 145-156.
[34] Pedlosky J. Geophysical Fluid Dynamics. 2nd Ed. Berlin: Springer-Verlag, 1987.
[35] Phillips O M. On the generation of waves by turbulent winds. J Fluid Mech, 1957, 2: 417-445.
[36] Da Prato G, Zabczyk J. Stochastic equations in infinite dimensions//Encyclopedia of Mathematics and its Application. Cambridge: Cambridge Univ Press, 1992.
[37] Da Prato G, Zabczyk J. Ergodicity for infinite dimensional systems//London Mathematical Society Lecture Note Series 229. Cambridge: Cambridge Univ Press, 1996.
[38] Rubenstein D. A spectral model of wind-forced internal waves. J Phys Oceanogr, 1994, 24: 819-831.
[39] Schmalfuss B. Backward cocycle and attractors of stochastic differential equations//Reitmann V, Riedrich T, Kokch N. International Seminar on Applied Mathematics-Nonlinear Dynamics: Attractor Approximation and Global Behavior. Dresden: Universität, 1992, 185-192.
[40] Temam. R. Infinite-dimensional dynamical systems in mechanics and physics. 2nd Ed. Appl Math Ser, Vol 68. New York: Springer-Verlag, 1997.
[41] Temam R. Navier-stokes equations: theory and numerical analysis. Rev Ed. Amsterdam: North-Holland, 1984.
[42] Samelson R, Temam R, Wang S. Some mathematical properties of the planetary geostrophic equations for large-scale ocean circulation. Appl Anal, 1998, 70 (1-2): 147-173.
[43] Temam R, Ziane M. Some mathematical problems in geophysical fluid dynamics. Handbook of Mathematical Fluid Dynamics, 2004.

# The Convergence of Non-Newtonian Fluids to Navier-Stokes Equations *

Guo Boling(郭柏灵) Guo Chunxiao(郭春晓)

**Abstract**

In this paper, the convergence of solutions for incompressible dipolar viscous non-Newtonian fluids is investigated. We obtain the conclusion that the solutions of non-Newtonian fluids converge to the solutions of Navier-Stokes equations in the sense of $L^2$-norm (resp. $H^1$-norm), as the viscosities tend to zero and the initial data belong to $H^1(\Omega)$ (resp. $H^2(\Omega)$). Moreover, we obtain $L^\infty$-norm convergence of solutions if the initial data belong to $H^2(\Omega)$.

**Keywords** non-Newtonian fluids; Navier-Stokes equations; convergence of solutions

## 1 Introduction

As is well known, the Navier-Stokes model of fluid restricts the linear relation between the stress tensor and the velocity gradient[16,19,20]. Fluids satisfying such constitutive relationship are called Newtonian fluids, e.g. water and air. However, for many fluid materials, such as molten plastics, synthetic fibers, biological fluids, paints and greases, etc., their flow behavior cannot be characterized by Newtonian relationship in real life. By relaxing the constraints of the Stokes hypothesis, the mathematical theory of viscous non-Newtonian fluids generalizes the usual Stokes model in three important respects: nonlinear constitutive relations between the viscous part of the stress tensor and velocity gradients, dependence of the viscous stress tensor on velocity gradients of order two or higher, and constitutive relations for higher order stress tensors which must be present in the balance of energy equation as soon as higher order velocity gradients are admitted into the theory[5,9].

The viscous non-Newtonian fluids which allow for nonlinearity in the constitutive theory have been extensively studied, see for example Ladyzhenskaya[10,11], Du and Gunzburger[7]. Particularly, Straughan in [18] obtained some results of three Ladyzhenskaya's models with dependence of viscosity on temperature. Bleustein and Green[4] developed dipolar fluids based on continuum thermodynamics. Further relevant work and details may be found in [15]. Regarding the viscoelastic part, the correct model was developed by Man and Sun[13] in 1987, where they applied viscoelastic fluid theory to study the generalized fluid of second grade in

* J Math Anal Appl, 2009, 357: 468-478.

their analysis of glacier flow, moreover, the model was capable of producing the effects of shear thickening and shear thinning, the details of this may be found in Section 4.2 of the book [17]. Franchi and Straughan in [8] investigated the stability and nonexistence of a fluid of second grade.

Many authors also have tremendous interest in equations which contain higher order velocity gradients than Navier-Stokes equation. Lions proposed to replace the negative Laplacian $-\Delta$ by the term $-\Delta+k(-\Delta)^{\frac{l}{2}}, l>2, k>0$ (in a way that is reminiscent of Taylor expansion), who considered such a modified problem in [12]. In summary, when the perturbations of Navier-Stokes equation are taken into account, higher order velocity gradients may be incorporated into the equation.

In this paper, we consider the isothermal dipolar incompressible viscous non-Newtonian fluids in 2D periodic domain as follows

$$u_t+u\cdot\nabla u-\nabla\cdot(\gamma(u)e(u))+\mu_1\Delta^2u-\nu\Delta u=-\nabla\pi^{n-N}+f(x),\quad (x,t)\in\Omega\times[0,T],\tag{1.1}$$

$$\nabla\cdot u(x,t)=0,\quad x\in\Omega,\tag{1.2}$$

$$u(x,0)=u_0(x),\tag{1.3}$$

subject to the spatial periodicity conditions

$$u(x,t)=u(x+L\chi_i,t),\quad \int_\Omega u(x,t)\mathrm{d}x=0,\quad t\geqslant 0,\tag{1.4}$$

where $\{\chi_i\}_{i=1}^2$ is the natural basis of $\mathbf{R}^2$,

$$\gamma(u)=2\mu_0|e(u)|^{p-2}.$$

Here $u$ is the velocity vector, $f$ is the external body force, $\nabla\pi^{n-N}$ denotes the pressure of non-Newtonian fluid system, $e(u)=(e_{ij}(u))$ is the symmetric deformations velocity tensor whose components are given

$$e_{ij}(u)=\frac{1}{2}\left(\frac{\partial u_i}{\partial x_j}+\frac{\partial u_j}{\partial x_i}\right),\quad |e(\cdot)|^2=\sum_{i,j=1}^2|e_{ij}(\cdot)|^2,$$

$\nu>0$ is constant, and $\mu_0,\mu_1>0$ are viscosities. The fluids are shear thinning in the case of $1<p<2$, and shear thickening in the case of $p>2$. See [1,9,14] for a detailed description of such model.

While $\mu_1=0$, the fluid is said to be monopolar, because the only first derivative of the velocity is involved in the stress tensor, such as the Ladyzhenskaya's model. Pokorny[14] investigated the Cauchy problem for monopolar and dipolar incompressible non-Newtonian fluids. Bellout et al.[2,3] studied the existence and uniqueness of weak solution of the dipolar viscous fluids for space-periodic version. Bloom and Hao[5,6] studied the dipolar viscous non-Newtonian fluids in an unbounded channel. While $\mu_0=\mu_1=0$, Eq. (1.1) turns out to be the famous Navier-Stokes equation[19,20]. Following this clue, we naturally ask the question what happens if the viscosities $\mu_0$ and $\mu_1$ tend to zero for non-Newtonian fluids, it is to this goal that the present paper is addressed.

We also consider the following Navier-Stokes equations

$$v_t + v\cdot\nabla v - \nu\Delta v = -\nabla\pi^{N-S} + f(x),\quad (x,t)\in\Omega\times[0,T], \tag{1.5}$$

$$\nabla\cdot v(x,t) = 0,\quad x\in\Omega, \tag{1.6}$$

$$v(x,0) = v_0, \tag{1.7}$$

subject to the spatial periodicity conditions

$$v(x,t) = v(x+L\chi_i,t),\quad \int_\Omega v(x,t)\mathrm{d}x = 0,\quad t\geqslant 0, \tag{1.8}$$

where $\{\chi_i\}_{i=1}^2$ is the natural basis of $\mathbf{R}^2$, $\nabla\pi^{N-S}$ denotes the pressure of Navier-Stokes system.

For the sake of convenience of the following contexts, first we set some notations.

$$\Omega = [0,L]^2,\quad L>0.$$

$L^p(\Omega)$ — the Lebesgue space with norm $\|\cdot\|_{L^p}$, particularly, $\|\cdot\|_{L^2} = \|\cdot\|$, and $\|u\|_{L^\infty} = \operatorname*{ess\,sup}_{x\in\Omega}|u(x)|$, for $p=\infty$.

$W^{\sigma,p}(\Omega)$ — the Sobolev space $\{u\in L^p(\Omega), D^k u\in L^P(\Omega), k\leqslant\sigma\}$ with norm $\|\cdot\|_{\sigma,p}$.

$\mathscr{C}(I,X)$ — the space of continuous functions from the interval $I$ to $X$.

$L^q(0,T;X)$ — the space of all measurable functions $u:[0,T]\to X$, with the norm

$$\|u\|_{L^q(0,T;X)}^q = \int_0^T \|u(t)\|_X^q \mathrm{d}t,$$

and when $q=\infty$, $\|u\|_{L^\infty(0,T;X)} \operatorname*{ess\,sup}_{t\in[0,T]}\|u(t)\|_X$.

Define a space of smooth functions that incorporates the periodicity with respect to $x$ and divergence free condition

$$\mathscr{V} = \left\{u\in C_{\mathrm{per}}^\infty(\Omega):\nabla\cdot u = 0, \int_\Omega u\mathrm{d}x = 0\right\},$$

where per represents that $u$ is periodic with respect to $x$.

$H$ = the closure of $\mathscr{V}$ in $L^2(\Omega)$ with norm $\|\cdot\|$, $(\cdot,\cdot)$ denotes the inner product in $H$.

$H^\sigma(\Omega)$ = the closure of $\mathscr{V}$ in $W^{\sigma,2}(\Omega)$ with norm $\|\cdot\|_\sigma (\sigma\geqslant 1)$.

Particularly, when $\sigma=2$, $V=H^2(\Omega)$, $V_p$ = the closure of $\mathscr{V}$ in $W^{1,p}(\Omega)$.

Since Navier-Stokes equation is the particular case of the non-Newtonian fluids for $\mu_0=\mu_1=0$, the aim of this paper is to investigate the difference between the solutions to non-Newtonian fluids and the solutions to Navier-Stokes equation, when the viscosities $\mu_0$ and $\mu_1$ tend to zero.

We discuss the following two cases of $p$ respectively.

**Case 1** $1<p<2$, we obtain the difference $\omega(t)$ between the solutions to non-Newtonian fluids and the solutions to Navier-Stokes equation, with the same initial data $\mu_0 = v_0\in H^1(\Omega)$, satisfying the following inequality for the finite time $t\in[0,T]$:

$$\|\omega(t)\|\leqslant C(\mu_0 t+\mu_1)^{\frac{1}{2}}.$$

**Case 2** $2<p<3$, we obtain the difference $\omega(t)$ between the solutions to non-Newtonian fluids and the solutions to Navier-Stokes equation, with the same initial data $u_0 = v_0\in H^1(\Omega)$, satisfying the following inequality for the finite time $t\in[0,T]$:

$$\|\omega(t)\|\leqslant C(\mu_0 t^{\frac{3-p}{2}}+\mu_1)^{\frac{1}{2}},$$

when $u_0 = v_0 \in V$ and the viscosities $\mu_0 = \mu_1$, we can further obtain

$$\|\nabla\omega(t)\| \leqslant C\mu_0^{\frac{1}{2}}\left[\left(1+(1+\mu_0^{\frac{p}{2}})t^{\frac{4-p}{2}}+\mu_0\right)\right]^{\frac{1}{2}},$$

$$\|\omega(t)\|_{L^\infty} \leqslant C\mu_0^{\frac{1}{4}}\left[(1+\mu_0^{\frac{1}{4}})(1+t^{\frac{3-p}{8}})\right].$$

Obviously, as $\mu_0, \mu_1 \to 0$, the above estimates of $\omega(t)$ tend to 0, which implies that the solutions of non-Newtonian fluids converge to the solutions of Navier-Stokes equation in the sense of different norms.

The paper is organized as follows. In Section 2, we recall some already known results. In Section 3, we obtain the conclusion that the solutions of non-Newtonian fluids converge to the solutions of Navier-Stokes equation in the sense of $L^2$-norm, as the viscosities tend to zero and the initial data belong to $H^1(\Omega)$. In Section 4, we obtain the conclusion that the solutions of non-Newtonian fluids converge to the solutions of Navier-Stokes equation in the sense of $H^1$-norm and $L^\infty$-norm, as the viscosities tend to zero and the initial data belong to $H^2(\Omega)$.

For notational simplicity, $C$ is a generic constant and may assume various values from line to line.

## 2 Preliminaries

We define a continuous trilinear form on $H^1(\Omega) \times H^1(\Omega) \times H^1(\Omega)$ as follows,

$$b(u,v,\omega) = \int_\Omega u_i \frac{\partial v_j}{\partial x_i}\omega_j \mathrm{d}x, \quad u,v,\omega \in H^1(\Omega),$$

which has the properties: $b(u,v,\omega) = -b(u,\omega,v)$, $b(u,v,v) = 0$.

$$(\nabla\cdot(\gamma(u)e(u)),\omega) = \int_\Omega \gamma(u)e_{ij}(u)e_{ij}(\omega)\mathrm{d}x, \quad u,\omega \in H^1(\Omega),$$

where $\gamma(u) = 2\mu_0|e(u)|^{p-2}$.

**Definition** 2.1 By a weak solution of the periodic problem (1.1)~(1.4) for $n = 2$, $1 < p < \infty$, we mean

$$u \in L^\infty(0,T;H) \cap L^p(0,T;V_p) \cap L^2(0,T;V), \quad \forall T > 0, \tag{2.1}$$

satisfying

$$\int_\Omega u(t)\psi(t)\mathrm{d}x - \int_0^t\int_\Omega u\frac{\partial\psi}{\partial s}\mathrm{d}x\mathrm{d}s + \int_0^t\int_\Omega u_i\frac{\partial u_j}{\partial x_i}\psi_j\mathrm{d}x\mathrm{d}s + \int_0^t\int_\Omega \gamma(u)e_{ij}(u)e_{ij}(\psi)\mathrm{d}x\mathrm{d}s$$
$$= \int_\Omega u_0\psi(0)\mathrm{d}x \tag{2.2}$$

a.e. $t \in [0,T]$ for every $\psi \in \mathscr{L}^1((0,T];H) \cap \mathscr{L}([0,T];V \cap V_p)$, and $\psi(x,T) = 0$.

By Galerkin approximations and some *a priori* estimates, one can obtain the existence and uniqueness of solutions, the method is classical and for more details, we refer to [2,3,5,9,14].

**Lemma 2.1** Let $u_0 \in H, f \in H$, and $p > 1, n = 2$, then there exists a unique weak solution of the system (1.1)~(1.4) in the sense of Definition 2.1.

**Lemma 2.2**[3] $u \in W^{s_1,p}(\Omega)$, $0 \leqslant s_2 \leqslant s \leqslant s_1 < \infty$, then there exists a constant $C$, such that

$$\| u \|_{s,p} \leqslant C \| u \|_{s_1,p}^{\alpha} \| u \|_{s_2,p}^{1-\alpha},$$

where $s = \alpha s_1 + (1-\alpha)s_2$.

**Lemma 2.3** (Gronwall Inequality) Let $f(t), g(t), h(t)$ be nonnegative continuous functions and satisfy the inequality

$$g(t) \leqslant f(t) + \int_0^t g(s)h(s)\mathrm{d}s, \quad \forall t > 0,$$

where $f'(t) \geqslant 0$, then

$$g(t) \leqslant f(t)\exp\left(\int_0^t h(s)\mathrm{d}s\right), \quad \forall t > 0.$$

# 3 The Convergence in $L^2$-norm

In this section, we obtain the conclusion that the solutions of non-Newtonian fluids converge to the solutions of Navier-Stokes equations in the sense of $L^2$-norm, we prove the main theorem with the aid of following lemmas. In other words, these lemmas make sure that the solution norm estimates don't tend to infinity, as $\mu_0, \mu_1 \to 0$.

**Lemma 3.1** For $u_0 \in H^1(\Omega)$, $f \in H$, in the case of $p > 1$, there exists constant $C$ such that

$$\|\nabla u(t)\| \leqslant C, \quad \int_0^t \|\Delta u(s)\|^2 \mathrm{d}s \leqslant C,$$

where the constant doesn't depend on the viscosities $\mu_0, \mu_1$.

**Proof** Taking the inner product of (1.1) with $-\Delta u$ in $H$, and noticing the spatial periodicity conditions, combining the divergence free condition, for $n = 2$, we can obtain $b(u, u, -\Delta u) = 0$ (see details in [16]), then we have

$$\frac{1}{2}\frac{\mathrm{d}}{\mathrm{d}t}\|\nabla u\|^2 + \nu\|\Delta u\|^2 + \mu_1\|u\|_3^2 + 2\mu_0\int_\Omega |e(u)|^{p-2} e_{ij}(u)e_{ij}(-\Delta u)\mathrm{d}x = (f, -\Delta u),$$

applying the ε-Young inequality,

$$\frac{1}{2}\frac{\mathrm{d}}{\mathrm{d}t}\|\nabla u\|^2 + \nu\|\Delta u\|^2 + \mu_1\|u\|_3^2 + 2\mu_0\int_\Omega |e(u)|^{p-2} e_{ij}(u)e_{ij}(-\Delta u)\mathrm{d}x \leqslant \frac{\|f\|^2}{2\nu} + \frac{\nu\|\Delta u\|^2}{2}. \tag{3.1}$$

Let $I = 2\mu_0\int_\Omega |e(u)|^{p-2} e_{ij}(u)e_{ij}(-\Delta u)\mathrm{d}x$, then

$$\begin{aligned} I &= 2\mu_0\int_\Omega |e(u)|^{p-2}\frac{\partial e_{ij}(u)}{\partial x_k}\frac{\partial e_{ij}(u)}{\partial x_k}\mathrm{d}x + 2\mu_0(p-2)\int_\Omega |e(u)|^{p-2}\frac{\partial e_{ij}(u)}{\partial x_k}\frac{\partial e_{ij}(u)}{\partial x_k}\mathrm{d}x \\ &= 2\mu_0(p-1)\int_\Omega |e(u)|^{p-2}\frac{\partial e_{ij}(u)}{\partial x_k}\frac{\partial e_{ij}(u)}{\partial x_k}\mathrm{d}x, \end{aligned} \tag{3.2}$$

owing to the assumed condition $p > 1$, then $I > 0$, we can transfer (3.1) into the following inequality,

$$\frac{1}{2}\frac{\mathrm{d}}{\mathrm{d}t}\|\nabla u\|^2+\frac{\nu}{2}\|\Delta u\|^2\leqslant\frac{\|f\|^2}{2\nu}. \tag{3.3}$$

Integrating both sides between 0 and $t$ in (3.3), then we can obtain

$$\|\nabla u(t)\|^2\leqslant\int_0^t\frac{\|f\|^2}{\nu}\mathrm{d}s+\|\nabla u(0)\|^2\leqslant C, \tag{3.4}$$

$$\int_0^t\|\Delta u(s)\|^2\mathrm{d}s\leqslant C. \tag{3.5}$$

□

**Lemma 3.2** For $v_0\in H^1(\Omega), f\in H$, there exists constant $C$ such that

$$\|\nabla v(t)\|\leqslant C,\quad \int_0^t\|\Delta v(s)\|^2\mathrm{d}s\leqslant C,$$

where the constant only depends on $v_0, f, \nu$.

**Proof** Taking the inner product of (1.5) with $-\Delta v$ in $H$, similarly, for $n=2$, we have $b(u,u,-\Delta u)=0$,

$$\frac{1}{2}\frac{\mathrm{d}}{\mathrm{d}t}\|\nabla v\|^2+\frac{\nu}{2}\|\Delta v\|^2\leqslant\frac{1}{2\nu}\|f\|. \tag{3.6}$$

Integrating both sides between 0 and $t$ in (3.6), then

$$\|\nabla v(t)\|^2\leqslant\int_0^t\frac{\|f\|^2}{\nu}\mathrm{d}s+\|\nabla v(0)\|^2\leqslant C,$$

$$\int_0^t\|\Delta v(s)\|^2\mathrm{d}s\leqslant C.$$

□

**Theorem 3.1** Consider the difference $\omega(x,t)=u(x,t)-v(x,t)$ between a solution of the non-Newtonian system and a solution of the Navier-Stokes system, and assume that these have the same initial value $u_0=v_0\in H^1(\Omega)$, for the finite time $t\in[0,T]$, then the difference $\omega(x,t)$ obeys following estimate:

**Case (1)** $1<p<2$,

$$\|\omega(t)\|\leqslant C(\mu_0 t+\mu_1)^{\frac{1}{2}}\exp(Ct).$$

**Case (2)** $2<p<3$,

$$\|\omega(t)\|\leqslant C(\mu_0 t^{\frac{3-p}{2}}+\mu_1)^{\frac{1}{2}}\exp(Ct).$$

**Proof** Obviously, $\omega(x,t)$ satisfies the following equality:

$$\omega_t+u\cdot\nabla\omega+\omega\cdot\nabla v-\nu\Delta\omega=\nabla\cdot(\gamma(u)e(u))-\mu_1\Delta^2u-(\nabla\pi^{n-N}-\nabla\pi^{N-S}), \tag{3.7}$$

taking the inner product of (3.7) with $\omega$ in $H$. For $n=2$, owing to the divergence free condition, we have $b(u,\omega,\omega)=0$,

$$\frac{1}{2}\frac{\mathrm{d}}{\mathrm{d}t}\|\omega\|^2+\nu\|\nabla\omega\|^2=-2\mu_0\int_\Omega|e(u)|^{p-2}e_{ij}(u)e_{ij}(\omega)\mathrm{d}x-\mu_1(\Delta^2u,\omega)-b(\omega,v,\omega).$$

We estimate the right hand side of the above equality term by term.

$$\mu_1|(\Delta^2u,\omega)|\leqslant\mu_1\|\Delta\omega\|\,\|\Delta u\|\leqslant\mu_1C(\|\Delta v\|^2+\|\Delta u\|^2). \tag{3.8}$$

From the divergence free condition, we easily obtain $b(\omega,v,\omega)=-b(\omega,\omega,v)$,

$$
\begin{aligned}
|b(\omega, v, \omega)| = |b(\omega, \omega, v)| &\leqslant \|\nabla \omega\| \|\omega\|_{L^4} \|v\|_{L^4} \\
&\leqslant C \|\nabla \omega\|^{\frac{3}{2}} \|\omega\|^{\frac{1}{2}} \|\nabla v\|^{\frac{1}{2}} \|v\|^{\frac{1}{2}} \\
&\leqslant C \|\nabla \omega\|^{\frac{3}{2}} \|\omega\|^{\frac{1}{2}},
\end{aligned} \tag{3.9}
$$

where we have employed the Gagliardo-Nirenberg inequality,

$$
\|\omega\|_{L^4} \leqslant C \|\omega\|^{\frac{1}{2}} \|\nabla \omega\|^{\frac{1}{2}},
$$

and the conclusion of Lemma 3.2.

According to (3.9), we can deduce the following inequality

$$
|b(\omega, v, \omega)| \leqslant \frac{\nu}{2} \|\nabla \omega\|^2 + \frac{C}{\nu} \|\omega\|^2, \tag{3.10}
$$

where we have used the $\varepsilon$ -Young inequality.

Finally, we estimate the nonlinear term $2\mu_0 \int_\Omega |e(u)|^{p-2} e_{ij}(u) e_{ij}(\omega) \mathrm{d}x$. Let

$$
M = 2\mu_0 \left| \int_\Omega |e(u)|^{p-2} e_{ij}(u) e_{ij}(\omega) \mathrm{d}x \right|.
$$

**Case (1)** $1 < p < 2$, then $0 < p - 1 < 1$,

$$
\begin{aligned}
M &\leqslant 2\mu_0 \int_\Omega |e(u)|^{p-1} |e_{ij}(\omega)| \mathrm{d}x \\
&\leqslant 2\mu_0 \|e(\omega)\| \left( \int_\Omega |e(u)|^{2(p-1)} \mathrm{d}x \right)^{\frac{1}{2}} \\
&\leqslant 2\mu_0 |\Omega|^{\frac{2-p}{2}} C \left( \int_\Omega |e(u)|^2 \mathrm{d}x \right)^{\frac{p-1}{2}} \\
&\leqslant \mu_0 C.
\end{aligned}
$$

where the third inequality is due to Holder inequality, and the conclusions of Lemmas 3.1 and 3.2.

Combining the above estimates, we can obtain the following inequality,

$$
\frac{1}{2} \frac{\mathrm{d}}{\mathrm{d}t} \|\omega\|^2 + \frac{\nu}{2} \|\nabla \omega\|^2 \leqslant \mu_0 C + C \|\omega\|^2 + C\mu_1 (\|\Delta u\|^2 + \|\Delta v\|^2).
$$

Applying the Gronwall inequality, then

$$
\begin{aligned}
\|\omega(t)\|^2 &\leqslant \mathrm{e}^{Ct} \|\omega(0)\|^2 + \int_0^t \mathrm{e}^{C(t-s)} [C\mu_0 + C\mu_1 (\|\Delta u(s)\|^2 + \|\Delta v(s)\|^2)] \mathrm{d}s \\
&\leqslant \mathrm{e}^{Ct} (\|\omega(0)\|^2 + C\mu_0 t + C\mu_1),
\end{aligned} \tag{3.12}
$$

where the last inequality is due to the conclusions of Lemmas 3.1 and 3.2.

Then for $1 < p < 2$, we obtain the estimate

$$
\|\omega(t)\| \leqslant C(\mu_0 t + \mu_1)^{\frac{1}{2}} \mathrm{e}^{Ct}.
$$

**Case (2)** $2 < p < 3$, then $1 < p - 1 < 2$,

$$
\begin{aligned}
M &\leqslant 2\mu_0 \int_\Omega |e(u)|^{p-1} |e_{ij}(\omega)| \mathrm{d}x \\
&\leqslant 2\mu_0 \|e(\omega)\| \left( \int_\Omega |e(u)|^{2(p-1)} \mathrm{d}x \right)^{\frac{p-1}{2(p-1)}}
\end{aligned}
$$

$$\leqslant 2\mu_0 C\|\Delta u\|^{p-1}, \tag{3.13}$$

where the third inequality is due to the Sobolev imbedding, for $n=2$,

$$H^2(\Omega) \hookrightarrow W^{1,q}(\Omega)(q = 2(p-1)).$$

Combining the above estimates, we can obtain the following inequality,

$$\frac{1}{2}\frac{\mathrm{d}}{\mathrm{d}t}\|\omega\|^2 + \frac{\nu}{2}\|\nabla\omega\|^2 \leqslant \mu_0 C\|\Delta u\|^{p-1} + C\|\omega\|^2 + C\mu_1(\|\Delta u\|^2 + \|\Delta v\|^2).$$

Applying the Gronwall inequality, then

$$\begin{aligned}\|\omega(t)\|^2 &\leqslant \mathrm{e}^{Ct}\|\omega(0)\|^2 + C\mu_1\int_0^t \mathrm{e}^{C(t-s)}(\|\Delta u(s)\|^2 + \|\Delta v(s)\|^2)\mathrm{d}s + \\ &\quad \mu_0 C\int_0^t \mathrm{e}^{C(t-s)}\|\Delta u(s)\|^{p-1}\mathrm{d}s \\ &\leqslant \mathrm{e}^{Ct}\left(\|\omega(0)\|^2 + C\mu_0 t^{\frac{3-p}{2}}\left(\int_0^t\|\Delta u(s)\|^2\mathrm{d}s\right)^{\frac{p-1}{2}} + C\mu_1\right) \\ &\leqslant \mathrm{e}^{Ct}(\|\omega(0)\|^2 + C\mu_0 t^{\frac{3-p}{2}} + C\mu_1),\end{aligned} \tag{3.14}$$

where the second inequality is due to the Holder inequality.

For $2 < p < 3$, we obtain the estimate

$$\|\omega(t)\| \leqslant C(\mu_0 t^{\frac{3-p}{2}} + \mu_1)^{\frac{1}{2}}\exp^{Ct}. \qquad \square$$

## 4 The Convergence in $H^1$-norm

In this section, we obtain the conclusion that the solutions of non-Newtonian fluids converge to the solutions of Navier-Stokes equations in the sense of $H^1$-norm and $L^\infty$-norm. Furthermore, we want to emphasize two points that differ from the above section. First, we can only obtain the $H^1$-norm convergence estimate in the case of $2<p<3$. For the case $1<p<2$, we cannot solve it with the present method. Second, these estimates are obtained under the added condition $\mu_0 = \mu_1$.

**Lemma 4.1** For $u_0 \in V, f \in H^1(\Omega), \mu_0 = \mu_1$, in the case of $2<p<3$, there exists constant $C$ such that

$$\|u(t)\|_2^2 \leqslant C(1+\mu_1), \quad \int_0^t \|u(s)\|_3^2\mathrm{d}s \leqslant C(1+\mu_1),$$

where the constant doesn't depend on the $\mu_1$.

**Proof** Taking the inner product of (1.1) with $\Delta^2 u$ in $H$,

$$\begin{aligned}&\frac{1}{2}\frac{\mathrm{d}}{\mathrm{d}t}\|\Delta u\|^2 + \mu_1\|u\|_4^2 + \nu\|u\|_3^2 \\ &= (f,\Delta^2 u) - 2\mu_0\int_\Omega |e(u)|^{p-2}e_{ij}(u)e_{ij}(\Delta^2 u)\mathrm{d}x - b(u,u,\Delta^2 u).\end{aligned}$$

We estimate the right hand side of the above equality respectively, obviously,

$$|(f,\Delta^2 u)| \leqslant \frac{\|f\|_1^2}{\nu} + \frac{\nu\|u\|_3^2}{4},$$

for the second term,

$$
\begin{aligned}
&-2\mu_0\int_\Omega |e(u)|^{p-2}e_{ij}(u)e_{ij}(\Delta^2 u)\mathrm{d}x\\
&=2\mu_0\int_\Omega |e(u)|^{p-2}\frac{\partial e_{ij}(u)}{\partial x_k}\frac{\partial^3 e_{ij}(u)}{\partial x_k^3}\mathrm{d}x+2\mu_0(p-2)\int_\Omega |e(u)|^{p-2}\frac{\partial e_{ij}(u)}{\partial x_k}\frac{\partial^3 e_{ij}(u)}{\partial x_k^3}\mathrm{d}x\\
&=2\mu_0(p-1)\int_\Omega |e(u)|^{p-2}\frac{\partial e_{ij}(u)}{\partial x_k}\frac{\partial^3 e_{ij}(u)}{\partial x_k^3}\mathrm{d}x. \qquad (4.1)
\end{aligned}
$$

Let $\mathcal{Q}=2\mu_0\left|\int_\Omega |e(u)|^{p-2}e_{ij}(u)e_{ij}(\Delta^2 u)\mathrm{d}x\right|$, then

$$
\begin{aligned}
\mathcal{Q}&=2\mu_0(p-1)\left|\int_\Omega |e(u)|^{p-2}\frac{\partial e_{ij}(u)}{\partial x_k}\frac{\partial^3 e_{ij}(u)}{\partial x_k^3}\mathrm{d}x\right|\\
&\leqslant 2\mu_0(p-1)\|\nabla e(u)\|_{L^\infty}\|e(u)\|_3\left(\int_\Omega |e(u)|^{2(p-2)}\mathrm{d}x\right)^{\frac{1}{2}}\\
&\leqslant 2\mu_0(p-1)C\|\nabla e(u)\|_{L^\infty}\|e(u)\|_3\left(\int_\Omega |e(u)|^{2}\mathrm{d}x\right)^{\frac{p-2}{2}}\\
&\leqslant 2\mu_0(p-1)C\|\nabla e(u)\|_{L^\infty}\|e(u)\|_3, \qquad (4.2)
\end{aligned}
$$

where due to $2<p<3, 0<p-2<1$, we can apply the Holder inequality, and the last inequality is due to the conclusion of Lemma 3.1.

Noticing that Gagliardo-Nirenberg inequality,

$$\|D^2 u\|_{L^\infty}\leqslant C\|u\|^{\frac{1}{4}}\|D^4 u\|^{\frac{3}{4}}.$$

Then

$$\mathcal{Q}\leqslant 2\mu_0(p-1)C\|u\|^{\frac{1}{4}}\|u\|_4^{\frac{7}{4}}\leqslant 2\mu_0(p-1)C\|u\|_4^{\frac{7}{4}}\leqslant\frac{\mu_1\|u\|_4^2}{2}+C\mu_1,$$

where we have used the assumed condition $\mu_0=\mu_1$, the conclusion of Lemma 3.1 and $\varepsilon$-Young inequality.

$$|b(u,u,\Delta^2 u)|\leqslant\|u\|_{L^\infty}\|\Delta u\|\|D^3 u\|+\|\nabla u\|_{L^4}^2\|D^3 u\|, \qquad (4.3)$$

we estimate the right hand side of (4.3) respectively.

As to the first term, applying Gagliardo-Nirenberg inequality,

$$\|u\|_{L^\infty}\leqslant C\|u\|^{\frac{1}{2}}\|\Delta u\|^{\frac{1}{2}},$$

$$\|u\|_{L^\infty}\|\Delta u\|\|D^3 u\|\leqslant C\|u\|^{\frac{1}{2}}\|\Delta u\|^{\frac{3}{2}}\|D^3 u\|\leqslant C\|\Delta u\|^{\frac{3}{2}}\|D^3 u\|.$$

From Lemma 2.2, we know

$$\|u\|_2\leqslant C\|u\|_1^{\frac{1}{2}}\|u\|_3^{\frac{1}{2}},$$

$$\|u\|_{L^\infty}\|\Delta u\|\|D^3 u\|\leqslant C\|u\|_1^{\frac{3}{4}}\|u\|_3^{\frac{7}{4}}\leqslant\frac{\nu}{8}\|u\|_3^2+C.$$

For the second term in the right hand side of (4.3), applying Gagliardo-Nirenberg inequality,

$$\|\nabla u\|_{L^4}\leqslant C\|\Delta u\|^{\frac{3}{4}}\|u\|^{\frac{1}{4}},$$

$$\|\nabla u\|_{L^4}^2\|D^3u\|\leqslant C\|\Delta u\|^{\frac{3}{2}}\|D^3u\|,$$

similarly, we can obtain

$$\|\nabla u\|_{L^4}^2\|D^3u\|\leqslant \frac{\nu}{8}\|u\|_3^2+C,$$

then

$$|b(u,u,\Delta^2u)|\leqslant \frac{\nu}{4}\|u\|_3^2+C. \tag{4.5}$$

Combining the above estimates, we have

$$\frac{1}{2}\frac{\mathrm{d}}{\mathrm{d}t}\|\Delta u\|^2+\frac{\nu}{2}\|u\|_3^2+\frac{\mu_1}{2}\|u\|_4^2\leqslant\frac{\|f\|_1^2}{\nu}+C+C\mu_1. \tag{4.6}$$

Integrating both sides between 0 and $t$ in (4.6), then

$$\|u(t)\|_2^2\leqslant C(1+\mu_1),$$

$$\int_0^t\|u(s)\|_3^2\mathrm{d}s\leqslant C(1+\mu_1).$$ □

**Lemma 4.2** For $v_0\in V, f\in H^1(\Omega)$, there exists constant $C$ such that

$$\|v(t)\|_2\leqslant C,\quad \int_0^t\|v(s)\|_3^2\mathrm{d}s\leqslant C,$$

where the constant only depends on $v_0, f, \nu$.

**Proof** Taking the inner product of (1.5) with $\Delta^2 v$ in $H$, then

$$\frac{1}{2}\frac{\mathrm{d}}{\mathrm{d}t}\|\Delta v\|^2+\nu\|v\|_3^2\leqslant\frac{\|f\|_1^2}{2\nu}+\frac{\nu\|v\|_3^2}{2}+|b(v,v,\Delta^2v)|,$$

and

$$|b(v,v,\Delta^2v)|\leqslant\frac{\nu\|v\|_3^2}{4}+C,$$

where the estimate is similar to (4.5).

$$\frac{1}{2}\frac{\mathrm{d}}{\mathrm{d}t}\|\Delta v\|^2+\frac{\nu\|v\|_3^2}{4}\leqslant\frac{\|f\|_1^2}{2\nu}+C. \tag{4.7}$$

Integrating both sides between 0 and $t$ in (4.7), then

$$\|v(t)\|_2\leqslant C,\quad \int_0^t\|v(s)\|_3^2\mathrm{d}s\leqslant C.$$ □

**Theorem 4.1** Consider the difference $\omega(x,t)=u(x,t)-v(x,t)$ between a solution of the non-Newtonian system and a solution of the Navier-Stokes system, and assume that these have the same initial value $u_0=v_0\in V, \mu_0=\mu_1$, for the finite time $t\in[0,T]$, then the difference $\omega(x,t)$ obeys the estimate:

$$\|\nabla\omega(t)\|\leqslant C\mu_0^{\frac{1}{2}}\left[(1+(1+\mu_0^{\frac{p}{2}})t^{\frac{4-p}{2}}+\mu_0)\right]^{\frac{1}{2}}\exp(C+Ct+C\mu_0),\quad 2<p<3.$$

**Proof** $\omega(x,t)$ satisfies the following equation:

$$\omega_t+u\cdot\nabla\omega+\omega\cdot\nabla v-\nu\Delta\omega=\nabla\cdot(\gamma(u)e(u))-\mu_1\Delta^2u-(\nabla\pi^{n-N}-\nabla\pi^{N-S}), \tag{4.8}$$

taking the inner product of (4.8) with $-\Delta\omega$ in $H$,

$$\frac{1}{2}\frac{\mathrm{d}}{\mathrm{d}t}\|\nabla\omega\|^2+\nu\|\Delta\omega\|^2\leqslant|(\mu_1\Delta^2u,\Delta\omega)|+2\mu_0\left|\int_\Omega|e(u)|^{p-2}e_{ij}(u)e_{ij}(\Delta\omega)\mathrm{d}x\right|+ |b(\omega,v,\Delta\omega)|+|b(u,\omega,\Delta\omega)|. \tag{4.9}$$

As to the estimate of the second term of the above inequality, let

$$\begin{aligned}
N&=2\mu_0\left|\int_\Omega|e(u)|^{p-2}e_{ij}(u)e_{ij}(\Delta\omega)\mathrm{d}x\right|\\
&=2\mu_0\left|\int_\Omega|e(u)|^{p-2}\frac{\partial e_{ij}(u)}{\partial x_k}\frac{\partial e_{ij}(\omega)}{\partial x_k}\mathrm{d}x+(p-2)\int_\Omega|e(u)|^{p-2}\frac{\partial e_{ij}(u)}{\partial x_k}\frac{\partial e_{ij}(\omega)}{\partial x_k}\mathrm{d}x\right|\\
&=2\mu_0(p-1)\left|\int_\Omega|e(u)|^{p-2}\frac{\partial e_{ij}(u)}{\partial x_k}\frac{\partial e_{ij}(\omega)}{\partial x_k}\mathrm{d}x\right|\\
&\leqslant2\mu_0(p-1)C\|e(u)\|_{L^\infty}^{p-2}\|\Delta u\|\|\Delta\omega\|\\
&\leqslant\mu_0C\|u\|_3^{p-2}(\|\Delta u\|^2+\|\Delta v\|^2)\\
&\leqslant C\mu_0(1+\mu_0)\|u\|_3^{p-2},
\end{aligned}$$

where the second inequality is due to the Sobolev imbedding, and the last inequality is the conclusions of Lemmas 4.1 and 4.2.

$$|(\mu_1\Delta^2u,\Delta\omega)|\leqslant\mu_1\|u\|_3\|\omega\|_3\leqslant\mu_1C(\|u\|_3^2+\|\omega\|_3^2)\leqslant C\mu_0(\|u\|_3^2+\|v\|_3^2),$$

where we have used the assumed condition $\mu_0=\mu_1$.

$$|b(u,\omega,\Delta\omega)|\leqslant\|\nabla u\,\nabla\omega\|\|\nabla\omega\|\leqslant C\|u\|_3\|\nabla\omega\|^2, \tag{4.10}$$

where we have used the divergence free condition and the Sobolev imbedding, for $n=2$, $H^2(\Omega)\hookrightarrow L^\infty(\Omega)$.

$$\begin{aligned}
|b(\omega,v,\Delta\omega)|&\leqslant C\|\nabla\omega\,\nabla v\|\|\nabla\omega\|+\|\omega D^2v\|\|\nabla\omega\|\\
&\leqslant C\|v\|_3\|\nabla\omega\|^2+C\|\omega\|_{L^4}\|D^2v\|_{L^4}\|\nabla\omega\|\\
&\leqslant C\|v\|_3\|\nabla\omega\|^2+C\|\omega\|_1\|D^2v\|_1\|\nabla\omega\|\\
&\leqslant C\|v\|_3\|\nabla\omega\|^2,
\end{aligned}\tag{4.11}$$

where we have used the divergence free condition, Holder inequality, and the Sobolev imbedding, for $n=2, H^2(\Omega)\hookrightarrow L^\infty(\Omega)$, and $H^1(\Omega)\hookrightarrow L^4(\Omega)$.

From the estimates (4.10) and (4.11), we have the following inequality

$$|b(\omega,v,\Delta\omega)|+|b(u,\omega,\Delta\omega)|\leqslant C(\|u\|_3+\|v\|_3)\|\omega\|_1^2.$$

After obtaining the estimates for the right hand side of (4.9) term by term, we have

$$\begin{aligned}
&\frac{1}{2}\frac{\mathrm{d}}{\mathrm{d}t}\|\nabla\omega\|^2+\nu\|\Delta\omega\|^2\\
\leqslant{}&C\mu_0(1+\mu_0)\|u\|_3^{p-2}+C\mu_0(\|u\|_3^2+\|v\|_3^2)+C(\|u\|_3+\|v\|_3)\|\omega\|_1^2.
\end{aligned}$$

Applying the Gronwall inequality, we deduce

$$\begin{aligned}
\|\nabla\omega(t)\|^2\leqslant{}&\exp^{\int_0^tC(\|u(s)\|_3+\|v(s)\|_3)\mathrm{d}s}\|\nabla\omega(0)\|^2+\\
&\int_0^t\exp^{\int_s^tC(\|u(\sigma)\|_3+\|v(\sigma)\|_3)\mathrm{d}\sigma}\left[C\mu_0(1+\mu_0)\|u(s)\|_3^{p-2}+C\mu_0(\|u(s)\|_3^2+\|v(s)\|_3^2)\right]\mathrm{d}s
\end{aligned}$$

$$\leqslant \exp^{\int_0^t C(\|u(s)\|_3^2+\|v(s)\|_3^2+1)ds}\Big[\|\nabla\omega(0)\|^2 + C\mu_0(1+\mu_0)$$

$$\int_0^t \|u(s)\|_3^{p-2}ds + C\mu_0(1+\mu_0) + C\mu_0\Big],$$

where we have used the conclusions of Lemmas 4.1 and 4.2.

Applying the Holder inequality, we have

$$\|\nabla\omega(t)\|^2 \leqslant \exp^{(C+Ct+C\mu_0)}(\|\nabla\omega(0)\|^2 + C\mu_0(1+\mu_0)t^{\frac{4-p}{2}}(1+u_0)^{\frac{p-2}{2}} + C\mu_0 + C\mu_0^2)$$

$$\leqslant (\|\nabla\omega(0)\|^2 + C\mu_0(1+\mu_0)^{\frac{p}{2}}t^{\frac{4-p}{2}} + C\mu_0 + C\mu_0^2)\exp(C+Ct+C\mu_0),$$

noticing

$$(a+b)^m \leqslant 2^m(a^m+b^m), \quad m \geqslant 1,$$

from the assumed condition $2 < p < 3$, we have $(1+\mu_0)^{\frac{p}{2}} \leqslant C(1+\mu_0^{\frac{p}{2}})$,

$$\|\nabla\omega(t)\| \leqslant \Big[\|\nabla\omega(0)\|^2 + C\mu_0(1+(1+\mu_0^{\frac{p}{2}})t^{\frac{4-p}{2}} + \mu_0)\Big]\exp(C+Ct+C\mu_0)$$

$$= \Big[C\mu_0(1+(1+\mu_0^{\frac{p}{2}})t^{\frac{4-p}{2}} + \mu_0)\Big]\exp(C+Ct+C\mu_0).$$

For $2 < p < 3$, we obtain the estimate

$$\|\nabla\omega(t)\| \leqslant C\mu_0^{\frac{1}{2}}\Big[(1+(1+\mu_0^{\frac{p}{2}})t^{\frac{4-p}{2}} + \mu_0)\Big]^{\frac{1}{2}}\exp(C+Ct+C\mu_0). \qquad \square(4.12)$$

**Theorem 4.2** Let $u_0 = v_0 \in V, \mu_0 = \mu_1$, then the difference $\omega(x, t)$ in Theorem 4.1 obeys the following estimate for the finite time $t \in [0, T]$:

$$\|\omega(t)\|_{L^\infty} \leqslant C\mu_0^{\frac{1}{4}}(1+\mu_0^{\frac{1}{4}})(1+t^{\frac{3-p}{8}}), \quad 2 < p < 3.$$

**Proof** Applying the Gagliardo-Nirenberg inequality,

$$\|\omega\|_{L^\infty} \leqslant C\|\omega\|^{\frac{1}{2}}\|\Delta\omega\|^{\frac{1}{2}},$$

then

$$\|\omega(t)\|_{L^\infty} = \|u(t)-v(t)\|_{L^\infty} \leqslant C\|u(t)-v(t)\|^{\frac{1}{2}}\|\Delta u(t)-\Delta v(t)\|^{\frac{1}{2}}, \tag{4.13}$$

from the conclusions of Lemmas 3.1, 3.2, 4.1, 4.2 and Theorem 3.1, we can easily deduce

$$\|u(t)-v(t)\|_{L^\infty} \leqslant C(\|u(t)\|_2^{\frac{1}{2}} + \|v(t)\|_2^{\frac{1}{2}})[\mu_0(1+t^{\frac{3-p}{2}})]^{\frac{1}{4}}$$

$$\leqslant (C(1+\mu_0)^{\frac{1}{4}} + C)[u_0(1+t^{\frac{3-p}{2}})]^{\frac{1}{4}}$$

$$\leqslant C\mu_0^{\frac{1}{4}}(1+\mu_0^{\frac{1}{4}})(1+t^{\frac{3-p}{8}}). \tag{4.14}$$

For $2 < p < 3$, we obtain the estimate

$$\|\omega(t)\|_{L^\infty} \leqslant C\mu_0^{\frac{1}{4}}(1+\mu_0^{\frac{1}{4}})(1+t^{\frac{3-p}{8}}). \tag{4.15}$$

The proof is completed. $\square$

**References**


[1] Bellout H, Bloom F, Nečas J. Phenomenological behavior of multipolar viscous fluids. Quart Appl Math, 1992, 50: 559 – 583.

[2] Bellout H, Bloom F, Nečas J. Weak and measure-valued solutions for non-Newtonian fluids. C R Acad Sci Paris, 1993, 317: 795 – 800.

[3] Bellout H, Bloom F, Nečas J. Young measure-valued solutions for non-Newtonian incompressible viscous fluids. Comm Partial Differential Equations, 1994, 19: 1763 – 1803.

[4] Bleustein J L, Green A E. Dipolar fluids. Internat J Engrg Sci, 1967, 5: 323 – 340.

[5] Bloom F, Hao W. Regularization of a non-Newtonian system in an unbounded channel: Existence and uniqueness of solutions. Nonlinear Anal, 2001, 44: 281 – 309.

[6] Bloom F, Hao W. Regularization of a non-Newtonian system in an unbounded channel: Existence of a maximal compact attractor. Nonlinear Anal, 2001, 43: 743 – 766.

[7] Du Q, Gunzburger M. Analysis of a Ladyzhenskaya model for incompressible viscous flow. J Math Anal Appl, 1991, 155: 21 – 45.

[8] Franchi F, Straughan B. Stability and nonexistence results in the generalized theory of a second grade fluid. J Math Anal Appl, 1993, 180: 122 – 137.

[9] Guo B, Lin G, Shang Y. Non-Newtonian fluids dynamical systems. National Defense Industry Press, 2006 (in Chinese).

[10] Ladyzhenskaya O A. The mathematical theory of viscous incompressible flow. New York: Gordon and Breach, 1969.

[11] Ladyzhenskaya O A. New equations for the description of the viscous incompressible fluids and solvability in the large of the boundary value problems for them// Boundary Value Problem of Mathematical Physics, Vol V. Providence, RI: Amer Math Soc, 1970.

[12] Lions J L. Quelques méthodes de résolution des problmés aux limites nonlinéaires. Paris: Dunod, 1969.

[13] Man C S, Sun Q K. On the significance of normal stress effects in the flow of glaciers. J Glaciol, 1987, 33: 268 – 273.

[14] Pokorny M. Cauchy problem for the non-Newtonian incompressible fluids. Appl Math, 1996, 41: 169 – 201.

[15] Quintanilla R, Straughan B. Bounds for some non-standard problems in porous flow and viscous Green-Naghdi fluids. Proc R Soc Lond Ser A, 2005, 461: 3159 – 3168.

[16] Robinson J C. Infinite-dimensional Dynamical Systems: An Introduction to Dissipative Parabolic PDEs and the Theory of Global Attractors. Cambridge Univ. Press, 2001.

[17] Straughan B. Explosive Instabilities in Mechanics. Heidelberg: Springer, 1998.

[18] Straughan B. Sharp global nonlinear stability for temperature dependent viscosity convection. Proc R Soc Lond Ser A, 2002, 458: 1773 – 1782.

[19] Temam R. Navier-Stokes Equations: Theory and numerical analysis. 3rd ed. North-Holland, 1984.

[20] Temam R. Infinite dimensional dynamical systems in mechanics and physics, 2nd ed. Berlin: Springer, 1997.

# Sharp Constant in Nonlocal Inequality and Its Applications to Nonlocal Schrödinger Equation with Harmonic Potential *

Chen Jianqing(陈建清) Guo Boling(郭柏灵) Han Yongqian(韩永前)

**Abstract**

This paper contains two parts. In the first part, we derive a variant of Gagliardo-Nirenberg interpolation inequality involving nonlocal nonlinearity and determine its best (smallest) constant. In the second part, we study two applications of this inequality and its best constant. In the first application, we use this best constant to establish a sharp criterion for the global existence and blow-up of solutions of the inhomogeneous Schrödinger equation with harmonic potential and nonlocal nonlinearity

$$i\varphi_t = -\Delta\varphi + |x|^2\varphi - \varphi|\varphi|^{p-2}\int \frac{|\varphi(y)|^p}{|x-y|^\alpha}dy$$

in the critical case $p = 2 + (2-\alpha)/N$. The result indicates that the existence of blow-up solutions not only depends on the mass of the initial data but also on the profile of the initial data. In the second application, we use this best constant to prove that when $2 + (2-\alpha)/N < p < (2N-\alpha)/(N-2)$, the solutions exist globally in time for one class of initial data whose norm can be as large as one wants.

**Keywords** nonlocal interpolation inequality; sharp constant; global solutions; blow-up; nonlocal Schrödinger equation

## 1 Introduction

This paper is concerned with the nonlinear Schrödinger equation with harmonic potential

$$i\varphi_t + \Delta\varphi - |x|^2\varphi + F(\varphi) = 0, \quad x \in \mathbf{R}^N, \quad t \geqslant 0, \tag{1.1}$$

where $\varphi := \varphi(x,t): \mathbf{R}^N \times \mathbf{R}_+ \to \mathbf{C}$ is a complex-valued function and $F(\varphi)$ is a nonlinearity (possibly nonlocal) satisfying suitable assumptions given later.

Equation (1.1) models a lot of physical phenomena. For example it is known as the Gross-Pitaevskii (GP) equation in the context of Bose-Einstein condensates (BEC) with parabolic traps. In fact, assuming a highly anisotropic trap, Kivshar et al[17] derived the GP-equation equation (1.1) with $F(\varphi) = \varphi|\varphi|^2$ as a model equation for the macroscopic dynamics of cooled atoms confined in a three dimensional parabolic potential created by a magnetic trap. Deconinck et al derive a three dimensional Schrödinger equation with potential and nonlocal nonlinearity; see [10, Equ. (5)].

Equation (1.1) with nonlocal nonlinearity $F(\varphi)$ has also appeared in other applications.

* Commun Math Sci, 2009, 7(3): 549–570.

For example, Kurth derived in [18] that Schrödinger equation with harmonic potential and nonlocal nonlinearity can be used to describe average pulse propagation in dispersion-managed fibers. Usually the nonlocal nonlinearity is assumed to be of Hartree type, i.e.,

$$F(\varphi) = (V * |\varphi|^2)\varphi,$$

where the kernel $V$ is in some weak $L^q$-space and $*$ denotes convolution[10, 14] and the references therein. It is of interest to understand the original nonlocal problem directly, although in most applications it is hard to get a clue on properties of the kernel.

These motivate us to study equation (1.1) directly with nonlocal nonlinearity of the following form

$$i\varphi_t = -\Delta\varphi + |x|^2\varphi - \varphi|\varphi|^{p-2}\int_{\mathbf{R}^N}\frac{|\varphi(y)|^p}{|x-y|^\alpha}dy, \quad t \geqslant 0, \quad x \in \mathbf{R}^N, \tag{1.2}$$

where $p \geqslant 2$ and $0 < \alpha < N$. The purpose here is to extend those results for the Schrödinger equation with local nonlinearity to the case of nonlocal nonlinearity and shed some new light for the Schrödinger with nonlocal nonlinearity. We remark that when $\alpha = 1, p = 2$ and $N = 3$, equation (1.2) is equivalent to the Schrödinger-Possion system with harmonic potential,

$$\begin{cases} i\varphi_t + \Delta\varphi = |x|^2\varphi - V(x)\varphi, & x \in \mathbf{R}^N \\ \Delta V = |\varphi|^2. \end{cases}$$

This equation typically arises in the mean field approximation of many body effects, modeled by the Possion equation with a confinement modeled by the quadratic potential of the harmonic oscillator. If the potential vanishes, then this equation models the classical limit of the field equations describing quantum mechanical non-relativistic many boson system[1]. For other variants of the Schrödinger equation, we refer the interested reader to [6] and the references therein.

Now we come back to equation (1.2). Given an initial data

$$\varphi(x,0) = \varphi_0(x), \tag{1.3}$$

we are interested in the following question: Under what conditions the solutions of equations (1.2)~(1.3) exist globally in time and under what conditions the solutions of equations (1.2)~(1.3) blow up in a finite time?

Noting that in the case of local nonlinearity $F(\varphi) = \varphi|\varphi|^{p-2}$, one can use the standard Gagliardo-Nirenberg inequality to study equation (1.1); we refer the interested readers to [11, 22, 25, 7, 8] for the stability of standing waves as well as global existence of equation (1.1) with various initial data. However, for the nonlocal nonlinearity like equation (1.2), the issue of whether or not a particular choice of initial data generates a blow-up solution of equation (1.2) is more subtle. In particular the arguments used in [7] cannot deduce a sharp condition on the existence of global solutions or blow-up solutions of equation (1.2), so we need to find other ways as we will see in the present paper.

In Section 2, we derive a variant of the Gagliardo-Nirenberg interpolation inequality involving nonlocal nonlinearity and determine the best (smallest) constant; see Theorem 2.3 and Theorem 2.7. We emphasize that the best constant determined here is not only of independent interest, but can also be used to study equations (1.2)~(1.3). In Section 3, we sketch some results on the Cauchy problem of equation (1.2). In Section 4, since we deal with the problem harmonic potential $|x|^2$, we use Theorem 2.7 to establish a sharp criterion on the solutions of equations (1.2)~(1.3) in the critical nonlinearity $p = 2 + (2-\alpha)/N$; see Theorem 4.1 and Theorem 4.3. The results indicate a sharp threshold on the existence of global solutions of equation (1.2). Furthermore, the results suggest the existence of blow-up solutions of equation (1.2) depends on both the mass of the initial data and the profile of the initial data. In Section 5, we give a sufficient condition to the existence of global solutions of equation (1.2) in the case where $2 + (2-\alpha)/N < p < (2N-\alpha)/(N-2)$. The interesting aspect is that equation (1.2) possesses global solutions even when one takes the initial data as large as one wants; see Theorem 5.3. The final section is an appendix where we outline the proof of Proposition 3.1.

To establish the results in view, we base our approach on an idea originated from [24,2]. We emphasize that the use of Theorem 2.7 is essential in the proofs of Theorem 4.1, Theorem 4.3, and Theorem 5.3. We point out that Theorem 2.7 can also be used to establish a sharp threshold to the existence and nonexistence of standing waves of NLS-equation with nonlocal nonlinearity as well as their stability. The results will be published elsewhere.

**Notations** Throughout this paper, $H^1(\mathbf{R}^N)$ denotes the standard Sobolev space with the standard norm. We denote the norm of the space $L^p(\mathbf{R}^n)$ by $\|\cdot\|_{L^q} 1 \leqslant q \leqslant \infty$, and denote the integral $\int_{\mathbf{R}^N} \mathrm{d}x$ simply by $\int$ unless stated otherwise. We also denote various positive constants by $C$ or $C_j$. $\bar{v}$ denotes the complex conjugate and Re denotes the real part.

## 2 Best (smallest) Constant

The goal of this section is to derive a variant of the Gagliardo-Nirenberg interpolation inequality (see Theorem 2.3) and determine its best constant (see Theorem 2.7). This inequality and the best constant will play an essential role in what follows. Firstly, we state some lemmas.

**Lemma 2.1** (Gagliardo-Nirenberg Inequality[5]) Let $1 < q < \frac{N+2}{N-2}$ when $N \geqslant 3$ and $1 < q < +\infty$ when $N = 1, 2$. Then there is a positive constant $C$ depending on $N$ and $q$ such that for any $u \in H^1(\mathbf{R}^N)$,

$$\int |u|^{q+1} \leqslant C\left(\int |\nabla u|^2\right)^{\frac{N(q-1)}{4}} \left(\int |u|^2\right)^{\frac{q+1}{2}-\frac{N(q-1)}{4}}.$$

**Lemma 2.2** [21, p. 31] Let $0 < \beta < N$ and $f \in L^q(\mathbf{R}^N)$, $h \in L^r(\mathbf{R}^N)$ with $\frac{1}{q} + \frac{1}{r} + \frac{\beta}{N} = 2$ and $1 < q, r < \infty$. Then

$$\int_{\mathbf{R}^N \times \mathbf{R}^N} \frac{|f(x)|\,|h(y)|}{|x-y|^\beta} \mathrm{d}x\mathrm{d}y \leqslant C(q,r,\beta,N) \|f\|_{L^q(\mathbf{R}^N)} \|h\|_{L^r(\mathbf{R}^N)}, \quad x, y \in \mathbf{R}^N,$$

where $C(q,r,\beta,N)$ is a positive constant depending on $q$, $r$, $\beta$, and $N$.

Next we use Lemma 2.1 and Lemma 2.2 to derive a variant of Gagliardo-Nirenberg interpolation inequality.

**Theorem 2.3** Let $0<\alpha<N$ and $(2N-\alpha)/N<p<2^*(\alpha)$, where $2^*(\alpha):=(2N-\alpha)/(N-2)$ when $N\geqslant 3$ and $(2N-\alpha)/N<p<+\infty$ when $N=1,2$. Then there is a positive constant $C(p,\alpha,N)$ depending on $p,\alpha$ and $N$ such that for any $u\in H^1(\mathbf{R}^N)$,

$$\int_{\mathbf{R}^N\times\mathbf{R}^N}\frac{|u(x)|^p|u(y)|^p}{|x-y|^\alpha}\mathrm{d}x\mathrm{d}y\leqslant C(p,\alpha,N)\left(\int|\nabla u|^2\right)^A\left(\int|u|^2\right)^B,\tag{2.1}$$

where $A=\dfrac{N(p-2)+\alpha}{2}$ and $B=\dfrac{2p-[N(p-2)+\alpha]}{2}$.

**Proof** Taking $|f(x)|=|u(x)|^p$, $|h(y)|=|u(y)|^p$, $q=r=\dfrac{2N}{2N-\alpha}$, and applying Lemma 2.2, we get that

$$\int_{\mathbf{R}^N\times\mathbf{R}^N}\frac{|u(x)|^p|u(y)|^p}{|x-y|^\alpha}\mathrm{d}x\mathrm{d}y\leqslant C\left(\int|u|^{\frac{2Np}{2N-\alpha}}\right)^{\frac{2N-\alpha}{N}}.\tag{2.2}$$

By the assumption of $(2N-\alpha)/N<p<2^*(\alpha)$ and Lemma 2.1, we obtain that

$$\int|u|^{\frac{2Np}{2N-\alpha}}\leqslant C\left(\int|\nabla u|^2\right)^{\frac{N(Np-2N+\alpha)}{2(2N-\alpha)}}\left(\int|u|^2\right)^{\frac{2Np-N(Np-2N+\alpha)}{2(2N-\alpha)}}.\tag{2.3}$$

It is now deduced from (2.2) and (2.3) that (2.1) holds. □

**Lemma 2.4** Let $N\geqslant 2$, $0<\alpha<N$ and $(2N-\alpha)/N<p<2^*(\alpha)$. If $u_n\rightharpoonup u$ weakly in $H^1_{\mathrm{radial}}(\mathbf{R}^N):=\{u\in H^1(\mathbf{R}^N); u(x)=u(|x|)\}$, then

$$\int_{\mathbf{R}^N\times\mathbf{R}^N}\frac{|u_n(x)|^p|u_n(y)|^p}{|x-y|^\alpha}\mathrm{d}x\mathrm{d}y\to\int_{\mathbf{R}^N\times\mathbf{R}^N}\frac{|u(x)|^p|u(y)|^p}{|x-y|^\alpha}\mathrm{d}x\mathrm{d}y.$$

**Proof** Firstly, for positive numbers $a,b,c$, and $d$, there holds

$$\begin{aligned}2|a^pb^p-c^pd^p|&=|(b^p+d^p)(a^p-c^p)+(a^p+c^p)(b^p-d^p)|\\&\leqslant(b^p+d^p)(a^{p-1}+c^{p-1})|a-c|+(a^p+c^p)(b^{p-1}+d^{p-1})|b-d|.\end{aligned}\tag{IE}$$

Putting $a=|u_n(x)|$, $b=|u_n(y)|$, $c=|u(x)|$, $d=|u(y)|$, we obtain from Lemma 2.2 with $\beta=\alpha$, $r=q=\dfrac{2N}{2N-\alpha}$, and the fact that $\{u_n\}$ is bounded in $H^1_{\mathrm{radial}}(\mathbf{R}^N)$ that

$$\begin{aligned}&\int_{\mathbf{R}^N\times\mathbf{R}^N}\frac{(|u_n(y)|^p+|u(y)|^p)(|u_n(x)|^{p-1}+|u(x)|^{p-1})||u_n(x)|-|u(x)||}{|x-y|^\alpha}\mathrm{d}x\mathrm{d}y\\
&\leqslant C_1\left[\int(|u_n(x)-u(x)||\,|u_n(x)|^{p-1}+|u(x)|^{p-1}|)^{\frac{2N}{2N-\alpha}}\right]^{\frac{2N-\alpha}{2N}}\times\\
&\quad\left(\int(|u_n(y)|^p+|u(y)|^p)^{\frac{2N}{2N-\alpha}}\right)^{\frac{2N-\alpha}{2N}}\\
&\leqslant C_2\left[\left(\int|u_n(x)-u(x)|^{\frac{2Np}{2N-\alpha}}\right)^{\frac{1}{p}}\left(\int[|u_n(x)|^{\frac{2Np}{2N-\alpha}}+|u(x)|^{\frac{2Np}{2N-\alpha}}]\right)^{\frac{p}{p-1}}\right]^{\frac{2N-\alpha}{2N}}\times\\
&\quad\left[\int(|u_n(y)|^{\frac{2Np}{2N-\alpha}}+|u(y)|^{\frac{2Np}{2N-\alpha}})\right]^{\frac{2N-\alpha}{2N}}\end{aligned}$$

$$\leqslant C_3\left(\int |u_n(x)-u(x)|^{\frac{2Np}{2N-\alpha}}\right)^{\frac{2N-\alpha}{2Np}}.$$

It is now deduced from the Strauss lemma, i.e.,

$u_n \rightharpoonup u$ weakly in $H^1_{\mathrm{radial}}(\mathbf{R}^N)$ implies that $u_n \to u$ strongly in $L^{\frac{2Np}{2N-\alpha}}(\mathbf{R}^N)$, that

$$\int |u_n(x)-u(x)|^{\frac{2Np}{2N-\alpha}} \to 0 \quad \text{as} \quad n\to\infty.$$

Therefore as $n\to\infty$, we have that

$$\int_{\mathbf{R}^N\times\mathbf{R}^N}\frac{(|u_n(y)|^p+|u(y)|)^p(|u_n(x)|^{p-1}+|u(x)|^{p-1})\,|\,|u_n(x)|-|u(x)|\,|}{|x-y|^\alpha}\mathrm{d}x\mathrm{d}y\to 0.$$

By a similar argument, we obtain that as $n\to\infty$,

$$\int_{\mathbf{R}^N\times\mathbf{R}^N}\frac{(|u_n(x)|^p+|u(x)|^p)(|u_n(y)|^{p-1}+|u(y)|^{p-1})\,|\,|u_n(y)|-|u(y)|\,|}{|x-y|^\alpha}\mathrm{d}x\mathrm{d}y\to 0.$$

Combining these with (IE), we easily obtain that

$$\int_{\mathbf{R}^N\times\mathbf{R}^N}\frac{|u_n(x)|^p|u_n(y)|^p}{|x-y|^\alpha}\mathrm{d}x\mathrm{d}y\to\int_{\mathbf{R}^N\times\mathbf{R}^N}\frac{|u(x)|^p|u(y)|^p}{|x-y|^\alpha}\mathrm{d}x\mathrm{d}y.$$

The proof is complete. □

Now we are going to determine the best (smallest) positive constant $C(p,\alpha,N)$ such that (2.1) holds. This will be done by minimizing the functional

$$J(u)=\frac{\left(\int|\nabla u|^2\right)^A\left(\int|u|^2\right)^B}{\int_{\mathbf{R}^N\times\mathbf{R}^N}\frac{|u(x)|^p|u(y)|^p}{|x-y|^\alpha}\mathrm{d}x\mathrm{d}y},\quad u\in H^1(\mathbf{R}^N),$$

where

$$A=\frac{N(p-2)+\alpha}{2},\quad B=\frac{2p-[N(p-2)+\alpha]}{2}.$$

According to inequality (2.1), the functional $J$ is well defined and $C^1$ in $H^1(\mathbf{R}^N)$. Moreover, we have that:

**Lemma 2.5** Let $N\geqslant 2$, $0<\alpha<N$, $(2N-\alpha)/N<p<(2N-\alpha)/(N-2)$ when $N\geqslant 3$, and $(2N-\alpha)/N<p<+\infty$ when $N=2$. Then

$$m=\inf\{J(u);u\in H^1(\mathbf{R}^N),u\neq 0\}$$

is attained by a function $\psi\in H^1(\mathbf{R}^N)$. Moreover $\psi$ is the minimal action solution of

$$-2A\Delta\psi+2B\psi-2mp\psi|\psi|^{p-2}\int_{\mathbf{R}^N}\frac{|\psi(y)|^p}{|x-y|^\alpha}\mathrm{d}y=0,\quad \psi>0 \text{ and } \psi\in H^1(\mathbf{R}^N) \tag{2.4}$$

and

$$m^{-1}=\int_{\mathbf{R}^N\times\mathbf{R}^N}\frac{|\psi(x)|^p|\psi(y)|^p}{|x-y|^\alpha}\mathrm{d}x\mathrm{d}y.$$

**Proof** First, it is easy to see that for any $0\neq u\in H^1(\mathbf{R}^N)$, $J(u)>0$. (2.1) implies that we can find a sequence $\{v_n\}\subset H^1(\mathbf{R}^N)$ such that

$$0 < m = \lim_{n\to\infty} J(v_n) < +\infty.$$

For any $\lambda, \mu > 0$, if we set $u^{\lambda,\mu}(x) = \lambda u(\mu x)$ then we have that

$$\int |\nabla u^{\lambda,\mu}(x)|^2 = \lambda^2 \mu^{2-N} \int |\nabla u|^2,$$

$$\int |u^{\lambda,\mu}(x)|^2 = \lambda^2 \mu^{-N} \int |u|^2$$

and

$$\int_{\mathbf{R}^N\times\mathbf{R}^N} \frac{|u^{\lambda,\mu}(x)|^p |u^{\lambda,\mu}(y)|^p}{|x-y|^\alpha} dx dy = \lambda^{2p} \mu^{\alpha-2N} \int_{\mathbf{R}^N\times\mathbf{R}^N} \frac{|u(x)|^p |u(y)|^p}{|x-y|^\alpha} dx dy.$$

It follows from the choice of $A$ and $B$ that

$$J(u^{\lambda,\mu}) = \frac{(\lambda^2\mu^{2-N}\int |\nabla u|^2)^A (\lambda^2 \mu^{-N}\int |u|^2)^B}{\lambda^{2p}\mu^{\alpha-2N}\int_{\mathbf{R}^N\times\mathbf{R}^N} \frac{|u(x)|^p |u(y)|^p}{|x-y|^\alpha} dx dy} = J(u).$$

Now let $\{u_n\} \subset H^1(\mathbf{R}^N)$ be a minimizing sequence of $m$. We can assume that $u_n > 0$ since $\int |\nabla | u_n \|^2 \leqslant \int |\nabla u_n|^2$. Furthermore, by symmetrization[3, 4,19], we can take $u_n(x) = n_n(|x|)$. Choosing

$$\mu_n = \frac{\| u_n \|_2}{\|\nabla u_n\|_2}, \quad \lambda_n = \frac{\| u_n \|_2^{\frac{N}{2}-1}}{\|\nabla u_n\|_2^{\frac{N}{2}}},$$

we obtain a sequence $\psi_n(x) = u_n^{\lambda_n,\mu_n}(x)$ satisfying the following properties:

$$\psi_n \in H^1(\mathbf{R}^N), \quad \psi_n(x) = \psi_n(|x|) \quad \text{and} \quad \psi_n \geqslant 0;$$

$$\| \psi_n \|_2^2 = 1 \quad \text{and} \quad \|\nabla \psi_n\|_2^2 = 1;$$

$$m = \lim_{n\to\infty} J(\psi_n).$$

Going if necessary to a subsequence, still denoted by $\{\psi_n\}$, we may assume that $\psi_n \rightharpoonup \psi$ weakly in $H^1_{\text{radial}}(\mathbf{R}^N)$. It is now deduced from Lemma 2.4 that

$$\int_{\mathbf{R}^N\times\mathbf{R}^N} \frac{|\psi_n(x)|^p |\psi_n(y)|^p}{|x-y|^\alpha} dx dy \to \int_{\mathbf{R}^N\times\mathbf{R}^N} \frac{|\psi(x)|^p |\psi(y)|^p}{|x-y|^\alpha} dx dy.$$

On the other hand, by weak convergence $\|\psi\|_2 \leqslant 1$ and $\|\nabla\psi\|_2 \leqslant 1$. Therefore

$$m \leqslant J(\psi) \leqslant \frac{1}{\int_{\mathbf{R}^N\times\mathbf{R}^N} \frac{|\psi(x)|^p |\psi(y)|^p}{|x-y|^\alpha} dx dy} = \lim_{n\to\infty} J(\psi_n) = m.$$

It follows that $(\int |\nabla \psi|^2)^A (\int |\psi|^2)^B = 1$ and hence $\|\psi\|_2 = 1$ and $\|\nabla\psi\|_2 = 1$. So $\psi_n \to \psi$ strongly in $H^1(\mathbf{R}^N)$ and

$$m = J(\psi) = \frac{(\int |\nabla\psi|^2)^{\frac{A}{2}} (\int |\psi|^2)^{\frac{B}{2}}}{\int_{\mathbf{R}^N\times\mathbf{R}^N} \frac{|\psi(x)|^p |\psi(y)|^p}{|x-y|^\alpha} dx dy} = \left(\int_{\mathbf{R}^N\times\mathbf{R}^N} \frac{|\psi(x)|^p |\psi(y)|^p}{|x-y|^\alpha} dx dy\right)^{-1}.$$

Noticing that

$$\left.\frac{\mathrm{d}}{\mathrm{d}\varepsilon}\right|_{\varepsilon=0} J(\psi+\varepsilon\eta)=0 \quad \text{for all} \quad \eta\in C_0^\infty(\mathbf{R}^N),$$

we have from a direct computation that $\psi$ satisfies (2.4). □

**Remark 2.6** Here and after, a minimal action solution of an equation is defined in the following sense. For example, for equation

$$-\Delta u+u-u|u|^{p-2}\int_{\mathbf{R}^N}\frac{|u(y)|^p}{|x-y|^\alpha}\mathrm{d}y=0,\quad u\in H^1(\mathbf{R}^N) \tag{2.5}$$

and functional

$$L(u)=\int\frac{|\nabla u|^2+|u|^2}{2}-\frac{1}{2p}\int_{\mathbf{R}^N\times\mathbf{R}^N}\frac{|u(x)|^p|u(y)|^p}{|x-y|^\alpha}\mathrm{d}x\mathrm{d}y,\quad u\in H^1(\mathbf{R}^N),$$

we give the following definitions. The set of solutions of equation (2.5) is denoted by $\Gamma$, namely

$$\Gamma=\{\phi\in H^1(\mathbf{R}^N);\phi\neq 0,L'(\phi)=0\}.$$

The set of minimal action solutions of equation (2.5) is denoted by $G$, namely

$$G=\{\phi\in\Gamma;L(\phi)\leqslant L(\psi)\quad\text{for any }\psi\in\Gamma\}.$$

**Theorem 2.7** Suppose the assumptions of Lemma 2.5 hold. The best constant $C(p,\alpha,N)$ in inequality (2.1) is exactly given as

$$C(p,\alpha,N)=\frac{p}{B}\left(\frac{B}{A}\right)^A\|w\|_2^{2-2p}$$

$$=\frac{2p}{2p-[N(p-2)+\alpha]}\left(\frac{2p-[N(p-2)+\alpha]}{N(p-2)+\alpha}\right)^{\frac{N(p-2)+\alpha}{2}}\|w\|_2^{2-2p},$$

where $w$ is the minimal action solution of

$$-\Delta\phi+\phi-\phi|\phi|^{p-2}\int_{\mathbf{R}^N}\frac{|\phi(y)|^p}{|x-y|^\alpha}\mathrm{d}y=0,\quad \phi>0 \text{ and } \phi\in H^1(\mathbf{R}^N). \tag{2.6}$$

**Remark 2.8** We emphasize that when $p=2+\frac{2-\alpha}{N}$ and $w$ is a minimal action solution of (2.6), then the following

$$\varphi(x,t)=(T-t)^{-\frac{N}{2}}\mathrm{e}^{-\frac{\mathrm{i}|x|^2}{4(T-t)}}w\left(\frac{x}{T-t}\right)\mathrm{e}^{\frac{\mathrm{i}t}{T(T-t)}}$$

is a solution of $\mathrm{i}\varphi_t+\Delta\varphi+\varphi|\varphi|^{p-2}(|x|^{-\alpha}*|\varphi|^p)=0$ and $\varphi$ blows up at finite time. But for general $p$, the blow-up derived by self-similarity is still **open** and we can not solve it at this moment. Actually, considering the Schrödinger equation with local nonlinearity $\mathrm{i}\varphi_t=-\Delta\varphi-|\varphi|^q\varphi$, one can get the blowup profile by self-similarity and pseudo-conformal invariance holds only for $q=\frac{4}{N}$. These are problems for further study.

**Proof of Theorem 2.7** Scaling

$$\psi(x)=(2pm)^{-\frac{1}{2p-2}}u(x)$$

we know from Lemma 2.5 that $u(x)$ satisfies

$$-2A\Delta\phi+2B\phi-\phi|\phi|^{p-2}\int_{\mathbf{R}^N}\frac{|\phi(y)|^p}{|x-y|^\alpha}\mathrm{d}y=0,\quad \phi>0 \text{ and } \phi\in H^1(\mathbf{R}^N).$$

A second scaling

$$u(x)=\left[2B\left(\frac{B}{A}\right)^{\frac{N-\alpha}{2}}\right]^{\frac{1}{2p-2}}w\left[\left(\frac{B}{A}\right)^{\frac{1}{2}}x\right]$$

shows that $w(x)$ is the ground state solution of equation (2.6). Now using

$$\psi(x)=(2pm)^{-\frac{1}{2p-2}}\left[2B\left(\frac{B}{A}\right)^{\frac{N-\alpha}{2}}\right]^{\frac{1}{2p-2}}w\left[\left(\frac{B}{A}\right)^{\frac{1}{2}}x\right]$$

and the fact that $\|\psi\|_2=1$, we obtain that

$$\begin{aligned}C(p,\alpha,N)&=m^{-1}=\frac{p}{B}\left(\frac{B}{A}\right)^{A}\|w\|_2^{2-2p}\\&=\frac{2p}{2p-[N(p-2)+\alpha]}\left[\frac{2p-(N(p-2)+\alpha)}{N(p-2)+\alpha}\right]^{\frac{N(p-2)+\alpha}{2}}\|w\|_2^{2-2p}.\end{aligned}$$

The proof is complete. □

**Remark 2.9** We emphasize that although we do not know whether the minimal action solution of equation (2.6) is unique or not, the best constant $C(p,\alpha,N)$ is independent of the choice of $w$. Indeed, denote $d=\inf\{L(u);u\in\Gamma\}$, then for a minimal action solution $w$ of equation (2.6), we have that $0<d=L(w)<+\infty$ and $d$ is independent of the choice of $w$[9]. On the other hand, since $w$ is a solution of equation (2.6) we have that

$$\int|\nabla w|^2-\frac{Np+\alpha-2N}{2p}\int_{\mathbf{R}^N\times\mathbf{R}^N}\frac{|w(x)|^p|w(y)|^p}{|x-y|^\alpha}\mathrm{d}x\mathrm{d}y=0$$

and

$$\int(|\nabla w|^2+|w|^2)-\frac{Np+\alpha-2N}{2p}\int_{\mathbf{R}^N\times\mathbf{R}^N}\frac{|w(x)|^p|w(y)|^p}{|x-y|^\alpha}\mathrm{d}x\mathrm{d}y=0.$$

Therefore

$$\int_{\mathbf{R}^N\times\mathbf{R}^N}\frac{|w(x)|^p|w(y)|^p}{|x-y|^\alpha}\mathrm{d}x\mathrm{d}y=\frac{2p}{2p-[N(p-2)+\alpha]}\int|w|^2$$

and

$$\int|\nabla w|^2=\frac{N(p-2)+\alpha}{2p-[N(p-2)+\alpha]}\int|w|^2.$$

It is deduced that

$$d=L(w)=\frac{p-1}{2p-[N(p-2)+\alpha]}\|w\|_{L^2}^2,$$

which is equivalent to

$$\|w\|_{L^2}=\left[\frac{2p-(N(p-2)+\alpha}{p-1}d\right]^{\frac{1}{2}}.$$

Thus $C(p,\alpha,N)$ is independent of the choice of $w$.

## 3 Cauchy Problem

In this section, we sketch some results on the existence of a local or global solution of equations (1.2)~(1.3). Define

$$\Sigma = \left\{ u \in H^1(\mathbf{R}^N); \int |x|^2 |u|^2 < +\infty \right\}.$$

Then $\Sigma$ is a Hilbert space under the inner product

$$\langle u, v\rangle_\Sigma = \mathrm{Re}\int (\nabla u \, \nabla \bar{v} + |x|^2 u\bar{v} + u\bar{v}).$$

The norm on $\Sigma$ is denoted by $\|u\|_\Sigma^2 = \int (|\nabla u|^2 + |x|^2|u|^2 + |u|^2)$. The following proposition is proved in the appendix for the readers convenience.

**Proposition 3.1** Let $0 < \alpha < \min\{N,4\}$ and $2 \leqslant p < 2^*(\alpha)$. For any $\varphi_0 \in \Sigma$, there is a $T = T(\|\varphi_0\|_\Sigma) > 0$ and a unique solution $\varphi$ of equation (1.2) with $\varphi \in C([0,T),\Sigma)$ and $\varphi(0) = \varphi_0$. Moreover, we have conservation of particle number

$$\int |\varphi|^2 \equiv \int |\varphi_0|^2 \tag{3.1}$$

and conservation of energy

$$E(\varphi) = \frac{1}{2}\int (|\nabla\varphi|^2 + |x|^2|\varphi|^2) - \frac{1}{2p}\int_{\mathbf{R}^N\times\mathbf{R}^N} \frac{|\varphi(x)|^p|\varphi(y)|^p}{|x-y|^\alpha}\mathrm{d}x\mathrm{d}y = E(\varphi_0) \tag{3.2}$$

for all $t \in [0,T)$, where either $T = +\infty$ or $T < +\infty$ and $\lim_{t\to T^-}\|\varphi\|_\Sigma = +\infty$.

**Theorem 3.2** Let $0 < \alpha < \min\{N,4\}$ and $2 \leqslant p < 2^*(\alpha)$.

- If $2 \leqslant p < 2 + (2-\alpha)/N$, then for any $\varphi_0 \in \Sigma$ the solution $\varphi(x,t)$ of equation (1.2) exists globally in time.
- If $2 \leqslant p = 2 + (2-\alpha)/N$, then the solution $\varphi(x,t)$ of equation (1.2) exists globally in time provided the initial data $\|\varphi_0\|_{L^2}$ sufficiently small.

**Proof** Let $\varphi(x,t) \in C([0,T),\Sigma)$ be the solution of equation (1.2) with initial data $\varphi_0$. By Proposition 3.1 and Theorem 2.3, we have that

$$\begin{aligned}
E(\varphi_0) &= E(\varphi) \\
&\geqslant \frac{1}{2}\int (|\nabla\varphi|^2 + |x|^2|\varphi|^2) - C\left(\int|\nabla\varphi|^2\right)^{\frac{N(p-2)+\alpha}{2}}\left(\int|\varphi|^2\right)^{\frac{2p-(N(p-2)+\alpha)}{2}} \\
&\geqslant \frac{1}{2}\int (|\nabla\varphi|^2 + |x|^2|\varphi|^2) - \\
&\quad C\left(\int(|\nabla\varphi|^2 + |x|^2|\varphi|^2)\right)^{\frac{N(p-2)+\alpha}{2}}\left(\int|\varphi|^2\right)^{\frac{2p-(N(p-2)+\alpha)}{2}}.
\end{aligned}$$

If $2 \leqslant p < 2 + (2-\alpha)/N$, then the Young inequality implies that there is $0 < \varepsilon < \frac{1}{2}$ and $C_\varepsilon$ such that

$$E(\varphi_0) \geqslant \left(\frac{1}{2} - \varepsilon\right)\int (|\nabla\varphi|^2 + |x|^2|\varphi|^2) - C_\varepsilon\left(\int|\varphi|^2\right)^{\frac{2p-[N(p-2)+\alpha]}{2-[N(p-2)+\alpha]}}.$$

Hence $\int (|\nabla\varphi|^2 + |x|^2|\varphi|^2)$ is bounded with respect to $t$. Proposition 3.1 implies that $\varphi(x,t)$ exists globally in time.

If $p=2+(2-\alpha)/N$, then

$$E(\varphi_0)\geqslant\frac{1}{2}\int(|\nabla\varphi|^2+|x|^2|\varphi|^2)-C\int|\nabla\varphi|^2\left(\int|\varphi|^2\right)^{p-1}$$
$$\geqslant\frac{1}{2}\int(|\nabla\varphi|^2+|x|^2|\varphi|^2)-C\int(|\nabla\varphi|^2+|x|^2|\varphi|^2)\left(\int|\varphi|^2\right)^{p-1}$$
$$\geqslant C_1\int(|\nabla\varphi|^2+|x|^2|\varphi_0|^2)$$

provided $\|\varphi^0\|_{L^2}$ sufficiently small. It follows that the solution $\varphi(x,t)$ of equation (1.2) exists globally in time when $p=2+(2-\alpha)/N$ and $\|\varphi_0\|_{L^2}$ is small enough. □

**Remark 3.3** It is observed that in Theorem 3.2, "$\|\varphi_0\|_{L^2}$ small enough" is vague. It is naturally to ask: how small? This is one of the goals of the next section.

## 4 Critical Mass for Critical Nonlinearity

In this section, we will use "our best constant" to give a sharp condition on the solution of equations (1.2)~(1.3) which exists globally in time or blows up in a finite time. In particular, we give an answer to the question: how small an initial data can ensure the existence of global solution of equations (1.2)~(1.3) in the case of $p=2+(2-\alpha)/N$? The answer is simple, as we see below.

**Theorem 4.1** Let $N\geqslant 2$, $0<\alpha<\min\{N,4\}$, and $p=2+(2-\alpha)/N$. If $\varphi_0\in\Sigma$ and

$$\|\varphi_0\|_{L^2}<\|w\|_{L^2},\tag{4.1}$$

where $w$ is a minimal action solution of equations (2.6), then equation (1.2)~(1.3) has a global solution $\varphi(x,t)\in C(\mathbf{R}_+,\Sigma)$.

**Proof** Let $\varphi(x,t)\in C([0,T),\Sigma)$ be a solution of equations (1.2)~(1.3) in the case of $p=2+(2-\alpha)/N$. Using Theorem 2.7, one has

$$\int_{\mathbf{R}^N\times\mathbf{R}^N}\frac{|\varphi(x)|^p|\varphi(y)|^p}{|x-y|^\alpha}\mathrm{d}x\mathrm{d}y\leqslant\frac{2N+2-\alpha}{N}\left(\frac{\int|\varphi|^2}{\int|w|^2}\right)^{\frac{N+2-\alpha}{N}}\int|\nabla\varphi|^2.\tag{4.2}$$

Combining (4.2) with the conservation of energy equation (3.2), we obtain that

$$E(\varphi_0)=\frac{1}{2}\int(|\nabla\varphi|^2+|x|^2|\varphi|^2)-\frac{N}{2(2N+2-\alpha)}\int_{\mathbf{R}^N\times\mathbf{R}^N}\frac{|\varphi(x)|^p|\varphi(y)|^p}{|x-y|^\alpha}\mathrm{d}x\mathrm{d}y$$
$$\geqslant\frac{1}{2}\left[1-\left(\frac{\int|\varphi|^2}{\int|w|^2}\right)^{\frac{N+2-\alpha}{N}}\right]\int|\nabla\varphi|^2+\frac{1}{2}\int|x|^2|\varphi|^2.\tag{4.3}$$

As

$$\int|\varphi|^2\equiv\int|\varphi_0|^2<\int|w|^2,$$

both $\int|\nabla\varphi|^2$ and $\int|x|^2|\varphi|^2$ are bounded for $t\in[0,T)$. It is deduced from Proposition 3.1

that $\varphi(x,t)$ exists globally in $t \in [0, +\infty)$. □

**Remark 4.2** We point out that we are studying the Schrödinger equation with harmonic potential, and the condition (4.1) is sharp in the sense of the following theorem.

**Theorem 4.3** Let $N \geqslant 2$, $0 < \alpha < \min\{N, 4\}$, and $p = 2 + (2-\alpha)/N$. If $\varphi_0 \in \Sigma$ with the form

$$\varphi_0(x) = c\lambda^{\frac{N}{2}} w(\lambda x),$$

where $\lambda > 0$, $w$ is the minimal action solution of (2.6), and $c$ is a complex number with $|c| \geqslant 1$, then

$$\|\varphi_0\|_{L^2} \geqslant \|w\|_{L^2}. \tag{4.4}$$

Moreover the solution $\varphi(x,t)$ of equations (1.2)~(1.3) must blow up in a finite time.

In order to prove Theorem 4.3, we need several lemmas. Firstly, we have the following virial identity which originated from Glassey[16].

**Proposition 4.4** Let $N \geqslant 2, 0 < \alpha < \min\{N, 4\}$. Let $\varphi_0 \in \Sigma$ and $\varphi \in C([0,T), \Sigma)$ be the solution of equations (1.2)~(1.3). If $h(t) = \frac{1}{2}\int |x|^2 |\varphi|^2$ and $p = 2 + (2-\alpha)/N$, then one has

$$h''(t) = 8E(\varphi_0) - 16h(t). \tag{4.5}$$

**Proof** We only prove equation (4.5) formally. Since $\varphi$ satisfies equation (1.2), we have that

$$\varphi_t = \mathrm{i}\left(\Delta\varphi - |x|^2\varphi + \varphi|\varphi|^{p-2}\int_{\mathbf{R}^N} \frac{|\varphi(y)|^p}{|x-y|^\alpha} \mathrm{d}y\right).$$

Therefore

$$h'(t) = \mathrm{Re}\int |x|^2 \bar{\varphi}\varphi_t = 2\mathrm{Im}\int \bar{\varphi} x \nabla\varphi$$

and

$$\begin{aligned} h''(t) &= 2\mathrm{Im}\int (\bar{\varphi}_t x \nabla\varphi + \bar{\varphi} x \nabla\varphi_t) \\ &= 2\mathrm{Im}\int \bar{\varphi}_t x \nabla\varphi - 2\mathrm{Im}\int \varphi_t (N\bar{\varphi} + x \nabla\bar{\varphi}) \\ &= -2\mathrm{Im}\int \varphi_t (N\bar{\varphi} + 2x \nabla\bar{\varphi}) \\ &= -2\mathrm{Re}\int (N\bar{\varphi} + 2x\nabla\bar{\varphi})\left(\Delta\varphi - |x|^2\varphi + \varphi|\varphi|^{p-2}\int_{\mathbf{R}^N} \frac{|\varphi(y)|^p}{|x-y|^\alpha}\mathrm{d}y\right). \end{aligned}$$

Direct computations show that

$$\mathrm{Re}\int (N\bar{\varphi} + 2x\nabla\bar{\varphi})\Delta\varphi = -2\int |\nabla\varphi|^2;$$

$$\mathrm{Re}\int (N\bar{\varphi} + 2x\nabla\bar{\varphi})|x|^2\varphi = -2\int |x|^2|\varphi|^2;$$

$$\mathrm{Re}\int (N\bar{\varphi} + 2x\nabla\bar{\varphi})\varphi|\varphi|^{p-2}\int_{\mathbf{R}^N} \frac{|\varphi(y)|^p}{|x-y|^\alpha}\mathrm{d}y$$

$$= N\int_{\mathbf{R}^N\times\mathbf{R}^N}\frac{|\varphi(x)|^p|\varphi(y)|^p}{|x-y|^\alpha}\mathrm{d}x\mathrm{d}y + \mathrm{Re}\int 2x\,\nabla\bar{\varphi}\varphi|\varphi|^{p-2}\int_{\mathbf{R}^N}\frac{|\varphi(y)|^p}{|x-y|^\alpha}\mathrm{d}y.$$

Since

$$\mathrm{Re}\int 2x\,\nabla\bar{\varphi}\varphi|\varphi|^{p-2}\int_{\mathbf{R}^N}\frac{|\varphi(y)|^p}{|x-y|^\alpha}\mathrm{d}y = \frac{\alpha-2N}{p}\int_{\mathbf{R}^N\times\mathbf{R}^N}\frac{|\varphi(x)|^p|\varphi(y)|^p}{|x-y|^\alpha}\mathrm{d}x\mathrm{d}y,$$

we obtain that

$$\mathrm{Re}\int(N\bar{\varphi}+2x\,\nabla\bar{\varphi})\varphi|\varphi|^{p-2}\int_{\mathbf{R}^N}\frac{|\varphi(y)|^p}{|x-y|^\alpha}\mathrm{d}y = \frac{N(p-2)+\alpha}{p}\int_{\mathbf{R}^N\times\mathbf{R}^N}\frac{|\varphi(x)|^p|\varphi(y)|^p}{|x-y|^\alpha}\mathrm{d}x\mathrm{d}y.$$

It is now deduced that

$$\begin{aligned} h''(t) &= 4\left(\int(|\nabla\varphi|^2-|x|^2|\varphi|^2)-\frac{N(p-2)+\alpha}{2p}\int_{\mathbf{R}^N\times\mathbf{R}^N}\frac{|\varphi(x)^p\varphi(y)|^p}{|x-y|^\alpha}\mathrm{d}x\mathrm{d}y\right)\\ &= 8E(\varphi)-16h(t). \end{aligned}$$

□

**Lemma 4.5** Let $N\geqslant 2$, $0<\alpha<\min\{N,4\}$, and $p=2+(2-\alpha)/N$. If $\varphi_0\not\equiv 0$ satisfies that

$$h(0)-\frac{1}{2}\int|x|^2|\varphi_0|\geqslant E(\varphi_0),$$

then the solution $\varphi$ of equations (1.2)~(1.3) blows up in a finite time.

**Proof** From Proposition 4.4, we have that

$$h(t)=\beta\sin(4t+\theta)+\frac{1}{2}E(\varphi_0), \tag{4.6}$$

where $\beta$ and $\theta$ are constants determined by $h(0)$ and $h'(0)$. Moreover,

$$\beta^2=\left(h(0)-\frac{1}{2}E(\varphi_0)\right)^2+\frac{1}{16}(h'(0))^2. \tag{4.7}$$

Thus if $h(0)\geqslant E(\varphi_0)$, (4.6) and (4.7) imply that there exists $T_0<\infty$ such that

$$\lim_{t\to T_0^-}h(t)=0.$$

It is deduced from[24]

$$\int|\varphi|^2\leqslant C\left(\int|x|^2|\varphi|^2\right)^{\frac{1}{2}}\left(\int|\nabla\varphi|^2\right)^{\frac{1}{2}}$$

that there exists $0<T<+\infty$ such that

$$\lim_{t\to T^-}\int|\nabla\varphi|^2=+\infty.$$

This proves that $\varphi(x,t)$ blows up in a finite time. □

**Proof of Theorem 4.3** For any positive constant $\lambda$ and complex number $c$ with $|c|\geqslant 1$, a direct computation yields that

$$\int|\varphi_0|^2=|c|^2\int|\lambda^{\frac{N}{2}}w(\lambda x)|^2\mathrm{d}x=|c|^2\int|w|^2\geqslant\int\|w\|^2.$$

On the other hand, since the function $w(x)$ makes the inequality (2.1) into an equality, one has that

$$\int_{\mathbf{R}^N\times\mathbf{R}^N}\frac{|w(x)|^p|w(y)|^p}{|x-y|^\alpha}\mathrm{d}x\mathrm{d}y=\frac{2N+2-\alpha}{N}\int|\nabla w|^2.$$

Therefore

$$E(\varphi_0)=\frac{1}{2}\int|\nabla\varphi_0|^2-\frac{N}{2(2N+2-\alpha)}\int_{\mathbf{R}^N\times\mathbf{R}^N}\frac{|\varphi_0(x)|^p|\varphi_0(y)|^p}{|x-y|^\alpha}dxdy+h(0)$$
$$=\frac{1}{2}(1-|c|^{\frac{2N+4-2\alpha}{N}})\lambda^2|c|^2\int|\nabla w|^2+h(0)$$
$$\leqslant h(0).$$

It follows from Lemma 4.5 that $\varphi(x,t)$ blows up in a finite time. The proof of Theorem 4.3 is complete. □

**Remark 4.6** From Theorem 4.1 and 4.3, we see that when $p=2+(2-\alpha)/N$, $\|w\|_{L^2}$ is the critical mass for the solutions of equations (1.2)~(1.3) which exist globally in time. The prescribed initial data in Theorem 4.3 also implies that the existence of blow-up solutions of equations (1.2)~(1.3) not only depends on the mass of the initial data but also on the profile of the initial data. So it is very reasonable to conjecture that for some class of initial data $\varphi_0$ with $\|\varphi_0\|_{L^2}\|w\|_{L^2}$, the solutions of equations (1.2)~(1.3) exist globally in time. In fact, this conjecture is true in the case of $2+(2-\alpha)/N<p<2^*(\alpha)$. Furthermore, we can prove that when $2+(2-\alpha)/N<p<2^*(\alpha)$, the solutions of equations (1.2)~(1.3) exist globally in time for a large class of initial data whose norm can be taken as large as one wants.

## 5 Global Solutions for Supercritical Nonlinearity

After developing the critical mass for the existence of global solutions and the blow-up solutions of equations (1.2)~(1.3) in the critical nonlinearity $p=2+(2-\alpha)/N$, attention is now focused on the existence of global solutions of equations (1.2)~(1.3) in the case of supercritical nonlinearity $2+(2-\alpha)/N<p<2^*(\alpha)$. An interesting aspect is that we can obtain global solutions for arbitrarily large data. We emphasize that the use of Theorem 2.7 is essential. First we need the following lemma from Begout[2].

**Lemma 5.1** Let $I\subset\mathbf{R}$ be an open interval, $s_0\in I$, $\theta>1$, $a>0$, $b>0$ and $\Phi(s)\in C(I,\mathbf{R}_+)$. Set $f(y)=a-y+by^\theta$ for any $y\geqslant0$. Define $y_*=(b\theta)^{-\frac{1}{\theta-1}}$ and $b_*=\frac{\theta-1}{\theta}y_*$. Assume that $\Phi(s_0)<y_*$, $a\leqslant b_*$, and $f\circ\Phi>0$. Then $\Phi(s)<y_*$ for any $s\in I$.

**Proof** Since $\Phi(s_0)<y_*$ and $\Phi$ is a continuous function, there exists a $\delta>0$ such that $\Phi(s)<y_*$ for any $s\in(s_0-\delta,\ s_0+\delta)\subset I$. If $\Phi(s)<y_*$ were not true for any $s\in I$, by continuity there would exist a $s_*\in I$ satisfying $\Phi(s_*)=y_*$. Then $f\circ\Phi(s_*)=f(y_*)=a-b_*\leqslant0$. However, this is impossible from $f\circ\Phi(s_*)>0$. Therefore $\Phi(s)<y_*$ for any $s\in I$. The proof is complete. □

Next, we define a real valued function $V(\lambda)$ as follows,

$$V(\lambda)=\left(\frac{A-1}{B}\right)^{\frac{A-1}{2B}}\|w\|_{L^2}^{\frac{p-1}{B}}\lambda^{-\frac{A-1}{2B}},\quad \lambda>0,$$

where $A=\frac{N(p-2)+\alpha}{2}$ and $B=\frac{2p-[N(p-2)+\alpha]}{2}$. Denote

$$S = \{u \in \Sigma; \|u\|_{L^2} \leqslant V(\|\nabla u\|_{L^2}^2 + \|xu\|_{L^2}^2)\}.$$

**Lemma 5.2** Let $N \geqslant 2, 0 < \alpha < \min\{N,4\}$, and $2 + (2-\alpha)/N < p < 2^*(\alpha)$. $S$ is an unbounded subset of $\Sigma$.

**Proof** For any $M > 0$, we choose $v \in \Sigma$, such that $\int |x|^2 |v|^2 > M$. Define $u_\lambda(x) = \lambda^{\frac{N+2}{2}} v(\lambda x)$ $(\lambda > 0)$, then we have from direct computations that

$$\int |x|^2 |u_\lambda(x)|^2 \mathrm{d}x = \int |x|^2 \lambda^{N+2} |v(\lambda x)|^2 \mathrm{d}x = \int |x|^2 |v|^2 \mathrm{d}x > M,$$

$$\int |\nabla u_\lambda(x)|^2 \mathrm{d}x = \lambda^4 \int |\nabla v|^2 \mathrm{d}x \quad \text{and} \quad \int |u_\lambda(x)|^2 \mathrm{d}x = \lambda^2 \int |v|^2 \mathrm{d}x.$$

Therefore for $\lambda$ small enough, we have

$$\begin{aligned}\|u_\lambda\|_{L^2}(\|\nabla u_\lambda\|_{L^2}^2 + \|xu_\lambda\|_{L^2}^2)^{\frac{A-1}{2B}} &= \lambda \|v\|_{L^2}(\lambda^4 \|\nabla v\|_{L^2}^2 + \|xv\|_{L^2}^2)^{\frac{A-1}{2B}} \\ &< \left(\frac{A-1}{B}\right)^{\frac{A-1}{2B}} \|w\|_{L^2}^{\frac{p-1}{B}}.\end{aligned}$$

According to the definition of $S$, we obtain that $u_\lambda \in S$ for $\lambda$ small enough. On the other hand,

$$\|u_\lambda\|_\Sigma^2 > \int |x|^2 |u_\lambda|^2 = \int |x|^2 |v|^2 \mathrm{d}x > M.$$

These prove that $S$ is unbounded in $\Sigma$. □

**Theorem 5.3** Let $N \geqslant 2$, $0 < \alpha < \min\{N,4\}$ and $2 + (2-\alpha)/N < p < 2^*(\alpha)$. If $\varphi_0 \in S$, then the solutions $\varphi(x,t)$ of equations (1.2)~(1.3) exist globally in $t \in [0, +\infty)$. Moreover for any $t \in (0,T)$ we have that

$$\|\varphi_0\|_{L^2}^{\frac{2B}{A-1}}(\|\nabla\varphi(t)\|_{L^2}^2 + \|x\varphi(t)\|_{L^2}^2) < \frac{A}{B}\|w\|_{L^2}^{\frac{2(p-1)}{A-1}}$$

and

$$\|\varphi(t)\|_\Sigma^2 \leqslant \frac{2N(p-2)+2\alpha}{N(p-2)+\alpha-2} E(\varphi_0) + \|\varphi_0\|_{L^2}^2. \tag{5.1}$$

**Proof** For any $t \in [0,T)$, applying Theorem 2.7 to $\varphi(t,x)$ and using the choice of $A$ and $B$, we obtain that

$$\int_{\mathbf{R}^N \times \mathbf{R}^N} \frac{|\varphi(x)|^p |\varphi(y)|^p}{|x-y|^\alpha} \mathrm{d}x\mathrm{d}y \leqslant C(p,\alpha,N) \|\varphi_0\|_{L^2}^{2B} \|\nabla\varphi\|_{L^2}^{2A}. \tag{5.2}$$

Denote $a = \int(|\nabla\varphi_0|^2 + |x|^2|\varphi_0|^2) > 0$. It is deduced from the energy identity and (5.2) that

$$\begin{aligned}\int(|\nabla\varphi|^2 + |x|^2|\varphi|^2) &= 2E(\varphi) + \frac{1}{p}\int_{\mathbf{R}^N\times\mathbf{R}^N} \frac{|\varphi(x)|^p|\varphi(y)|^p}{|x-y|^\alpha}\mathrm{d}x\mathrm{d}y \\ &= 2E(\varphi_0) + \frac{1}{p}\int_{\mathbf{R}^N\times\mathbf{R}^N} \frac{|\varphi(x)|^p|\varphi(y)|^p}{|x-y|^\alpha}\mathrm{d}x\mathrm{d}y\end{aligned}$$

$$
\begin{aligned}
&< a + \frac{1}{p}\int_{\mathbf{R}^N\times\mathbf{R}^N}\frac{|\varphi(x)|^p|\varphi(y)|^p}{|x-y|^\alpha}\mathrm{d}x\mathrm{d}y\\
&\leqslant a + \frac{C(p,\alpha,N)}{p}\|\varphi_0\|_{L^2}^{2B}\|\nabla\varphi\|_{L^2}^{2A}\\
&\leqslant a + \frac{C(p,\alpha,N)}{p}\|\varphi_0\|_{L^2}^{2B}\left(\int(|\nabla\varphi|^2+|x|^2|\varphi|^2)\right)^A. \qquad (5.3)
\end{aligned}
$$

Let

$$
b := \frac{C(p,\alpha,N)}{p}\|\varphi_0\|_{L^2}^{2B} = \frac{1}{B}\left(\frac{B}{A}\right)^A\|w\|_{L^2}^{2-2p}\|\varphi_0\|_{L^2}^{2B},
$$

$$
\theta = A = \frac{N(p-2)+\alpha}{2} > 1
$$

and

$$
\Phi(t) = \int(|\nabla\varphi(t)|^2 + |x|^2|\varphi(t)|^2).
$$

Obviously $\Phi(0) = a$. At the same time, we define $f(y) = a - y + by^\theta$. Then (5.2) implies that

$$
0 < a - y + by^\theta, \quad \text{where} \quad y = \Phi(t).
$$

Denote

$$
y_* = (b\theta)^{-\frac{1}{\theta-1}}, \quad b_* = \frac{\theta-1}{\theta}y_*.
$$

Then $b_* < y_*$. By direct computation and the exact value of $C(p,\alpha,N)$ we have that

$$
y_* = \frac{A}{B}\|w\|_{L^2}^{\frac{2(p-1)}{A-1}}\|\varphi_0\|_{L^2}^{-\frac{2B}{A-1}}
$$

and

$$
b_* = \frac{A-1}{B}\|w\|_{L^2}^{\frac{2(p-1)}{A-1}}\|\varphi_0\|_{L^2}^{-\frac{2B}{A-1}}. \qquad (5.4)
$$

Moreover, using $\|\varphi_0\|_{L^2} \leqslant V(\|\nabla\varphi_0\|_{L^2}^2 + \|x\varphi_0\|_{L^2}^2) = V(a)$, we know that

$$
\|\varphi_0\|_{L^2} \leqslant \left(\frac{A-1}{B}\right)^{\frac{A-1}{2B}}\|w\|_{L^2}^{\frac{p-1}{B}}a^{-\frac{A-1}{2B}},
$$

which implies that

$$
a \leqslant \frac{A-1}{B}\|w\|_{L^2}^{\frac{2(p-1)}{A-1}}\|\varphi_0\|_{L^2}^{-\frac{2B}{A-1}} \qquad (5.5)
$$

because of $2 + (2-\alpha)/N < p < 2^*(\alpha)$.

Now using (5.4), (5.5) and Lemma 5.1, we obtain that $\Phi(t) < y_*$ for any $t \in [0, T)$. It follows from $\int|\varphi|^2 \equiv \int|\varphi_0|^2$ that $\|\varphi(t)\|_\Sigma^2$ is bounded from above uniformly with respect to $t \in [0, T)$. In other words, the solutions of equations (1.2)~(1.3) with $\varphi_0$ satisfying (5.1) exist globally in $t \in [0, +\infty)$.

Since $\Phi(t) < y_*$ for any $t \in [0, T)$, we obtain that

$$\|\varphi_0\|_{L^2}^{\frac{2B}{A-1}}(\|\nabla\varphi(t)\|_{L^2}^2+\|x\varphi(t)\|_{L^2}^2)<\frac{A}{B}\|w\|_{L^2}^{\frac{2(p-1)}{A-1}}.$$

Next, for the solutions obtained in the above, we give an explicit upper bound on $\|\varphi(t)\|_{\Sigma}^2$. First we have that

$$\begin{aligned}E(\varphi_0)=E(\varphi)&=\frac{1}{2}\int(|\nabla\varphi|^2+|x|^2|\varphi|^2)-\frac{1}{2p}\int_{\mathbf{R}^N\times\mathbf{R}^N}\frac{|\varphi(x)|^p|\varphi(y)|^p}{|x-y|^\alpha}dxdy\\ &\geqslant\frac{1}{2}\int(|\nabla\varphi|^2+|x|^2|\varphi|^2)-\frac{C(p,\alpha,N)}{2p}\|\varphi\|_{L^2}^{2B}\|\nabla\varphi\|_{L^2}^{2A}\\ &\geqslant\frac{1}{2}\int(|\nabla\varphi|^2+|x|^2|\varphi|^2)-\frac{C(p,\alpha,N)}{2p}\|\varphi\|_{L^2}^{2B}\left(\int|\nabla\varphi|^2+|x|^2|\varphi|^2)\right)^A\\ &=\frac{1}{2}\int(|\nabla\varphi|^2+|x|^2|\varphi|^2)\left[1-\frac{C(p,\alpha,N)}{p}\|\varphi\|_{L^2}^{2B}\times\right.\\ &\left.\left(\int(|\nabla\varphi|^2+|x|^2|\varphi|^2)\right)^{A-1}\right]\\ &=\frac{1}{2}\int(|\nabla\varphi|^2+|x|^2|\varphi|^2)\left(1-\frac{1}{A}\left[\left(\frac{AC(p,\alpha,N)}{p}\|\varphi_0\|_{L^2}^{2B}\right)^{-\frac{1}{A-1}}\times\right.\right.\\ &\left.\left.\left(\int|\nabla\varphi|^2+|x|^2|\varphi|^2\right)^{-1}\right]^{1-A}\right).\end{aligned}$$

Since $\Phi(t)<y_*$, we have that

$$\left(\frac{AC(p,\alpha,N)}{p}\|\varphi_0\|_{L^2}^{2B}\right)^{-\frac{1}{A-1}}\left(\int(|\nabla\varphi|^2+|x|^2|\varphi|^2\right)^{-1}>1,$$

by using the exact value of $C(p,\alpha,N)$ obtained in Theorem 2.7. Therefore

$$\left[\left(\frac{AC(p,\alpha,N)}{p}\|\varphi_0\|_{L^2}^{2B}\right)^{-\frac{1}{A-1}}\left(\int(|\nabla\varphi|^2+|x|^2|\varphi|^2)\right)^{-1}\right]^{1-A}<1$$

Because of $2+(2-\alpha)/N<p<\dfrac{2N-\alpha}{N-2}$. It follows that

$$E(\varphi_0)\geqslant\frac{1}{2}\int(|\nabla\varphi|^2+|x|^2|\varphi|^2)\left(1-\frac{1}{A}\right),$$

which yields that

$$\int(|\nabla\varphi|^2+|x|^2|\varphi|^2)\leqslant\frac{2N(p-2)+2\alpha}{N(p-2)+\alpha-2}E(\varphi_0).$$

Therefore

$$\|\varphi(t)\|_{\Sigma}^2\leqslant\frac{2N(p-2)+2\alpha}{N(p-2)+\alpha-2}E(\varphi_0)+\|\varphi_0\|_{L^2}^2.$$

The proof is complete. □

**Remark 5.4** By Lemma 5.2, we get that equations (1.2)~(1.3) possesses global solutions for a large class of initial data whose norm can be as large as we want. On the other hand, from the definition of $V(\lambda)$ and Theorem 5.3 we know that $V(\lambda)\to\|w\|_{L^2}$ as $p\to 2+(2-\alpha)/N$. So we obtain the sharp condition for global existence in the case of initial data $\| \varphi_0 \|_{L^2} < \| w \|_{L^2}$, which coincides with Theorem 4.1. In the case of critical nonlinearity $p = 2 + (2 - \alpha)/N$, the condition (4.1) is sharp. However, we do not know whether or not the condition (5.1) is sharp in the case of supercritical nonlinearity $2 + (2 - \alpha)/N < p < 2^*(\alpha)$.

## Appendix

In this appendix, we outline the proof of Proposition 3.1 by using similar arguments [20]. Consider the Schrödinger operator in $\mathbf{R}^N$,

$$A = -\Delta + |x|^2.$$

Let $S(t)$ be the propagator of $A$ and $k(t, x, y)$ be its Schwartz kernel. It is known from [20] that

$$k(t, x, y) = \left(\frac{-\mathrm{i}}{2\pi \sin t}\right)^{N/2} \mathrm{e}^{\mathrm{i}\{\cos t(|x|^2 + |y|^2) - 2x \cdot y\}/2\sin t}.$$

**Proposition A.1**[20] For any $q \geqslant 2$ and $0 < |t| \leqslant T_0$, $S(t)$ is a bounded operator from $L^{q'}$ to $L^q$, $1/q + 1/q' = 1$, and the map $t \to S(t)$ is strongly continuous to $B(L^{q'}, L^q)$. Moreover, for $0 < |t| \leqslant T_0$, we have

$$\| S(t) u \|_q \leqslant \left(\frac{2\pi |t|}{C_1}\right)^{N/q - N/2} \| u \|_{q'}, \tag{A.1}$$

where

$$C_1 = \sup_{x, y, |t| \leqslant T_0} \left| \frac{t}{\sin t} \right|^{1/2}.$$

**Definition A.2** (Admissible Pair) The pair $(r, q)$ is admissible if $2/r = N(1/2 - 1/q)$, $2 \leqslant q \leqslant \infty$ for $N = 1$, $2 \leqslant q < \infty$ for $N = 2$, and $2 \leqslant q < 2N/(N-2)$ for $N \geqslant 3$.

**Lemma A.3** Let $(r, q)$ be any admissible pair.

(Ⅰ) For any $0 < T \leqslant T_0$ and $\psi \in L^2(\mathbf{R}^N)$, we have $S(t)\psi \in L^r(-T, T; L^q)$, and there exists a constant $C$ such that

$$\| S(t)\psi \|_{L^r(-T;T;L^q)} \leqslant C \| \psi \|_{L^2}. \tag{A.2}$$

(Ⅱ) For any $0 < T \leqslant T_0$ and $g \in L^{r'}(-T, T; L^{q'}(\mathbf{R}^N))$, we have $\int_0^t S(t-\tau) g(\tau) \mathrm{d}\tau \in L^r(-T;T;L^q) \cap C([-T, T]; L^2)$, and there exists a constant $C$ such that

$$\left\| \int_0^t S(t-\tau) g(\tau) \mathrm{d}\tau \right\|_{L^r(-T,T;L^q)} \leqslant C \| g \|_{L^{r'}(-T,T;L^{q'})}, \tag{A.3}$$

$$\left\| \int_0^t S(t-\tau) g(\tau) \mathrm{d}\tau \right\|_{C([-T,T];L^2)} \leqslant C \| g \|_{L^{r'}(-T,T;L^{q'})}. \tag{A.4}$$

**Proof** For any $\psi \in L^2(\mathbf{R}^N)$ and $f \in L^{r'}(-T, T; L^{q'}(\mathbf{R}^N))$, we obtain

$$\left| \int_{-T}^T \int_{\mathbf{R}^N} S(t)\psi \bar{f}(t) \mathrm{d}x \mathrm{d}t \right| = \left| \int_{-T}^T \int_{\mathbf{R}^N} \psi \bar{S}(-t) \bar{f}(t) \mathrm{d}x \mathrm{d}t \right|$$
$$\leqslant \| \psi \|_{L^2_x} \left\| \int_{-T}^T S(-t) f(t) \mathrm{d}t \right\|_{L^2_x}, \tag{A.5}$$

$$\left\| \int_{-T}^T S(-t) f(t) \mathrm{d}t \right\|_{L^2_x}^2 = \int_{-T}^T \int_{\mathbf{R}^N} f(t, x) \int_{-T}^T \bar{S}(t-\tau) \bar{f}(\tau) \mathrm{d}\tau \mathrm{d}x \mathrm{d}t$$

$$\leqslant \|f\|_{L^{r'}(-T,T;L^{q'})}\left\|\int_{-T}^{T}S(t-\tau)f(\tau)\mathrm{d}\tau\right\|_{L^{r}(-T,T;L^{q})}. \tag{A.6}$$

By using equation (A.1) and Hardy-Littlewood-Sobolev inequality[23, 117-120], we obtain

$$\left\|\int_{-T}^{T}S(t-\tau)f(\tau)\mathrm{d}\tau\right\|_{L^{r}(-T,T;L^{q})}\leqslant C\left\|\int_{-T}^{T}\|t-\tau\mid^{N/q-N/2}\mid f(\tau)\|_{L_x^{q'}}\mathrm{d}\tau\right\|_{L^{r}(-T,T)}$$

$$\leqslant C\|f\|_{L^{r'}(-T,T;L^{q'})}. \tag{A.7}$$

From equations (A.5)~(A.7) we have equation (A.2).

For any $f\in L^{r'}(-T,T;L^{q'}(\mathbf{R}^N)$, by using equation (A.2) and equation (A.7) we obtain

$$\left|\int_{-T}^{T}\int_{\mathbf{R}^N}\int_0^t S(t-\tau)g(\tau)\mathrm{d}\tau\bar{f}(t,x)\mathrm{d}x\mathrm{d}t\right|$$

$$=\left|\int_0^t\int_{\mathbf{R}^N}g(\tau,x)\int_{-T}^{T}\bar{S}(\tau-t)\bar{f}(t)\mathrm{d}t\mathrm{d}x\mathrm{d}\tau\right|$$

$$\leqslant \|g\|_{L^{r'}(-T,T;L^{q'})}\left\|\int_{-T}^{T}S(t-\tau)f(\tau)\mathrm{d}\tau\right\|_{L^{r}(-T,T;L^{q})}$$

$$\leqslant C\|g\|_{L^{r'}(-T,T;L^{q'})}\|f\|_{L^{r'}(-T,T;L^{q'})},$$

$$\left\|\int_0^t S(t-\tau)g(\tau)\mathrm{d}\tau\right\|_{L^{r}(-T,T;L^{q})}\leqslant C\|g\|_{L^{r'}(-T,T;L^{q'})}, \tag{A.8}$$

$$\left\|\int_0^t S(t-\tau)g(\tau)\mathrm{d}\tau\right\|_{L_x^2}^2=\int_0^t\int_{\mathbf{R}^N}g(\tau,x)\bar{S}(\tau-t)\int_0^t\bar{S}(t-s)\bar{g}(\tau)\mathrm{d}s\mathrm{d}x\mathrm{d}\tau$$

$$\leqslant \|g\|_{L_r^{r'}(-T,T;L^{q'})}\left\|S(\tau-t)\int_0^t S(t-s)g(s)\mathrm{d}s\right\|_{L_r^{r}(-T,T;L^{q})}$$

$$\leqslant C\|g\|_{L_r^{r'}(-T,T;L^{q'})}\left\|\int_0^t S(t-s)g(s)\mathrm{d}s\right\|_{L_x^2},$$

$$\left\|\int_0^t S(t-\tau)g(\tau)\mathrm{d}\tau\right\|_{C([-T,T];L^2)}\leqslant C\|g\|_{L^{r'}(-T,T;L^{q'})}. \tag{A.9}$$

This completes the proof of Lemma A.3. □

Now applying Lemma A.3, we can prove the local existence of solution for equations (1.2)~(1.3) and obtain some conservation laws with $\varphi_0\in\Sigma$. We consider the integral equation associated to equation (1.2)

$$\varphi(t)=S(t)\varphi_0-\mathrm{i}\int_0^t S(t-\tau)F(\varphi(\tau))\mathrm{d}\tau,\quad F(\psi):=\psi|\psi|^{p-2}\int_{\mathbf{R}^N}\frac{|\psi(y)|^p}{|x-y|^{\alpha}}\mathrm{d}y. \tag{A.10}$$

For getting some conservation laws precise, we also consider regularized versions of equation (A. 10), as in [20, 15]

$$\varphi(t)=S(t)h*(g\varphi_0)-\mathrm{i}\int_0^t S(t-\tau)h*\{gF(h*\varphi(\tau))\}\mathrm{d}\tau, \tag{A.11}$$

where we let $h$ belong to the Schwarz space and be an even function with $\|h\|_{L^1}=1$, and $g$ be a $C^{\infty}$ cut-off function with compact support. From now on, we denote $\phi=(g,h)$ for notational convenience. Let $\delta$ denote the Dirac delta function. If $\phi = (1,\delta)$, then equation (A.11) is equivalent to equation (A.10).

Let

$$\frac{1}{\rho} = \begin{cases} \dfrac{2N-\alpha+2p-2}{2pN}, & N \geqslant 3 \\ \dfrac{1}{2}\left(\dfrac{N-\alpha}{2N} + \max\left\{\dfrac{1}{2} - \dfrac{1}{pN}, \dfrac{2N-\alpha}{2pN}\right\}\right), & N = 1,2 \end{cases} \tag{A.12}$$

and $2/\gamma = N(1/2 - 1/\rho)$. It is obvious that $(\gamma, \rho)$ is an admissible pair.

**Lemma A.4** Let $0 < \alpha < N$, $2 \leqslant p < 2^*(\alpha)$, and $\rho$ be defined in equation (A.12). For any $\psi \in H_\rho^1(\mathbf{R}^N)$, we have

$$\|F(\psi)\|_{H_{\rho'}^1} \leqslant C\|\psi\|_{H_\rho^1}^{2p-1}. \tag{A.13}$$

**Proof** By using the Hölder inequality, we obtain

$$\|F(\psi)\|_{H_{\rho'}^1} \leqslant C\|\psi\|_{H_\rho^1}\|\psi\|_{L^{\rho^*}}^{p-2}\| |\psi|^p * |y|^{-\alpha}\|_{L^{q_1}} + C\|\psi\|_{L^{\rho^*}}^{p-1}\|\{|\psi|^{p-2}(\bar{\psi}\partial_y\psi + \psi\partial_y\bar{\psi})\} * |y|^{-\alpha}\|_{L^{q_2}},$$

where

$$\frac{1}{\rho^*} = \frac{1}{2p-2}\left(1 - \frac{2}{\rho} + \frac{N-\alpha}{N}\right), \quad \frac{1}{q_1} = \frac{p}{\rho^*} - \frac{N-\alpha}{N}, \quad \frac{1}{q_2} = \frac{1}{\rho} + \frac{p-1}{\rho^*} - \frac{N-\alpha}{N}. \tag{A.14}$$

By the Hardy-Littlewood-Sobolev inequality[23, 117 120], we obtain

$$\|F(\psi)\|_{H_{\rho'}^1} \leqslant C\|\psi\|_{H_\rho^1}\|\psi\|_{L^{\rho^*}}^{2p-2}. \tag{A.15}$$

From (A.12), we have

$$\frac{1}{\rho} \geqslant \frac{1}{\rho^*} \geqslant \frac{1}{\rho} - \frac{1}{N}.$$

By applying the Sobolev imbedding theorem and equation (A.15), we obtain equation (A.13). □

**Lemma A.5** Let $0 < T \leqslant T_0$, $0 < \alpha < N$, $2 \leqslant p < (2N-\alpha)/(N-2)$, $\rho$ be defined in equation (A.12), and $(\gamma,\rho)$ be the admissible pair. For any $u, v \in L^\gamma(-T,T;H_\rho^1(\mathbf{R}^N))$, we have

$$\left\|\int_0^t S(t-\tau)h * \{gF(h*u) - gF(h*v)\}\mathrm{d}\tau\right\|_{L^\gamma(-T,T;L^\rho)}$$
$$\leqslant CT^{1-2p/\gamma}\left\{\|u\|_{L^\gamma(-T,T;H_\rho^1)}^{2p-2} + \|v\|_{L^\gamma(-T,T;H_\rho^1)}^{2p-2}\right\}\|u-v\|_{L^\gamma(-T,T;L^\rho)}, \tag{A.16}$$

$$\left\|\int_0^t S(t-\tau)h * \{gF(h*u)\}\mathrm{d}\tau\right\|_{L^\gamma(-T,T;H_\rho^1)} \leqslant CT^{1-2p/\gamma}\|u\|_{L^\gamma(-T,T;H_\rho^1)}^{2p-1}, \tag{A.17}$$

where the constant $C$ is independent of $\phi = (g,h)$.

**Proof** By using equation (A.3), the Hölder inequality, and the Sobolev imbedding theorem, we have

$$\begin{aligned}
&\left\|\int_0^t S(t-\tau)h*\{gF(h*u)-gF(h*v)\}\mathrm{d}\tau\right\|_{L^\gamma(-T,T;L^\rho)}\\
&\leqslant C\|F(h*u)-F(h*v)\|_{L^{\gamma'}(-T,T;H^1_{\rho'})}\\
&\leqslant C\left(\int_{-T}^T\left\{\|(u-v)(t)\|_{L^\rho}^{\gamma'}\left(\|u(t)\|_{L^{\rho*}}^{(p-2)\gamma'}+\|v(t)\|_{L^{\rho*}}^{(p-2)\gamma'}\right)\|\,|h*u|^p*|y|^{-\alpha}\|_{L^{q_1}}^{\gamma'}+\right.\right.\\
&\left.\left.\|v(t)\|_{L^{\rho*}}^{(p-1)\gamma'}\|(|h*u|^p-|h*v|^p)*|y|^{-\alpha}\|_{L^{q_2}}^{\gamma'}\right\}\mathrm{d}t\right)^{1/\gamma'}\\
&\leqslant C\left[\int_{-T}^T\|(u-v)(t)\|_{L^\rho}^{\gamma'}\left(\|u(t)\|_{L^{\rho*}}^{(2p-2)\gamma'}+\|v(t)\|_{L^{\rho*}}^{(2p-2)\gamma'}\right)\mathrm{d}t\right]^{1/\gamma'}\\
&\leqslant CT^{1-2p/\gamma}\left\{\|u\|_{L^\gamma(-T,T;H^1_\rho)}^{2p-2}+\|v\|_{L^\gamma(-T,T;H^1_\rho)}^{2p-2}\right\}\|u-v\|_{L^\gamma(-T,T;L^\rho)},
\end{aligned}$$

where $\rho_*$, $q_1$ and $q_2$ are defined in equation (A.14), constant $C$ is independent of $\phi=(g,h)$.

From Lemmas A.3 and A.4, we obtain that

$$\begin{aligned}
&\left\|\int_0^t S(t-\tau)h*\{gF(h*u)\}\mathrm{d}\tau\right\|_{L^\gamma(-T,T;H^1_\rho)}\leqslant C\|F(h*u)\|_{L^{\gamma'}(-T,T;H^1_{\rho'})}\\
&\leqslant C\left(\int_{-T}^T\|u(t)\|_{H^1_\rho}^{(2p-1)\gamma'}\mathrm{d}t\right)^{1/\gamma'}\leqslant CT^{1-2p/\gamma}\|u\|_{L^\gamma(-T,T;H^1_\rho)}^{2p-1},
\end{aligned}$$

where the constant $C$ does not depend on $\phi=(g,h)$. □

**Lemma A.6** Let $0<\alpha<N$ and $2\leqslant p<2^*(\alpha)$. For any $\varphi_0\in H^1(\mathbf{R}^N)$, there exists a $0<T=T(\|\varphi_0\|_{H^1})\leqslant T_0$ and a unique solution $\varphi$ of equations (A.11) and (A.12) such that $\varphi\in C([-T,T);H^1(\mathbf{R}^N))\cap L^\gamma(-T,T;H^1_\rho(\mathbf{R}^N))$, where $T$ does not depend on $\phi$, $\rho$ is defined in equation (A.12), and $(\gamma,\rho)$ is admissible pair.

**Proof** Let constants $T>0$ and $R_0>0$ be selected later, and

$$E=E(T,R_0)=\left\{\varphi\mid\varphi\in L^\gamma(-T,T;H^1_\rho(\mathbf{R}^N)),\|\varphi\|_{L^\gamma(-T,T;H^1_\rho(\mathbf{R}^N))}\leqslant R_0\right\}.$$

Note that $E$ is never empty by Lemma A.3. Endowed with the metric

$$\mathrm{dist}(u,v)=\|u-v\|_{L^\gamma(-T,T;L^\rho)},$$

$E$ is a complete metric space. Indeed, since $L^\gamma(-T,T;H^1_\rho(\mathbf{R}^N))$ is reflexive, the closed ball of radius $R_0$ is weakly compact. We wish to find conditions on $T$ and $R_0$ such that the map $M:\varphi\to M\varphi$ given by

$$M\varphi=S(t)h*(g\varphi_0)-\mathrm{i}\int_0^t S(t-\tau)h*\{gF(h*\varphi(\tau))\}\mathrm{d}\tau\quad\text{and}$$

$$F(\psi):=\psi|\psi|^{p-2}\int_{\mathbf{R}^N}\frac{|\psi(y)|^p}{|x-y|^\alpha}\mathrm{d}y$$

is a strict contraction on $E$. For any $\varphi\in E$, from Lemma A.3 and A.5 we have

$$\|M\varphi\|_{L^\gamma(-T,T;H^1_\rho(\mathbf{R}^N))}\leqslant C\|\varphi_0\|_{H^1}+CT^{1-2p/\gamma}\|\varphi\|_{L^\gamma(-T,T;H^1_\rho(\mathbf{R}^N))}^{2p-1},$$

where the constant $C$ is independent of $\phi=(g,h)$. Let us take $R_0=2C\|\varphi_0\|_{H^1}$. Note that $1-2p/\gamma>0$. Thus there exists $T>0$, which does not depend on $\phi=(g,h)$, such that for any $\varphi \in E$ we have

$$CT^{1-2p/\gamma}\|\varphi\|^{2p-1}_{L^{\gamma}(-T,T;H^1_p(\mathbf{R}^N))} \leqslant R_0/2.$$

This implies that $M: E \to E$. For any $u, v \in E$, it follows from Lemma A.5 that

$$\|Mu - Mv\|_{L^{\gamma}(-T,T;L^p)}$$
$$\leqslant CT^{1-2p/\gamma}\left\{\|u\|^{2p-2}_{L^{\gamma}(-T,T;H^1_p)} + \|v\|^{2p-2}_{L^{\gamma}(-T,T;H^1_p)}\right\}\|u - v\|_{L^{\gamma}(-T,T;L^p)},$$

where the constant $C$ is independent of $\phi = (g, h)$. Then there exists $0 < T = T(\|\varphi_0\|_{H^1})$, which does not depend on $\phi = (g, h)$, such that

$$\|Mu - Mv\|_{L^{\gamma}(-T,T;L^p)} \leqslant \frac{1}{2}\|u - v\|_{L^{\gamma}(-T,T;L^p)}.$$

So $M: E \to E$ is strictly contractive. There exists a unique fixed point $\varphi \in E$. From Lemma A.3, $\varphi \in C([-T,T];H^1(\mathbf{R}^N))$. Lemma A.6 is proved. □

**Lemma A.7** Let $0 < \alpha < N$, $2 \leqslant p < 2^*(\alpha)$, $\varphi_0 \in \Sigma$ and $\varphi \in C([-T,T];H^1(\mathbf{R}^N))$ be the solution of equation (A. 12). Then $\varphi \in C([-T,T];\Sigma)$ and for all $t \in [-T,T]$, $\varphi$ satisfies the equalities

$$\int_{\mathbf{R}^N}|\varphi(t)|^2\mathrm{d}x = \int_{\mathbf{R}^N}|\varphi_0|^2\mathrm{d}x, \tag{A.18}$$

$$E(\varphi(t)) = E(\varphi_0). \tag{A.19}$$

**Proof** Following the proof of Theorem 4.7 in [20], we can prove this lemma. □

**Proof of Proposition 3.1** Applying Lemma A.6 and Lemma A.7, we obtain Proposition 3.1. □

## References


[1] Baym G, Pethick C J. Ground state properties of magnetically trapped Bose-condensed rubidium gas. Phys Rev Lett, 1996, 76: 6 – 9.

[2] Bégout P. Necessary conditions and sufficient conditions for global existence in the nonlinear Schrödinger equation. Adv Math Sci Appl, 2002, 12: 817 – 829.

[3] Brascamp H J, Lieb E H, Luttinger J M. A general rearrangement inequality for multiple integrals. J Funct Anal, 1974, 17: 227 – 237.

[4] Burchard A, Guo Y. Compactness via symmetrization. J Funct Anal, 2004, 214: 40 – 73.

[5] Caffarelli L, Kohn R, Nirenberg L. First order interpolation inequalities with weights. Compositio Math, 1984, 53: 259 – 275.

[6] Cao Y, Musslimani Z H, Titi E S. Nonlinear Schrödinger-Helmholtz equation as numerical regularization of the nonlinear Schrödinger equation. Nonlinearity, 2008, 21: 879 – 898.

[7] Cazenave T. An introduction to nonlinear Schrödinger equations. Rio de Janeiro: Textos de Metodos Matematicos, 1989: 22.

[8] Chen G, Zhang J. Remarks on global existence for the supercritical nonlinear Schrödinger equation with a harmonic potential. J Math Anal Appl, 2006, 320: 591 – 598.

[9] Chen J, Guo B. Strong instability of standing waves for a nonlocal Schrödinger equation. Phys D, 2007, 227:

142 - 148.

[10] Deconinck B, Kutz J N. Singular instability of exact stationary solutions of the nonlocal Gross-Pitaevskii. e-print cond-mat/0208441.

[11] Fukuizumi R. Stability and instability of standing waves for the nonlinear Schrödinger equation with harmonic potential. Discrete Contin Dyn Syst, 2001, 7: 525 - 544.

[12] Fujiwara D. A construction of the fundamental solution for the Schrödinger equation. J Analyse Math, 1979, 35: 41 - 96.

[13] Fujiwara D. Remarks on convergence of the Feynman path integrals. Duke Math J, 1980, 47: 559 - 600.

[14] Garcia-Ripoll J J, Konotop V V, Malomed B, et al. A quasi-local Gross-Pitaevskii for Bose-Einstein condensates. Math Compt Simulation, 2003, 62: 21 - 30.

[15] Ginibre J, Velo G. On a class of nonlinear Schrödinger equation, I: the Cauchy problem. J Funct Anal, 1979, 32: 1 - 32.

[16] Glassey R T. On the blowing-up of solutions to the Cauchy problem for the nonlinear Schrödinger equation. J Math Phys, 1977, 18: 1794 - 1797.

[17] Kivshar Y S, Alexander T J, Turitsyn S K. Nonlinear modes of a macroscopic quantum oscillator. Phys Lett, A, 2001, 278: 225 - 230.

[18] Kurth M. On the existence of infinitely many modes of a nonlocal nonlinear Schrödinger equation related to dispersion-managed solitons. SIAM J Math Anal, 2004, 36: 967 - 985.

[19] Lieb E H, Loss M. Analysis, AMS graduate studies in mathematics. 2nd Edition. Providence AMS, 2001: 13.

[20] Oh Y G. Cauchy problem and Ehrenfest's law of nonlinear Schrödinger equations with potentials. J Diff Equ, 1989, 81: 255 - 274.

[21] Reed M, Simon B. Methods of modern mathematical physics, Vols II, IV. Elsevier(Singapore) Pte Ltd., 2003.

[22] Rose H A, Weinstein M I. On the bound states of the nonlinear Schrödinger equation with linear potential. Phys D, 1988, 30: 207 - 218.

[23] Stein E M. Singular integrals and differentiability properties of functions. Princeton University Press, 1970.

[24] Weinstein M I. Nonlinear Schrödinger equations and sharp interpolation estimates. Commun Math Phys, 1983, 87: 567 - 576.

[25] Zhang J. Stability of attractive Bose-Einstein condensates. J Stat Phys, 2000, 102: 731 - 746.

# Orbital Stability of Standing Wave Solution for a Quasilinear Schrödinger Equation *

Guo Boling(郭柏灵) Chen Jianqing(陈建清)

**Abstract**

Via minimization arguments and the Concentration Compactness Principle, we prove the orbital stability of standing wave solutions for a class of quasilinear Schrödinger equation arising from physics.

**Keywords** standing wave solution; orbital stability; quasilinear Schrödinger equation

## 1 Introduction

This paper is motivated by recent interest in the following type of quasilinear Schrödinger equation:

$$i\partial_t z = -\Delta z + V(x)z - k(\Delta|z|^2))z - \theta|z|^{p-2}z, \qquad (1.1)$$

where $i$ is the imaginary unit, $p > 2$ and $k, \theta \in \mathbf{R}_+$, $\Delta z = \sum_{j=1}^{N} \partial^2 z/\partial x_j^2$ is the standard Laplacian operator, $z := z(x,t) : \mathbf{R}^N \times \mathbf{R}_+ \to \mathbf{C}$ is a complex-valued function.

Problems of this kind arise naturally from various domains of mathematical physics and have been derived as models of several physical phenomena in the theory of superfluid films and in dissipative quantum mechanics (see e. g. Kurihura[8], Nakamura[14]). For more physical motivations and more references dealing with applications, we refer the interested readers to Lange et al.[9], Poppenberg et al.[15] and the references therein.

Via critical point theory, finding a standing wave of the form $z(x,t) = e^{i\mu t}u(x)$ of problem (1.1) is equivalent to solving the following elliptic equation:

$$-\Delta u + (V(x) + \mu)u - k(\Delta(|u|^2))u - \theta|u|^{p-2}u = 0. \qquad (1.2)$$

Under the basic assumption of $\inf_{x\in\mathbf{R}^N} V(x) + \mu > 0$, the existence of nontrivial solutions of (1.2) has been studied in the last two decades. Some of them are due to Strauss[17] and Rabinowitz[16] in the case of $k = 0$ and Ambrosetti-Liu-Wang-Wang[2, 11, 12, 13] in the case of $k \neq 0$ with some additional assumptions on $V(x)$ and $p$. Using the basic function space $H^s(\mathbf{R}^N) = W^{s,2}(\mathbf{R}^N)$ and setting

$$X = \left\{u \in H^1(\mathbf{R}^N) \,\middle|\, \int_{\mathbf{R}^N} |u|^2 |\nabla u|^2 dx < \infty, \int V(x)|u|^2 < \infty\right\},$$

* Quarterly of Applied Mathemaics, 2009, LXVII(4): 781 - 791.

we can define functionals:

$$F_1(u)=\frac{1}{2}\int(|\nabla u|^2+V(x)|u|^2),\quad F_2(u)=\frac{1}{2}\int|u|^2,$$

$$F_3(u)=\frac{1}{4}\int|\nabla|u|^2|^2,\quad F_4(u)=\frac{1}{p}\int|u|^p,$$

$$F_5(u)=F_1(u)+\mu F_2(u)+kF_3(u)-\theta F_4(u),$$

$$F_6(u)=\int(|\nabla u|^2+(V(x)+\mu)|u|^2+k|\nabla|u|^2|^2-\theta|u|^p).$$

Problem (1.2) has a formal variational structure. For any $\phi\in\mathscr{D}(\mathbf{R}^N)$, $F_j, j=1,\cdots,6$, have directional derivatives at $u$ in the direction $\phi$, denoted by $\langle F'_j(u),\phi\rangle$. We say that $u$ is a weak solution of (1.2) if and only if for any $\phi\in\mathscr{D}(\mathbf{R}^N)$ and some $\mu,\theta$, it follows that

$$\langle F'_1(u)+\mu F'_2(u)+kF'_3(u)-\theta F'_4(u),\phi\rangle=0. \tag{1.3}$$

Up to now, we have three methods to study the existence of standing wave solutions for problems of this kind.

Method Ⅰ: Study the minimization problem

$$m_{\mathrm{I}}=\inf\{F_1(u)+\mu F_2(u)+kF_3(u)\mid u\in X, F_4(u)=\lambda_{\mathrm{I}}>0\}. \tag{1.4}$$

If $m_{\mathrm{I}}$ is achieved, then the minimizer is a solution of (1.2) for some $\theta$.

Method Ⅱ: Consider the minimization problem

$$m_{\mathrm{II}}=\inf\{F_1(u)+kF_3(u)-\theta F_4(u)\mid u\in X, F_2(u)=\lambda_{\mathrm{II}}>0\}. \tag{1.5}$$

If $m_{\mathrm{II}}$ is achieved, then again the minimizer is a solution of (1.2) for some $\mu$.

Method Ⅲ: Study the minimization problem

$$m_{\mathrm{III}}=\inf\{F_5(u)\mid 0\neq u\in X, F_6(u)=0\}. \tag{1.6}$$

If $m_{\mathrm{III}}$ is achieved, then the minimizer is a solution of (1.2).

When $k=0$, the three methods mentioned above are almost equivalent in the sense that they can be changed from one to another by scaling. However, when $k\neq 0$, the term introduced by $F_3(u)$ makes the problem substantially different from the case of $k=0$ since $F_3$ is nonconvex and the scaling argument does not work. Method Ⅲ has been used in [11, 12, 13] to study the existence of solutions of (1.2). But it seems that it is too difficult to be used to study orbital stability (see the precise definition in Section 3) of standing waves of (1.1) since $F_6$ usually does not satisfy any conservative laws (see Lemma 2.1). Hence our purpose here is to take Method Ⅱ to study the existence and orbital stability of standing waves of (1.1) under the basic assumption $\inf_{x\in\mathbf{R}^N}V(x)\geqslant 0$ and $\lim_{|x|\to\infty}V(x)=+\infty$.

We recall that in the case of $k=0$, the existence and orbital stability of a standing wave solution of (1.1) has been studied by Cazenave and Lions[6] via the concentration compactness principle. However, the presence of the quasilinear term $(\Delta(|z|^2))z$ makes the problem substantially different from the semilinear case. For example, when $k=0$, one can use $\left\{u\mid\int|u|^2=\text{const.}\right\}$ or $\left\{u\mid\int|u|^p=\text{const.}\right\}$ as a constraint to study a corresponding minimization problem. Then the study of orbital stability can proceed after a standard scaling argument. But the term introduced by $(\Delta(|z|^2))z$ makes the scaling arguments fail. Hence we need to restrict the arguments to a function space which consists of functions with radial symmetry. The main results are Theorem 3.2 and Theorem 3.4, which are contained in Section 3 after some preliminaries given in Section 2. In Section 4, we study problem (1.1) in one spatial dimension. Some improvements of Theorem 3.2 and Theorem 3.4 are obtained; see Theorem 4.6 and Theorem 4.8.

## 2 Preliminaries

Throughout this paper, all integrals are taken over $\mathbf{R}^N$ (or $\mathbf{R}$, depending upon the context) unless stated otherwise. All $dx$ in the integrals are omitted. $o(1)$ will denote a generic infinitesimal as $n \to \infty$. $\to$ denotes the strong convergence and $\rightharpoonup$ the weak convergence. $B(x, R)$ will denote a ball centered at $x$ with radius $R$, and $B(0, R)$ is simply denoted by $B_R$. To continue, we list some lemmas which will be useful in what follows. First

**Lemma 2.1**(Conservative Laws) Let $z(x, t)$ be a solution of (1.1) with initial value $z_0$. Then it follows that

$$\int |z(x,t)|^2 = \int |z_0|^2 \equiv \text{const. (Conservation of "mass")}, \tag{2.1}$$

$$\begin{aligned}&\int\left[\frac{1}{2}(|\nabla z|^2 + V|z|^2) + \frac{k}{4}|\nabla|z|^2|^2 - \frac{\theta}{p}|z|^p\right]\\ =&\int\left[\frac{1}{2}(|\nabla z_0|^2 + V|z_0|^2) + \frac{k}{4}|\nabla|z_0|^2|^2 - \frac{\theta}{p}|z_0|^p\right]\\ \equiv&\text{const. (Conservation of "energy")}.\end{aligned} \tag{2.2}$$

**Proof** This kind of result is known[9]. □

**Lemma 2.2**[10] Let $(\rho_n)$ be bounded in $L^1(\mathbf{R})$. By extracting a subsequence, we may assume that $(\rho_n)$ satisfies one of the following two possibilities:

(i)(Vanishing) $\lim\limits_{n\to\infty}\sup\limits_{y\in\mathbf{R}}\int_{y+B_R}\rho_n(x)dx = 0$ for all $0 < R < +\infty$.

(ii)(Nonvanishing) There exist $\alpha > 0, R < +\infty$ and $y_n \subset \mathbf{R}$ such that

$$\lim_{n\to\infty}\int_{y_n+B_R}\rho_n(x)dx \geqslant \alpha > 0.$$

**Lemma 2.3**[10] Let $(u_n)$ be bounded in $H^1(\mathbf{R})$. Assume that for some $q > 2$ and $R > 0$,

$$\sup_{y\in\mathbf{R}}\int_{y+B_R}|u_n|^q dx \to 0 \quad \text{as} \quad n \to \infty.$$

Then $u_n \to 0$ in $L^\beta(\mathbf{R})$ for any $\beta \geqslant 2$.

**Lemma 2.4**([15, Lemma 2 on page 333]) If $u_n \rightharpoonup u$ in $H^1(\mathbf{R})$ and $u_n \to u$ a.e. in $\mathbf{R}$, then

$$\liminf_{n\to\infty}\int|(|u_n|^2)'|^2 \geqslant \int|(|u|^2)'|^2 + \liminf_{n\to\infty}\int|(|u_n - u|^2)'|^2.$$

Let $H_r^1(\mathbf{R}^N) = \{u \in H^1(\mathbf{R}^N) \mid u(x) = u(|x|)\}$ and $H = H_r^1 \cap X$ and $\|u\|^2 = \int(|\nabla u|^2 + V(x)|u|^2 + |u|^2)$. Then we have the following lemma from Strauss[17].

**Lemma 2.5** Let $N \geqslant 2$. Then the following embedding is compact:

$$H \hookrightarrow L^q(\mathbf{R}^N), \quad 2 \leqslant q < 2^*,$$

where $2^* = \frac{2N}{N-2}$ for $N \geqslant 3$ and $\infty$ for $N = 2$.

**Lemma 2.6** ([15, Lemma 13 on page 340]) If $(u_n) \subset H$ are such that $u_n \rightharpoonup u$ in $H^1(\mathbf{R}^N)$, then

$$\liminf_{n\to\infty} \int |\nabla| u_n|^2|^2 \geqslant \int |\nabla| u|^2|^2.$$

# 3 The Case of $N \geqslant 2$

In this section, we assume $V(x) = V(|x|)$. We will first follow the line of Method II to study the existence of a standing wave of (1.1) in the case of $\theta > 0$. Then we will study the orbital stability of the standing wave. Now for any $\lambda > 0$ fixed, consider the following minimization problem:

$$m_r = \inf\{E(u) = F_1(u) + kF_3(u) - \theta F_4(u) \mid u \in H, F_2(u) = \lambda\}. \tag{3.1}$$

**Lemma 3.1** Let $(u_n) \subset H$ be a minimizing sequence of $m_r$. If $2 < p < 2 + \frac{4}{N}$, then $\| u_n \|$ and $\int |\nabla| u_n|^2|^2$ are uniformly bounded with respect to $n$.

**Proof** Since $(u_n) \subset H$ is a minimizing sequence of $m_r$, i.e.,

$$m_r + o(1) = F_1(u_n) + kF_3(u_n) - \theta F_4(u_n); \quad F_2(u_n) = \lambda, \tag{3.2}$$

by Sobolev and interpolation inequalities, we deduce that for any $u \in H$ and

$$s = \left(\frac{1}{p} - \frac{1}{2^*}\right) \Big/ \left(\frac{1}{2} - \frac{1}{2^*}\right),$$

$$|u|_p^p \leqslant |u|_2^{ps} |u|_{2^*}^{p(1-s)} \leqslant M |u|_2^{ps} |u|^{p(1-s)}.$$

The Young inequality implies that

$$|u|_p^p \leqslant \varepsilon \| u \|^2 + A_\varepsilon |u|_2^{2ps/(2-p+ps)}.$$

In here we use the assumption $2 < p < 2 + \frac{4}{N}$. It follows that

$$m_r + 1 \geqslant F_1(u_n) + \frac{k}{4} \int |\nabla| u_n|^2|^2 - \frac{\varepsilon}{p} \| u_n \|^2 - \frac{A_\varepsilon}{p} |u_n|_2^{2ps/(2-p+ps)}. \tag{3.3}$$

We obtain from choosing $\varepsilon$ small enough and using $|u_n|_2^2 = 2\lambda$ that $m_r \nrightarrow -\infty$ and $\| u_n \|$ and $\int |\nabla| u_n|^2|^2$ are uniformly bounded with respect to $n$. □

**Theorem 3.2** Suppose $N \geqslant 2$, $2 < p < 2 + \frac{4}{N}$ and $k > 0$. Then for any fixed $\theta > 0$ and $\lambda > 0$, $m_r$ is achieved at some $u_0$, i.e., $u_0(x) = u_0(|x|)$, $F_2(u_0) = \lambda$, $F_1(u_0) + kF_3(u_0) - \theta F_4(u_0) = m_r$.

**Proof** Let $(u_n) \subset H$ be a minimizing sequence of $m_r$. Lemma 3.1 implies that $\| u_n \|$ and $\int |\nabla| u_n|^2|^2$ are uniformly bounded with respect to $n$. Going if necessary to a subsequence, we can assume that $u_n \rightharpoonup u_0$ in $H^1(\mathbf{R}^N)$ and $u_n \to u_0$ a.e. in $\mathbf{R}^N$. The Brezis-Lieb Lemma[3] implies that

$$F_1(u_n) = F_1(u_n - u_0) + F_1(u_0) + o(1). \tag{3.4}$$

Combining this with (3.1), Lemma 2.5 and Lemma 2.6 we have that

$$\begin{aligned} m_r + o(1) &= F_1(u_n) + \frac{k}{4}\int |\nabla |u_n|^2|^2 - \frac{\theta}{p}|u_n|_p^p \\ &\geqslant F_1(u_0) + kF_3(u_0) - \theta F_4(u_0) + \lim_{n\to\infty} F_1(u_n - u_0) \\ &\geqslant F_1(u_0) + kF_3(u_0) - \theta F_4(u_0) \geqslant m_r. \end{aligned}$$

It follows that $m_r = F_1(u_0) + kF_3(u_0) - \theta F_4(u_0)$ and $\lim_{n\to\infty} F_1(u_n - u_0) = 0$. Thus we obtain from $|u_n - u_0|_2^2 \to 0$ that $\|u_n - u_0\| \to 0$, i.e., $u_n \to u_0$ in $H^1(\mathbf{R}^N)$. □

The method for studying the minimization problem (3.1) is not new, but it has the merit of studying the orbital stability of the standing wave of (1.1) in the spirit of Cazenave and Lions[6]; see also Albert[1]. Roughly speaking, a set of solutions of (1.1) is said to be stable if any solution of (1.1) remains near the set whenever it starts near the set.

**Definition 3.3** A set $S \subset H$ is $H$-stable with respect to (1.1) if for any $\varepsilon > 0$, there is $\delta > 0$ such that for any $z_0 \in H \cap H^s(\mathbf{R}^N)\left(s \geqslant \frac{N}{2} + 7\right)$ with

$$\inf_{v\in S}\left(\|z_0 - v\| + \left|\int |\nabla |z_0|^2|^2 - \int |\nabla |v|^2|^2\right|\right) < \delta,$$

the solution $z(t,x)$ of (2.1) with initial data $z(0,x) = z_0$ can be extended to a global solution in $C([0,\infty), H^s(\mathbf{R}^N))$ and

$$\sup_{0\leqslant t<\infty}\inf_{v\in S}\left(\|z(\cdot,t) - v\| + \left|\int |\nabla |z(\cdot,t)|^2|^2 - \int |\nabla |v|^2|^2\right|\right) < \varepsilon.$$

Otherwise $S$ is called $H$-unstable.

Now for any $\lambda > 0$, we know from Theorem 3.2 that the set of minimizers, denoted by $S_\lambda$, of the minimization problem (3.1) is not empty. Then for any $u_0 \in S_\lambda$, $\int |\nabla |u_0|^2|^2$ is finite. Hence the integral $\int \nabla |u_0|^2 \nabla(u_0 v)$ exists for any $v \in \mathscr{D}(\mathbf{R}^N)$ because

$$\int |u_0||\nabla u_0|^2 + |u_0|^2|\nabla u_0| \leqslant \int (1 + |u_0|^2)|\nabla u_0|^2 + |u_0|^2(1 + |\nabla u_0|^2) < +\infty. \tag{3.5}$$

Therefore, the standard proof of the Ljusternik Theorem on Lagrange multipliers[4] implies that there exist $\gamma$ such that $u_0$ is a weak solution of (1.2) for $\mu = -\gamma$. It follows that $z(x,t) = e^{i\mu t}u_0(x)$ is a standing wave of (1.1). Thus $e^{i\mu t}u_0(\cdot)$ is the orbit of $u_0$. Moreover, for any $t \geqslant 0$, if $u \in S_\lambda$, then $e^{i\mu t}u(x) \in S_\lambda$. Our orbital stability result is

**Theorem 3.4** Assume that $N \geqslant 2$ and $2 < p < 2 + \frac{4}{N}$. Then $S_\lambda$ is $H$-stable with respect to (1.1) in the sense of Definition 3.3.

**Proof** Suppose the conclusion is false. Then there exists a number $\varepsilon_0 > 0$, such that for any $\frac{1}{n} > 0$, there is $\psi_n \in H \cap H^s(\mathbf{R}^N)\left(s \geqslant \frac{N}{2} + 7\right)$ with

$$\inf_{g\in S_\lambda}\left(\|\psi_n-g\|+\left|\int|\nabla|\psi_n|^2|^2-\int|\nabla|g|^2|^2\right|\right)<\frac{1}{n}$$

but

$$\sup_{t>0}\inf_{g\in S_\lambda}\left(\|z_n(\cdot,t)-g\|+\left|\int|\nabla|z_n(\cdot,t)|^2|^2-\int|\nabla|g|^2|^2\right|\right)\geqslant\varepsilon_0,$$

where $z_n(x,t)$ solves (1.1) with $z_n(x,0)=\psi_n$. Hence we can pick the first time $t_n$ such that

$$\inf_{g\in S_\lambda}\left(\|z_n(\cdot,t_n)-g\|+\left|\int|\nabla|z_n(\cdot,t_n)|^2|^2-\int|\nabla|g|^2|^2\right|\right)=\varepsilon_0.$$

Then since $\psi_n\to S_\lambda$ in $H^1(\mathbf{R}^N)$ and $m_r=E(g)$ for all $g\in S_\lambda$ and $F_2(g)=\lambda$, we have $E(\psi_n)\to m_r$ and $F_2(\psi_n)\to\lambda$ as $n\to\infty$. Thus we can find a sequence $\beta_n\to 1$ such that $F_2(\beta_n\psi_n)=\lambda$ for all $n$. It follows from Lemma 2.1 that the sequence $q_n=\beta_n z_n(\cdot,t_n)$ satisfies $F_2(q_n)=\lambda$ and

$$\lim_{n\to\infty}E(q_n)=\lim_{n\to\infty}E(z_n(\cdot,t_n))=\lim_{n\to\infty}E(\psi_n)=m_r$$

and is therefore a minimizing sequence for $m_r$. The proof of Theorem 3.2 implies that $q_n\to q_0$ strongly in $H^1(\mathbf{R}^N)$ and

$$\lim_{n\to\infty}\int|\nabla|q_n|^2|^2=\int|\nabla|q_0|^2|^2.$$

Hence $q_0\in S_\lambda$ and for $n$ large, we have that

$$\begin{aligned}\varepsilon_0&\leqslant\|z_n(\cdot,t_n)-q_0\|+\left|\int|\nabla|z_n(\cdot,t_n)|^2|^2-\int|\nabla|q_0|^2|^2\right|\\&=\left|\frac{1}{\beta_n}-1\right|\|q_n\|+\|q_n-q_0\|+\\&\quad\left|\frac{1}{\beta_n^4}-1\right|\int|\nabla|q_n|^2|^2+\left|\int|\nabla|q_n|^2|^2-\int|\nabla|q_0|^2|^2\right|\to 0\end{aligned}$$

as $n\to\infty$, which is a contradiction. □

**Remark 3.5** Please note that in the stating of this theorem, we have tacitly assumed that the potential $V(x)$, $p$ and the initial data $z_0$ belong to a class in which a unique solution of the initial-value problem for (1.1) exists for all $t\in[0,T^*)$ and some $T^*<\infty$ or $T^*=\infty$. In view of local well-posedness results for this kind of problems studied by Lange-Poppenberg[9] and the arguments developed by Cazenave[5], we will continue to make these assumptions without any comments. Moreover, if $z_0$ is radially symmetrical with respect to $x$, then so is $z(x,t)$.

**Remark 3.6** Note that in Theorem 3.2 and Theorem 3.4, $V(x)$ can cover the case of $V(x)=|x|^2$. Problem (1.1) in the case of $V(x)=|x|^2$ and $k=0$ describes the Bose-Einstein condensate with attractive interparticle interactions under a magnetic trap as the Gross-Pitaevski equation with a harmonic potential term[18]. Hence Theorem 3.4 generalizes [19, Theorem 3.2].

**Remark 3.7** As pointed out in [1, page 4], although the concentration compactness method for proving the orbital stability of a standing wave has the advantage of requiring less analysis than others (see, e.g., [7] and the references therein), it produces a weak result in that it only demonstrates stability of a set of minimizing solutions without providing information on the structure of the set, or distinguishing among its different members. This kind of possible confusion will be clarified in the case of $N = 1$.

## 4 The Case of $N = 1$

In this section, we will give a stronger orbital stability of the standing wave of problem (1.1) in the case of $V(x) \equiv 0$ and $N = 1, k \geqslant 0, \theta > 0$, At this time, we write (1.1) as

$$i\partial_t z = - z'' - k(|z|^2)''z - \theta|z|^{p-2}z \tag{4.1}$$

and denote by $''$ (resp. $'$) the second (resp. first) order spatial derivatives. Finding a standing wave of (4.1) of the form $e^{i\mu t}u(x)$ is equivalent to solving the associated elliptic problem

$$- u'' + \mu u - k((|u|^2)'')u - \theta|u|^{p-2}u = 0. \tag{4.2}$$

Using the basic function space $H^1(\mathbf{R}) = W^{1,2}(\mathbf{R})$ with the standard norm $\|u\|^2 = \int(|\nabla u|^2 + |u|^2)$ and the continuous embedding $H^1(\mathbf{R}) \hookrightarrow L^\infty(\mathbf{R})$, we know that the functional

$$I(u) = \int\left(\frac{1}{2}|u'|^2 + \frac{k}{4}|(|u|^2)'|^2 - \frac{\theta}{p}|u|^p\right) \tag{4.3}$$

is well defined on $H^1(\mathbf{R})$. Moreover, we have

**Lemma 4.1**([15, Lemma 1]) $I$ is of class $C^1$ on $H^1(\mathbf{R})$.

For any $\lambda > 0$ fixed, we define

$$\mathscr{M} = \left\{u \in H^1(\mathbf{R}) \mid \frac{1}{2}\int_{\mathbf{R}}|u|^2 = \lambda\right\}, \quad m = \inf_{u\in\mathscr{M}} I(u). \tag{4.4}$$

**Lemma 4.2** If $2 < p < 6$, then $-\infty < m < 0$.

**Proof** Suppose $\psi(x) \in \mathscr{M}$. Then so is $\xi^{\frac{1}{2}}\psi(\xi x)$ for $\xi > 0$. Hence

$$\begin{aligned} I(\xi^{\frac{1}{2}}\psi(\xi x)) &= \int_{\mathbf{R}}\left(\frac{1}{2}\xi^3|\psi'(\xi x)|^2 + \frac{k}{4}\xi^4|(|\psi(\xi x)|^2)'|^2 - \frac{\theta}{p}\xi^{\frac{p}{2}}|\psi(\xi x)|^p\right) \\ &= \int\left(\frac{1}{2}\xi^2|\psi'|^2 + \frac{k}{4}\xi^3|(|\psi|^2)'|^2 - \frac{\theta}{p}\xi^{\frac{p}{2}-1}|\psi|^p\right). \end{aligned}$$

It follows from $2 < p < 6$ and $m \leqslant I(\xi^{\frac{1}{2}}\psi(\xi x))$ that $m < 0$ provided $\xi > 0$ small enough. Now for any $u \in \mathscr{M}$, we can use Sobolev and interpolation inequalities such that for some $0 < \alpha < 1$,

$$|u|_p^p \leqslant |u|_2^{p\alpha}|u|_\infty^{p(1-\alpha)} \leqslant d|u|_2^{p\alpha}\|u\|^{p(1-\alpha)}. \tag{4.5}$$

The Young inequality implies that

$$|u|_p^p = \varepsilon\|u\|^2 + A_\varepsilon|u|_2^{2p\alpha/(2-p+p\alpha)}. \tag{4.6}$$

In here, we use the assumption $2 < p < 6$. By choosing some $u \in \mathscr{M}$ such that

$$m + 1 \geqslant \frac{1}{2}\|u\|^2 - \lambda + \frac{k}{4}\int_{\mathbf{R}}|(|u|^2)'|^2 - \frac{\theta}{p}\varepsilon\|u\|^2 - \frac{\theta}{p}A_\varepsilon|u|_2^{2p\alpha/(2-p+p\alpha)} \tag{4.7}$$

and $\varepsilon$ small enough, we can get that $m > -\infty$. □

**Lemma 4.3** Let $\{u_n\}$ be a minimizing sequence of $m$. Then there exist $w \neq 0$ and $(y_n) \subset \mathbf{R}$ such that $u_n(\cdot + y_n) \rightharpoonup w$ in $H^1(\mathbf{R})$.

**Proof** Let $\{u_n\}$ be a minimizing sequence of $m$. Then from the proof of the second part of Lemma 4.2, we know that $(u_n)$ is bounded in $H^1(\mathbf{R})$. Using the fact that $\frac{1}{2}\int_{\mathbf{R}} |u_n| = \lambda > 0$ and Lemma 2.3, we know that the vanishing case does not occur for $\rho_n(x) = |u_n(x)|^2$. Therefore, there exist $\varepsilon > 0$ and $R > 0$ such that

$$\liminf_{n\to\infty} \sup_{y\in\mathbf{R}} \int_{y-R}^{y+R} |u_n(x)|^2 \geqslant \varepsilon > 0. \tag{4.8}$$

We may assume that there are $(y_n) \subset \mathbf{R}$ and $R > 0$ such that

$$\liminf_{n\to\infty} \int_{y_n-R}^{y_n+R} |u_n(x)|^2 \geqslant \frac{\varepsilon}{2} > 0. \tag{4.9}$$

Hence $w_n(x) = u_n(x + y_n)$ satisfies $I(w_n) = I(u_n)$ and $\frac{1}{2}\int_{\mathbf{R}} |w_n|^2 = \lambda$; i.e., the sequence $(w_n)$ is also a minimizing sequence which satisfies

$$\liminf_{n\to\infty} \int_{-R}^{R} |w_n(x)|^2 \geqslant \frac{\varepsilon}{2} > 0. \tag{4.10}$$

$(w_n)$ is also bounded in $H^1(\mathbf{R})$. Going if necessary to a subsequence, we have $w_n \rightharpoonup w$ in $H^1(\mathbf{R})$ and thus $w_n \to w$ in $L^2_{\mathrm{loc}}(\mathbf{R})$. (4.10) implies that $w \neq 0$. □

**Lemma 4.4** Let $w_n \rightharpoonup w$ in $H^1(\mathbf{R})$ and $w_n \to w$ a.e. in $\mathbf{R}$. Then

$$\lim_{n\to\infty} (|w_n|_q^q - |w_n - w|_q^q) = |w|_q^q, \quad q \geqslant 2; \tag{4.11}$$

$$\lim_{n\to\infty} (|w_n'|_2^2 - |(w_n - w)'|_2^2) = |w'|_2^2. \tag{4.12}$$

**Proof** These are direct consequence of the Brezis-Lieb Lemma[3]. □

**Proposition 4.5** Suppose $4 \leqslant p < 6$. Then $m$ is achieved by some nonnegative $w$; i.e., there is $0 \neq w \in \mathcal{M}$ with $I(w) = m$ and $w \geqslant 0$.

**Proof** Let $(u_n)$ be a minimizing sequence of $m$. Lemma 4.3 implies that there are $(y_n)$ and $w \neq 0$ such that $w_n(x) = u_n(x + y_n)$ and $w_n \rightharpoonup w \neq 0$ in $H^1(\mathbf{R})$. To conclude the proof, we have to show that $w \in \mathcal{M}$, i.e., $\frac{1}{2}|w|_2^2 = \lambda$.

Arguing by a contradiction, we may assume $0 < \frac{1}{2}|w|_2^2 < \lambda$. For $v_n = w_n - w$, we may assume from Lemma 4.4 that $\frac{1}{2}|v_n|_2^2 < \lambda$. Denote

$$\tilde{w} = w \Big/ \left(\frac{1}{2\lambda}\right)^{\frac{1}{2}} |w|_2 \quad \text{and} \quad \tilde{v} = v_n \Big/ \left(\frac{1}{2\lambda}\right)^{\frac{1}{2}} |v_n|_2.$$

We have that

$$\begin{aligned} m &\geqslant \int_{\mathbf{R}} \left(\frac{1}{2}|w_n'|^2 + \frac{k}{4}|(|w_n|^2)'|^2 - \frac{\theta}{p}|w_n|^p\right) \\ &\geqslant \int_{\mathbf{R}} \left(\frac{1}{2}|w'|^2 + \frac{k}{4}|(|w|^2)'|^2 - \frac{\theta}{p}|w|^p\right) + \end{aligned}$$

$$\int_{\mathbf{R}}\left(\frac{1}{2}|v_n'|^2-\frac{\theta}{p}|v_n|^p\right)+\frac{k}{4}\liminf_{n\to\infty}\int_{\mathbf{R}}|(|v_n|^2)'|^2$$
$$=\left(\sqrt{\frac{1}{2\lambda}}|w|_2\right)^2\int_{\mathbf{R}}\frac{1}{2}|\tilde{w}'|^2+\frac{k}{4}\left(\sqrt{\frac{1}{2\lambda}}|w|_2\right)^4\int_{\mathbf{R}}|(|\tilde{w}|^2)'|^2-$$
$$\left(\sqrt{\frac{1}{2\lambda}}|w|_2\right)^p\frac{\theta}{p}\int_{\mathbf{R}}|\tilde{w}|^p+\left(\sqrt{\frac{1}{2\lambda}}|v_n|_2\right)^2\int_{\mathbf{R}}\frac{1}{2}|\tilde{v}_n'|^2-$$
$$\left(\sqrt{\frac{1}{2\lambda}}|v_n|_2\right)^p\frac{\theta}{p}\int_{\mathbf{R}}|\tilde{v}_n|^p+\frac{k}{4}\liminf_{n\to\infty}\left(\sqrt{\frac{1}{2\lambda}}|v_n|_2\right)^4\int_{\mathbf{R}}|(|\tilde{v}_n|^2)'|^2$$
$$>\left(\sqrt{\frac{1}{2\lambda}}|w|_2\right)^4\int_{\mathbf{R}}\left(\frac{1}{2}|\tilde{w}'|^2+\frac{k}{4}|(|\tilde{w}|^2)'|^2-\frac{\theta}{p}|\tilde{w}|^p\right)(\text{since } p\geqslant 4)+$$
$$\liminf_{n\to\infty}\left(\sqrt{\frac{1}{2\lambda}}|v_n|_2\right)^4\int_{\mathbf{R}}\left(\frac{1}{2}|\tilde{v}_n'|^2+\frac{k}{4}|(|\tilde{v}_n|^2)'|^2-\frac{\theta}{p}|\tilde{v}_n|^p\right)$$
$$=\left(\sqrt{\frac{1}{2\lambda}}|w|_2\right)^4 I(w)+\liminf_{n\to\infty}\left(\sqrt{\frac{1}{2\lambda}}|v_n|_2\right)^4 I(\tilde{v}_n)$$
$$\geqslant m\left(\left(\sqrt{\frac{1}{2\lambda}}|w|_2\right)^4+\liminf_{n\to\infty}\left(\sqrt{\frac{1}{2\lambda}}|v_n|_2\right)^4\right)$$
$$\geqslant m\left(\left(\sqrt{\frac{1}{2\lambda}}|w|_2\right)^2+\liminf_{n\to\infty}\left(\sqrt{\frac{1}{2\lambda}}|v_n|_2\right)^2\right)(\text{since } m<0)$$
$$=m, \tag{4.13}$$

which is a contradiction. In here the last equality follows from using Lemma 4.4 and $\frac{1}{2}|w_n|_2^2=\lambda$. Up to now we have proved that $\frac{1}{2}|w|_2^2=\lambda$, i.e., $w_n\to w$ in $L^2(\mathbf{R})$. Since $(w_n)$ is bounded in $H^1(\mathbf{R})$, the standard interpolation and Sobolev inequalities imply that $w_n\to w$ in $L^p(\mathbf{R})$. We now obtain from $(w_n)$ being a minimizing sequence and Lemma 2.4 that

$$m\geqslant\int_{\mathbf{R}}\left(\frac{1}{2}|w_n'|^2+\frac{k}{4}|(|w_n|^2)'|^2-\frac{\theta}{p}|w_n|^p\right)$$
$$\geqslant\int_{\mathbf{R}}\left(\frac{1}{2}|w'|^2+\frac{k}{4}|(|w|^2)'|^2-\frac{\theta}{p}|w|^p\right)+\lim_{n\to\infty}\int_{\mathbf{R}}\frac{1}{2}|(w_n-w)'|^2$$
$$\geqslant m+\lim_{n\to\infty}\int_{\mathbf{R}}\frac{1}{2}|(w_n-w)'|^2. \tag{4.14}$$

Hence $\int_{\mathbf{R}}|(w_n-w)'|^2\to 0$. Combining this with $w_n\to w$ in $L^2(\mathbf{R})$ we get that $w_n\to w$ in $H^1(\mathbf{R})$. Note that if $(w_n)$ is a minimizing sequence of $m$, then so is $(|w_n|)$. Therefore the conclusion follows. □

**Theorem 4.6** Assume $4\leqslant p<6$. Then there exists $\gamma<0$ such that (4.2) admits a positive solution $u\in H_0^1(\mathbf{R})$ for any $\theta>0$ and $\mu=-\gamma$.

**Proof** With the help of Proposition 4.5 and an argument in front of Theorem 3.4, we know that there exists a nonnegative $u\in H^1(\mathbf{R})$ with $|u|_2^2=2\lambda$ and a Lagrange multiplier $\gamma\in\mathbf{R}$ such that

$$-\int u''v-k\int(u^2)''uv-\theta|u|^{p-2}uv=\gamma\int uv, \tag{4.15}$$

for all $v \in H^1(\mathbf{R})$. Putting $v = u$ in (4.15), we get that

$$\gamma|u|_2^2 = |u'|_2^2 + k\int |(|u|^2)'|^2 - \theta|u|_p^p.$$

It follows from $0 > m = \frac{1}{2}|u'|_2^2 + \frac{k}{4}\int |(|u|^2)'|^2 - \frac{\theta}{p}|u|_p^p$ and $p \geqslant 4$ that $\gamma < 0$. Hence $u \in H_0^1(\mathbf{R})$ is a nonnegative solution of (4.2) for any $\mu = -\gamma$. The proof of positivity of $u$ is the same as [15, Theorem 1] and is omitted here. The proof is completed. □

Just as was pointed out in Remark 3.5, we need to use the following basic assumption to study the orbital stability.

(A) For some $p \geqslant 4, k > 0$ and some $z_0$, the Cauchy problem (4.1) with initial value $z(x,0) = z_0 \in H^s(\mathbf{R})\left(s \geqslant \frac{1}{2} + 7\right)$ has a solution $z(x,t)$ in $[0, T^*)$ with $T^* < \infty$ or $T^* = \infty$.

Denote by $S_\lambda$ the set of minimizers of problem (4.4). Then with the help of Lemma 4.3 and Proposition 4.5, we can check in detail the proof of Theorem 3.4 and get directly the following orbital stability result.

**Proposition 4.7** Suppose (A) holds and $4 \leqslant p < 6, \theta > 0$. Then $S_\lambda$ is $H$-stable with respect to (4.1) in the sense of Definition 3.3.

We are now in a position to state the following stronger orbital stability result.

**Theorem 4.8** Suppose (A) holds and $4 \leqslant p < 6, \theta > 0$. Then for any $\varepsilon > 0$, there exists $\delta > 0$ such that if $z_0 \in H^s(\mathbf{R})$ and

$$\inf_{\eta,\xi\in\mathbf{R}}\left(\|z_0 - e^{i\eta}u_0(\cdot + \xi)\| + \left|\int |\nabla|z_0|^2|^2 - \int |\nabla|u_0|^2|^2\right|\right) < \delta, \tag{4.16}$$

then the solution $z(x,t)$ of (4.1) with initial value $z(x,0) = z_0$ can be extended to a global solution and for all existence times $t$, it follows that

$$\inf_{\eta,\xi\in\mathbf{R}}\left(\|z(\cdot,t) - e^{i\eta}u_0(\cdot + \xi)\| + \left|\int |\nabla|z|^2|^2 - \int |\nabla|u_0|^2|^2\right|\right) < \varepsilon. \tag{4.17}$$

**Proof** It suffices to prove that $S_\lambda = \{e^{i\eta}u_0(x + \xi)\}$. If $u \in S_\lambda$, then $e^{i\eta}u(x) \in S_\lambda$ and an argument presented in the proof of Theorem 4.6 implies that $u$ being a solution of (4.2) with some $\mu$. Note that $u(x)$ being a positive solution of (4.2) is equivalent to $w(x) = \theta^{\frac{1}{p-2}}\mu^{\frac{1}{2-p}}u(\mu^{-\frac{1}{2}}x)$ being a solution of

$$-w'' + w - k\theta^{\frac{2}{2-p}}\mu^{\frac{1}{p-2}}(w^2)''w = w^{p-1}, \quad w > 0, \quad w \in H^1(\mathbf{R}). \tag{4.18}$$

By a result of Ambrosetti-Wang[2], we know that (4.18) has a unique positive solution $w_0(x)$ up to a translation. It follows that $u_0(x) = \theta^{\frac{1}{p-2}}\mu^{\frac{1}{p-2}}w_0(\mu^{\frac{1}{2}}x)$ is the unique positive solution of (4.2) up to a translation. The proof is complete. □

**References**

[1] Albert J P. Concentration compactness and the stability of solitary-wave solutions to nonlocal equations.

Contemporary Mathematics, 1999, 221: 1 – 29.

[2] Ambrosetti A, Wang Z Q. Positive solutions to a class of quasilinear elliptic equations on **R**. Disc Contin Dynam Syst, 2003, 9: 55 – 68.

[3] Brezis H, Lieb E. A relation between pointwise convergence of functions and convergence of functionals. Proc Amer Math Soc, 1983, 88: 486 – 490.

[4] Browder F E. Variational methods for nonlinear elliptic eigenvalue problems. Bull Amer Math Soc, 1965, 71: 176 – 183.

[5] Cazenave T. Stable solutions of the logarithmic Schrödinger equation. Nonlinear Anal TMA, 1983, 7: 1127 – 1140.

[6] Cazenave T, Lions P L. Orbital stability of standing waves for some nonlinear Schrödinger equations. Comm Math Phys, 1982, 85: 549 – 561.

[7] Grillakis M, Shatah J, Strauss W. Stability theory of solitary waves in the presence of symmetry, Ⅰ. J Functional Analysis, 1987, 74: 160 – 197; Ⅱ. J Functional Analysis, 1990, 94: 308 – 348.

[8] Kurihura S. Large amplitude quasi-solitons in superfluid film. J Phys Soc Japan, 1981, 50: 3262 – 3267.

[9] Lange H, Poppenberg M, Teismann H. Nash-Moser methods for the solution of quasilinear Schrödinger equations. Comm Partial Diff Equations, 1999, 24: 1399 – 1418.

[10] Lions P L. The concentration-compactness principle in the calculus of variations. The locally compact case, Parts 1 and 2. Ann Inst Henri Poincaré Anal Non Linéaire, 1984, 1: 109 – 145; 223 – 283.

[11] Liu J Q, Wang Z Q. Soliton solutions for quasilinear Schrödinger equations, I. Proc Amer Math Soc, 2003, 131: 441 – 448.

[12] Liu J Q, Wang Y, Wang Z Q. Soliton solutions for quasilinear Schrödinger equations, Ⅱ. J Differential Equations, 2003, 187: 473 – 493.

[13] Liu J Q, Wang Y, Wang Z Q. Solutions for quasilinear Schrödinger equations via the Nehari method. Comm Partial Diff Equations, 2004, 29: 879 – 901.

[14] Nakamura A. Damping and modification of exciton solitary waves. J Phys Soc Japan, 1977, 42: 1824 – 1835.

[15] Poppenberg M, Schmitt K, Wang Z Q. On the existence of soliton solutions to quasilinear Schrödinger equations. Cal Var and PDEs, 2002, 14: 329 – 344.

[16] Rabinowitz P. On a class of nonlinear Schrödinger equations. Z Angew Math Phys, 1992, 43: 270 – 291.

[17] Strauss W. Existence of solitary waves in higher dimensions. Comm Math Phys, 1977, 55: 149 – 162.

[18] Tsurumi T, Wadati M. Collapses of wavefunctions in multi-dimensional nonlinear Schrödinger equations under harmonic potential. J Phys Soc Japan, 1997, 66: 1 – 8.

[19] Zhang J. Stability of standing waves for nonlinear Schrödinger equations with unbounded potentials. Z Angew Math Phys, 2000, 51: 498 – 503.

# 后　　记

在华南理工大学李元元校长、出版社范家巧社长、理学院吴敏院长以及河南理工大学数学与信息计算科学学院原保全院长的大力支持下,《郭柏灵论文集》第七卷、第八卷出版了,文集收录了2005至2009年发表的论文。

《郭柏灵文集》第七卷、第八卷的出版,经费上得到河南理工大学数学与信息计算科学学院的大力资助。华南理工大学数学系偏微分方程方向的研究生对文集出版的校对工作付出了辛勤劳动,在此一并表示衷心感谢。

华南理工大学理学院

2010年9月19日